VENUS AERONOMY

VENUS AERONOMY

Edited by

C. T. RUSSELL

Institute of Geophysics and Planetary Physics,
University of California at Los Angeles.

Reprinted from Space Science Reviews, Volume 55, Nos. 1–4, 1991

SPRINGER SCIENCE+BUSINESS MEDIA, B.V.

Library of Congress Cataloging-in-Publication Data

Venus aeronomy / [edited] by C.T. Russell.
      p.   cm.
    ISBN 978-0-7923-1091-4   ISBN 978-94-011-3300-5 (eBook)
    DOI 10.1007/978-94-011-3300-5
    1. Venus (Planet)--Atmosphere.   I. Russell, C. T.
QB621.V47   1991
551.5'0999'22--dc20   90-26481

Printed on acid-free paper

TABLE OF CONTENTS

INTRODUCTION

For almost three decades since Mariner 2 flew by the planet in December 1962, Venus has been the subject of intense investigation by both the Soviet and American space programs. Since the intrinsic magnetic field of Venus is exceedingly weak, if it exists at all, we expect many phenomena of the upper atmosphere and ionosphere of Venus to differ from their terrestrial counterparts. While flybys and landings of the many Venus missions provided useful data on these phenomena, orbital missions were needed for their detailed investigation. Such orbital missions were provided by the Soviet program with Veneras 9 and 10 in October 1975 and by the United States with the Pioneer Venus Orbiter in December 1978. Originally designed for a prime mission of only 243 days, the Pioneer Venus Orbiter is still functioning over a decade later, providing data nearly 24-hours a day through one of the most active solar cycles to date. We expect these transmissions to continue until September 1992 when gravitation perturbations will lower the periapsis of the PVO orbit so that the spacecraft will be lost to the atmosphere.

The Venera 9 and 10 and the Pioneer Venus observations have led to an explosion of knowledge about the upper atmosphere and ionosphere of Venus and their interaction with the solar wind. The availability of data over a full solar cycle has led to greater understanding than had been planned prior to launch and has resolved many of the initial differences between the Venera and the early PVO observations which were obtained at different phases of the solar cycle.

The seven articles which follow in this special issue attempt to capture this explosion in our understanding of Venus. We have divided the task into seven topical areas. We begin with the solar wind interaction. In the paper "The Magnetosheath and Magnetotail of Venus", J. L. Phillips and D. J. McComas explain how the planet slows and diverts the solar wind flow with the formation of a bow shock and show how the magnetotail is formed. In the paper "The Structure of the Venus Ionosphere", L. H. Brace and A. J. Kliore combine the results of *in-situ* and radio occultation investigations to reveal the density, temperature and composition of the ionosphere and how it varies over the solar cycle. This is followed by K. L. Miller's and R. C. Whitten's manuscript, "Ion Dynamics in the Venus Ionosphere" which describes and explains the flows in the Venus ionosphere.

One of the puzzles of the early PVO observations was the variation in the state of magnetization of the ionosphere. Some days the ionosphere was field free and at other times it was strongly magnetized. Thus behavior was difficult to understand if the current systems in the ionosphere were directly driven by the solar wind. The interaction was more subtle than this simple picture predicted. In their paper "Magnetic Fields in the Ionosphere of Venus", J. G. Luhmann and T. E. Cravens review the observations and the theoretical explanation of this behavior.

One of the important diagnostics of the physical processes occurring in a plasma is the waves it produces. Pioneer Venus included a simple plasma wave instrument measuring wave power in four narrow bands. R. J. Strangeway in his paper "Plasma Waves at

Venus" reviews the results of this investigation from the bow shock to lowest altitudes in the night ionosphere. One of the earliest interpretations of these latter signals was that they were caused by electrical discharges, or lightning, in the upper atmosphere. F. L. Scarf, who was invited to be a co-author on this article before his untimely death in July 1988, was one of the strongest proponents of this view. Strangeway examines the arguments for and against this interpretation.

In a paper solicited independently of this collection of papers, but included here because of its appropriateness, I review the totality of evidence for lightning on Venus from the Venera landers, the Venera 9 orbiter and the Pioneer Venus Orbiter. Included in this review is a discussion of the properties of terrestrial lightning so that we may know what we should expect at another planet. While alternate explanations are still being examined for these waves, the atmospheric electric source is still the strongest candidate explaining simply most of the observed properties. If this explanation is indeed correct, then lightning on Venus is a very prevalent phenomenon, probably much more so than on Earth.

In the last paper of the series, "The Structure, Luminosity and Dynamics of the Venus Atmosphere", J. L. Fox and S. W. Bougher review observations and models related to the chemical and thermal structures, airglow and auroral emissions and dynamics of the Venus atmosphere. This discussion includes a treatment of the extended exospheres of hydrogen and oxygen that surround Venus as well as phenomena such as the unexpectedly cold nightside thermosphere. Finally they review the major aspects of the circulation and dynamics of the thermosphere: subsolar to anti-solar convection, superrotation and turbulent processes.

These articles represent the present state of our understanding of the upper atmosphere, ionosphere and solar wind interaction with Venus. Much effort has been expended in preparing these reviews and we thank the authors for their exhaustive reviews. We are also grateful for the assistance of many reviewers, especially W. C. Knudsen, who also gave unselfishly of their time to assist in ensuring the quality and accuracy of these papers. We caution the reader that knowledge, like the Venus ionosphere, is dynamic, and that our understanding may continue to evolve and improve. If readers have any questions about these papers, I am confident that the authors would like to hear from them. Please do not hesitate to begin a dialogue.

C. T. RUSSELL
March 1990

THE MAGNETOSHEATH AND MAGNETOTAIL OF VENUS

JOHN L. PHILLIPS and DAVID J. McCOMAS

Los Alamos National Laboratory, Los Alamos, NM 87545, U.S.A.

Abstract. We describe the observational history and assess the current understanding of the magnetosheath and magnetotail of Venus, stressing recent developments. We make recommendations for research that can be done using existing observations, as well as desirable trajectory and instrumentation characteristics for future spacecraft missions.

Table of Contents

1. Introduction

The subjects to be discussed in this review, the Venus magnetosheath and magnetotail, make up a major portion, but by no means the entirety, of the interaction of the solar wind with Venus. For discussions of other aspects of the Venus atmosphere and the solar wind interaction the reader is referred to the companion review papers in this volume. Previous reviews of these subjects have been presented by Breus (1979), Russell and Vaisberg (1983), and Luhmann (1986a). Collections of original papers can be found in a special Pioneer Venus issue of *Journal of Geophysical Research* (December 1980),

Space Science Reviews **55**: 1–80, 1991.
© 1991 *Kluwer Academic Publishers.*

two special Venus issues of *Icarus* (August and December 1982), and a Venus atmosphere issue of *Advances in Space Research* (Volume 5, No. 9, 1985). Additionally, the 1983 University of Arizona volume entitled *Venus* features a comprehensive series of articles on the solid planet, the atmosphere, and the solar wind interaction.

Of the solar-planetary interactions thus far explored, all are unique in some respect. Each planet has a combination of atmospheric characteristics, magnitude and orientation of the planetary magnetic field, and the presence or absence of satellites which defines its surrounding plasma environment. At the magnetized planets, the primary obstacle to the solar plasma is the planetary magnetic field itself, and the region dominated by the planetary field is called the magnetosphere. A standing bow shock wave is formed upstream of the obstacle, and the region of shocked solar wind plasma surrounding the magnetosphere is called the magnetosheath. The planetary field of Venus, if one exists at all, has been shown to be inconsequential in diverting the solar wind (e.g., Phillips and Russell, 1987). Our current understanding is that the obstacle to the solar wind consists primarily of shielding currents carried by the ionosphere, modified by a comet-like pickup process involving exospheric and ionospheric ions. Thus, there is no distinct boundary, or magnetopause, separating regions dominated by planetary and solar magnetic fields. Some authors have used the term 'magnetosphere' to describe certain portions of the interaction region. We will avoid this term, as it implies the existence of a planetary magnetic field, and thus is a misnomer in the case of Venus. The 'ionopause' has been defined in various ways by various authors, but generally can be considered to be an altitude above which the ion density decreases sharply. While some authors have referred to the region of shocked solar plasma and lesser amounts of planetary plasma, lying between the bow shock and the ionopause, as the 'ionosheath', we will use the more common term of 'magnetosheath'.

Just as the obstacle to the solar wind at Venus is fundamentally different from those of the magnetized planets thus far explored by spacecraft, so too the magnetotail is unique. It resembles a draped cometary tail more than it does those at the magnetized planets: Mercury, Earth, Jupiter, Saturn, Uranus, and Neptune. While the polarities of the tail lobes at the magnetized planets are controlled primarily by the planetary fields, the lobes at Venus respond directly to changes in the orientation of the interplanetary magnetic field (IMF). The presence of a neutral exosphere which extends far into the regions dominated by the solar wind plasma adds an additional comet-like aspect to the interaction. The ionization and subsequent assimilation by the shocked solar wind of the exospheric particles produces observable effects in both the magnetosheath and the magnetotail. Figure 1 contains a schematic of the Venus–solar wind interaction, with the major plasma regimes and typical locations indicated.

Before proceeding to the detailed discussions of phenomenology, we will digress briefly to describe the basic numerical simulation types applicable to modeling the Venus–solar wind interaction. Modeling results will be mentioned often in subsequent sections, so it is important that the reader be aware of the physics included in, and excluded from, various model types. At present, three types of models have been used to study the Venus magnetosheath. The first is the gas dynamic model developed by

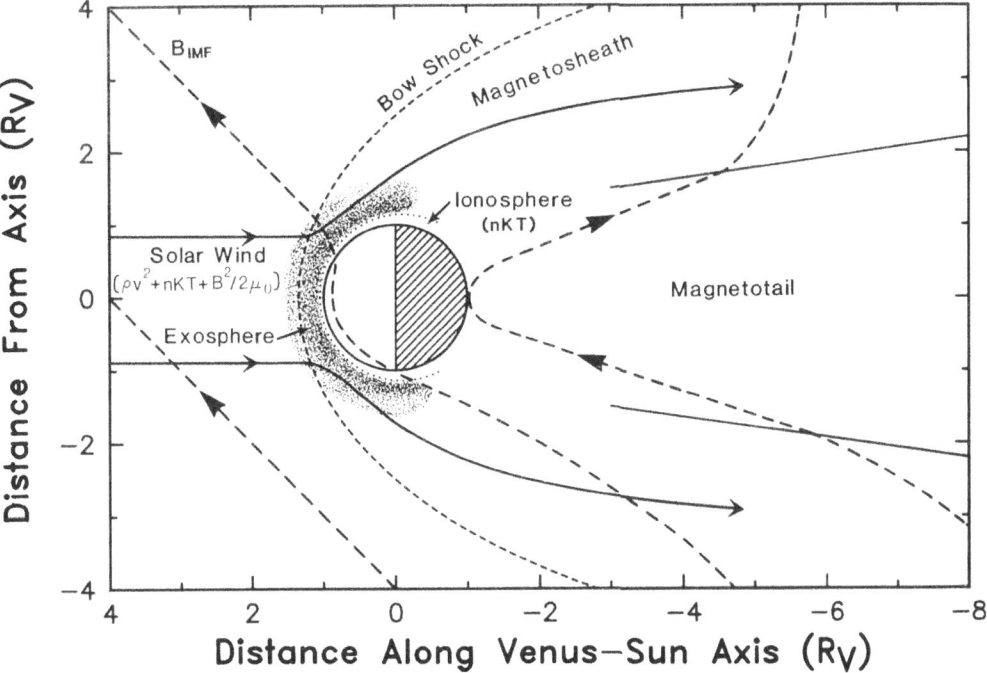

Fig. 1.   Schematic of the solar wind interaction with Venus for a typical IMF 'toward sector' configuration, in a plane defined by the solar wind **V** and **B** vectors. Aberration due to planetary motion is not incorporated. The momentum flux of the solar wind, including inertial, thermal and magnetic contributions, is withstood by the thermal pressure of the ionosphere. Unlabeled solid and dashed lines represent streamlines and magnetic field lines, respectively. The connection of the magnetotail to the near-planet interaction region is not yet well understood, and so is not shown here.

Spreiter and Stahara (1980). In this model the fluid equations for hypersonic flow around an impenetrable obstacle are solved. The magnetic field is not explicitly included in calculation of the model flow field, but instead is determined subsequently based on the distortion of the fluid elements and thus has no effect on the flow parameters. A variation of this model which has been used to assess the behavior of planetary pickup ions and their effects consists of the gas dynamic model with test particle planetary ions. The trajectories of newborn ions are determined by the background electric and magnetic fields calculated for the magnetosheath. Such a model is not self consistent in that it does not include the reaction of the flow parameters to the mass addition, but it gives a first-order picture of pickup ion behavior (cf. Luhmann *et al.*, 1985; Phillips *et al.*, 1987). A third type is the 'mass-loaded' gas dynamic model where a source term added to the continuity equation allows assessment of the effects of the addition of a heavy, cold photoion 'fluid' at low altitudes (cf. Berlotserkovskii *et al.*, 1987; Stahara *et al.*, 1987). Such a simulation, while treating the planetary ion source self-consistently, does not take finite gyroradius effects into account and, therefore, can underestimate planetary ion effects at higher altitudes. We note that none of the afore-mentioned simulations incorporate realistic tail properties, and that, therefore, these models are applicable only to the upstream and near-terminator regions.

Two model types which can answer certain questions left by those mentioned above are an MHD model and a hybrid model. We are currently aware of progress in simulating the interaction with both of these techniques, though published results are not yet available. The former method adds the magnetic terms to the fluid equations and thus describes self-consistently the effects of the magnetic field on the plasma flow. Observed and predicted asymmetries due to MHD effects in the interaction region should be obtainable with such a model. Test particle pickup ions or a fluid mass loading source term can be added to an MHD simulation in the same manner, and with the same limitations, as for a gasdynamic simulation. A hybrid model, which would treat ions as individual particles and electrons as a fluid, would incorporate not only magnetic effects but also the finite Larmor radius effects not available with a fluid mass loading treatment. Such a simulation could presumably model asymmetries related to ion pickup in a self-consistent manner.

2. Observational History

Since the beginning of the spacecraft era, Venus has been visited by a total of 20 spacecraft missions, 15 from the Soviet Union and 5 from the United States. Another U.S. spacecraft, the Magellan radar mapper, is enroute to Venus at the time of this writing. The trajectories and/or instrument packages for many of the Soviet missions, as well as for the Pioneer Venus multiprobe, were optimized for atmospheric or surface studies, and therefore have not contributed significantly to our knowledge of the Venus magnetosheath, wake and magnetotail. The remainder of this review will be concerned primarily with observations by the following spacecraft: Mariner 5 and 10; Venera 4, 6, 9, and 10; Pioneer Venus Orbiter. Other spacecraft whose measurements have provided indirect evidence concerning the Venus–solar wind interaction include Mariner 2 and Venera 11, 12, 15, and 16. Contributions of these latter spacecraft will be mentioned as part of the historical development, but will not be described in detail.

The first observational constraint on the Venus–solar wind interaction was established by the Mariner 2 spacecraft, which flew by Venus on December 14, 1962. Closest approach was at 41000 km from the center of the planet, just sunward of the terminator plane. The lack of any planetary perturbation of the interplanetary magnetic field (IMF) led the observers (Smith *et al.*, 1963, 1965) to conclude that the spacecraft had never passed through a planetary bow shock. Using theoretical predictions for solar wind flow around a magnetized obstacle (Spreiter and Jones, 1963), they then concluded that the planetary magnetic moment must be less than one-tenth of that of Earth. Similar results, that is no perturbation of the interplanetary medium, were observed by the Mariner 2 ion spectrometer (Neugebauer and Snyder, 1965). Both magnetometer and plasma experimenters estimated that the subsolar bow shock, if it existed, was 4 R_V or less from the center of the planet. Additionally, the absence of energetic protons and electrons, which would have suggested magnetospheric trapping, led the energetic particle experimenters (Frank et al., 1963) to conclude that the spacecraft had never entered a magnetosphere and that therefore the planetary magnetic moment must be smaller than the terrestrial moment.

The next spacecraft to encounter Venus were Venera 4, which impacted the planetary night side on October 18, 1967, and Mariner 5, which flew by one day later, approaching to within 4100 km from the planetary surface and roughly 2500 km from the optical shadow. Important observational milestones established by both missions included the first observations of the planetary bow shock by magnetic field and plasma experiments (Bridge et al., 1967; Dolginov et al., 1968; Gringauz et al., 1968). Venera 4 crossed the shock inbound at a distance of $\sim 4\,R_V$ from the planetary center at a solar zenith angle (SZA), that is the angle between the planet-centered position vector and the Venus–Sun line, of $\sim 110°$. The shock crossing was signaled by increased ion flux, interpreted as increased plasma density, as well as by enhanced magnetic field magnitude. Gringauz et al. (1968) estimated the subsolar shock position, based on the observed solar wind conditions near Venus and on this nightside crossing, to be 2000 km above the planetary surface. Mariner 5 encountered the bow shock twice, at $8.25\,R_V$ from planet center and $136°$ SZA inbound, and at $2.0\,R_V$ and $59°$ outbound (Greenstadt, 1970). The plasma data were later reanalyzed by Shefer et al. (1979), who noted multiple shock crossings along the outbound trajectory.

Other findings of the Venera 4 mission included low (less than $\sim 10^3$ cm^{-3}) nightside ionospheric density (Gringauz et al., 1968) and the apparent lack of a planetary magnetic field, which Dolginov et al. (1968) estimated to be smaller than the terrestrial field by a factor of at least 3000. As during the Mariner 2 encounter, Mariner 5 observed no plasma or magnetic field signature indicating crossing of a magnetopause (Bridge et al., 1967), and the energetic particle investigators again observed no energetic electrons and protons, and again concluded that the spacecraft had not penetrated a planetary magnetosphere (Van Allen et al., 1967). Subsequently, however, Russell (1976a) reinterpreted the Venera 4 magnetometer observations and concluded that the measurements allowed a possible planetary field of roughly $\frac{1}{1000}$th of the terrestrial field. Russell (1976b) also reinterpreted the Mariner 5 magnetometer observations and postulated the existence of an intrinsic magnetotail similar to that of Earth.

Two further discoveries from the Mariner 5 mission provided vital links in the overall picture of the Venus–solar wind interaction. First, a 'plasmapause', or transition between solar and ionospheric plasmas, was found by the radio occultation experimenters (Mariner Stanford Group, 1967), to exist at an altitude of about 500 km on the planetary day side. This transition, now known as the 'ionopause', appears as a sudden drop in ion density from near 10^4 cm^{-3} to values similar to those in interplanetary space. Occultation on the nightside ionosphere, however, revealed an ionosphere that was extended in height and had no clear upper boundary, though the peak ion density was much lower than on the dayside. Second, an extended exosphere of atomic hydrogen was discovered through measurement of the Lα airglow via ultraviolet photometry (Barth et al., 1967). The measurements were analyzed further by Barth (1968), who concluded that a mixture of atomic and molecular hydrogen was present, with total densities of $10-1000$ cm^{-3} at altitudes of a few thousand kilometers. Other findings of the Mariner 5 particle and field experiments include (Bridge *et al.*, 1967): (1) fluctuating magnetic fields within the magnetosheath, and (2) a region just outside the optical shadow of the planet, characterized by magnetic field nearly aligned with the Sun–Venus line and by low plasma density and flow speed. Entry of the latter region, bounded on the outside during the inbound (downstream) crossing by a magnetic field discontinuity, may have constituted the first direct observation of the magnetotail.

Thus at the conclusion of the analysis of the Venera 4 and Mariner 5 encounters the basic observational building blocks for the Venus–solar wind interaction were in place. The bow shock, the dayside ionopause, the lack of an intrinsic magnetic obstacle, the extended neutral exosphere, and a nightside tail-like structure had all been discovered. However, the existence of an induced magnetotail had not been unambiguously demonstrated, and the nature of the upper boundary of the nightside ionosphere, and the nightside ionization source, were unknown. Mariner 5 ionopause measurements, coupled with the plasma and field measurements inside the bow shock, led quickly to formulation of an interaction model in which the highly conducting ionosphere is able to divert the solar plasma via shielding currents. This ionospheric shielding leads to pile-up of the interplanetary magnetic flux in an 'induced magnetosphere' or magneto-sheath (Mariner Stanford Group, 1967; Dessler, 1968; Johnson and Midgley, 1969).

The next spacecraft to visit Venus with suitable particle or field instrumentation was Venera 6, which impacted the planet on May 17, 1969 (Venera 5 had arrived at Venus on the previous day and carried plasma instrumentation, but its low telemetry rate precluded significant measurements in the near-planet environment). The Venera 6 plasma measurements (Gringauz *et al.*, 1970) served primarily to support the earlier findings that an observable bow shock, crossed by this spacecraft near 130° SZA at an altitude of ~ 30000 km, extended downstream from the planet.

The Mariner 10 spacecraft, which encountered Venus on February 5, 1974 enroute to Mercury, flew sunward within an 8.5° half-angle cone centered on the nightside Venus–Sun axis (but outside the optical shadow) for a distance of $\sim 750\,R_V$ prior to closest approach at 5700 km altitude near the terminator plane. The bow shock, crossed near the terminator, was not a sharp, precisely identifiable boundary, but instead was characterized by magnetic fluctuations and a broad transition region, suggesting a quasi-parallel alignment between the IMF and the local shock normal (Ness *et al.*, 1974). The shock was observed somewhat closer to the planet than expected, based on modeling of the Venera 4, Venera 6, and Mariner 5 downstream crossings. This varia-tion was suggested by Ness *et al.* to result from changes in the composition of the ionosphere, the solar wind speed, or the IMF. Similarly, Bridge *et al.* (1974) noted that the solar wind pressure was higher for Mariner 10 than for Mariner 5 and suggested that this might be a factor in controlling the shock location. Although modulation of the bow shock position by the solar cycle was not suggested until much later (Slavin *et al.*, 1979b), the Mariner 10 observations provided the first observational suggestion of such modulation (the previous missions had encountered Venus near solar cycle maximum, while the Mariner 10 flyby was near solar minimum). In its passage up the Venus wake, the magnetometer experiment (Ness *et al.*, 1974) measured intermittent field rotations and fluctuations which suggested multiple crossings between the downstream magnetosheath and a disturbed wakelike region. Lepping and Behannon (1978) concluded that the field in the wake region did not agree with predictions either (1) for a steady comet-like draping configuration (e.g., Alfvén, 1957), or (2) for a magnetotail based on a planetary field (cf. Russell, 1976a, b).

A further finding by the Mariner 10 plasma experiment (Bridge *et al.*, 1974) was that the energetic electron density was somewhat depleted within the magnetosheath. The investigators suggested that this was due to evacuation of the magnetic flux tubes which were compressed while passing near the ionopause. In the downstream region the plasma data (Yeates *et al.*, 1978) indicated intermittent regions of low plasma density and bulk speed in the same regions where magnetic fluctuations had been observed, and additional regions of high density and speed in the magnetically quieter regions which had been described by Lepping and Behannon (1978) as 'sheath'. Yeates *et al.* (1978) suggested that the low density and speed regions, which generally contained magnetic field oriented transversely to the flow, could be related to plasma instabilities generated by pickup of planetary ions, while the high density regions were more typical of the expected behavior for a viscous interaction between solar and planetary plasmas.

The next missions to encounter Venus were the twin orbiters Venera 9 and Venera 10, which entered orbit on October 22 and October 25, 1975, respectively. Both spacecraft had periapsis altitudes of ~ 1600 km, apoapsis altitudes near 113 000 km, and periods of roughly 48 hours. Based on downstream magnetic field measurements indicating a steady magnetotail with two lobes of opposite magnetic polarities, Dolginov *et al.* (1978) concluded that Venus had a small but measurable magnetic moment, and that the spacecraft had penetrated a magnetosphere in transiting the near-planet nightside. These findings were contested by Yeroshenko (1979), who showed that the Venera 4, 9, and 10 observations were consistent with draping of the IMF around the planetary obstacle; that author also postulated that the draped field closed in the planetary umbra. A unified view of the Venus magnetotail was later presented by Dolginov *et al.* (1981), who concluded that a comet-like draping configuration was indeed supported by the Venera 9 and 10 observations.

Venera 9 and 10, as the first Venus orbiters, presented the first opportunities for systematic mapping of the planetary bow shock. Romanov (1978) and Romanov *et al.* (1978) demonstrated that the shock location had a measurable asymmetry. The shock was found to be at lowest altitude in the plane through the planetary center defined by the solar wind **V** and **B** vectors, and highest in the orthogonal flow-aligned plane. These authors attributed the effect to the different velocities of magnetohydrodynamic (MHD) waves propagating parallel and perpendicular to the magnetic field, as had been suggested by Cloutier (1976).

An important milestone achieved with Venera 9 and 10 wide-angle plasma experiments (Gringauz *et al.*, 1976; Verigin *et al.*, 1978) was the measurement of the plasma properties of the planetary umbra and penumbra. These authors found that a conical umbra, containing a stable electron population and barely discernible ion fluxes, extended $3-4\,R_V$ downstream from the planet. The umbra was surrounded by a transition region, or penumbra, containing plasma which was less dense, slower, and hotter, than in the dayside magnetosheath. The nightside interaction region was investigated by Vaisberg *et al.* (1976) and Romanov *et al.* (1978), using the narrow-angle electrostatic plasma experiment. These authors noted the existence of a boundary layer, or 'rarefaction region', which extended as far as $5\,R_V$ downstream from the planet and apparently contained a mixture of solar and planetary plasmas. Figure 2 is a schematic of the interaction based on the Vaisberg *et al.* (1976) findings.

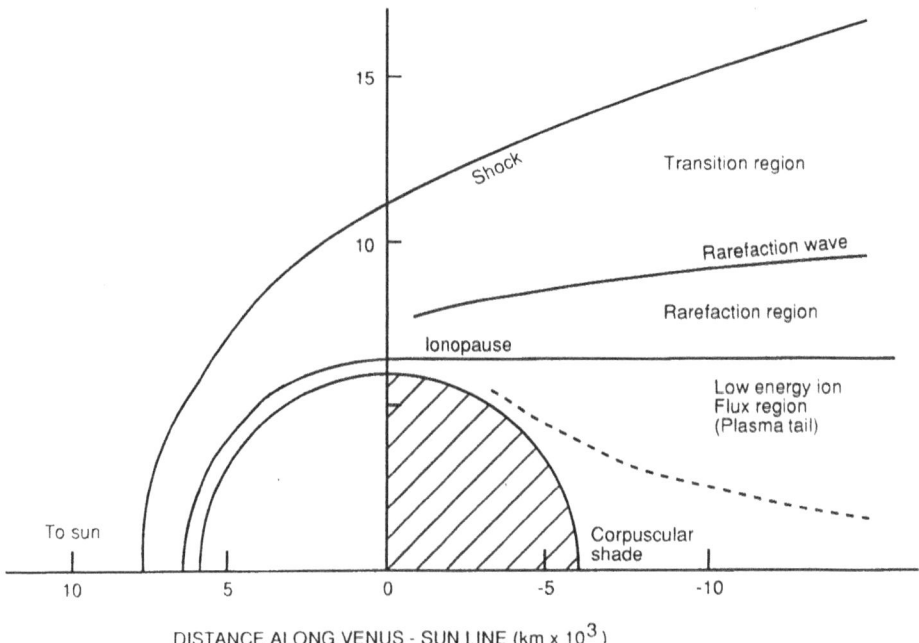

DISTANCE ALONG VENUS - SUN LINE (km x 10^3)

Fig. 2.   Axisymmetric schematic of the near-planet interaction region, based on Venera 9 and 10 terminator and nightside observations. The connection of the tail boundary and the rarefaction wave to near-planet plasma structures is unknown. Adapted from Vaisberg *et al.* (1976).

A further result of the Venera 9 and 10 missions was a re-evaluation of the Lα emissions by Bertaux *et al.* (1978), who had the advantage of linewidth, as well as intensity, measurements. These researchers found that their observations could be accounted for by a two-temperature atomic hydrogen exosphere, rather than the atomic plus molecular hydrogen model of Barth (1968).

The next suitably equipped spacecraft to encounter Venus was the Pioneer Venus Orbiter (PVO), which entered orbit on December 4, 1978. PVO, which continues to return measurements over a decade later, is responsible for many milestones in our understanding of the Sun–Venus interaction. These include: (1) observation of the complexity of the nightside and near-terminator ionosphere (e.g., Brace *et al.*, 1980,

1987; Cravens *et al.*, 1982) and analysis of possible ionization sources (e.g., Knudsen *et al.*, 1980, 1987); (2) systematic mapping and description of the bow shock (e.g., Russell *et al.*, 1981, 1985a; Tatrallyay *et al.*, 1983); (3) systematic mapping and description of the ionopause (e.g., Elphic *et al.*, 1980c; Knudsen *et al.*, 1982; Phillips *et al.*, 1988); (4) measurement of the flow field and magnetic characteristics of the magnetosheath (e.g., Mihalov *et al.*, 1980; Phillips *et al.*, 1986b); (5) Identification of picked up planetary ions in the near-planet magnetosheath (e.g., Mihalov and Barnes, 1981) and in the downstream region (e.g., Intriligator, 1982); (6) description and interpretation of the variable magnetization of the ionosphere (e.g., Luhmann *et al.*, 1980; Shinagawa and Cravens, 1988; Cloutier *et al.*, 1987); (7) Observation of an extended exosphere of suprathermal oxygen (Nagy *et al.*, 1981); (8) identification, mapping, and description of the magnetotail (e.g., Russell *et al.*, 1981; Saunders and Russell, 1986; McComas *et al.*, 1986); (9) observation and analysis of solar cycle effects (e.g., Knudsen *et al.*, 1987; Russell *et al.*, 1988).

Additionally, the rich PVO data set has promoted fruitful collaborations between observers, theoreticians, and numerical modelers which has contributed significantly to our understanding of global aspects of the interaction (e.g., Spreiter and Stahara, 1980; Zelenyi and Vaisberg, 1982; Mihalov *et al.*, 1982; Slavin *et al.*, 1983; Luhmann *et al.*, 1986; Breus *et al.*, 1987b; Stahara *et al.*, 1987). As much of the current understanding of the magnetosheath and magnetotail is based on PVO observations, discussion of these observations and interpretations will be left to the topical sections of this review.

The Venera 11 and 12 flyby/probe missions arrived on December 25 and December 21, 1978, respectively. The main contributions of Venera 11 and 12 to our knowledge of the Venus magnetosheath and magnetotail were improved ultraviolet photometry measurements of the Lα airglow and exospheric modeling based on these observations (Bertaux *et al.*, 1981, 1982). Subsequent missions to Venus have included Venera 13 through 16 and VEGA 1 and 2. The optimization of these spacecraft for different tasks has resulted in little new knowledge of the Venus–solar wind interaction. However, radio occultation measurements with Venera 15 and 16 (e.g., Osmolovskii and Samoznaev, 1987) have provided new data on the ionosphere in the waning portion of the solar cycle.

3. Boundaries and Building Blocks

3.1. THE SOLAR WIND AT VENUS

The solar wind plasma at 0.72 AU, while highly variable and structured, typically has a convection speed of ~ 400 km s^{-1} and number density of ~ 10–15 cm^{-3}. The frozen-in magnetic field is on average about 11 nT and is typically oriented roughly $40°$ from the Sun–Venus line in the proper sense for an Archimedean spiral of either inward or outward polarity and prograde solar rotation. The ram or dynamic pressure of the solar wind, ρV^2, which averages 4 to 5.5 nanopascals, constitutes roughly 98% of the total solar wind pressure, while the magnetic field and the plasma thermal pressure each contribute roughly 1% of the total. Sonic, Alfvén, and magnetosonic Mach numbers are typically 6.6, 6.1, and 4.7, respectively (Phillips *et al.*, 1986b), based on ion parameters and magnetic fields observed by PVO near solar maximum and assuming a constant solar wind electron temperature of 1.5×10^5 K. The reader is referred to Russell *et al.* (1988) for histograms of solar wind dynamic pressure and magnetosonic Mach number representing a wide range of solar activity. The motion of Venus results in an average 5° offset of the solar wind velocity vector from the Sun–Venus line in the planetary frame, such that the subflow point is typically offset 5° locally westward from the subsolar point, and such that the magnetotail lags the planet azimuthally by the same angle. As the rotation of Venus is retrograde, the dusk side leads planetary motion and the dawn side lags, so that the subflow point is offset duskward.

3.2. THE BOW SHOCK

The solar wind is deflected around the planetary obstacle by a fast magnetosonic shock wave. This bow shock, which appears as a standing (though variable) structure in the planetary reference frame, also slows and heats the solar plasma. Two aspects of the bow shock, its strength and its position, will be reviewed here, as they provide information about the nature of processes occurring in the magnetosheath and near the ionopause.

The strength of the shock, defined here as the ratio of upstream to downstream field and particle parameters, has been described by Russell *et al.* (1979c) and Mihalov *et al.* (1980). Both studies concluded that the shock was weaker than was predicted by gasdynamic modeling (Stahara and Spreiter, 1976; Spreiter *et al.*, 1970) for the observed upstream conditions. These observations provided part of the impetus for theorizing that a small fraction of the solar wind may be absorbed by the ionosphere (Gombosi *et al.*, 1980), or that charge exchange with the neutral exosphere may more significantly deplete the solar plasma flux (Gombosi *et al.*, 1980, 1981), thus providing momentum and energy loss mechanisms on the downstream side of the shock. However, subsequent analysis by Tatrallyay *et al.* (1984) showed that incorporation of MHD effects and the use of the appropriate upstream Mach number results in predicted magnetic field shock jumps which are in reasonable agreement with the observations. Figure 3 demonstrates this agreement. While further work remains to be done on this subject (for example, the lack of electron temperature measurements for the PVO shock crossings requires the assumption of a temperature model), it appears at present that there is no overriding discrepancy in shock jump predictions and observations which would require that non-MHD processes be at work inside the shock.

The position of the bow shock has been extensively measured, and much modeling and analysis has been devoted to explaining the shock shape and to inferring the shape of the underlying obstacle. By far the largest data set of shock crossings is that of PVO, which includes subsolar, near-terminator, and downstream measurements. The near-terminator crossings have been examined by PVO investigators in a series of papers (Tatrallyay *et al.*, 1983, 1984; Alexander and Russell, 1985; Alexander *et al.*, 1986; Russell *et al.*, 1979c, 1988); findings of interest here are those concerning modulation

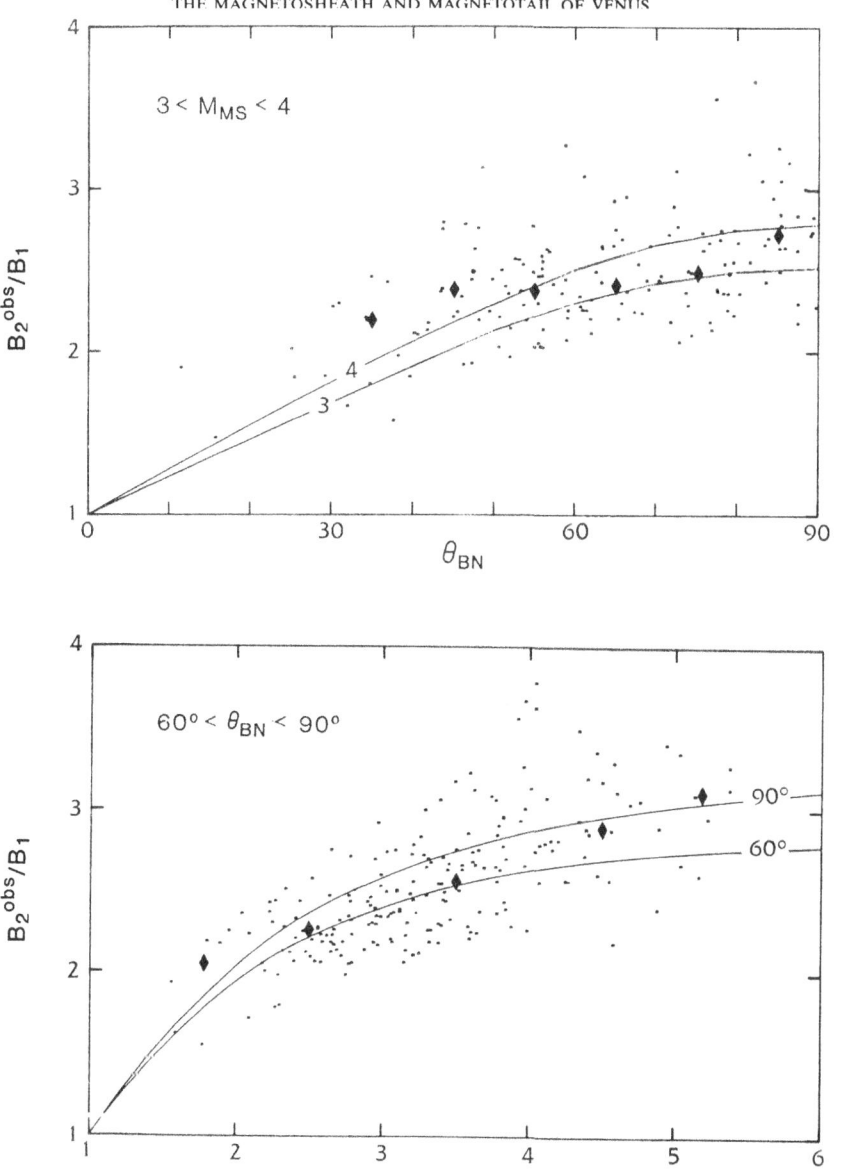

Fig. 3. (*Top*) PVO observations (dots) of jumps in total magnetic field at the bow shock versus the angle θ_{BN} between the shock normal and the upstream magnetic field, for magnetosonic Mach number of 3–4. Diamonds are medians of θ_{BN} bins. Lines are calculated MHD shock jumps for M_{MS} of 3 and 4, for $\gamma = 1.85$. (*Bottom*) Observed shock jumps in magnetic field versus M_{MS} for quasiperpendicular shock geometry (θ_{BN} between 60° and 90°). Lines show calculated MHD shock jumps for θ_{BN} of 60° and 90°. From Tatrallyay *et al.* (1984).

of the shock location by (1) solar cycle and solar EUV flux, (2) upstream solar wind parameters, and (3) orientation of the IMF. The conclusions of these papers have been updated and synthesized in the most recent study (Russell *et al.*, 1988), which found that the terminator shock position varies dramatically with the solar cycle. At solar minimum, when Venera 9 and 10 observed the shock, it was much closer to the planet than at solar maximum, just after orbital insertion of PVO. Much later in the PVO mission, nearing solar minimum, the shock position had returned to the Venera 9 and 10 location; this solar cycle variation is shown in Figure 4. Alexander and Russell (1985) and Russell *et al.* (1988) attributed this effect to modulation of the neutral atmospheric scale height, and the thus of the mass-loading rate, by solar EUV flux. The terminator bow shock position has also been shown to respond to solar wind parameters such as magnetosonic Mach number and dynamic pressure (Russell *et al.*, 1988); Figure 5 illustrates these responses.

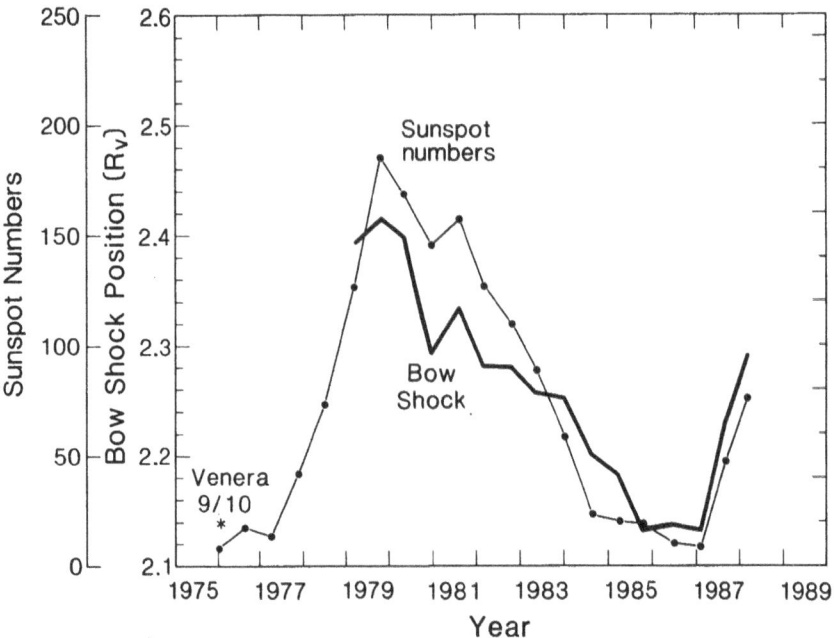

Fig. 4.   The location of the bow shock, normalized to the terminator, measured by PVO through 1987 (heavy trace). Lighter trace shows the sunspot number averaged over each PVO observing season of roughly 100 Earth days. The asterisk at lower left marks the bow shock location observed by Venera 9 and 10. Figure courtesy of C. T. Russell.

Effects on the bow shock of the orientation of the IMF are somewhat more subtle. Alexander *et al.* (1986) and Russell *et al.* (1988) demonstrated shock asymmetries which appear to result from the pickup of planetary ions by the solar plasma. These studies found that the shock is on average closer to the planet when upstream **B** and **V** are aligned than when they are crossed, and is closer to the planet in the hemisphere of

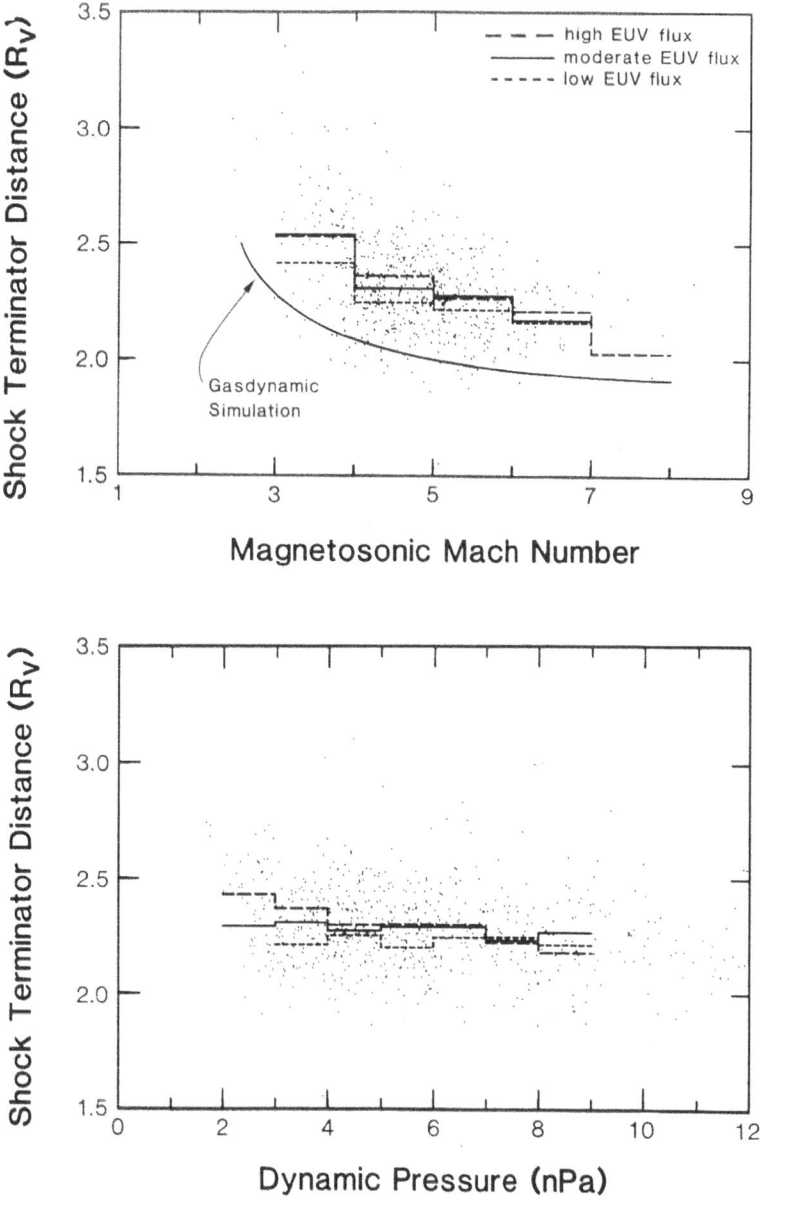

Fig. 5.   The location of the bow shock, normalized to the terminator, measured by PVO through April 1987, as functions of magnetosonic Mach number (*top*) and solar wind dynamic pressure (*bottom*). Each panel has three traces, corresponding to conditions of high, moderate, and low measured EUV fluxes, as shown in top panel legend. The smooth curve in the top panel shows the shock position for a gasdynamic simulation with $\gamma = 1.67$. From Russell *et al.* (1988).

locally downward convective electric field ($\mathbf{E} = -\mathbf{V} \times \mathbf{B}$) than in the upward $\mathbf{E}$
hemisphere. The first of these two magnetic field-related asymmetries is shown in the
top panel of Figure 6. Terminator shock position is plotted versus the cosine of the IMF
cone angle, or the angle betwen the upstream $\mathbf{V}$ and $\mathbf{B}$, for polar and equatorial shock

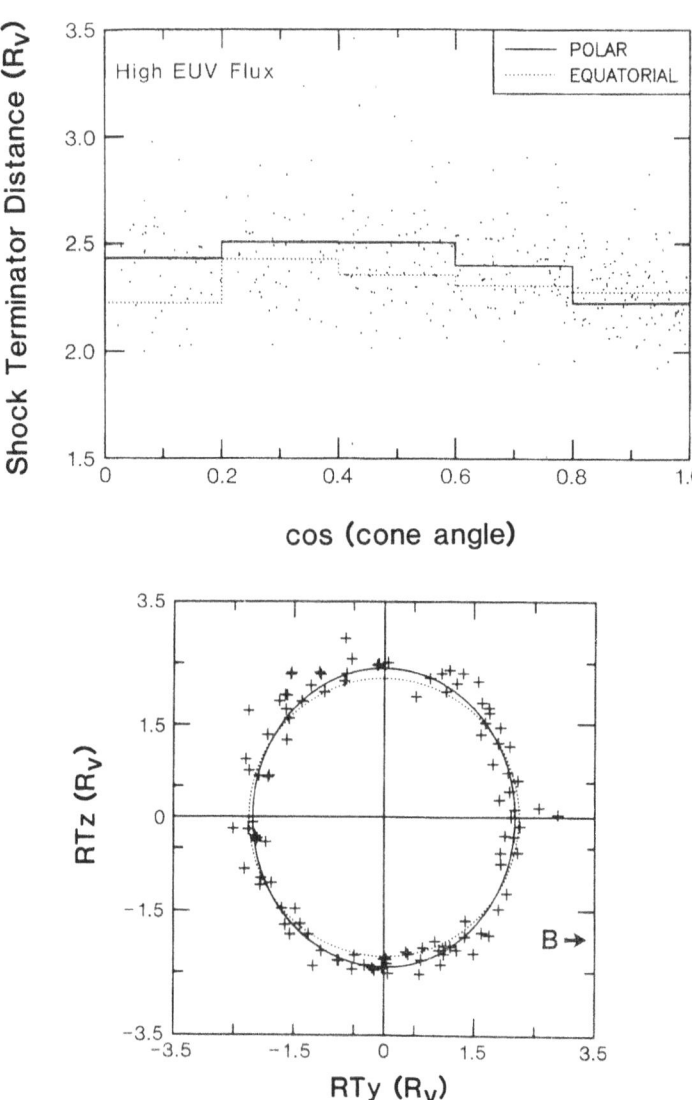

Fig. 6. Asymmetries in the location of the bow shock, as observed by PVO through April 1987. (*Top*)
Terminator shock position during periods of high EUV flux as a function of the cosine of the IMF cone angle,
that is the acute angle between the solar wind streamlines and the IMF field lines. (*Bottom*) Terminator shock
position (crosses) projected on the terminator plane, in a coordinate system with the cross-flow component
of the IMF oriented in the + Y direction. Data are restricted to crossings with cone angle of at least 78.5°.
The solid ellipse is a fit to the plotted observations, while the dotted circle is a similar cross-section for shock
crossings (not shown) with low cone angle. From Russell *et al.* (1988).

crossings during periods of high EUV flux. Note that the polar shock altitude varies directly with IMF cone angle. This trend is less pronounced during periods of lower EUV flux. The cone angle effect was attributed by Alexander *et al.* (1986) and Russell *et al.* (1988) to the increased efficiency of ion pickup when **B** and **V** are mutually perpendicular and thus the motional electric field is maximum, and conversely to the relatively inefficient ion pickup for a quasi-parallel (low cone angle) geometry.

A consequence of the varying speed of propagation of a magnetosonic wave, with fastest propagation perpendicular to the magnetic field, is that the cross section of the bow shock should be somewhat elliptical, with the major axis in the $V \times B$ direction (Cloutier, 1976). This effect was seen initially in the Venera 9 and 10 observations by Romanov (1978), and has been confirmed in the PVO crossings by Russell *et al.* (1988). The bottom panel of Figure 6 shows the shock cross section for large IMF cone angles in a coordinate system for which the upstream cross-flow magnetic field points to the right. The crosses represent observed shock crossings, while the solid curve is a best fit ellipse. Note that the shock cross section is elliptical, with the major axis alignment as predicted above. The dotted trace, showing average shock position for low IMF cone angles, is circular, indicating that the asymmetry vanishes when the upstream **B** and **V** are aligned. As noted previously, the shock altitude is also lowest for low cone angles. Russell *et al.* (1988) postulated that the small and nearly circular cross section for coaligned **B** and **V** is due to the smallness or lack of ion pickup effects for these conditions.

Smirnov *et al.* (1981) observed an additional morning-evening asymmetry in Venera 9, 10, and PVO shock crossings, and postulated that the entire shock was aberrated by $\sim 7°$ (in addition to the $\sim 5°$ aberration created by planetary motion). Such an aberration had been predicted by Walters (1964) for the terrestrial bow shock as a result of the prevailing IMF spiral orientation and its effect on shock jump relations. For typical IMF configurations the Walters effect would be observable as a lower near-terminator shock altitude on the leading side of the planet (quasi-parallel geometry) than on the lagging side (quasi-perpendicular geometry). Subsequent analysis of PVO shock crossings (Tatrallyay *et al.*, 1983; Russell *et al.*, 1988) has not included a test to isolate such an effect while excluding other asymmetries. It appears at this time that the Walters asymmetry, while not obviously supported by PVO observations, has not been conclusively disproven.

The overall shape of the shock, that is its standoff distance and flaring angle, can provide some clues about the nature of the effective planetary obstacle. This notion has been explored in detail by Slavin *et al.* (1979a, b, 1980, 1983) and Slavin and Holzer (1981), who generally found that the shock was blunter, or more flared, than predicted by gasdynamic modeling. Although the near-terminator shock had been well sampled, the subsolar shock position had not been directly measured until the evolution of the PVO orbit caused the spacecraft to graze or cross the shock near periapsis on the dayside. Russell *et al.* (1985a) found that the prevailing standoff altitude for the subsolar shock was 2280 km. When this altitude is used as an input for modeling the overall shock shape use of the observed ionopause surface as the obstacle to the flow results in calculation of a near-terminator shock position which is considerably lower than the observed position, with greatest altitude discrepancy near solar maximum. The authors inferred that the effective obstacle in the near terminator region was about 800 km and 2500 km above the ionopause near solar minimum and solar maximum, respectively.

The position of the downstream bow shock was studied by Russell *et al.* (1981) using multiple crossings by PVO and Venera 9 and 10, and single shock observations by Mariner 5 and Venera 4 and 6. Figure 7 shows the observations used in this study, as well as two conic fits based solely on the PVO data. The study found that the flaring angle of the downstream shock is about 22°, corresponding to a somewhat low magnetosonic Mach number of 2.7. Thus it is not surprising that Mariner 10 observed no distinct shock crossings in its pre-encounter transit up the planetary wake.

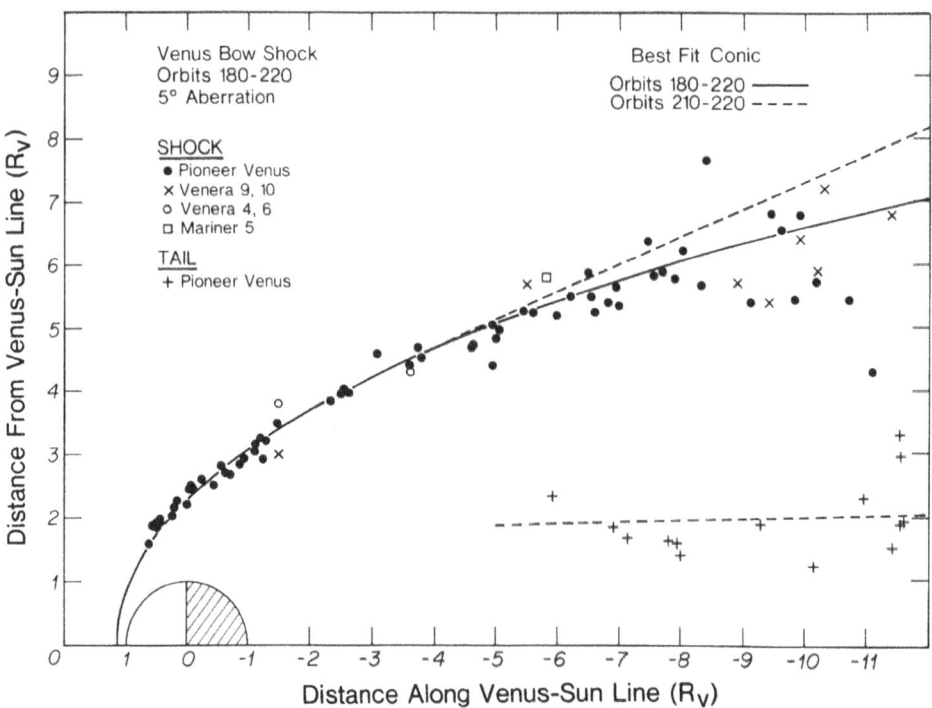

Fig. 7. Axisymmetric display of the downstream bow shock crossings observed by various spacecraft, as listed in legend, as well as PVO tail crossings. Solid curve is a fit to all PVO crossings shown, while the dashed curve is a fit to a restricted set of crossings. From Russell *et al.* (1981).

The possibility that the solar wind may be modified by planetary effects upstream of the bow shock has received recent attention. Brace *et al.* (1985) and Brace (1987) noted that the PVO electron temperature probe observed increases in detector current upstream of the shock and suggested that mass loading by newly created planetary ions could result in formation of a shock precursor region. However, Knudsen *et al.* (1989)

found no evidence for upstream mass loading, based on analysis of PVO retarding potential analyzer observations. These authors concluded that the apparent shock precursor was actually a result of variation in the density of the electron sheath trapped by the electrostatic potential of the spacecraft. Moore *et al.* (1989) noted the existence of a suprathermal ion population just outside of the bow shock, and evaluated the possibility that planetary pickup processes were modifying the solar wind upstream of the shock. They concluded that the observed ions were unlikely to be newly created planetary ions, and that the most likely explanation for the anomalous population was shock-related field aligned beams of solar wind ions similar to those observed at Earth, or possibly solar or planetary ions which had leaked upstream from the magnetosheath.

Winske (1986) used numerical simulations to test the hypothesis that low frequency magnetic field fluctuations observed downstream of the shock were caused by instabilities, possibly occurring upstream of the shock, related to beams of newly ionized planetary oxygen. That study concluded that the observed fluctuations were most likely caused by the shock itself, and not by ion pickup effects, but that the presence of an unusually high upstream population of planetary O^+ could cause broadening of the shock transition. Such a pickup effect on the shock has so far not been supported observationally. Thus, the overall perturbation of the upstream solar wind by the planet appears to be at most a minor effect and to be most likely due to shock-related processes rather than to upstream pickup of planetary ions.

3.3. THE IONOPAUSE

The upper boundary of the ionosphere, or ionopause, has been defined in many ways and is often difficult to resolve. Concepts such as the boundary layer, the plasma mantle, and the magnetic barrier make identification of a boundary between solar and planetary plasmas somewhat difficult. However, for the purpose of describing the basic shape of the ionosphere we will initially consider the ionopause to be an impenetrable surface separating the shocked convecting solar plasma of the magnetosheath and the partially ionized planetary ionosphere. A distinct ionopause was first noted in the Mariner 5 radio occultation observations (Mariner Stanford Group, 1967). While on the dayside the plasma density dropoff at the ionopause is generally abrupt and well-defined, on the nightside it is often ragged and filamentary in nature, with rays of ionospheric plasma extending downstream (e.g., Brace *et al.*, 1987). There has been a continuing controversy concerning the magnetization of the ionosphere, and the ramifications for the morphology of the ionospheric shielding current system (cf. Shinagawa and Cravens, 1988; Cloutier *et al.*, 1987). We will defer discussion of this issue to the later section on the Magnetic Barrier.

The basic shape of the ionopause, as defined by gradients in plasma density measured by the PVO Electron Temperature Probe and Retarding Potential Analyzer experiments near solar maximum (e.g., Brace *et al.*, 1980; Knudsen *et al.*, 1982), is a blunt conic, with altitude of ~ 300 km near the subsolar point, rising to $\sim 900-1000$ km near the terminator. The variability of the ionopause position generally increases with increasing solar zenith angle. Figure 8 shows the median observed ionopause position, as a

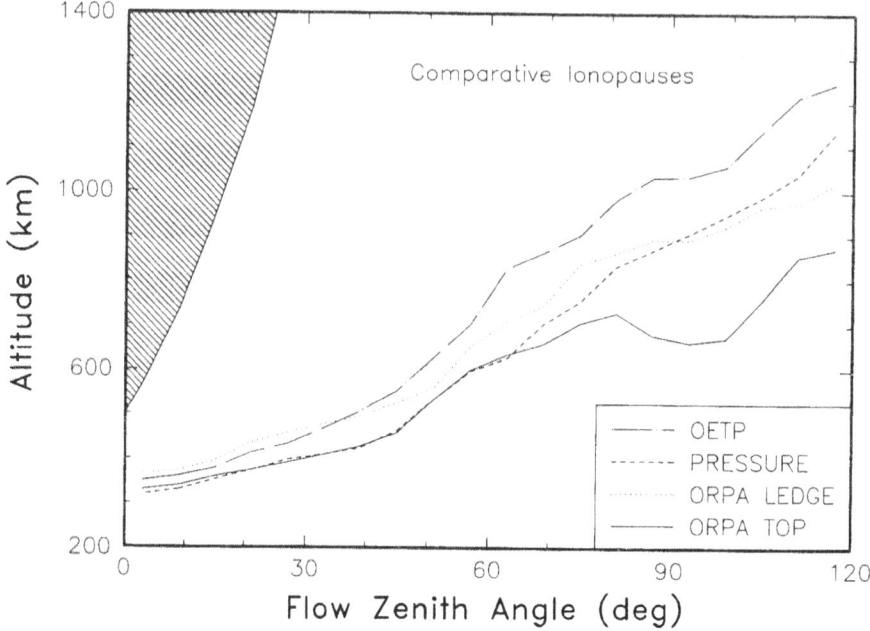

Fig. 8.   Contours in altitude-flow zenith angle space showing 50% probability of the PVO spacecraft being above or below the ionopause. Four ionopause definitions were used: (1) the altitude where the PVO Electron Temperature Probe (OETP) measured an electron density of 100 cm^{-3}; (2) where the thermal pressure and magnetic pressure were equal; (3) where the PVO Retarding Potential Analyzer (ORPA) measured an ion density of 100 cm^{-3} ('ledge'); (4) where the ion density measured by the ORPA began to deviate from that expected for hydrostatic equilibrium ('*top*'). Hatched area at upper left indicates no spacecraft coverage. From Phillips *et al.* (1988).

function of 'flow zenith angle', which is similar to SZA but incorporates aberration due to planetary motion. The ionopause altitude near is highest when upstream solar wind pressure is low, and decreases with increasing solar wind pressure to a lower limit near 200 km in the subsolar region (e.g., Elphic *et al.*, 1980b). The reader is referred to studies by Luhmann *et al.* (1987b) and Mahajan *et al.* (1989) for details of the low altitude ionopause configuration. Phillips *et al.* (1988) found that no significant dawn-dusk asymmetries exist in the ionopause shape, but that there is an easily observable asymmetry related to pickup of planetary ions; this effect will be discussed in a subsequent section.

   The long time history of *in situ* and remote observation of the Venus ionosphere has enabled analysis of solar cycle effects. Knudsen *et al.* (1987) and Knudsen (1988) found, via comparison of PVO retarding potential analyzer and radio occultation measurements, that the ionopause was much lower (250 km rather than 625 km) at solar minimum than at solar maximum for SZA of 65°. This effect is shown schematically in Figure 9. They inferred that the ionopause variation was caused by reduction in peak ionospheric thermal pressure to a level below the prevailing solar wind dynamic pressure, and that the reduced ionopause altitude should prevail throughout the dayside.

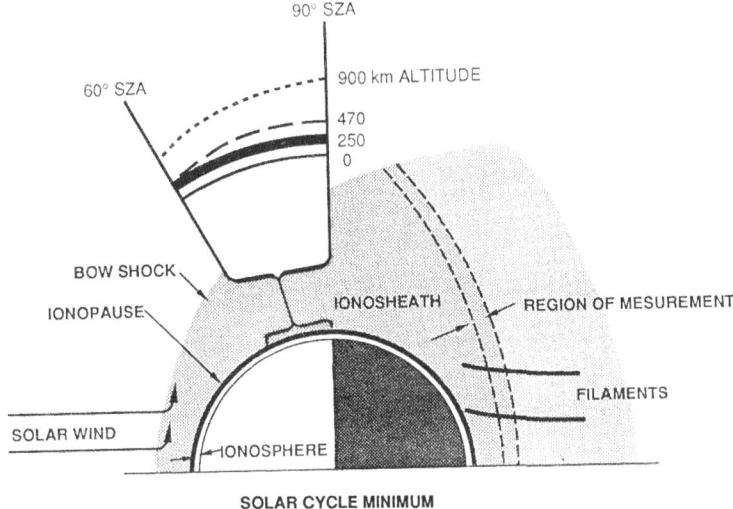

Fig. 9.   Schematic representation of Venus ionospheric morphology at solar minimum (*top*) and maximum (*bottom*). The expanded view in the top panel illustrates the prevailing ionopause position between 60° and 90° SZA at solar minimum (heavy solid trace), near the middle of the waning portion of the solar cycle (dashed trace), and solar maximum (dotted trace). From Knudsen (1988).

Thus, while the ionopause has not been extensively sampled *in situ* for solar minimum (due to rising PVO periapsis altitude), it is reasonable to conclude that the typical altitudes listed in the previous paragraph are valid only for periods of high solar activity.

As will be addressed in the later discussion on the magnetic barrier, it is also likely that the nature of the magnetosheath-ionosphere interface is different near solar minimum, and perhaps is similar to the boundary observed at solar maximum during periods of unusually high solar wind pressure. Knudsen *et al.* (1987) found a solar cycle-related variation in the nature of the plasma transition at the ionopause, based once again on PVO radio occultation and retarding potential analyzer results. At solar maximum, the ionopause, at least for low to moderate solar wind pressure, was observed as an abrupt cutoff in ionospheric plasma density. At solar minimum, however, the ionopause was generally poorly defined. This change is similar to that between the occultation measurements by Mariner 5 near the previous solar maximum (Mariner Stanford Group, 1967), when the ionopause was observed as a clean transition, and those by Mariner 10 near solar minimum (Fjeldbo *et al.*, 1975), when the ionospheric density profile was observed to be much more ragged.

3.4. THE NEUTRAL EXOSPHERE

Prior to the PVO and Venera 11 and 12 missions, Venus was known to have a hydrogen exosphere with a scale height larger than expected for thermal atomic hydrogen, based

Hot Oxygen Densities at Venus

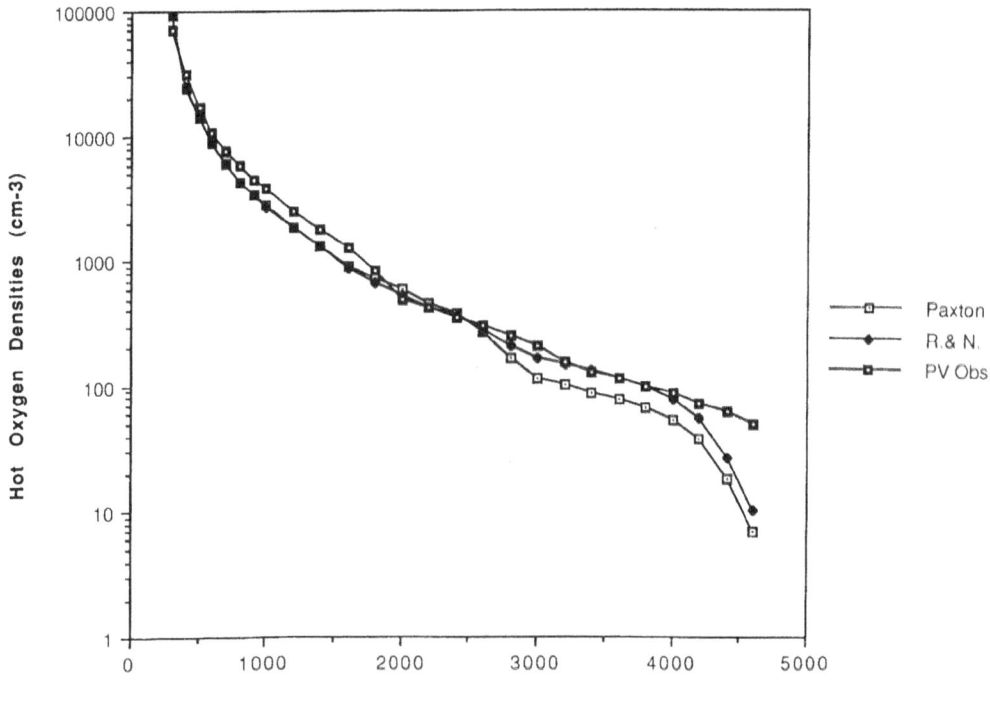

Altitude (km)

Fig. 10.   Density profile of hot exospheric atomic oxygen, showing values for solar maximum based on (1) calculations using branching ratios for dissociative recombination of O_2^+ adopted from Paxton (1983); (2) calculations using branching ratios from Rohrbaugh and Nisbet (1973); (3) PVO ultraviolet spectrometer observations reported by L. J. Paxton and A. I. F. Stewart. From Nagy and Cravens (1988).

on Lyman-α measurements from earlier Mariner and Venera spacecraft (Barth *et al.*, 1967; Bertaux *et al.*, 1978). Subsequent analysis of Venera 11 and 12 (Bertaux *et al.*, 1982, 1985) determined that the nonthermal component was most likely atomic hydrogen with a temperature of roughly 10^3 K and a density of 10^3 cm^{-3} at a 250 km exobase. Cravens *et al.* (1980) found that the major source of the hot hydrogen on the dayside was an exothermic reaction involving molecular hydrogen (e.g., Kumar *et al.*, 1981), while on the nightside charge exchange of H$^+$ with atomic oxygen and hydrogen (Hodges and Tinsley, 1981) contributed significantly. A more significant result in terms of modification of the solar wind interaction was the discovery of hot exospheric oxygen (Nagy *et al.*, 1981), based on PVO ultraviolet spectrometer observations at 1304 Å. The main source of the hot oxygen was shown to be dissociative recombination of O$_2^+$ ions. Atomic oxygen is the main constituent of the exosphere up to perhaps 3000 km, above which hydrogen dominates, though the far greater mass of the oxygen suggests that it is the dominant constituent in exchanging momentum and energy with the solar wind. Figure 10 (Nagy and Cravens, 1988) shows revised calculations for the exospheric oxygen density.

4. The Dayside and Near-Terminator Magnetosheath

4.1. MACROSCOPIC PROPERTIES

Direct observations of the plasma and field properties of the magnetosheath have been made by Venera 9, 10, and PVO. For the most part, the trajectory and instrumental characteristics of PVO, as well as the duration of its mission, have resulted in superior observations in the dayside and high-altitude magnetosheath, while the Venera 9 and 10 measurements are superior in some respects in the umbra and just above the terminator ionopause. The prevailing magnetic configuration of the magnetosheath was explored by Phillips *et al.* (1986b), using a large set of PVO observations from near solar maximum. This study found that the magnetic field magnitude within the magnetosheath responds most directly to the dynamic pressure of the solar wind, as shown in Figure 11 (top). The compression of the field, however, relative to the upstream IMF magnitude, is primarily controlled by upstream magnetosonic Mach number, as shown in Figure 11 (bottom). The draping of the magnetic field is observable from the bow shock all the way down to the ionosphere itself, which supports the concept of an induced magnetosphere, as had been previously established by various researchers (e.g., Russell *et al.*, 1979b; Elphic *et al.*, 1980b; Marubashi *et al.*, 1985; Luhmann *et al.*, 1986). Figure 12 shows the average field draping pattern for data rotated into planes defined by the solar wind **V** and **B** vectors. A further finding of Phillips *et al.* (1986a, b) was that the average magnitude of the magnetosheath field was greater in the hemisphere of locally upward convective electric field than in the opposite hemisphere. This was interpreted as evidence for greater slowing of the solar wind flow in the upward **E** hemisphere as a result of greater mass loading by planetary ions; this concept will be explored further in a later section.

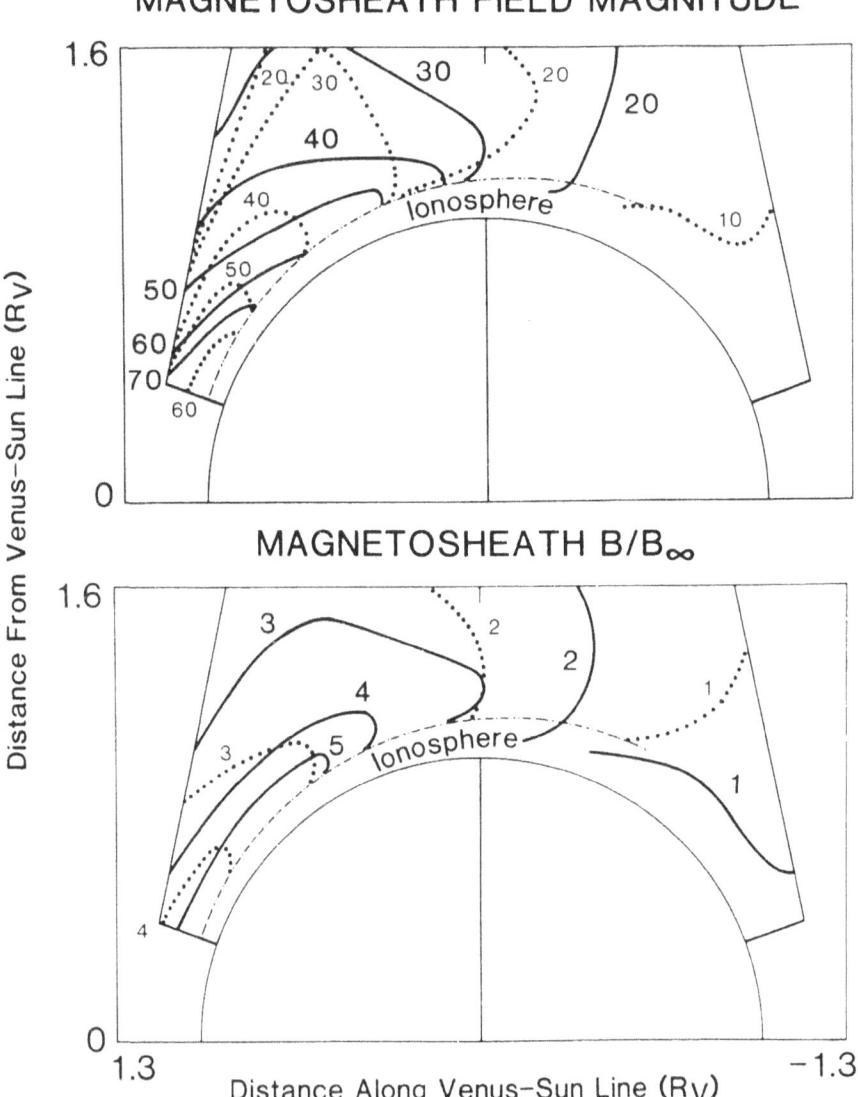

Fig. 11.   (*Top*) Contours of magnetosheath magnetic field magnitude, in an axisymmetric projection, from PVO observations near solar maximum. Solid lines represent intervals of high solar wind dynamic pressure (> 5.2 nPa), while dotted lines are for low pressure (< 3.5 nPa). (*Bottom*) Similar display of magnetic field compression, B/B_∞, for intervals of $M_{MS} > 5.5$ (solid lines) and $M_{MS} < 4.0$ (dotted lines). From Phillips *et al.* (1986b).

The magnetosheath flow field has been mapped by Venera 9 and 10 and PVO researchers. While the PVO plasma analyzer requires long integration times to generate a single ion energy per charge (*E*/*q*) spectrum, and hence is hampered by time aliasing due to the spacecraft motion through the magnetosheath, the large number of PVO

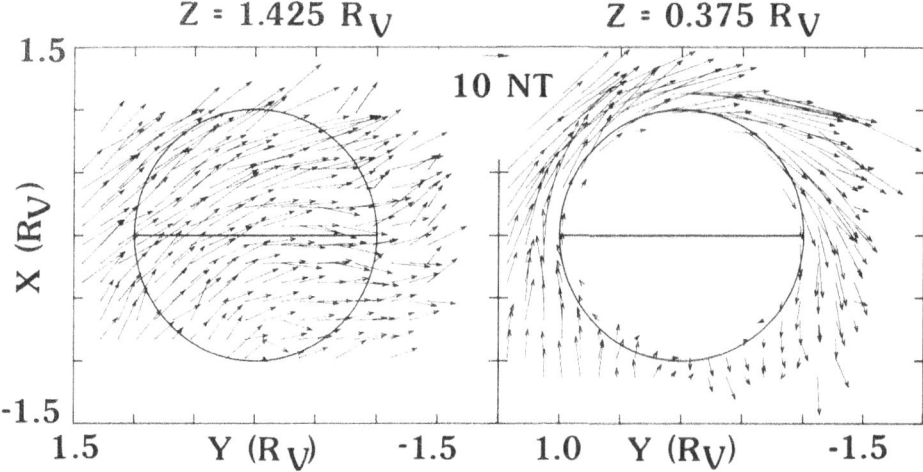

Fig. 12. Averaged magnetic field vectors observed by PVO near solar maximum, in a coordinate system in which the upstream **V** and cross-flow **B** vectors are in the $-X$ and $-Y$ directions, respectively. Data are folded into a single $+Z$ hemisphere; each panel represents a Z position range of $\pm 0.075\,R_V$ from the position indicated at top. From Phillips *et al.* (1986a).

orbits partly compensates for this shortcoming. Figure 13 shows the magnetosheath flow field as measured by the PVO plasma analyzer (Mihalov *et al.*, 1980) in an axisymmetric projection. These authors noted only minimal suggestion of deflection of the sheath flow into the planetary wake, and also noted that the density of the solar plasma was frequently below the instrument sensitivity threshold for observations downstream from the terminator but well outside the optical shadow of the planet.

Systematic mapping of plasma density and temperature within the dayside magneto-sheath has not been done, and indeed may be impossible with the available observations due to the long integration times required by the PVO plasma analyzer. A limited model-observation comparison for near-terminator magnetosheath plasma parameters was carried out by Mihalov *et al.* (1982) using the gasdynamic code of Spreiter and Stahara (1980). This study found that the simulation provided reasonably reliable predictions of ion velocity, density, and temperature, but that the upstream Mach number had to be artificially lowered to reproduce the observed shock position. Another result was that the observed flow tended to be slightly slower than the modeled flow near the ionopause, perhaps suggesting a type of viscous boundary layer between the two plasma regimes.

The limitations of the available observations of bulk plasma properties within the magnetosheath have focused the comparison of observations and models primarily on the magnetic field and on the location of the bow shock for given ionospheric characteristics. As mentioned previously, modelers have consistently found that the pronounced flaring of the shock suggests effects in the magnetosheath that have thus far not been incorporated into the current gasdynamic simulations. These effects might include

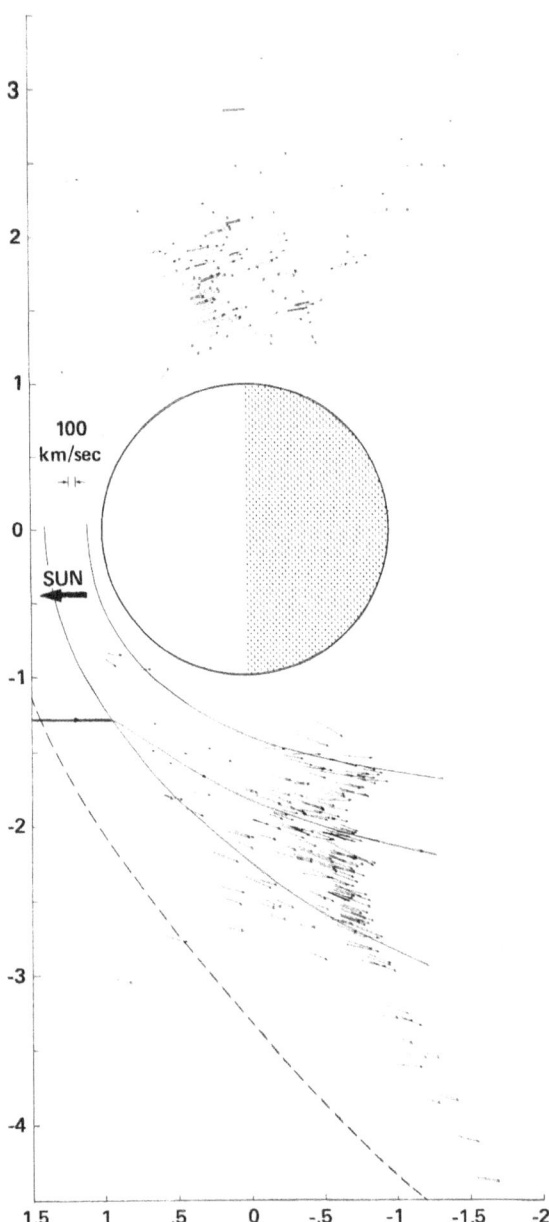

Fig. 13.   Magnetosheath flow field measured by the PVO Plasma Analyzer near solar maximum. The projection is axisymmetric, with each velocity vector projected onto the plane through the planetary center defined by the measurement position vector and the Sun. The upper (lower) half of the figure contains vectors measured north (south) of the Venus orbital plane. The three solid traces represent a model ionopause shape and the corresponding bow shock shape and a representative streamline for Mach 8. The dashed trace shows the modeled Mach 2 bow shock. From Mihalov *et al.* (1980).

MHD processes, which can also create asymmetries ordered by the magnetic field, as well as pickup of planetary ions by the magnetosheath plasma. Within these limits, Luhmann *et al.* (1986) have demonstrated that a gasdynamic simulation, with magnetic field derived from the distortion of fluid elements, can be quite successful in predicting the draping of the magnetosheath field in both time-stationary and simple time-dependent scenarios. Figure 14 shows an example of the observed and modeled magnetosheath field configurations from this study. Although the gasdynamic simulations cannot reproduce all the pertinent physics of the interaction, their success in explaining the observations is such that these models remain a valuable tool in enhancing our understanding of magnetosheath fluid characteristics.

4.2. TURBULENCE AND WAVES

The presence of fluctuating magnetic fields in the magnetosheath has been observed by virtually all the spacecraft missions penetrating this region. Greenstadt (1970) noted that the field was relatively steady when Mariner 5 entered the magnetosheath but was much more disturbed prior to the outbound bow shock crossing. Based on the observed field orientations, he concluded that the inbound crossing had occurred in a quasi-perpendicular portion of the shock while the outbound crossing occurred where **B** was nearly aligned with the shock normal, and that fluctuations were probably associated with the quasiparallel shock region. Lepping and Behannon (1978) noted field fluctuations in the Mariner 10 observations which appeared to be modulated by the magnetic field orientation, and suggested that the sheath itself was relatively quiet, while the wake was more disturbed. However, they acknowledged the possibility that the observed fluctuations had been convected downstream from the quasi-parallel portion of the shock, or that the spacecraft had sampled regions adjacent to the quasi-parallel regions of the downstream shock as a result of flapping of the interaction region. Plasma observations from the same region (Yeates *et al.*, 1978) also indicated downstream turbulence in regions characterized by low bulk speeds, low densities and magnetic field aligned transverse to the flow. These authors suggested that the fluctuations could result from instabilities associated with the ion pickup process.

Systematic mapping of the disturbed regions of the near-planet magnetosheath was performed for PVO magnetic field data on a case study basis by Luhmann *et al.* (1983), and for a large statistical data set by Phillips *et al.* (1986b). The former study noted that the fluctuations were typically left-hand polarized hydromagnetic waves with periods of 10–40 s. Luhmann *et al.* concluded that the fluctuating fields were indeed due to waves generated at the quasiparallel portions of the bow shock. Figure 15 shows examples of the field fluctuations and the corresponding shock geometry from a subsequent work by Luhmann *et al.* (1987a). The Phillips *et al.* (1986b) study showed that the fluctuations were on average more pronounced when the upstream velocity and magnetic field vectors were aligned, and that for these low cone angle cases the fluctuations peaked in the subsolar magnetosheath and showed no obvious field-controlled asymmetry in location. However, when the IMF cone angle was higher, the fluctuating fields were much more prevalent on streamlines connected to the quasiparallel part of the shock

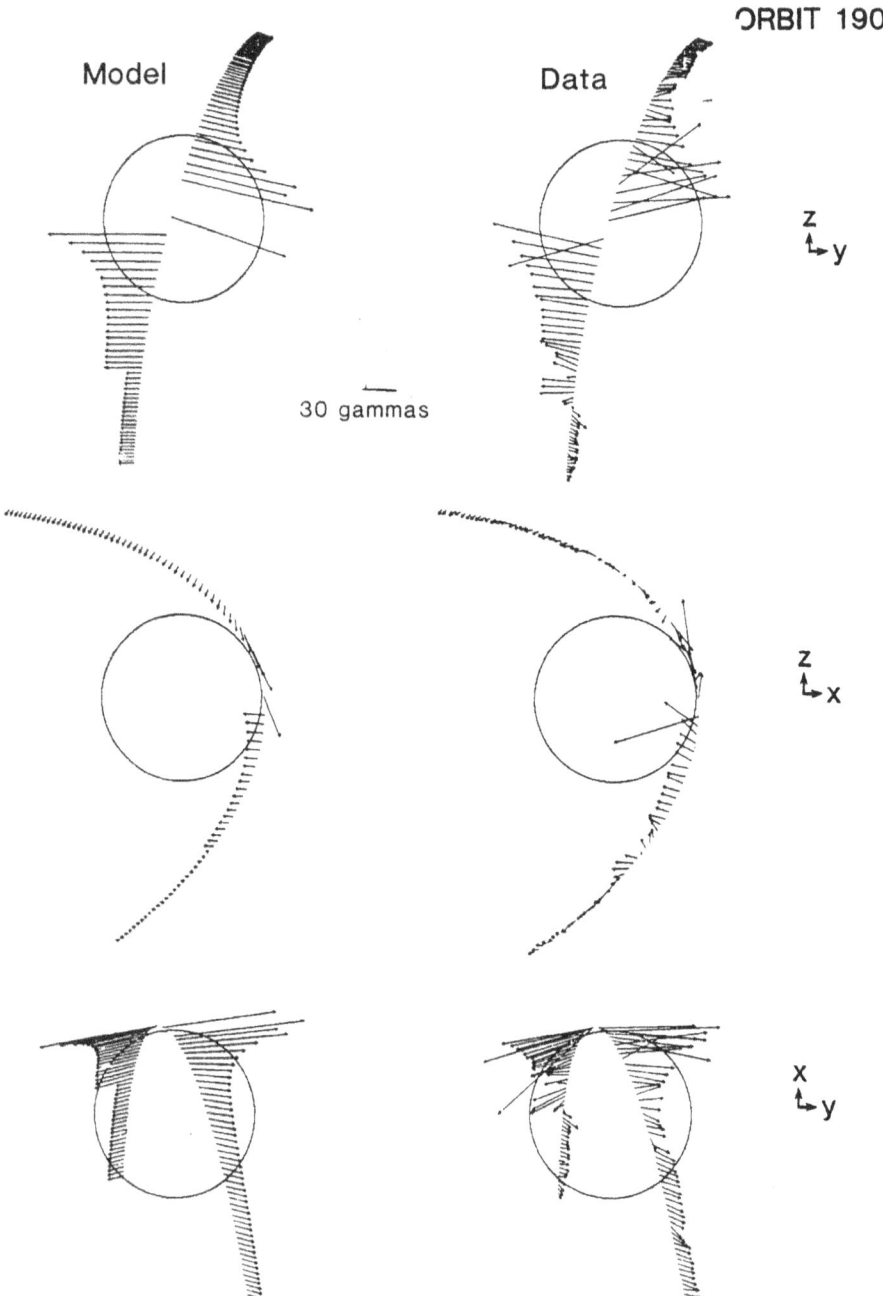

Fig. 14. Observed (right) and simulated (left) magnetic field draping pattern along the PVO trajectory for orbit 190. Three views are shown: from the Sun (*top*), noon-midnight (*middle*) and from orbital North (*bottom*). The simulated field vectors are from a gasdynamic model, with magnetic field calculated from the distortion of fluid elements. A reversal in the IMF occurred while the spacecraft was near periapsis. This figure illustrates the validity of the gasdynamic approximation even for the case of temporally varying IMF. From Luhmann *et al.* (1986).

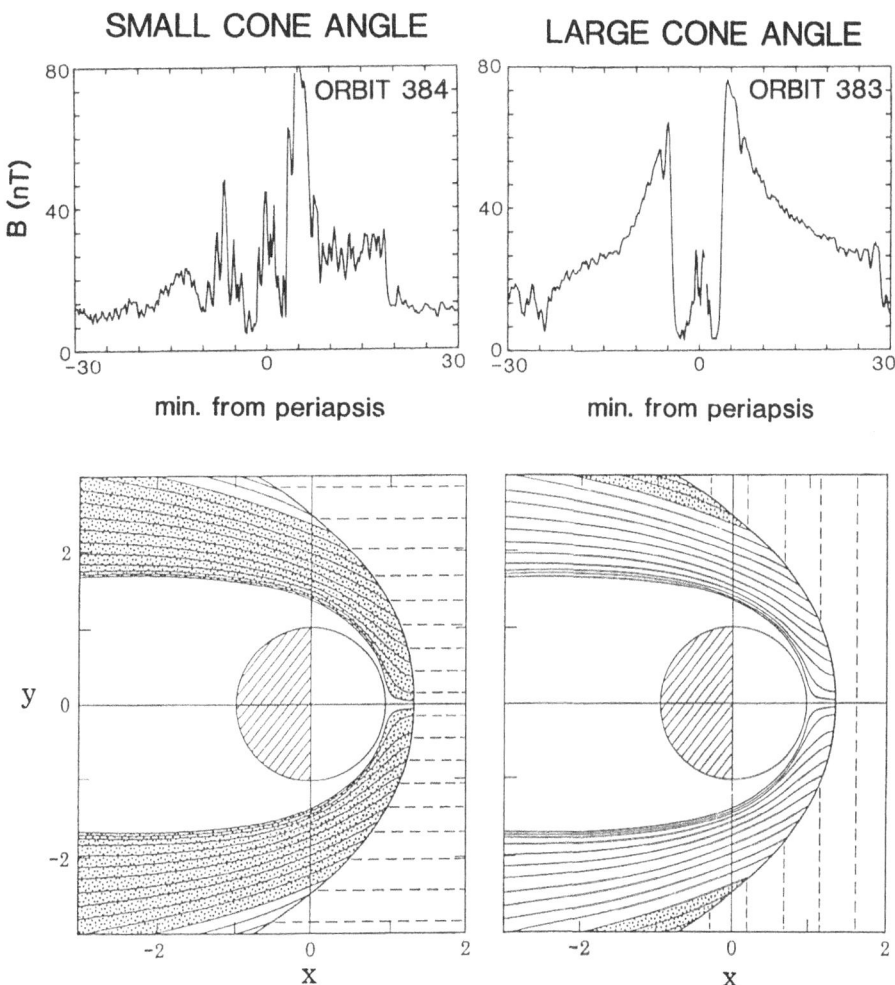

Fig. 15. (Top panels) magnetic field magnitudes vs minutes from periapsis for two PVO orbits in the subsolar magnetosheath. Low field magnitudes near periapsis are due to transit through the ionosphere. Note that the magnetosheath magnetic field is steady for the orbit with large IMF cone angle (solar wind **V** and **B** are crossed) but that fluctuations are pronounced when the cone angle is low (**V** and **B** are nearly parallel or antiparallel. This tendency was attributed to the quasiparallel subsolar shock geometry for low cone angles, shown schematically (bottom panels). The stippling shows regions downstream from the quasiparallel portions of the bow shock for upstream **B** parallel (left) and perpendicular (right) to the upstream **V**. From Luhmann *et al.* (1987).

than on the opposite side of the planet where the shock was quasiperpendicular. Note, however, that in the limit of 90° cone angle quasiparallel shock geometry exists on both flanks of the magnetosheath, as shown in Figure 15, and no spatial asymmetry in the fluctuations should be present.

To resolve the relative importance of generation of hydromagnetic waves by (1) the shock, and (2) the ion pickup process, Winske (1986) conducted a numerical simulation and concluded that the shock was the most likely source of the large-amplitude fluctuations. The case for shock generated waves appears convincing at the present time, while turbulence generated by the pickup process has thus far not been unambiguously demonstrated in the near planet region.

4.3. THE MAGNETIC BARRIER

The concept of a magnetic barrier at Venus, that is a region in which solar wind ram pressure is transmitted to the ionosphere via conversion into enhanced magnetic field and thus contains the ionosphere via shielding currents, received impetus from two early observations. First, the radio occultation measurements of the Mariner Stanford Group (1967) revealed that the dayside ionosphere had an abrupt upper boundary, while the nightside ionosphere extended to greater altitudes and apparently lacked such a boundary. This led to formulation of a model in which the resistance of the highly conductive ionosphere to the passage of the solar wind magnetic field resulted in the enhancement of that magnetic field above or within the ionosphere such that the solar wind plasma was completely deflected (Dessler, 1968; Johnson and Midgley, 1969). A second observation contributing to the magnetic barrier picture was made by the electron analyzer aboard Mariner 10, which measured a pronounced depletion in electrons with energies of 100 eV or greater in the inner magnetosheath. While this observation was originally attributed to an interaction with the exosphere, subsequent theoretical work by Zwan and Wolf (1976) suggested that plasma depletion might be a ubiquitous feature of the solar wind interaction with an obstacle. In the Zwan–Wolf model, a magnetic flux tube which drapes over an obstacle is depleted of plasma; flux tubes passing closest to the nose of the obstacle are most depleted. This scenario results in minimum density at the stagnation point, whereas a gasdynamic interaction would result in maximum density at the stagnation point. Thus one might expect the magnetosheath to be nearly devoid of plasma at just above the ionopause. In such a model, the streaming pressure of the solar wind would increasingly give way to magnetic pressure in the magnetosheath, which would in turn be withstood by plasma pressure (perhaps augmented by magnetic fields) in the ionosphere.

   Initial PVO observations (Russell *et al.*, 1979a) showed that the dayside ionosphere was quite often devoid of large-scale magnetic fields. Based on these findings, Johnson and Hanson (1979) devised a model for a shielding current system in which the currents flowed along the ionopause, but not within the ionosphere itself, and closed in the magnetosheath. This model, shown in Figure 16, contains distinct regions in which plasma energy is converted to electrical current, thus slowing the flow, and other regions in which the draped magnetic field reaccelerates the plasma. Brace *et al.* (1980) and Elphic *et al.* (1980a, b) confirmed that the total pressure at the ionopause matched the incident solar wind pressure quite well and that the ionopause was generally marked by an abrupt transition between rarefied, low-β solar plasma and high-β ionospheric plasma. This revised magnetic barrier model was refined analytically by Pivovarov *et al.* (1982), who found that their predictions for magnetic field draping agreed with PVO observations. Their analytical model predicted a dayside thickness for the magnetic

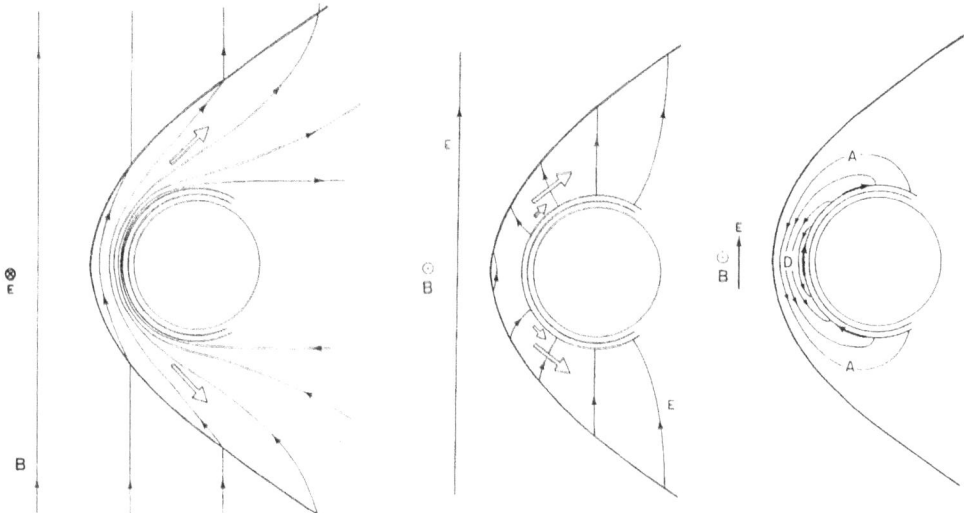

Fig. 16.   Schematic of the magnetic field draping (left), electric field (middle), and the shielding current system (right) postulated for the idealized case of an impenetrable ionosphere. Upstream **B** and **V** are assumed to be perpendicular; flow vectors are indicated by large hollow arrows, while **B** and **E** are shown vectorially by solid traces. Note in the right panel that the magnetosheath currents close at the ionopause and do not flow through the ionosphere itself. In region D of the right panel, **E** · **J** is negative and energy is transferred from the decelerating plasma to electrical currents. In region A, **E** · **J** is positive and the **J** × **B** force associated with the shielding currents accelerates the flow. From Johnson and Hanson (1979).

barrier, defined as the magnetosheath region within which the gasdynamic approximation is invalid, of ~ 600 km.

Under conditions in which the ionosphere is not magnetized, subsequent PVO observations still support this model. Figure 17 shows the relationship between upstream solar wind dynamic pressure, modified to include a small component of thermal pressure, and the sum of magnetic and ionospheric thermal pressure measured at the ionopause, for the complete PVO set of ionopause crossings (all crossings occurred near solar maximum). The ionopause was identified in this data set as the altitude at which the magnetic and thermal pressures were equal. It should be noted here that the lack of high time resolution plasma measurements make it nearly impossible to conclusively confirm the Zwan–Wolf hypothesis for the Venus magnetosheath, but interpretation of the available plasma measurements, plus the agreement demonstrated in Figure 17, strongly suggest that this model is correct.

The issue of ionospheric magnetization is somewhat beyond the scope of this review; the interested reader is referred to the companion paper (*SSR*, this issue) on that subject. However, we will pursue one aspect of current research on ionospheric magnetization, as it involves the nature of the magnetosheath current system. Several researchers (Russell *et al.*, 1983; Luhmann *et al.*, 1984; Cravens *et al.*, 1984; Phillips *et al.*, 1984; Shinagawa *et al.*, 1987; Shinagawa and Cravens, 1988), have interpreted ionospheric magnetization as a time-dependent process involving convection and diffusion of

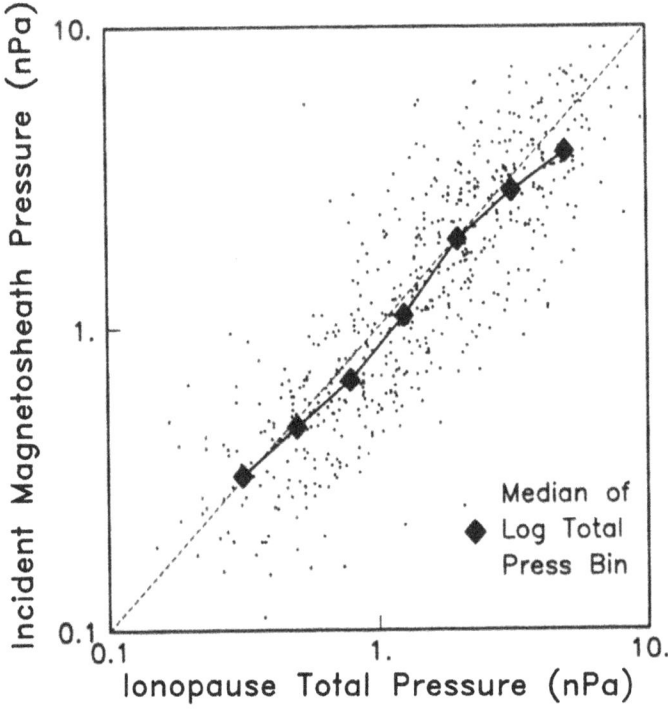

Fig. 17. Incident magnetosheath pressure vs. total pressure at the ionopause, defined to be the altitude where plasma β is unity, for PVO observations near solar maximum. Incident magnetosheath pressure is based on the normally incident component of observed upstream dynamic pressure, plus a small thermal component. Heavy diamonds are magnetosheath pressure medians, binned by log ionopause total pressure. From Phillips *et al.* (1988).

magnetic field from the overlying magnetosheath. Alternative views have been presented by Cloutier and coworkers. Cloutier and Daniell (1973) postulated a steady-state interaction model in which ionospheric current systems are driven by pressure gradients and by electric fields induced in the ionosphere by solar wind penetration. This model, in which the characteristic ionospheric magnetic field signatures obsrved by PVO were later interpreted as resulting from the spacecraft trajectory through a standing current system, was developed further in a series of papers (e.g., Cloutier *et al.*, 1983). The current system proposed in this model is shown in Figure 18 (left panel); note that the magnetosheath currents close partly along the ionopause and partly within the ionosphere itself.

Cloutier *et al.* (1987) subsequently noted that the steady state model could not account for PVO magnetic field observations at high SZA during intervals of low solar wind pressure, and also found that the ionopause, based on observed ion density gradients, is not necessarily colocated with the current layer signaled by magnetic field gradients in the upper ionosphere. Concluding that the physics of ionopause formation and of the magnetic field gradient must be different, whereas in the magnetic barrier scenario they

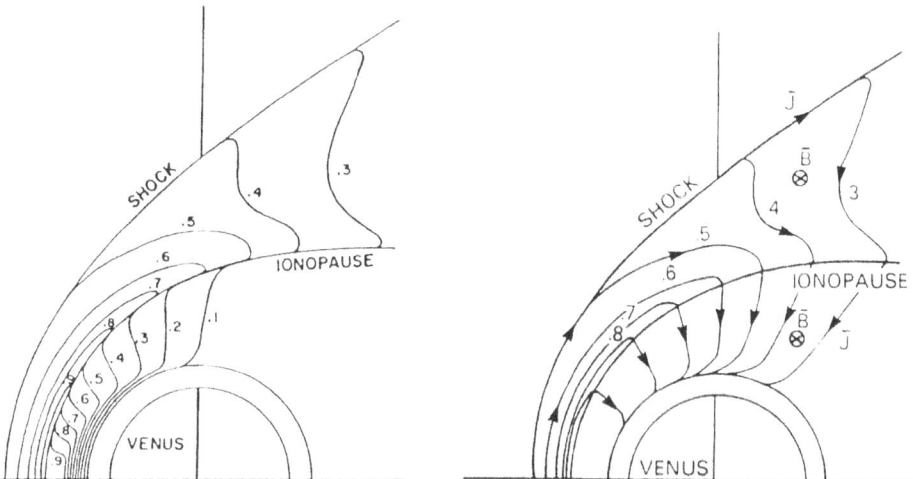

Fig. 18. Current systems postulated to exist in the magnetosheath and ionosphere in the solar wind interaction models of Cloutier and coworkers. The ionopause altitude is exaggerated for clarity. The labeled contours indicate constant magnetic field magnitude, with one-tenth of the total current flowing between adjacent contours. The left panel shows the large scale standing current system which closes not only at the ionopause but also within the ionosphere itself due to pressure gradients and to electric fields caused by the absorption of a fraction of the solar wind by the ionosphere. The right panel shows a subsequent model with no current sheet at the ionopause, such that the magnetosheath currents close entirely within the ionosphere. From Cloutier *et al.*, 1983 (left) and 1987 (right).

are the same, Cloutier *et al.* (1987) developed a different model based on the one-dimensional mass, momentum, and energy equations. In this later model, the large-scale current system in the magnetosheath crosses the ionopause and closes in the ionosphere. This current system is shown in Figure 18 (right panel); note that this differs from the preceding system (left panel) in that there is no current sheet at the ionopause.

It appears at this time that the preponderance of observational evidence favors the formation of a magnetic barrier which transmits the solar wind pressure to the ionosphere. When ionospheric thermal pressure exceeds the upstream solar wind pressure, the ionopause approaches the idealization of a tangential discontinuity between the low-β magnetosheath and the high-β ionosphere. At solar minimum, when peak ionospheric pressure is typically less than the solar wind ram pressure (e.g., Knudsen *et al.*, 1987), or when solar wind ram pressure is unusually high, the magneto-sheath shielding currents may close within the ionosphere. This could be the result of ionospheric absorption of the solar wind, perhaps in a manner similar to the earlier models of Cloutier and coworkers (e.g., Cloutier *et al.*, 1983). Note, however, that the diffusion-convection models (e.g., Shinagawa and Cravens, 1988) also link the magneto-sheath current system with an ionospheric system when the ionosphere is magnetized, without an explicit requirement for solar wind penetration of the ionosphere. A key feature of the later Cloutier *et al.* (1987) current system model (Figure 18, right panel) is the downward acceleration of ionospheric ions to sonic speeds near 150 to 160 km altitude. This prediction is thus far unsupported by observations or corroborated by other researchers, and we conclude that this more recent model is unconvincing.

4.4. ION PICKUP THEORY AND OBSERVATIONS

As noted previously, the excess flaring and solar cycle dependence of the Venus bow shock, as compared to simulation results, suggests that another mechanism is at work in the magnetosheath. The probable candidate for this mechanism is the assimilation of planetary ions by the shocked solar wind (e.g., Elco, 1969). The increased flaring of the near-terminator bow shock position near solar maximum has also been interpreted in this context; that is, due to a comet-like pickup process which is most pronounced during periods of high solar activity (e.g., Alexander et al., 1986). The neutral exosphere, composed primarily of hydrogen, oxygen and possibly helium, is subject to ionization by solar photons, charge exchange with solar wind ions, and possibly by other mechanisms such as the critical ionization velocity (CIV) phenomenon (e.g., Formisano et al., 1982).

Once ionized, a planetary particle is accelerated by the electric and magnetic fields of the magnetosheath. If one ignores fluctuating electromagnetic fields and considers only a steadily convecting magnetic field in the magnetosheath, and the resultant motional electric field $E = -V \times B$, then the initial trajectory of a newborn ion in the planetary frame should be a cycloid consisting of gyration at the ambient plasma speed superimposed on a drift at the same speed. In the limit of spatially uniform magnetosheath magnetic and velocity fields, newly ionized particles would appear as a gyrating ring distribution in velocity space in the reference frame of the background flow. In the planetary frame, particle motion would be parallel to the ambient flow (but at twice the speed) at the top of the cycloidal trajectories, and particles would be stationary at the bottom of the cycloids. Calculations of such pickup trajectories were performed by Wallis (1972) and by Cloutier et al. (1974) and were pursued further by Luhmann et al. (1985) and Phillips et al. (1987) using the gasdynamic simulation of Spreiter and Stahara (1980); sample trajectories are shown in Figure 19.

The pickup process described above becomes ineffective when V and B are nearly aligned due to the small motional electric field. However, another type of ion pickup, stochastic assimilation due to magnetic field fluctuations convecting through the magnetosheath, may in fact have its maximum efficiency for V parallel to B. Such a process, proposed by Brinca et al. (1984), was assessed by Luhmann et al. (1987a), who found that the observed fluctuations could accelerate planetary ions to the ambient flow speeds in a time comparable to the period of the fluctuations (15 s in that study). As the magnetic fluctuations are most pronounced downstream of the quasiparallel portion of the bow shock, which has its greatest extent for lowest cone angles (Luhmann et al., 1983; see Figure 15), this mechanism may be the prevalent pickup method for aligned solar wind B and V. The Luhmann et al. (1987a) study also concluded that low-frequency waves provide the most efficient coupling of ions to the background flow; fluctuations faster than the pickup ion gyrofrequency are ineffective. These findings were supported by numerical simulations (Winske, 1986), which indicated that O^+ ions can be assimi-

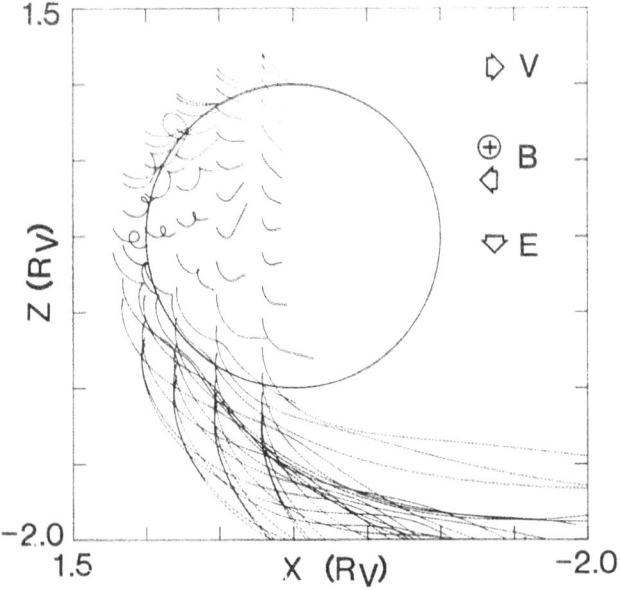

Fig. 19.   Test particle trajectories for newborn O^+ ions, launched in a spherical shell at 1400 km altitude, using the magnetosheath flow field from the gasdynamic simulation of Spreiter and Stahara (1980). Magnetic field was calculated from the distortion of fluid elements, and the electric field is purely motional. A 'toward sector' IMF configuration is used; in the displayed coordinate system, $\mathbf{X}$, $\mathbf{Y}$, and $\mathbf{Z}$ are antiparallel to the solar wind $\mathbf{V}$, cross-flow $\mathbf{B}$, and motional $\mathbf{E}$, respectively. Trajectories which terminate abruptly represent particle entry into the ionosphere, which is assumed to be unmagnetized. From Phillips *et al.* (1987).

lated into the magnetosheath flow in less than one oxygen gyroperiod even when the angle between the velocity and magnetic field vectors is very small (Figure 20).

Observations by Taylor *et al.* (1981) of correlation between suprathermal ion fluxes and low-frequency electric field fluctuations near the ionopause suggest that a relationship may exist between pickup of planetary ions and plasma waves or instabilities. The microphysics of the ion pickup has been explored by Curtis (1981) in the context of electromagnetic and electrostatic instabilities which might be excited by the presence of a fast beam of newborn ions in the magnetosheath flow frame. Curtis concluded that the Alfvén wave instability is most effective in picking up ions when $\mathbf{V}$ and $\mathbf{B}$ are crossed, and that both electrostatic and electromagnetic processes are ineffective when $\mathbf{V}$ and $\mathbf{B}$ are aligned, suggesting that the residence times of planetary ions near their points of ionization could be quite long (> 10 s) under conditions of field-aligned flow. Note, however, that this study did not consider the existence of a convecting wave field in the magnetosheath which Winske (1986) and Luhmann *et al.* (1987a) found to be an effective pickup mechanism as described above.

Daniell (1981) performed a similar analysis to that of Curtis (1981) with somewhat different initial assumptions, and concluded that conditions near the ionopause resulted

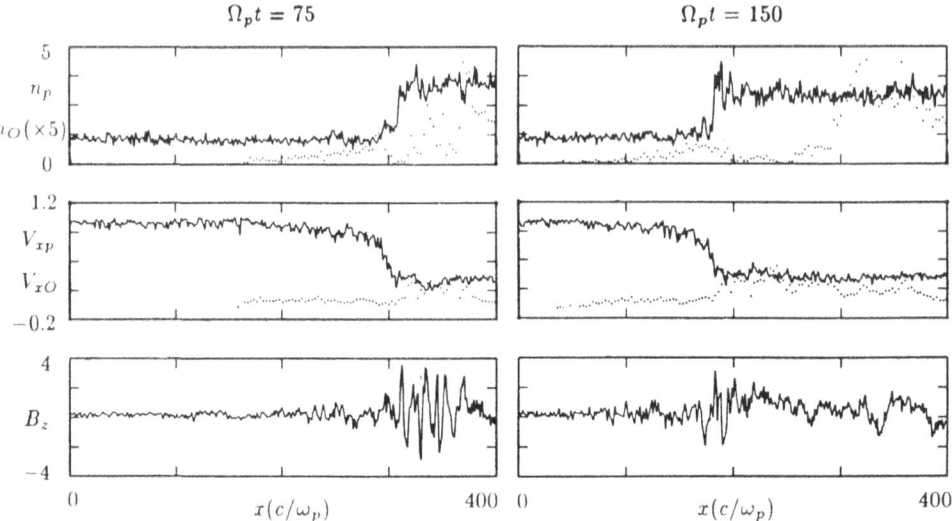

Fig. 20.   Results of a numerical simulation showing assimilation of O^+ ions into a convecting magneto-sheath flow field containing hydromagnetic fluctuations. The three quantities plotted are density (*top*), velocity (*middle*), and transverse magnetic field (*bottom*), with solid traces in the top two panels representing solar wind protons, and dotted traces representing picked up oxygen. The abcissa of each panel is a spatial coordinate, and the left and right columns show two different simulation times. The bow shock, which is generated prior to injection of the pickup ions, is easily visible, with solar wind at left and magnetosheath at right. Oxygen is added at time $\Omega_p t = 50$; note that the newborn ions are quickly accelerated nearly to ambient magnetosheath speed. From Winske (1986).

in large wave growth rates primarily for the electrostatic Bernstein mode. Another finding of the Daniell (1981) study was that a ring distribution of ions picked up by steady electric and magnetic fields would be rapidly thermalized, and that downstream of the planet the ions would probably be streaming along with the ambient flow. Thus, even in this idealized case one might expect to observe heavy ions downsteam from the planet with the same velocities as the background solar plasma. In the case of pickup due to fluctuating fields, full assimilation of the planetary particles is intrinsic to the pickup process, although the efficiency of the particle acceleration is sensitive to variety of factors (e.g., Luhmann *et al.*, 1987a). Accordingly, observations that downstream planetary ion fluxes are roughly aligned with the background flow (e.g., Perez-de-Tejada *et al.*, 1982) do not favor either pickup mechanism to the detriment of the other.

Returning to the initial pickup of newborn ions by steady magnetosheath magnetic and electric fields, one can easily see an asymmetry in the calculated pickup configuration shown in Figure 19. O^+ ions picked up on the side of the planet with locally upward **E** execute upward cycloidal trajectories, while those picked up in regions of downward **E** precipitate downward. In the study featuring these calculated trajectories (Phillips *et al.*, 1987), downward precipitating ions were presumed lost to the interaction. Such asymmetric pickup, resulting from the fact that the initial Larmor radii of heavy ions in the near-Venus region are comparable to the planetary radius, is similar in some respects but also involves fundamental differences, when compared to a pickup scenario originally proposed by Cloutier *et al.* (1974). This earlier study showed via analytical calculations that the finite Larmor radius effects in the pickup process could create an asymmetry in the altitude profiles of planetary ions in the magnetosheath. However, in the Cloutier *et al.* (1974) model, such an asymmetry does not ultimately result in loss of ions downstream preferentially on one side of the planet. First, in that model the ionosphere is magnetized and thus pickup ions entering the ionosphere at shallow angles continue to gyrate back into the magnetosheath. Second, loss of ions which penetrate the ionosphere deeply enough to collide with planetary neutrals is suppressed by accumulation of positive charge in the planetary hemisphere of downward magnetosheath electric field. As a result, although in the near planet environment the pickup ion distributions and their resultant detectability by plasma instruments may be very different in one hemisphere than in the other, once thermalized downstream from the planet the ion fluxes should be the same in each hemisphere. Observational assessment of the validity of this model is difficult, due in part to instrumental factors. However, the observations by Intriligator (1989) and Slavin *et al.* (1989), which will be discussed in more detail in a subsequent section, of asymmetries in heavy ion fluxes far downstream from the planet, suggest that downward precipitating ions are in fact lost to the interaction.

While double-peaked E/q spectra suggesting pickup of planetary ions were observed in the near-terminator magnetosheath by Venera 10 (Vaisberg *et al.*, 1976), subsequent observations by PVO appear to be more convincing. Direct observations by the PVO plasma analyzer of picked up planetary ions in the dayside or near-terminator magnetosheath were first reported by Mihalov and Barnes (1981). The instrument, which is capable of detecting O^+ only at speeds up to 310 km s^{-1}, measured heavy (high E/q) ion fluxes on 12 of the first 250 orbits. In general, the secondary (heavy ion) E/q peak occurred at lowest energy per charge for observations closest to the planet. The authors interpreted the measured ion fluxes as singly ionized atomic oxygen in the process of acceleration up to the ambient magnetosheath bulk speed.

In a subsequent study, Phillips *et al.* (1987) used similar PVO observations, integrated with magnetic field observations and numerical calculations, to assess the effects of the finite Larmor radii of picked up ions. This study identified 10 ion pickup events in the magnetosheath; a sequence of E/q spectra for PVO transit from the solar wind into the magnetosheath, including a pickup event, is shown in Figure 21. They noted that all 10 events occurred in the hemisphere of upward electric field, and had velocities appropriate for newborn ions observed during the initial portions of initial cycloidal trajectories. Figure 22 shows the positions and velocities associated with the ten pickup events. Note that the oxygen velocities are roughly aligned with the proton velocities, but are sometimes displaced in the locally upward direction, and that the oxygen speeds are generally equal to or slower than the corresponding proton speeds. The observed hemispherical asymmetry in pickup ion locations is consistent with the notion of ions in the hemisphere of locally downward **E** precipitating into the ionosphere and being lost to the interaction, but does not rule out a scenario such as that of Cloutier (1974).

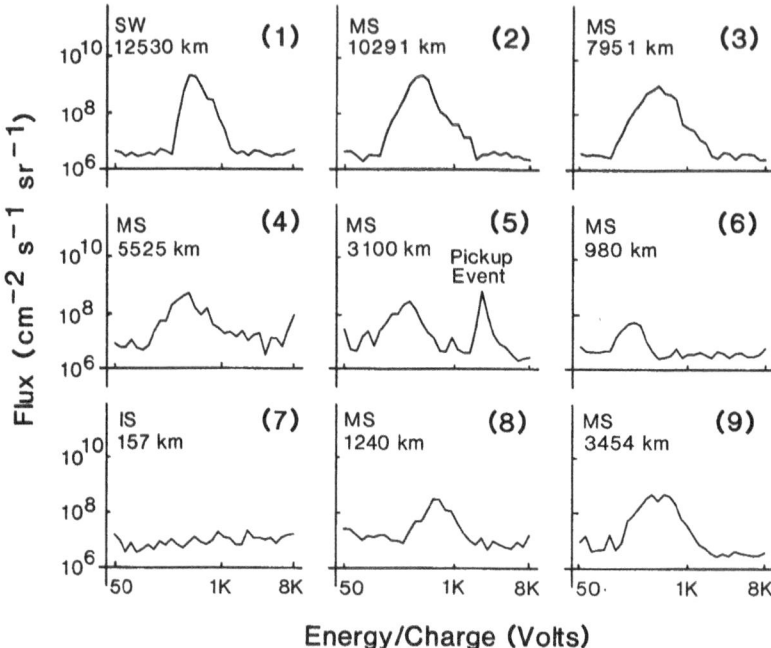

Fig. 21. Sequence of PVO Plasma Analyzer spectra representing transit of spacecraft during orbit 131 from solar wind (SW) to magnetosheath (MS) to ionosphere (IS) and then back to magnetosheath. The spatial region and altitude of the center of the scanning period are indicated for each spectrum. The center panel contains a pickup event, visible as a second peak at high E/Q; panel (4) also contains an enhancement at high E/Q which may represent heavy ions. From Phillips *et al.* (1987).

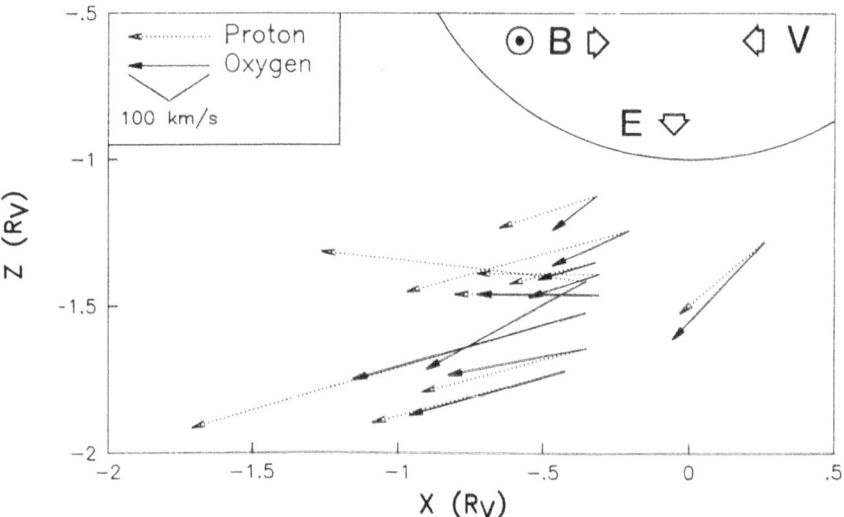

Fig. 22. Positions and velocity vectors for protons (dotted traces) and pickup ions, assumed to be O^+ (solid traces), for ten ion spectra measured by the PVO plasma analyzer in the magnetosheath. Data are displayed in a coordinate system ordered by the upstream **B**, **V**, and **E** vectors, shown at upper right. From Phillips *et al.* (1987).

Phillips *et al.* then examined the prevailing magnetic field draping pattern in planes defined by the solar wind **B** and **V** vectors and offset from the planetary center in the plus or minus **E** directions, and concluded that the field was more sharply draped in the upward **E** hemisphere (Figure 23). They then performed a numerical experiment in which newborn ions were launched a test particles in a model magnetosheath. Treating the test ions as a current, they demonstrated that such a current could account for the difference in field draping in the two hemispheres. In a related paper, Phillips *et al.*

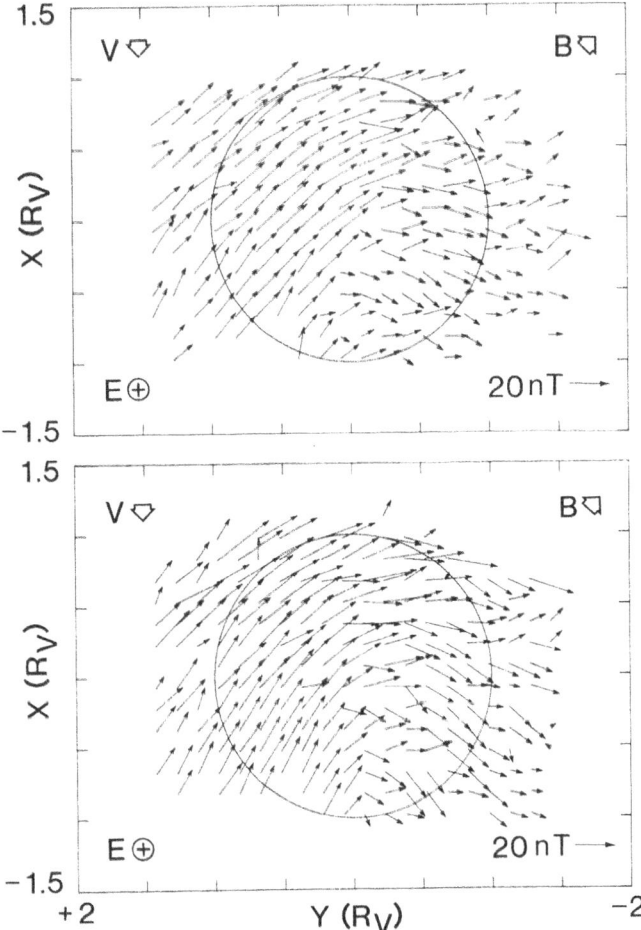

Fig. 23. Spatially-averaged magnetic field vectors, based on PVO observations near solar minimum, for slabs of data aligned with the solar wind **V** and **B** vectors. Each slab is 200 km thick and is displaced from the point of tangency to the planetary surface by 1000 km. The top panel shows the slab displaced in the direction opposite the motional electric field and thus represents part of the hemisphere of locally downward **E**, while the bottom slab is displaced parallel to the electric field and thus represents a region of upward **E**. Note the more pronounced draping of the magnetic field for locally upward **E** (bottom panel), which was postulated to result from an upward (i.e., into the plane of the figure) current of pickup ions. From Phillips *et al.* (1987).

(1988) showed that the momentum transferred to the solar wind via the ion pickup process could account for the observed difference in ionopause altitude in the $+ \mathbf{E}$ and $- \mathbf{E}$ hemispheres. This altitude asymmetry takes the form of a displacement of the ionopause in the direction opposite to the convective electric field of the solar wind such that the near-terminator ionopause is $\sim 200–300$ km higher on the side of the planet with locally downward $\mathbf{E}$ than on the opposite side. This effect was attributed by Phillips *et al.* (1988) to momentum transfer between the solar wind plasma and newly born ions in the exosphere; Figure 24 illustrates the asymmetry. This effect on ionopause position was shown to be most pronounced when $\mathbf{V}$ and $\mathbf{B}$ are crossed, creating a large electric field, and to vanish when the field and flow are aligned and stochastic pickup, with no finite Larmor radius effect, should prevail.

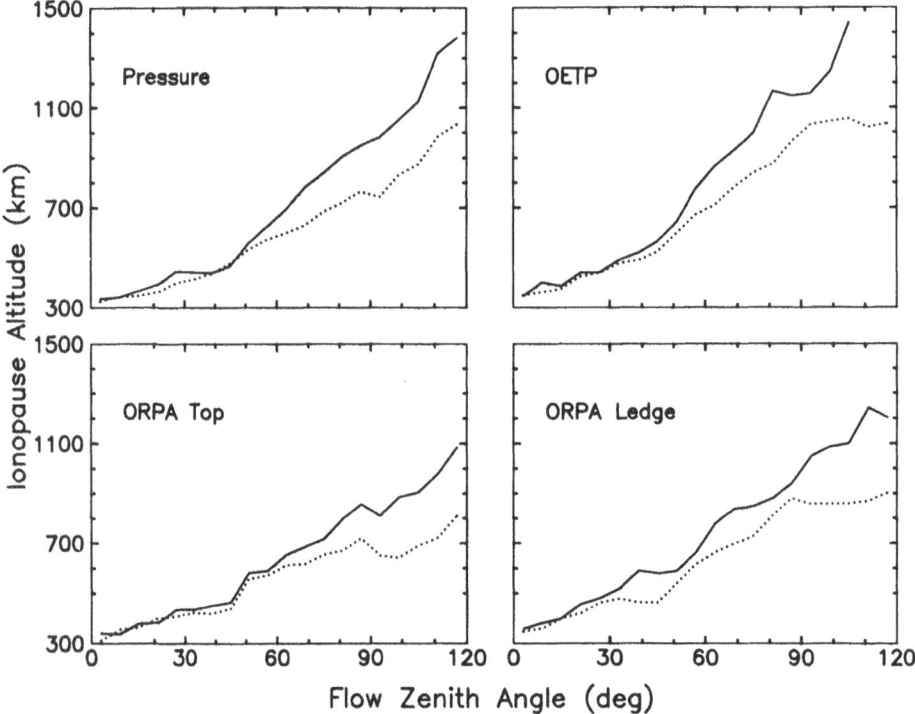

Fig. 24. Contours in altitude-flow zenith angle space showing 50% probability of PVO being above or below the ionopause, for the four ionopause definitions shown in Figure 8. Solid and dotted traces correspond to hemispheres of locally downward and upward motional electric field, respectively. From Phillips *et al.* (1988).

Assimilation of planetary ions should constitute an important aspect of the overall interaction, particularly near solar maximum, as indicated by the variation of the bow shock position (e.g., Russell *et al.*, 1988). Non-self-consistent modeling of the influence of the ion pickup process on the magnetosheath bulk properties, such as that described above (Phillips *et al.*, 1987), suggests the need for incorporation of ion pickup self-consistently into global models. This problem has been considered by Breus and colleagues (Krymskii and Breus, 1986; Breus, 1986; Breus *et al.*, 1987a, b; Belotserkovskii *et al.*, 1987) and by Stahara *et al.* (1987). In the first of these studies, Krymskii and Breus (1986) used analytical calculations to assess pickup effects on the interaction, and concluded that the shock should move away from the planet, and that the peak magnetic field in the barrier region should decrease, as a result of ion pickup. Breus (1986), Breus *et al.* (1987b), and Belotserkovskii *et al.* (1987) reported the results of a global gasdynamic simulation, with magnetic field derived from fluid parameters as in the Spreiter and Stahara (1980) model, but incorporating a mass source term representing fluid pickup of planetary ions. Momentum and energy source terms were not included (pickup ions have zero initial velocity, but subsequently affect the momentum and energy equations through their incorporation into mass density), nor were finite Larmor radius effects. They found that the incorporation of mass-loading caused the magnetosheath to thicken substantially, particularly near the terminator, and concluded that their model provided a greatly improved prediction of shock position and magnetic field draping.

The mass-loaded gasdynamic model of Stahara *et al.* (1987) produced conflicting results. Using the same upstream conditions and the same 400 km hot oxygen scale height that were used by Breus (1986) and Belotserkovskii *et al.* (1987), that study found that mass-loading affected the bow shock location very little. Figure 25 shows the bow shock shape with and without mass-loading for the two differint models; hot oxygen parameters and mass source term are shown in the figure. While unable to explain the disparity between the two computational results, Stahara *et al.* (1987) found in their model that the effect of a 400 km scale height for mass-loading was to create a boundary layer of enhanced density above the ionopause and to have little effect at higher altitudes. They were able to duplicate the observed terminator bow shock position only by increasing this scale height to 3200 km, or alternatively by increasing the oxygen density at the exobase. Figure 26 shows the bow shock position for a range of mass-loading scale heights. The enhanced scale height which provides the best fit with the observations suggests a connection with the Larmor radius of pickup ions. FLR effects, at least on the side of the planet with upward electric field, would have the effect of stretching upward the regions of energy and momentum exchange between solar plasma and planetary ions.

In addition to the unresolved differences between the simulations of Belotserkovskii *et al.* (1987) and Stahara *et al.* (1987), certain simplifications employed by both studies suggest the need for further computational efforts. First and foremost is the gasdynamic approximation, which produces an axisymmetric flow and density field and of course cannot duplicate MHD effects such as a Zwan and Wolf (1976) depletion layer. Second is the fluid treatment of pickup ions, in which a newly created ion is assumed to be immediately assimilated into the local flow. The Larmour radius of the newborn ions is roughly comparable to the planetary radius (e.g., Phillips *et al.*, 1987), and the results shown in Figure 26 suggest that incorporation of such a scale may be necessary for

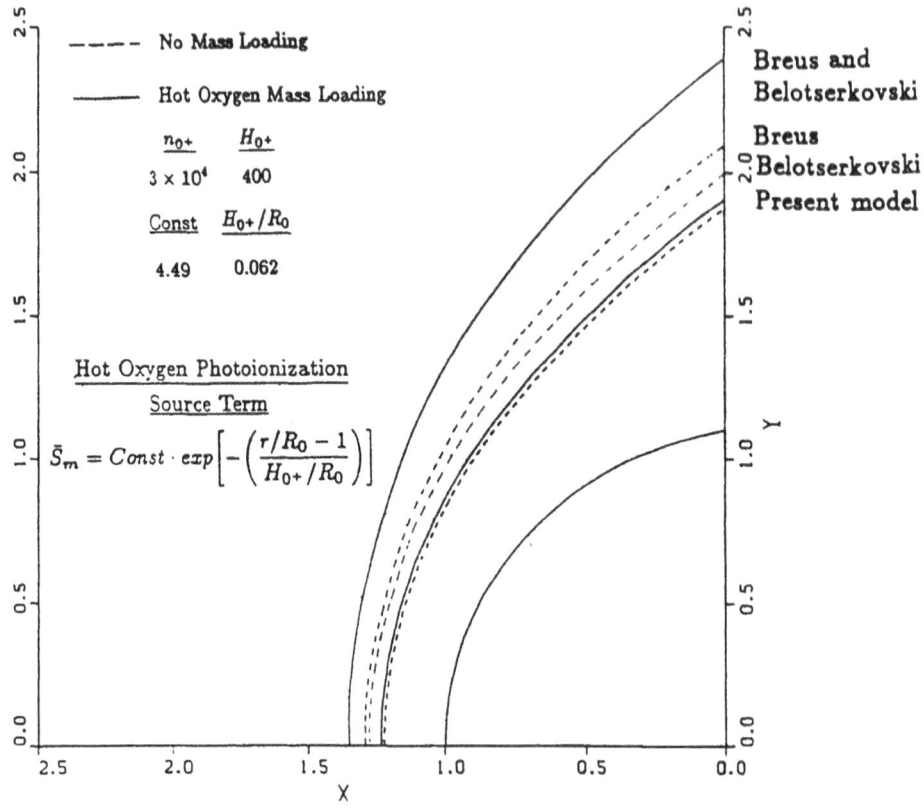

Fig. 25.   Axisymmetric depiction of differing results of massloading a gasdynamic simulation for the solar wind-Venus interaction. The innermost solid trace represents the obstacle shape, while the dashed and outer solid traces represent the modeled bow shock for Mach 3, $\gamma = \frac{5}{3}$, and appropriate upstream conditions for PVO orbit 582, without and with mass-loading, respectively. Results shown are for the Breus (1986) and Belotserkovskii *et al.* (1987) simulations, in which mass-loading produced a substantial elevation of the terminator bow shock, and for the Stahara *et al.* (1987) model, in which the same mass source term produced only a minor effect on bow shock location. The mass source term is shown in the figure. From Stahara *et al.* (1987).

proper modeling. Note, however, that in the limit of parallel or antiparallel solar wind **B** and **V**, where FLR effects should be minor, the fluid pickup approximation may be satisfactory (e.g., Luhmann *et al.*, 1987a). Comprehensive simulation of the pickup process will probably require a hybrid MHD approach with fluid solar wind and discrete pickup ions, or perhaps an iterative fluid approach incorporating test particles to generate the mass, momentum and energy source terms.

4.5. THE MAGNETOSHEATH BOUNDARY LAYER

One of the least understood aspects of the Venus–solar wind interaction, in part due to limitations of PVO instrumentation, concerns the plasma characteristics of the region just above the ionopause. We note here that our identification of this region as a 'magnetosheath boundary layer' differs somewhat from that of Vaisberg and coworkers

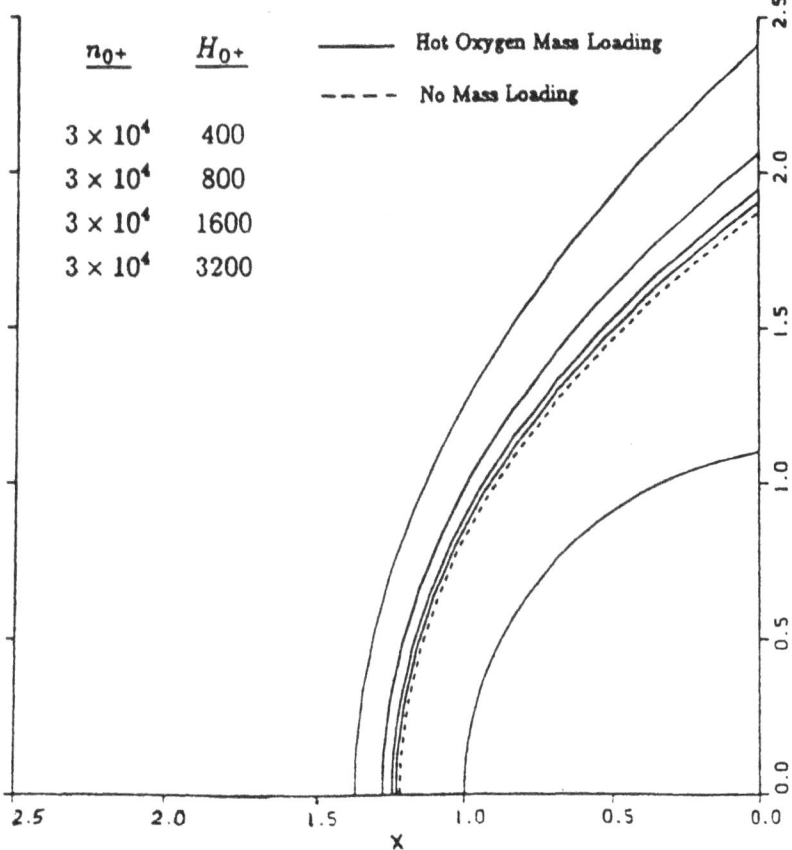

Fig. 26. Results of varying the hot oxygen scale height in the simulation of Stahara *et al.* (1987). The innermost solid trace represents the ionospheric obstacle, the dashed trace is the computed bow shock location without mass-loading, and the four outer solid traces show the mass-loaded bow shock shape corresponding to the four oxygen scale heights listed in the figure. Higher (lower) bow shock altitudes correspond to larger (smaller) scale heights. From Stahara *et al.* (1987).

(e.g., Russell and Vaisberg, 1983), who reserved this term for a rarefaction region observed in the downstream magnetosheath. This rarefaction region (see Figure 2) will be discussed further in subsequent sections. The boundary layer between the ionopause and magnetosheath was called the 'mantle' by Spenner *et al.* (1980), who identified it as a unique layer in which plasmas of ionospheric and magnetosheath characteristics could be observed. That study concluded that the mantle, which varied in thickness from a few hundred km in the subsolar region to ~ 1500 km near the terminator, was in fact the magnetic barrier. Observed characteristics of the mantle included high magnetic field, low plasma density, and electron energy spectra intermediate between those observed in the magnetosheath and the ionosphere.

The morphology of the mantle, shown in Figure 27, suggests a mapping nightward to the 'low energy ion flux' region (Figure 2) of Vaisberg *et al.* (1976), such that the upper

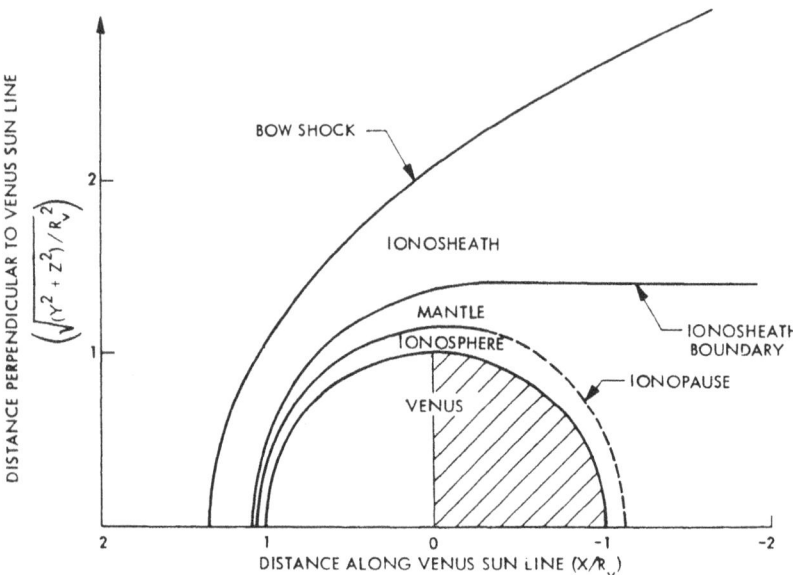

Fig. 27. Axisymmetric schematic of the Venus-solar wind interaction region, featuring a 'mantle' region with plasma characteristics between those of the ionosphere and magnetosheath (called the 'ionosheath' here). Compare with Figure 2. From Spenner *et al.* (1980).

edge of the mantle eventually forms the boundary between the magnetotail and the magnetosheath. The source of the plasma in the mantle region is still unclear; various authors have attributed it to upward transport from the ionosphere, photoionization of planetary neutrals within the mantle, and possible to anomalous processes such as CIV (e.g., Zelenyi and Vaisberg, 1982). The contribution of the mantle plasma in balancing ionospheric thermal pressure is unclear. Vaisberg *et al.* (1980) estimated it to be $\frac{1}{4}$ to $\frac{1}{3}$ of the magnetic pressure, based on an observed deficit in magnetic field above the ionopause. However, Knudsen *et al.* (1982) and Phillips *et al.* (1988) mapped the observed upstream solar wind pressure into the magnetosheath and inferred that the magnetic barrier thermal pressure required to produce the observed ionopause shape was of the order of 12% or less of the total pressure.

   Another aspect of the interaction of solar and planetary plasma near the ionopause concerns the viscous transfer of energy and momentum between these two components, as investigated by Perez-de-Tejada and coworkers in a series of papers (e.g., Perez-de-Tejada and Dryer, 1976; Perez-de-Tejada, 1986a, b). These studies were motivated by several observations. First, plasma of planetary origin is observed in the mantle, as discussed above. Second, the plasma fluxes above the near-terminator and nightside ionopause appear to be weaker than in the outer magnetosheath (e.g., Gringauz *et al.*, 1976; Vaisberg *et al.*, 1976; Mihalov *et al.*, 1980). Third, a generally antisolar ionospheric convection pattern, with supersonic convection speeds near the terminator, has been observed (Knudsen *et al.*, 1980), and has been interpreted as consistent with the deficit in magnetosheath momentum flux (Perez-de-Tejada, 1982, 1986a). Fourth, there is some evidence for closure of the magnetosheath flow in the planetary wake; this will be discussed in more detail in a subsequent section. Finally, low altitude magnetosheath plasma temperatures appear anomalously high near and downstream of the terminator (e.g., Verigin *et al.*, 1978). Figure 28 shows the increasing temperature of ion distributions with decreasing altitude in the near-terminator magnetosheath (Perez-de-Tejada *et al.*, 1985). All angular spectra are scaled relative to their respective peaks; analysis of ion fluxes indicates a density of 15 cm^{-3} for the outer magnetosheath (scan II) and only 3 cm^{-3} for the inner magnetosheath (scan I).

While the viscous interaction theory appears to have some observational support, most of the evidence mentioned above has alternative explanations. First, mixing of

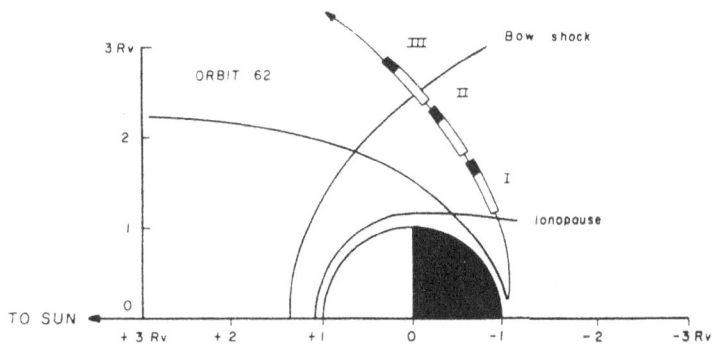

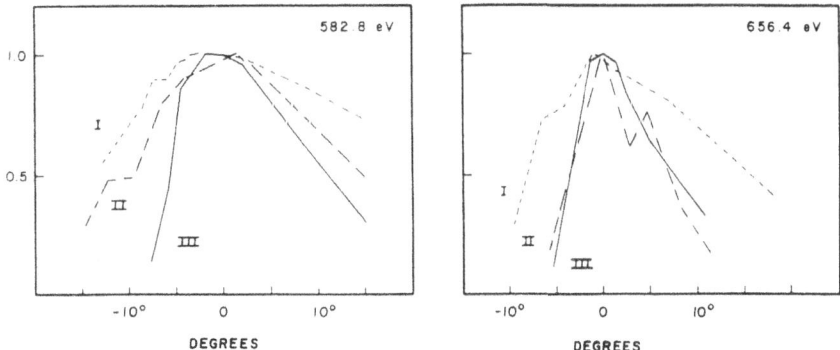

Fig. 28. (*Top*) Axisymmetric projection of the PVO trajectory for orbit 62; heavy black regions mark the angular scans of the plasma analyzer for three numbered angular spectra in the lower magnetosheath (I), outer magnetosheath (II), and solar wind (III). (Bottom) Azimuthal distributions of peak plasma fluxes at 582.8 and 656.4 eV measured in the three numbered scans. Note the broadening of the angular distributions for azimuthal scan I relative to scan II, suggesting a heating process above the ionopause. All fluxes are normalized to the peak flux for a particular energy and scan; peak fluxes were actually substantially lower for azimuthal scan I than for scan II at the depicted energies. From Perez-de-Tejada *et al.* (1985).

planetary and solar plasmas might occur for a variety of reasons not involving viscosity, such as ionization of exospheric neutrals or outward leakage of ionospheric particles along magnetic field lines penetrating the dayside ionosphere. Second, plasma depletion is expected, without invoking viscous mechanisms, as a result of the Zwan and Wolf (1976) effect. Third, the observed antisunward ionospheric convection is explainable simply as the result of thermal pressure gradients, with little or no momentum input from the magnetosheath (Knudsen et al., 1981; Elphic et al., 1984; Whitten et al., 1984). Fourth, Knudsen et al. (1982) have shown that the closure of the magnetosheath flow, as evidenced by the ionopause altitude nightward of the terminator, is appropriate for isentropic expansion into the wake of an obstacle. The final argument in favor of a viscous interaction, the high magnetosheath temperatures measured by Venera 9 and 10, so far has not been countered by other observational evidence, and thus is the most compelling of the pro-viscosity evidence. We note here that Knudsen (1989) has demonstrated that interactions with the hot oxygen component of the atmosphere may heat the upper ionosphere; similar arguments might be invoked for heating of the lower magnetosheath. Additionally, a recent theoretical study by Linker et al. (1989) demonstrated that, for some flow conditions, assimilation of cold particles can heat the background plasma. A fundamental weakness of the viscosity hypothesis is that its proponents have provided no specific physical description, based on first principles, of a viscous process. The question of precisely what mechanism (if any) transfers momentum and energy across the sheared velocity field at the ionopause remains unanswered. Until a specific model is proposed and tested against the observational evidence, and it is demonstrated that the observations support a specific viscous mechanism and exclude other explanations, the viscosity hypothesis is unconvincing.

Another observational finding may have important ramifications for the physics of the low altitude magnetosheath. In the previous discussion on ion pickup, we have focused on pickup of single newborn ions by the convective electric field, or by fluctuating electric and magnetic fields, in the magnetosheath. However, observations by the PVO electron temperature probe (Brace et al., 1982a) indicate that detached clouds and attached streamers of ionospheric plasma are common in the vicinity of the dayside, near-terminator and nightside ionopause. These bodies of plasma may produce a local magnetic draping effect, with magnetic stresses serving to accelerate the ionospheric plasma downstream, as in the single case study by Russell et al. (1982). While Wolff et al. (1980) suggested that detached plasma clouds, as well as ionospheric flux ropes, might be generated by the Kelvin–Helmholtz instability at the ionopause, existence of this instability at Venus has not been verified observationally. In summary, the generation mechanism, and the overall impact on the magnetosheath flow, of bulk scavenging of ionospheric plasma has not been determined.

5. The Near-Planet Wake and Nightside Magnetosheath

5.1. FLOW AND MAGNETIC FIELD CONFIGURATION

The near-planet wake of Venus is a region of remarkable complexity. Though measurements by the PVO and Venera spacecraft have revealed many intriguing field and particle phenomena, thus far no unified picture of this region has emerged. At low altitudes, PVO observations show that the ionosphere is highly structured, with no easily definable ionopause. Filaments of ionospheric plasma, as well as detached clouds, persist to an undetermined distance downstream. A study of the nightside plasma between 2000 and 2500 km altitude has shown that the major ion constituent is suprathermal (9–16 eV) O^+, with little thermal plasma remaining (Brace *et al.*, 1987). An even more energetic (40–200 eV) ion population, generally flowing tailward but also having a large flow component perpendicular to the orbital plane, is also occasionally observed (Kasprzak *et al.*, 1987). The magnetic field is observed to be radially directed in regions between the plasma rays (Brace *et al.*, 1987). Prior to the evolution of the PVO orbit such that this filamentary structure could be observed, lower altitude measurements had revealed the presence of ionospheric 'holes', or regions of depleted density and radially aligned magnetic field (e.g., Brace *et al.*, 1982b; Luhmann *et al.*, 1982). The magnetic polarities within these holes were shown to be appropriate for enhanced draping of the IMF (Luhmann and Russell, 1983; Marubashi *et al.*, 1985; Phillips and Russell, 1987), leading to a concept of hole formation such as that shown in Figure 29. However, the results of the Brace *et al.* (1987) study indicate that the actual dynamic behavior of the nightside magnetic field is much more complex, possible involving multiple tail rays which form near the terminator, converge into the wake, and coalesce into a single larger ray.

Evidence for convergence of the magnetosheath flow into the wake, which might facilitate ray coalescence or hole formation in the simpler scenario of Figure 29, is strong but not conclusive. Limited direct observations by the PVO plasma analyzer suggest that the flow convergence may be such as to close behind the planet somewhere near 5 R_V downstream (Intriligator *et al.*, 1979). Perez-de-Tejada *et al.* (1983) demonstrated that the draping of the magnetic field in some cases supports such convergence; examples of this draping are shown in Figure 30. Additional evidence for convergence of the magnetic field, based on a statistical study of PVO magnetometer measurements in the magnetosheath and upper ionosphere, was demonstrated by Marubashi *et al.* (1985). The reader is cautioned, however, that convergence of the magnetic field does not necessarily indicate convergence of the flow field. While a certain amount of flow deflection should be expected for simple isentropic expansion of the magnetosheath flow, without any viscous interaction with the ionosphere (Knudsen *et al.*, 1982), Perez-de-Tejada (1986b) maintained that such expansion is insufficient to explain the observed nightside ionopause locations. In the Perez-de-Tejada scenario, an unspecified viscous interaction between the magnetosheath and ionospheric plasmas converts some of the magnetosheath kinetic energy into thermal energy. The heated magnetosheath

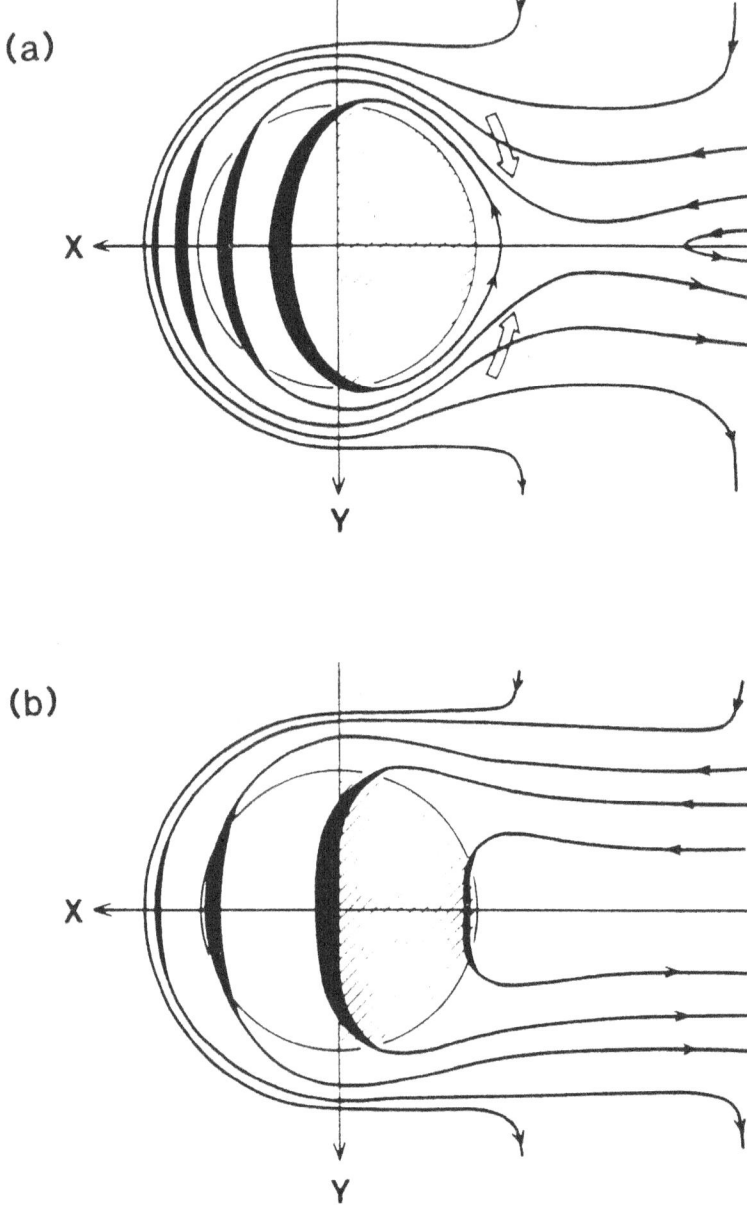

Fig. 29.   Schematic of the magnetic field configuration during periods when the magnetosheath plasma flow (open arrows) converges strongly into the wake (*top*) and when such convergence is weak (*bottom*). The former scenario was postulated to result in formation of radially aligned magnetic fields and ionospheric holes on the nightside. Field reconnection, as shown for the lowest altitude field line in the top panel, is not essential for this hole formation scenario. Recent studies facilitated by higher PVO periapsis suggest that this picture, though perhaps qualitatively correct, is oversimplified. From Marubashi *et al.* (1985).

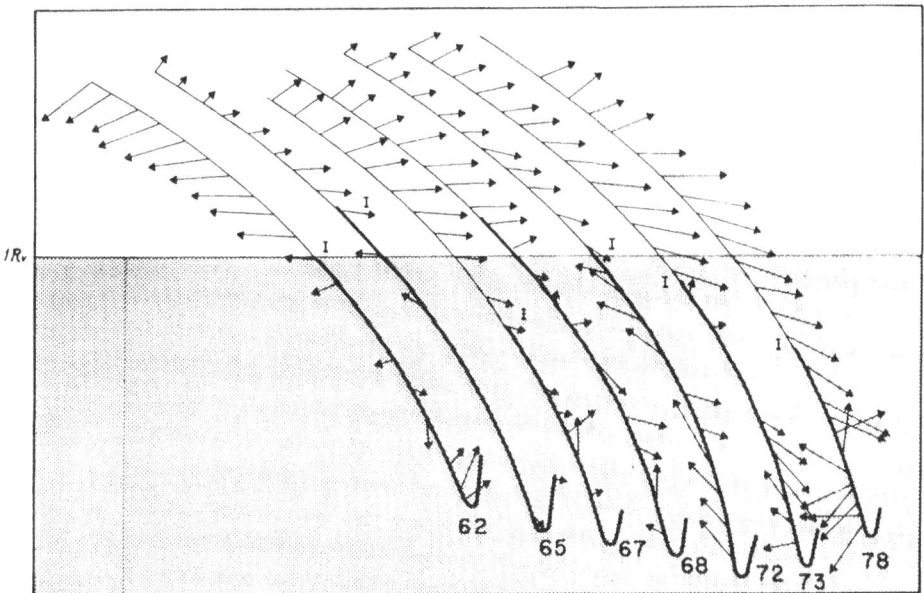

Fig. 30.   Composite of PVO magnetic field measurements (orbit numbers are shown at bottom), with the shaded area representing the planetary umbra. The ionopause, marked with the letter 'I', can be seen as a reduction in magnetic field magnitude and a general disruption of the orderly draping pattern in the magnetosheath. Orbits such as 78, with the ionopause located deep within the umbra, show a field draping pattern which suggests flow convergence into the wake. Such convergence is less apparent for high ionopause orbits such as 65. From Perez-de-Tejada *et al.* (1983).

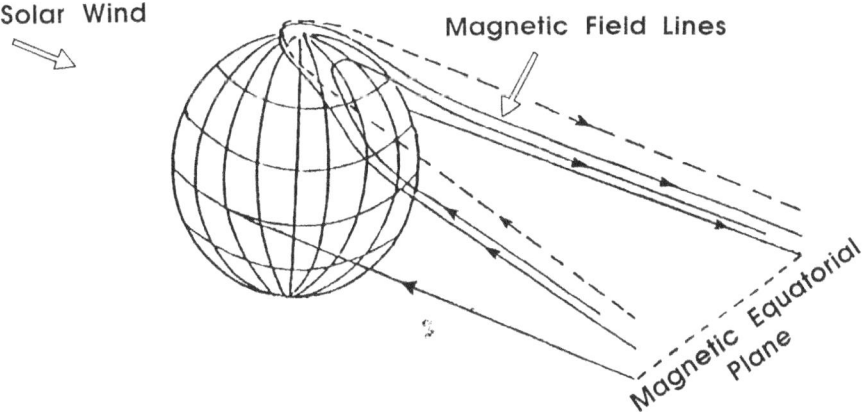

Fig. 31.   Schematic of the magnetic field draping suggested for the scenario of viscous heating of the magnetosheath plasma near the ionopause. The draped magnetic field lines (solid curves) sink into the umbra due to the expansion of the plasma behind the planet. The dashed curves shows the field line geometry expected for an inviscid interaction. From Perez-de-Tejada (1986b).

plasma then expands into the nightside umbra, deflecting the magnetic field into the wake at an angle somewhat greater than for the inviscid case. Figure 31 illustrates this postulated field draping.

Aside from questions of flow closure behind the planet and localized effects such as ionospheric holes, it appears that the overall draping of the magnetosheath magnetic field on the planetary nightside is fully consistent with the notion of an induced magnetotail which responds to the IMF orientation. Figure 32 shows the field configuration produced by a laboratory terrella experiment (Dubinin *et al.*, 1978) and the field measured by Venera 9 and 10 (Dolginov *et al.*, 1981). The draping of the field is readily apparent in the near-planet nightside; some differences between the model and observations may be due to the presence of a radial IMF component, whereas in the laboratory model the field is purely transverse to the flow. Effects on the draping pattern of a spiral IMF orientation were noted in the PVO measurements by Phillips *et al.* (1986b) for the dayside and near-terminator magnetosheath and by McComas *et al.* (1986) for the deep magnetotail, but have thus far not been demonstrated in the Venera 9 and 10 nightside observations.

5.2. Plasma observations

Although the low-altitude planetary nightside has been extensively explored by PVO, spacecraft periapsis has generally been in or just above the ionosphere. At higher altitudes (but well below the PVO apoapsis region and the Mariner 10 tail transit), the Venera 9 and 10 observations, augmented by the single Mariner 5 flyby, have provided the majority of the information about the various downstream plasma regimes. The Venera measurements were made using a sunward looking wide-angle ion Faraday cup with proton energy per charge range of 0–4400 eV, an antisunward looking electron retarding potential analyzer with a 0–300 V range, retarding potential electron analyzers with 0–300 V range, and several narrow-angle electrostatic ion analyzers with various look directions and a combined E/q range of 50–19800 eV. As noted previously, Figure 2 shows schematically the plasma regions and boundaries postulated by Vaisberg *et al.* (1976) based on narrow-angle ion measurements. A slightly different picture, shown in Figure 33, was drawn by Gringauz (1983) based on wide-angle measurements. We will attempt to describe the distinct plasma regions thus far discovered by starting in the optical shadow of the planet.

Near the antisolar region, Venera 9 and 10 transited a region in which the Faraday cup ion instrument measured no clearly defined ion fluxes, though sporadic fluxes were measured at energies up to the instrument limit (4.4. keV). The electron analyzer measured total densities of less than 1 cm^{-3}. Verigin *et al.* (1978) estimated a closure point for this 'corpuscular umbra' at $\sim 4\,R_V$ downstream, similar to the value derived from PVO plasma analyzer observations (Intriligator *et al.*, 1979). It is tempting to speculate that the high energy ion fluxes occasionally observed in the umbra were due to picked up planetary ions or possibly to crossing of extended ionospheric tail rays (cf. Brace *et al.*, 1987), but the observations are too sparse to support this hypothesis convincingly. In fact, the Venera 9 and 10 narrow-angle ion analyzer, looking roughly

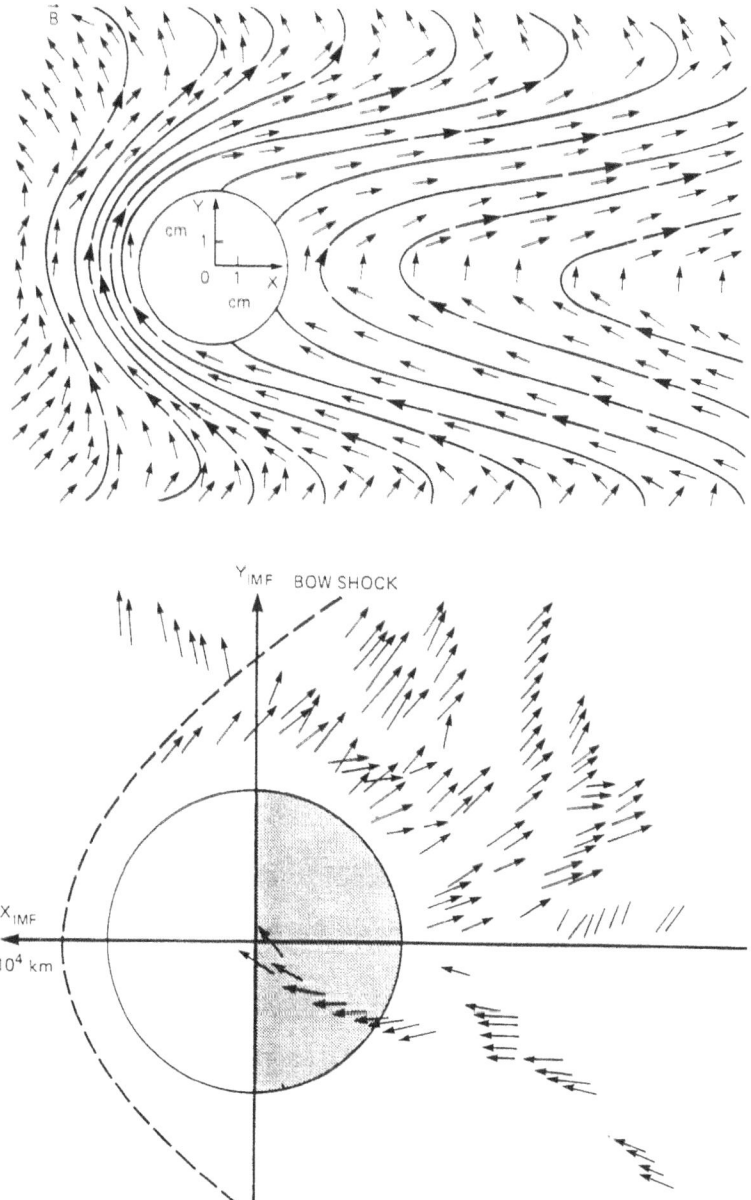

Fig. 32.   A comparison of the magnetic field from a laboratory terrella experiment (*top*) by Dubinin *et al.*
(1978) and that observed by Venera 9 and 10 (Dolginov *et al.*, 1981) in the near-terminator and nightside
magnetosheath (*bottom*). Observations are projected onto a plane defined by the Venus–Sun axis and the
transverse IMF direction. From Sagdeev *et al.* (1983).

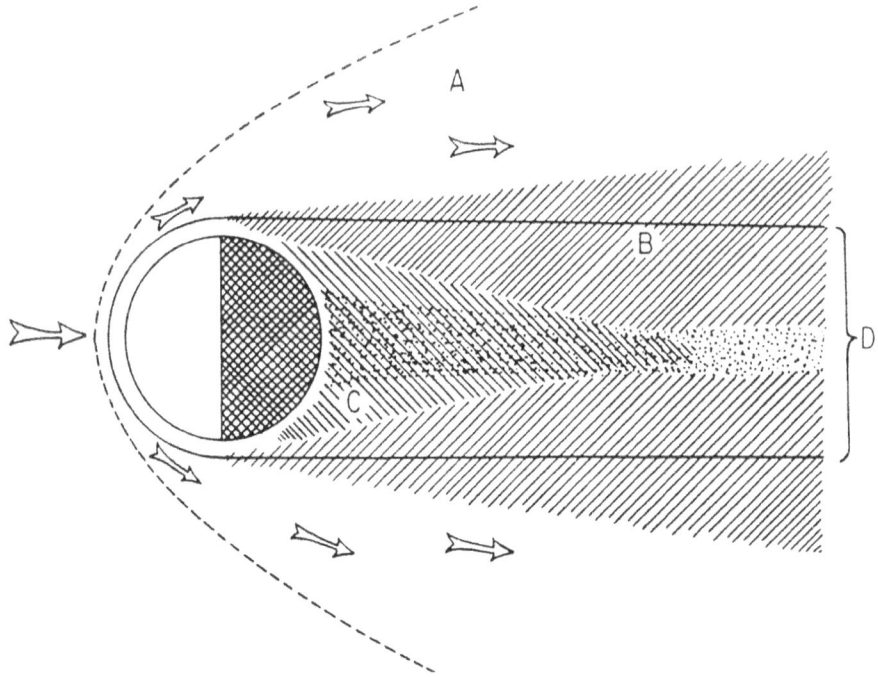

Fig. 33.   Schematic of the plasma regions observed by the Venera 9 and 10 wide-angle ion and electron experiments: (A) transition region (magnetosheath), (B) corpuscular penumbra, and (C) corpuscular umbra. Region D is the 'wake of the magnetosphere', similar conceptually to the magnetotail, and the dotted region along the antisolar axis is the plasma sheet separating the magnetotail lobes. From Gringauz (1983).

sunward, recorded ion fluxes in the near-planet wake only at energies of 50 eV and below (Vaisberg *et al.*, 1976), and the sunward-looking Faraday cup on board Mariner 5 had observed only marginally detectable ion fluxes while transiting this region (Shefer *et al.*, 1979).

Venera 9 and 10 experimenters observed the umbra to be enclosed by a distinct plasma region, labeled by Gringauz *et al.* (1976) as the 'corpuscular penumbra' (Figure 33). This region may in fact be the same as that transited by Mariner 5 (Shefer *et al.*, 1979), as instrumental characteristics for that mission make differentiation of the umbra and penumbra questionable. The penumbra (e.g., Gringauz *et al.*, 1976) is characterized by lower ion convection speeds (~ 100–200 km s^{-1}, based on the assumption that the ions are protons) and lower densities (a few cm^{-3}) than in the outer magnetosheath. Verigin *et al.* (1978) noted that contours of constant electron density, normalized to the upstream solar wind density, were generally aligned with the Sun–Venus axis; Figure 34 shows such contours. Based on the observed density gradients, Verigin *et al.* concluded that there must be enhanced magnetic field in the wake of the ionopause (region *D* in Figure 33) and that this region represented the magnetotail. Another finding of that study was that the penumbra grows in width with increasing distance downstream from the terminator.

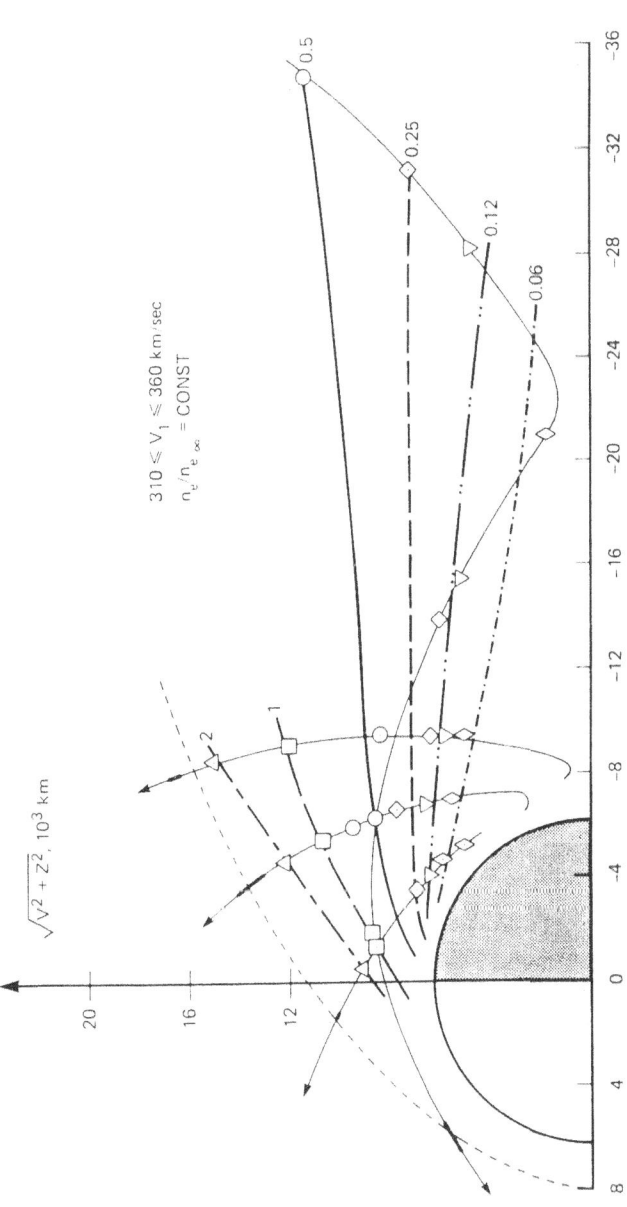

Fig. 34.   Contours of constant electron number density, relative to the solar wind density, in the magnetosheath, umbra and penumbra of Venus (Verigin *et al.*, 1978). Contours are based on Venera 9 and 10 retarding potential analyzer observations, restricted to electron energies greater than roughly 10 eV. From Sagdeev *et al.* (1983).

Using Venera 9 and 10 narrow-angle plasma measurements, Vaisberg *et al.* (1976) arrived at a different picture for the plasma regimes in or near the planetary wake, as shown in Figure 2. Note that Gringauz *et al.*'s penumbra, shown as region *B* in Figure 33, appears to include both the rarefaction region and the low energy ion flux region of Figure 2. Vaisberg *et al.* identified a boundary, interpreted as the extension of the ionopause, separating these two plasma regimes. Just below the boundary, particle fluxes of two distinct energies and flow directions were observed, with increasing domination by slower and cooler ions as the spacecraft approached the antisolar axis. Vaisberg *et al.* interpreted the cooler population as accelerated ionospheric ions. Above the extended ionopause, in the 'rarefaction region' of Figure 2, ion convection speeds and temperatures were substantially higher, suggesting that the plasma in this region was thermalized solar wind. The Venera investigators (e.g., Romanov *et al.*, 1978) also interpreted the upper boundary of this region of heated solar plasma, or the lower limit of the magnetosheath proper (or 'transition region'), as a rarefaction wave. Figure 35

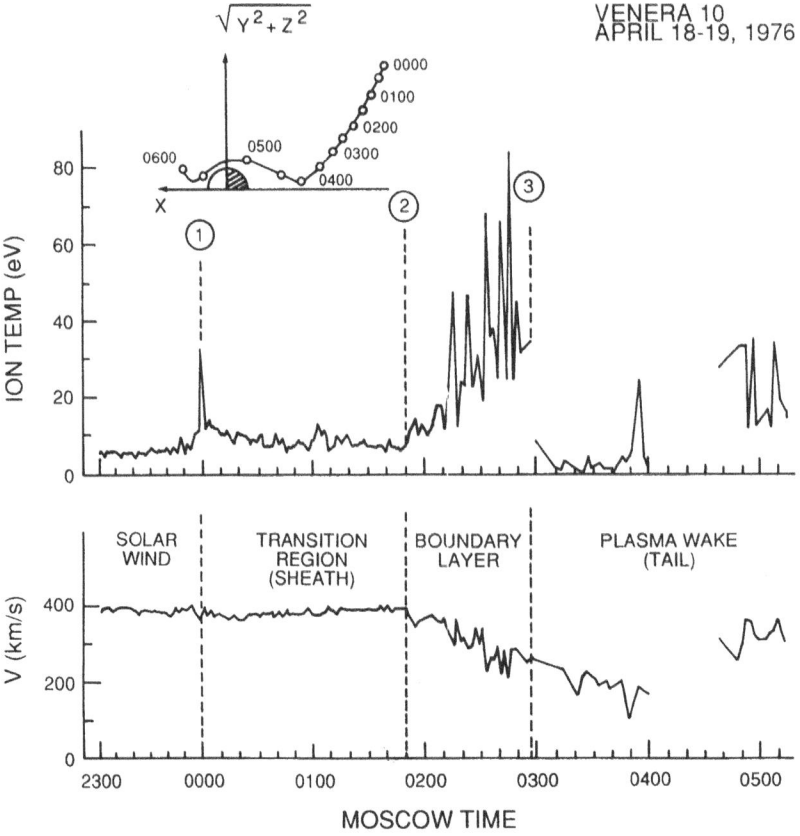

Fig. 35.   Time series of ion temperature and flow speed measured by Venera 10 on April 18–19, 1976. Trajectory is shown as a solid line at top. Labeled points refer to (1) the bow shock crossing; (2) entry into the boundary layer, seen a velocity decrease and temperature increase, and (3) entry into the low energy ion region. Adapted from Romanov *et al.* (1978).

is a time series of Venera 10 ion temperature and flow speed measurements showing transit through the magnetosheath, the rarefaction region, and then into the low energy ion region, or plasma wake. An 'intermediate region' possessing plasma characteristics similar to those of the Venera rarefaction region, that is low density, high temperature, and bulk speed declining with approach to the planet, were also noted during the single Mariner 5 flyby by Shefer *et al.* (1979).

One other phenomenon noted by the Venera observers is the presence of bursts of energetic (a few keV) ions within the umbra (Gringauz *et al.*, 1976) and near the extended ionopause (Vaisberg *et al.*, 1976), but not in the transition region (outer magnetosheath). The origin of these bursts, which are often accompanied by magnetic fluctuations, is presently unknown. Romanov *et al.* (1978) referred to the extended ionopause as the magnetotail boundary, and noted that bursts of high-energy ions and magnetic fluctuations were generally observed while crossing this boundary. As discussed previously in the context of the magnetosheath, the location of picked up planetary ions is sensitive to both the IMF clock and cone angles. Thus it is reasonable to postulate that the burstiness of the ion fluxes observed by Venera 9 and 10 could be due to rotations of the upstream magnetic field such that the spacecraft intermittently transited regions favorable for pickup ion observation.

Outside of Vaisberg's rarefaction region, Gringauz' penumbra, Shefer's intermediate region, and Romanov's boundary layer (all similar in definition, at least in terms of their upper boundaries), observations support the predicted inviscid flow of the shocked, but not anomalously heated, solar wind. In this outer magnetosheath, or transition region (Figure 2), velocities are lower, and density and temperature higher, than in the undisturbed solar wind. It appears at this time that the scenario of Vaisberg *et al.* (1976), as shown in Figure 2, is fundamentally correct, although the use of the term 'ionopause' is somewhat different from the common usage. While some observations have been interpreted in support of the viscous interaction model of Perez-deTejada and co-workers, so far no model for the underlying physical processes which might heat and slow the solar plasma within the rarefaction region has been demonstrated convincingly. Within the context of the wake plasma regions the definition of the near-planet magnetotail is somewhat arbitrary; the tail may be bounded by the rarefaction wave or perhaps by the extended ionopause of Figure 2.

6. The Magnetotail

6.1. FORMATION AND CONFIGURATION

While magnetic field draping of the upstream IMF plays an important role in the morphology of the dayside and near-terminator magnetosheath regions as well as the near-planet wake, as described in previous sections, the role of draping is even more fundamental to the formation and morphology of the Venus magnetotail. Figure 36 shows the basic tail draping concept first introduced by Alfvén (1957) to explain cometary tails. In Figure 36(a), straight IMF lines approach the cometary obstacle.

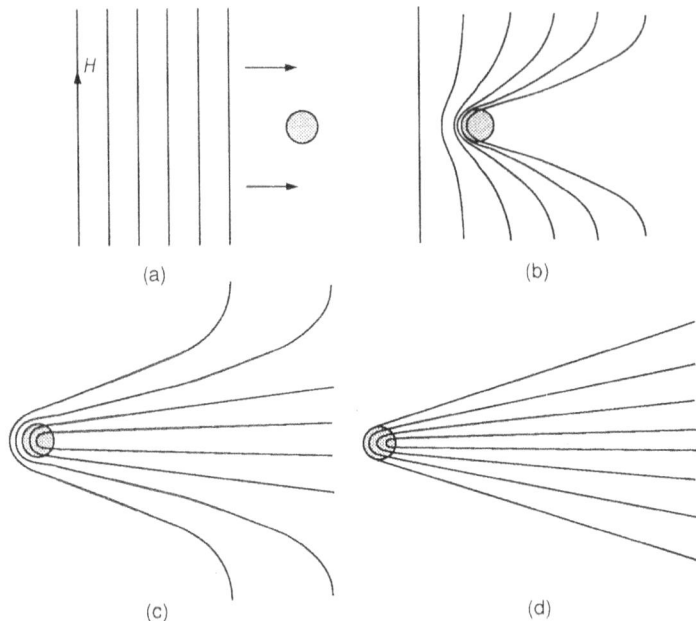

Fig. 36. Original schematic of cometary tail draping process. In panels (a) through (d) the IMF carried by the solar wind becomes increasingly bent about the obstacle and draped back into a taillike configuration. From Alfvén (1957).

Those portions of the flow which pass sufficiently close to the obstacle become mass loaded owing to pick-up of cometary ions and, due to conservation of momentum, become slowed. The velocity shear between streamlines which pass close to the obstacle and are mass loaded, and those further away which do not experience substantial mass loading, bends or drapes the imbedded IMF about the obstacle. Figures 36(b)–(d) show increasingly greater draping of the IMF with time into a magnetotail-like configuration. Eventually, even the slowed portions of the field will pull through the mass loading region and be released down the tail as highly kinked, draped field lines.

The real tail formation process at Venus is, of course, far more complicated than this simple cartoon. In addition to mass loading of the magnetosheath flow by exospheric planetary ions, the conductive Venus ionosphere provides an obstacle to the magnetized magnetosheath flow by supporting shielding currents which largely exclude this flow, as discussed previously. The flow path for plasma around the obstacle is appreciably greater than along streamlines away from the obstacle, which are not diverted, and again the imbedded field becomes draped. Several authors have considered the relative importance of mass loading versus simple diversion of the flow (e.g., Luhmann *et al.*, 1985; Russell *et al.*, 1985b; McComas *et al.*, 1987a), and some argue that this relative importance varies with the exospheric scale height variations with solar cycle (Alexander and Russell, 1985; Russell *et al.*, 1988). In general, it appears that diversion of the flow is primarily responsible for formation of the Venus magnetotail, although mass loading effects clearly affect the details of the interaction. The three-dimensional interaction between the solar wind flow and Venus is displayed schematically in Figure 37. A flow streamline which would intersect the planet (dashed) is diverted over the obstacle in a direction perpendicular to the upstream IMF lines. In addition, the flow along this streamline is slowed due to mass loading with cold ionospheric material. The plasma flow behind Venus is generally tailward, with small perturbations to allow the diverted plasma to flow back into the rarefied near-Venus wake region.

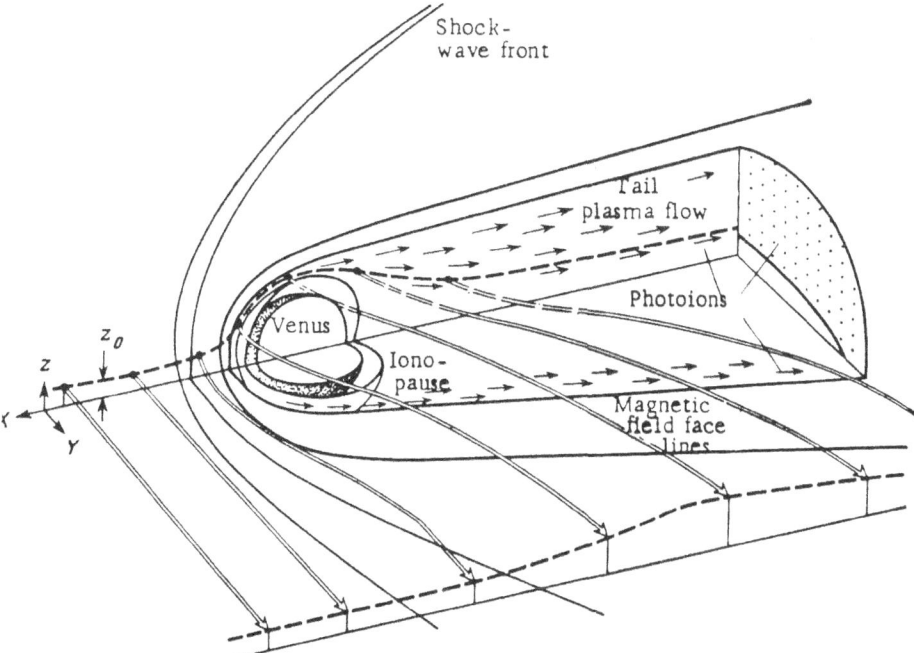

Fig. 37. Schematic diagram of the plasma convection and magnetic field configuration around Venus. Small black arrows behind the planet indicate plasma flowing into and refilling the near-tail region while the long tubes pointing out of the page indicate the draped and diverted magnetic field. From Zelenyi and Vaisberg (1982).

At Earth, and other planets which possess substantial intrinsic magnetic fields, the orientations of the tail's internal structures are dominated by the orientation of the planetary field. For example the terrestrial tail lobes (top panel of Figure 38) are positioned northward and southward of a central east–west plasma sheet. At Venus, on the other hand, the magnetotail field arises purely from draped IMF, and the orientation of this tail's internal configuration varies with the upstream IMF direction (bottom panel of Figure 38). For the nominal case of the IMF lying in the ecliptic plane, the draped tail lobes lie to the east and west of a cross-tail, field reversing current sheet (often called the plamsa sheet by analogy with the terrestrial case) which is oriented north–south. For convenience in subsequent descriptions of tail phenomena, we here

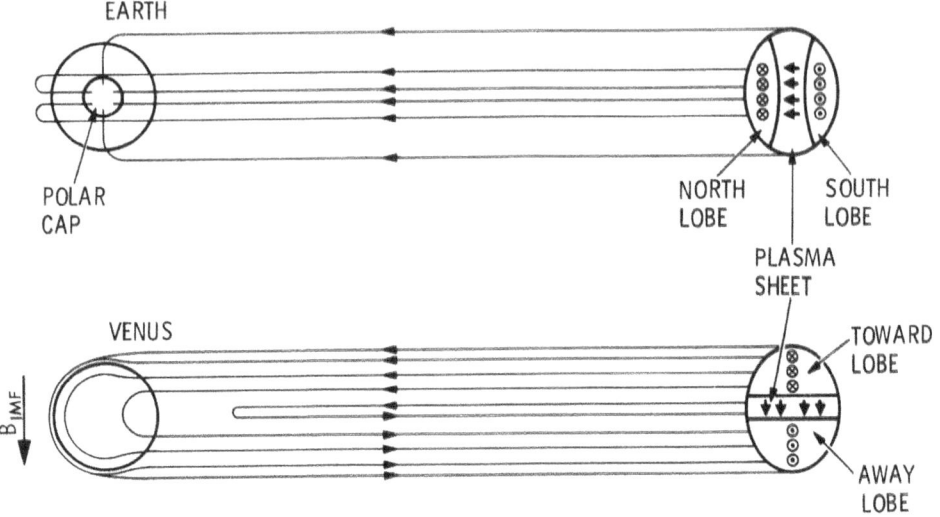

Fig. 38. Comparison of the global tail configuration between Earth (top panel) and Venus (bottom panel). Both tails are comprised of two magnetic lobes separated by a field reversing cross-tail current sheet. At Earth, where the tail configuration is dominated by the internal planetary field, the lobes lie to the north and south of an east–west oriented current sheet. At Venus, however, the orientation of the tail structures is determined solely by the upstream IMF orientation. From Slavin *et al.* (1984).

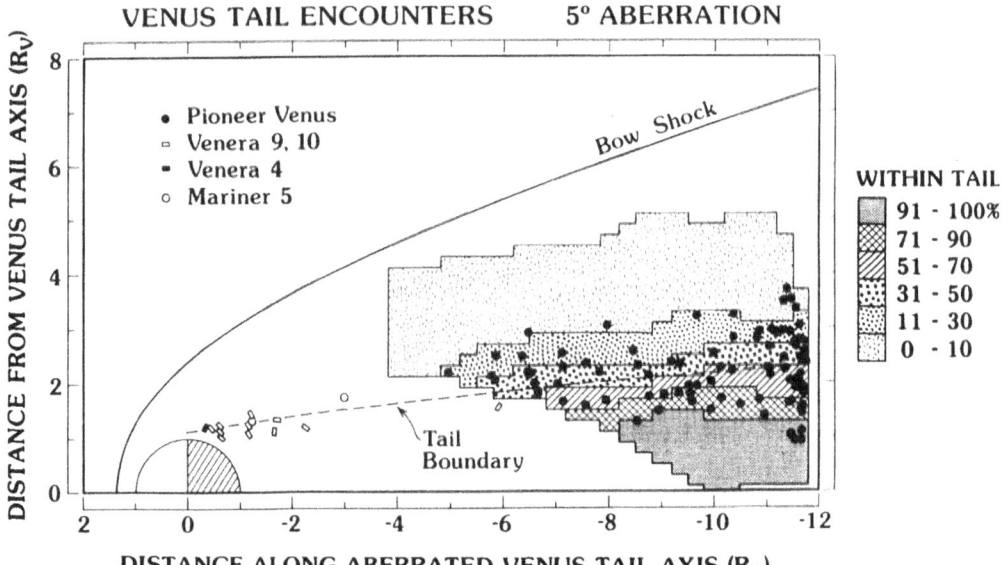

Fig. 39. The average cross-sectional dimension of the Venus magnetotail as determined by tail boundary crossings from Venera 4, 9, and 10, Mariner 5, and PVO. Contours of fractional time spent within the tail by PVO are superposed. From Saunders and Russell (1986).

define the 'central tail plane' as the plane through planetary center containing the upstream solar wind **V** and **B** vectors. Note that this plane is orthogonal to the plasma sheet.

The Venus magnetotail average dimension (Saunders and Russell, 1986) is shown in Figure 39. This figure combines early tail boundary crossings near the planet from Venera 4, 9, and 10 and Mariner 5 with a statistical set of PVO tail crossings from ~ 5 to $12\,R_V$ downstream. The cross-sectional shape of the magnetotail is harder to determine due the fact that direct spacecraft crossings between the magnetosheath and the cross-tail current sheet are less well defined in the PVO magnetic field data than are transits between the sheath and the tail lobes. Consequently, the average location of the tail boundary is less well established along its north and south edges, where the current sheet intersects the magnetosheath, than near the equatorial plane where a relatively sharp lobe-sheath boundary prevails (see Figure 38). The best available statistical cross section of the tail is shown in Figure 40.

6.2. Magnetic field observations

Early observations of the Venus magnetotail from Mariner 10 (Ness *et al.*, 1974) indicated lobe-like regions in which the magnetic field was enhanced in magnitude and

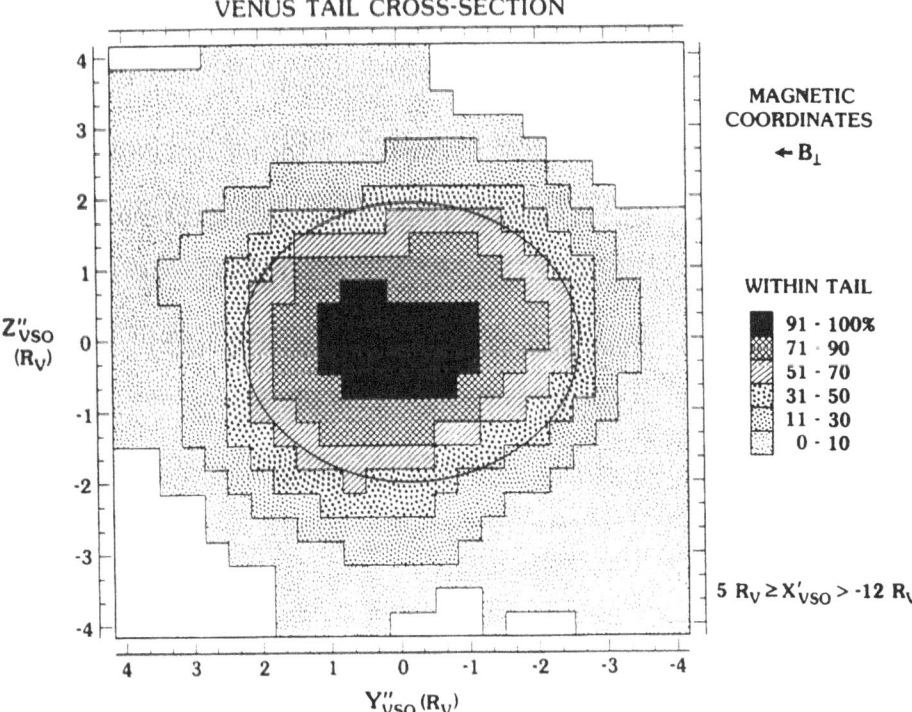

Fig. 40.   Venus tail cross section based on a large statistical PVO data set between 5 and $12\,R_V$ downstream. The coordinate system is such that the average cross-flow magnetic field points along $+ Y''$ (VSO) and the motional electric field points in $+ Z''$ (VSO). From Saunders and Russell (1986).

was directed roughly parallel and anti-parallel to the Venus–Sun line. In these regions, high-frequency magnetic field fluctuations were less pronounced than in the magneto-sheath. The similarities between the field configurations in draped and intrinsic magnetotails as discussed above (two lobes and a central current sheet) made it impossible to unambiguously resolve the nature of a planetary magnetotail from single spacecraft transits such as that of Mariner 5 (cf. Bridge *et al.*, 1967; Russell, 1976b). Even after the more extensive observations by Venera 9 and 10 observations, Dolginov *et al.* (1978) maintained that the Venus tail was due to an intrinsic planetary field.

Careful analyses of the Mariner 10 (Lepping and Behannon, 1978) and Venera 9 and 10 (Yeroshenko, 1979) magnetometer results finally resolved this issue in favor of the induced magnetotail formed by IMF draping. Figure 41 summarizes the Venera results. In this figure the radial (x) component of the magnetic field observed in the Venus umbra is displayed in a plane normal to the Sun–Venus axis, with position and **B** vectors rotated about that axis such that the cross-flow magnetic field is toward the left. This rotation clearly orders the B_x observations into two lobes with polarities appropriate for draped IMF.

VENERA 9 AND 10

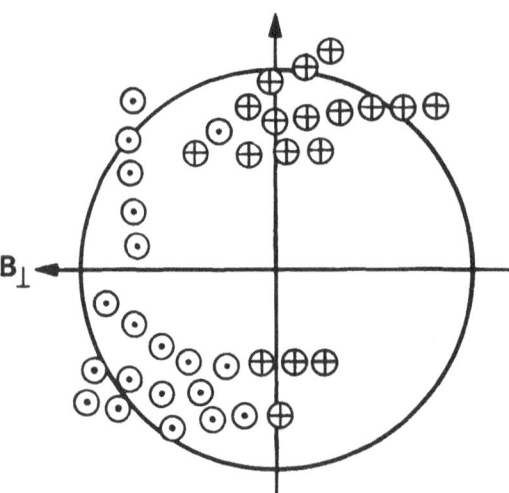

Fig. 41.   Summary of Venera 9 and 10 magnetic field draping ordered by the cross-flow magnetic field component. While the B_x components into and out of the page fell on both sides of the tail in inertial coordinates, they are clearly separated into two lobes in magnetic coordinates. From Yeroshenko (1979).

The most complete picture of Venus magnetotail physics has been derived from the PVO magnetic field measurements, owing to the numerous spacecraft transits of the tail during the prolonged Pioneer mission. As a result of the polar orbital orientation, PVO generally passes through the tail only when periapsis is near local noon and apoapsis is near local midnight. Apoapsis, where the spacecraft is moving most slowly and therefore spends the most time, is near $12\,R_V$. Consequently, most magnetotail observations cover the range from ~ 7 to $\sim 12\,R_V$, with a median downstream distance of roughly $10.5\,R_V$.

PVO magnetometer and plasma wave data were examined for several tail orbits by Russell *et al.* (1981). The 5.4 kHz plasma wave data and magnetic field components for a typical tail pass (orbit 191) are shown in Figure 42. Tail entry near 09:00 UT is signaled by a reduction in the plasma wave activity, enhancement of B_x, and increased variability in magnetic field magnitude and orientation. The strongly x-aligned field is indicative of entry directly into one of the draped tail lobes. Exit from the opposite tail lobe near 17:35 UT is indicated by the reduction in the magnetic field, rotation of the field away from the strongly x-directed lobe configuration, and decrease in the field variability. Reversal of B_x coincident with a large reduction in field magnitude indicates that the cross-tail current sheet was transited between 13:45 and 17:00 UT. Earlier brief encounters (incomplete traversals) with this current sheet occur at $\sim 11:15$ UT and 12:15 to 12:45 UT. This study, as well as that by Russell *et al.* (1985b) also showed

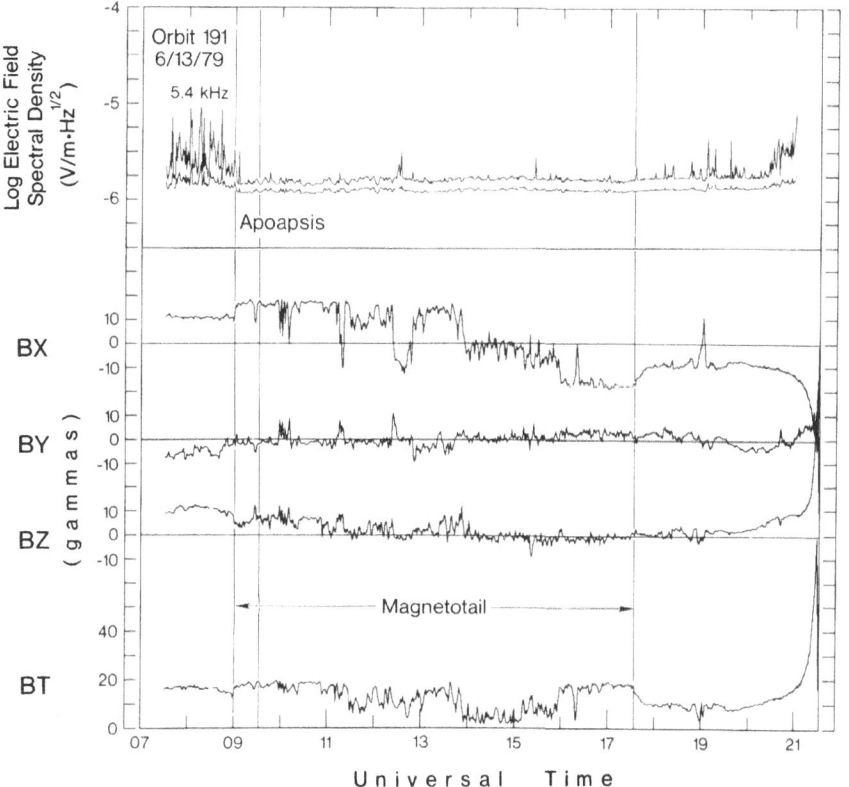

Fig. 42.   PVO magnetometer and 5.4 kHz plasma wave data for a typical tail pass (orbit 191). Tail entry and exit are indicated with vertical lines, as is spacecraft apoapsis. From Russell *et al.* (1981).

that there are several MWb of magnetic flux contained in the Venus tail; this is far more than found in the dayside or nightside ionosphere. A more complete statistical study (Saunders and Russell, 1986) estimated an average crosstail field of 2 nT, yielding roughly 3 MWb of tail flux. The average cross-tail field finding was later revised upward to 4 nT by McComas *et al.* (1986), indicating an even greater flux content. The consequences of this large amount of flux in the tail are that (1) most of the field lines observed in the tail must close across the tail, rather than in the ionosphere or dayside magnetosheath, and that (2) the solar wind can resupply the entire magnetotail flux in only a few minutes, and thus the field lines must transit the near-planet region in a similar time.

Inside the Venus magnetotail two magnetic field signatures, lobe and cross-tail current sheet, are typically observed. This was as shown for a single tail crossing in Figure 42 and is demonstrated statistically in Figure 43. This latter figure plots ~ 9400 one-minute averaged PVO magnetic field measurements from 38 tail orbits. The field magnitude is plotted versus the draping angle of the field in the plane through the planetary center defined by the average magnetic field and plasma flow vectors. Orientations of ± 90° correspond to field vectors parallel to the Venus–Sun line while

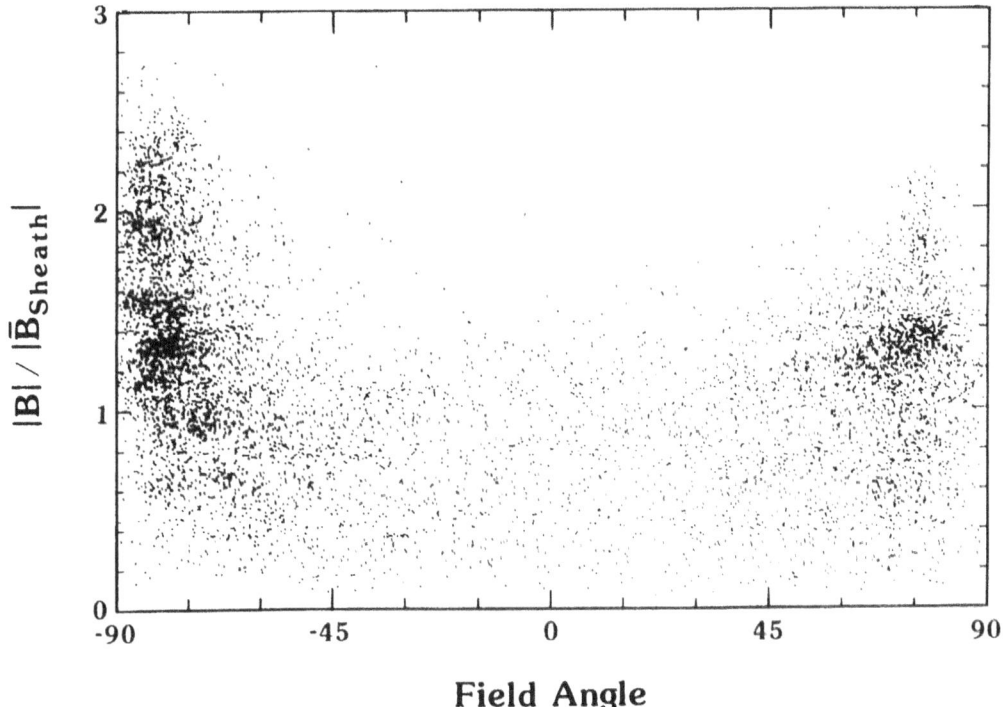

Field Angle

Fig. 43.   Relative magnetic field magnitude plotted versus field draping angle measured from the cross-flow field direction (positive angles measured toward the planet). Tail lobes are indicated by regions of relatively large field magnitudes and highly draped field directions (near ± 90°) while the cross-tail current sheet is characterized by smaller field magnitudes and angles. From McComas *et al.* (1986).

angles near $0°$ represent cross-tail field alignment. The field magnitude is normalized to the average adjacent magnetosheath magnitude observed along each orbit. While there is a large amount of variability in Figure 43, the tail lobes are characterized by enhanced field magnitudes and nearly radial alignments, while current sheet intervals have reduced field magnitudes and **B** vectors pointing across the tail. The relative paucity of current sheet sampling is indicative of the much larger spatial extent of the tail lobes. Figure 44 shows the same angle, for the same statistical sampling as in the previous figure, plotted versus location across the tail along the direction of the average cross-flow magnetic field. The current sheet is encountered at all locations across the tail, as are both tail lobe polarities. However, the tailward pointing lobe (negative angles) is preferentially found on the $+ Y$ side, while sunward pointing **B** (positive angles) is found on the opposite side.

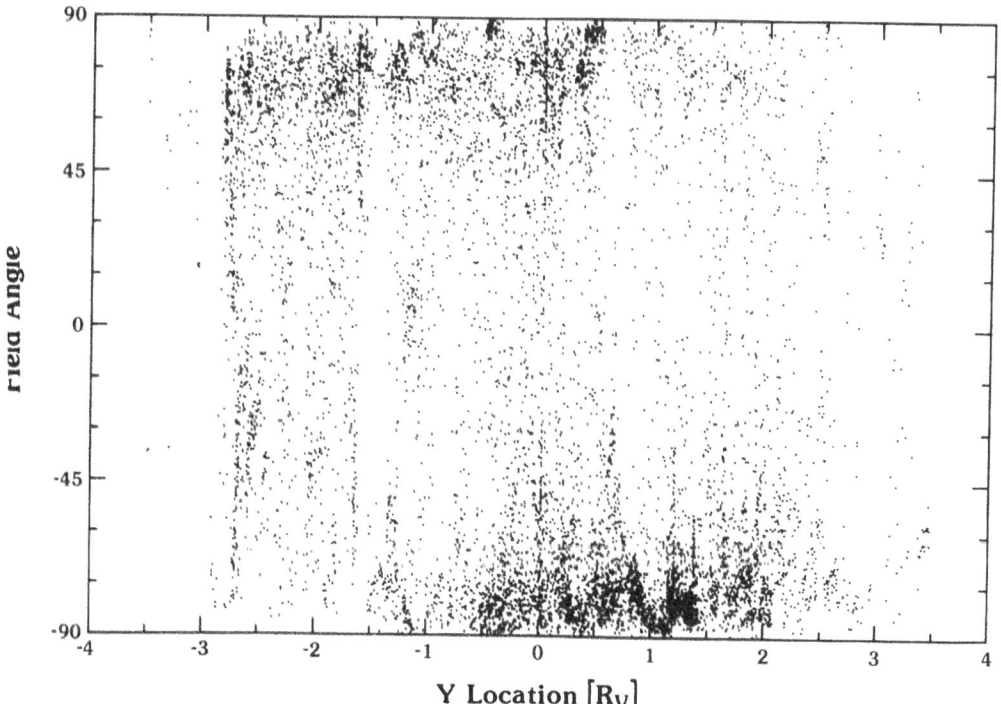

Fig. 44. Same magnetic field draping angle as in Figure 43, plotted versus the cross-tail location in the **B – V** plane. While either lobe and current sheet are observed at all cross-tail locations, the tailward (sunward) pointing lobe is preferentially found on $+ Y(- Y)$ side, consistent with IMF draping. From McComas *et al.* (1986).

While the PVO magnetotail magnetic field data are clearly ordered by draping into a cross-tail current sheet and two tail lobes, there is a tremendous amount of field variability on an orbit-by-orbit basis. Such variability is in part due to flapping of the entire tail structure caused by variations in the upstream solar wind velocity and by rotations in the upstream IMF. In addition, and perhaps more importantly, variations in the upstream IMF spiral angle cause the cross-tail current sheet to flap within the tail itself (McComas *et al.*, 1986). The combination of these effects makes it very difficult to separate the spatial configuration of the tail from temporal variations on a single orbit basis. Thus recent studies of the Venus tail magnetic configuration have centered on statistical analyses (e.g., Saunders and Russell, 1986; McComas *et al.*, 1986). On average, the mapping of the upstream spiral field configuration into the tail causes the

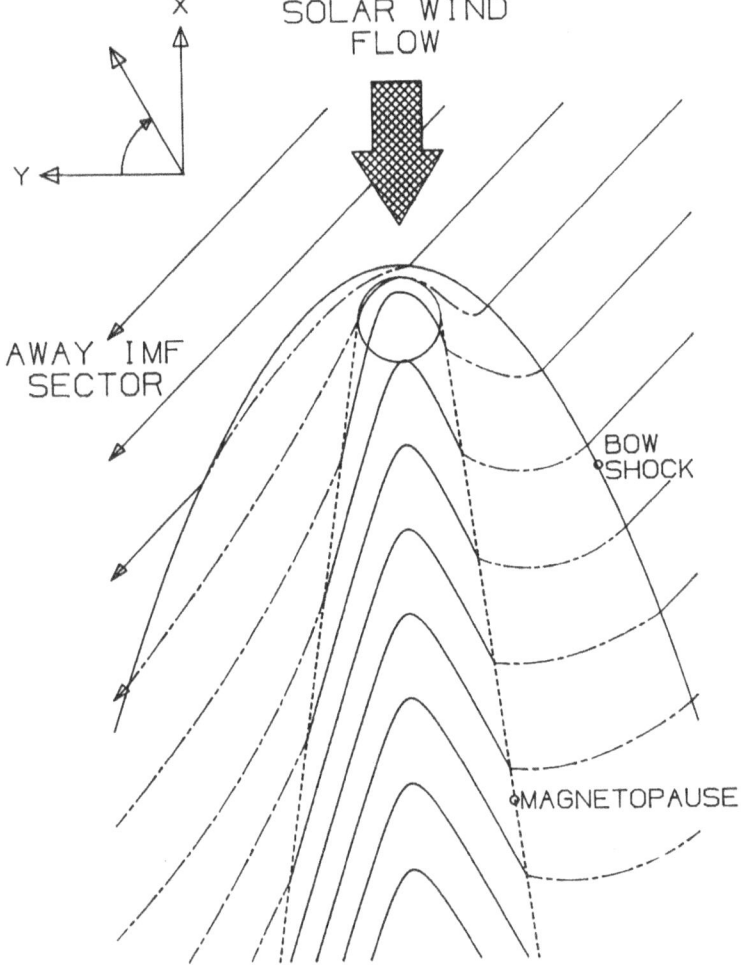

Fig. 45.   Schematic representation of the draped magnetic field regions around and behind Venus in the plane of the upstream solar wind velocity and IMF. The tail boundary (magnetopause) is dashed as are the magnetosheath portions of the field which interconnect the IMF with the tail field. Note that the mapping of the normal IMF spiral configuration into the tail requires that there be more magnetic flux in the lobe which lags planetary motion than in the leading lobe at any given downtail location. From McComas *et al.* (1986).

tail lobes to be uneven in size and the cross-tail current sheet to be offset towards the $-Y$ side of the tail as shown schematically in Figure 45. This figure summarizes the draping induced (east–west) asymmetries of the magnetotail. The characteristic thickness of the current sheet is approximately the same as the planetary diameter (McComas *et al.*, 1986), which suggests that the direct interaction between the flowing plasma and ionospheric obstacle plays a major role in tail formation (McComas *et al.*, 1987a).

In addition to east–west draping effects, with asymmetry sense determined by IMF sector polarity, the magnetic field data also display a substantial north–south (that is, convective electric field directed away or toward the central tail plane) asymmetry. Unfortunately, there is no coordinate system which can combine both toward and away IMF sector data and maintain both east–west and north–south tail asymmetries. For example, the coordinate transformations used by McComas *et al.* (1986) retain the east–west (draping) asymmetries displayed in Figure 45, but mix electric field polarities and thus wash out any north–south asymmetries. For this reason, as well as to improve the statistics, the sample space was compressed into the central tail plane by those authors. In contrast, the coordinate transformations used by Saunders and Russell (1986) conserve the polarity of the motional electric field and hence north–south asymmetries, but mix any east–west effects. Figure 46 shows the north–south asymmetry of the cross-tail magnetic field component. The cross-tail magnetic field magnitude is larger on average in the $+Z$ ('north') half of the tail, in which the upstream **E** points away from the central tail plane. This field asymmetry has the same sense as that noted in the magnetosheath and attributed to ion pickup by Phillips *et al.* (1987); the role of the electric field in the ordering of magnetotail plasma observations will be discussed in the next section. The reader is cautioned that any apparent east–west asymmetry in this figure may be spurious due to combining of IMF sectors and thus mixing the draping patterns shown on the left and right halves of Figure 45.

The issue of magnetic field reconnection in the Venus magnetotail has been raised by several authors. For example, Kivelson and Russell (1983) found evidence in one PVO orbit for a field geometry favorable for reconnection across the tail current sheet. Marubashi *et al.* (1985) suggested that reconnection could occur in the near wake region as a result of convergence of the magnetosheath flow behind the planet. While no direct evidence of reconnection dynamics has been observed, note in Figure 46 that some of the spatial bins contain cross-tail magnetic field oriented opposite (negative B_Y) to the average field direction. McComas *et al.* (1986) statistically examined the likelihood of tail reconnection by comparing the variability of large magnetotail and adjacent magnetosheath magnetic field data sets, and set an upper bound of 5% for such oppositely directed magnetic flux. While the existence and the importance of magnetotail reconnection has not been conclusively resolved, it appears at present that it plays at most a minor role in tail dynamics.

6.3. PLASMA AND PLASMA WAVE OBSERVATIONS

In contrast to the extensive study of magnetic field observations, there has been far less analysis of plasma measurements in the Venus magnetotail. This is largely due to the

VENUS CROSS-TAIL MAGNETIC FIELD

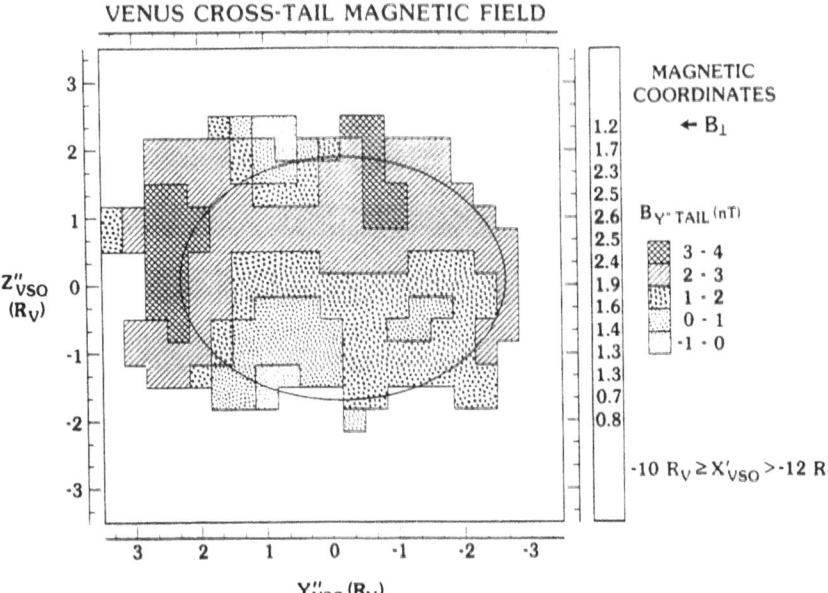

Fig. 46.　Cross-tail magnetic field component as a function of location in the tail when ordered by the field orientation. The field component crossing the tail is nearly twice as large on average in the half where the motional electric field is pointing away from the planet (*top*) than in the other (*bottom*). From Saunders and Russell (1986).

fact that the PVO plasma instruments were designed for solar wind or ionospheric sampling, and were not well-suited for the tail plasma environment. Only limited observations with the electrostatic plasma analyzer onboard PVO have proven useful for probing the tail properties, and such analysis is problematic as described below. Much of the current quantitative knowledge of the tail plasma is based on measurements from earlier missions. The earliest plasma observations were from the Venera 9 and 10 and Mariner 10 spacecraft. Yeates *et al.* (1978) reported two characteristic types of electron distributions observed during the single Mariner 10 flyby within 80 R_V downstream of Venus. These distributions were marked by differences in the orientation of the electron anisotropy and flow speed, and by differing levels of magnetic fluctuations. Electron densities were similar to those observed in the solar wind near Venus ($\sim 10\text{--}15\text{ cm}^{-3}$).

　　The most interesting Venera plasma observations within the magnetotail were taken with Venera 10 on the April 17 and April 19, 1976 passes at $\sim 0.5\text{--}6\,R_V$ downtail (Romanov *et al.*, 1978). The April 19 results were shown previously as Figure 35. In that figure, ions measured within the tail (after $\sim 03:00$ Moscow time) contain a very cold component traveling most slowly nearest the center of the tail, with a minimum speed of roughly half of that of the solar wind (based on the assumption that the ions are protons). Those authors suggested that these plasma observations as well as the relatively sharp boundaries between the various plasma regions are indicative of a

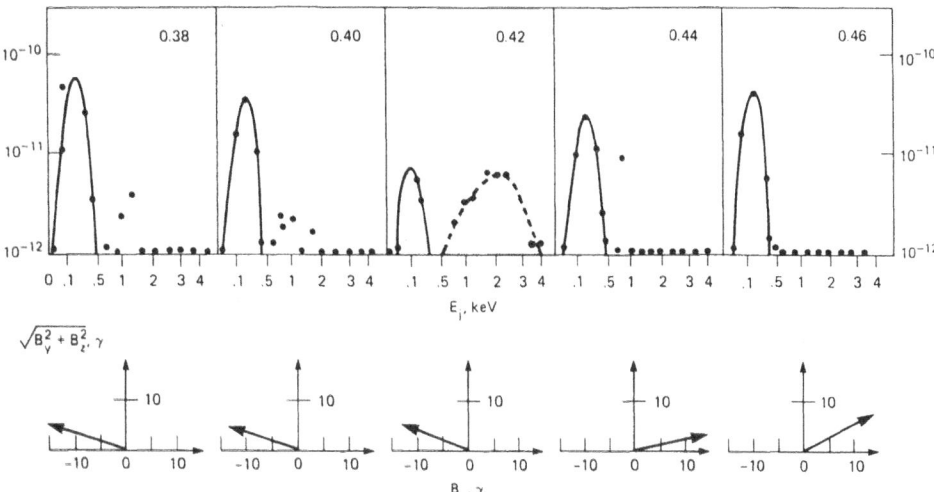

Fig. 47.   Ion spectra and magnetic field observations from the April 19, 1976 Venera 10 pass down the near-Venus tail. Intense flows of energetic ions (shaded) were observed coincidently with the reversal of the B_x field component. From Sagdeev *et al.* (1983).

plasma tail of ionospheric origin. Ion spectra (top panels) and magnetic field observations (bottom panels) from the same Venera 10 pass are displayed in Figure 47. The local magnetic field is shown in a cylindrical projection defined by the Venus–Sun line (X-axis). The reversal of the field between the third and fourth panel columns indicates crossing of the cross-tail current sheet. At the same time, an additional (shaded) peak is observed in the ion spectra. This higher energy ion peak, centered near 2 keV and with a temperature of $\sim 10^3$ eV, was interpreted by Verigin *et al.* (1978) as plasma sheet ions.

At distances greater than $\sim 5\,R_V$ downtail, the only other plasma observations available are those from Pioneer Venus. However, the PVO plasma analyzer was designed for solar wind observations and is inadequate in several respects for magnetotail measurements. The typical PVO plasma spectrum measured in the tail region has lower peak E/q than in the magnetosheath, suggesting slowed solar wind plasma, although some of the time the ion fluxes are sufficiently low that no signal is observed above the instrumental noise level. On rare occasions a second peak is simultaneously present in the PVO ion spectra at higher energy per charge (Intriligator, 1982; Mihalov and Barnes, 1982). Reliable plasma moments cannot be routinely derived from the ion counts in the magnetotail, although a best guess sort of calculation for some of these double peaked ion spectra indicates an O^+ ion temperature of a few $\times 10^5$ K (Mihalov and Barnes, 1982).

Figure 48 displays three double peaked events from three different PVO orbits. Note that the E/q ratios of the two peaks are consistent with the interpretation of singly or doubly ionized oxygen (higher peaks) convecting with the slowed solar wind hydrogen and alphas (lower peaks). For this reason, these signatures have been interpreted as

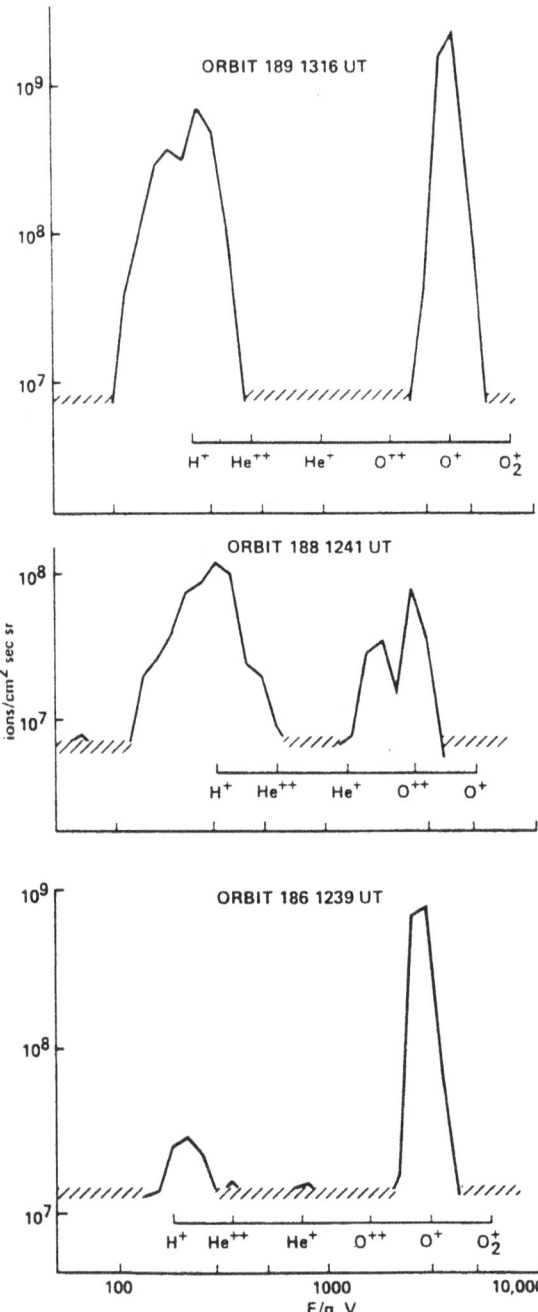

Fig. 48. Energy/charge spectra from three PVO orbits which display second peaks at the appropriate locations for singly or doubly ionized oxygen convecting with the same speed as the protons (primary peaks). From Mihalov and Barnes (1982).

direct evidence of planetary ions which have been picked up by the flowing solar plasma. Such ions could either be directly scavenged from the upper ionosphere (e.g., Wolff et al., 1980; Intriligator, 1982) or could represent initially neutral exospheric material which became ionized above the ionopause and was accelerated by the solar wind flow (e.g., Cloutier et al., 1974; Phillips et al., 1987). Comparison of the Venus and Comet Giacobini–Zinner magnetotails (McComas et al., 1987a) indicated that, although the pickup of newly ionized exospheric material contributes to the bending of the magnetic field lines, it is the direct interaction of the ionospheric obstacle with the flowing plasma which creates the strongly draped magnetotail. Consequently, the majority of the planetary material found in the Venus magnetotail has probably been assimilated along those streamlines which pass closest to the planet. In any case, the planetary ions are found flowing in essentially the same downtail direction as the bulk solar wind independent of local magnetic field direction (Intriligator et al., 1982, Perez-de-Tejada et al., 1982), suggesting that the assimilation process is nearly complete after a few R_V of downstream travel.

The only statistical study to date of pickup ions in the tail (Intriligator, 1989) shows that the locations of pick-up ion observations are well ordered by the orientation of the solar wind motional electric field ($\mathbf{E} = -\mathbf{V} \times \mathbf{B}$) just as they are nearer to the planet. Figure 49 summarizes the locations of 167 ion pickup events observed by the PVO

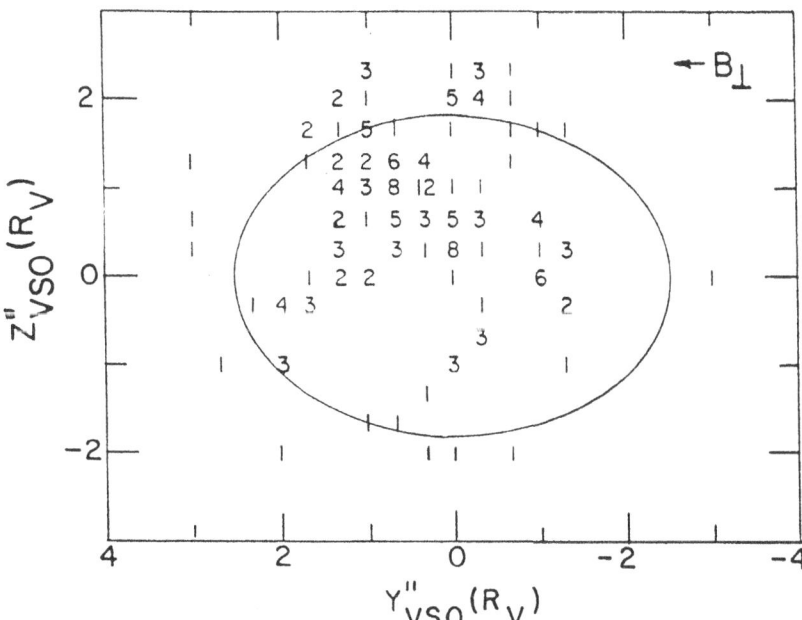

Fig. 49.   Locations of 167 PVO E/q spectra containing pickup ion signatures in the region downstream from Venus. Data are plotted in magnetic coordinates in which the average cross-flow field component points to the left and the motional electric field points upward. More than four times as many of these pickup events are found in the half plane where the motional electric field is pointing away from the planet. From Intriligator (1989).

plasma analyzer. The events are displayed in a coordinate system in which the upstream motional electric field is pointing toward the top of the page. Clearly these observations preferentially occur in the tail regions in which $\mathbf{E}$ points away from the central tail plane, which is consistent with ions gyrating initially parallel to the electric field and then downtail. (Similarly to Figure 46, any apparent east–west asymmetry in Figure 49 may be spurious due to the coordinate rotations used here.) Slavin et al. (1989) also found this electric field related asymmetry in the location of O^+ observations for 12 carefully selected PVO orbits. Additional results of this study were that the greatest concentration of the planetary ions was observed in the vicinity of the cross tail current sheet and that essentially no O^+ was observed outside the tail. However, the authors noted that, since the plasma flow speed is characteristically slowest in the center of the tail and fastest outside the tail, and the PVO plasma analyzer can only detect O^+ traveling less than 310 km s^{-1}, this may simply be a selection effect.

In addition to the in situ plasma measurements described above, McComas et al. (1986) used the observed average magnetic field configuration of the Venus magnetotail and the ideal MHD equations to infer the consistent average plasma properties of the tail. These calculations provided average downtail flow speeds which vary from ~ 250 km s^{-1} at $8 R_V$ downtail to ~ 470 km s^{-1} at $12 R_V$. The calculated plasma densities, ion temperatures, and plasma-β for the Venus tail lobes and cross-tail current sheet are given in Table I for both the proton and oxygen ions. The derived current sheet β of 12 is similar, given the large experimental uncertainties, with the value of 8.5 previously calculated by Slavin et al. (1984). It is interesting to note that these derived average plasma properties do not fall within the parameter range measurable with the PVO plasma analyzer (McComas et al., 1987b). In addition, broadband plasma wave noise is considerably higher overall in the Venus tail than in the Earth's tail, indicating generally higher levels of turbulence downstream from Venus (Intriligator and Scarf, 1984). This observation is consistent with the unusually high derived temperatures shown in Table I. Integration over the appropriate tail cross section of the number flux

TABLE I

Summary of the average Venus tail plasma parameters derived from the average magnetic field configuration using the MHD equations. From McComas et al. (1987b)

	Average consistent plasma properties	
	P^+	O^+
ρ_{lobe}	$\sim 1.2 \times 10^{-22}$ kg m^{-3}	
n_{lobe}	~ 0.07 cm^{-3}	~ 0.005 cm^{-3}
ρ_{CS}	$\sim 1.6 \times 10^{-21}$ kg m^{-3}	
n_{CS}	0.9 cm^{-3}	~ 0.06 cm^{-3}
$T_{isothermal}$	$\sim 6 \times 10^6$ K	$\sim 9 \times 10^7$ K
	$\beta_{lobe} \simeq 0.08$	$\beta_{CS} \simeq 12$

down the tail calculated by this technique gives a mass loss rate of roughly 10^{26} amu s^{-1} or $\sim 6 \times 10^{24}$ O$^+$ ions per second, in excellent agreement with the previous estimate by Mihalov and Barnes (1982) of $\lesssim 10^{25}$ ion s^{-1}. Finally, because these derived properties are based on MHD considerations such as pressure balance, they do not rule out another coexisting plasma population in the tail which supplies very little pressure (e.g., extremely cold ionospheric material).

Enhanced plasma wave activity, apparently associated with changing ion distributions, has been observed by the PVO electric field instrument in the Venus tail (Intriligator and Scarf, 1982, 1984). These plasma waves have been interpreted as Doppler-shifted ion-acoustic waves. The coincident signatures in the electric field and in the ion distributions has been suggested to be indicative of the tail boundary region, yielding boundary locations which can be quite different from those identified on the basis of magnetic field measurements alone (Intriligator and Scarf, 1984). This point highlights the great difficulty in definitively separating data taken in the Venus magnetotail from that in the surrounding magnetosheath, given the limitations of the PVO plasma analyzer and the large amount of variability in the magnetic field measurement. Slavin *et al.* (1989) found that the tail boundary is a relatively sharp 'magnetopause like' current layer over the hemisphere which exhibits less mass loading (E toward the planet) but is broader and resembles a slow mode expansion fan over the more heavily mass loaded hemisphere. This asymmetry, like that observed in the location of picked

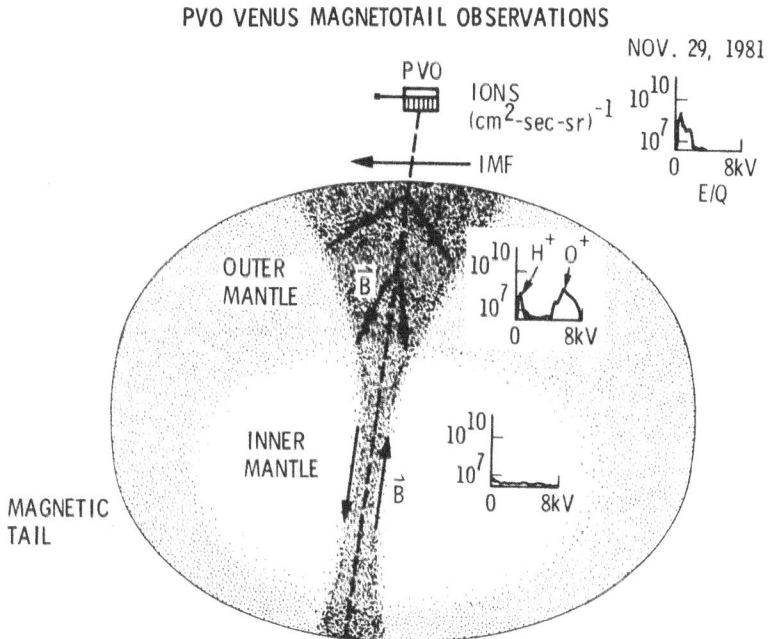

Fig. 50.   Schematic diagram of the magnetic field and plasma conditions in the Venus tail. From Slavin *et al.* (1989).

up heavy ions, was interpreted as being due to the more efficient pickup in the hemisphere of locally upward **E**. Slavin *et al.* have synthesized this result with north–south hemispherical asymmetries in the location of pickup ions and in the magnitude of the cross tail magnetic field component (Figure 46) to derive the cross-sectional tail schematic shown in Figure 50. These authors suggest that the draped field in the magnetotail may be preferentially supplied over the hemisphere with the locally outward electric field.

7. Conclusions

7.1. Current understanding

The last few years of research into the interaction of the solar wind with Venus have been productive and enlightening. Much of the research can be categorized by one or more of the following general descriptions: (1) analysis of large data sets of PVO observations in order to describe the average characteristics of the interaction; (2) theoretical and observational work concerning the effects of pickup of planetary ions by the solar wind; (3) study of solar cycle effects on various aspects of the interaction; (4) improved numerical simulations. We will attempt to summarize below the current state of knowledge concerning the magnetosheath and magnetotail, emphasizing recent findings.

The boundaries of the magnetosheath, that is the bow shock and ionopause, both exhibit solar cycle effects and asymmetries ordered by the IMF. The bow shock is observed to be farthest from the planet during periods of high solar EUV flux, as at solar maximum, and to have an elliptical cross section caused by the differing speeds of propagation of a fast magnetosonic shock wave parallel and perpendicular to **B**. The shock is at lowest altitudes when the upstream **V** and **B** are aligned, is higher when **V** and **B** are crossed, and is slightly farther from the planet in the hemisphere of locally upward motional **E** (e.g., Russell *et al.*, 1988). The shock is somewhat blunter (higher at the terminator) than predicted by gasdynamic simulations, suggesting an enhanced planetary obstacle (Russell *et al.*, 1985), MHD effects, or perhaps inflation of the magnetosheath by planetary ions (cf. Belotserkovskii *et al.*, 1987; Stahara *et al.*, 1987).

The ionopause was shown to be much lower at solar minimum than at solar maximum (e.g., Knudsen *et al.*, 1987), presumably due to reduction in peak ionospheric pressure to a value insufficient to stand off the solar wind. This finding suggests that a magnetized ionosphere may be the prevalent condition near solar minimum, though the evolution of the PVO orbit has so far precluded *in situ* confirmation of this notion. The ionopause near solar maximum was shown by Phillips *et al.* (1988) to be lower in the hemisphere of upward electric field than in the opposite hemisphere, an asymmetry attributed to momentum exchange between newborn planetary ions and the magnetosheath plasma. Although most published literature supports the notion of a magnetic barrier and accompanying plasma depletion layer above the dayside ionopause, an alternative shielding current system has been proposed by Cloutier *et al.* (1987). The disagreement over the nature of the shielding currents is essentially equivalent to the ongoing controversy concerning the magnetization of the dayside ionosphere (cf. Cloutier *et al.*, 1987; Shinagawa and Cravens, 1988). However, we feel at this time that the observational evidence and the most convincing theoretical and computational analysis support the magnetic barrier picture, with large-scale ionospheric magnetic fields generated by downward diffusion and convection.

The magnetosheath has been the subject of recent attention from PVO researchers. Phillips *et al.* (1986b) characterized the prevailing magnetic field magnitude and draping configuration, and additionally found that the magnetic field was enhanced in the hemisphere with locally upward motional electric field. In a subsequent study, Phillips *et al.* (1987) found that PVO observations of pickup ions occurred preferentially in the upward **E** hemisphere, and used a numerical experiment to show that the initial gyration of newborn ions constituted a current which was appropriate to account for the enhanced field magnitude. The issues of hydromagnetic fluctuations in the magnetosheath were addressed by Winske (1986) and Luhmann *et al.* (1987a), who found that the source of the observed fluctuations was most likely the quasiparallel portions of the bow shock. The viscous or inviscid nature of the magnetosheath-ionosphere plasma interface remains unclear. While the viscous scenario has received support from observations of heated solar wind plasma above the terminator ionopause (Perez-de-Tejada *et al.*, 1985), the evidence is far from conclusive and the nature of the postulated viscous mechanism is unknown.

Large statistical studies of PVO magnetic field observations have clarified the average magnetic configuration of the magnetotail from ~ 8 to $12\,R_V$ downstream. The tail consists of draped interplanetary magnetic field, similar in general characteristics to the cometary tails predicted by Alfvén (1957) and containing typically 3 MWb of magnetic flux (Saunders and Russell, 1986). The specifics of the magnetic draping were explored by McComas *et al.* (1986) using coordinate transformations designed to eliminate the effects of tail flapping and to isolate draping-related asymmetries. That study found that east–west asymmetric draping effects were easily observable, with the tail lobe which lags planetary motion containing significantly more magnetic flux at any given downtail distance for a typical spiral IMF of either polarity. One of the findings of the Saunders and Russell (1986) study was that the magnetic field crossing the magnetotail was enhanced where the motional electric field pointed away from the central tail plane. This enhancement suggested a mechanism involving pickup of planetary ions, as was later demonstrated for the magnetosheath (Phillips *et al.*, 1987). Such a correlation was discovered observationally by Intriligator (1989), who found that a large majority of tail E/q spectra containing pickup ion signatures were found in the upward **E** sections of the tail or downstream magnetosheath. Thus the effects of ion pickup in the near-planet region appear to persist well downtail. Similar results were demonstrated by Slavin *et al.* (1989) using case studies of PVO magnetic field and plasma observations. Thus far, however, the average plasma parameters in the tail have only been inferred from the magnetic field observations (McComas *et al.*, 1986, 1987b). These properties, shown in Table I, await observational confirmation or repudiation.

Primarily due to the lack of PVO observations in the 2–6 R_V downstream range, little new knowledge concerning the mapping of the magnetotail to the well-defined near-planet regions has become available. One key observational fact related to this mapping is that the cross-tail 'plasma sheet' appears to be roughly one planetary diameter across (e.g., McComas *et al.*, 1986; Slavin *et al.*, 1989). This suggests a connection with the planetary ionosphere, similar to that shown in Figure 2 (Vaisberg *et al.*, 1986). Further, the enhanced magnetic field magnitude observed in the tail lobes may indicate a connection to the similarly enhanced magnetic barrier region in the low altitude magnetosheath, perhaps as shown in Figure 27 (Spenner *et al.*, 1980). This interpretation, in which the plasma sheet is populated by ionospheric plasma scavenged on the nightside while the magnetosheath and tail lobes contain the more diffuse and faster moving exospheric pickup ions, is but one of the many possible connections of the downstream topological regions to the near-planet environment. Saunders *et al.* (1986) proposed a similar idea, suggesting that the interface between the mantle and the magnetosheath probably forms the outer boundary of the tail. Definitive resolution of this mapping is one of the major unanswered questions concerning the Venus-solar wind interaction.

7.2. RECOMMENDATIONS FOR NEAR-TERM RESEARCH

Many opportunities exist for further research using currently available particle and field observations and modern numerical simulations. As noted above, the connection of the deep tail to the near-planet environment remains unknown, and should clearly be a focus of ongoing observational research. One relevant question which should be easily answerable concerns the rarefaction wave noted by Vaisberg *et al.* (1976) downstream of the planet (Figure 2). Is this feature consistently (or sporadically) observable in the near-planet magnetosheath? Thorough analysis of PVO plasma observations in and near the tail, despite the shortcomings of such observations, may provide key links in the overall interaction scenario. We are aware of one such study in progress (K. R. Moore, unpublished manuscript), and are encouraged by preliminary results which we will not review here.

A spatial region which is observed by PVO but which has received little attention thus far is the downstream magnetosheath. Magnetic field and plasma observations there could provide estimates of the degree of mixing of solar and planetary plasmas and perhaps help resolve the viscous heating vs isentropic expansion issue. Another subject which deserves more analysis is the generation and transport of bulk scavenged ionospheric plasma, observable in the near-planet environment as clouds and streamers (e.g., Brace *et al.*, 1987). Analysis of upstream conditions which modulate the occurrence rate or location of these features, and perhaps a search of observations farther downstream for indirect signatures, could provide constraints on the importance of bulk scavenging, as opposed to diffuse exospheric pickup, of planetary ions. This subject is of particular interest due to a possible analogy with cometary rays.

Much potential remains for further development in numerical simulation of the Venus-solar wind interaction. A three-dimensional MHD simulation of the basic interaction should be within the current state of the art. Such a simulation should incorporate realistic downstream free boundary conditions, rather than an impenetrable wake, so as to produce a realistic magnetotail and downstream shock and sheath. The next modeling step should be to include realistic mass, momentum, and energy source terms due to planetary ions. This might be done by iterative experimentation with test particle arrays, mapping the regions of mass, momentum and energy exchange between the test particles and the ambient plasma and using the results as source terms for the next iteration. Ideally, however, a hybrid simulation with fluid solar plasma and discrete test particles should be used. While we recognize that such a global hybrid simulation may exceed the limits of today's techniques and computational resources, it should nevertheless be a goal for future models.

7.3. RECOMMENDATIONS FOR FUTURE SPACECRAFT

The successes and shortcomings of the various spacecraft missions to Venus have resulted in evolution of our notions of the key unanswered questions and the spacecraft and instrument characteristics required to resolve these questions. We will attempt to outline a desirable orbit and field and particle instrument suite for the next Venus orbiter, focusing entirely on magnetosheath and magnetotail science and ignoring other scientific objectives. The primary measurements required to resolve sheath and tail objectives are (1) magnetic field and (2) plasma observations with appropriate energy, mass, time, and angular resolution. Other instruments, perhaps including a plasma wave detector and an energetic particle sensor, may also be helpful but are not absolutely essential, and so will not be considered here. The performance characteristics of the PVO magnetometer have been entirely satisfactory, and would probably suffice admirably for future missions. The plasma instrumentation, however, requires major improvements relative to all previous missions.

We recommend a baseline configuration of a spinning spacecraft with spin plane roughly aligned with the Venus orbital plane and period on the order of 10 s, similar to PVO. To answer the unresolved scientific questions in the magnetosheath and tail using plasma instrument technology available today would require two separate instruments. First and most important would be a 'hot' plasma analyzer incorporating both E/q and mass per charge (m/q) resolution, perhaps using time of flight technology. We suggest the following instrument parameters: (1) E/q range of 100 eV to 40 keV (this upper limit corresponds to a speed of nearly 700 km s^{-1} for O$^+$); (2) m/q resolution up to at least 32 amu per charge, corresponding to O$_2^+$; (3) capability of producing a complete three dimensional spectrum spanning most of 4π steradians in velocity space in one spacecraft spin. Complementing this analyzer would be a 'cold' plasma analyzer with similar m/q range, angular resolution, and integration time, and the following additional characteristics: (1) E/q range of 1 eV to 300 eV; (2) ability to resolve densities of less than 1 cm^{-3}.

The two plasma analyzers described above, optimized for magnetosheath and magnetotail science, would together provide a minimal capability for solar wind analysis. In order to provide continuous upstream solar wind monitoring, we suggest two identical spacecraft, with similar lines of apsides but out of phase by 180°. Desirable periapsis would be ~ 200 km, while apoapsis would be in the 10–$15 \, R_V$ range. A low inclination orbit would provide sampling of the tail at variable downstream distances during seasons with periapsis on the dayside but not near local noon. An alternative method would be to have a single orbiter with the plasma instrumentation described above, and a second simpler spacecraft, located at the upstream Lagrange point, equipped only with a magnetometer and a specialized solar wind instrument. This upstream monitoring capability, achieved either with twin spacecraft or with a dedicated monitor, should be considered essential for the next generation of Venus orbiters.

Acknowledgements

We thank K. R. Moore for helpful discussions and for critical reading of the manuscript, J. G. Luhmann for valuable suggestions, and R. N. Robichaud for redrafting of several figures. We also thank F. S. Johnson, W. C. Knudsen, J. G. Luhmann, J. D. Mihalov, A. F. Nagy, C. T. Russell, J. A. Slavin, S. S. Stahara, and D. Winske for providing original figures. We appreciate the careful and constructive reviews of the three referees. This review was prepared at Los Alamos National Laboratory under the auspices of the United States Department of Energy.

References

Alexander, C. J. and Russell, C. T.: 1985, 'Solar Cycle Dependence of the Location of the Venus Bow Shock', *Geophys. Res. Letters* **12**, 369.

Alexander, C. J., Luhmann, J. G., and Russell, C. T.: 1986, 'Interplanetary Field Control of the Venus Bow Shock: Evidence for Comet-Like Ion Pickup', *Geophys. Res. Letters* **13**, 917.

Alfvén, H.: 1957, 'On the Theory of Comet Tails', *Tellus* **9**, 92.

Barth, C. A.: 1968, 'Interpretation of the Mariner 5 Lyman Alpha Measurements', *J. Atmospheric Sci.* **25**, 564.

Barth, C. A., Pearce, J. B., Kelly, K. K., Wallace, L., and Fastie, W. G.: 1967, 'Ultraviolet Emissions Observed near Venus from Mariner V', *Science* **158**, 1675.

Belotserkovskii, O. M., Breus, T. K., Krymskii, A. M., Mitnitskii, V. Ya., Nagy, A. F., and Gombosi, T. I.: 1987, 'The Effect of Hot Oxygen Corona on the Interaction of the Solar Wind with Venus', *Geophys. Res. Letters* **14**, 503.

Bertaux, J. L., Blamont, J. E., Marcelin, M., Kurt, V. G., Romanova, N. N., and Smirnov, A. S.: 1978, 'Lyman-Alpha Observations of Venera 9 and 10, I. The Non-Thermal Hydrogen Population in the Exosphere of Venus', *Planetary Space Sci.* **26**, 817.

Bertaux, J. L., Blamont, J. E., Lepine, V. M., Kurt, V. G., Romanova, N. N., and Smirnov, A. S.: 1981, 'Venera 11 and Venera 12 Observations of EUV Emissions from the Upper Atmosphere of Venus', *Planetary Space Sci* **29**, 149.

Bertaux, J. L., Lepine, V. M., Kurt, V. G., and Smirnov, A. S.: 1982, 'Altitude Profile of H in the atmosphere of Venus from Lyman α Observations of Venera 11 and Venera 12 and Origin of the Hot Exospheric Component', *Icarus* **52**, 221.

Bertaux, J. L., Chassefiere, E., and Kurt, V. G.: 1985, 'Venus EUV Measurements of Hydrogen and Helium from Venera 11 and Venera 12', *Adv. Space Res.* **5**, 119.

Brace, L. H.: 1987, 'Additional Pioneer Venus Observations of the Precursor to the Venus Bow Shock (Abstract)', *EOS Trans. AGU* **68**, 342.

Brace, L. H., Theis, R. F., Hoegy, W. R., Wolfe, J. H., Mihalov, J. D., Russell, C. T., Elphic, R. C., and Nagy, A. F.: 1980, 'The Dynamic Behavior of the Venus Ionosphere in Response to Solar Wind Interactions', *J. Geophys. Res.* **85**, 7663.

Brace, L. H., Theis, R. F., and Hoegy, W. R.: 1982a, 'Plasma Clouds Above the Ionopause of Venus and Their Implications', *Planetary Space Sci.* **30**, 29.

Brace, L. H., Theis, R. F., Mayr, H. G., Curtis, S. A., and Luhmann, J. G.: 1982b, 'Holes in the Nightside Ionosphere of Venus', *J. Geophys. Res.* **87**, 199.

Brace, L. H., Curtis, S. A., Russell, C. T., and Scarf, F. L.: 1985, 'A Precursor to the Venus Bow Shock (Abstract)', *EOS Trans. AGU* **66**, 294.

Brace, L. H., Kasprzak, W. T., Taylor, H. A., Theis, R. F., Russell, C. T., Barnes, A., Mihalov, J. D., and Hunten, D. M.: 1987, 'The Ionotail of Venus: Its Configuration and Evidence for Ion Escape', *J. Geophys. Res.* **92**, 15.

Breus, T. K.: 1979, 'Solar Wind Interaction', *Space Sci. Rev.* **23**, 253.

Breus, T. K.: 1986, 'Mass-Loading at Venus: Theoretical Expectations', *Adv. Space Res.* **6**, 167.

Breus, T. K., Krymskii, A. M., and Mitnitskii, V. Ya.: 1987a, 'Effect of an Extended Neutral Atmosphere on the Interaction of the Solar Wind and the Nonmagnetic Bodies of the Solar System. I. Venus', *Kosmich. Issled.* **25**, 124.

Breus, T. K., Krymskii, A. M., Mitnitskii, V. Ya., Gombosi, T., and Nagy, A.: 1987b, 'Role of the Hot Oxygen Corona in the Interaction of Venus with the Solar Wind', *Kosmich. Issled.* **25**, 626.

Bridge, H. S., Lazarus, A. J., Snyder, C. W., Smith, E. J., Davis, L., Jr., Coleman, P. J., Jr., and Jones, D. E.: 1967, 'Mariner V: Plasma and Magnetic Fields Observed near Venus', *Science* **158**, 1669.

Bridge, H. S., Lazarus, A. J., Scudder, J. D., Ogilvie, K. W., Hartle, R. E., Asbridge, J. R., Bame, S. J., Feldman, W. C., and Siscoe, G. L.: 1974, 'Observations at Venus Encounter by the Plasma Science Experiment on Mariner 10', *Science* **183**, 1293.

Brinca, A. L.: 1984, 'On the Coupling of Test Ions to Magnetoplasma Flows through Turbulence', *J. Geophys. Res.* **89**, 115.

Cloutier, P. A.: 1976, 'Solar Wind Interaction with Planetary Ionospheres', in N. Ness (ed.), *Solar Wind Interaction with Planets Mercury, Venus and Mars*, NASA Spec. Publ. SP-397, p. 111.

Cloutier, P. A. and Daniell, R. E., Jr.: 1973, 'Ionospheric Currents Induced by Solar Wind Interaction with Planetary Atmospheres', *Planetary Space Sci.* **21**, 463.

Cloutier, P. A., Daniell, R. E., Jr., and Butler, D. M.: 1974, 'Atmospheric Ion Wakes of Venus and Mars in the Solar Wind', *Planetary Space Sci.* **22**, 967.

Cloutier, P. A., Tascione, T. F., Daniell, R. E., Taylor, H. A., and Wolff, R. S.: 1983, in D. H. Hunten, L. Colin, and T. M. Donahue (eds.), 'Physics of the Interaction of the Solar Wind with the Ionosphere of Venus', *Venus*, Univ. of Arizona, Tucson, pp. 941–979.

Cloutier, P. A., Taylor, H. A., Jr., and McGary, J. E.: 1987, 'Steady State Flow/Field Model of Solar Wind Interaction with Venus: Global Implications of Local Effects', *J. Geophys. Res.* **92**, 7289.

Cravens, T. E., Nagy, A. F., and Gombosi, T. I.: 1980, 'Hot Hydrogen in the Exosphere of Venus', *Nature* **283**, 178.

Cravens, T. E., Brace, L. H., Taylor, H. A., Jr., Russell, C. T., Knudsen, W. L., Miller, K. L., Barnes, A., Mihalov, J. D., Scarf, F. L., Quenon, S. J., and Nagy, A. F.: 1982, 'Disappearing Ionospheres in the Nightside of Venus', *Icarus* **51**, 271.

Cravens, T. E., Shinagawa, H., and Nagy, A. F.: 1984, 'The Evolution of Large-Scale Magnetic Fields in the Ionosphere of Venus', *Geophys. Res. Letters* **11**, 267.

Curtis, S. A.: 1981, 'Solar Wind Pickup of Ionized Venus Exosphere Atoms', *J. Geophys. Res.* **86**, 4715.

Daniell, R. E., Jr.: 1981, 'A Source of Plasma Turbulence at the Ionopause of Venus', *J. Geophys. Res.* **86**, 10094.

Dessler, A. J.: 1968, in J. C. Brandt and M. B. McElroy (eds.), 'Ionizing Plasma Flux in the Martian Upper Atmospheres', *The Atmosphere of Venus and Mars*, Gordon and Breach, New York, pp. 241–250.

Dolginov, Sh. Sh., Yeroshenko, Ye. G., and Zhugov, D. N.: 1968, 'Magnetic Field Investigation with Interplanetary Station "Venera-4"', *Kosmich. Issled.* **6**, 561.

Dolginov, Sh. Sh., Zhugov, L. N., Sharova, V. A., and Buzin, V. B.: 1978, 'Magnetic Field and Magnetosphere of the Planet Venus', *Kosmich. Issled.* **16**, 827.

Dolginov, Sh. Sh., Dubinin, E. M., Yeroshenko, Ye. G., Izrailevich, P. L., Podgorny, I. M., and Shkol'nikova, S. I.: 1981, 'Field Configuration in the Magnetic Tail of Venus', *Kosmich. Issled.* **19**, 624.

Dubinin, E. M., Podgorny, I. M., Potanin, Ya. N., and Shkol'nikova, S. I.: 1978, 'Determining the Magnetic Moment of Venus by Magnetic Measurements in the Tail', *Kosmich. Issled.* **16**, 870.

Elco, R. A.: 1969, 'Interaction of the Solar Wind with Planetary Atmospheres', *J. Geophys. Res.* **74**, 073.

Elphic, R. C., Russell, C. T., Slavin, J. A., Brace, L. H., and Nagy, A. F.: 1980a, 'The Location of the Dayside Ionopause of Venus: Pioneer Venus Magnetometer Observations', *Geophys. Res. Letters* **7**, 561.

Elphic, R. C., Russell, C. T., Slavin, J. A., and Brace, L. H.: 1980b, 'Observations of the Dayside Ionopause and Ionosphere of Venus', *J. Geophys. Res.* **85**, 7679.

Elphic, R. C., Russell, C. T., Luhmann, J. G., Scarf, F. L., and Brace, L. H.: 1980c, 'The Venus Ionopause Current Sheet: Thickness, Length Scale and Controlling Factors', *J. Geophys. Res.* **86**, 11430.

Elphic, R. C., Mayr, H. G., Theis, R. F., Brace, L. H., Miller, K. L., and Knudsen, W. C.: 1984, 'Nightward Ion Flow in the Venus Ionosphere: Implication of Momentum Balance', *Geophys. Res. Letters* **11**, 1007.

Fjeldbo, G., Seidel, B., Sweetman, D., and Howard T.: 1975, 'The Mariner 10 Radio Occultation Measurements of the Ionosphere of Venus', *J. Atmospheric Sci.* **32**, 1232.

Formisano, V., Galeev, A. A., and Sagdeev, R. Z.: 1982, 'The Role of the Critical Ionization Velocity Phenomenon in Production of Inner Coma Cometary Plasma', *Planetary Space Sci.* **30**, 491.

Frank, L. A., Van Allen, J. A., and Hills, H. K.: 1963, 'Mariner 2: Preliminary Reports on Measurements of Venus, Charged Particles', *Science* **139**, 905.

Gombosi, T. I., Cravens, T. E., Nagy, A. F., Elphic, R. C., and Russell, C. T.: 1980, 'Solar Wind Absorption by Venus', *J. Geophys. Res.* **85**, 7747.

Gombosi, T. I., Horanyi, M., Cravens, T. E., Nagy, A. F., and Russell, C. T.: 1981, 'The Role of Charge Exchange in the Solar Wind Absorption by Venus', *Geophys. Res. Letters* **8**, 1265.

Greenstadt, E. W.: 1970, 'Dependence of Shock Structure at Venus and Mars on Orientation of the Interplanetary Magnetic Field', *Cosmic Electro* **1**, 380.

Gringauz, K. I.: 1983, in D. M. Hunten, L. Colin, T. M. Donahue, and V. I. Moroz (eds.), 'The Bow Shock and the Magnetosphere of Venus According to Measurements from Venera 9 and 10 Orbiters', *Venus*, Univ. Arizona, Tucson, pp. 980–993.

Gringauz, K. I., Bezrukikh, V. V., Musatov, L. S., and Breus, T. K.: 1968, 'Plasma Measurements in the Vicinity of Venus by the Space Vehicle "Venera-4"', *Kosmich. Issled.* **6**, 411.

Gringauz, K. I., Bezrukikh, V. V., Volkov, G. I., Musatov, L. S., and Breus, T. K.: 1970, 'Interplanetary Plasma Disturbances Near Venus Determined from "Venera-4" and "Venera-6" Data', *Kosmich. Issled.* **8**, 431.

Gringauz, K. I., Bezrukikh, V. V., Breus, T. K., Gombosi, T., Remizov, A. P., Verigin, M. I., and Volkov, G. I.: 1976, 'Preliminary Results of the Plasma Measurements with the Wide-Angle Sensors Aboard Venera-9 and Venera-10', *Kosmich. Issled.* **14**, 839.

Hodges, R. R. and Tinsley, B. A.: 1981, 'Charge Exchange in the Venus Ionosphere as the Source of Hot Exospheric Hydrogen', *J. Geophys. Res.* **86**, 7649.

Intriligator, D. S.: 1982, 'Observations of Mass Addition to the Shocked Solar Wind of the Venusian Ionosheath', *Geophys. Res. Letters* **9**, 727.

Intriligator, D. S.: 1989, 'Results of the First Statistical Study of Pioneer Venus Orbiter Plasma Observations in the Distant Venus Tail: Evidence for a Hemispheric Asymmetry in the Pickup of Ionospheric Ions', *Geophys. Res. Letters* **16**, 167.

Intriligator, D. S. and Scarf, F. L.: 1982, 'Plasma Turbulence in the Downstream Ionosheath of Venus', *Geophys. Res. Letters* **9**, 1325.

Intriligator, D. S. and Scarf, F. L.: 1984, 'Wave-Particle Interactions in the Venus Wake and Tail', *J. Geophys. Res.* **89**, 47.

Intriligator, D. S., Collard, H. R., Mihalov, J. D., Whitten, R. C., and Wolfe, J. H.: 1979, 'Electron Observations and Ion Flow from the Pioneer Venus Orbiter Plasma Analyzer Experiment', *Science* **205**, 116.

Johnson, F. S. and Hanson, W. B.: 1979, 'A New Concept for the Daytime Magnetosphere of Venus', *Geophys. Res. Letters* **6**, 581.

Johnson, F. S. and Midgley, J. E.: 1969, 'Induced Magnetosphere of Venus', *Space Res.* **9**, 760.

Kasprzak, W. T., Niemann, H. B., and Mahaffy, P.: 1987, 'Observations of Energetic Ions on the Nightside of Venus', *J. Geophys. Res.* **92**, 291.

Kivelson, M. G. and Russell, C. T.: 1983, 'The Interaction of Flowing Plasmas with Planetary Ionospheres: A Titan–Venus Comparison', *J. Geophys. Res.* **88**, 49.

Knudsen, W. C.: 1988, 'Solar Cycle Changes in the Morphology of the Venus Ionosphere', *J. Geophys. Res.* **93**, 8756.

Knudsen, W. C.: 1990, 'Role of Hot Oxygen in Venusian Ionospheric Ion Energetics and Supersonic Anti-sunward Flow', *J. Geophys. Res.* **95**, 1097.

Knudsen, W. C., Spenner, K., Miller, K. L., and Novak, V.: 1980, 'Transport of Ionospheric O$^+$ ions Across the Venus Terminator and Implications', *J. Geophys. Res.* **85**, 7803.

Knudsen, W. C., Spenner, K., and Miller, K. L.: 1981, 'Anti-Solar Acceleration of Ionospheric Plasma Across the Venus Terminator', *Geophys. Res. Letters* **8**, 241.

Knudsen, W. C., Miller, K. L., and Spenner, K.: 1982, 'Improved Venus Ionopause Altitude Calculation and Comparison with Measurement', *J. Geophys. Res.* **87**, 2246.

Knudsen, W. C., Kliore, A. J., and Whitten, R. C.: 1987, 'Solar Cycle Changes in the Ionization Sources of the Nightside Venus Ionosphere', *J. Geophys. Res.* **92**, 13391.

Knudsen, W. C., Luhmann, J. G., Russell, C. T., and Scarf, F. L.: 1989, 'The Venus Precursor: An Environmental Effect on the Pioneer Venus Spacecraft', *J. Geophys. Res.* **94**, 197.

Krymskii, A. M. and Breus, T. K.: 1986, 'Some Aspects of the Flow of the Solar Wind over Venus', *Kosmich. Issled* **24**, 778.

Krymskii, A. M. and Breus, T. K.: 1988, 'Magnetic Fields in the Venus Ionosphere: General Features', *J. Geophys. Res.* **93**, 8459.

Kumar, S., Hunten, D. M., and Taylor, H. A.: 1981, 'H_2 Abundance in the Atmosphere of Venus', *Geophys. Res. Letters* **8**, 237.

Lepping, R. P. and Behannon, K. W.: 1978, 'Mariner 10 Magnetic Field Observations of the Venus Wake', *J. Geophys. Res.* **83**, 3709.

Linker, J. A., Kivelson, M. G., and Walker, R. J.: 1989, 'The Effect of Mass Loading on the Temperature of a Flowing Plasma', *Geophys. Res. Letters* **16**, 763.

Luhmann, J. G.: 1986, 'The Solar Wind Interaction with Venus', *Space Sci. Rev.* **44**, 241.

Luhmann, J. G. and Russell, C. T.: 1983, 'Magnetic Fields in the Ionospheric Holes of Venus: Evidence for an Intrinsic Field?' *Geophys. Res. Letters* **10**, 409.

Luhmann, J. G., Elphic, R. C., Russell, C. T., Mihalov, J. D., and Wolfe, J. D.: 1980, 'Observations of Large-Scale Steady Magnetic Fields in the Dayside Venus Ionosphere', *Geophys. Res. Letters* **7**, 917.

Luhmann, J. G., Russell, C. T., Brace, L. H., Taylor, H. A., Jr., Knudsen, W. C., Scarf, F. L., Colburn, D., and Brace, A.: 1982, 'Pioneer Venus Observations of Plasma and Field Structure in the Near Wake of Venus', *J. Geophys. Res.* **87**, 9205.

Luhmann, J. G., Tatrallyay, M., Russell, C. T., and Winterhalter, D.: 1983, 'Magnetic Field Fluctuations in the Venus Magnetosheath', *Geophys. Res. Letters* **10**, 655.

Luhmann, J. G., Russell, C. T., and Elphic, R. C.: 1984, 'Time Scales for the Decay of Induced Large-Scale Magnetic Fields in the Venus Ionosphere', *J. Geophys. Res.* **89**, 362.

Luhmann, J. G., Russell, C. T., Spreiter, J. R., and Stahara, S. S.: 1985, 'Evidence for Mass-Loading of the Venus Magnetosheath', *Adv. Space Res.* **5**, 307.

Luhmann, J. G., Warniers, R. J., Russell, C. T., Spreiter, J. R., and Stahara, S. S.: 1986, 'A Gas Dynamic Magnetosheath Model for Unsteady Interplanetary Fields: Applications to the Solar Wind Interaction with Venus', *J. Geophys. Res.* **91**, 3001.

Luhmann, J. G., Russell, C. T., Phillips, J. L., and Barnes, A.: 1987, 'On the Role of the Quasi-Parallel Bow Shock in Ion Pickup: A Lesson from Venus?', *J. Geophys. Res.* **92**, 2544.

Luhmann, J. G., Russell, C. T., Scarf, F. L., Brace, L. H., and Knudsen, W. C.: 1987b, 'Characteristics of the Mars-Like Limit of the Venus-Solar ind Interaction', *J. Geophys. Res.* **92**, 8545.

Mahajan, K. K., Mayr, H. G., Brace, L. H., and Cloutier, P. A.: 1989, 'On the Lower Altitude Limit of the Venusian Ionopause', *Geophys. Res. Letters* **16**, 759.

Mariner Stanford Group, Venus: 1967, 'Ionosphere and Atmosphere as Measured by Dual-Frequency Radio Occultation of Mariner V', *Science* **158**, 1678.

Marubashi, K., Grebowsky, J. M., Taylor, H. A., Jr., Luhmann, J. G., Russell, C. T., and Barnes, A.: 1985, 'Ionosheath Plasma Flow in the Wake of Venus and the Formation of Ionospheric Holes', *J. Geophys. Res.* **90**, 1385.

McComas, D. J., Spence, H. E., Russell, C. T., and Saunders, M. A.: 1986, 'The Average Magnetic Field Draping and Consistent Plasma Properties of the Venus Magnetotail', *J. Geophys. Res.* **91**, 7939.

McComas, D. J., Gosling, J. T., Russell, C. T., and Slavin, J. A.: 1987a, 'Magnetotails at Unmagnetized Bodies: Comparison of Comet Giacobini–Zinner and Venus', *J. Geophys. Res.* **92**, 10111.

McComas, D. J., Spence, H. E., and Russell, C. T.: 1987b, in Q. T. Y. Lui (ed.), 'The Average Configuration of the Induced Venus Magnetotail', *Magnetotail Physics*, Johns Hopkins University Press, pp. 389–392.

Mihalov, J. D. and Barnes, A.: 1981, 'Evidence for the Acceleration of Ionospheric O$^+$ in the Magnetosheath of Venus', *Geophys. Res. Letters* **8**, 1277.

Mihalov, J. D. and Barnes, A.: 1982, 'The Distant Interplanetary Wake of Venus: Plasma Observations from Pioneer Venus', *J. Geophys. Res.* **87**, 9045.

Mihalov, J. D., Wolfe, J. H., and Intriligator, D. S.: 1980, 'Pioneer Venus Plasma Observations of the Solar Wind–Venus Interaction', *J. Geophys. Res.* **85**, 7613.

Mihalov, J. D., Spreiter, J. R., and Stahara, S. S.: 1982, 'Comparison of Gas Dynamic Model with Steady Solar Wind Flow around Venus', *J. Geophys. Res.* **87**, 10363.

Moore, K. R., McComas, D. J., Russell, C. T., and Mihalov, J. D.: 1989, 'Suprathermal Ions Observed Upstream of the Venus Bow Shock', *J. Geophys. Res.* **94**, 3743.

Nagy, A. F. and Cravens, T. E.: 1988, 'Hot Oxygen Atoms in the Upper Atmospheres of Venus and Mars', *Geophys. Res. Letters* **15**, 433.

Nagy, A. F., Cravens, T. E., Yee, J. H., and Stewart, A. I. F.: 1981, 'Hot Oxygen Atoms in the Upper Atmosphere of Venus', *Geophys. Res. Letters* **8**, 629.

Ness, N. F., Behannon, K. W., Lepping, R. P., Whang, Y. C., and Schatten, K. H.: 1974, 'Magnetic Field Observations Near Venus: Preliminary Results from Mariner 10', *Science* **183**, 1301.

Neugebauer, M. and Snyder, C. W.: 1965, 'Solar-Wind Measurements Near Venus', *J. Geophys. Res.* **70**, 1587.

Osmolovskii, I. K. and Samoznaev, L. N.: 1987, 'Electron Density Distribution in the Nightside Venusian Atmosphere Using Radiographic Data', *Kosmich. Issled.* **25**, 292.

Paxton, L. J.: 1983, 'Atomic Carbon in the Venus Thermosphere, Observations and Theory', Ph.D. Thesis, University of Colorado.

Perez-de-Tejada, H.: 1982, 'Viscous Dissipation at the Venus Ionopause', *J. Geophys. Res.* **87**, 7405.

Perez-de-Tejada, H.: 1986a, 'Fluid Dynamic Constraints of the Venus Ionospheric Flow', *J. Geophys. Res.* **91**, 6765.

Perez-de-Tejada, H.: 1986b, 'Distribution of Plasma and Magnetic Fluxes in the Venus Near Wake', *J. Geophys. Res.* **91**, 8039.

Perez-de-Tejada, H. and Dryer, M.: 1976, 'Viscous Boundary Layer for the Venusian Ionopause', *J. Geophys. Res.* **81**, 2023.

Perez-de-Tejada, H., Intrilligator, D. S., and Russell, C. T.: 1982, 'Orientation of Planetary O$^+$ fluxes and Magnetic Field Lines in the Venus Wake', *Nature* **299**, 325.

Perez-de-Tejada, H., Dryer, M., Intrilligator, D. S., Russell, C. T., and Brace, L. H.: 1983, 'Plasma Distribution and Magnetic Field Orientation in the Venus Near Wake: Solar Wind Control of the Nightside Ionopause', *J. Geophys. Res.* **88**, 9019.

Perez-de-Tejada, H., Intrilligator, D. S., and Scarf, F. L.: 1985, 'Plasma Measurements of the Pioneer Venus Orbiter in the Venus Ionosheath: Evidence for Plasma Heating Near the Ionopause', *J. Geophys. Res.* **90**, 1759.

Phillips, J. L. and Russell, C. T.: 1987, 'Upper Limit on the Intrinsic Magnetic Field of Venus', *J. Geophys. Res.* **92**, 2253.

Phillips, J. L., Luhmann, J. G., Russell, C. T., and Alexander, C. J.: 1986a, 'Interplanetary Magnetic Field Control of the Venus Magnetosheath Field and Bow Shock Location', *Adv. Space Res.* **5**, 179.

Phillips, J. L., Luhmann, J. G., and Russell, C. T.: 1986b, 'Magnetic Configuration of the Venus Magneto-sheath', *J. Geophys. Res.* **91**, 7931.

Phillips, J. L., Luhmann, J. G., Russell, C. T., and Moore, K. R.: 1987, 'Finite Larmor Radius Effect on Ion Pickup at Venus', *J. Geophys. Res.* **92**, 9920.

Phillips, J. L., Luhmann, J. G., Knudsen, W. C., and Brace, L. H.: 1988, 'Asymmetries in the Location of the Venus Ionopause', *J. Geophys. Res.* **93**, 3927.

Pivovarov, V. G., Erkaev, N. V., Volokitin, A. S., Breus, T. K., and Ivanova, S. V.: 1982, 'Problem of a Magnetic Barrier at Venus', *Kosmich. Issled.* **20**, 97.

Rohrbaugh, R. P. and Nisbet, J. S.: 1973, 'Effect of Energetic Oxygen Atoms on Neutral Density Models', *J. Geophys. Res.* **78**, 6788.

Romanov, S. A.: 1978, 'Asymmetry of the Region of Interaction of the Solar Wind with Venus According to the Data of the Venera-9 and Venera-10 Spacecraft', *Kosmich. Issled.* **16**, 318.

Romanov, S., Smirnov, V., and Vaisberg, O.: 1978, 'Interaction of the Solar Wind with Venus', *Kosmich. Issled.* **16**, 746.

Russell, C. T.: 1976a, 'The Magnetic Moment of Venus, Venera-4 Measurements Reinterpreted', *Geophys. Res. Letters* **3**, 125.

Russell, C. T.: 1976b, 'The Magnetosphere of Venus: Evidence for a Boundary Layer and a Magnetotail', *Geophys. Res. Letters* **3**, 589.

Russell, C. T. and Vaisberg, O.: 1983, in D. H. Hunten, L. Colin, and T. M. Donahue (eds.), 'The Interaction of the Solar Wind with Venus', *Venus*, Univ. of Arizona, Tucson, pp. 873–940.

Russell, C. T., Elphic, R. C., and Slavin, J. A.: 1979a, 'Initial Pioneer Venus Magnetic Field Results: Dayside Observations', *Science* **203**, 745.

Russell, C. T., Elphic, R. C., and Slavin, J. A.: 1979b, 'Initial Pioneer Venus Magnetic Field Results: Nightside Observations', *Science* **205**, 114.

Russell, C. T., Elphic, R. C., and Slavin, J. A.: 1979c, 'Pioneer Magnetometer Observations of the Venus Bow Shock', *Nature* **282**, 815.

Russell, C. T., Luhmann, J. G., Elphic, R. C., and Scarf, F. L.: 1981, 'The Distant Bow Shock and Magnetotail of Venus: Magnetic Field and Plasma Wave Observations', *Geophys. Res. Letters* **8**, 843.

Russell, C. T., Luhmann, J. G., Elphic, R. C., Scarf, F. L., and Brace, L. H.: 1982, 'Magnetic Field and Plasma Wave Observations in a Plasma Cloud at Venus', *Geophys. Res. Letters* **9**, 45.

Russell, C. T., Luhmann, J. G., and Elphic, R. C.: 1983, 'The Properties of the Low Altitude Magnetic Belt in the Venus Ionosphere', *Adv. Space Res.* **2**, 13.

Russell, C. T., Luhmann, J. G., and Phillips, J. L.: 1985a, 'The Location of the Subsolar Bow Shock of Venus: Implications for the Obstacle Shape', *Geophys. Res. Letters* **12**, 627.

Russell, C. T., Saunders, M. A., and Luhmann, J. G.: 1985b, 'Mass-Loading and the Formation of the Venus Tail', *Adv. Space Res.* **5**, 177.

Russell, C. T., Chou, E., Luhmann, J. G., Gazis, P., Brace, L. H., and Hoegy, W. R.: 1988, 'Solar and Interplanetary Control of the Location of the Venus Bow Shock', *J. Geophys. Res.* **93**, 5461.

Sagdeev, R. Z., Moroz, V. I., and Breus, T.: 1983, *Results of Soviet Studies of Venus*, NASA SP-461, Washington, pp. 169–190.

Saunders, M. A. and Russell, C. T.: 1986, 'Average Dimension and Magnetic Structure of the Distant Venus Magnetotail', *J. Geophys. Res.* **91**, 5589.

Saunders, M. A., Russell, C. T., and Luhmann, J. G.: 1986, 'Interactions with Planetary Ionospheres and Atmospheres: A Review, in *Comparative Study of Magnetospheric Systems*, Cepadues Edition, Paris, pp. 131–147.

Shefer, R. E., Lazarus, A. J., and Bridge, H. S.: 1979, 'A Re-examination of Plasma Measurements from the Mariner 5 Venus Encounter', *J. Geophys. Res.* **84**, 2109.

Shinagawa, H., Cravens, T. E., and Nagy, A. F.: 1987, 'A One-Dimensional Time-Dependent Model of the Magnetized Ionosphere of Venus', *J. Geophys. Res.* **92**, 7317.

Shinagawa, H. and Cravens, T. E.: 1988, 'A One-Dimensional Multispecies Magnetohydrodynamic Model of the Dayside Ionosphere of Venus', *J. Geophys. Res.* **93**, 11263.

Slavin, J. A. and Holzer, R. E.: 1981, 'Solar Wind Flow about the Terrestrial Planets, 1. Modeling Bow Shock Position and Shape', *J. Geophys. Res.* **86**, 11401.

Slavin, J. A., Elphic, R. C., Russell, C. T., Wolfe, J. H., and Intriligator, D. S.: 1979a, 'Position and Shape of the Venus Bow Shock: Pioneer Venus Orbiter Magnetometer Observations', *Geophys. Res. Letters* **6**, 901.

Slavin, J. A., Elphic, R. C., and Russell, C. T.: 1979b, 'A Comparison of Pioneer Venus and Venera Bow Shock Observations: Evidence for a Solar Variation', *Geophys. Res. Letters* **6**, 905.

Slavin, J. A., Elphic, R. C., Russell, C. T., Scarf, F. L., Wolfe, J. H., Mihalov, J. D., Intriligator, D. S., Brace, L. H., Taylor, H. A., and Daniell, R. E.: 1980, 'The Solar Wind Interaction with Venus: Pioneer Venus Observations of Bow Shock Location and Structure', *J. Geophys. Res.* **85**, 7625.

Slavin, J. A., Holzer, R. E., Spreiter, J. R., Stahara, S. S., and Chausee, D. S.: 1983, 'Solar Wind Flow about the Terrestrial Planets. 2. Comparison with Gasdynamic and Implications for Solar Wind Planetary Interactions', *J. Geophys. Res.* **88**, 19.

Slavin, J. A., Smith, E. J., and Intriligator, D. S.: 1984, 'A Comparative Study of Distant Magnetotail Structure at Venus and Earth', *Geophys. Res. Letters* **11**, 1074.

Slavin, J. A., Intriligator, D. S., and Smith, E. J.: 1989, 'Pioneer Venus Orbiter Magnetic Field and Plasma Observations in the Venus Magnetotail', *J. Geophys. Res.* **94**, 2383.

Smirnov, V. N., Vaisberg, O. L., Romanov, S. A., Slavin, J. A., Russell, C. T., and Intrilligator, D. S.: 1981, 'Three-Dimensional Shape and Position of the Shock Wave at Venus', *Kosmich. Issled.* **19**, 613.

Smith, E. J., Davis, L., Coleman, P. J., and Sonett, C. P.: 1963, 'Mariner II: Preliminary Reports on Measurements of Venus, Magnetic Field', *Science* **139**, 909.

Smith, E. J., Davis, L., Coleman, P. J., and Sonett, C. P.: 1965, 'Magnetic Measurements Near Venus', *J. Geophys. Res.* **70**, 1571.

Spenner, K., Knudsen, W. C., Miller, K. L., Novak, V., Russell, C. T., and Elphic, R. C.: 1980, 'Observation of the Venus Mantle, the Boundary Region between Solar Wind and Ionosphere', *J. Geophys. Res.* **85**, 7655.

Spreiter, J. R. and Jones, W. P.: 1963, 'On the Effect of a Weak Interplanetary Magnetic Field on the Interaction between the Solar Wind and the Geomagnetic Field', *J. Geophys. Res.* **68**, 3555.

Spreiter, J. R. and Stahara, S. S.: 1980, 'Solar Wind Flow Past Venus: Theory and Comparisons', *J. Geophys. Res.* **85**, 7715.

Spreiter, J. R., Summers, A., and Rizzi, A.: 1970, 'Solar Wind Flow Past Non-Magnetic Planets: Venus and Mars', *Planetary Space Sci.* **18**, 1281.

Stahara, S. S. and Spreiter, J. R.: 1976, *Calculation of Solar Wind Flows About Terrestrial Planets*, Nielsen Engineering and Research Report, Mountain View, Calif.

Stahara, S. S., Molvik, G. A., and Spreiter, J. R.: 1987, 'A New Computational Model for the Prediction of Mass Loading Phenomena for Solar Wind Interactions with Cometary and Planetary Ionospheres', *AIAA* **87-1410**, Amer. Inst. of Aero. and Astronautics, New York.

Tatrallyay, M., Russell, C. T., Mihalov, J. D., and Barnes, A.: 1983, 'Factors Controlling the Location of the Venus Bow Shock', *J. Geophys. Res.* **88**, 5613.

Tatrallyay, M., Russell, C. T., Luhmann, J. G., Barnes, A., and Mihalov, J. D.: 1984, 'On the Proper Mach Number and Ratio of Specific Heats for Modeling the Venus Bow Shock', *J. Geophys. Res.* **89**, 7381.

Taylor, H. A., Daniell, R. E., Hartle, R. E., Brinton, H. C., Bauer, S. J., and Scarf, F. L.: 1981, 'Dynamic Variations Observed in Thermal and Superthermal Ion Distributions in the Dayside Ionosphere of Venus', *Adv. Space Res.* **1**, 247.

Vaisberg, O. L., Romanov, S. A., Smirnov, V. N., Karpinsky, I. P., Khazanov, B. I., Polenov, B. V., Bogdanov, A. V., and Antonova, N. M.: 1976, 'The Structure of the Field of Interaction Between the Solar Wind and Venus According to Measurements of the Characteristics of the Ion Flux Aboard the Automatic Stations Venera-9 and Venera-10', *Kosmich. Issled.* **14**, 827.

Vaisberg, O. L., Intriligator, D. S., and Smirnov, V. N.: 1980, 'An Empirical Model of the Venusian Outer Environment. 1. The Shape of the Dayside Solar Wind-Atmosphere-Interface', *J. Geophys. Res.* **85**, 7642.

Van Allen, J. A., Krimigis, S. M., Frank, L. A., and Armstrong, T. P.: 1967, 'Venus: An Upper Limit on Intrinsic Magnetic Dipole Moment Based on Absence of a Radiation Belt', *Science* **158**, 1673.

Verigin, M. I., Gringauz, K. I., Gombosi, T., Breus, T. K., Bezrukikh, V. V., Remizov, A. P., and Volkov, G. I.: 1978, 'Plasma Near Venus from the Venera 9 and 10 Wide-Angle Analyzer Data', *J. Geophys. Res.* **83**, 3721.

Wallis, M. K.: 1972, 'Comet-Like Interaction of Venus with the Solar Wind, I', *Cosmic Electro.* **3**, 45.

Walters, G. K.: 1964, 'Effect of Oblique Interplanetary Magnetic Field on Shape and Behavior of the Magnetosphere', *J. Geophys. Res.* **69**, 1769.

Whitten, R. C., McCormick, P. T., Merritt, D., Thompson, K. W., Brynsvold, R. R., Eich, C. J., Knudsen, W. C., and Miller, K. L.: 1984, 'Dynamics of the Venus Ionosphere: A Two-Dimensional Model Study', *Icarus* **60**, 317.

Winske, D.: 1986, 'Origin of Large Magnetic Fluctuations in the Magnetosheath of Venus', *J. Geophys. Res.* **91**, 11951.

Wolff, R. S., Goldstein, B. E., and Yeates, C. M.: 1980, 'The Onset and Development of Kelvin–Helmholtz Instability at the Venus Ionopause', *J. Geophys. Res.* **85**, 7697.

Yeates, C. M., Ogilvie, K. W., and Siscoe, G. L.: 1978, 'Plasma Electron Observations in the Wake of Venus', *J. Geophys. Res.* **83**, 1524.

Yeroshenko, Ye. G.: 1979, 'Unipolar Induction Effects in the Magnetic Tail of Venus', *Kosmich. Issled.* **17**, 93.

Zelenyi, L. M. and Vaisberg, O. L.: 1982, 'Formation of the Plasma Mantle in the Venus Magnetosphere', *Kosmich. Issled.* **20**, 604.

Zwan, B. J. and Wolf, R. A.: 1976, 'Depletion of Solar Wind Plasma Near a Planetary Boundary', *J. Geophys. Res.* **81**, 1636.

THE STRUCTURE OF THE VENUS IONOSPHERE

L. H. BRACE*

Laboratory for Atmospheres, Goddard Space Flight Center, Greenbelt, MD 20771, U.S.A.

and

A. J. KLIORE

Jet Propulsion Laboratory, California Institute of Technology, Pasadena, CA 91103, U.S.A.

Abstract. Our current knowledge of the spatial structure of the Venus ionosphere and its temporal behavior is reviewed, with emphasis on the more recent Pioneer Venus measurements and analysis not covered in earlier reviews. We will stress the ionosphere structure, since other papers in this issue deal with its dynamics, and its magnetic properties. We also discuss some of the limitations that the orbit has placed on the spatial and temporal coverage of the ionosphere. For the benefit of future users of the data some of the factors which affect the measurement accuracies are discussed in an Appendix.

Table of Contents

* Currently at Space Physics Research Laboratory, University of Michigan, Ann Arbor, MI 48109, U.S.A.

Space Science Reviews **55**: 81–163, 1991.
© 1991 *Kluwer Academic Publishers.*

1. Introduction

The ionosphere of Venus is the ionized component of its upper atmosphere above about 100 km where ions and electrons are produced locally by photo-ionization and, on the nightside, by a combination of ion flow from the dayside and local ion production by suprathermal electron impact. The ionosphere is bounded above by a sharp gradient in the electron density known as the ionopause, a region of transition between the cold planetary plasma and the post-shock solar wind plasma. This interface is a complex, 3-dimensional surface that surrounds the planet and extends several thousand kilometers downstream where it bounds a complex ray-like region known as the ionotail.

The theoretical aspects of the Venus ionosphere have been reviewed by Nagy *et al.* (1983), including ionization and photochemical processes, electron and ion heating and cooling processes and certain effects of plasma transport. We will not repeat a review of these processes here, since little has changed. The major structural characteristics of the ionosphere below 1000 km or so have been described extensively earlier reviews (Schunk and Nagy, 1980; Brace *et al.*, 1983, Breus *et al.*, 1985). These reviews were based primarily on the early Pioneer Venus Orbiter (PVO) measurements that were made near solar maximum (1979–1982) and the Venera 9 and 10 radio occultation measurements made near solar minimum. We will not discuss these earlier results extensively but will briefly outline the major findings and then move on to describe the very extensive measurements made since that time, primarily by PVO. These recent results represent an important extension of our earlier knowledge because they cover much higher altitudes, especially in the nightside ionosphere, and a wide range of solar activity. Mahajan and Kar (1988) discussed some of these more recent results as part of a broader review of planetary ionospheres.

We will focus on the structure of the ionosphere, including its global variations and smaller scale features. Temporal variations will also be discussed, including solar cycle and solar wind dynamic pressure effects. Other papers in this issue deal with the magnetic properties of the ionosphere (Luhmann and Cravens, 1991) and the dynamics of the ionosphere, including the role of ion transport in the maintenance of the nightside ionosphere (Miller and Whitten, 1991). Still another review covers our current knowledge of the upper atmosphere, including the thermosphere from which the ionosphere is formed (Fox and Bougher, 1991). Since these topics are intimately related, the reader will find some overlap in their content.

As we go along, we will also note some questions about ionospheric behavior that have not yet been adequately answered. Some of these may be addressed using existing data, some may be answered by future PVO measurements, and still others must await new missions. For the benefit of future users of the PVO data base, we include an Appendix which describes the types of data that are available and some of the unique difficulties that were encountered in making the *in situ* observations. We also mention reported discrepancies among the measurements from different instruments and note possible sources of error. Much of the Pioneer Venus data is available from the National Space Science Data Center at Goddard Space Flight Center, Greenbelt, Maryland.

It is important to say at the outset that, although a great deal has been learned about the Venus ionosphere, our understanding remains rather shallow in most respects. Venus, like Earth, is a complex world. A simple exploratory mission like PVO will not reveal all of the important aspects of the ionospheric structure, nor can such a mission unravel all of the competing processes that produce its great variability. But a few things have been learned about how the Venus ionosphere works, and much of this is due to the many experimental and theoretical tools that have been developed in the course of exploring our own ionosphere (see references in Nagy *et al.*, 1983).

2. Historical Context

Many spacecraft have been sent to Venus since the beginning of the space age. Figure 1 indicates their arrival dates in relation to recent sunspot cycles. Not all of these missions provided information on the ionosphere. The first such measurements were the radio

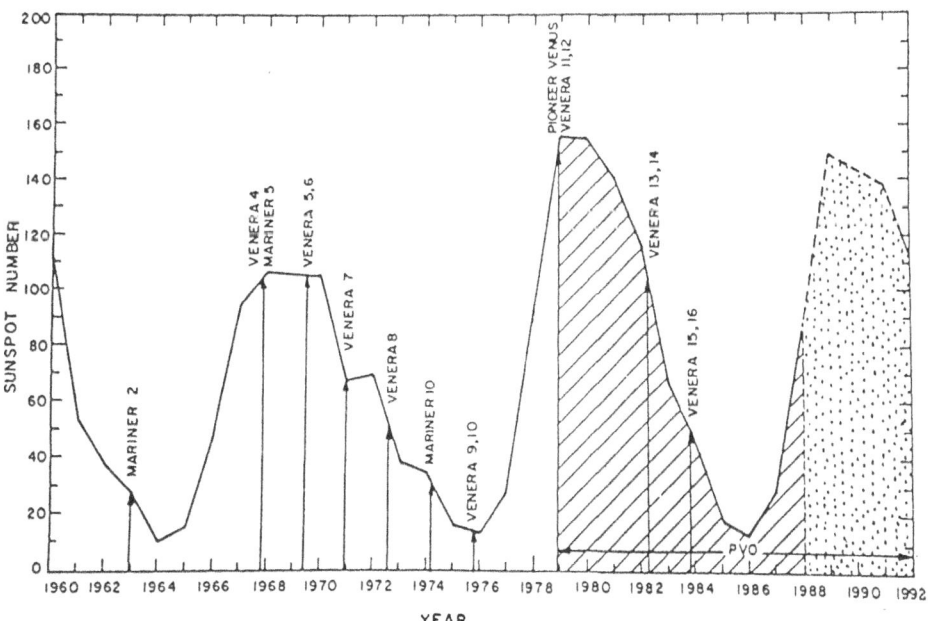

Fig. 1.   Solar activity and the arrival times of Venus missions (from Mahajan and Kar, 1988).

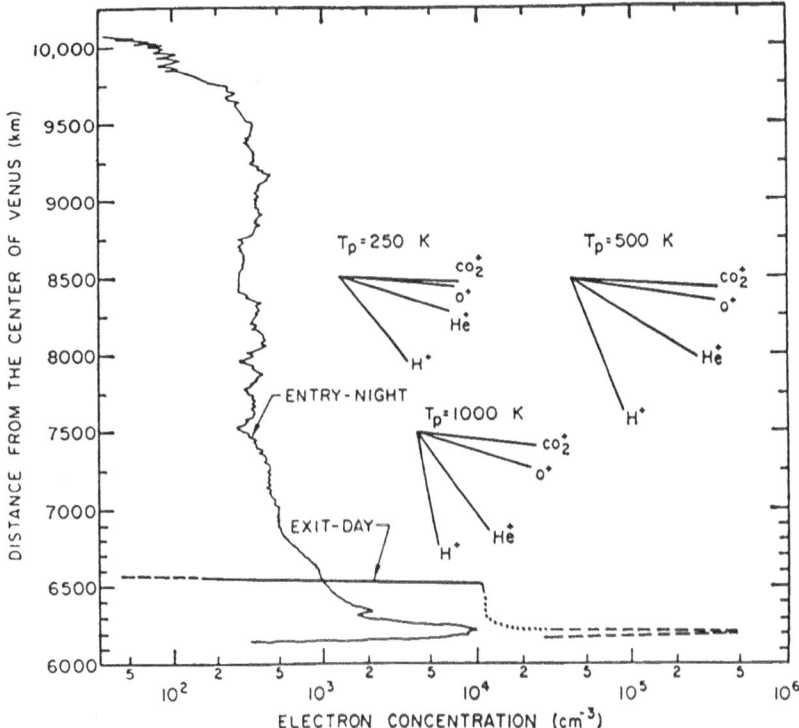

Fig. 2.   Mariner 5 N_e profiles showing a dense but thin ionosphere on the dayside and a weaker but more extended ionosphere on the nightside. The ion scale heights for various ion species at various temperatures are inset at the right to illustrate the ambiguity of height profiles of N_e (from the Mariner Stanford Group, 1967).

occultation profiles of the electron density, N_e, at the entry and exit points of Mariner 5 (Mariner Stanford Group, 1967; Kliore *et al.*, 1967; Fjeldbo and Eshleman, 1968). These profiles, shown in Figure 2, sparked extensive speculation about what kinds of ions and neutrals were present in the upper atmosphere of Venus and what the electron, ion, and neutral temperatures were. Unfortunately, density profiles cannot provide unique information about these things, as evidenced by the range of ion masses and plasma temperatures that were offered by early investigators. In diffusive equilibrium, the plasma density scale height, H, is given by

$$H = k(T_e + T_i)/m_i g,\tag{1}$$

where k is the Boltzmann constant, T_e and T_i are the electron and ion temperatures, m_i is the mean ion mass, and g is the acceleration of gravity. From Equation (1) it is clear that many fundamentally different ionospheres could have the same scale height. This fact is also evident in the insets in Figure 2, which depict the scale heights for various combinations of plasma temperatures and mass. A cold ionosphere with light ions (H^+, H_2^+, or He^+) may have a topside scale height that is identical to one that is populated by heavier ions (O^+, O_2^+, or CO_2^+) at much higher temperatures. For example, McElroy and Strobel (1969) envisioned the extended nightside ionosphere of Mariner 5 as consisting of H_2^+ and He^+ ions at a temperature of about 650 K. PVO later found that the major ion in this region was O^+ at temperatures of order of 3000 K (Miller *et al.*, 1980). The dayside profile extended in altitude to only 500 km and had a smaller scale height. Photochemical calculations for an assumed neutral atmosphere of 75% CO_2 and 25% N_2 suggested that CO_2^+ would be the major ion (McElroy, 1969). PVO later showed that O_2^+ is the major ion below 200 km, while O^+ dominates at higher altitudes (Taylor *et al.*, 1980).

Mariner 10 flew by Venus in 1974 and its radio occulation measurements (Howard *et al.*, 1974) revealed a dayside ionosphere that was highly compressed, as shown in Figure 3. The topside had two ledges. Bauer and Hartle (1974) interpreted the lower ledge as a region dominated by O^+ ions that had been compressed by downward ion transport induced by the interaction of the solar wind with the ionosphere. PVO *in situ* measurements later confirmed this interpretation (Hartle *et al.*, 1980). In the absence of information on the composition and temperatures, however, this was only one of several potentially viable interpretations of the profile.

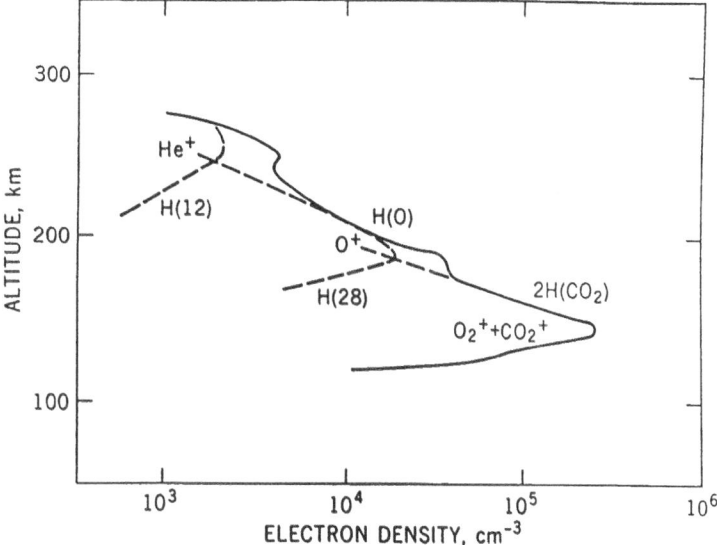

Fig. 3. Dayside N_e profile from Mariner 10 with ions proposed by Bauer and Hartle (1974) to fit the profile.

The lesson provided by these early analyses is that electron density profiles do not provide reliable information about the major ion and neutral species or their temperatures. In addition, the interpretation of these profiles may be further complicated by ion transport, vertical and horizontal gradients in the temperatures, and the existence of ionospheric magnetic fields which inhibit the transport of mass and energy.

When the Venera 9 and 10 satellites began their radio occulation measurements in 1975, hundreds of additional density profiles were obtained (Aleksandrov *et al.*, 1976a, b). The repeated measurement of both dayside and nightside profiles allowed the dynamics of the ionosphere to be appreciated for the first time. These data also permitted the solar zenith angle (SZA) variations of the density and its peak altitude to be determined at solar minimum, but they offered no solution to the fundamental questions concerning the composition and temperature of the ionosphere. Further progress would await the arrival of PVO which would begin direct measurements of these parameters 3 years later.

3. The Pioneer Venus Mission

The Pioneer Venus mission consisted of an orbiter spacecraft (PVO) and an entry bus which carried four entry probes to measure various characteristics of the lower atmosphere (Colin, 1983). The ionospheric goals of the mission were described by Bauer *et al.* (1977). We will be concerned here primarily with the results from the orbiter, although the bus also carried both neutral and ion mass spectrometers which made measurements during its single passage through the ionosphere and thermosphere (Taylor *et al.*, 1980; von Zahn *et al.*, 1980). But before reviewing the PVO contributions, we will describe the limitations that the orbit has placed on its *in situ* observations.

3.1. THE ORBIT AND ITS EVOLUTION

The Orbiter was injected into a highly eccentric orbit having an inclination of 105°, a period of 24 hours, with periapsis at 17° N. Apoapsis was at an altitude of about 72 000 km. Periapsis was initially placed at an altitude of 378 km, but it was quickly lowered to the vicinity of 150 km, where it was maintained for the next 19 months by the occasional use of onboard propulsion to counter the orbit-perturbing effects of solar gravitation. About 80% of the fuel had been expended in this way by the summer of 1980, when a decision was made to save the remaining fuel for future use. For the next 6 years periapsis spiraled upward through the low latitude ionosphere, reaching its maximum altitude in 1986. This is illustrated in Figure 4, which shows how the periapsis altitude and the $F10.7$ solar index (Covington, 1969) have changed thus far in the mission. The expected changes between now and re-entry in 1992 (Schatten, 1988) are also shown. The altitude of periapsis is now decreasing, and the spacecraft will re-enter the thermosphere where it is expected to operate for several months prior to final entry and incineration late in 1992.

The rise in periapsis has uniquely determined the regions of the Venus ionosphere which could be observed *in situ*. Since these changes in the orbit are occurring on a time-scale of the order of a solar cycle, it is clear that *in situ* measurements will not be available for all combinations of altitude and solar activity. It is likely, however, that periapsis will return to the lower ionosphere during a period of distinctly lower levels of solar activity than existed in 1979–1980.

The spatial coverage that will have been obtained during the PVO lifetime is illustrated

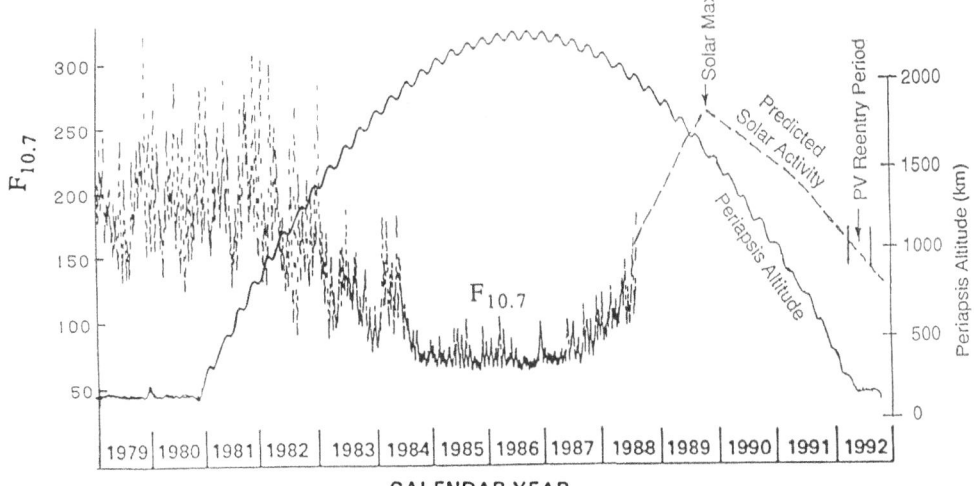

Fig. 4. The PVO altitude of periapsis and the daily $F_{10.7}$ index of solar activity. The values beyond 1988 are predicted. Low altitude *in situ* measurements were only possible near solar maximum (1979–1980), while only higher altitude measurements were possible at solar minimum. Periapsis will return to the lower ionosphere in 1992, perhaps at a time of moderate solar activity.

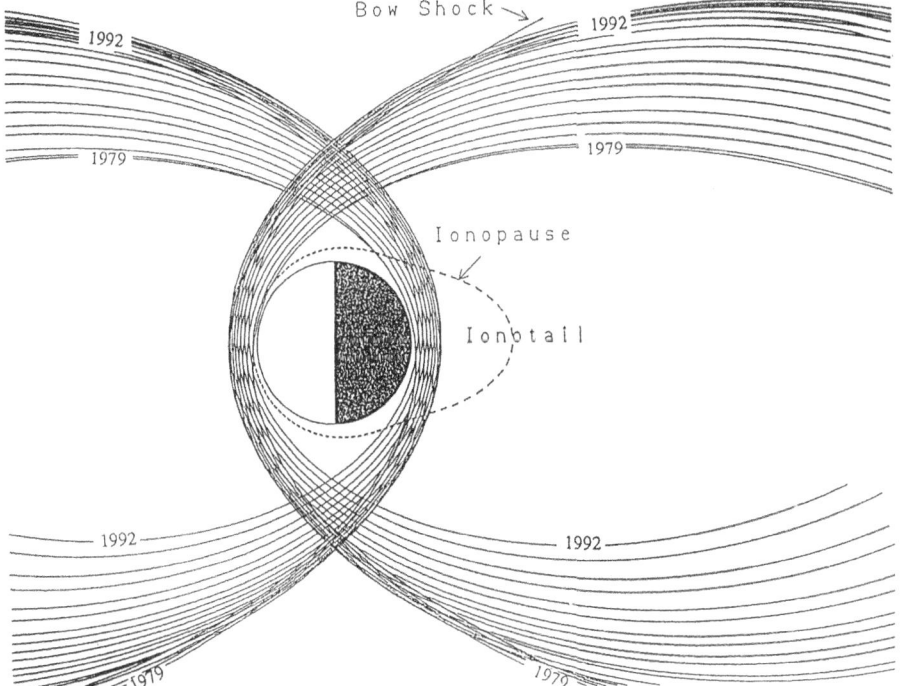

Fig. 5. Another view of the evolution of the PVO orbit between 1979 and 1992, showing average bow shock and ionopause locations. Periapsis passages covered the nightside ionosphere (the ionotail) to altitudes of 2300 to 3000 km.

in Figure 5 which shows the noon and midnight orbits for each Venus year projected against the dusk sky. These represent only 2 of the 224 orbits per Venus year, but they illustrate how the sampling geometry is changing during the mission. The interval from 1983–1989 has been useful primarily for measuring the high altitude fringes of the nightside ionosphere, a region known as the ionotail (Section 4.12).

Returning to Figure 5, the period of low periapsis (1979–1980) provided quasi-altitude profiles of the ionosphere almost every day. With periapsis initially at $17°$ N, the inbound profiles tended to represent conditions at low northern latitudes, while the outbound profiles reflected the structure of the equatorial ionosphere. Periapsis has continued to drift southward at the rate of about $2°$ per Earth year, so it will be near $10°$ S at re-entry in September 1992.

Full diurnal coverage was provided by the annual motion of Venus about the Sun, which caused a local time drift of 0.108 h orbit^{-1}. The first 600 days provided 2.5 sweeps through all local times before periapsis began to rise. The coverage obtained during the first 6 Venus years (≈ 1400 passages) is illustrated in Figure 6, which shows

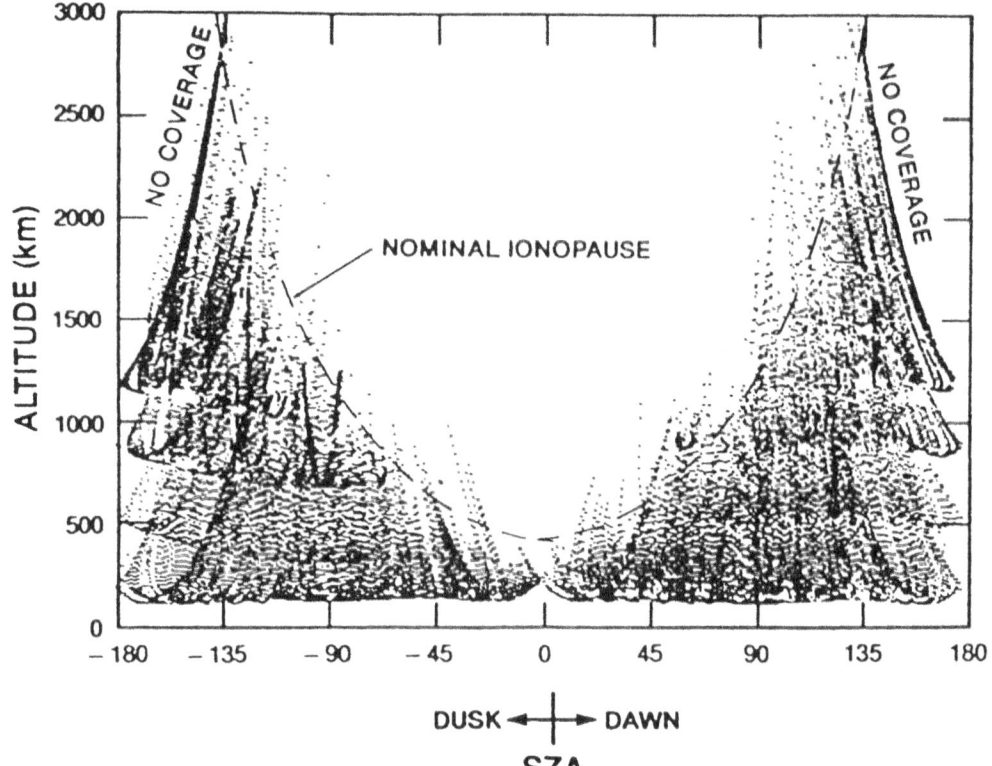

Fig. 6. Altitude and SZA coverage within the ionosphere from the first 6 Venus years (1979–1983) as periapsis spiralled upward around the planet, remaining at low latitudes throughout this period (from Theis *et al.*, 1984).

the spacecraft altitude and SZA at 12 s intervals along each of the approximately 1400 orbits from this period.

3.2. THE PLASMA MEASUREMENT TECHNIQUES

For our purposes, the PVO instruments of primary interest are the Ion Mass Spectrometer (OIMS), the Retarding Potential Analyzer (ORPA), the Electron Temperature Probe (OETP), the Plasma Analyzer (OPA), and the Neutral Mass Spectrometer (ONMS). These and the other PVO instruments were described in detail elsewhere (*IEEE Trans. Geosci. and Remote Sens.*, 1980). Their heritage lies in Earth satellites such as the OGO, ISIS, Atmosphere Explorer, and Aeros. Together these instruments measure the densities of the individual ions, and the total ion and electron densities (N_i and N_e) and temperatures (T_i and T_e), the ion drift velocities, v_i, suprathermal electron fluxes, the solar wind velocity and density, and the neutral gas densities. The ONMS also measures energetic ions with energies > 40 eV. N_e and N_i are assumed to be equal, although observations of them by different instruments may disagree at times, for reasons discussed in the Appendix. The total solar extreme ultraviolet flux (V_{EUV}) was deduced from measurements of the photoelectron emission current from one of the OETP sensors (Brace *et al.*, 1988a). In addition to these *in situ* measurements, the radio occultation method (ORO) provided N_e profiles during many occultation seasons throughout the mission thus far (Kliore and Mullen, 1989a, b). These data complement the *in situ* measurements because they extend to altitudes well below the main peak which are not reached by the spacecraft, and because they provided measurements of

TABLE I

Parameters measured by PVO[a]

Instruments	Parameter (range)
Ion mass spectrometer (OIMS)	N_i (16 pre-selected cold ions) Superthermal ions (9–16 eV)
Neutral mass spectrometer (ONMS)	N_n (2–46 amu) Energetic ions (> 40 eV)
Retarding potential analyzer (ORPA)	N_i, N_e (10–10^7 cm^{-3}) T_i (100–$10\,000$ K) T_e (300–$25\,000$ K) v_i (0–5 km s^{-1}) Suprathermal electrons (< 50 eV)
Electron temperature probe (OETP)	N_e (10^1–10^5 cm^{-3}) N_i (10^2–10^7 cm^{-3}) T_e (300–$25\,000$ K) V_{EUV} ($> 1 \times 10^{11}$ ph cm^{-2} s^{-1})
Radio occultation (ORO)	N_e profiles ($> 10^3$ cm^{-3})
Plasma analyzer (OPA)	Solar wind density, temperature and velocity

[a] For detailed instrument capabilities, see papers in *IEEE Trans.* (1980).

the solar cycle variations of N_e in the lower ionosphere after periapsis rose, precluding continued *in situ* measurements at those altitudes. The ORO profiles also provided a valuable check upon the accuracy of the *in situ* measurements. Table I lists the measured parameters and the approximate dynamic range of some of the measurements.

4. The Ionosphere Structure

Nearly all parameters of the Venus ionosphere were completely unknown when PVO arrived in 1978. As noted earlier, some aspects of its N_e structure had been measured by the Mariner and Venera radio occultation experiments. PVO would resume the occultation measurements but would also measure the composition and temperature of the ions, the electron temperature, the ion drift velocity, and the magnetic and plasma wave properties of the ionosphere. The thermospheric neutrals would also be measured, and the gas temperature would be inferred from the neutral scale heights. The eccentricity of the orbit would allow altitude variations to be examined, as well as smaller scale structures in the ionosphere. The total solar EUV and the solar wind parameters also would be measured. Finally, the apparently complex global structure of the ionosphere could be investigated as the orbit swept through all local times.

Since the Venera and Pioneer Venus measurements made prior to 1983 have been the topic of several reviews (referenced earlier), we will not concentrate on them in this review. Instead, we will note a few of the earlier results to provide the background for our discussion of more recent measurements and analysis. Because of the rising periapsis altitude of PVO most of these recent results refer to higher altitudes, but some of them deal with the effects of solar activity, so data from the entire PVO mission are included.

The global structure of the Venus ionosphere is controlled by ion production, loss, and transport processes. These processes were described in the review paper by Nagy *et al.* (1983), and the references therein. The companion review of the dynamics of the ionosphere by Miller and Whitten (1989) will update some of the theoretical work that has been done since the Nagy *et al.* review, particularly with regard to the role of the nightward ion flow in maintaining the nightside ionosphere.

Large temporal variations occur on many time-scales in the Venus ionosphere. Some of this variability may arise from the dynamics of the thermosphere itself, but most of it can be traced to the fact that the ionosphere is the primary obstacle to the highly variable solar wind. The companion review by Phillips and McComas (1991) will discuss the magnetospheric aspects of solar wind interactions. The following is a review of what has been learned about the structure and variability of the Venus ionosphere, again with primary emphasis on the recent PVO measurements.

To provide a framework for the review of a very complex topic, it may be helpful to introduce the sketch shown in Figure 7. The features illustrated there will be discussed separately in later sections. Brace *et al.* (1983) constructed an earlier version of this drawing to illustrate many of the features of the Venus ionosphere that had been discovered up to that time. Since then, the measurements at higher altitudes have permitted the upper nightside ionosphere to be examined, and these results have been

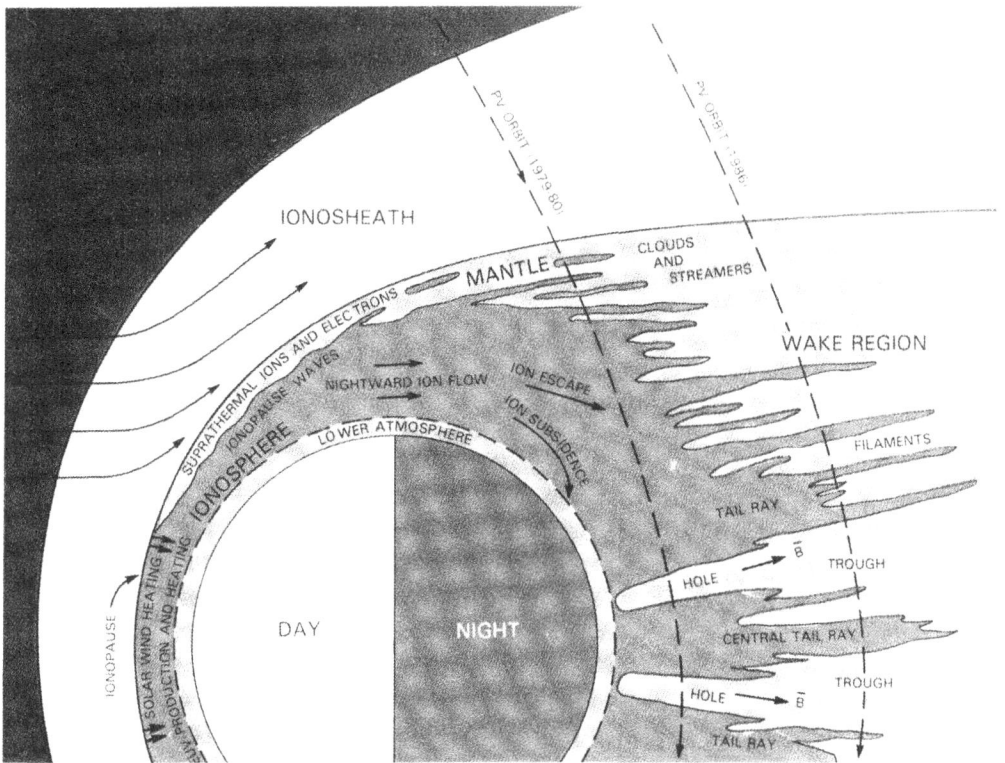

Fig. 7.   A drawing which illustrates the complex global configuration of the Venus ionosphere (altitude scale expanded by a factor two relative to planet). The ionosphere is produced by solar EUV radiation and is heated and shaped by solar wind interactions at the ionopause, producing cometlike features on the nightside and planetary ion escape. The two PVO orbits (shown dashed) illustrate how the rising altitude of periapsis between 1980 and 1986 has allowed the ionotail to be examined (after Brace *et al.*, 1987).

added to the cartoon. To permit annotation, the scale of the ionosphere is expanded by a factor of 2 or 3 relative to the planet and the bow shock. The nightside orbit tracks for 1979–1980 and 1986 are also drawn to this expanded scale.

The dayside ionosphere is created by solar EUV radiation, and some of these ions flow nightward to help form the nightside ionosphere, a region of great structure and variability. Some of the ions carried by this nightward flow are heated or accelerated enough to escape down the Venus wake. Large holes form in the nightside ionosphere, and these contain strong radial magnetic fields of unknown origin. An ionopause forms the boundary between the hot solar wind plasma and the cold ionospheric plasma. The ionopause hugs the planet at the nose, expands outward toward the terminator, and becomes an increasingly complex surface as it bounds the highly dynamic nightside ionosphere. Just above the dayside ionopause is a region called the mantle, which contains a mixture of shocked solar wind plasma and plasma of ionospheric origin. Plasma clouds and streamers form above the terminator ionopause, and these extend an unknown distance downstream. Tail rays form in the umbra, in a region known as the ionotail, and these extend at least a few thousand km downstream.

It is important to caution the reader that Figure 7 is only a cartoon. It shows some of the most persistent features and phenomena that have been found and examined during the PVO mission. It is not intended to be a snapshot of reality at Venus at any particular instant. The *in situ* measurements on which this picture is based cannot provide a global image of the Venus plasma environment. Changes in the orbit have not allowed the altitudinal structure and the solar cycle variations to be separated uniquely. There is no assurance that the holes shown in the nightside ionosphere or the ray structure of the ionotail exist throughout the solar cycle. These, and the other features, are simply shown in the region where they were most commonly observed.

4.1. EARLY RADIO OCCULTATION PROFILES OF N_e

The early radio occultation measurements from PVO, such as those shown in Figures 8 and 9 (Kliore *et al.*, 1979a, b), largely confirmed the Venera 9 and 10 results concerning the diurnal variations of the ionosphere and its nightside variability. The dayside peak density was nearly twice as large as the densities observed in the same region at solar minimum by Venera 9 and 10. This difference was the first indication that N_e varied significantly with solar activity, although such a variation was to be expected (we will return to this topic in Section 4.9). The height variations of the main peak are shown in Figure 10 (Cravens *et al.*, 1981). The peak was well behaved on the dayside at an average altitude of about 140 km, consistent with the Venera and Mariner results that are also shown. But the peak height increased to over 150 km at 100° SZA, then varied widely at larger SZA. The rise in the peak altitude just beyond the terminator is consistent with a Chapman layer in which the altitude of maximum ion production moves upward at sundown (Garvik *et al.*, 1980; Cravens *et al.*, 1981). This rise is partly

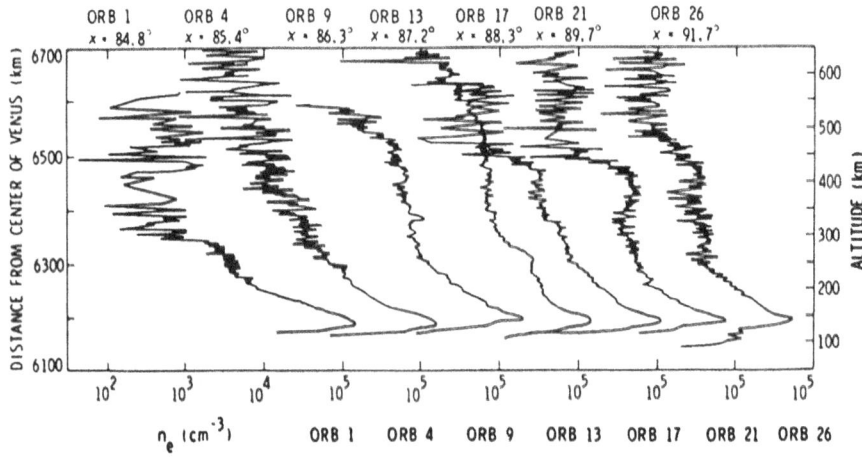

Fig. 8.   Dusk region N_e profiles derived from PVO radio occultation measurements (from Kliore *et al.*, 1979a).

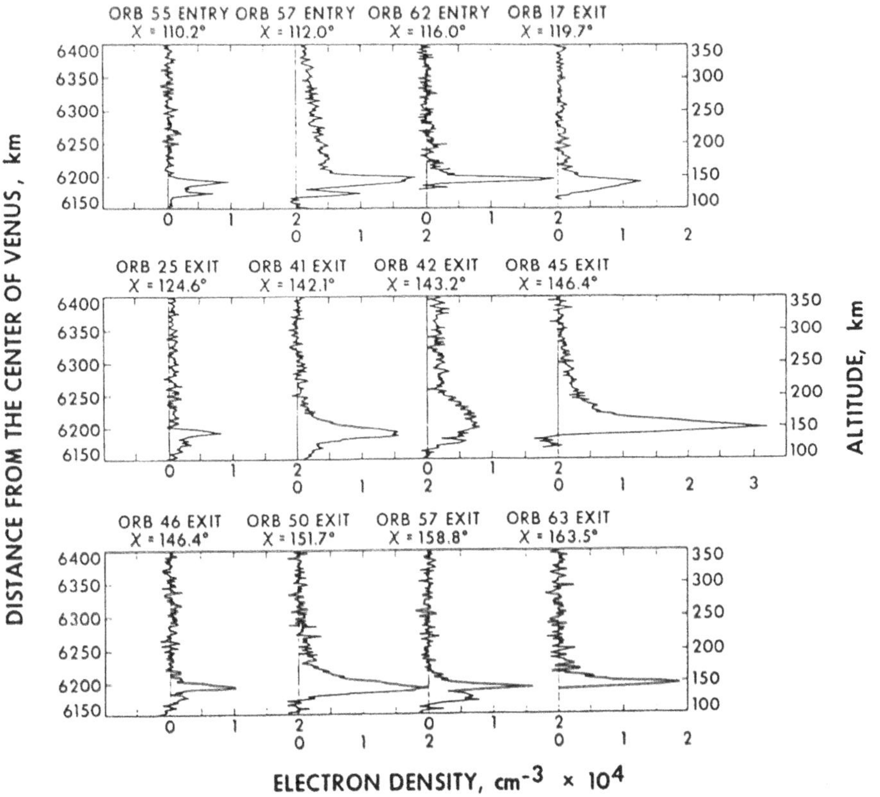

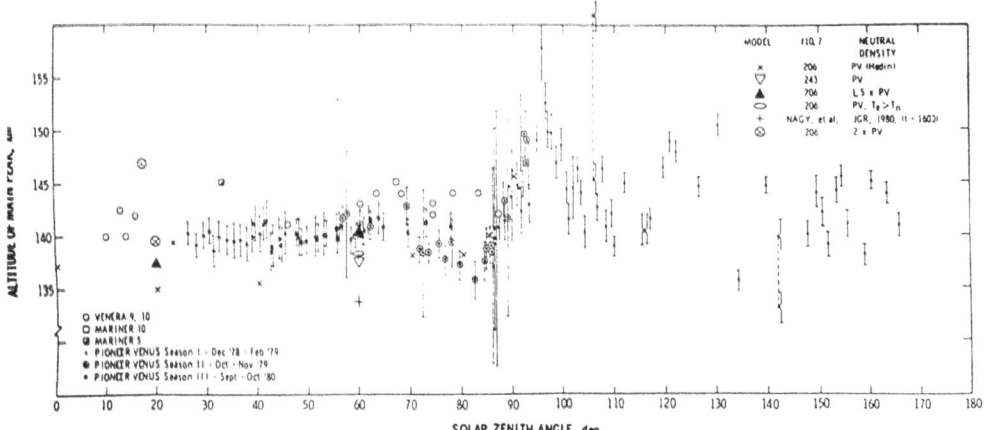

Fig. 9.   Nightside N_e profiles from the PVO radio occultation measurements showing its great variability (frm Kliore *et al.*, 1979b).

Fig. 10.   Variations in the height of the main peak observed by PVO, Veneras 9 and 10, and Mariners 5 and 10. Theoretical peak heights from Nagy *et al.* (1980) are also shown (from Cravens *et al.*, 1981).

opposed by the decline in the density of the ionizable constituents as the thermosphere temperature declines with increasing SZA (Niemann *et al.*, 1980).

The nightside ionosphere was more variable than the dayside. The density exhibits a main peak at an altitude of about 145 km, and a weaker peak is often observed below the main peak at 125 km or so. Gringauz *et al.* (1979) attributed the main peak to the precipitation of energetic electrons that they measured directly at higher altitudes in the Venus wake. The density of the main peak often changed by more than a factor of 3 from orbit to orbit. It was argued that this variability was produced by temporal fluctuations in the energetic particle precipitation (Keldish, 1977) measured nearly simultaneously from the same spacecraft. We will see later that variations in ion transport from the dayside ionosphere may provide an equally viable explanation for the nightside variability.

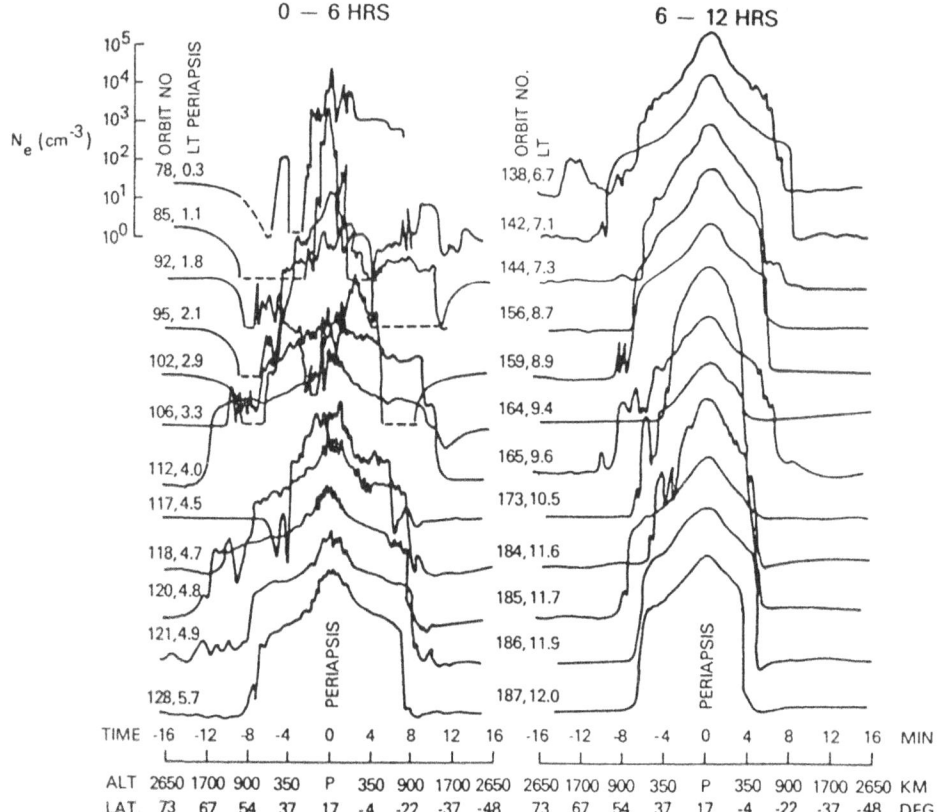

Fig. 11. A series of N_e profiles contrasting the highly structured nightside ionosphere (*left*) with the smoother dayside ionosphere (*right*). The periapsis altitude (*P*) was typically between 150 and 160 km during this period. Abrupt changes in N_e are evident at the ionopause crossings. Surface waves on the ionopause are evident at many of the dayside crossings, as in orbits 138, 173, and 184 (after Brace *et al.*, 1980).

4.2. EARLY PVO *IN SITU* MEASUREMENTS

The *in situ* measurements provided further evidence for a highly dynamic ionosphere. Figure 11 shows the N_e profiles measured by the OETP during a series of passages (Brace *et al.*, 1980) which illustrate some of the major features. The nightside passes are at the left, and the dayside passes are at the right. Periapsis (P) was near 150 km in all of these orbits. In each pass a sharply defined ionopause is marked by an abrupt drop in N_e by about a factor of 100 at the top of the ionosphere. The dayside ionopause

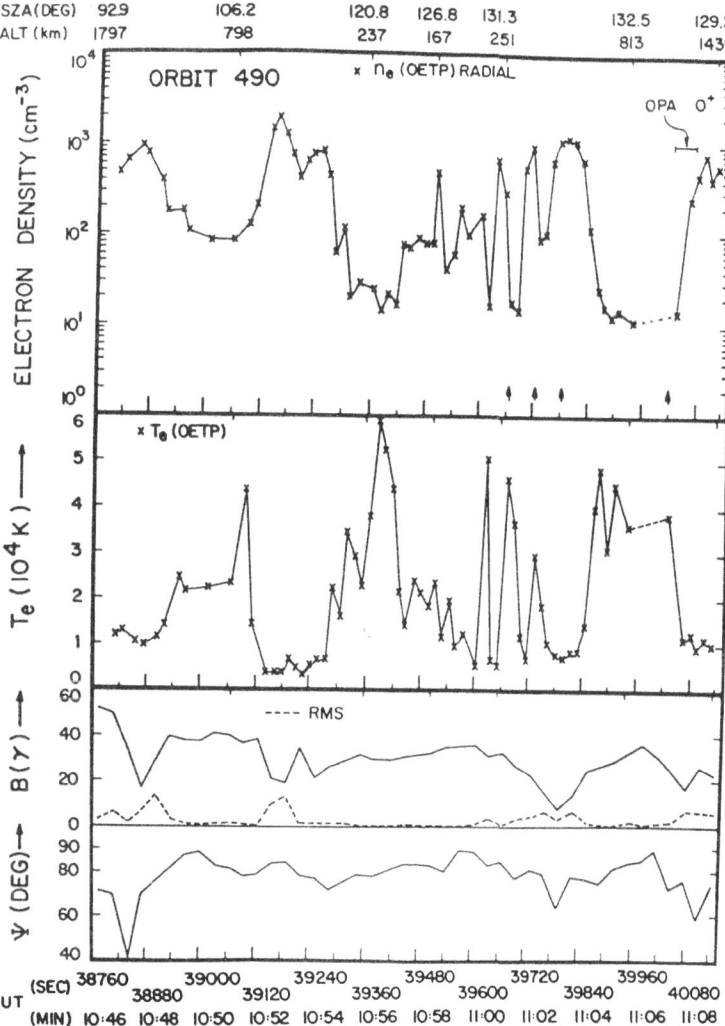

Fig. 12.   Plasma and magnetic field measurements through a 'disappearing ionosphere' on orbit 490. N_e was unusually low and T_e very high, and a strong, largely horizontal magnetic field, B, permeates the ionosphere. The OPA observed energetic O^+ ions flowing southward and tailward in the region shown. Superthermal ions (not shown) were observed by the OIMS and ONMS (after Cravens *et al.*, 1982).

crossings often exhibit wavelike structures; e.g., orbits 165, 173, and 184. These are believed to be surface waves on the ionopause which are made particularly visible by the nearly tangential nature of the crossings. These surface waves are illustrated in Figure 7, and they will be discussed in greater detail in the section on the ionopause (4.11). On the dayside, the underlying ionosphere is quite smooth, with a larger scale height above about 200 km and a smaller scale height below. A single smooth N_e maximum is seen as the spacecraft passes through periapsis, usually a few kilometers above the height of the N_e peak.

Unlike the dayside ionosphere, the nightside is highly structured and varies greatly from orbit to orbit, as illustrated by the series of pass plots at the left of Figure 11. Each profile is viewed as a snapshot of the density along the trajectory of the spacecraft, reflecting a mixture of structure in both latitude and altitude. Brace *et al.* (1980) attributed much of this structure to horizontal stratification. This layering was suggested by the occurrence of similar density signatures at the same altitudes on the inbound and outbound legs of the orbit. These structures have not been examined systematically. More recently, evidence has been set forth that much of this structure represents waves propagating horizontally across the nightside ionosphere. The evidence for this explanation is reviewed in Section 4.7.

The extreme variability of the nightside ionosphere is its most notable characteristic. The great orbit to orbit variations shown in Figure 11 demonstrate that large global changes can occur on time-scales shorter than the 24 hour interval between passages. Figure 11 understates this variability, however. Cravens *et al.* (1982) reported many instances in which the nightside ionosphere almost vanished. Figure 12 shows the plasma and magnetic field profiles obtained on one such occasion. The ionosphere was heavily depleted and magnetized at this time, and the ions and electrons were much warmer than usual. The ions in some regions were flowing generally southward and tailward. Since these 'disappearing ionospheres' tended to occur at times of very high solar wind dynamic pressure, Cravens *et al.* attributed them to solar wind driven depressions of the dayside ionopause height and a resulting constriction of the nightward ion flow which maintains the nightside ionosphere. Some of the more energetic nightside plasma at these times was attributed to enhanced charge exchange between solar wind protons and thermospheric atomic oxygen. Other manifestations of solar wind interactions are discussed in greater detail in Sections 4.11 and 4.12.

4.3. THE MEAN STRUCTURE OF THE IONOSPHERE

The mean structure of the ionosphere below 1000 km was summarized in an empirical model (Bauer *et al.*, 1985) that is included in the Venus International Reference Atmosphere, VIRA (Keating *et al.*, 1985). This model still represents the most comprehensive tabulation of the mean density, temperature and composition of the ionosphere as a function of altitude and SZA, although it is incomplete or inaccurate in several respects noted below.

Several PVO instruments contributed to the VIRA ionosphere model. The N_e and T_e tabulations are based on the OETP measurements, as modelled in analytical form by

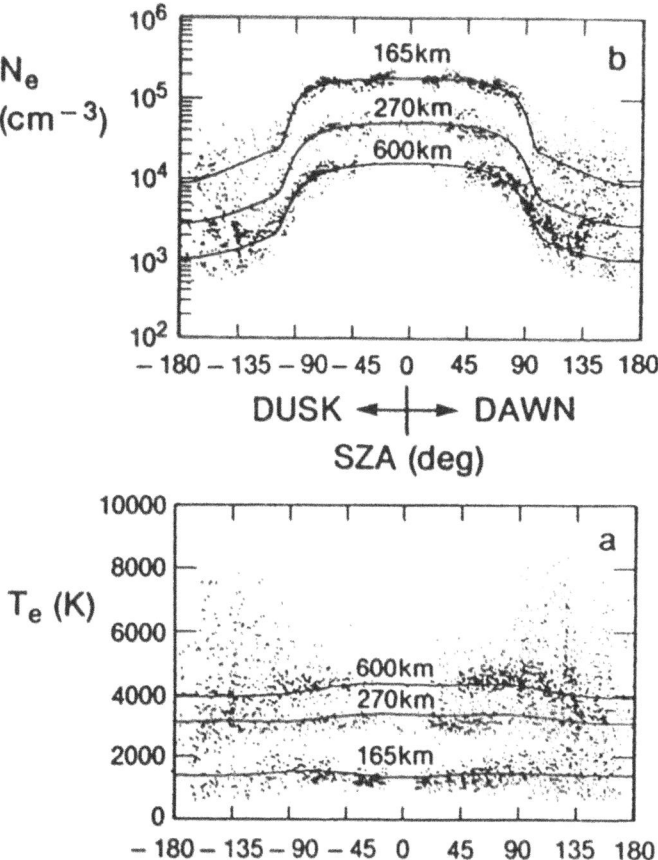

Fig. 13. N_e and T_e model values at 165, 270, and 600 km, and the corresponding PVO measurements by the OETP. The scatter in the data illustrates the relatively small temporal variability of the dayside ionosphere and the larger nightside variations (after Theis *et al.*, 1984).

Theis *et al.* (1984), with N_e normalized at 150 km to the radio occultation measurements. Figure 13 shows the OETP measurements and the Theis model results for selected altitudes. Although the data are generally well represented by the models, spatial structures and temporal variability cause very large scatter, especially on the nightside. The nightside N_e at 165 km is lower than its midday value by a factor of about 30, in agreement with the diurnal variation evident in the occultation measurements. Surprisingly, T_e exhibits little diurnal variation, a point that we will return to in discussions of the thermal balance of the ionosphere.

The ORPA measurements are the basis for T_i in the VIRA model (Miller *et al.*, 1984). The ORPA measures N_i and T_e, but with poorer spatial resolution than the OETP. This instrument also measures the ion velocity, v_i (Knudsen *et al.*, 1981), but this parameter is not included in the VIRA model. Figure 14 shows the ORPA median values of N_i, T_i, and T_e within 6 SZA ranges (Miller *et al.*, 1980, 1984). The diurnal variation in N_i

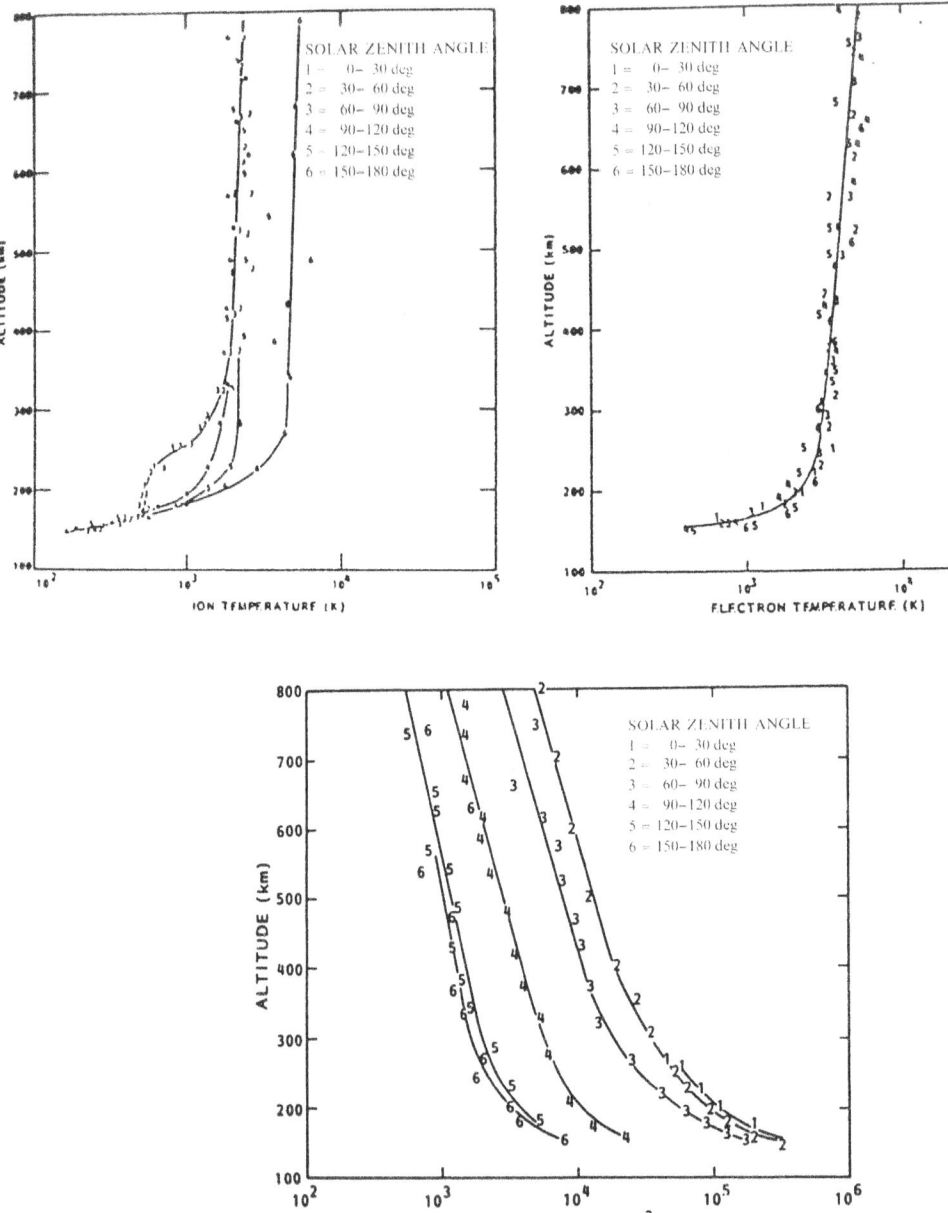

Fig. 14. ORPA median T_i, T_e, and N_i height profiles for 30° intervals of SZA, measured by the ORPA. There is little diurnal variation in the temperature, except that T_i is extraordinarily high in the antisolar region (SZA > 150°) (after Miller *et al.*, 1980).

agrees well with that found in the OETP measurements of N_e. T_e exhibits little variation with SZA, also in agreement with the OETP measurements. T_i increases with SZA, actually exceeding T_e in the anti-solar region. The causes of these high ion temperatures are unknown, but the dynamics of convergent ion flow in that region may be a factor. We will see in later sections that this nightward ion flow is also a major factor in the maintenance of the nightside ionosphere. These effects of nightward ion flow are discussed in greater detail in the companion review on ionosphere dynamics by Miller and Whitten (1990).

The ion composition tables given in the VIRA model are based on the measurements of the OIMS (Taylor *et al.*, 1980, 1985). Figure 15 shows the diurnal variation of the ion densities at an altitude of 200 km. For simplicity, the dawn-dusk asymmetries evident in the H^+ measurements were not included in the VIRA model. These asymmetries, which are most important for the ion composition, are discussed in greater detail in Section 4.6.

From the above discussion, it is clear that the 1985 VIRA ionosphere model has important shortcomings, perhaps the most important being the lack of solar cycle effects. The excursion of periapsis to high altitudes after 1980 prevented PVO from measuring the solar cycle variations of the lower ionosphere, at least for the temperature, composition, and ion velocity, all of which require *in situ* measurements. The Venera radio occulation measurements, made in 1975, and the PVO measurements, made since 1983, show that the solar cycle effects are very large indeed. These effects, to the extent that they are known, are described in Section 4.9.

The *in situ* measurements are necessarily heavily biased toward the low latitude of periapsis, so latitudinal structure, if present, is difficult to detect in the PVO measurements. However, the radio occultation measurements of N_e at the peak suggest that the SZA variation with latitude is similar to the SZA variations with local time. Searches for latitudinal and seasonal variations are recounted in Section 4.8.

Finally, improvements in measurement accuracy can be expected to be reflected in future VIRA models. Ongoing comparisons of the plasma measurements have revealed areas of disagreement which require further investigation. Some of these discrepancies, and possible sources of error, are discussed in the Appendix.

4.4. THERMAL BALANCE OF THE IONOSPHERE

The thermal structure of an ionosphere is determined by the balance between the electron and ion local heating and cooling rates, with local differences in these rates made up by heat transport, primarily to lower altitudes where cooling to the neutrals is more efficient. The ultimate energy sources are solar EUV and the solar wind. In the following sections, heating of the dayside and nightside ionosphere are discussed separately.

Dayside Thermal Structure

Heat balance calculations for the dayside ionosphere that had been made prior to PVO (Nagy *et al.*, 1975, Cravens *et al.*, 1978). These calculations suggested that T_e and T_i

L. H. BRACE AND A. J. KLIORE

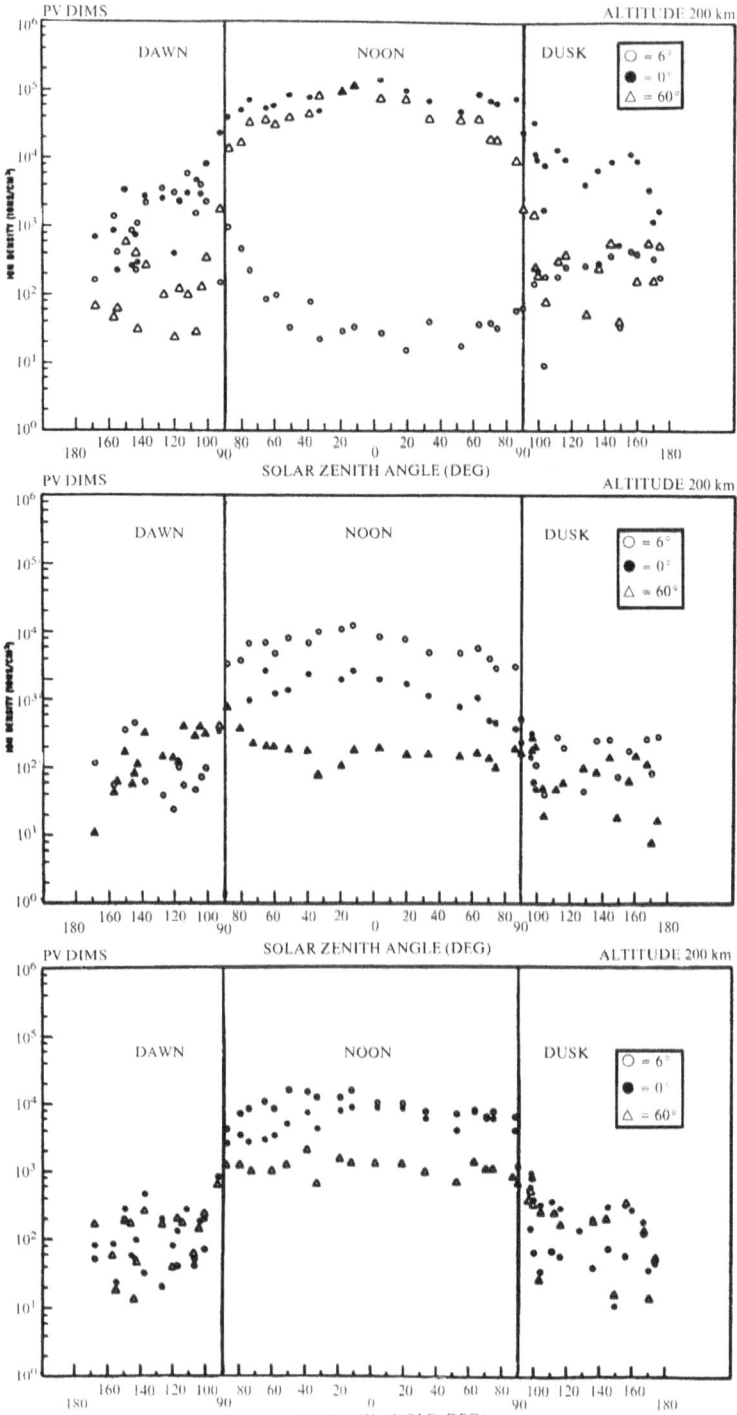

Fig. 15.   The diurnal variation of ion composition at 200 km measured by the OIMS. The major ions are O⁺ and O₂⁺ on the dayside and O⁺ and H⁺ on the nightside. A predawn bulge in H⁺ is the most obvious asymmetry in the diurnal variation (from Taylor *et al.*, 1980).

would be significantly elevated above the neutral gas temperature. Photoelectrons were assumed to be the major heat source for the ionosphere electrons, although most of their energy goes into heating and dissociating the neutrals. The ions were assumed to be heated primarily by the ionosphere electrons and cooled by collisions with the neutrals.

Cravens *et al.* had noted that the Viking 1 and 2 Lander T_i measurements at Mars could be modelled well only by introducing additional heat at the top of the ionosphere. The same model was used to predict the temperatures at Venus, with the solar EUV heating scaled up appropriately to the smaller solar distance. The top panel of Figure 16 shows the resulting T_e and T_i profiles, both with and without additional heating at the top. They used an ion heat source of 2.2×10^8 eV cm^{-2} s^{-1} and an electron heat source of 3×10^{10} eV cm^{-2} s^{-1}. These profiles turned out to be remarkably similar to those obtained later by PVO. The middle and bottom panels show the PVO measurements at SZA = 60° and two theoretical profiles based on unmagnetized and magnetized ionospheres (Cravens *et al.*, 1980). Maintenance of the observed high temperatures in an unmagnetized ionosphere required an electron heat flux of 3×10^{10} eV cm^{-2} s^{-1}, the same as that used by Cravens *et al.* (1978). The required ion heat flux was 5×10^7 eV cm^{-2} s^{-1}, a factor of 4 smaller than Cravens *et al.* (1978). Similarly high electron heat fluxes had been proposed earlier by Brace *et al.* (1979) and Knudsen *et al.* (1979b) based on initial PVO temperature measurements.

Uncomfortable with these large heat inputs from above. Cravens *et al.* (1980) introduced a small magnetic field ($B = 10\gamma$) and a reduced electron mean free path ($\lambda = 10$ km) to lower the vertical heat conductivity. This assumption was justified by the magnetometer measurements which often found largely horizontal magnetic fields and small scale magnetic structures which have been called flux ropes (Elphic *et al.*, 1980b). The resulting profiles (lower panel) agree equally well with the measurements while requiring a downward electron heat flux of only 5×10^9 eV cm^{-2} s^{-1} and an ion heat flux of 3×10^7 eV cm^{-2} s^{-1}. A small internal chemical heat source was introduced to achieve better agreement with the T_i measurements below 220 km. Rohrbaugh *et al.* (1979) had shown earlier that ion-neutral reactions of O^+ and CO_2^+ create energetic O_2^+ ions which helped to explain the higher than expected measurements of T_i in the Martian ionosphere. The same chemical heat source works well at Venus.

The argument that solar wind heating was required to explain the elevated ionosphere temperatures prompted Taylor *et al.* (1979) to look for evidence of electron heating by plasma waves which could be observed by the PVO Electric Field Detector. They noted an abrupt decline in the wave amplitude at 100 Hz as the spacecraft entered the ionosphere. This decline was attributed to Landau damping of whistler waves by the ionospheric electrons, a process that heats the electrons. The waves were assumed to have been generated at the bow shock or in the dayside magnetosheath. The observed wave energy was large enough to produce the observed temperatures, so a magnetized ionosphere may not be required to achieve agreement with the observed temperatures. This assumption is troublesome anyway, because it requires the ionosphere thermal structure to depend on the magnetic state of the ionosphere, an effect that is not observed (Elphic *et al.*, 1984a), as will be discussed shortly.

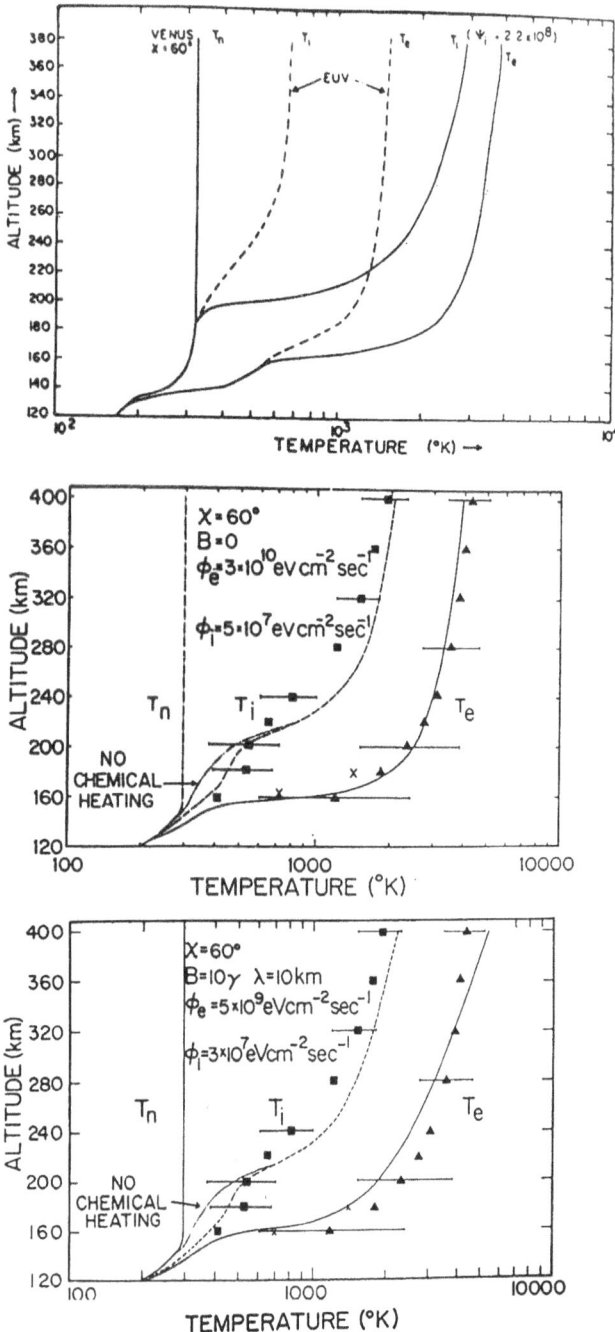

Fig. 16. Calculations of T_e and T_i before PVO (upper panel) and after PVO (middle and bottom). The dashed lines in the top panel assumed only solar EUV heating, while the solid lines assumed downward heat fluxes suggest by Viking Lander results at Mars. The middle and bottom panels represent an attempt to fit the PVO temperature measurements with and without introducing an ionospheric magnetic field. Smaller heat fluxes are resuired when the magnetized nature of the ionosphere is considered (after Cravens et al., 1978, 1979).

The first attempts to study the effects of solar EUV variations on the ionosphere occurred at solar maximum, so only the EUV changes associated with solar rotation could be examined. Taylor *et al.* (1982a) reported dayside ion and neutral density variations that were well correlated with Earth satellite measurements of the solar EUV flux and ground based measurements of the $F_{10.7}$ cm flux. The ion density variations extended into the pre-dawn ionosphere, a result that supported the belief that nightward ion flow was a major factor in the maintenance of the nightside ionosphere (Spenner *et al.*, 1981). See the companion review by Miller and Whitten (1991) for more details on the roles of nightward flow and energetic electrons in the formation of the nightside ionosphere.

Elphic *et al.* (1984a) employed the OETP measurements from the same period to examine the response of the lower dayside ionosphere to solar rotation. Figure 17 shows the ratios of individual orbit-average N_e and T_e below 200 km to empirical model average values. The results are shown for four solar rotation periods near solar maximum. The bottom panel shows an index of the total solar EUV flux based on the photoelectron current from the axial OETP sensor. The amplitude of its 28.5 day variation was scaled

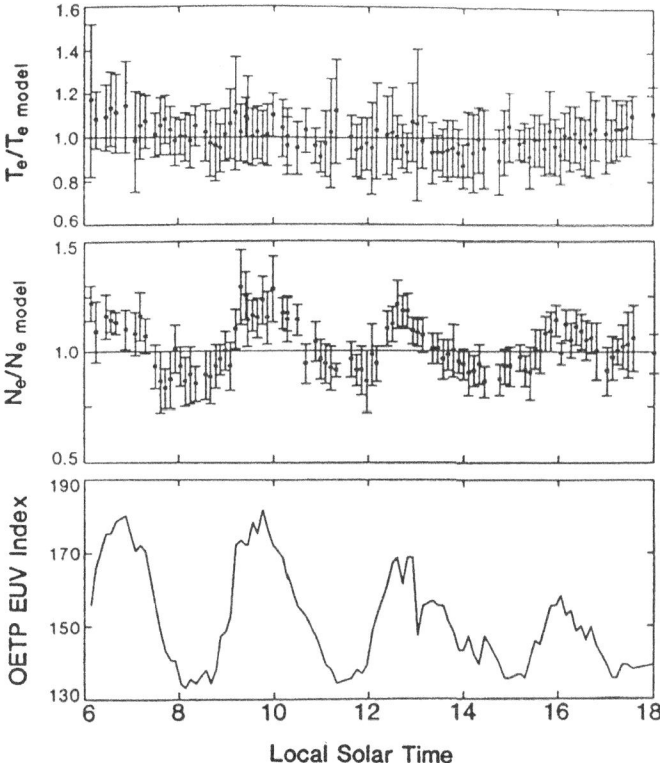

Fig. 17. The effects of EUV variations (with solar rotation) on T_e and N_e below 200 km. The ratios of the OETP measurements to modelled average values are shown. Only N_e responds to the variations the total solar EUV flux (bottom panel) also measured by the OETP (after Elphic *et al.*, 1984).

approximately to that of the $F_{10.7}$ index, but the OETP solar index had the advantage that it represented the EUV variations experienced by the Venus ionosphere. Elphic *et al.* concluded that N_e (below 200 km) was highly responsive to the EUV, but T_e was not. This result is consistent with the ORPA photoelectron measurements of Knudsen *et al.* (1980a), who found that photoelectron heating accounts for only about 20% of the total electron heat input to the dayside ionosphere.

To see if solar wind interactions might be responsible for the day-to-day (orbit-to-orbit) scatter in T_e and N_e, Elphic *et al.* removed the solar EUV contributions and correlated the residuals with the magnetic state of the ionosphere, which they took as an indicator of solar wind dynamic pressure. They considered two states; the flux rope state which occurs at times of low solar wind pressures, and the fully magnetized state which occurs at times of high solar wind pressures (Luhmann *et al.*, 1987). They found that N_e and T_e at altitudes below 200 km are not significantly affected by the magnetic state of the ionosphere.

This is not to say that solar wind heating has no effect on the ionosphere. The sharp rise in T_e at the ionopause reflects solar wind heating. Luhmann *et al.* (1987) and Mahajan *et al.* (1988) have shown that T_e increases above 250 km at times of high solar wind dynamic pressure, but these increases may simply reflect the movement of the ionopause to these altitudes at these times. A real increase in the heat flux would have been detected by Elphic *et al.* in the T_e measurements made below 200 km.

All of this leaves our understanding of the dayside thermal balance in a dilemma. It appeas that T_e does not respond to changes in either the EUV flux or the solar wind, which are the main heat sources for the ionosphere. The absence of a T_e response to EUV variations could arise from opposing changes in the electron cooling rate caused by solar EUV-driven increases in N_e. But the absence of a T_e response to solar wind variations is curious, because solar wind heating has been invoked specifically to explain the observed high temperatures. These results make it clear that the thermal balance of the dayside ionosphere is not yet understood.

Nightside Thermal Structure

The thermal structure of the nightside ionosphere is perhaps even less well understood. Miller and Whitten (1991) cover this topic in greater detail in a companion review. The two most obvious heat sources for the nightside are heat conduction from the dayside and heating at the ionopause. Energetic electron precipitation and convective heat flow could also be important (Knudsen *et al.*, 1980). Hoegy *et al.* (1980) employed the empirical T_e model of Theis *et al.* (1980) to examine the relative importance of horizontal heat conduction across the terminator and downward conduction of heat introduced at the top of the ionosphere. Figure 18 illustrates the heat conduction paths which provided the best fit with the T_e model. This conduction pattern suggests that the nightside ionosphere below about 300 km is heated primarily by conduction from the dayside, while the upper ionosphere is heated primarily from above. The total nightside heating was more than an order of magnitude less than the dayside heating, but this represents an upper limit because no account was taken of the smaller saturated heat

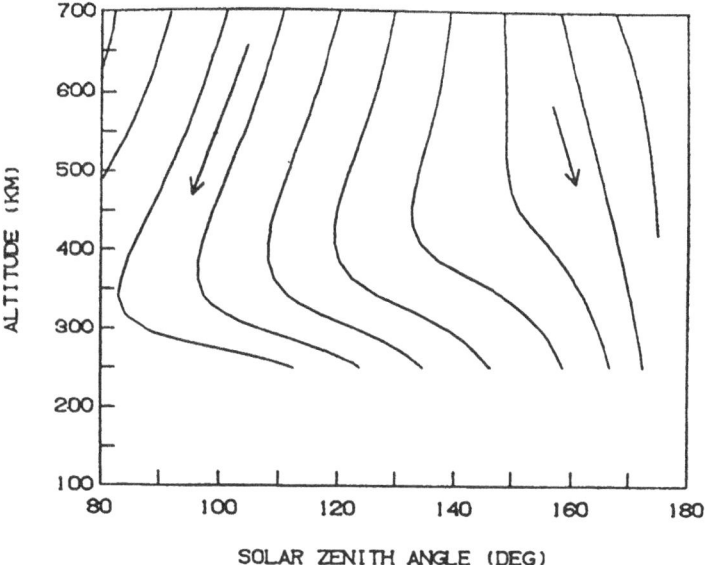

Fig. 18.   Model heat flux lines (magnetic field lines) that satisfy the electron energy balance for the Venus nightside T_e model of Theis *et al.* (1980). The arrows indicate the paths of heat flow from both the dayside ionosphere and from above (after Hoegy *et al.*, 1980).

conductivity that should apply at the low densities of the nightside ionosphere (Merritt and Thompson, 1980).

Since electron heat conduction proceeds primarily parallel to the magnetic field, the lines in Figure 18 can be viewed as the average magnetic field configuration in the nightside ionosphere. The nightside measurements themselves (Russell *et al.*, 1980) reveal a magnetic field that is highly chaotic, but no attempt to produce an average nightside magnetic field model has been reported.

The nightside ion temperature is even less well explained than is the electron temperature. Knudsen *et al.* (1980b) had suggested that these elevated temperatures may be due to a reconvergent shock in which the supersonic nightward flow is converted into thermal energy. Bougher and Cravens (1984) examined the ion energy balance using a two-dimensional model which takes into account the horizontal and vertical bulk transport of heat. They were successful in explaining the T_i variations up to 150° SZA without introducing a topside heat source or any magnetic limitations on the ion heat conductivity. But they could not reproduce the higher temperatures reported by Miller *et al.* (1980) for the antisolar region.

4.5. ION COMPOSITION

The composition of the ionosphere turned out to be complex and highly variable. The diurnal variation of the major ions at an altitude of 200 km was shown earlier in Figure 15. Quasi-vertical ion profiles from a single subsolar passage are shown in Figure 19 (Taylor *et al.*, 1980). O^+ is the major ion above 190 km, while O_2^+ dominates

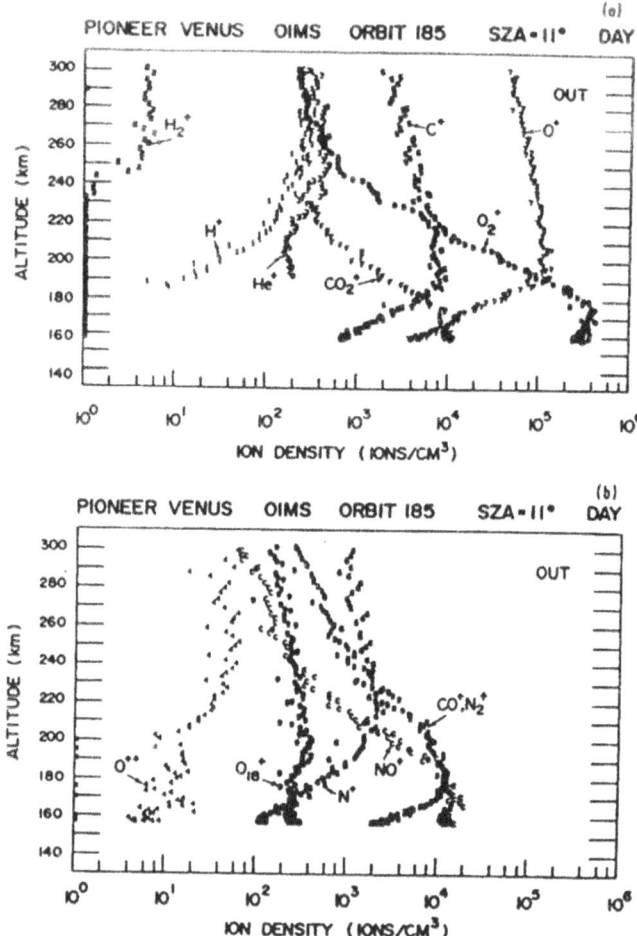

Fig. 19.   An example of the ion composition measurements from a single PVO passage through the subsolar ionosphere. These profiles illustrate the rich variety of ions that are present in the Venus ionosphere (after Taylor *et al.*, 1980).

below that altitude. It is interesting that CO_2^+ remains a minor ion at low altitudes, in spite of the fact that CO_2 is the major neutral constituent. H^+ is a major constituent at high altitudes on the nightside, but the origin of the mass 2 ion was uncertain. Its concentration was about 1% of that of H^+. From an analysis of the OIMS measurements, Hartle and Taylor (1983) showed this ion to be deuterium, D^+, rather than H_2^+. This supported a conclusion of McElroy *et al.* (1982) that the preferential escape of H should have produced a 100-fold enrichment of D relative to H. Similar conclusions were reached by Donahue *et al.* (1982) based on mass spectrometer measurements in the lower atmosphere by the Pioneer Venus Large Entry Probe.

The Dayside Composition

The review of ionospheric theory by Nagy *et al.* (1983) describes the early attempts to derive the ion composition at Venus from N_e profiles using ionosphere theory. Most of these efforts were largely unsuccessful because too little was known about the composition and temperature of the thermosphere. In particular, there was no information on the concentrations of atomic oxygen and molecular nitrogen. Small amounts of these constituents cause very large increases in the concentrations of the molecular ions O_2^+,

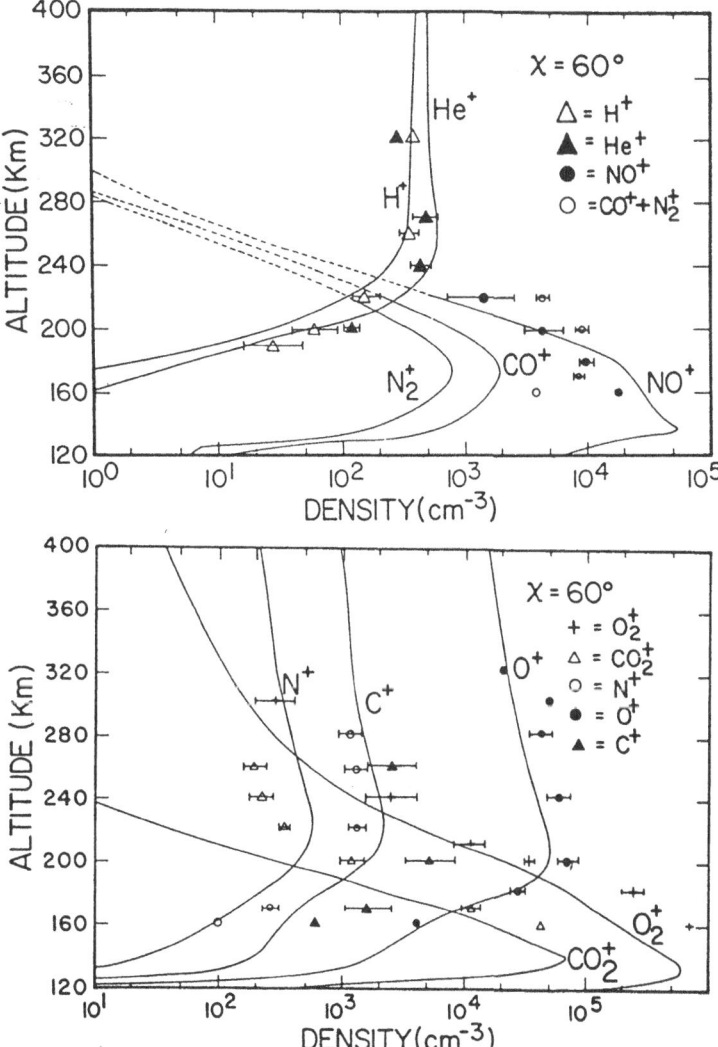

Fig. 20. Comparison of theoretical ion densities (solid) with the OIMS ion density measurements. Reasonable agreement was obtained except for the molecular ions N_2^+ and mass 28 ($CO^+ + N_2^+$) (after Nagy *et al.*, 1980).

NO^+, N_2^+, CO^+, and CO_2^+, the ions that turned out to be the major constituents in the lower ionosphere of Venus. However, once PVO measured the thermospheric composition (Niemann *et al.*, 1980; Hedin *et al.*, 1983), the ion chemistry was well enough known to largely reproduce the ion composition measurements. Figure 20 illustrates an early attempt by Nagy *et al.* (1980) to match the OIMS ion densities measured at $60°$ SZA. The agreement was adequate for most of the major constituents, considering the questionable accuracy of the measurements that was implied by differences in the ion composition measured by the OIMS and ORPA at altitudes below about 200 km. The agreement with theory was poor for the minor ions N^+, N_2^+, and CO^+ whose ion chemistry is significantly altered by the presence of metastable species,

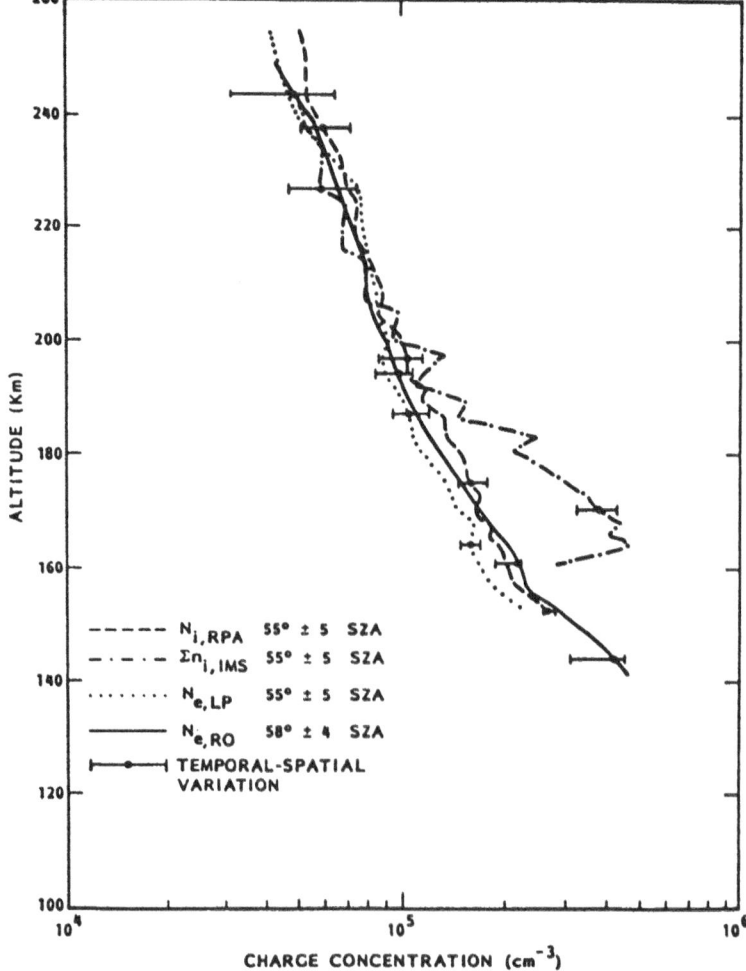

Fig. 21.   Comparison of the total plasma density profiles near SZA = 55° from the Pioneer Venus ORPA, OIMS, OETP (LP), and the ORO. The agreement is good, except below 200 km where the OIMS total density exceed that measured by other methods by as much as a factor of 2.5 (after Miller *et al.*, 1984).

particularly $O^+(^2D)$, $O^+(^2P)$, $N(^2D)$, and $N(^2P)$. The inclusion of the appropriate metastable ion chemistry brought the minor ion calculations into general agreement with the observations (Fox, 1982).

Returning to the question of measurement accuracy, the ORPA was able to obtain median height profiles of the major ions O^+, CO_2^+, and an ion group consisting of the sum of NO^+, N_2^+, and CO^+ (Miller *et al.*, 1984). These results were similar to those of the OIMS above 200 km, but they differed in some important respects at lower altitudes. The OIMS found a systematic peak in O_2^+ near 175 km that was not evident in the ORPA dayside profiles. This peak can be seen for example in the upper panel in Figure 19. Miller *et al.* (1984) specifically noted that the total ion density obtained by the OIMS on the dayside was up to a factor of 2.5 larger at 165 km than that measured by the ORPA. This difference is evident in their comparison of the total plasma density profiles from the various PVO instruments, shown in Figure 21 for solar zenith angles near 55°. Breus *et al.* (1985) also noted that neither the radio occultation N_e profiles (ORO) or theoretical profiles based on ion photochemistry are consistent with the OIMS measurements of a peak in the total ion density some 25 to 30 km above the main peak. These discrepancies in the ion composition and density measurements have not been resolved, but in the Appendix we note some potential sources of measurement error that could be involved.

The Nightside Ion Composition

The existence of a substantial nighttime ionosphere was a bit surprising to early investigators. The planet was known to rotate very slowly, with the period of surface darkness lasting 58 Earth days (Colin, 1983), so any ionization that was produced in the dayside thermosphere had been expected to recombine soon after sunset. This expectation did not consider a number of factors, however. First, the atmosphere at the cloud tops, and probably the overlying thermosphere, rotates much more rapidly than the planet itself (Schubert *et al.*, 1980; Mayr *et al.*, 1980), reducing the time that a particular sector of the thermosphere is in darkness to about 4 Earth days. Secondly, and more important, Venus does not have an intrinsic magnetic field to restrict the nightward flow of ions (Russell *et al.*, 1980). And finally, energetic, or suprathermal, electrons precipitate into the nightside of Venus to produce ionization locally.

The relative importance of ion transport and local ion production has been a matter of considerable controversy. The earliest interpretation of the Mariner 5 nightside profile (Figure 2) invoked the nightward transport of H_e^+ or H_2^+ ions (McElroy, 1968; McElroy and Strobel, 1969). These attempts to reproduce the density profiles were largely successful, but they did not get the ion composition right because too little was known about the thermosphere composition to determine the appropriate ion chemistry. After Venera 9 and 10 detected energetic electrons in the Venus wake, several groups showed that the fluxes were adequate to produce the densities observed at the nightside peak (Gringauz *et al.*, 1977, 1979; Chen and Nagy, 1978). The case for an energetic electron source was strengthened by the good correlation that existed between the variations in the energetic electron fluxes and the peak densities.

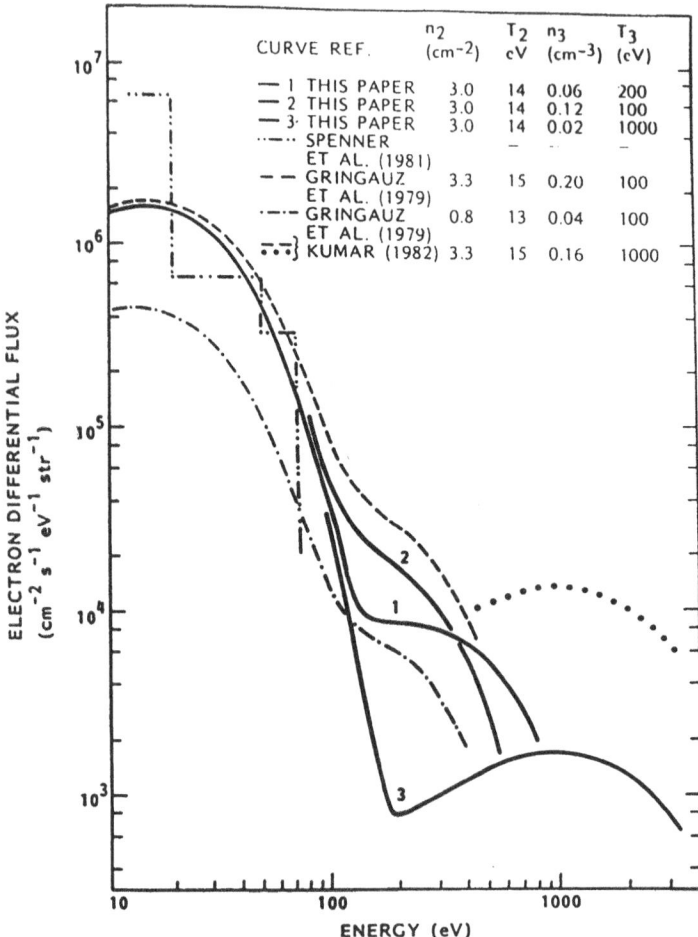

Fig. 22.   Differential electron energy fluxes in the Venus umbra from PVO and Venera measurements. n_2 and T_2 are the density and temperature of the primary suprathermal component, and n_3 and T_3 refer to the same parameters for a more energetic electron component. These fluxes are adequate to produce a substantial fraction of the nightside ionization observed near the main peak of the ionosphere (after Knudsen and Miller, 1985).

Knudsen and Miller (1985) reported PVO observations of differential electron energy distributions that were similar to the Venera results Gringauz *et al.* Both are shown in Figure 22. A PVO spectrum used by Spenner *et al.* (1981) and one used by Kumar (1982) are also shown. Two suprathermal electron populations were present in both the PVO and Venera spectra. The main population in the PVO measurements had an average temperature of 14 eV and a density of 3 cm^{-3}. The second, more energetic, electron component had an energy of several hundred eV and much smaller densities. The PVO integral energy flux was about a factor of 2 larger than the Venera flux, but they are otherwise very similar. The differences could be due to the different instrument look angles in an anisotropic medium, or the fact that the PVO measurements were made at solar maximum, while the Venera fluxes were measured at solar minimum.

Perhaps a more important difference in the Venera and PVO suprathermal electrons was that the Venera fluxes were highly variable from orbit to orbit, while the PVO fluxes were less variable, making it more difficult to explain the variability of the nightside ionosphere in this way. This made nightward transport a more likely source of the nightside ionosphere, at least at solar maximum. Cravens *et al.* (1983) were able to reproduce most aspects of the nightside ion composition using nightward ion flow in a two-dimensional model. Since then, Knudsen *et al.* (1986) have assembled profiles of the mean density of the major ions in the central nightside ionosphere between 145 km and 900 km based on ORPA measurements. They compared their O^+, O_2^+, and total ion density profiles with the theoretical profiles of Cravens *et al.* (1983), which separately considered the effects of nightward ion transport and local electron precipitation. Their results suggest that nightward transport is almost totally responsible for the O^+ in the central nightside ionosphere, while the O_2^+ is produced almost equally by both nightward ion transport and electron precipitation. The measured molecular ion densities exceeded the theoretical values by about a factor of 2, suggesting that the model overestimates the total nightward ion flow.

4.6. DAWN-DUSK ASYMMETRY IN THE ION COMPOSITION

Some particularly surprising features of the nightside ion composition were the high concentrations of H^+ and its large dawn-dusk asymmetry (Taylor *et al.*, 1979a), effects that were shown earlier in Figure 15. This asymmetry has been attributed to the super-rotation of the upper atmosphere (Mayr *et al.*, 1980). H^+ is more than an order of magnitude more abundant on the nightside than on the dayside because of wind-induced diffusion and exospheric flow of the neutral atomic hydrogen from which it is formed by the charge exchange reaction,

$$O^+ + H \leftrightarrow H^+ + O. \tag{2}$$

Since H^+ can be assumed to be in chemical equilibrium below the exobase, its concentration is determined by charge exchange, a relationship which Brinton *et al.* (1980) used to derive the H concentration from the PVO measurements of O, O^+, and H^+ made below 160 km. ONMS cannot measure the H concentration itself because it is too chemically reactive to be observed before it recombines in the mass analyzer.

Taylor *et al.* (1985) employed the OIMS data from the first three Venus years to provide an empirical model of the relative ion composition for the subsolar and antisolar regions. These are shown in Figure 23. The relative ion composition on the dayside and nightside is actually quite similar, except that the lighter ion H^+ is greatly enhanced at night because of the greater nightside H concentration. In fact, H^+ and O^+ have similar concentrations in the predawn bulge. This is illustrated in the single outbound profile through that sector shown in Figure 24 (Taylor *et al.*, 1985). It is notable that the scale height of O^+ at high altitude is nearly as great as that of H^+. In diffusive equilibrium, the scale heights of these ions would be different by a factor of 16. The combination

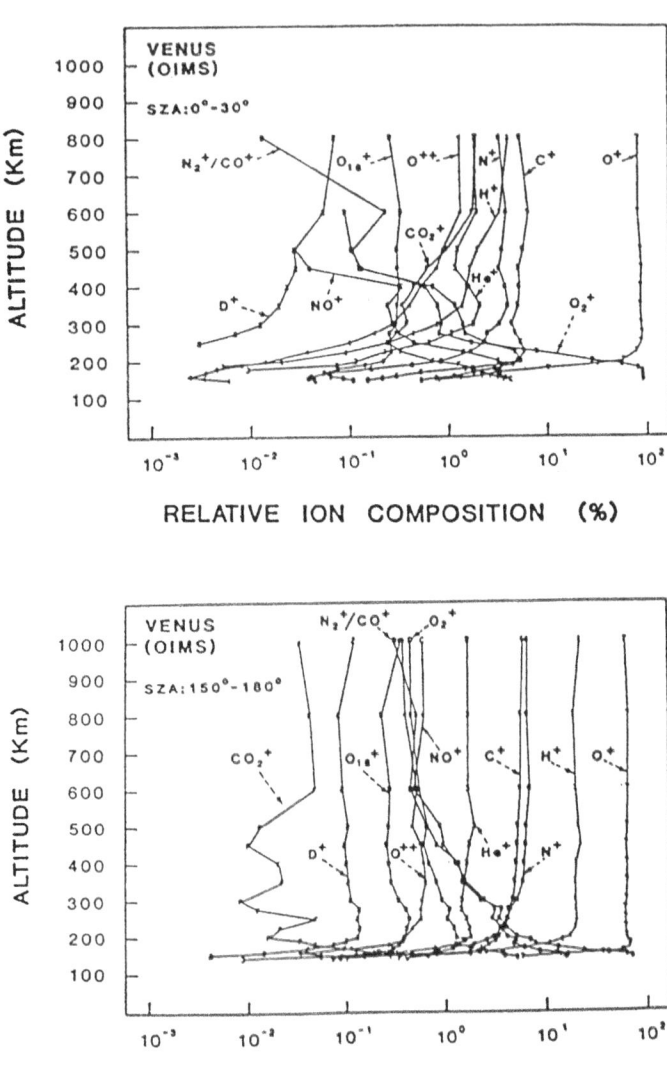

Fig. 23.   Relative ion composition for the subsolar ionosphere (upper panel) and antisolar ionosphere (lower) derived from OIMS measurements from many orbits. These data were used as the basis for the ion composition in the VIRA ionosphere model (after Taylor *et al.*, 1985).

of large and equal H^+ and O^+ scale heights suggests that the predawn region has either strong upward ion flow, or that this region contains unobserved higher energy ion and/or electron populations that affect the polarization electric fields which determine the ion scale heights.

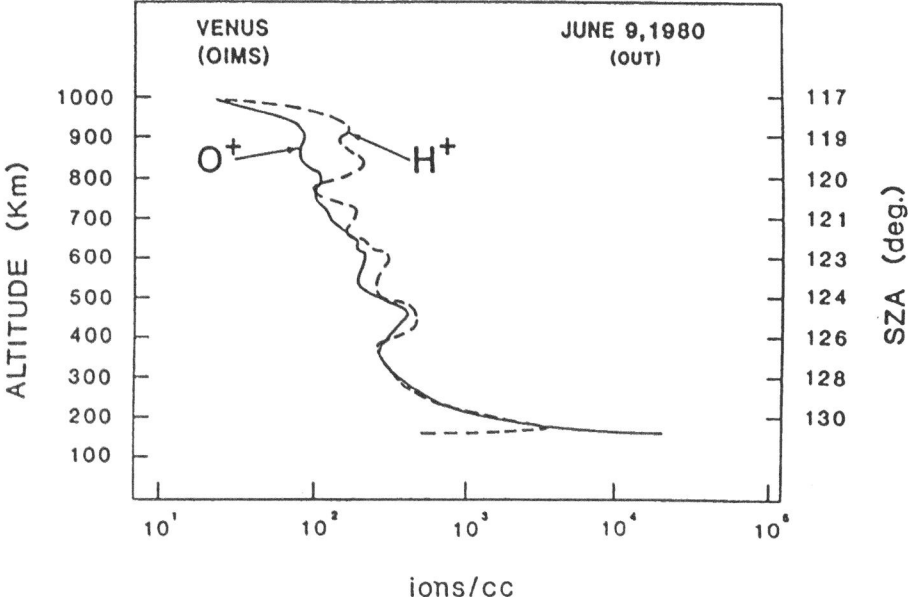

Fig. 24.   Profiles of O^+ and H^+ in the predawn ionosphere. The similar scale heights of the two ions suggest that this region is not in diffusive equilibrium. This result could be caused by upward ion flow out of the ionosphere or the presence of energetic ions and/or electrons (after Taylor *et al.*, 1985).

4.7. SMALL-SCALE SPATIAL STRUCTURE ON THE NIGHTSIDE

Nearly all parameters of the nightside ionosphere exhibit a lot of small scale spatial features, even during periods of low and steady solar wind pressure. The most common and largest amplitude nightside features are discussed in the next 3 sections.

4.7.1. *Post-Terminator Waves*

The lower ionosphere just downstream of the terminator characteristically exhibits large amplitude wavelike structures in N_e and T_e and the magnetic field. Brace *et al.* (1983b) called these structures post-terminator waves, and found that they occur at altitudes below about 175 km and at SZA between about 90° and 120°. The magnetic field exhibits spatial structure that is coherent with the N_e and T_e structure. The east–west component of the field, B_E, exhibited the greatest wave amplitudes. Figure 25 shows an example of these waves in a single southward passage on an orbit whose periapsis was at an altitude of 155 km. The bottom panel is an expanded view of the 2 min about periapsis, an interval when the spacecraft altitude changed by only about 15 km while it travelled 1200 km horizontally. Clearly the horizontal structure of the waves dominates these profiles. N_e, T_e, and B_E exhibit the same fundamental period, with a wavelength along the orbit of the order of 200 km (20 s at the spacecraft velocity of 10 km s^{-1}). This is not their true wavelength because the orbit is inclined 15° to the terminator. If one assumes that the waves are generated near the terminator and are propagating

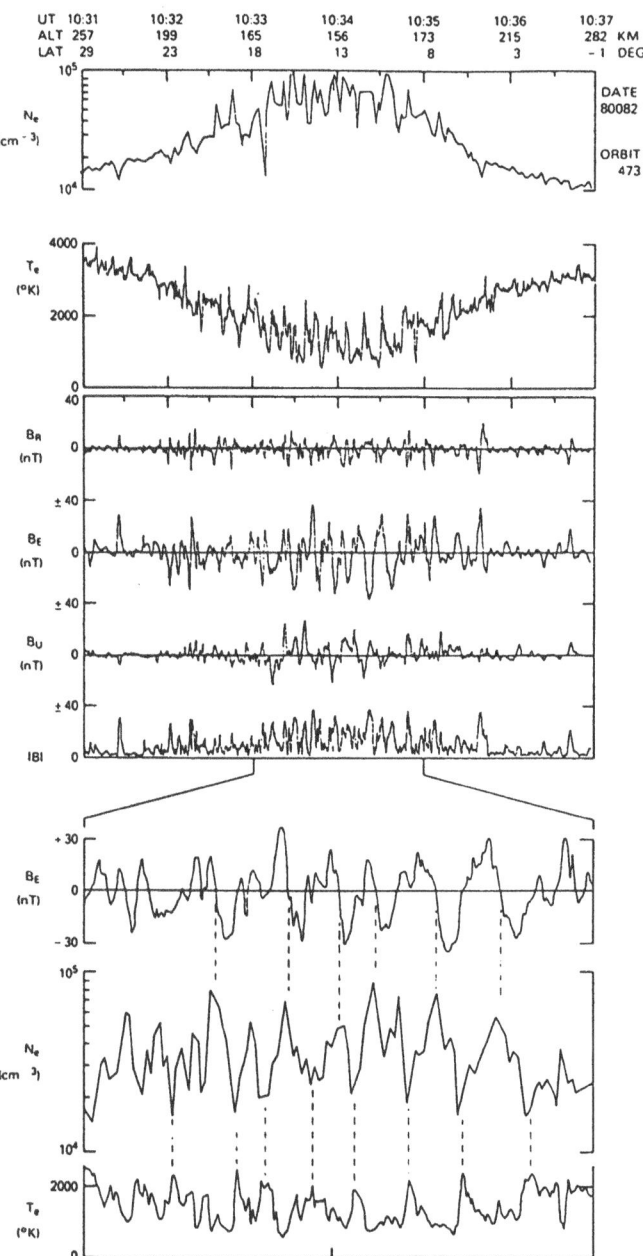

Fig. 25.   Wave-like structure observed in the plasma and the magnetic fields of the post-terminator ionosphere below 175 km. The waves in N_e and T_e are 180° out of phase, while the B_E waves are shifted 90°. The energy source for the waves is likely to be the nightward ion flow, but the wave generation process is uncertain (after Brace *et al.*, 1983b).

toward the antisolar point, their longitudinal wavelengths would be about 50 km. The wavelike nature of these structures extends downstream about 30° from the terminator, where they become less regular, as discussed in the next section. The post-terminator waves are seen at both the dawn and dusk terminators, but the nearly fixed latitude of periapsis ($\approx 15°$ N) has made them observable only in the range of 10° to 20° north latitude. They are assumed to extend downstream from the entire terminator, however.

Another interesting feature of the post-terminator waves is the unique phase relationships among the plasma and magnetic parameters. N_e and T_e are out of phase, and the B_E waves are shifted by 90° from both, as can be seen from the dashed lines in Figure 25. In each case, magnetic field reversals from east to west are associated with an N_e peak, while magnetic reversals from west to east are associated with N_e troughs.

No generation mechanism for the post-terminator waves has been clearly identified. Brace et al. (1983b) suggested wave production by a shear instability driven by ion-neutral collisions at altitudes below the exobase, which is near the periapsis altitude in this local time sector. Knudsen et al. (1980b) reported high ion flow velocities in this sector. Spenner et al. (1981) showed the resulting nightward ion flux to be a major factor in maintaining the nightside ionosphere. Since the thermospheric gas is moving nightward at much lower velocities (Niemann et al., 1980), one imagines that the waves may be signatures of the overturning of ionospheric plasma as it accelerates into the nightside, its bulk velocity increasing with altitude due to ion-neutral drag below the exobase. The resulting tumbling motion of this high beta plasma could be expected to wind up the entrained field into structures having the same scale sizes as the plasma. But this is only speculation. No quantitative explanation for the post-terminator waves has been offered. We suspect that the PVO data base contains much more information about these waves than has been examined thus far.

4.7.2. Waves in the Antisolar Region

Large amplitude ionospheric structure is not limited to the post-terminator region. Hoegy et al. (1986, 1990) recently extended the study of these waves into the antisolar region in an attempt to understand the complex thermal structure of the nightside ionosphere, particularly at altitudes below 175 km where high amplitude structure is very common. They found that N_e, T_e, and B_E continue to exhibit small-scale variations at larger SZA, but that these variations are more chaotic and have less reliable phase relationships than do the post-terminator waves. Figure 26 shows an example of the structures that were encountered near periapsis (154 km) during orbit 514. The local time of periapsis was 23 hours, and the spacecraft was below 170 km during the entire 2-min interval (1200 km horizontal distance) that is shown. Hoegy et al. also discovered that structures of similar sizes were often present in the neutral particle densities, and that these structures tended to have specific phase relationships with the N_e irregularities. The phases suggest that at least some of the ionospheric structure in the antisolar region is induced by thermospheric gravity waves.

The source of the electron heating in these structures is unknown. The temperature

Orbit 514

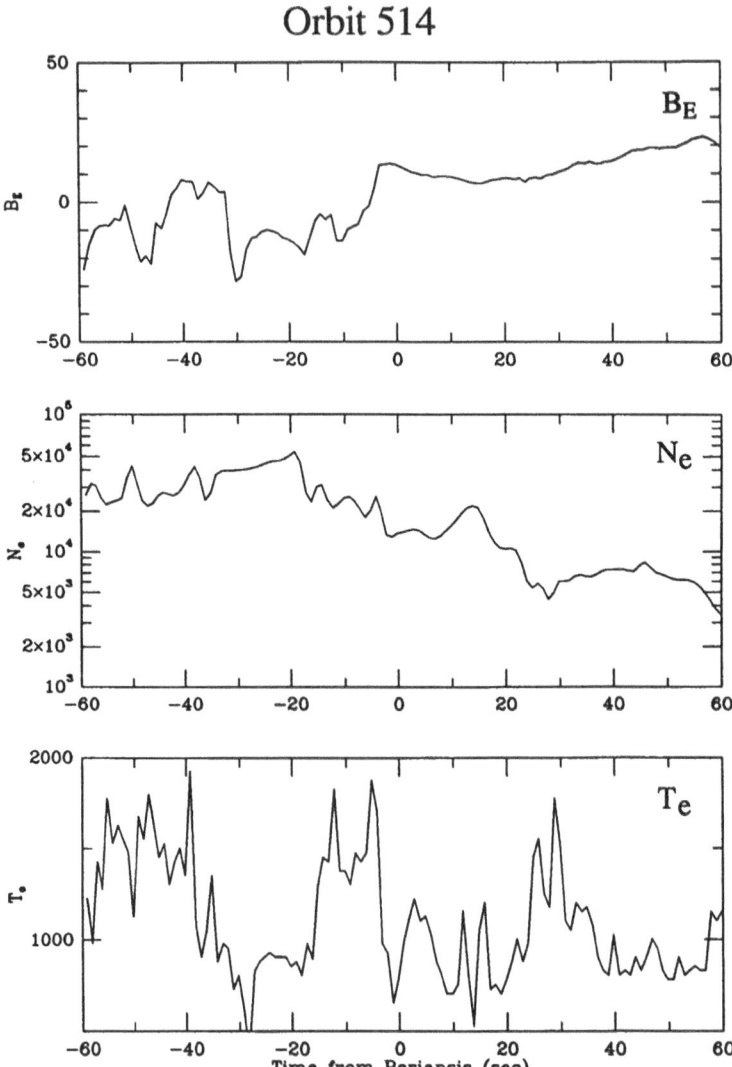

Fig. 26. B_E, N_e, and T_e structure in the antisolar ionosphere. These structures are larger, less wavelike, and less well phase correlated than the post-terminator waves (after Hoegy *et al.*, 1990).

peaks are clearly not signatures of adiabatic heating, because they are out of phase with the N_e peaks. One possibility is electron heat conduction from the overlying ionopause or from the dayside ionosphere. Photoelectrons, originating on the dayside, are another possible heat source, but it seems unlikely that they could survive the long, torturous path along the apparently tangled magnetic field lines of the post-terminator region (e.g., Figure 25).

Local electron cooling by neutral collisions should also affect T_e at these altitudes. The electrons are even more strongly cooled at altitudes only one or two scale heights below periapsis, so one would expect to find cooler regions where the dip angle of the magnetic field was large. A large magnetic dip angle should permit the electrons to conduct energy into the cooler regions below, but Hoegy *et al.* found no systematic correlation between the dip angle and T_e in individual PVO passes. The source of these structures, and reason for their well ordered phase relationships, are among the most important unanswered questions about the nightside ionosphere.

4.7.3. *Ionospheric Holes*

Another major surprise in the nightside ionosphere came with the discovery of large holes in N_e in the antisolar region (Brace *et al.*, 1980, 1982b). Taylor *et al.* (1980) measured the same structures in the ions and referred to them as plasma depletions. The density in these holes is lower than that in the surrounding ionosphere by one or two orders of magnitude. Figure 27 is an N_e profile from a single passage in which two such holes were encountered; one centered at about 29° north latitude and the other at 24° south latitude. Figure 28 shows the behavior of a variety of ionosphere parameters during the passage through another pair of ionospheric holes, including the plasma wave amplitudes (top panel), the magnetic field components (second panel), T_e (third panel), N_e and O^+ (fourth panel), T_i (bottom panel). Strong quasi-vertical magnetic fields exist within the holes ($B_x > B_y$ or B_z), while the magnetic field in the surrounding ionosphere is usually weak and more nearly horizontal (Luhmann *et al.*, 1982). This vertical orienta-

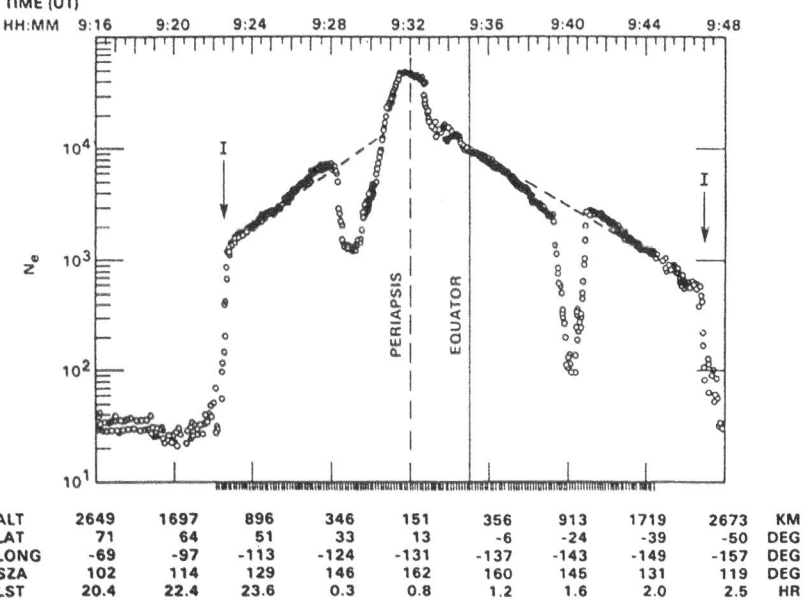

Fig. 27. Ionospheric holes in an otherwise smooth nightside N_e profile. The dashed lines represent the smooth ionosphere without holes, and the ionopause crossings (*I*) are identified (from Brace *et al.*, 1982b).

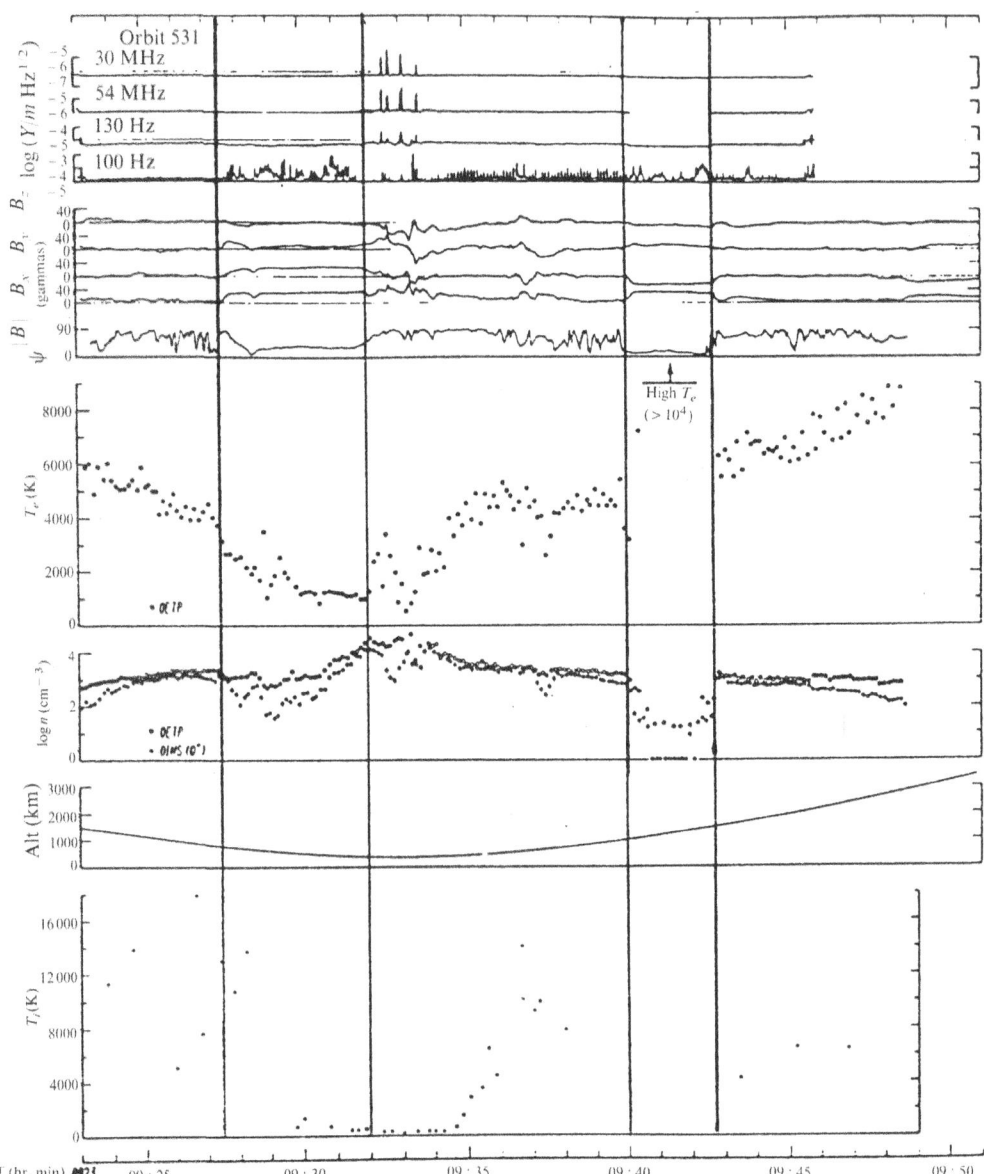

Fig. 28. Plasmawave, magnetic field, T_e, N_e, O^+, and T_i measurements through a pair of ionosphere holes in the antisolar region. The magnetic field is nearly vertical within the holes. All other parameters are perturbed in the holes (after Luhmann *et al.*, 1982).

tion of the magnetic field in the holes provides the primary evidence that the holes themselves are configured vertically as shown earlier in Figure 7. Additional evidence for their vertical orientation is the observations of similar holes at higher altitudes in the antisolar region (Brace *et al.*, 1987).

The plasma in the holes is complex and significantly different than that found in the surrounding ionosphere. Two electron populations were present in the outbound hole, which was traversed between 09:40 UT and 09:43 UT. One component was cold and one was hot. The cold component has a temperature of about half that of the surrounding ionosphere, while the hot component is an order of magnitude warmer. The hot component is observed only at higher altitudes in the holes, where N_e is low. T_i is also much lower in the holes than in the surrounding ionosphere (bottom panel of Figure 28). The ion composition in the holes is also quite different than in the surrounding ionosphere. Although only O^+ was shown for this pass, H^+ often becomes a major ion in the holes, while O^+ is the major ion outside (Grebowsky et al., 1983).

An examination of all of the nightside passages of PVO has suggested that holes are very common in the antisolar region, and they may be a nearly permanent feature (Brace et al., 1982b). This is illustrated in the series of N_e profiles shown in Figure 29, all taken during a single season of nightside transits in 1980. One or two holes are observed in most of the passes that occur within an hour or so of midnight, so they are assumed to be a more or less steady-state feature. Their changing latitudes shows that they do move about from orbit to orbit, but there is no way to be sure that the same hole is being observed on consecutive passages. More often than not the holes are seen in north–south pairs, but the absence of a hole encounter during any given passage could simply mean that they did not happen to lie in the spacecraft path. Their north–south dimension along the orbit, which is measured directly, is of the order of 500 to 1000 km. If the east–west dimension of the holes was small at the time of a PVO passage, either or both of them could be missed by a spacecraft in a nearly polar orbit, leading to N_e profiles showing either one hole or no holes. Brace et al. argued that their average east–west dimension must be somewhat larger than 1000 km to explain their high occurrence rate as single holes, and as north–south pairs.

The depth of penetration of the holes into the ionosphere is not well known. Since the holes are more or less centered on the equator, the slight northward offset of periapsis caused PVO to cross the inbound holes at generally lower altitudes (200–500 km) than the outbound holes (300–1000 km). Inbound hole crossings were reported as low as 160 km (Brace et al., 1982b), but this could happen only on those rare occasions when a northern hole has moved equatorward to the vicinity of periapsis (15–17 N). Luhmann et al. (1982) suggested that one PVO passage may have observed this situation in which strong quasi-vertical magnetic fields were present at periapsis, but no hole was seen. The spacecraft is assumed to have passed under a hole on that occasion. Surprisingly, an ion drift velocity of over 7 km s^{-1} was measured in the region below the hole, with a downward vertical component of about half of the total velocity. It is not clear what such high downward ion velocities at these altitudes implies about the hole formation process, or about the dynamics of the nightside ionosphere in general.

Although solar wind interactions are generally believed to be involved in creating the holes, the process is uncertain. Brace et al. (1982b) attributed the holes to the emergence of magnetic flux which enters the dayside ionosphere and is convected nightward by the bulk ion flow. In that scenario, the stagnation of ion flow in the antisolar region allows magnetic flux loops to emerge from the ionosphere, creating the holes by removing plasma by $\mathbf{B} \times \mathbf{v}$ acceleration. Once formed, a hole is maintained by its radial magnetic field which inhibits ion transport from the surrounding unmagnetized ionosphere, while

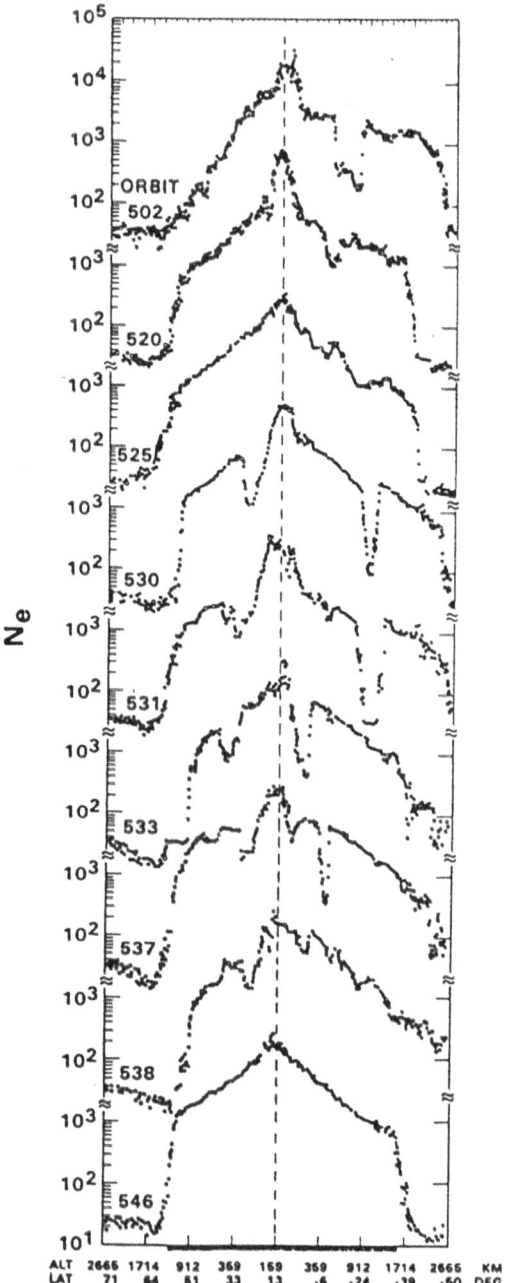

Fig. 29.   Series of N_e profiles through the antisolar region showing how common the ionospheric holes can be. They are most often seen as north–south pairs, but may be seen alone, or not at all. The outbound crossings occur a higher altitude and exhibit lower densities (from Brace *et al.*, 1982).

at the same time allowing the ions within the holes to flow vertically into the underlying thermosphere where recombination proceeds rapidly.

This scenario also explains why the electrons and ions in the holes are cold. The radial field thermally isolates the hole from the hot ionosphere around it, while permitting field-aligned electron heat conduction from the upper ionosphere into the colder lower ionosphere. Once the hole exists, it can be invaded by field aligned transport of hot solar wind electrons from the wake, thus accounting for its higher temperature component. It is also possible that the hot component represents ionospheric electrons that have been accelerated in the ongoing hole formation and maintenance process.

Grebowsky and Curtis (1981) proposed an alternate formation scenario in which the holes are created by parallel electric fields that are generated by one or more processes that can occur in the overlying plasma sheet. In this model, the parallel electric field also accelerates plasma sheet electrons into the holes to produce the hot electrons measured there. Later Grebowsky *et al.* (1983) used he plasma fluid equations to demonstrate that nightward ion convection could produce most of the known characteristics of the holes, more or less in the manner suggested by Brace *et al.* (1982b). They showed that the nightward ion flow can produce the sharp gradient in the magnetic field strength at the edge of a hole, and that this gradient would allow the hole to be formed by ion subsidence followed by chemical recombination, once ion transport from the surrounding iono- sphere has been cut off.

The question arises as to what one can say about the geographical distribution of the holes. Brace *et al.* (1982b) found that the holes tend to form in two zones of latitude which more or less straddle the Venus equator. The two hole pattern is offset to the north by a few degrees and shifted toward dawn by about an hour. No explanation for the northward offset was suggested. It was argued that the shift toward dawn tended to support the involvement of the solar wind in hole formation, although the observed shift of 15° is larger than the 5° solar wind aberration caused by the orbital motion of Venus.

Marubashi *et al.* (1985) studied the hole distribution using OIMS measurements. Assuming that the holes are a manifestation of solar wind interactions, they attempted to better order their global pattern by using the orientation of the IMF to define the magnetic equator at the time of each passage. Figure 30 shows the PVO orbits segments lying within the holes both before and after rotation by the IMF. The left panel is the distribution in the conventional Venus–Sun–Earth coordinates, VSE, and the right panel shows the same holes in solar wind magnetic coordinates, SWM. The SWM hole distribution was more randomly distributed about the antisolar point. But the magnetic field polarities in the holes were remarkable well ordered by the polarity of the IMF. This ordering of the polarities is consistent with IMF draping around the planet, and it confirms the suspicion that the holes are formed by solar wind interactions with the ionosphere.

The existence of such complex structures as the holes and the waves makes it quite clear that the nightside ionosphere is a highly dynamic region. Its temporal variability and spatial structure are somehow linked to variations in the solar wind, but the exact processes have not been identified with any precision. The small scale structures also

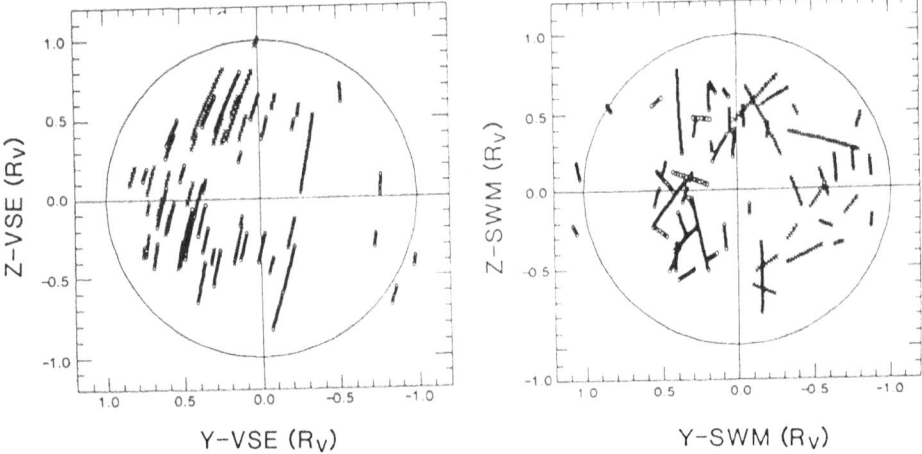

Fig. 30.   PVO orbit segments lying within the ionospheric holes, shown in solar ecliptic coordinates, VSE, (left) and rotated into solar wind magnetic coordinates, SWM, (right). The 0 and × symbols represent the polarity of the magnetic field, toward or away from Venus, respectively. The polarity of the field is essentially random in VSE coordinates, but is well ordered in SWM coordinates, consistent with IMF draping (after Marubashi *et al.*, 1985).

should drive home the point that *in situ* profiles of the nightside ionosphere have limited validity as height profiles. Horizontal structure can easily be far larger than the height variations.

4.8. LATITUDINAL AND SEASONAL VARIATIONS

The Venus ionosphere would not be expected to exhibit large seasonal or interhemi-spherical differences. The geographic pole is tilted only 3° from the ecliptic pole, and the Venus orbit is nearly circular. Latitudinal variations associated with SZA would be expected, and additional latitude effects might be produced by the solar wind through its effects upon the magnetic properties of the ionosphere (Luhmann and Cravens, 1991). The IMF is aligned preferentially parallel to the ecliptic plane, so magnetic effects on the ionosphere would tend to produce latitudinal structure, either by inhibiting or promoting north–south ion drift. Since the ionosphere is heavily magnetized only at times of high solar wind pressure (Luhmann *et al.*, 1980), magnetically induced latitu-dinal structure, if present, should be expected only occasionally. No detailed search for this effect has been reported.

The great eccentricity of the PVO orbit ($e = 0.82$) and the initial northward offset of periapsis from the equator have combined to make latitudinal gradients and inter-hemispherical differences rather difficult to observe, at least through *in situ* measure-ments. The large altitude changes within each passage tend to emphasize the altitudinal structure which is already dominant on the dayside.

The most direct detection of latitudinal structure should come from comparing the measurements from inbound transits (middle latitudes) and outbound transits (low latitudes) during the same PVO passage. One such comparison is illustrated in the excellent agreement of the inbound and outbound ion density profiles shown in Figure 31 (Taylor *et al.*, 1979a). The inbound and outbound measurements agreed very well where they overlapped at 800 km. This agreement comes in spite of the fact that the latitude was at 50° N on the inbound leg and 20° S on the outbound leg. This would appear to be typical, since Taylor *et al.* (1980) report no evidence of systematic latitudinal variations in the ion composition.

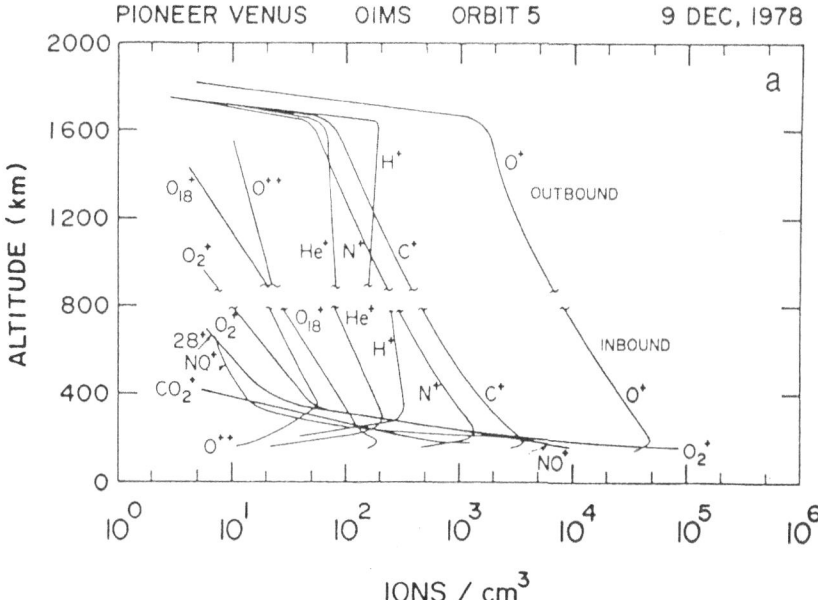

Fig. 31.   The absence of latitudinal structure is evident in the continuity between the inbound (50° N) and outbound (20° S) OIMS ion profiles on a single PVO afternoon passage (after Taylor *et al.*, 1979a).

Several authors have looked for latitude variations statistically. Theis *et al.* (1980) used their empirical model of N_e and T_e behavior with altitude and SZA to look for statistically significant differences between the inbound and outbound OETP measurements. They found none. Figure 32 shows the ratios of the individual N_e and T_e measurements to the corresponding empirical model values computed for the location of each of the measurements. All local times are included. Since the empirical model made no distinction between inbound and outbound data, real latitudinal structure would produce inbound and outbound differences. Although the scatter is great, largely because of the variability and structure of the nightside ionosphere. The OETP measurement reveal no evidence of latitude variations. Perhaps a similar analysis of the same data base for limited local time ranges would provide a more sensitive test for latitudinal structure.

Miller *et al.* (1984) took a different approach in their search for latitudinal and local

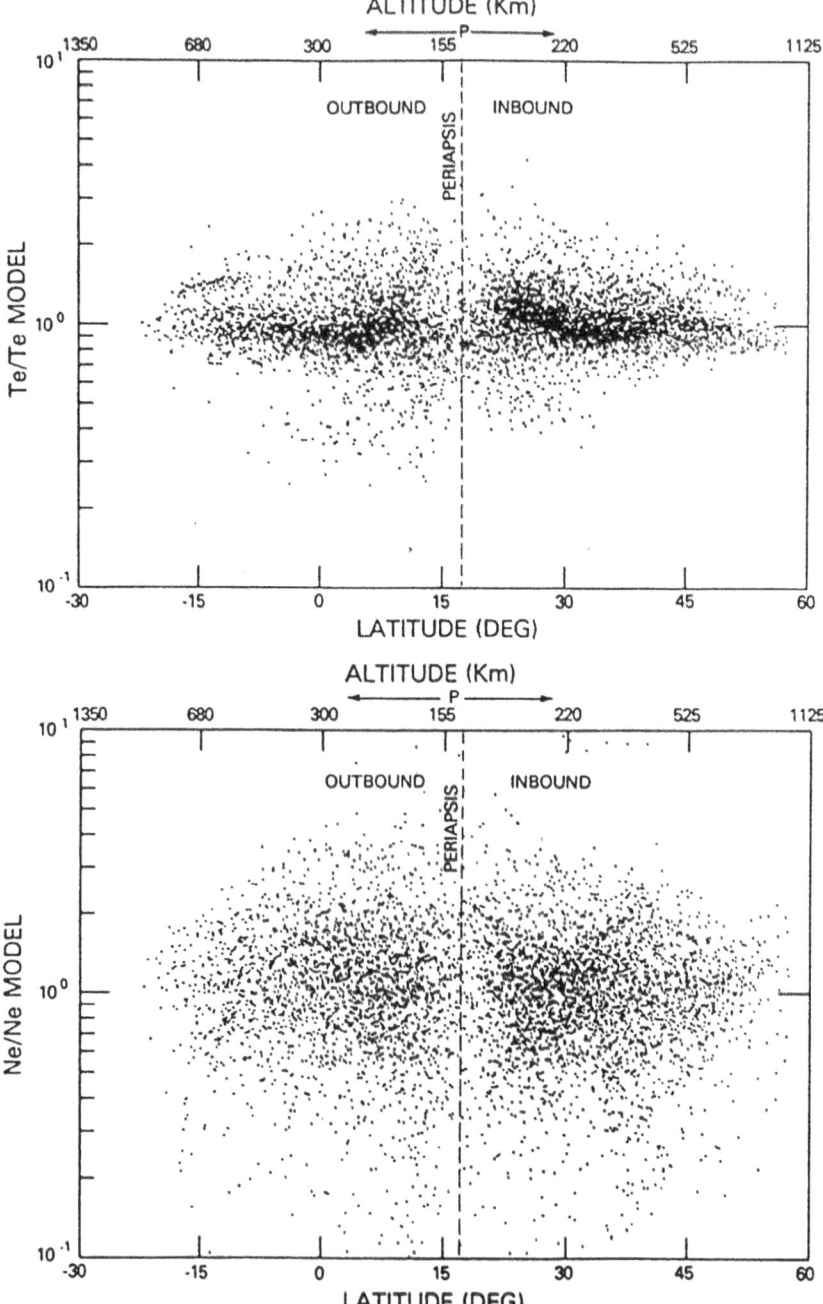

Fig. 32.   Ratios of T_e and N_e to the average model values, T_e model and N_e model, with data from all SZA included. A ratio of unity indicates exact agreement of the measurements. Significant latitudinal or inter-hemispherical gradients in these parameters would produce a pattern of inbound/outbound asymmetry in the ratios. The great scatter represents primarily the temporal and spatial variability of the nightside ionosphere (after Theis *et al.*, 1980).

time structure in the dayside ionosphere. They modeled the inbound and outbound ORPA measurements separately. They attempted to remove the very large altitudinal variations of N_i and T_i by normalizing the data to median daytime profiles of these parameters. Then they removed the local time variations in various ways depending upon the parameter being modelled. For example, they assumed a Chapman layer SZA variation of N_i at the peak. The resulting global N_i pattern is shown in Figure 33 as contours in latitude (or altitude) versus SZA. If one interprets the inbound-outbound differences in these contours as latitudinal or interhemispherical in origin, differences of between 10% and 30% are present at SZA < 60°. Larger differences occur near the dusk terminator.

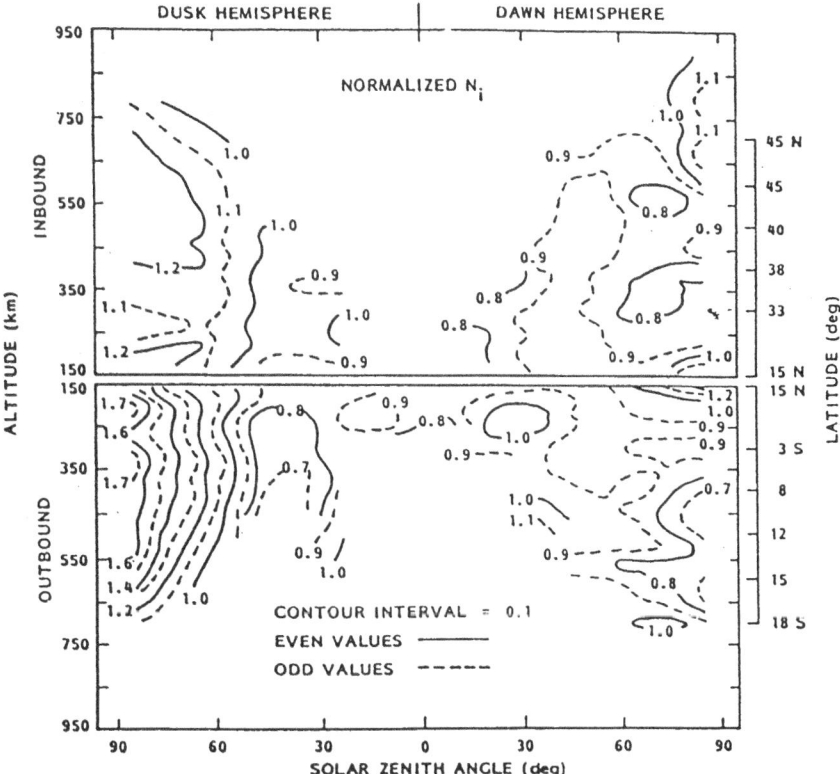

Fig. 33.   Contours of N_i normalized for SZA and altitude variations. A dawn-dusk asymmetry is clearly shown, but inbound–outbound differences are small, except near the dusk terminator (after Miller *et al.*, 1984).

We suspect, however, that much of the apparent SZA and latitudinal structure shown in Figure 33 lies within the uncertainties of the altitudinal and local time corrections that were applied in normalizing the N_i measurements. Local deviations of the measurements from the mean altitude and SZA models employed in the normalization could produce a coherent pattern of apparent latitudinal, interhemispherical, or SZA variations.

Statistical variations due to solar EUV and solar wind heating and ionization effects, if not successfully removed, could also introduce false global structure. We doubt that these effects are well enough understood to be removed with sufficient precision to clearly reveal what must be very small latitudinal and seasonal variations.

The latitude coverage provided by the radio occultation method is not constrained to the low latitude of the PVO periapsis. ORO profiles of N_e have been obtained at essentially all latitudes and seasons. Neither seasonal or latitudinal effects have been reported. From this, and the *in situ* evidence examined thus far, we conclude that the dayside ionosphere exhibits little latitudinal structure that is not related to the SZA, nor are interhemispherical structure or seasonal differences detectable. There are indeed important dawn-dusk asymmetries in the upper ionosphere, however. These appear primarily in the ion composition, as discussed earlier in Section 4.6. The nightside ionosphere, on the other hand, is so highly structured and dynamic (Knudsen *et al.*, 1979b; Taylor *et al.*, 1980; Brace *et al.*, 1980) that weak global trends are very difficult to isolate.

4.9. SOLAR CYCLE EFFECTS IN THE IONOSPHERE

Direct measurements of the lower ionosphere (< 500 km) were obtained only during the solar maximum conditions of 1979–1980. The subsequent rise in periapsis made it impossible to observe the response of the main ionosphere to the ensuing decline of solar activity, except through the use of the radio occultation technique. However, Knudsen *et al.* (1986) pointed out that the PVO nightside *in situ* measurements at solar maximum gave an order of magnitude higher densities than were obtained by the Venera 9 and 10 radio occulation measurements at solar minimum. They attributed this difference to a higher level of nightward O^+ transport at solar maximum. Later, Knudsen *et al.* (1987) used the PVO radio occultation measurements to examine the differences in the average dayside N_e profiles for 1980 and 1986. The solar maximum and minimum profiles are shown in Figure 34. N_e decreased at all altitudes at solar minimum, a behavior supported by the Venera 15 and 16 radio occultation profiles at the terminator just prior to solar minimum (Osmolovskii and Samoznaev, 1987).

Kliore and Mullen (1989) recently examined the solar cycle variations of the dayside ionosphere in greater detail. They employed 105 N_e profiles obtained by PVO between 1979 and 1986 and 11 profiles from Venera 9 and 10 measurements made in 1975 to examine the solar cycle variations of the peak density. The results are shown in Figure 35. The solar minimum data from PVO and Venera 9 and 10 (1984–1985 versus 1975) agree very well, thus confirming an important solar cycle effect. The peak density in the subsolar ionosphere declined by about 33% from solar maximum to solar minimum. However, a more recent and more extensive analysis of the PVO observations from 1984–1987 that are as yet unpublished shows an even greater solar cycle variation in the peak density. The dayside peak density decreased by approximately 50% from solar maximum to solar minimum, going from 20.3 (± 8.3) $\times 10^3$ cm^{-3} to 10.1 (± 4.3) $\times 10^3$ cm^{-3}. The average altitude of the main peak decreased from 142.2 (± 6.3) km at solar maximum to 136.8 (± 6.4) km at solar minimum.

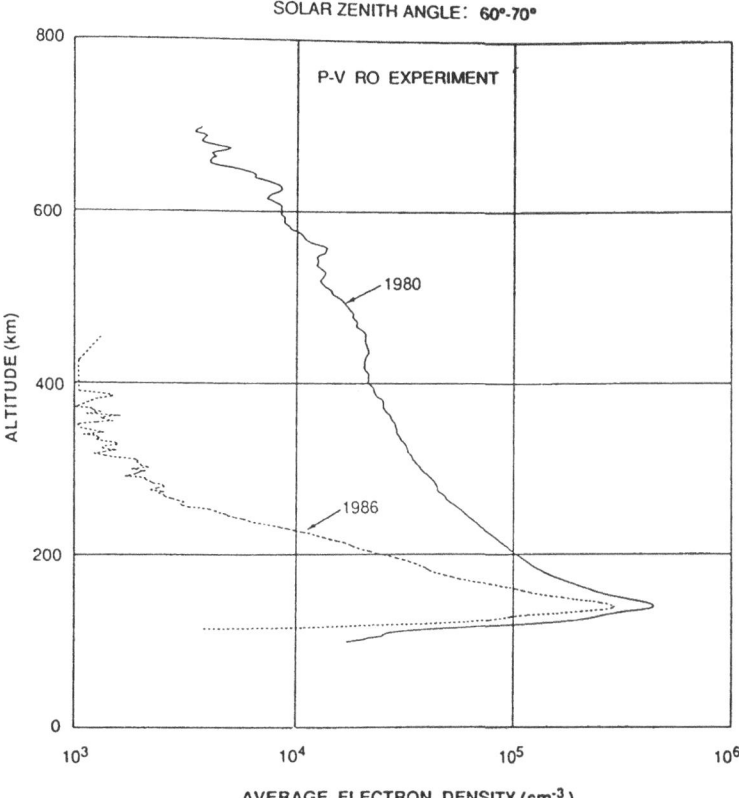

Fig. 34. N_e profiles obtained by PVO radio occultation at solar maximum (1980) and solar minimum (1986). The profiles show a great depletion of the dayside upper ionosphere at solar minimum (after Knudsen *et al.*, 1987).

Kliore and Mullen (1989) also found that the solar cycle behavior of the dayside N_e peak is consistent with the behavior of a Chapman layer (Chapman, 1931); i.e., a layer in which ion production by solar EUV radiation is balanced by chemical recombination. Employing the Hinteregger *et al.* (1981) formula for estimating the EUV flux from the $F10.7$ cm index, Kliore and Mullen showed that the log of N_e at the peak tracks the log of the solar EUV flux with a slope of 0.376, as shown in Figure 36. This is in excellent agreement with the value of 0.36 derived by Bauer (1983) who employed earlier PVO measurements of the SZA variations of the peak to show that the dayside of the Venus ionosphere behaves like a simple Chapman layer.

The solar cycle variation of the peak density may also have implications for the variation of T_e and the neutral gas temperature, T_n. When Kliore and Mullen (1989) calculated the solar cycle variation in the CO_2^+ production rate, they were able to obtain the expected Chapman layer behavior of the peak N_e only by introducing a solar cycle variation in T_e and T_n, assuming $T_e = T_n$ at the peak. Their calculations employed the

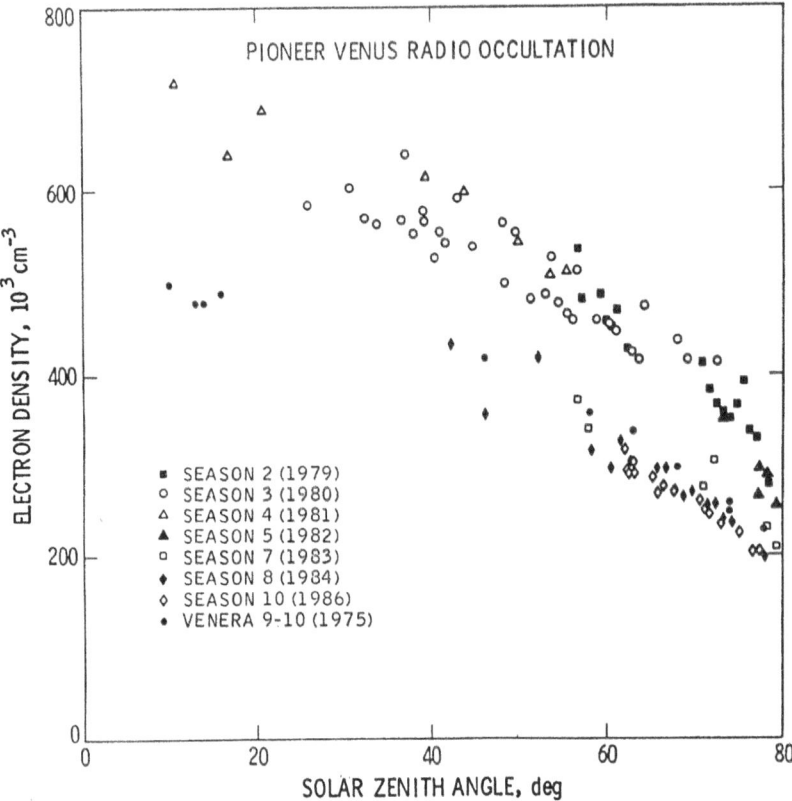

Fig. 35. Peak N_e vs SZA from PVO (1979–1986) and Venera (1975) radio occultation measurements. The peak density is smaller by about 33% at solar minimum (from Kliore and Mullen, 1989).

incident EUV flux (Hinteregger, 1981), the absorption and ionization cross-sections (Torr *et al.*, 1979), and the VIRA upper atmosphere models (Keating *et al.*, 1985; Hedin *et al.*, 1983), with solar-cycle corrections applied. A Chapman layer would be expected to exhibit the relationship $\Delta \log(N_{peak}/T_e^{0.275})/\Delta \log P_{CO_2^+} = 0.5$. However, in order to obtain the desired 0.5 slope, Kliore and Mullen had to assume that T_e (and T_n) decreased by about 25% from solar maximum to solar minimum. This is the best current information on the solar cycle variations of T_e and T_n at the peak altitude.

The ionosphere above the peak was far more responsive to solar activity, however (Kliore and Mullen, 1990). The change in the average N_e scale height in the SZA range of 55° to 75° was determined from averaged ORO data from 1979–1980 and 1984–1986. The results are shown in Figure 37. The scale height decreased from solar maximum to solar minimum by a factor of about 1.5 at 200 km, and by about a factor of 3 at 300 km. Assuming that diffusive equilibrium holds in this region, the effective plasma temperature, $(T_e + T_i)$, is inferred to have dropped by a factor of 3 from maximum to minimum at 200 km.

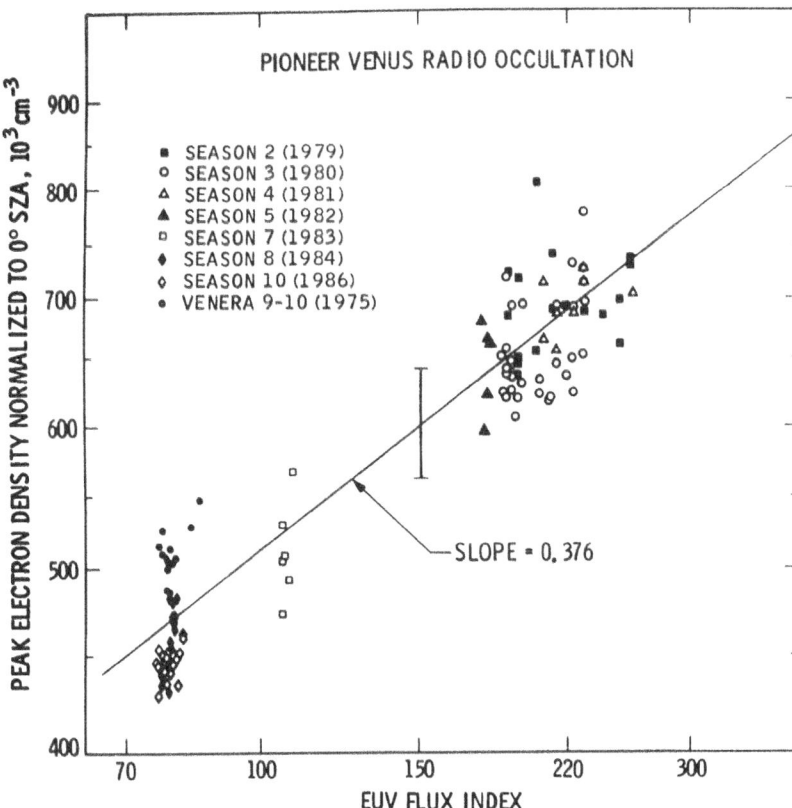

Fig. 36. Solar cycle variations in the peak N_e in the subsolar ionosphere plotted vs the solar EUV index based on the $F_{10.7}$ cm flux measured at PVO and Venera radio occultation results are shown. The slope of 0.376 is consistent with an ionosphere that responds to solar EUV flux variations much like a Chapman layer would (from Kliore and Mullen, 1989).

Since a factor of 3 solar cycle variation in $T_e + T_i$ at 200 km seems unlikely, the observed change in scale height suggests that the dayside ionosphere at these altitudes (200–300 km) does not remain in diffusive equilibrium at solar minimum. As shown by Knudsen *et al.* (1987), the dayside ionopause moves into this altitude range at solar minimum. Thus the ion scale height is reduced by $\mathbf{j} \times \mathbf{B}$ forces (Hartle *et al.*, 1980), and by ion pickup by the magnetosheath magnetic field convecting tailward across the top of the ionosphere (Luhmann, 1986; Phillips *et al.*, 1988).

Thus our knowledge of the solar cycle variation of the temperature of the main ionosphere temperature is very indirect. The same is true for the composition of the ionosphere and thermosphere, since it can only be measured *in situ*. Perhaps PVO will resolve some of these changes during its re-entry period in August and September 1992, probably at lower levels of solar activity.

Solar cycle variations on the nightside are much larger than those on the dayside. Knudsen *et al.* (1987) explained this in terms of large solar cycle changes in the night-

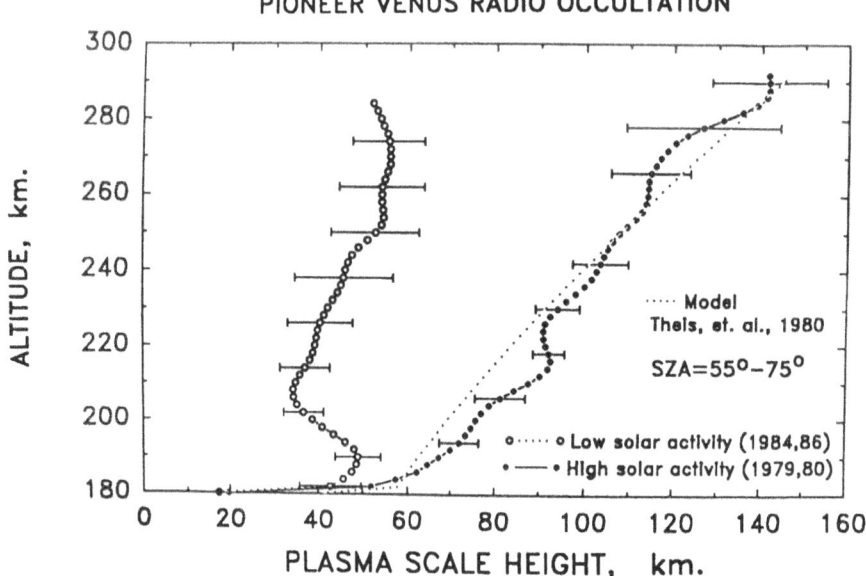

Fig. 37.   The plasma scale height in the dayside ionosphere from radio occultation measurements at solar
maximum and minimum. It has been argued that this change may be caused by very much lower plasma
temperatures at solar minimum, but no *in situ* measurements of this effect have been made (after Kliore and
Mullen, 1990).

ward ion flux. The ORPA measurements of ion velocity at solar maximum suggested
that 85% of the nightward ion flow occurs above 250 km. Therefore, the removal of
most of the upper ionosphere at solar minimum should reduce the nightward flow
significantly, leaving a much weaker nightside ionosphere. This is a major topic of the
companion review by Miller and Whitten (1991), so we will not review all of its
implications here.

4.10. SUPRATHERMAL ELECTRONS AND SUPERTHERMAL IONS

Nonthermal ions and electrons are observed in many regions of the near-Venus environment. Some authors call these electrons superthermal rather than suprathermal, and
sometimes the ions are called energetic rather than superthermal. We will use the terms
used by the authors in each case.

As noted earlier, Brace *et al.* (1982b) reported a hot electron component within the
ionospheric holes. Knudsen *et al.* (1980a) measured suprathermal electrons in the
dayside ionosphere and identified them as photoelectrons transported from the dayside.
Spenner *et al.* (1980) reported suprathermal electrons in the mantle, the region (shown
in Figure 7) which marks the boundary between the shocked solar wind plasma and the
ionosphere. Gringauz *et al.* (1979) and Spenner *et al.* (1981) have reported suprathermal
or energetic electrons on the nightside of Venus, and both showed that they were an
important ionizing source.

The OIMS on PVO consistently observed superthermal O^+ ions at the ionopause and above, and found them within the nightside ionosphere, particularly associated with the ionospheric holes (Taylor et al., 1980). Brace et al. (1987), using the OIMS measurements, have shown that essentially all of the nightside ions at altitudes in the vicinity of 2000 km are superthermal, most with energies in the range of 9 to 16 eV, at least near solar minimum. Cold ions were generally absent at these altitudes, while they are known to dominate at lower altitudes in the ionosphere. The altitude of transition from thermal to superthermal ions, if a unique one exists, has not been reported. It would appear possible to locate this transition using the existing OIMS measurements. Superthermal H^+ ions may also be present at high altitudes on the nightside, but these ions would not be observed by the OIMS because of an instrumental effect in which their higher energies produce an apparent shift below the mass range of the instrument (Brace et al., 1987).

The ONMS on PVO has measured energetic ions (> 40 eV), mainly above 1500 km on the nightside (Kasprzak et al., 1987). Ions having lower energies are not normally measured when the instrument is in its neutral composition mode. The energetic ions have about the same composition as that found in the underlying ionosphere, so they are undoubtedly of ionospheric origin. These ions, while more energetic than the OIMS superthermals, were also seen at the ionopause at all SZA (Kasprzak et al., 1982). The source of both of these superthermal ion populations is unknown, and it is not clear whether they represent different parts of the same superthermal population or whether they represent distinctly different ion populations. Clearly, the presence of superthermal ions of ionospheric composition shows that ion acceleration processes are operating both at the ionopause, and within the nightside ionosphere. We will return to this topic in Section 4.12 on the Venus ionotail.

4.11. THE IONOPAUSE

The ionopause is perhaps the most readily identified feature in the cold plasma measurements. It is marked by a steep gradient in plasma density at the top of the ionosphere, where N_e (and N_i) change by a factor of the order of 100. It is a reliable feature that can be identified both inbound and outbound in almost every ionosphere passage (e.g., Figures 11, 27, 29). Its altitude is usually taken arbitrarily as the point within the ionopause density gradient at which N_e or N_i passes through a value of 100 cm^{-3} (Brace et al., 1980; Knudsen et al., 1980b). Taylor et al. (1980) took the ionopause at the low altitude boundary of the region of superthermal ions measured by the OIMS. Elphic et al. (1980a) placed the ionopause at the altitude where the ionospheric plasma pressure ($N_e k T_p$) equals the magnetic pressure ($B^2/8\pi$), where $T_p = T_e + T_i$. The pressure balance ionopause typically occurs a few kilometers below the density ionopause (Phillips et al., 1988). The ionopause density gradient is usually to steep, however, so that the differences in the ionopause altitudes determined in these ways are typically much smaller than the orbit to orbit changes. These changes have been quite clearly tied to solar wind variations. The ionopause is a product of the solar wind interactions.

In the absence of an intrinsic planetary magnetic field, the ionosphere of Venus is the primary obstacle to the solar wind (e.g., Johnson and Midgely, 1969; Bauer *et al.*, 1970; Elphic *et al.*, 1980a; Phillips *et al.*, 1985). Solar wind ions and electrons, and the entrained IMF, impact the ionosphere directly, driving currents that produce a magnetic barrier to the further penetration of the solar wind. The magnetic barrier can also be viewed as a compression of the IMF as it drapes around the ionosphere. This compression squeezes the solar wind plasma out of the region just above the ionopause, an effect described by Zwan and Wolf (1976), causing most of the shocked solar wind to flow around the planet after having applied its dynamic pressure to the ionosphere via the induced magnetic barrier.

Figure 38 illustrates the solar wind interaction process schematically (Luhmann,

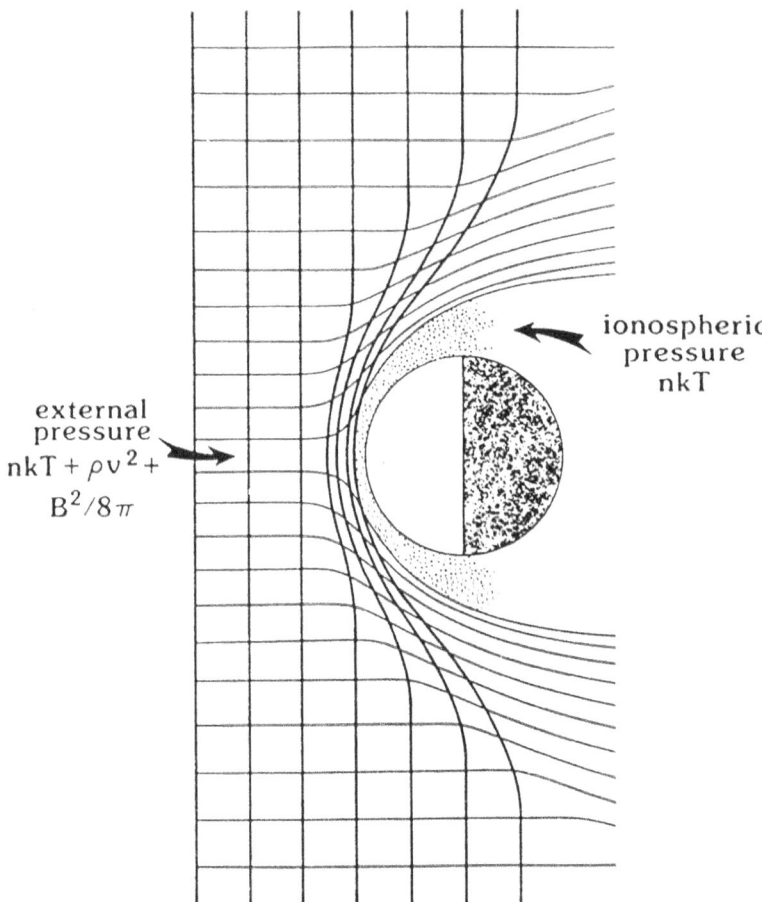

Fig. 38.   Diagram suggesting how the solar wind applies pressure to the ionosphere via a build up in the magnetic field as it drapes around the planet. The solar wind dynamic pressure (ρv^2) greatly exceeds the solar wind plasma pressure (nkT) and the IMF pressure $(B^2/8\pi)$. An ionopause forms at the altitude where the ionospheric plasma pressure (nkT) balances the radial component of these external pressures (after Luhmann, 1986).

1986). The external pressure includes the sum of the solar wind dynamic pressure, ρv^2, its thermal pressure, nkT, and the magnetic pressure of the IMF, $B^2/8\pi$. The dynamic pressure is by far the largest component, except at the terminator where the other terms may dominate. The internal pressure of the ionosphere is simply its thermal pressure. In general, the external and internal pressures are balanced across the ionopause, as illustrated in Figure 39 which shows the pressures calculated from PVO measurements

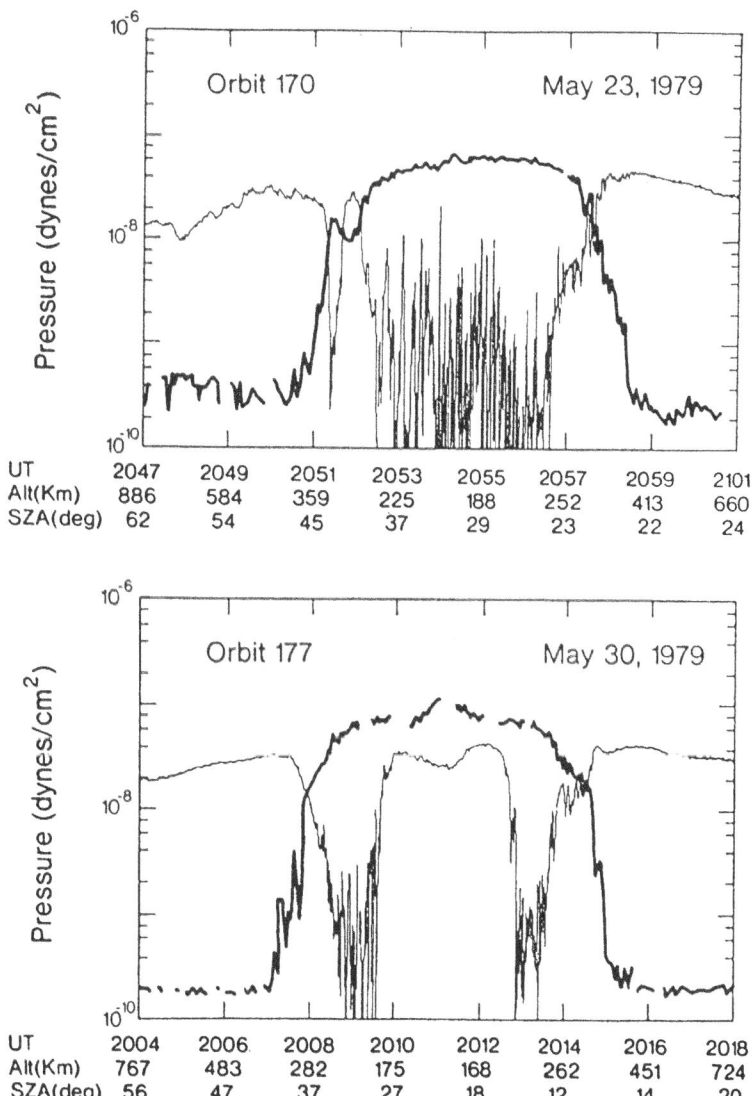

Fig. 39.   Two orbits which illustrate the pressure continuity across the ionopause, based on PVO measurements. The magnetic pressure outside the ionopause (light line) is in equilibrium with the plasma pressure (heavy line) of the underlying ionosphere (after Elphic *et al.*, 1980b).

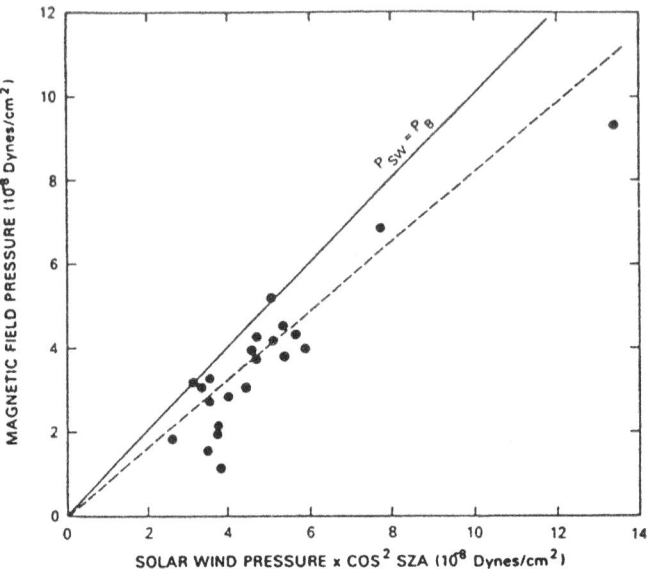

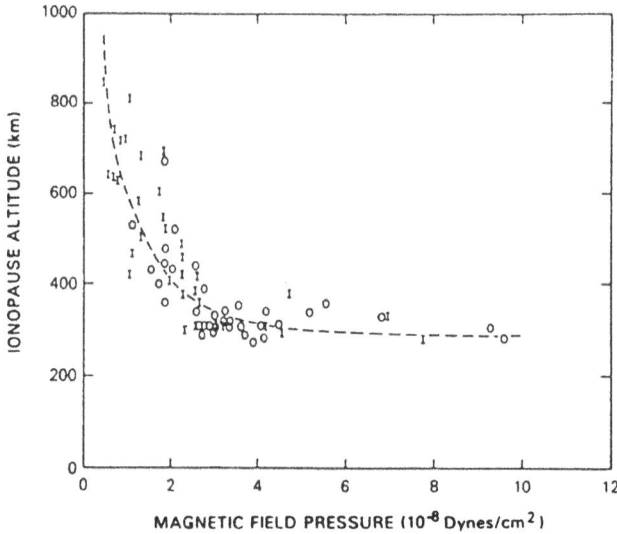

Fig. 40.   PVO measurements of the pressure in the magnetic barrier just above the ionopause plotted versus the component of the solar wind dynamic pressure that is normal to the ionopause (upper panel). The dashed line represents perfect agreement. The general agreement for a wide range of solar wind dynamic pressures supports the idea that the magnetic cushion conveys the solar wind pressure to the ionosphere. The lower panel shows the response of the ionopause altitude, inbound (I) and outbound (O), to the orbit to orbit changes in magnetic pressure. The ionopause altitude declines with increasing solar wind pressure, but levels off above about 4×10^{-8} dyne cm^{-2} (after Brace et al., 1980).

made during two PVO passages through the dayside ionosphere (Elphic *et al.*, 1980b). The magnetic pressure (light line) dominates outside the ionosphere and the plasma pressure (heavy line) dominates inside. The total pressure is essentially continuous across the ionopause.

The PVO measurements have been used to demonstrate statistically that the solar wind dynamic pressure is converted to magnetic pressure in the region just above the ionopause. This conversion is illustrated in the top panel of Figure 40 by the near agreement between the dynamic pressure of the solar wind and the pressure in the magnetic barrier (Brace *et al.*, 1979, 1980; Elphic *et al.*, 1980a), for ionopause crossing at SZA $< 40°$. The solar wind pressure scale has been multiplied by $\cos^2$ SZA to include only the component of dynamic pressure acting normal to the ionopause surface. As a result of this pressure balance, the height of the ionopause is sensitive to changes in the solar wind. The bottom panel of Figure 40 shows that the ionopause moves in and out in response to changes in the magnetosheath magnetic pressure, which are driven by changes in the solar wind pressure. Inbound (I) crossings and outbound crossings (O) are shown separately for ionopauses at SZA $< 67°$. The ionopause expands when less pressure is applied and contracts when higher pressures are applied. However, the ionopause altitude reaches a lower limit of about 300 km when the external pressure exceeds about 4×10^{-8} dyne cm^{-2}. Mahajan *et al.* (1989) have argued that this saturation occurs when the ionopause is driven so deeply into the thermosphere that photo-ion production loads down the solar wind interaction. When this occurs, further increases in solar wind dynamic pressure cause very little further decreases in the ionopause altitude, an affect that is evident in Figure 40. The downward diffusion of the large number of photo-ions at these times forms a new kind of ionopause which Mahajan *et al.* called a photodynamical ionopause. The plasma scale height in the photodynamical ionopause matches that of the main ionizable constituent, atomic oxygen. This effect was noted earlier by Bauer and Hartle (1974) in the Mariner 5 dayside N_e profile, and by Hartle *et al.* (1980) in early ion profiles from PVO.

Since the scale height of the photodynamical ionopause is determined by the atomic oxygen scale height, the ionopause is thicker at times of high solar wind pressure. Elphic *et al.* (1981) examined the thickness of the ionopause and the current sheet within it. They concluded that the ionopause thickness is only a few ion gyro radii at low solar wind dynamic pressures, i.e., when the ionopause forms at high altitude. When the ionopause moves to very low altitudes, at times of high dynamic solar pressure, its thickness expands to several tens of ion gyroradii.

4.11.1. *Global Configuration of the Ionopause*

The global configuration of the ionopause has been difficult to characterize, both because of its great variability and because of inherent limitations imposed by the PVO orbit. While its overall configuration must be somewhat elliptical, as shown in the cartoon in Figure 7, its surface is quite irregular on smaller scales. Nearly tangential traversals of the dayside ionopause by PVO have shown that it is often distorted by surface waves having scale lengths of several hundred km along the orbit (Brace *et al.*,

1980; Russell *et al.*, 1986). Streamers, tail rays and filaments of ionospheric plasma extend downstream from the ionopause for several thousand km on the nightside to form an extended ionotail (Brace *et al.*, 1982a, 1987), a region that will be discussed in detail in Section 4.12.

A significant characteristic of the ionopause is its apparent global coherence. Brace *et al.* (1980) found that the inbound and outbound ionopause altitudes tended to move in and out together. They concluded that this correlation was caused by its global scale interaction with the solar wind; i.e., the entire dayside ionopause responds simultaneously to changes in the solar wind dynamic pressure. Small differences in inbound and outbound ionopause heights were observed, but these could have been produced by a skewed IMF configuration, or by real temporal changes in the solar wind pressure during the few minutes it took for the spacecraft to pass through the underlying ionosphere.

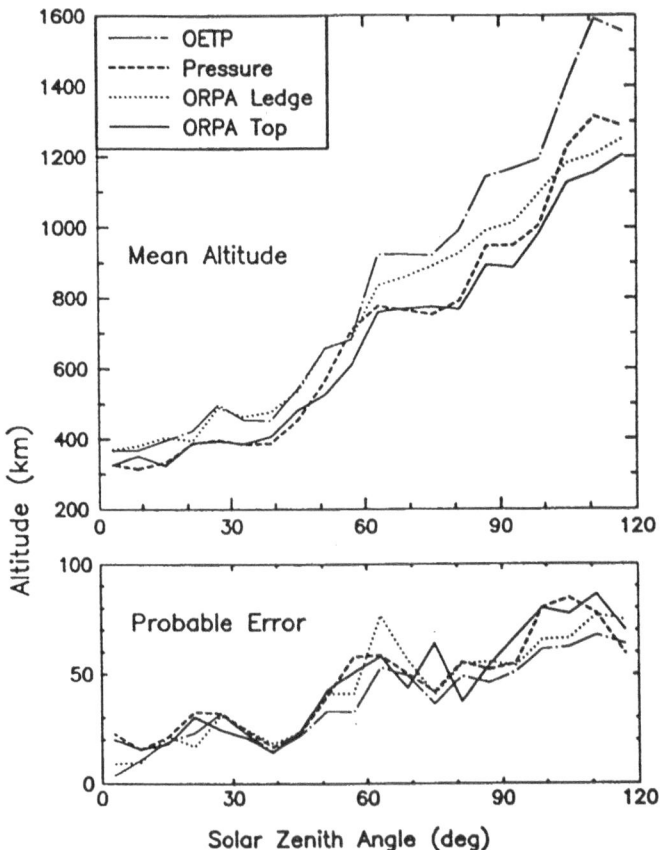

Fig. 41.   The SZA variations of the ionopause height using ionopause definitions based on various PVO measurements. All definitions show similar SZA variation of the ionopause height (after Phillips *et al.*, 1988).

The most extensive study of the SZA behavior of the ionopause has been conducted by Phillips *et al.* (1988). They examined the ionopause as defined independently by the OETP measurements, the ORPA measurements, and by the pressure balance criterion. The SZA effects are shown in Figure 41 for solar maximum conditions. Phillips *et al.* used two different ORPA-defined ionopauses; the Top ionopause (taken at an altitude just below the ionopause density gradient) and the Ledge ionopause (taken at the 100 cm^{-3} density level). The SZA variations are quite similar by all of these definitions. The ionopause rises from about 350 km in the subsolar region to over 1000 km at 120° SZA. This increase occurs because the normal component of the solar wind dynamic pressure goes to zero near the terminator, leaving only the thermal and magnetic pressure of the magnetosheath to balance the ionospheric plasma pressure. These authors attributed an observed dawn-dusk asymmetry in the ionopause height to flow aberration and sampling biases in ionopause data base. They also found an additional asymmetry which was correlated with the orientation of the IMF and which they attributed to asymmetric pickup of planetary ions. In an earlier study, Knudsen and Miller (1982) successfully modelled the SZA variation of the ionopause height observed by the ORPA by including only the pressure exerted by the ionosheath plasma.

The height of the ionopause appears to change a great deal with solar activity. The PVO *in situ* measurements described above only refer the ionopause behavior at solar maximum, since the rise of periapsis which began in 1980 has precluded *in situ* measurements of the dayside ionopause since that time. However, continued measurements of the very high altitude nightside ionopause have been possible throughout the mission. These results will be discussed in Section 4.12 in the context of the ionotail and its solar cycle variations.

4.11.2. *Stability of the Ionopause*

The question arises as to why the ionopause seems to be such a stable boundary; i.e., why does a distinct ionopause exist at every PVO entry and exit of the ionosphere? Why is there not sometimes a gradual transition between the ionosphere and the magnetosheath? Johnson and Hanson (1979) suggested that the curvature of the magnetic field as it drapes around the ionosphere could cause the ionopause to be unstable to the flute instability, thus allowing ropes of magnetic field to be pulled downward into the ionosphere. Alternatively, Dubinin *et al.* (1980) found that the magnetic curvature at the ionopause could give rise to an interchange instability which could cause the empty flux tubes just above the ionopause to become embedded in the ionosphere and permit flutes of ionospheric plasma to float outward into the mantle flow.

Elphic and Ershkovich (1984) examined the stability of the ionopause in the light of gravitational and curvature effects. They found that the buoyancy of the plasma stabilizes the ionopause against the flute instability, and that the Kelvin–Helmholtz mode is the dominant instability over the dayside ionopause. They predicted that the K–H instability should generate surface waves in the ionopause that will affect the electrodynamic coupling between the solar wind and ionospheric plasmas, particularly at larger solar zenith angles where the shear velocity is greater. Examples of such surface waves were evident in some of the dayside N_e profiles of Figure 11. Whatever their cause, the breaking of the ionopause waves in the terminator region of the ionopause may be responsible for the formation of the ionospheric clouds and streamers that extend great distances from the ionosphere at the flanks (Brace *et al.*, 1982a).

Russell *et al.* (1986), suggested that the mass loading of magnetosheath flux tubes just above the subsolar ionopause may overcome the inherent stability of the ionopause in that region. The low altitude of the ionopause in the subsolar region (typically 300–400 km) enhances the photo-ion production rate within these flux tubes, producing greater mass loading there than elsewhere along the ionopause. These new ions slow the convection of magnet flux toward the poles, and this provides time for the further production of ions that further inhibit the poleward flow. Cumulatively, this process may make the tubes less buoyant against gravity, allowing them to sink into the dayside ionosphere to form the magnetic flux ropes reported by Elphic and Russell (1983), as well as the more fully-magnetized ionospheres discussed for example by Shinagawa and Cravens (1988) and Cloutier *et al.* (1987). We will not review the large-scale magnetic structure of the ionosphere here, since this is the topic of one of the companion reviews in this issue (Luhmann and Cravens, 1991).

4.12. THE IONOTAIL

The rise of periapsis after the summer of 1980 permitted the nightside upper ionosphere of Venus to be measured *in situ* at ever increasing altitudes. The highly structured nature of the ionotail, illustrated earlier in Figure 7, is evident in the series of PVO profiles shown in Figure 42. These profiles are from north–south transits of the antisolar ionosphere at altitudes between 2000 and 3000 km. Densities of the order of 10^3 cm^{-3} were present in many of the enhanced regions which Brace *et al.* (1987) called tail rays. N_e was lower by one or two orders of magnitude in the troughs between the rays. Similar measurements through many seasons of nightside passages have made it clear that the nightside ionosphere extends at least several thousand kilometers downstream from the planet. These tail rays are reminiscent of the plasma clouds observed earlier in the mission above the terminator ionopause (Brace *et al.*, 1982a). The ionopause configuration shown in Figure 7 was drawn in a way to suggest a connection between the troughs in the ionotail and the holes in the underlying ionosphere (Figures 27–29). This connection is made physical by the magnetometer measurements of largely radial magnetic fields that couples these regions by allowing vertical ion transport and electron heat conduction.

4.12.1. *A Detailed Look at an Ionotail Passage*

A closer look at what was going on in the ionotail at solar minimum is shown in Figure 43 (Brace *et al.*, 1987). Various PVO measurements made during a single ionotail passage are shown to illustrate its complex behavior. Although this passage was called typical, there was great diversity from orbit to orbit. Tail rays can be seen in the N_e measurements near 00:25 UT and 00:35 UT, where the maximum densities are about 600 cm^{-3} (second panel). A weaker ray may be seen near 00:40 UT. The OIMS

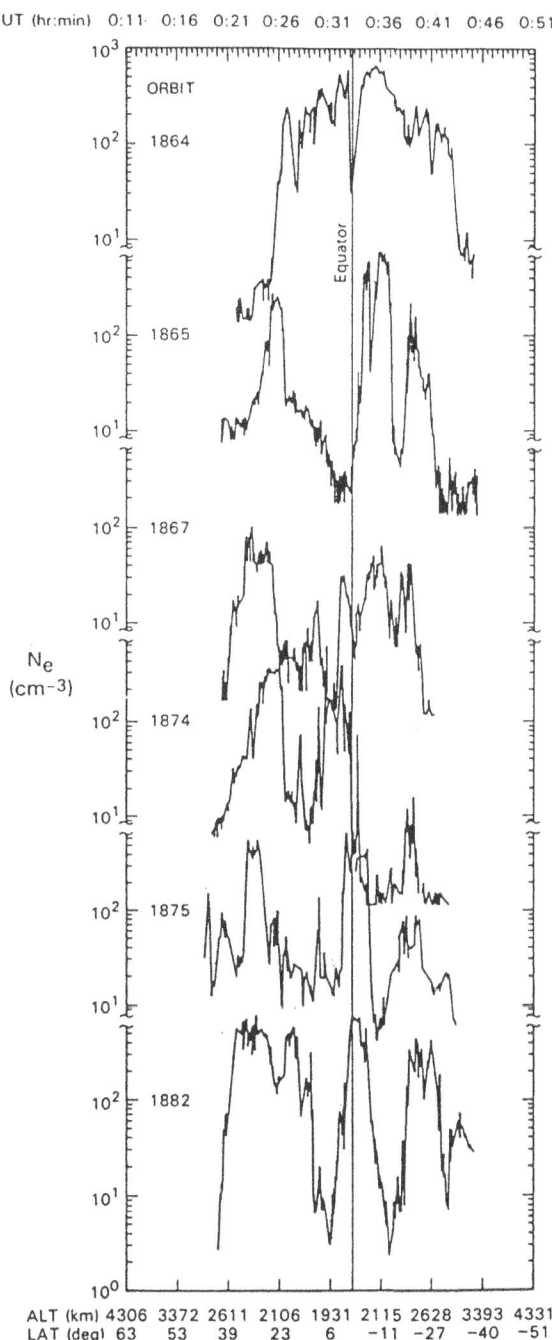

Fig. 42.   A sequence of N_e profiles of the ionotail at 2000 to 3000 km altitude. Large enhancements, separated by deep troughs, are typical of the antisolar region at these altitudes. Measurements like these are the basis for the comet like ionotail configuration shown earlier in Figure 7 (after Brace *et al.*, 1987).

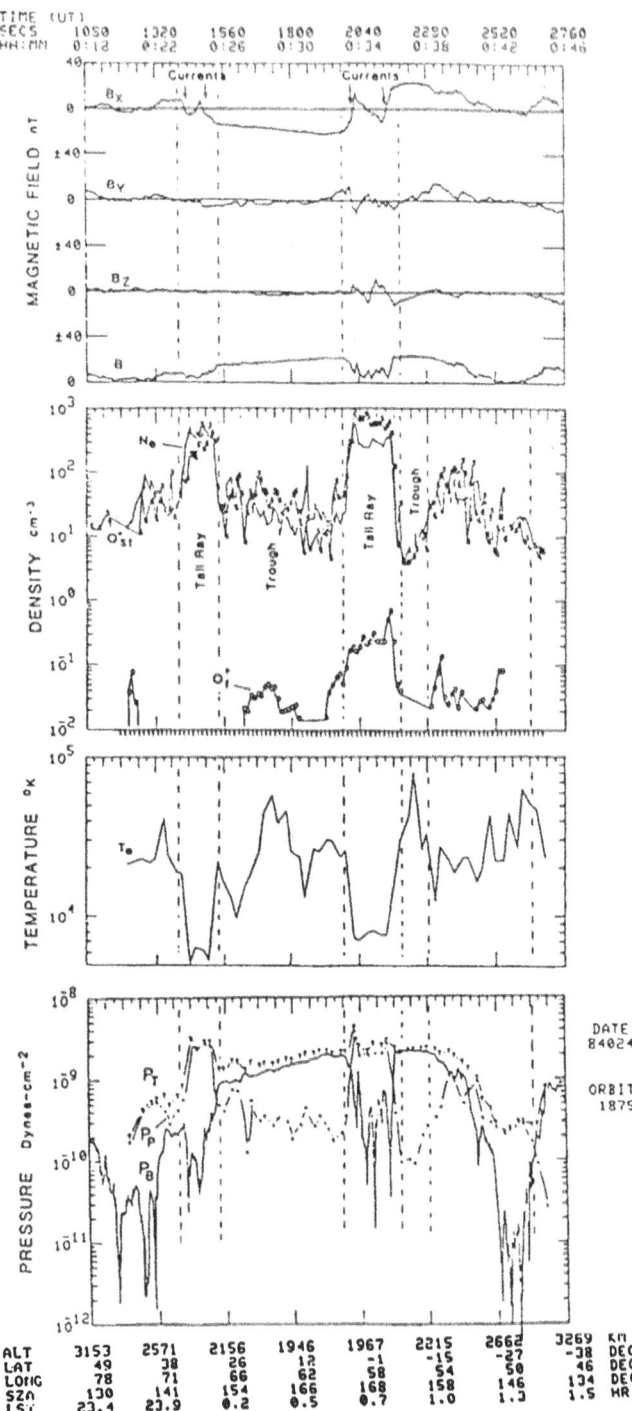

Fig. 43. Magnetic and plasma measurements in a single ionotail crossing, showing the tail rays composed of superthermal ions, O_{st}^+, magnetic field reversals indicating the presence of currents at the boundaries of the rays, high T_e in the troughs between rays, and the pressure balance conditions across the rays. Fast ions (> 40 eV), O_f^+, may be seen in some of the tail rays and also in some of the troughs (after Brace *et al.*, 1987).

observed the same tail rays as superthermal oxygen ions, O_{st}^+, having energies in the range of 9 to 16 eV. No cold ions were observed in this pass, and their absence is also typical of the other ionotail passages from this interval that have been examined thus far. The N_e and O_{st}^+ densities agree reasonably well, and this is consistent with the absence of cold ions, or significant numbers of superthermals of other species.

The electrons in the denser tail rays are much cooler than those in the surrounding trough plasma (third panel), but T_e is similar to that found in the underlying ionosphere (<1 eV). T_e is higher in the troughs and in the weaker tail ray at $00:40$ UT. This thermal structure suggests that the tail-ray boundaries mark the transition between two different plasma populations. The source of the trough electrons has not been identified, however. They may originate in the magnetotail, or they may be ionospheric electrons that have been heated and transported into the troughs. It has been suggested that the sharp B_x reversals which were observed at the edges of the two larger rays (top panel) indicated the presence of intense current sheets there. These currents were viewed by Brace et al. (1987) as signatures of plasma acceleration at the ray boundaries. A larger scale B_x reversal across each entire tail ray implies a draped magnetic field configuration, such as that proposed by Russell et al. (1982) for the magnetic field about an ionospheric plasma cloud near the terminator. Thus the tail rays and plasma clouds could be similar structures seen at different SZAs.

The second panel of Figure 43 also shows measurement by the ONMS of fast oxygen ions, O_f^+ (>40 eV), within the central tail ray. Smaller O_f^+ fluxes were found in the troughs on either side. The conversion of the O_f^+ fluxes to absolute densities cannot be done very accurately, but it is clear that the fast ion densities are at least two orders of magnitude lower than N_e (or O_{st}^+). The O_f^+ densities, while larger in the central tail ray, were not present in the ray at $00:25$ UT. Thus the fast ions do not necessarily represent simply a high energy tail of the 9–16 eV O_{st}^+ population, but that may sometimes be the case. The $26°$ off axis orientation of the ONMS aperature has made it possible to show that the O_f^+ ions usually have a bulk velocity which has a large tailward component (Kasprzak et al., 1987; Brace et al., 1987). This is considered evidence for ion escape from the Venus ionotail.

The pressure balance, or imbalance, across the tail rays in passages such as this one has provided clues as to the dynamics of the ionotail. The bottom panel of Figure 43 shows the magnetic pressure, P_B, the plasma pressure, P_p, and the total pressure, P_T, calculated from the other measurements. Note that P_p tends to dominate inside the tail rays, and P_B dominates in the troughs between the rays. The central tail ray ($00:35$ UT) appeared to be in equilibrium with the magnetized region surrounding it, with P_T nearly constant across the ray boundaries. This suggests a pressure balance at the boundaries of that ray that is similar to that found at the dayside ionopause, a situation illustrated earlier in Figure 39. It is significant, however, that a pressure balance would not have occurred if the ions had been assumed to be cold. This is considered evidence that the ion temperature is indeed very high in the ionotail, as deduced from the OIMS measurements.

The tail ray near $00:25$ UT appears not to be in equilibrium with the magnetic pressure on either side of the ray; i.e., P_T is discontinuous at both edges of this ray. The plasma pressure in this ray is factor of 3 to 5 times larger than the magnetic pressure outside. This would appear to be an unstable situation for the ray, and it is not clear what is maintaining its sharp boundary. The weaker ray near 00:40 UT exhibited a similarly weak magnetic signature in which P_p was approximately equal to P_B. This could be the signature of a ray whose plasma is being dissipated by its interaction with the magnetosheath magnetic field as it convects into the Venus tail. It is interesting that the current sheets mentioned above tend to occur at the edges of the rays, where the plasma pressure of the tail ray equals the magnetic pressure of the region surrounding the ray.

The acceleration or heating processes that produce the superthermal ions in the ionotail (O_{st}^+ and O_f^+) are not understood. Nearly all of the ions observed in the lower ionosphere are cold (Taylor *et al.*, 1980); and nearly all of ions in the ionotail are superthermal (Brace *et al.*, 1987). The acceleration region(s) has not been located. Superthermal ions were found routinely in the vicinity of the ionospheric holes even at low altitudes, suggesting that ion acceleration occurs over a wide altitude range. The current sheets at the edges of the tail rays at 2000 km altitude suggest that this acceleration might occur along the tail ray surfaces which may extend deep into the ionosphere. It is interesting, however, that the superthermal ions exist throughout the rays, not just at the boundaries. This implies that ion acceleration occurs not only at the tail ray boundaries but also within the ionosphere itself. A detailed survey of the OIMS data should be conducted to locate these acceleration regions by systematically identifying the transitions from cold to hot ions.

4.12.2. *Implications for Ion Escape*

The very existence of hot ions in the ionotail has implications for planetary ion escape. The ion mass spectrometer cannot distinguish between ions with high bulk velocities and ions having very high temperatures. But in either case the 9 to 16 eV ions observed by the OIMS have velocities which exceed the escape velocity at 2000 km altitude. Brace *et al.* (1987) have estimated a total planetary O^+ escape rate from the ionotail of 5×10^{25} s^{-1}. Superthermal H^+ may also be present in the ionotail, but these ions could not be measured by the OIMS because they would be mass shifted below the range of the instrument. Brace *et al.* pointed out that a hot H^+ population having a density one half that of the hot O^+ would produce an H^+ escape rate of the same order as that caused by charge exchange between hot H^+ ions and cold atomic hydrogen in the exosphere (Kumar *et al.*, 1983). The latter has been thought to be the most important escape process for hydrogen. The generally tailward bulk velocity of O_f^+ measured by the ONMS also supports the conclusion that planetary ions are being lost from the ionotail. These more enegetic ions (> 40 eV) do not contribute significantly to the total loss rate because their density is typically less than 1% of the total population, but they do show that at least one component of the ionotail is moving tailward.

Additional planetary ion escape may also occur by bulk acceleration of the plasma clouds or streamers mentioned earlier (Brace *et al.*, 1982a). If these features are detached from the ionopause, much of their plasma seems likely to be accelerated down the tail by magnetic pick up. Russell *et al.* (1982) examined the magnetic field which was draped around one cloud and found that the convecting field should accelerate the cloud tailward at velocities well above the escape velocity. A similar process will occur if and when ion tail rays are detached in the tailward flow of the magnetosheath.

4.12.3. *Solar Cycle and Shortterm Variations of the Ionotail*

The ionotail exhibits large solar cycle and orbit to orbit variations. Brace *et al.* (1987) concluded that the shortterm variations were caused by solar wind induced magnetic fields which push the tail rays about while removing plasma around the edges. Because of this dynamic behavior, one should consider the ionotail configuration shown in Figure 7 as merely a snapshot of a highly dynamic region rather than an accurate model of the region. The observations that influenced this drawing were all made near solar minimum (1983–1986), however, so it does not represent the ionotail conditions at the higher levels of solar activity that have occurred recently.

To understand the solar cycle variations of the ionotail, one must consider the solar cycle behavior of the N_e in the dayside ionosphere (Section 4.9) which has been assumed to be its main source. The solar cycle control of the dayside density (and ionopause altitude) was demonstrated in Figures 35, 36, and 37. Knudsen (1988) also employed ORPA measurements to infer the presence of a strong solar cycle variation in the central nightside ionosphere up to 2000 km. Figure 44 shows the density measurements at solar minimum (low altitudes) and solar maximum (higher altitudes). The discontinuity between the low altitude and high altitude ion densities was attributed to a large solar cycle effect. Knudsen concluded from this result, and from earlier radio occulation results (Knudsen, 1986), that the entire ionosphere above about 250 km is removed at solar minimum. He attributed the decline in the ionopause height to a reduction in the dayside thermal pressure. The ionopause height strongly controls the nightward ion flow, since most of the transport occurs in the O^+ layer which lies above 200 km where ion-neutral drag forces are sufficiently weak to permit large flow velocities (Knudsen, 1988). When the ionosphere above 200 km nearly disappears at solar minimum, this scenario predicts that the main ion source for the nightside ionosphere and ionotail is cutoff. Thus one expects the nightside densities to decline at solar minimum to a level that can be maintained by local ion production by energetic or suprathermal electrons (Gringauz *et al.*, 1979; Spenner *et al.*, 1981). The solar cycle variations of this source, if any, have not been reported.

This situation is illustrated in Figure 45, which shows the ionotail configurations that might be expected at solar maximum and solar minimum. A typical 1986 transit of PVO through the antisolar region is also shown. The ionopause height and the average ionotail density measured during such a transit should show the effects of solar EUV and solar wind variations. When the solar EUV flux increases at solar maximum both the nightside N_e and ionopause height would be expected to increase. The reverse would be expected at solar minimum. Wide variations in the solar wind dynamic pressure,

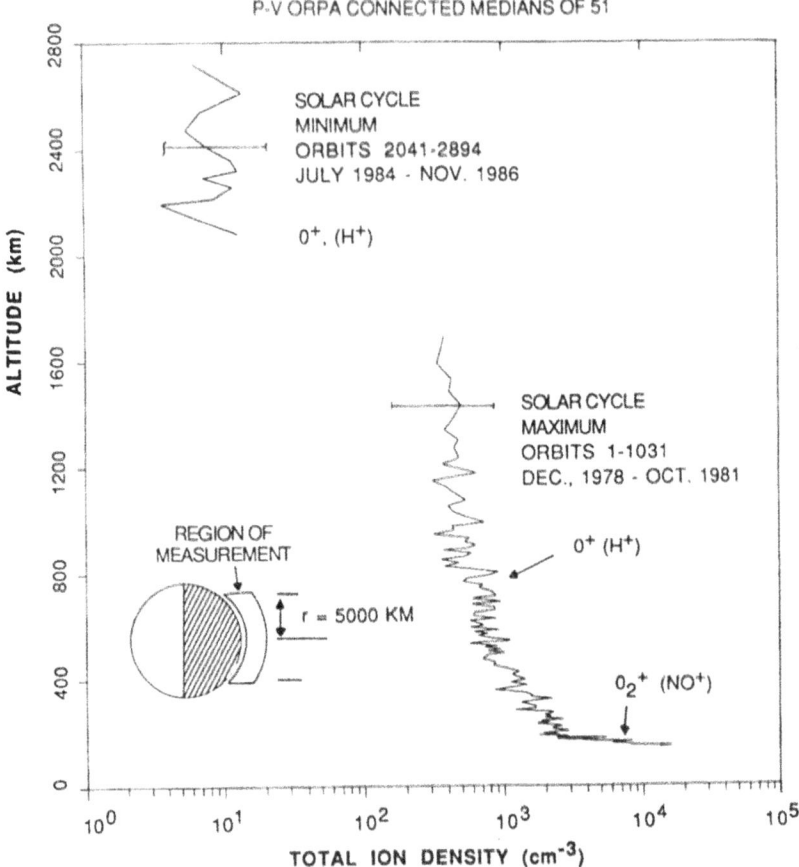

Fig. 44.   Median ion density in the nightside ionosphere between 2000 and 2700 km (near solar minimum) and between 150 and 1600 km (near solar maximum). These data strongly suggested a large solar cycle variation in the nightside ionosphere density (after Knudsen, 1988).

P_{SW}, occur throughout the solar cycle, and the effects of these changes on the ionopause and the ionotail density are superposed on the solar cycle effects.

To investigate the above scenario in greater detail, Brace *et al.* (1990) examined the PVO measurements of the P_{SW}, the intensity of the solar EUV flux, and the nightside N_e. The goal was to see how the ionotail density responded to the solar cycle variations in the EUV and orbit to orbit changes in P_{SW}. The PVO measurements made between late 1982 and early 1989 provided a wide range of variations in both parameters at altitudes above 1400 km. This period included solar minimum, a significant portion of the declining phase of solar cycle 21, and the initial rise in solar cycle 22. P_{SW} was calculated from the OPA measurements of the solar wind density and velocity. The V_{EUV} and N_e were measured by the OETP. To suppress the orbit to orbit variations, the average V_{EUV} and the average ionotail density, $\underline{N}_e$, for each of the 11 nightside transit

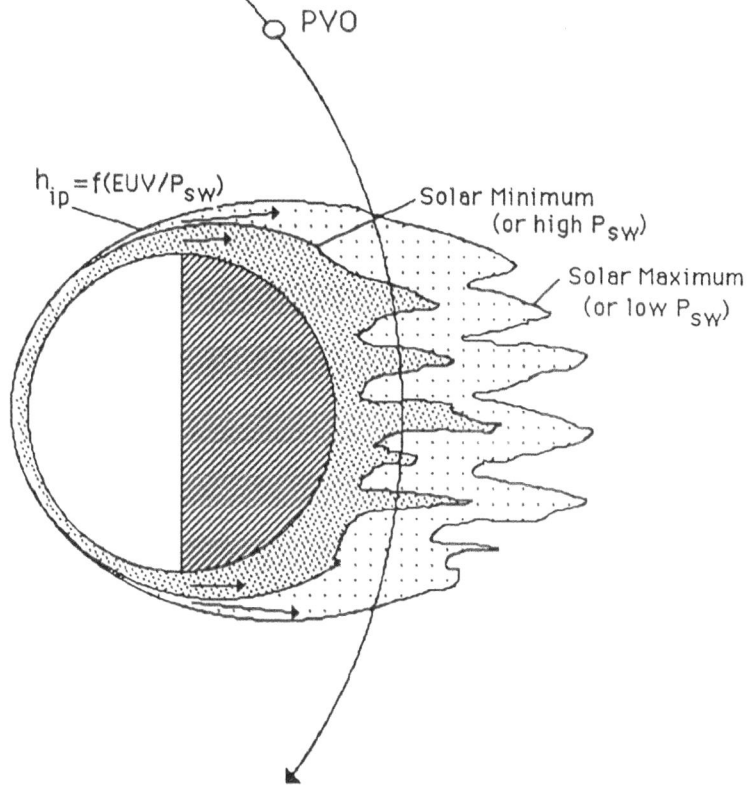

Fig. 45.   A Cartoon illustrating how the solar cycle related changes in the EUV flux may produce changes in the height of the ionopause near the terminator, causing variations in the nightward ion flow and the extent of the ionotail. At any given level of solar activity, solar wind dynamic pressure variations also change the ionopause height, thus modulating the density of the nightside ionosphere (after Brace *et al.*, 1989).

seasons was calculated, and these are shown in Figure 46. N_e included all of the N_e measurements made anywhere in the umbra during each of the 11 Venus years. $\underline{N}_e$ is intended to be a proxy for the total electron content of the ionotail, which we have no way to measure directly. From this result, Brace *et al.* concluded that the ionotail is greatly depleted at solar minimum. $\underline{N}_e$ was lower by about a factor of 20 at solar minimum than it was in December of 1982 and in January of 1989 when the EUV flux was 30 to 40% higher. This variation is consistent with the solar cycle variation inferred by Knudsen *et al.* (1988) from the ORPA measurements of N_i in the central nightside ionosphere (Figure 44).

The orbit to orbit variations of the ionotail were also examined by Brace *et al.* (1989), primarily to look for the effects of changes in P_{SW}. About 200 nightside orbits that passed within 2 hr of local midnight were examined. The N_e measurements were averaged over each entire umbra passage, not just the portions of eah passage that contained measurable densities ($N_e > 2$ cm^{-3}). The orbit-averaged densities, $\overline{N}_e$, were

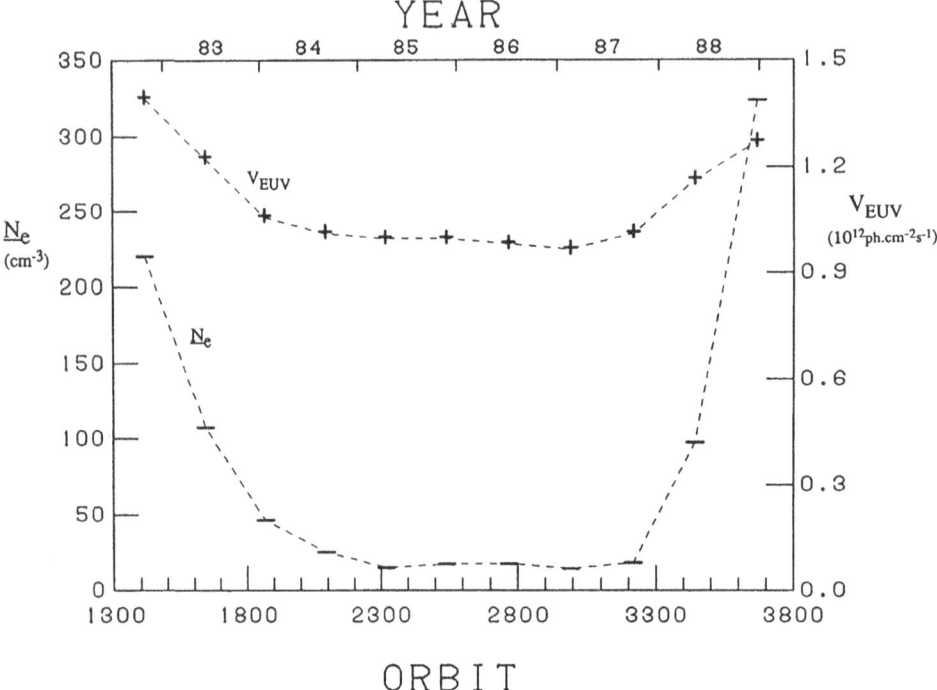

Fig. 46.   The seasonally averaged N_e response of the ionotail to variations in the solar extreme ultraviolet flux, V_{EUV}. Results are shown from all 11 Venus seasons from 1983–1988. The ionotail density varies by a factor of 20 or more for a 30% variation of the solar EUV flux (after Brace *et al.*, 1990).

then compared individually with the daily V_{EUV} and P_{SW} measurements that were taken an hour our two before periapsis. The results are shown in Figures 47 and 48. Each point represents the conditions at the time of one passage through the ionotail. To reduce the scatter due to EUV variations, Figure 48 shows only the orbits in which V_{EUV} was between 0.9 and 1.1×10^{12} ph cm^{-2} s^{-1}. The effect of this was to focus on the short-term variability of the ionotail at solar minimum. From these data it is clear that $\overline{N}_e$ in the ionotail varies directly with the solar EUV flux and inversely with P_{SW}.

These orbit to orbit changes are consistent with the suggestion of Knudsen *et al.* (1988) that the ionopause height on the dayside exercises strong control over the ionotail density. Since the ionopause height is determined by a balance between the EUV flux and P_{SW}, both of these factors affect the ionotail density, and perhaps independently. However, the EUV variations clearly dominate the solar cycle behavior of the ionotail, while P_{SW} variations produce the orbit to orbit changes, at least at solar minimum. Continuing PVO measurements should soon permit the ionotail behavior at solar maximum to be examined.

It is clear from the scatter in Figures 47 and 48 that the ionotail responds to other factors as well. Brace *et al.* (1990) concluded from a cursory look at the magnetic field measurements during these orbits that the IMF had little if any influence on the ionotail

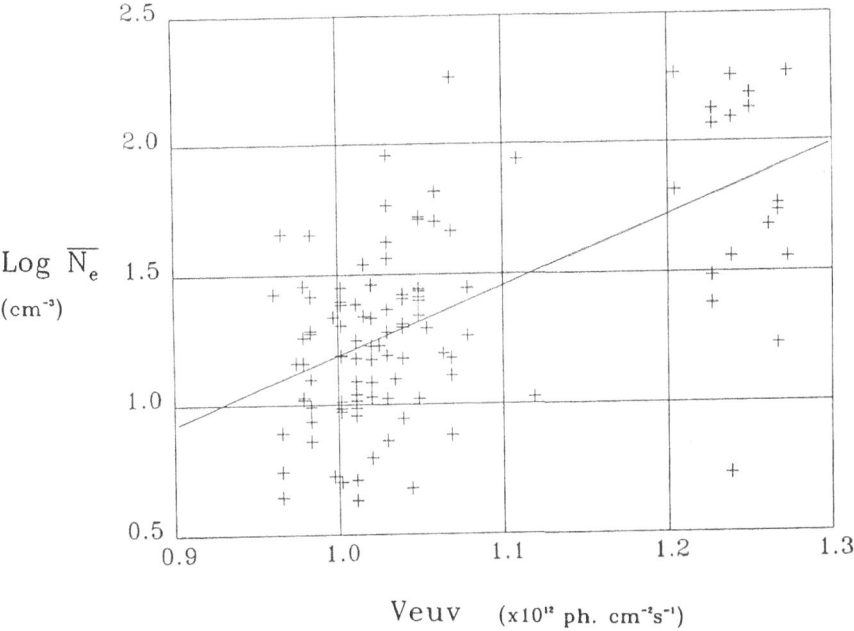

Fig. 47.   The ionotail average N_e from many individual orbits through the antisolar region plotted against the solar EUV measurements. V_{EUV} was measured 1–2 hours earlier in each orbit. The scatter arises in part from the wide range of P_{SW} conditions that are represented in the data base (after Brace *et al.*, 1990).

average density. It is also possible that some of the scatter arises from the lack of simultaneity between the solar wind and ionosphere measurements. At this time we can only conclude that solar EUV and solar wind variations produce much of the variability observed in the ionotail.

5. Contrasts with the Ionospheres of Earth and Mars

In the above discussion we have stressed the fact that much of the unique structure and behavior of the Venus ionosphere arises from the absence of a planetary magnetic field. To emphasize this point further, it may be useful to recall some of the ways in which the magnetic field of Earth dominates its ionospheric structure and to speculate on what the situation may be at Mars.

The primary effect of the geomagnetic field is to constrain the transport of ions and electrons to occur primarily parallel to the field, except perhaps at high latitudes where magnetospheric electric fields drive the ions across magnetic field lines. At middle and low latitudes, ions created on the dayside are constrained to move nightward with the magnetic field lines as they corotate with the planet. This geomagnetic control produces a terrestrial ionosphere that has very different SZA variations than that of Venus, specifically one in which latitudinal structure dominates. This magnetic control affects

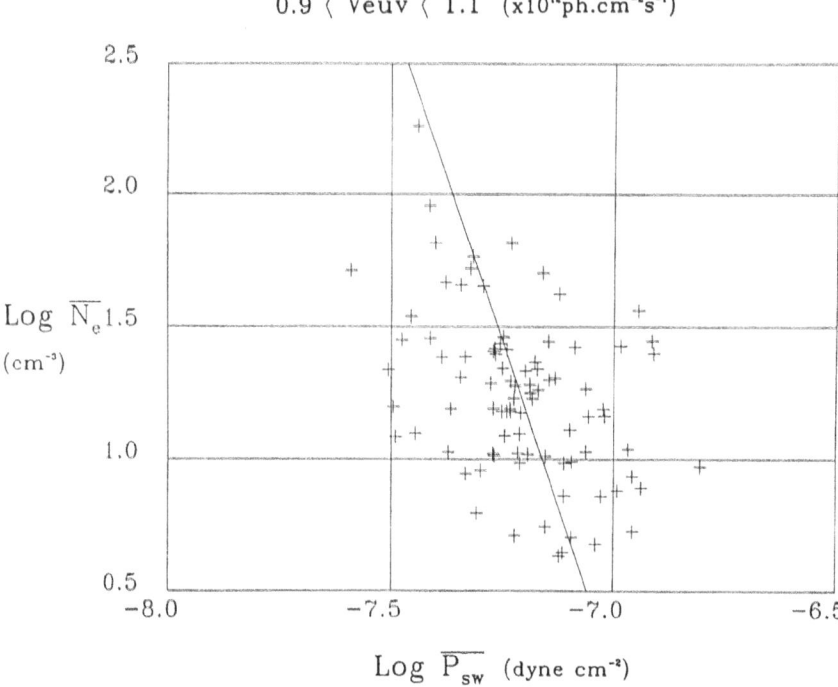

Fig. 48.   The day to day response of the ionotail average N_e to P_{SW} changes in the period of low activity between 1984 and 1987. P_{SW} was measured 1–2 hours earlier on the same orbits, so they only approximately represent the conditions at the time of the ionotail measurements (after Brace *et al.*, 1990).

all ionospheric parameters, including N_e, T_e, T_i and the ion composition and velocities. For brevity, however, we will only mention the effects on N_e and T_e.

Magnetic Control of N_e

The magnetic control on N_e is imposed in several ways. First, the motions of the thermosphere and ionosphere are coupled by ion-neutral drag forces. The ions are anchored to the geomagnetic field, so that poleward (antisunward) circulation of the neutrals produces a field-aligned downward force on the ions, causing a broad N_e trough to form at middle latitudes. The equatorward winds on the nightside drive ions upward to form broad N_e maxima at middle latitudes in the upper F-region. Zonal winds also tend to drive ions up or down the magnetic field at longitudes where the declination of the magnetic field is large. These wind induce large longitudinal variations in the ionosphere. The absence of an intrinsic magnetic field at Venus appears to have produced an ionosphere that is largely free of these kinds of latitudinal and longitudinal structure. Instead, SZA effects apparently dominate everywhere in the Venus dayside ionosphere of Venus, although the orbit of PVO has only allowed the low latitude ionosphere to be examined directly.

The role of a planetary magnetic field in shielding the planet from the solar wind also affects its ionosphere. The Earth's magnetic field, for example, interacts with the solar wind far above the surface. The geomagnetic field stores some of the solar wind energy in the magnetosphere, and later channels some of it into the upper atmosphere in the form of energetic particles. Particle precipitation produces high amplitude, small scale structure in the density and temperature of the high latitude thermosphere and ionosphere.

Another effect of the Earth's magnetic field is to provide a huge high altitude magnetic reservoir for the storage of cold ions (H^+ and He^+) that originate in the underlying F-region. This region, known as the plasmasphere, is a donut shaped shell which extends from the F-region out to a distance of 4 to 7 Earth radii at the magnetic equator. This reservoir is filled during the day by field-aligned upward diffusion of these light ions, and it is depleted at night by downward ion transport. This downward ion flux, together with equatorward winds which lift the F-region, help to maintain the ionosphere throughout the night against recombination. These effects reduce the amplitude of the day to night variation of the ionosphere to a much smaller amplitude than that observed on Venus which does not have such a reservoir. Instead, the dayside Venusian ionosphere acts as a reservoir from which nightward ion flow can maintain the nightside ionosphere against chemical recombination and escape. This nightward flow becomes weaker at solar minimum, however, so the amplitude of the diurnal variation in the nightside N_e is increased.

Magnetic Control of T_e

The temperature structure of an ionosphere is also affected by the magnetic properties of the planet. The nightward transport of heat by conduction is inhibited in the Earth's ionosphere by the largely vertical geomagnetic field. However, this inhibition is partially compensated by thermal energy that is stored in the plasmasphere on the dayside and later returned to the nightside ionosphere by downward heat conduction. This heat reservoir is filled during the day by photoelectron heating. The heat capacity of the longer flux tubes (middle latitudes) is sufficient to maintain a substantial heat conduction flux into the middle latitude ionosphere all night. The lower electron content of the shorter flux tubes provides a smaller heat capacity, so the F-region below about 40° geomagnetic latitude cools rapidly after sunset. The magnetosphere itself supplies an additional nighttime heat source of the middle latitude ionosphere. Ions in the magnetospheric ring current heat the thermal electrons near the plasmapause, adding to the photoelectron heat reservoir. It sometimes provides enough additional electron heating to excite thermospheric atomic oxygen to produce a band 6300 A airglow across the middle latitude sky. These events, which are observed from the ground, are known as a sub-auroral red arcs, or Sar arcs. They usually occur during the recovery phase of a large geomagnetic storm.

Thus the absence of a strong intrinsic magnetic field at Venus makes the thermal structure of its ionosphere much simpler than that of Earth. Solar wind heating at the ionopause is believed to provide the main source of energy for the upper Venus dayside ionosphere, although Section 4.4 brought out some questions about this, and the exact heating process is not yet clear. Photoelectron heating is more important in the lower dayside ionosphere. The nightside ionosphere appears to be heated by conduction both from above and horizontally from the dayside ionosphere. Nightward plasma convection may also be important, particularly for the ion temperature. Acceleration processes within the nightside ionosphere also play a role in the energetics of this region, and they may contribute significantly to planetary ion escape.

Implications for Mars

The ionosphere of Mars is much less well understood. No satellite missions have been conducted to examine its thermosphere and ionosphere and their interactions with the solar wind. Most of what is known about the Martian ionosphere comes from radio occultation measurements, and the *in situ* measurements made from the two Viking Landers which provided quasi-vertical profiles of the neutral and ion composition, density, and temperature. The Mariner 9 orbiter and the Viking 1 and 2 orbiters made radio occultation measurements of N_e, but the nightside occultation results have not been widely reported, apparently because the nightside ionosphere was generally too weak to measure reliably. Mariner 9 and Viking did not carry magnetometers to investigate the planetary magnetic field. This deficiency may be rectified by the magnetometer instrument on the planned Mars Observer mission which, inexplicably, will have no plasma diagnostics. Investigations of bowshock crossing distances observed during early flyby missions suggest that Mars may have a weak magnetic field (Slavin *et al.*, 1981), but these observations may also be explained by mass loading (Luhmann *et al.*, 1987; Phillips and McComas, 1991).

If Mars is weakly magnetized it would be truly unique among the planets which have atmospheres. The major question is whether its magnetic field will be found strong enough to affect the structure of its ionosphere. If not, its ionosphere can be expected to look more like that of Venus, with an ionopause and direct solar wind heating. It is not clear from the radio occultation or Viking Lander measurements whether the Martian ionosphere has an ionopause, but this is a real possibility at solar maximum when the plasma pressure will exceed the dynamic pressure of the solar wind, at least at times of low solar wind pressure. Luhmann *et al.* (1987) have pointed out that the Martian dayside ionosphere is much weaker than that of Venus, in part because of its greater distance from the Sun. Thus, even a weak intrinsic magnetic field could become the primary obstacle to the solar wind.

Such a magnetic field would have important ionospheric effects, as it would inhibit horizontal ion transport, thus affecting the latitudinal and local time structure. If strong enough, an intrinsic field could even form a weak plasmasphere which would help to maintain and heat the nightside ionosphere. The answer to these questions, and many others, will only be obtained by a mission which is dedicated to the Martian upper atmosphere and ionosphere and their interactions with the solar wind.

6. Summary

The absence of an intrinsic magnetic field allows the solar wind to interact directly with the Venus ionosphere. These interactions cause a magnetic barrier to form just above the dayside ionopause, a cushion which isolates the shocked solar wind plasma from the cold ionospheric plasma. The dynamic pressure of the solar wind is conveyed to the ionosphere via this magnetic cushion, and the ionopause forms at the altitude where the normal component of the solar wind pressure equals the ionosphere plasma pressure. This pressure balance forms a complex three dimensional ionopause which envelops the planet. Its surface is often undulating and wavelike on the dayside and stretches out on the nightside to form the outer boundary of a long ionotail that is cometlike in many respects. When the solar wind dynamic pressure is especially high, the ionopause in the subsolar region may be driven to quite low altitudes where photo-ion production is sufficient to substantially change the nature of the ionopause pressure balance. At these times $\mathbf{j} \times \mathbf{B}$ forces induce a strong downward ion flow within the ionopause which limits its further descent with increasing solar wind dynamic pressure. Some of the photo-ions produced in the ionopause density gradient are convected away in the magnetosheath flow, a process which mass loads the flow and probably further limits the downward pressure on the ionosphere. At these times of high solar wind dynamic pressure, the underlying dayside ionosphere becomes heavily magnetized.

The dayside ionosphere of Venus is less variable in many ways than the ionosphere of Earth. Its dayside peak behaves much like a Chapman layer in which local ion production by photoionization is approximately in equilibrium with ion recombination. The variations of the ionopause height with solar wind pressure do not seem to affect the underlying ionosphere very much. The temperature structure on the dayside is also quite stable, with only small day to day (orbit to orbit) variations. Solar wind related heating at the ionopause is believed to be the source of most of the dayside heating, at least at solar maximum. (*In situ* measurements of the lower ionosphere have only been made at that time in the cycle.) Locally produced photoelectrons supply only about 20% of the energy that is required to explain the observed temperatures. The average electron temperature in the nightside ionosphere is almost the same as that on the dayside. Electrons in the nightside ionosphere appear to be heated primarily by conduction and convection from the dayside and by conduction from the overlying ionopause and ionotail. The precipitation of energetic electrons also plays a role in the energy budget. The ion temperature is somewhat higher on the nightside than the dayside, especially at SZA > 150°, an effect that apparently cannot be entirely explained by convergence of the supersonic ion flow into the antisolar ionosphere.

The ionotail density is highly variable with solar activity, varying by more than a factor of 20 from solar minimum to solar minimum. The ionotail is also highly dynamic on very short time-scales, its density variations apparently related to changes in the solar wind dynamic pressure. Most of the ions in the ionotail are superthermal or energetic, at least during conditions of low or moderate solar activity for which the PVO data on energetic ions have been examined. The energy distribution of these ions has not been measured but present evidence suggests that most are in the range 9–16 eV. A small percentage of the superthermal ions exceed 40 eV. Significant planetary ion escape rates would be expected at these energies.

The main ion source for the nightside ionosphere at solar maximum appears to be ion flow from the dayside ionosphere. This nightward flow is reduced at times of high solar wind dynamic pressure and low solar EUV flux, largely because of their combined effect upon the height of the ionopause. Much of the upper ionosphere is removed at solar minimum, greatly reducing nightward ion transport and depleting the nightside ionosphere and ionotail. The decrease in nightward flow at solar minimum probably increases the relative importance of energetic electron precipitation in maintaining the nightside ionosphere at that time.

Comparisons of the ionospheres of Earth and Venus illustrate the fundamental importance of an intrinsic magnetic field to the nature of a planetary ionosphere. The magnetic field of Earth diverts the solar wind at altitudes well above the ionosphere, so direct solar wind heating is small. Solar wind energy is stored in the magnetosphere, however, and some of this energy is deposited later in the high latitude ionoshere. Its intrinsic magnetic field also affects the Earth's ionospheric structure by inhibiting the nightward ion transport and by providing a reservoir which supplies cold ions and electrons to the nightside ionosphere. At Venus, the absence of an intrinsic magnetic field allows the solar wind to come into virtual contact with the ionosphere, removing plasma, configuring the ionopause into a comet-like shape, heating the underlying ionosphere globally, and producing much of the observed nightside complexity in ways that are not yet understood.

Acknowledgements

The authors are grateful to many members of the Pioneer Venus Science team for extensive discussions and helpful comments. We are especially grateful to Professor W. B. Hanson for his insightful comments, and to one other referee.

Appendix. The PVO Data Base and Sources of Measurement Error

This Appendix contains a more detailed discussion of the PVO data base, the spatial resolution and accuracy of the measurements, and some of the discrepancies in N_i and N_e that have been reported. In general, it can be said that the measurement errors have not fundamentally limited PVO investigations of the ionosphere. The same statement cannot be made for the limited breadth of the PVO science payload, particularly the lack of energetic particle instruments which would have aided investigations of solar wind interaction processes. Instrumentation for high-resolution measurements of vector ion and neutral wind velocities and temperatures was also lacking. But PVO was, after all, a low cost mission with broad exploratory goals and a playload weight and data rate that were too low to permit many additional valuable instruments to be carried.

A.1. THE UNIFIED ABSTRACT DATA SYSTEM (UADS)

To aid in the exchange and interpretation of the PVO measurements, an online Unified Abstract Data System (UADS) was established by the PV Project. The UADS combined the various *in situ* measurements that were averaged in a way to assure simultaneity and a uniform spatial resolution. Entries occur at specific times defined at 12 s intervals during each passage. The entries are limited to 30 min either side of periapsis. The actual measurements may have been made more or less frequently than 12 s, depending upon the particular instrument, its measurement mode, and the spacecraft telemetry bit rate being used. The radio occultation profiles are not included in the UADS data base, since they do not fit into the 12 s data format. These, and other remote measurements, are available in other forms.

The UADS on line sytem was discontinued in 1981 and the data then available were submitted via magnetic tapes to the National Space Science Data Center at Goddard Space Flight Center for further analysis by interested investigators. These tapes contain data for most of the first 1000 orbits. In 1981, cost constraints made it infeasible to continue to maintain and expand the online UADS data base. Since then, each investigator has independently submitted the measurements from his own instrument as they become available, mainly in the common 12 s format of the original UADS.

The UADS files are not static. As new data are acquired or older data are reprocessed with improved algorithms, new UADS tapes are generated to replace earlier versions. In this way the investigators plan to meet their goal of assembling the most complete and accurate data base possible by the end of the mission. This approach inevitably means that later versions of the UADS files, while more complete and more accurate, may differ in detail from earlier versions.

Many studies require data with higher spatial resolution than is provided by the UADS files. Most instruments are capable of providing this higher resolution data, and these are submitted to the NSSDC as High Frequency Data Files, or they may be obtained directly from the appropriate PV investigators.

A.2. MEASUREMENT CAPABILITIES

As noted in Section 3.2, PVO carries three instruments which make *in situ* measurements within the ionosphere. These are the OIMS, ORPA, and OETP. A fourth instrument, the ONMS, measures the neutral gas densities that are needed for understanding many aspects of ionosphere behavior. The ONMS also detects superthermal ions having energies greater than about 40 eV, the retarding potential of an outer shield that was intended to exclude cold ionospheric ions while the instrument was measuring the concentrations of the neutral thermospheric constituents. The ONMS also has a cold ion composition mode in which the outer shield is grounded and the filament is off, but this mode excluded the measurement of neutral densities, so it has seldom been used. The radio occultation experiment, ORO, provides height profiles of N_e both above and below the periapsis altitude. These measurements are available only during the occultation seasons, however.

Each of these instruments has its own strengths and limitations. For detailed instrument descriptions we refer the reader to the instrument descriptions presented in the special issue of *IEEE Trans. Geoscience and Remote Sensing* (1980). The OIMS reports the concentration of all ion species present, including minor ion species, and has an inflight data selection system which permits high spatial resolution. The OETP provides similarly high resolution in N_e, N_i, and T_e. The measurements of the photoemission current from the OETP sensors can be employed as a measure of the total EUV flux from the Sun (Brace *et al.*, 1988a). The ORPA measures the individual concentrations of the major ions, the ion temperature, T_i. However, the ORPA is not able to distinguish between ions of similar mass (O_2^+ and CO^+, or O^+ and N^+) or can it detect the presence of minor ions that represent less than a few percent of the total. In its electron mode, the ORPA measures T_e and the integral electron flux for energies up to about 50 eV. Its sensor head was tilted off the spin axis at an angle of $25°$ to permit it to look nearly directly into the velocity vector once per spin. Thus the spatial resolution of the ORPA measurements is limited to the spacecraft spin period of approximately 13 s. The ORPA also measures the ion drift velocity when the total density exceeds about 10^3 cm^{-3}, but with poorer spatial resolution.

The actual spatial resolution of the ORPA measurements may be poorer than one value per spin period, because the instrument has several modes of operation in which it measures different parameters. These modes are mutually exclusive, so the thermal electrons and ions are not measured simultaneously with the superthermal electrons. The measurement mode can be alternated throughout a passage to obtain sequential measurements of these parameters, with a corresponding loss of spatial resolution. The OETP and OIMS, on the other hand, provide continuous high resolution measurements on every passage through the ionosphere. This high resolution (order of 1 s^{-1} at typical telemetry rates), is attained through the use of adaptive inflight servo systems. Spin modulation was minimized in the OIMS measurements by mounting the sensor parallel to the spin axis, an approach that further increases the effective spatial resolution. The OETP radial sensor is mounted perpendicular to the spin axis for the same reason, but shadowing of that sensor by the spinning spacecraft introduces measurement errors at certain spin angles. The resulting reductions of spin effects turned out to be very valuable for resolving ionospheric structure having much smaller spatial scales than 120 km, the distance the spacecraft travels in one spin period. The nightside ionosphere has many such small scale features, and the ionopause crossings often occur within one or two spacecraft spin periods.

Thus the apparently overlapping capability of the various PVO instruments does not necessarily make their measurements redundant. Instead, their differences in spatial resolution, sensitivity, and sources of measurement error combine to make their measurements complementary. It will be important for future users of data to take these factors into account when choosing data sets for a particular study.

A.3. Discrepancies among the Measurements

The high degree of measurement redundancy among the PVO instruments has naturally lead to inconsistencies, particularly in the meamsurements of the ion composition and the ion and electron densities, N_i and N_e, which can be obtained in at least four ways. The OIMS and ORPA measure N_i. The OETP also measures N_i at high densities and N_e at low densities, with overlap between 1×10^3 and $1 \times 10^5\,cm^{-3}$. The radio occultation analysis yields height profiles of N_e down to densities of the order of $10^3\,cm^{-3}$. These profiles have been useful in checking the absolute accuracy of the *in situ* measurements. However, these profiles can be compared only statistically with the direct measurements because they usually represent a region quite remote from the satellite. They also represent N_e averaged over a relatively long horizontal path through the ionosphere, so small-scale structure is not resolved.

Systematic comparisons of the early UADS data base, conducted by Miller *et al.* (1984), uncovered a number of discrepancies, particularly in the ion composition and density measurements. They found that the OIMS values of N_i were larger than those from the ORPA by a factor of between 1.5 and 2.5 in portions of the dayside ionosphere, with the largest discrepancies in the afternoon at altitudes between 170 and 200 km. Miller *et al.* attributed this to OIMS O_2^+ concentration that were about a factor of 3 higher in that region. The density measurements above 200 km, where O^+ dominates, are in better agreement. The ORPA measurements of N_i were in essential agreement with the OETP measurements of N_e, except at the lowest altitudes on the nightside, where N_e values were a bit higher. The ORO density profiles agreed well statistically at these altitudes with the ORPA and OETP densities, and generally gave lower densities than those given by the OIMS. The T_e values from the OETP were in general agreement with those from the ORPA, except on the dayside below 200 km, where the OETP values were several percent higher.

It is important to recognize in making such comparisons that the UADS files are continually being updated and corrected as the investigators receive new data and increase their understanding of the sources of measurement error. These changes can be expected to alter the remaining discrepancies among the various measurements to some degree, but the major differences described above are likely to remain until a consensus is reached on how to, and whether to, attempt to make the entire ionosphere data base more internally consistent. In the next section we will discuss some possible sources of measurement error.

A.4. Possible Sources of Measurement Error

The accuracy of the PVO measurements varies with the parameter in question and depends upon the conditions present. In most situations the accuracy is believed by the investigators to be adequate to resolve the major features of ionospheric behavior, particularly considering the wide dynamic range of these parameters. However, this mission offered new challenges that were not encountered in the Earth missions, so it is not surprising to find significant measurement errors under some conditions. The very low data rate afforded by the spacecraft required the instruments to include onboard data processing schemes to transform their raw spectra or volt-ampère curves into physical parameters. Without onboard processing the small scale structure of the ionosphere could not have been resolved. These systems were limited in sophistication by strict payload weight limitations; most of the instruments weighed only a kilogram or two. Part of the challenge was to recover representative raw experimental data (mass spectra, energy spectra, or volt-ampère curves) to verify the accuracy of the onboard processing scheme. In general, these systems performed well, but the discrepancies among the measurements show that undetected errors are still present in some cases.

In addition to the above problems, the complexity of the Venus ionosphere itself may have presented conditions that are beyond the capabilities of the simple PVO instruments. The following section gives some examples.

A.4.1. *Suprathermal Electron Effects*

Suprathermal, or energetic, electrons are common in the nightside ionosphere. They cause errors in the measurements of cold ionospheric electrons by making the spacecraft potential so negative that the thermal electrons cannot reach the sensors. This situation occurs when the spacecraft is in darkness and spacecraft photoelectron emission is not available to prevent charging of the spacecraft to high negative potentials. This electrostatic shielding is most important for the ORPA measurements because its sensor is mounted on the spacecraft surface so the cold electrons must overcome the full spacecraft potential to enter the sensor. The OETP is less affected by the spacecraft sheath because its sensors are mounted 40 cm (axial probe) and 100 cm (radial probe) from the spacecraft surface where the local plasma potential is closer to that of the undisturbed ionosphere. In general, only the radial probe measurements are reported for regions of low density in darkness, where spacecraft charging has been most severe.

A.4.2. *Non-Maxwellian Electron Effects*

Another example of an unexpected complexity that may cause measurement problems is the case of non-Maxwellian electrons in the nightside ionosphere. Brace *et al.* (1980) noted that thermal and superthermal electrons often exist together there with similar densities, making it difficult to characterize the electron temperature accurately. The OETP volt-ampère curves obtained in these regions are not well fitted for a single value T_e. Often a good fit can be obtained by assuming a two temperature distribution, but these fits are not performed routinely. Such distributions are common in the ionospheric holes (Brace *et al.*, 1982b) and in the lower nightside ionosphere in the vicinity of small-scale N_e structure (Hoegy *et al.*, 1989). Even in the absence of superthermals, however, strong spatial gradients in T_e cause the electron energy distribution to be non-Maxwellian, thus causing poor curve fits and clouding the definition of temperature. The UADS file did not envision this kind of complexity, so the T_e measurements in these regions do not properly describe the thermal energy of the electrons. Usually the temperature that is entered in the file is that of the cold component, but the higher temperature component may be given when no cold electrons are present.

A.4.3. *Superthermal Ion Effects*

Superthermal ions may affect the accuracy of the various PVO measurements of total ion density. Taylor *et al.* (1980) observed superthermal ions at the ionopause and within the nightside ionosphere. At higher altitudes on the nightside the OIMS measured increasingly larger percentages of superthermal ions, but the response of the instrument to these ions is not well understood (Brace *et al.*, 1987). The evidence for their existence is a shift in the apparent mass of O^+ from 16 to 14 amu, a shift that corresponds to ion energies in the range of 9 eV to 16 eV. Superthermal H^+ ions are not observed because H^+ at these energies would fall below the mass range of the OIMS. Thus the OIMS total density may be underestimated when superthermal ions are present because the H^+ ions are not included.

The N_i measurements by the OETP radial probe are also affected by superthermal ions if they represent a significant fraction of the total density. The calculation of N_i assumes that the ion flux to the collector is produced by the velocity of the collector through the ionosphere (10 km s^{-1}). If the thermal velocity of the ions is comparable to the spacecraft velocity, additional ion current is collected and N_i is overestimated.

A.4.4. *Ion Drift Effects*

High ion velocities in the ionosphere can introduce important errors into the ion measurements. Knudsen *et al.* (1980b) and Taylor *et al.* (1980) have shown that very high ion drift velocities are present, particularly at high altitudes near the terminator. The cold plasma instrument designs generally assume that the ion drift velocity will be small compared to the spacecraft velocity, thus requiring the ion sensors to be mounted to look into the velocity vector. Ion drift produces changes in the ion velocity into the sensor and changes in the angle of approach, both of which affect the ion fluxes that reach the collector of the instrument. The ORPA and the OIMS are mounted to provide small angles of attack near periapsis, but the minimum angle of attack tends to increase with altitude. The ORPA measures the ion drift component normal to the sensor, so its N_i measurements are affected by ion drift only to the extent that the correction for the assumed angle of arrival may be incorrect. The OETP measurements of N_i depend linearly upon knowledge of the ion drift velocity, which is assumed in the data processing to be the satellite velocity. The angle of arrival is unimportant because the maximum in the spin modulated ion flux to its cylindrical collector is used to derive N_i, and this always occurs at the two points in the spin cycle where the velocity vector is perpendicular to the probe axis. These ion measurements are used only at low altitudes, however, where the ion drift velocities are small. The N_e measurements are used at the higher altitudes. Off-axis ion drift velocities cause errors in the OIMS measurements because they reduce the transmission efficiency of the analyzer, an effect that is particularly important for the heavier ions. Thus ion drift leads to an underestimate of the density and an apparent change in the relative ion composition, but these effects should be limied to high altitudes where the ion drift velocities may be a significant fraction of the spacecraft velocity. It is interesting that both of these factors, high ion drift velocities and large angles of attack, tend to be important to the measurement accurately at higher altitudes, but the discrepancies among the PVO density measurements are greater at lower altitudes.

A.4.5. *Spacecraft Photoelectron Effects*

Spacecraft photoelectrons represent another source of error in the *in situ* electron measurements. When the spacecraft is in sunlight, its sunlit side is surrounded by a cloud of photoelectron whose thickness depends upon the spacecraft potential, which itself depends upon the ionospheric plasma density (Brace *et al.*, 1988b). At densities greater than a few hundred cm^{-3}, the instruments operate in an electron environment that is dominated by the ionospheric electrons. Spacecraft photoelectrons dominate when the ambient density is lower, but the lower limit for reliable measurements depends on the mounting location of the particular sensor and its sensitivity to the photoelectron background. The OETP radial sensor is least affected because of its greater distance from the spacecraft, and because one can select the measurements taken only when the collector is on the dark side of the spacecraft where the spacecraft photoelectron background is lower (Brace *et al.*, 1988b). The ORPA is mounted on the spacecraft surface where much larger photoelectron fluxes are available, and this limits the density range over which cold ionospheric electrons can be measured when the spacecraft is sunlit. The photoelectrons have energies of only a few eV, however, so more energetic ambient electron populations can still be measured (Knudsen *et al.*, 1980).

A.4.6. *Periapsis Effects*

An entire class of instrumental errors can occur at very low altitudes due to spacecraft-atmosphere interactions. These errors can be grouped under the general term 'periapsis effects'. These effects are not well understood, and their importance varies with altitude, with the parameter being measured, and with the location and type of sensor. The main known effects are described below.

Perhaps the most well established periapsis effect arises from impact ionization (Hanson *et al.*, 1981; Whipple *et al.*, 1983; Curtis *et al.*, 1985). The spacecraft, traveling through the dense lower thermosphere at high velocities, behaves somewhat like a meteorite. At the PVO periapsis velocity of 10 km s^{-1}, the impact energy ($\frac{1}{2}mv^2$) for CO_2 is 23 eV. This is enough energy to ionize a small fraction of the CO_2 that the spacecraft encounters and produce a measurable cloud of 1 to 2 eV secondary electrons above the leading surface of the spacecraft. Lighter molecules (O_2 and CO) contribute less impact ionization because their impact energies are only slightly greater than their ionization potentials.

The OETP axial sensor is mounted on the ram end of the spacecraft, so it is ideally located to observe impact electrons. Easily measurable fluxes of these electrons are seen when the spacecraft is below about 165 km on the dayside and 150 km on the nightside. The impact electron density at periapsis may be as high as 10% of the ambient N_e at the nightside peak (order of 10^4 cm^{-3}). Their density has been shown proportional to the thermospheric CO_2 concentration, one of the neutral gas parameters that is measured by the ONMS (Whipple *et al.*, 1983). Since their temperature is more than 10 times that of the ionosphere at 150 km, the two electron components are easily distinguishable in the electron retardation region of the volt-ampère curves from the axial sensor. The radial OETP sensor observes no secondary electrons at its location, probably because of it is mounted further from the spacecraft (1 m) and views only lateral spacecraft surfaces which recieve only a small fraction of the ram flux of CO_2.

Impact ionization is also seen in other PVO instruments. Miller *et al.* (1984) suggested that impact electrons may be responsible for the anomalous increase in T_e in the ORPA measurements below about 167 km in the daytime ionosphere. The ORPA is mounted on the forward looking surface of the spacecraft where this effect is greatest. Plasma waves in the vicinity of 100 Hz have been observed consistently when periapsis is very low. These waves have also been attributed to the impact process (Curtis *et al.*, 1985).

Impact ionization may indirectly produce errors in the ion measurements as well. A byproduct of the impact ionization process is the creation of a dc electric field upstream of the spacecraft (Parker and Holeman, 1980). This electric field is produced by the difference in mobility of the sputtered ions and electrons. The resulting charge separation creates a region of positive space charge near the surface and a region of negative spacecharge farther ahead. Ambient ions must pass through these electric fields to reach the ion sensors, so the composition and energy of the measured ions may be perturbed. No analysis of this effect on the PVO ion measurement techniques has been reported.

An analogous periapsis effect involves the thermospheric neutrals that are not ionized by impact with the spacecraft. These neutrals tend to be thermalized at the spacecraft surface and re-emitted at much lower velocities, thus creating a dense cloud of neutrals just upstream of the spacecraft. This enhancement is caused by the low departure velocity of the gas that has become thermalized by collision(s) with the ram surface. If complete thermal accomodation occurs, the maximum density of the ram cloud is more than a hundred times greater than the ambient density at that point in the thermosphere. At typical periapsis altitudes (about 150 km), the ambient neutral density is greater than 10^{10} cm^{-3}, and the ram cloud density at the surface may be of the order of 10^{12} cm^{-3}. At this density the ion and neutral mean free paths that are comparable to the size of the cloud, which extends several spacecraft diameters upstream (a few meters). Some of the incoming ions and neutrals will experience collisions with the outflowing neutrals in the cloud, changing their energies and velocities in the reference frame of the space-craft. The effect of the ram cloud upon the measurements may be difficult to separate from impact ionization effects since they occur together near periapsis.

In summary, these 'periapsis effects' affect the various PVO instruments differently, but no quantitative analysis of the resulting errors has been reported. From various signatures in their data the investigators can often recognize when these effects are present. We suspect that, in assembling data for their own analyses, or for submission to the NSSDC, the investigators have deleted the measurements that have been most obviously affected. However, one cannot be sure that all of these effects have been recognized and removed.

A.5. VIRA ATTEMPTS TO DEAL WITH MEASUREMENT ERRORS

In a first attempt to mitigate the discrepancies in the density measurements by different instruments, Bauer *et al.* (1985) assembled the VIRA ionosphere model with these differences in mind. A series of tables were presented listing the altitude variations of each major ionosphere parameter. The global model by Theis *et al.* (1984), based on the OETP measurements, was adopted for the VIRA electron density and temperature. The densities were normalized at 150 km to the average radio occultation density measurements, which are believed to be more reliable at the high densities usually present at the peak. The OIMS results were employed only to establish the relative ion composition rather than the absolute ion densities. Since the ORPA measures only the major ions, its data could not be the basis for a complete ion composition model. The ORPA measurements were used to define the ion temperature in the VIRA model, and to provide the pattern of global ion drift velocities.

In conclusion, the accuracy of the PVO measurements is difficult to assess. While the periapsis effects can be expected to cause errors at the lowest altitudes, the largest disagreements occur in the afternoon near 175 km, well above periapsis. The investigators, aware of these disagreements, have been comparing measurements to further identify areas of disagreement, and to correct errors where possible. These efforts are expected to lead to continuing improvements in the accuracy of the UADS data and of future VIRA models. In the meantime, the choice of PVO data for a particular investigation will fall to the user.

References

Aleksandrov, Yu. N., Vasil'ev, M. B., Vysholof, A. S., Dolbezhev, G. G., Dubrovin, V. M. Saitsev, A. L., Kolosov, M. A., Petrov, G. M., Savich, N. A., Samovol, V. Z., Samoznaev, L. H., Siborenki, A. Ai., Khasyanov, A. F., and Shtern, D. Ya.: 1976a, *Kosmich. Issled.* **14**, 824.
Aleksandrov, Yu. N., Vasil'ev, M. B., Vysholof, A. S., Dubrovin, V. M., Zaitsev, A. L., Kolosov, M. A., Krymov, A. A., Makovoz, G. I., Petrov, G. M., Savich, N. A., Samovol, V. Z., Samoznaer, L. N., Siborenkio, A. I., Khasyanov, A. F., and Shtern, D. Ya.: 1976b, *Kosmich. Issled.* **14**, 819.
Bauer, S. J.: 1983, *Ostereichische Akademie der Wissenschaften, Sitzungsberichte*, **II**, **192**, Nos. 8–10, 309.
Bauer, S. J. and Hartle, R. E.: 1974, *Geophys. Res. Letters* **1**, 7.
Bauer, S. J., Brace, L. H., Hunten, D. M., Intriligator, D. S., Knudsen, W. C., Nagy, A. F., Russell, C. T., Scarf, F. L., and Wolfe, J. H.: 1977, *Space Sci. Rev.* **20**, 413.
Bauer, S. J., Brace, L. H., Taylor, H. A., Breus, T. K., Kliore, A. J., Knudsen, Nagy, A. F., Russell, C. T., and Savich, N. A.: 1985, *Adv. Space Res.* **5**, 233.
Bauer, S. J., Hartle, R. E., and Herman, J. R.: 1970, *Nature* **225**, No. 5232, 533.
Bougher, S. W. and Cravens, T. E.: 1984, *J. Geophys. Res.* **89**, 3837.
Brace, L. H., Theis, R. F., Krehbiel, J. P., Nagy, A. F., Donahue, T. M., McElroy, M. B., and Pedersen, A.: 1979, *Science* **203**, 763.
Brace, L. H., Theis, R. F., Hoegy, W. R., Wolfe, J. H., Mihalov, J. D., Russell, C. T., Elphic, R. C., and Nagy A. F.: 1980, *J. Geophys. Res.* **85**, 7663.
Brace, L. H., Theis, R. F., and Hoegy, W. R.: 1982a, *Planetary Space Sci.* **30**, 29.
Brace, L. H., Theis, R. F., Mayr, H. G., Curtis, S. A., and Luhmann, J. G.: 1982b, *J. Geophys. Res.* **87**, 199.
Brace, L. H., Gombosi, T. I., Kliore, A. J., Knudsen, W. C., Nagy, A. F., Taylor, H. A.: 1983a, in L. Colin and D. Hunten (eds.), *Venus*, Univ. of Arizona Press, Tucson, Chapter 23, p. 779.
Brace, L. H., Elphic, R. C., Curtis, S. A., and Russell, C. T.: 1983b, *Geophys. Res. Letters* **10**, 1116.
Brace, L. H., Kasprzak, W. T., Taylor, H. A., Theis, R. F., Russell, C. T., Barnes, A., Mihalov, J. D., and Hunten, D. M.: 1987, *J. Geophys. Res.* **92**, 15.

Brace, L. H., Hoegy, W. R., and Theis, R. F.: 1988a, *J. Geophys. Res.* **93**, 7282.

Brace, L. H., Theis, R. F., Curtis, S. A., and Parker, L. W.: 1988b, *J. Geophys. Res.* **93**, 12735.

Brace, L. H., Theis, R. F., Luhmann, J. G., and Mihalov, J. D.: 1989, *EOS* **70**, No. 15, 386.

Brace, L. H., Theis, R. F., Mihalov, J. D.: 1990, *J. Geophys. Res.* **95**, 4075.

Breus, T. K., Gringauz, K. I., and Verigin, M. I.: 1985, *Adv. Space Res.* **5**, 9, 145.

Brinton, H. C., Taylor, H. A., Niemann, H. B., Mayr, H. G., Nagy, A. F., Cravens, T. E., and Strobel, D. F.: 1980, *Geophys. Res. Letters* **7**, 865.

Chapman, S.: 1931, *Proc. Phys. Soc.* **43**, 26.

Chen, R. H. and Nagy, A. F.: 1978, *J. Geophys. Res.* **83**, 1133.

Cloutier, P. A., Taylor, H. A., Jr., and McGary, J. E.: 1987, *J. Geophys. Res.* **92**, 7289.

Colin, L.: 1983, in D. M. Hunten, L. Colin, T. M. Donahue, and V. I. Moroz (eds.), *Venus*, Univ. of Arizona Press, Tucson, Chapter 2, p. 10.

Covington, A. E.: 1969, *J. Roy. Astron. Soc. Can.* **63**, 125.

Cravens, T. E. and Nagy, A. F.: 1983, *Rev. Geophys. Space Phys.* **21**(2), 263.

Cravens, T. E., Nagy, A. F., Chen, R. H., and Stewart, A. I.: 1978, *Geophys. Res. Letters* **5**, 613.

Cravens, T. E., Nagy, A. F., Brace, L. H., Chen, R. H., and Knudsen, W. C.: 1979, *Geophys. Res. Letters* **6**, 341.

Cravens, T. E., Gombosi, T. I., Kozyra, J., Nagy, A. F., Brace, L. H., and Knudsen, W. C.: 1980, *J. Geophys. Res.* **85**, 7778.

Cravens, T. E., Kliore, A. J., Kozyra, J., and Nagy, A. F.: 1981, *J. Geophys. Res.* **86**, 11323.

Cravens, T. E., Brace, L. H., Taylor, H. A., Russell, C. T., Knudsen, W. C., Miller, K. L., Barnes, A., Mihalov, J. D., Scarf, F. L., Quenon, S. J., and Nagy, A. F.: 1982, *Icarus* **51**, 271.

Cravens, T. E., Crawford, S. L., Nagy, A. F., and Gombosi, T. I.: 1983, *J. Geophys. Res.* **88**, 5595.

Curtis, S. A., Brace, L. H., Niemann, H. B., and Scarf, F. L.: 1985, *J. Geophys. Res.* **90**, 6631.

Donahue, T. M., Hoffman, J. H., Hodges, R. D., Jr., and Watson, A. J.: 1982, *Science* **261**, 630.

Dubinin, E. M., Israelevich, P. L., and Podgorny, I. M.: 1980, *Kosmich. Issled.* **18**, 470.

Elphic, R. C. and Ershkovich, A. I.: 1984, *J. Geophys. Res.* **89**, 997.

Elphic, R. C. and Russell, C. T.: 1983, *J. Geophys. Res.* **88**, 2993.

Elphic, R. C., Russell, C. T., Slavin, J. A., Brace, L. H., and Nagy, A. F.: 1980a, *Geophys. Res. Letters* **7**, 561.

Elphic, R. C., Russell, C. T., Slavin, J. A., and Brace, L. H.: 1980b, *J. Geophys. Res.* **85**, 7679.

Elphic, R. C., Russell, C. T., Luhmann, J. G., Scarf, F. L., and Brace, L. H.: 1981, *J. Geophys. Res.* **86**, 11430.

Elphic, R. C., Brace, L. H., Theis, R. F., and Russell, C. T.: 1984a, *Geophys. Res. Letters* **11**, 124.

Elphic, R. C., Mayr, H. G., Theis, R. F., Brace, L. H., Miller, K. L., and Knudsen, W. C.: 1984b, *Geophys. Res. Letters* **11**, 1007.

Elphic, R. C., Brace, L. H., and Russell, C. T.: 1985, *Adv. Space Res.* **5**, No. 4, 313.

Fjeldbo, G. and Eshleman, V. R.: 1968, *Planetary Space Sci.* **70**, 123.

Fox, J. L.: 1982, *Icarus* **51**, 248.

Fox, J. L. and Bougher, S. W.: 1991, *Space Sci. Rev.* **55**, 357.

Garvik, A. L., Ivanov-Kholodny, G. S., Mihalov, G. S., Savich, N. A., and Samoznaev, L. N.: 1980, *Space Research XX*, Pergamon Press, Oxford, p. 231.

Grebowsky, J. M. and Curtis, S. A.: 1981, *Geophys. Res. Letters* **8**, 1273.

Grebowsky, J. M., Mayr, H. G., Curtis, S. A., and Taylor, H. A.: 1983, *J. Geophys. Res.* **88**, 3005.

Gringauz, K. I., Verigin, M., Breus, T. K., and Gombosi, T.: 1977, *Dokl. Acad. Nauk. SSR* **232**, 1039.

Gringauz, K. I., Verigin, M., Breus, T. K., and Gombosi, T.: 1979, *J. Geophys. Res.* **84**, 2123.

Hanson, W. B., Sanatani, S., and Hoffman, J. H.: 1981, *J. Geophys. Res.* **86**, 11, 350.

Hartle, R. E. and Taylor, H. A., Jr.: 1983, *Geophys. Res. Letters* **10**, 965.

Hartle, R. E., Taylor, H. A., Bauer, S. J., Brace, L. H., Russell, C. T., and Daniell, R. E.: 1980, *J. Geophys. Res.* **85**, 7739.

Hedin, A. E., Niemann, H. D., Kasprzak, W. T., and Seiff, A.: 1983, *J. Geophys. Res.* **88**, 73.

Hinteregger, H. E., Fukui, and Gilson, B. R.: 1981, *Geophys. Res. Letters* **8**, 1147.

Hoegy, W. R., Brace, L. H., Theis, R. F., and Mayr, H. G.: 1980, *J. Geophys. Res.* **85**, 7811.

Hoegy, W. R., Brace, L. H., and Russell, C. T.: 1986, *EOS* **67**, No.16, 299.

Hoegy, W. R., Brace, L. H., Kasprzak, W. T., and Russell, C. T., 1990, *J. Geophys. Res.* **95**, 4085.

Howard, H. T., Tyler, G. L., Fjeldbo, G., Kliore, A. J., Levy, G. S., Brunn, D. L., Dickinson, R., Edelson, R. E., Martin, W. L., Postal, R. B., Seidel, B. Desplaukis, T. T., Shirley, D. L., Shirley, D. L., Stelzreid,

C. T., Sweetnam, D. N., Sygielbaum, A. E., Esposito, P. B., Anderson, J. D., Shapiro, I. I., and Reasenberg, R. D.: 1974, *Science* **183**, 1297.

IEEE Trans. on Geosci. and Remote Sens.: 1980, **GE-18**, 1.

Johnson, F. S. and Hanson, W. B.: 1979, *Geophys. Res. Letters* **6**, 581.

Johnson, F. S. and Midgely, T. E.: 1969, *Space Res.* **9**, 760.

Kasprzak, W. T., Taylor, H. A., Brace, L. H., Niemann, H. B., and Scarf, F. L.: 1982, *Planetary Space Sci.* **30**, 1107.

Kasprzak, W. T., Niemann, H. B., and Mahaffy, P.: 1987, *J. Geophys. Res.* **92**, 291.

Keating, G. M. *et al.*: 1985, *Adv. Space Res.* **5**, 117.

Keldish, M. V.: 1977, *Icarus* **30**, 605.

Kliore, A. J. and Mullen, L.: 1989, *J. Geophys. Res.* **94**, 13339.

Kliore, A. J. and Mullen, L.: 1990, *Adv. Space Res.* (in press).

Knudsen, W. C.: 1986, *J. Geophys. Res.* **91**, 11,936.

Knudsen, W. C.: 1987, *J. Geophys. Res.* **92**, 7308.

Knudsen, W. C.: 1988, *J. Geophys. Res.* **93**, 8756.

Knudsen, W. C. and Miller, K. L.: 1985, *J. Geophys. Res.* **90**, 2695.

Knudsen, W. C., Spenner, K., Whitten, R. C., Spreiter, J. R., Miller, K. L., and Novak, V.: 1979a, *Science* **203**, 757.

Knudsen, W. C., Spenner, K., Whitten, R. C., Spreiter, J. R., Miller, K. L., and Novak, V.: 1979b, *Science* **205**, 105.

Knudsen, W. C., Spenner, K., Whitten, R. C., and Miller, K. L.: 1980, *Geophys. Res. Letters* **7**, 1045.

Knudsen, W. C., Spenner, K., Michelson, P. F., Whitten, R. C., Miller, K. L., and Novak, V.: 1980a, *J. Geophys. Res.* **85**, 7754.

Knudsen, W. C., Spenner, K., Miller, K. L., and Novak, V.: 1980b, *J. Geophys. Res.* **85**, 7803.

Knudsen, W. C., Spenner, K., and Miller, K. L.: 1981, *Geophys. Res. Letters* **8**, 241.

Knudsen, W. C., Miller, K. L., and Spenner, K.: 1982, *J. Geophys. Res.* **87**, 2246.

Knudsen, W. C., Kliore, A. J., and Whitten, R. C.: 1987, *J. Geophys. Res.* **92**, 13, 392.

Kumar, S.: 1982, *Geophys. Res. Letters* **9**, 595.

Kumar, S., Hunten, D. M., and Pollack, J. B.: 1983, *Icarus* **55**, 365.

Luhmann, J. G.: 1986, *Space Sci. Rev.* **44**, 241.

Luhmann, J. G. and Cravens, T. E.: 1991, *Space. Sci. Rev.* **55**, 201.

Luhmann, J. G., Elphic, R. C., Russell, C. T., Mihalov, J. D., and Wolfe, J. H.: 1980, *Geophys. Res. Letters* **7**, 917.

Luhmann, J. G., Russell, C. T., Brace, L. H., Taylor, H. A., Knudsen, W. C., Colburn, D. S., Scarf, F. L., and Barnes, A.: 1982, *J. Geophys. Res.* **87**, 9205.

Luhmann, J. G., Russell, C. T., Scarf, F. L., Brace, L. H., and Knudsen, W. C.: 1987, *J. Geophys. Res.* **92**, A8, 8545.

Mahajan, K. K. and Kar, J.: 1988, *Space. Sci. Rev.* **47**, 303.

Mahajan, K. K., Paul, R., and Kar, J.: 1988, *Ind. J. Radio Space Phys.* **17**, 50.

Mahajan, K. K., Mayr, H. G., Brace, L. H., and Cloutier, P. A.: 1989, *Geophys. Res. Letters* **16**, 759.

Mariner Stanford Group: 1967, *Science* **158**, 167.

Marubashi, K., Grebowsky, J. M., Taylor, H. A., Jr., Luhmann, J. G., Russell, C. T., and Barnes, A.: 1985, *J. Geophys. Res.* **90**, 1385.

Mayr, H. G., Harris, I., Niemann, H. B., Brinton, H. C., Spencer, N. W., Taylor, H. A., Hartle, R. E., Hoegy, W. R., and Hunten, D. M.: 1980, *J. Geophys. Res.* **85**, 7841.

McElroy, M. B.: 1968, *J. Geophys. Res.* **73**, 1513.

McElroy, M. B.: 1968, *J. Atmos. Sci.* **25**, 574.

McElroy, M. B.: 1969, *J. Geophys. Res.* **74**, 29.

McElroy, M. B. and Strobel, D. F.: 1969, *J. Geophys. Res.* **74**, 1118.

McElroy, M. B., Prather, M. J., and Rodriques, J. M.: 1982, *Science* **216**, 1614.

Merritt, D. and Thompson, K.: 1980, *J. Geophys. Res.* **85**, 6778.

Miller, K. L. and Whitten, K. L.: 1991, *Space Sci. Rev.* **55**, 165.

Miller, K. L., Knudsen, W. C., Spenner, K., Whitten, R. C., and Novak, V.: 1980, *J. Geophys. Res.* **85**, 7759.

Miller, K. L., Knudsen, W. C., and Spenner, K.: 1984, *Icarus* **57**, 386.

Nagy, A. F., Lui, S. C., Donahue, T. M., and Atreya, S. K.: 1975, *Geophys. Res. Letters* **2**, 83.

Nagy, A. F., Cravens, T. E., Smith, S. G., Taylor, H. A., and Brinton, H. C.: 1980, *J. Geophys. Res.* **85**, 7795.

Nagy, A. F., Cravens, T. E., and Gombosi, T. I.: 1983, in D. M. Hunten, L. Colin, T. M. Donahue, and V. I. Moroz (eds.), Univ. Arizona Press, Chapter 24, p. 841.

Niemann, H. B., Kasprzak, W. T., Hedin, A. E., Hunten, D. M., and Spencer, N. W.: 1980, *J. Geophys. Res.* **85**, 7817.

Osmolovskii, I. K. and Samoznaev, L. J.: 1987, *Kosm. Issled.* **25**, 292.

Parker, L. W., and Holeman, E. G.: 1980, *European Space Agency Report SP-155*, ESTEC, Noordwijk, The Netherlands.

Phillips, J. L. and McComas, D. J.: 1991, *Space. Sci. Rev.* **55**, 1.

Phillips, J. L., Luhmann, J. G., and Russell, C. T.: 1985, *Adv. Space Res.* **9**, 173.

Phillips, J. L., Luhmann, J. G., Knudsen, W. C., and Brace, L. H.: 1988, *J. Geophys. Res.* **93**, 3927.

Rohrbaugh, R. P., Nisbet, J. S., Bleuler, E., and Herman, J. R.: 1979, *J. Geophys. Res.* **84**, 3327.

Russell, C. T.: 1979, in Kennel *et al.* (eds.), *Solar System Plasma Physics*, North Holland Publ. Co., Chapter II.5, Vol. 2, p. 208.

Russell, C. T., Elphic, R. C., and Slavin, J. A.: 1980, *J. Geophys. Res.* **85**, A13, 8319.

Russell, C. T., Luhmann, J. G., Elphic, R. C., Scarf, F. L., and Brace, L. H.: 1982, *Geophys. Res. Letters* **9**, 45.

Russell, C. T., Singh, R. N., Luhmann, J. G., Elphic, R. C., and Brace, L. H.: 1986, *Adv. Space Res.* **7**(12), 115.

Savich, N. A., Osmolovsky, I. K., and Samonznaev, L. N.: 1982, *Proc. 13th Int. Symp. Space Tech. Sci.*, No. 1533, AGNE, Tokyo, Japan.

Schatten, K. H.: 1988, *EOS* **69**, 1662.

Schubert, G., Covey, C., Del Genio, A., Elson, L. S., Keating, G., Seiff, A., Young, R. E., Apt, J., Counselman, C. C., Kliore, A. J., Limaye, S. S., Revercomb, H. E., Sromovsky, L. A., Suomi, V. E., Taylor, F., Woo, R., and von Zahn, U.: 1980, *J. Geophys. Res.* **85**, 8007.

Schunk, R. W. and Nagy, A. F.: 1980, *Rev. Geophys. Space Phys.* **18**, 813.

Shinagawa, H. and Cravens, T. E.: 1988, *J. Geophys. Res.* **93**, 11,263.

Slavin, J. A. and Holzer, R. E.: 1981, *J. Geophys. Res.* **86**, 11401.

Spenner, K., Knudsen, W. C., Miller, K. L., Novak, V., Russell, C. T., and Elphic, R. C.: 1980, *J. Geophys. Res.* **85**, 7655.

Spenner, K., Knudsen, W. C., Whitten, R. C., Michelson, P. F., Miller, K. L., and Novak, V.: 1981, *J. Geophys. Res.* **86**, 9170.

Taylor, H. A., Jr., Brinton, H. C., Bauer, S. J., Hartle, R. E., Donahue, T. M., Cloutier, P. A., Daniell, R. E., and Blackwell, B. H.: 1979a, *Science* **203**, 752.

Taylor, H. A., Jr., Brinton, H. C., Bauer, S. J., Hartle, R. E., Cloutier, P. A., Daniell, R. E., and Donahue, T. M.: 1979b, *Science* **205**, 96.

Taylor, H. A., Brinton, H. C., Bauer, S. J., Hartle, R. E., Cloutier, P. A., and Daniell, R. E.: 1980, *J. Geophys. Res.* **85**, 7765.

Taylor, H. A., Mayr, H., Brinton, H., Niemann, H., Hartle, R., and Daniell, R. E.: 1982a, *Icarus* **52**, 211.

Taylor, H. A., Hartle, R. E., Niemann, H. B., Brace, L. H., Daniell, R. E., Bauer, S. J., and Kliore, A. J.: 1982b, *Icarus* **51**, 283.

Taylor, H. A., Mayr, H. G., Niemann, H. B., and Larson, J.: 1985, *Adv. Space Res.* **9**, 157.

Taylor, W. W. L., Scarf, F. L., Russell, C. T., and Brace, L. H.: 1979, *Science* **205**, 112.

Theis, R. F., Brace, L. H., and Mayr, H. G.: 1980, *J. Geophys. Res.* **85**, 7787.

Theis, R. F., Brace, L. H., Elphic, R. C., and Mayr, H. G.: 1984, *J. Geophys. Res.* **89**, 1477.

Torr, M. R., Torr, D. G., Ong, R. A., and Hinteregger, H. E.: 1979, *Geophys. Res. Letters* **6**, 771.

von Zahn, U., Fricke, K. H., Hunten, D. M., Krankowsky, D., Mauersberger, K., and Nier, A. O.: 1980, *J. Geophys. Res.* **85**, 7829.

Whipple, E. C., Brace, L. H., Parker, L. W.: 1983, *Proc. 17th ESLAB Symp. Spacecraft Interactions*, 127, ESA Report SP-198.

Wolff, R. S., Goldstein, B. E., and Yeates, C. M.: 1980, *J. Geophys. Res.* **85**, 7697.

Zwan, B. J. and Wolf, R. A.: 1976, *J. Geophys. Res.* **81**, 1636.

ION DYNAMICS IN THE VENUS IONOSPHERE

K. L. MILLER

Center for Atmospheric and Space Sciences, Utah State University, Logan, UT 84322, U.S.A.

and

R. C. WHITTEN

Space Science Division, NASA Ames Research Center, U.S.A.

Abstract. Dynamics play an important role in defining the characteristics of the Venus ionosphere. The absence of a significant internal magnetic field at Venus allows the ionization to respond freely to gradients in the plasma pressure. The primary response to a gradient in plasma pressure is the nightward flow of the ionization away from a photoionization source on the dayside. The flow is approximately symmetric about the Sun–Venus axis and provides the source of O^+ that maintains the nightside ionosphere during solar maximum. Modelling efforts have generally been successful in describing the average nightward ion velocity. Asymmetric and temporally-variable flow is measured, but is not well described by the models. Departures from axially-symmetric flow described in this paper include ionospheric superrotation at low altitudes and an enhanced flow at high altitude at the dawn terminator. Variability that is the result of changes in the ionopause height induced by changes in solar wind dynamic pressure is especially strong on the nightside. Ion flow to the nightside is also reduced during solar minimum because of a depressed ionopause.

1. Introduction

Beginning with Mariner 2 in 1962, 19 missions have been launched to study Venus (Colin, 1983; Mahajan and Kar, 1988). Until Venera 9 and 10 in 1975, all were fly-bys. The later Venera spacecraft orbited at altitudes above the ionosphere or probed the lower atmosphere. Most carried probes that descended through the atmosphere to the surface. Much was learned about the ionosphere from radio occultation experiments on the early probes. But until the *in situ* measurements of Pioneer Venus, the dynamics of the ionosphere could only be inferred from the shapes of the radio occultation profiles.

Since its insertion into orbit in December 1978, Pioneer Venus has provided an unprecedented opportunity to study the dynamics of the ionosphere of a planet with a substantial ionosphere, but with no strong internal magnetic field. Analysis of the data from the complement of instruments on board the Pioneer–Venus orbiter has broadened our understanding of the plasma processes in the ionosphere of Venus and of the effects of solar radiation and solar wind. Increased availability of data has led to theoretical studies of the physical and chemical properties of the ionosphere. It is evident from these studies that ion dynamics play a key role in defining the characteristics of the ionosphere, especially on the nightside of the planet.

This paper presents a summary of measurements of the ion velocity in the ionosphere of Venus, and a discussion of the mathematical theories that help us understand the measurements. The majority of the ion velocity measurements presented here made by the Retarding Potential Analyzer (ORPA) on board the Pioneer–Venus orbiter space-

Space Science Reviews **55**: 165–199, 1991.
© 1991 *Kluwer Academic Publishers*.

craft. Inferences of superthermal ion flow have also been made by the ion mass spectrometer (OIMS).

Pioneer Venus was inserted into orbit around Venus in December 1978. The periapsis of the orbiter was maintained near 150 km altitude during the first 600 orbits allowing measurements throughout the topside ionosphere. Ionospheric measurements were also made on several later orbits, during the period while periapsis altitude was rising but below the ionopause, and on the nightside, when the ionosphere extended to altitudes above periapsis. Nearly all of the information on ion dynamics at Venus comes from the first 3.5 Venus years of the Pioneer–Venus mission. Since the Pioneer–Venus orbiter is in the ionosphere only near periapsis, the primary data set includes measurements during three dayside and four nightside seasons. The rising periapsis meant that the dayside ionosphere was not sampled after this time. Ion velocity measurements made on the nightside after the fourth season were made only sporadically.

Data from Venus have shown that its ionosphere is strongly influenced by dynamic processes. There is no strong internal magnetic field to constrain the motion of the ionospheric plasma, as there is on Earth. The ionosphere is thus able to respond freely to internal pressure gradients which cause the ionization to flow across the terminator into the nightside.

The absence of a strong planetary magnetic field and the lack of planetary-scale ionospheric rotation lead to very different conditions within the ionosphere of Venus than experienced on the Earth. Large day-night differences are observed in the ionospheric quantities. The influence of the solar wind is strong at the upper boundary of the ionosphere. Characteristics of the ionosphere itself are dominated by solar electromagnetic radiation on the dayside, and its absence on the nightside. As a result, most of the measured quantities exhibit near-axial symmetry about the Sun–Venus axis. Statistical treatments of ionospheric densities by Miller *et al.* (1980, 1984) show the ion density at low altitudes to be characterized by three regions (Figure 1). The ion dynamics in each of these regions also have unique characteristics. The dayside, extending to a solar zenith angle of 80°, is characterized by an ion density which is controlled by photoionization, and decreases monotonically with increasing solar zenith angle. In this region, the ions are accelerated toward the nightside to speeds approaching the thermal velocity. Nightward of approximately 110°, the ion density is nearly constant with solar zenith angle. Here the supersonic nightward flow slows and becomes chaotic. Separating these region is a 30°-wide region at the therminator where the ion density drops by about an order of magnitude and where ion velocities are greatest.

The ionization source on the dayside is provided by solar photoionization. The ion density near the ionization peak varies with solar zenith angle in the manner expected for local production in a Chapman layer (Cravens *et al.*, 1981). That is, the ion density is, to a first approximation, inversely proportional to the square root of the solar zenith angle. The ion density at higher altitudes is controlled primarily by diffusion (Nagy *et al.*, 1980). On the nightside, however, solar photoionization does not contribute locally to the ionization, and the maintenance of the ionosphere at the observed level requires another source of ionization. In addition, the median ion density being nearly constant

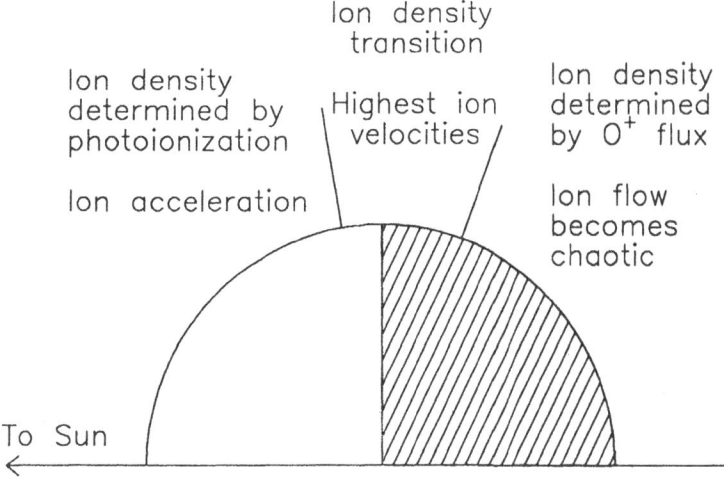

Fig. 1. Approximate location of three regions with similar characteristics of ion density and velocity.

with solar zenith angle requires this source to be acting nearly uniformly on the average throughout the nightside.

The thermal plasma is accelerated to super-sonic velocities as it crosses the terminator region. The flow then slows as it converges on the nightside. The flow of the ionospheric plasma abruptly becomes chaotic at what may be the signature of a recompression shock in the vicinity of 140° solar zenith angle. Although the mean flow is approximately axially-symmetric about the Sun–Venus axis, asymmetries exist between the dawn and dusk hemispheres which influence the characteristics of the ionosphere.

As with many other ionospheric properties that have been measured at Venus, a dominant characteristic of ion velocity in the ionosphere of Venus is variability. There are significant variations in the measured velocity on a time-scale less than the orbital period of the spacecraft, and even within the tens of minutes that Pioneer Venus is in the ionosphere near periapsis. This variability is reflected in the variability of ion density, especially on the nightside of Venus.

The first objective of this paper is to describe the physical relationships that can be formulated mathematically to describe the ion dynamics. Second, the mean characteristics of ion flow in the Venus ionosphere will be described. This mean state as determined by observations may not necessarily correspond to an average of perturbed states of the ionosphere. It is, however, valuable in that it represents conditions that are most likely to occur in the ionosphere. Third will be a description of the modelling studies that have addressed the question of Venus ion dynamics. Fourth, some of the main asymmetries and deviations from the mean flow will be described. This includes the observed dawn-dusk asymmetry in the average ion velocity, and the temporal variability that results from changes in solar activity and solar wind pressure.

2. Theory

2.1. CONSERVATIONS EQUATIONS

In treating ionospheric dynamics, it is necessary to investigate the physical laws which govern the production, loss, heating, and transport of ions and electrons; that is, the laws of momentum, energy and mass conservation, of which Equations (1)–(3) are mathematical statements for steady-state conditions.

Momentum

$$\rho_i \mathbf{v}_i \cdot \nabla \mathbf{v}_i + \rho_i \nu_i \mathbf{v}_i + 2\rho_i \boldsymbol{\Omega} \times \mathbf{v}_i + \nabla \cdot \eta_i \nabla \mathbf{v}_i +$$

$$+ N_i e \mathbf{E} + \kappa \alpha_i N_i \nabla T_i + \mathbf{g} \rho_i = -\nabla p_i . \tag{1}$$

Energy

$$\nabla \cdot \kappa \nabla T_i - C_{vi} [\nabla \cdot (\mathbf{v}_i T_i) - \frac{T_i}{3} \nabla \cdot \mathbf{v}_i] = -Q_i + L_i . \tag{2}$$

Mass

$$\nabla \cdot \boldsymbol{\Phi}_i = q_i - l_i \tag{3}$$

with the flux, $\boldsymbol{\Phi}_i$, given by

$$\boldsymbol{\Phi}_i = N_i \mathbf{v} . \tag{4}$$

Because $\mathbf{v}$ is the solution of Equation (1), diffusion has been automatically included in the definition of the flux. Simple in appearance, the solution of these equations can, in general, be obtained only by numerical integration and even then only with many simplifying approximations. For example, because of the very small mass, we can simplify Equation (1) for electrons to

$$\nabla p_e = -e \mathbf{E} N_e . \tag{5}$$

The symbols contained in Equations (1)–(3) have the following meanings:

ρ_i = mass density of the ith charged particle species;
$\mathbf{v}_i$ = bulk flow velocity of the ith species;
ν_i = collision frequency of the ith species;
Ω = planetary angular velocity;
η_i = dynamic viscosity of the ith species;
p_i = gas kinetic pressure of the ith species;
N_i = concentration of the ith species;
T_i = gas kinetic temperature of the ith species;
m_i = mass of the ith ionic species;
k = Boltzmann constant;
g = gravitational acceleration (positive sense if downward);
e = electronic charge;

$\mathbf{E}$ = electric field, including but not limited to, charge separation electric fields;

α_i = thermal diffusion coefficient (e.g., see Schunk and Walker, 1969; Nakada and Sullivan, 1980);

κ = thermal conductivity (see Merritt and Thompson, 1980; Whitten and Knudsen, 1980);

C_{v_i} = specific heat capacity at constant volume of the ith species;

Q_i = heating rate;

L_i = rate of energy loss;

D_i = diffusion coefficient;

$\hat{\mathbf{r}}$ = (inward) radial unit vector;

$\hat{\Theta}$ = tangential unit vector;

H_{p_i} = plasma scale height (ith species);

q_i = production rate of the ith ionic species;

l_i = loss rate of the ith ionic species.

Equations (1)–(3) contain within themselves a formulation of an important mechanism for ion acceleration and flow from the dayside to the nightside: that due to the pressure gradient. Neglecting thermal diffusion and assuming the same flow velocities for all ions (that of the major ion species) and the approximate equality of the number density of the major ion species to the electron density, Equation (1) for the major ion species can be written in two dimensions (we assume axial symmetry about the subsolar–anti-solar axis) as

$$\frac{1}{R}\frac{\partial u^2/2}{\partial \Theta} + w\frac{\partial u}{\partial z} + \frac{wu}{R} + vu + \frac{1}{\rho R}\frac{\partial}{\partial z}\,\eta\,\frac{\partial u}{\partial z} = -\frac{1}{\rho R}\frac{\partial p}{\partial \Theta}\,, \tag{6}$$

$$\frac{\partial w^2/2}{\partial z} + \frac{u}{R}\frac{\partial w}{\partial \Theta} - \frac{u^2}{R} + vw + \frac{1}{\rho R}\frac{\partial}{\partial z}\,\eta\,\frac{\partial w}{\partial z} = -\frac{1}{\rho R}\frac{\partial p}{\partial z}\,, \tag{7}$$

where u and w are, respectively, the horizontal and vertical components of velocity, p_e is the electron gas pressure, η is the coefficient of viscosity given by $\eta = \eta_0 T^{5/2}$ (Braginskii, 1958), and Θ is the solar zenith angle. This equation would hold, approximately, for a non-rotating planet like Venus. The coupled mass and momentum equations, Equations (3), (6), and (7), have been numerically integrated by Whitten $et\ al.$ (1984) (also see McCormick $et\ al.$, 1987) to yield ion velocities. These equations proved to be very difficult to integrate because of computational instabilities. To render the solution more tractable, although simultaneously considerably less accurate than that obtained by Whitten and coworkers, Singhal and Whitten (1987) ignored the terms in Equation (6) containing the vertical velocity component, w. The horizontal was then easily computed frm the modified form

$$\frac{\partial u^2/2}{\partial \mu} = -R(1-\mu^2)^{-1/2}\left[vu + \frac{1}{\rho}\frac{\partial}{\partial z}\,\eta\,\frac{\partial u}{\partial z}\right] - \frac{1}{\rho}\frac{\partial}{\partial \mu}\,(p_e + p_i)\,, \tag{8}$$

where $\mu = \cos\Theta$. To crudely simulate the large-SZA attenuation, Singhal and Whitten

(1987) multiplied the velocity obtained from Equation (8) by a factor sin Θ or equivalently $(1 - \mu^2)^{1/2}$, which provides a crude approximation to the u's computed from the two-dimensional model of Whitten *et al.* (1984). If we now ignore viscosity and assume that the plasma temperature is only weakly dependent upon SZA, Equation (8) can be integrated to yield (see Singhal and Whitten, 1987, for a derivation and discussion)

$$
v_i = (1 - \mu^2)^{1/2} \left\{ \frac{2k}{m} \left[T_p(1) \ln N_i(1) - T_p(\mu) \ln N_i(\mu) \right] - 2R \int_0^{\text{SZA}} v_i v \, d\Theta \right\}^{1/2},
$$
(9)

where T_p is the plasma temperature (the sum of the ion and electron temperatures) and $\mu = \cos(\text{SZA})$. Because the computed bulk velocity acquires no vertical component when approximation (8) is used, it is not attenuated at large SZA. As a consequence, it shows a maximum value at the anti-solar point, a physical impossibility. An approximate resolution of this problem, obtained by introducing a factor $\sin \Theta = (1 - \mu^2)$ in Equation (9), is justified in Section 3.1.

In the application of the momentum equation (Equation (1)), the inertial acceleration $\mathbf{v} \cdot \nabla \mathbf{v}$ needs to be resolved into its components in spherical geometry. Since we consider in this paper only the horizontal components of the velocity, v_i, the inertial acceleration is approximated by

$$
\mathbf{v} \cdot \nabla \mathbf{v} = \hat{\Theta} \, \frac{v}{R} \, \frac{\partial v}{\partial \Theta} + \hat{\mathbf{r}} \, \frac{v^2}{R} \,,
$$
(10)

where $\hat{\Theta}$ is a unit vector in the horizontal direction. Only the first term on the right-hand side of Equation (10) is included in computing v_i; the second term may be significant in increasing the plasma scale height, H_p. The inverse plasma scale height then takes the form

$$
\frac{1}{H'_{p_i}} = \frac{1}{H_{p_i}} \left(1 - \frac{v_i^2}{gR} \right).
$$
(11)

Additional dayside ionization will result if the solar wind particles interact directly with and are absorbed by the atmosphere (Luhmann *et al.*, 1987). In this case, the solar wind ions will undergo charge transfer to the absorbing neutral particles (mainly O and CO_2). The energetic neutral particles so formed can again become ionized via charge transfer, and so on, until they are eventually absorbed in the atmosphere. However, Kar and Mahajan (1987), using PVO magnetometer data, have shown that there is a strong linear correlation between ionospheric magnetization and solar wind dynamic pressure. That is, there appears to be a linear relationship between the solar wind pressure and the induced magnetic field strength. Cloutier *et al.* (1987) have arrived at similar conclusions by constructing a model of ionospheric motions on the dayside; the model was based on the integration of the plasma flow equations. They showed that the steady-state solar wind interaction requires a downward flow of the ionospheric plasma together with the downward transport of the interplanetary magnetic field into the ionospheric region. Hence, one can expect the dayside ionosphere to be magnetized, the intensity depending upon the solar wind dynamic pressure. The result is a considerable degree of screening of the ionosphere from direct interaction with the solar wind particles. Shinagawa and coworkers (Shinagawa et al., 1987; and Shinagawa and Cravens, 1988) have developed one-dimensional models of the magnetization by solving the relevant Maxwell equations. However, their work is outside the scope of the present paper (see Luhman and Cravens, 1991).

Some studies by Whitten and coworkers (Whitten et al., 1984; McCormick et al., 1987) have employed models which included vertical flow. They yielded decreasing velocities with increasing SZA on the nightside because of the inclusion of the vertical component. More detailed studies are in progress and will be reported elsewhere.

Suprathermal electron streams have been observed on the nightside by Gringauz et al. (1979, 1983) and by Knudsen and coworkers (Spenner et al., 1981; Knudsen and Miller, 1985). The source of these streams is unclear although they are undoubtedly partly of solar wind origin. They are sufficient to produce enough ionization to make an observable ionosphere when ion transport from the dayside is shut off by a low ionopause at the terminator (Knudsen et al., 1987; Knudsen, 1988).

2.2. HEATING

Ion motions are, of course, driven by thermal energy sources. These include photo-electron heating, Joule heating, solar wind-induced heating via wave interactions, precipitating electrons (on the nightside) chemical heating of the ions via the exothermic reactions of CO_2^+ and O^+ (Rohrbaugh et al., 1979), and elastic collisions with the hot oxygen component (Gombosi et al., 1980; Knudsen, 1989).

All of them undoubtedly contribute to electron and ion heating in varying degrees which are still unclear. McCormick et al. (1976) computed the photoelectron heating rate by solving a one-dimensional form of the Boltzmann equation. The resulting electron temperatures at high altitude were about 2000 K, roughly one-half to two-thirds that observed by the Pioneer Venus spacecraft (e.g., Miller et al., 1980). If one assumes that the ion gas is heated only by collisions with thermal electrons, the computed ion temperatures are less than 2000 K, again in disagreement with observations (e.g., Miller et al., 1980). Obviously, additional energy sources must be present. This conclusion is reinforced by the observed high temperatures of the nocturnal ion and electron gases, which require the presence of heat sources (Whitten et al., 1986; Singhal and Whitten, 1986). In this context, chemical heating is undoubtedly small and influences the thermal structure of the ion gas only at rather low altitudes ($\sim$ 200 to 250 km on the dayside). The relative contribution of Joule heating, precipitating electrons, and heating via wave interactions is not known. The problem of the missing ion heat source on the dayside may have been solved in part by some recent work by Knudsen (1989). He evaluated the heat supply by collisions of O^+ with hot oxygen atoms (Nagy et al., 1981; Nagy and

Cravens, 1988) and obtained a thermal flux supply of 1×10^8 eV cm^{-2} s^{-1} which he claims to be sufficient to account for the observed dayside ion temperature. The hot oxygen component appears to be an insufficient heat source for the nightside ion gas, however. Because the topic of ion energetics is properly the subject of a separate review (Brace and Kliore, 1991), it is not discussed further here.

2.3. IONIZATION

The principal source of ionization for any planetary atmosphere is solar EUV radiation, although streams of ionizing particles of solar origin may also contribute in a major way. The solar ultraviolet radiation can itself vary by a factor of four over the solar cycle, thus changing the upper ionospheric ion density by a similar ratio since the major neutral species there is atomic oxygen. These variations have a pronounced influence on the nightside ionosphere especially, because they control the height of the ionopause and thus the size of the annulus through which the plasma can flow from dayside to nightside (Knudsen et al., 1987; see Section 5.3).

Numerous observations of the solar EUV spectrum have been made in recent years (e.g., Heroux and Hinteregger, 1978; Hinteregger, 1979, 1981, are representative). As mentioned previously, Oppenheimer et al. (1981) have also inferred EUV variations over solar cycle 21 from observations of He$^+$ in the Earth's upper ionosphere. The principal result of their work is the determination that the variation of EUV irradiance from solar minimum to maximum was by a factor of about four.

2.4. NEUTRAL ATMOSPHERE

The structure of the neutral atmosphere is very important to ionospheric dynamics because of the drag exerted on horizontally moving ions and the strong influence on vertical diffusion. As a result of numerous observations made by the Pioneer Venus Orbiter (PVO) (Keating et al., 1980; Niemann et al., 1980) knowledge of the structure of the upper atmosphere of Venus during periods of solar maximum is quite complete. These observations have been supplemented by observations of the Pioneer Venus entry bus (Von Zahn et al., 1980) during morning conditions near solar maximum. Composition measurements of Niemann et al. (1980) and of Von Zahn et al. (1980) were made with the aid of neutral ion mass spectrometers, while the atmospheric density measurements of Keating and coworkers were obtained from observed orbital changes caused by collisions with thermospheric particles. Unfortunately, there exists no reliable data for solar minimum conditions.

Hedin et al. (1983) have constructed a detailed model of the thermosphere (corresponding to solar maximum) based on neutral mass spectrometer observations of Niemann et al. (1980). The model was constructed with the aid of a spherical harmonic expansion; since the series was truncated at the fifth order, the rapid changes near the terminator are not faithfully represented. Figure 2 shows concentration profiles of the principal constituents at noon and at midnight as predicted by the Hedin model. Figure 3 portrays the approximate variation of the exospheric temperature with solar zenith angle; the 'ripples' in the curve are an artifact of the truncation of the series

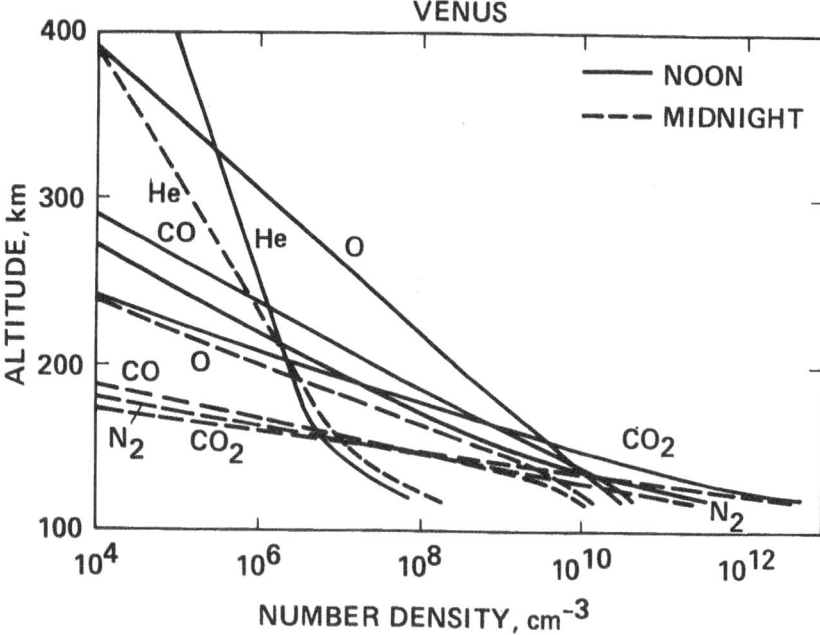

Fig. 2.   The day- and nightside neutral constituent number densities on Venus as predicted by the model
of Hedin *et al.* (1983).

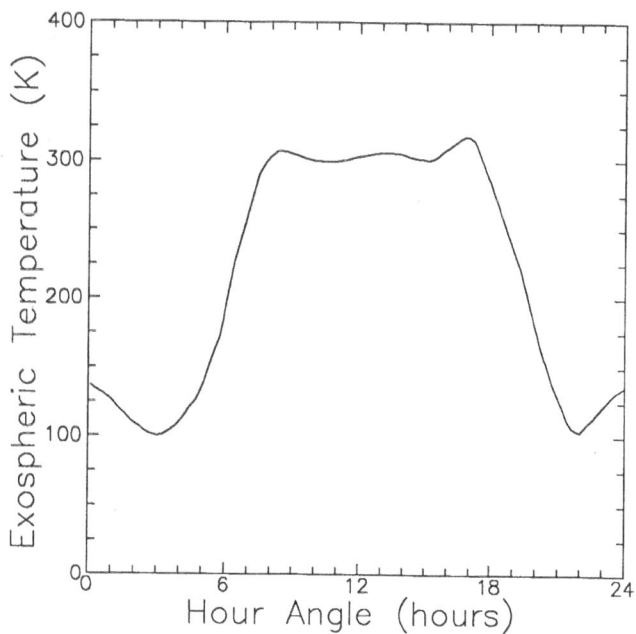

Fig. 3.   The Venus exospheric temperature as a function of solar zenith angle (after Hedin *et al.*, 1983).

expansion. McCormick *et al.* (1987) have developed a slightly different model atmosphere based on ionospheric observations coupled with a two-dimensional ionospheric
model. However, their results are somewhat ambiguous because the presence of a
magnetic field is expected to affect ion diffusion in a manner similar to that of an
increased atmospheric density. The Keating model (Keating *et al.*, 1980) based on
orbital observations and dynamics differs from that of Hedin and coworkers in that the
thermospheric temperatures at the terminator that are predicted by the former are close
to the mean nighttime values. Since there is no good theoretical basis for such variation,
it appears that the Hedin *et al.* (1983) thermospheric diurnal temperature variation is
to be preferred.

Nagy and coworkers (Nagy *et al.*, 1981; Nagy and Cravens, 1988) have suggested that
energetic oxygen atoms are present in the high atmosphere of Venus as a result of the
dissociative recombination of O_2^+. Their early model calculations have proved to be
about an order of magnitude too large, but their more recent values (Nagy and Cravens,
1988) ($\sim 10^3$ cm^{-3}) are very close to the values deduced from Pioneer Venus Orbiter
EUV observations (Paxton and Stewart, 1988). As Knudsen (1989) has shown, the
smaller values are still large enough to substantially influence the dayside ionospheric
thermal balance.

3. Observations

3.1. METHOD OF ION VELOCITY MEASUREMENTS

The method by which the Pioneer Venus ORPA measures the velocities of the ions is
described by Knudsen *et al.* (1979, 1980c). Basically, it is the measurement of the mean
O^+ kinetic energy in three directions on three successive spacecraft rotations. The mean
energy is determined by a least-squares fit of a travelling Maxwellian distribution
function to the measured current/voltage curve.

The ORPA is mounted at an angle of 25° from the spacecraft spin axis. In its most
common mode, it is programmed to record data when the instrument is in the spacecraft
ram direction in one spin period, 45° earlier in the next spin period, and 45° later in
the third spin period. The line-of-sight of the second and third measurements are thus
separated from the ram measurement line-of-sight by about 18.6°. The resolution of
these three measurements into a Cartesian coordinate system provides the vector sum
of the spacecraft and ion velocities. After subtracting the velocity of the spacecraft, the
resultant vector formed from the three line-of-sight velocity measurements corresponds
to the ion velocity of the ambient ions.

The spacecraft speed of approximately 9 km s^{-1} at periapsis insures that the flow is
directed into the ORPA for measurements when the spacecraft is in the ionosphere.
Since the spacecraft spin period is about 12 s, this orbital velocity also means that the
spacecraft has traveled about 100 km of horizontal distance between measurements. In
spite of approximately 200 km of horizontal distance separating the first and last
measurement, the vectors formed from three successive measurements are usually

self-consistent and in a nightward direction. The results suggest that there is a primary flow of ions that is global in scale, in spite of significant small-scale spatial and temporal variations.

The other instrument on board the Pioneer Venus Orbiter that is sensitive to ion velocity is the ion mass spectrometer (OIMS) (Taylor *et al.*, 1980a). The instrument is a Bennett RF spectrometer, and performs a mass analysis by an initial RF voltage that increases the kinetic energy of resonant ions, followed by a retarding voltage that rejects all ions that have not achieved the maximum energy. A servo system is used to tune the instrument for changes in ion-ram velocity and spacecraft potential. Monitoring this servo loop gives a measure of the ram energy of the ions, from which the component of ion drift in the ram direction can be derived.

In addition to the measurement of ionospheric ram velocities by the OIMS, Taylor *et al.* (1979, 1980b, 1981) have also reported the detection of superthermal ions at the top of the ionosphere. The OIMS cannot discriminate between ions of different masses at these velocities, but rather infers the presence of superthermal ions by detecting an ion component with kinetic energies that are too high for the OIMS to measure.

3.2. MEAN ION VELOCITIES

This section is a statistical summary of the ion velocity measurements. A coordinate system is used to present the majority of the data in which all positions and velocities have been projected onto a plane containing the Sun, Venus, and the spacecraft. The two coordinates of this projection are solar zenith angle (SZA) and altitude. This is a useful representation since most ionospheric quantities are nearly axially-symmetric about the Sun–Venus axis.

The method of presentation of ion velocities for use in studies of the Venus ionosphere can itself influence the conclusions drawn from the data. Presenting data in coordinates of SZA and altitude is equivalent to assuming axial symmetry about the Sun–Venus axis. Although this is apparently the predominant symmetry in the Venus ionosphere, non-negligible asymmetries exist. The most apparent is between dawn and dusk hemispheres.

The ionopause and any phenomena that are strongly affected by the solar wind would, because of the motion of the planet along its orbit, tend to have an axis of symmetry that is aberrated by about 5° toward the dusk terminator. But, since most of the processes that affect, or are affected by, ion dynamics occur below the ionopause, the different incident directions of the solar electromagnetic radiation and the solar wind are not considered in the discussion.

The ion velocity, like most of the quantities that have been measured in the Venus ionosphere, is variable on a time-scale shorter than the orbital period of the spacecraft. Figure 4 shows ion velocities measured on orbits 448 and 449, separated by only about 24 hours. Periapsis of the two orbits were at solar zenith angles of 61° and 62°, respectively. Both show nearly constant nightward flow to an altitude of about 400 km. Above 400 km, velocities from orbit 449 deviate from this nightward velocity in a manner that is probably not realistic. The indication is that there are major changes in the direction of flow within the 200-km interval of the velocity measurement. The

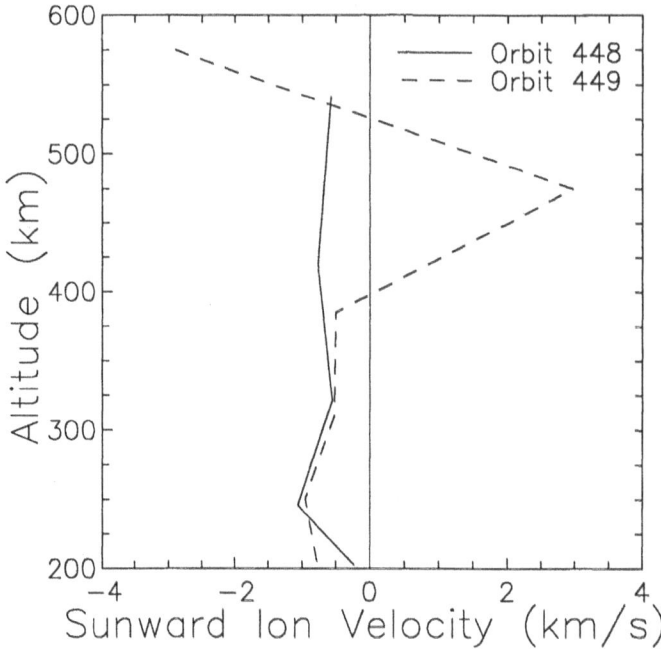

Fig. 4. Horizontal component of O$^+$ velocities in orbits 448 and 449, possibly showing the effect of changes or shears in the ion velocity on a scale that is shorter than that required for making a vector measurement by the ORPA. SZA at periapsis is approximately 62°.

velocities in orbit 448, on the other hand, progress smoothly from the periapsis to ionopause.

Figure 5 shows average ion velocities measured in the outbound leg of the Pioneer–Venus orbit during the first 3.5 years of the Pioneer–Venus mission (Knudsen *et al.*, 1982a). The ionization is flowing generally in an anti-Sunward direction with average speeds that approach 4 km s^{-1} near the terminator. Although significant variability is observed in non-averaged measurements from individual orbits, the average ion velocity varies smoothly at solar zenith angles less than 140°.

Measured line-of-sight velocity components can be quite large on the nightside, and the resolved velocity vectors are not as consistently anti-Sunward as at smaller SZA. Averaging such large, but randomly directed vectors results in a vector field with small magnitudes. Such is the case with ion velocities within 40° of the anti-solar point which appear to be small in Figure 5, but, because of more chaotic conditions which prevail there, can actually represent averages of large velocities. The large velocities are themselves probably in error, since any velocity shear occurring within the 200-km interval of the three measurements would invalidate the result.

The velocity vectors shown in Figure 5 are from the outbound legs of the Pioneer Venus orbit. The ion velocity measured by the ORPA has a consistent downward component on the inbound leg of the orbit, and is nearly horizontal on the outbound

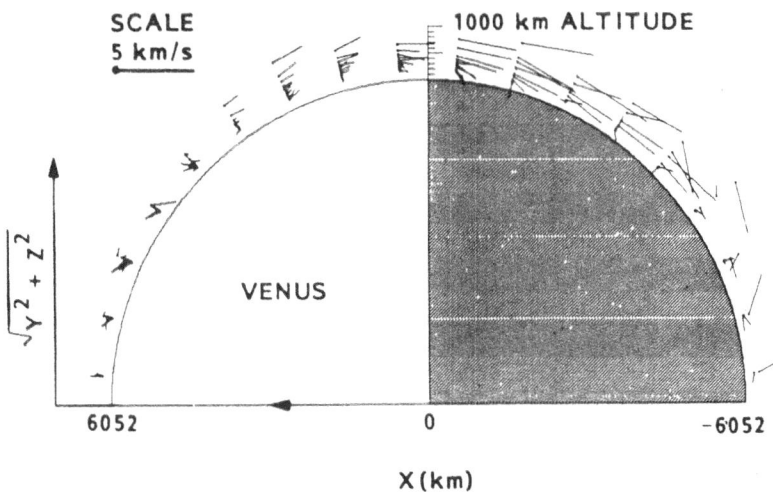

Fig. 5.   Average ion velocities measured in the outbound leg of the orbit during the first 3.5 years of the Pioneer Venus mission (after Knudsen *et al.*, 1982a).

leg. Some downward flux is expected on the basis of theory (Cravens *et al.*, 1984; Cloutier *et al.*, 1987), but the measured velocities are much larger than expected. Although no systematic error has been determined from the ORPA data, apparent large vertical velocities may well be in error as a result of the large angles of attack of the instrument on the inbound leg. The angles of attack are smaller and are within acceptable limits on the outbound leg. For this reason, data from the outbound leg only are used for this statistical presentation. The use of outbound data has the additional advantage that it is from the vicinity of the equator, which reduces the ambiguity when asymmetries in the flow are isolated.

Averages of the horizontal anti-Sunward component of the O^+ velocity field are shown in Figure 6, together with theoretical results. These values are averages of all the outbound data measured by the Pioneer–Venus ORPA. The data have been smoothed to remove random variations which are probably related to velocity shears. The amount of smoothing used for the velocity measurements shown in Figure 6 are represented by 'error bars' shown at selected data points. The error bars indicate the range within which two-thirds of the measurements fall. This is representative of the measurements on the dayside, and to about 140° SZA on the nightside. Nightward of 140°, the velocity vectors become randomly oriented and cannot be accurately represented by average vectors.

In addition to the axially-symmetric bulk flow of the ionosphere, the OIMS (Taylor *et al.*, 1980b) measured regions of superthermal ion flow. The superthermal flow region begins, in most cases, in the region of the ionopause and extends to higher altitudes. Figure 7 shows an example from the outbound leg of an orbit on the nightside of Venus.

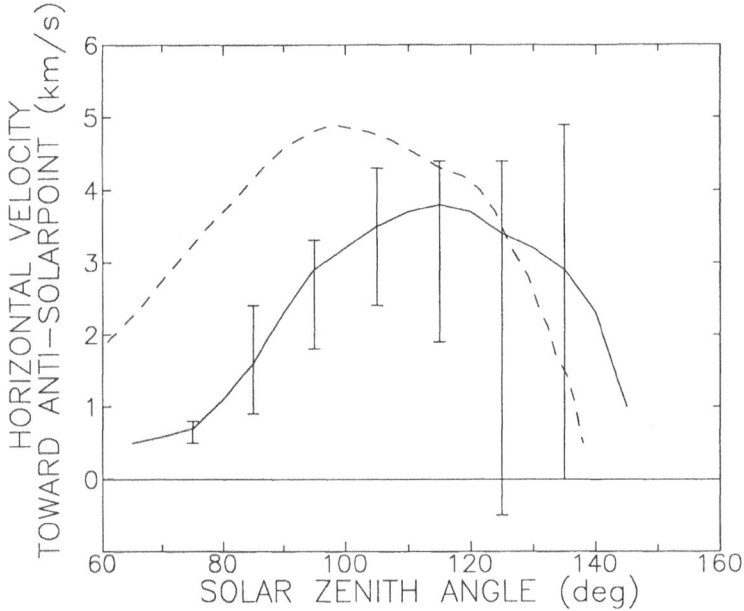

Fig. 6.   Observed and calculated horizontal ion velocities toward the anti-solar point at 400 km altitude. The dashed curve was obtained with the simplified model of Singhal and Whitten (1987). The observed data (obtained from the ORPA experiment on PVO) are the average values, while the error bars show the standard deviation from the mean.

V_f is the component of ion velocity parallel to the spacecraft axis and directed into the instrument, as measured by the OIMS. Superthermal ion flow, marked by the shading, is observed mainly above the ionopause, but is also seen in isolated regions at lower altitudes. Taylor *et al.* (1980b) also reported detached regions of superthermal ionospheric ions above the ionopause, presumably associated with patches of detached ionization or streamers.

3.3. MAGNETIC FIELD EFFECTS

While Venus has quite convincingly been shown to possess no intrinsic magnetic field (e.g., Russell *et al.*, 1980), induced magnetic fields certainly are present within the ionosphere. In fact, the field intensities sometimes exceed 100 nT (Luhmann *et al.*, 1987). Such fields must influence plasma motions because the ratio of thermal plasma pressure to magnetic pressure (usually referred to as 'beta') may be substantially less than unity. In fact, plasma motions and the magnetic field are closely coupled and should be treated as a single system, at least near the subsolar point (Cloutier *et al.*, 1987; Shinagawa *et al.*, 1987; Shinagawa and Cravens, 1988). A review of the characteristics of the magnetic field and its effects on the Venus ionosphere is presented elsewhere (Luhmann and Cravens, 1991). Under quiescent solar wind conditions the ionosphere of Venus is nearly field-free and the modeling techniques described in this paper are thoroughly valid. Under disturbed conditions, however, the plasma motions

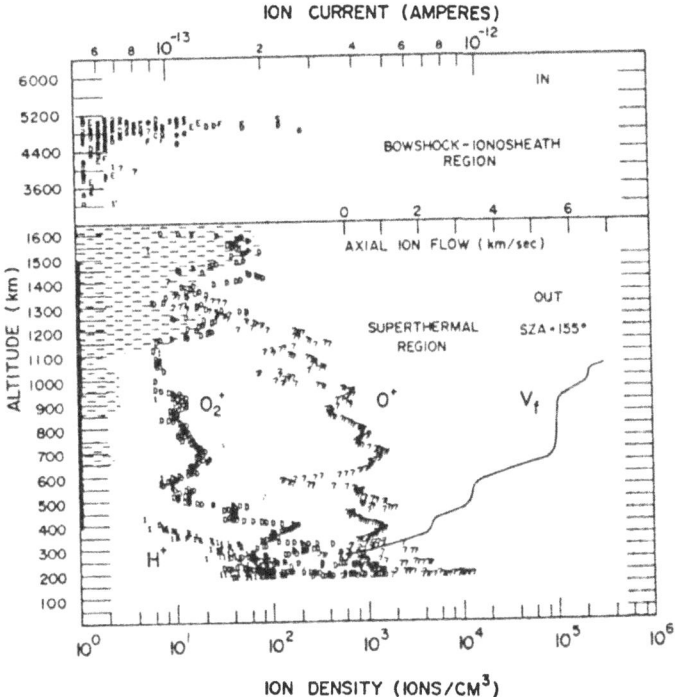

Fig. 7.   Measurements by the OIMS during an orbit on the nightside of Venus. The energetic ion currents in the ionosheath have energies greater than 100 eV. Regions of superthermal ion flow (energy = 30–80 eV) are shown by shading. The width of the shading indicates an uncalibrated relative intensity. V_f is the axial component of the ion flow (after Taylor *et al.*, 1980b).

are substantially altered in the subsolar region (e.g., Shinagawa *et al.*, 1987) and two-dimensional modeling techniques are less applicable. As one approaches the terminator, the induced fields tend to drape themselves around the planet, allowing free horizontal flow, but inhibiting the vertical transport of the plasma. This influence was pointed out by McCormick *et al.* (1987) in trying to reconcile the Pioneer Venus retarding potential analyzer measurements of ion density with the neutral atmospheric structure obtained from the neutral mass spectrometer experiment.

4. Theoretical Results

4.1. Ion velocities and densities

The trans-terminator plasma pressure gradient of the Venus ionosphere may be sufficient to accelerate the plasma to the observed supersonic speeds (Knudsen *et al.*, 1981; Spenner *et al.*, 1981; Whitten *et al.*, 1982, 1984; Elphic *et al.*, 1984b; McCormick *et al.*, 1987; Singhal and Whitten, 1987). Crude two-dimensional simulations of the dynamics were first reported by Chen (1977) and by Chen and Nagy (1978), who extended a one-dimensional model to examine horizontal plasma transport as far as the terminator (SZA = 90°). Whitten *et al.* (1982) used an appropriately modified one-dimensional model to look behind the terminator to SZA = 102.5°. Both sets of investigators found theoretical evidence for very substantial transport of ions from the dayside to the nightside hemisphere as a result of the transterminator plasma pressure gradient.

The next step was the development of genuine two-dimensional models which would predict ion densities far into the nightside hemisphere. Cravens *et al.* (1983) constructed a model which predicted the densities of ion species (O^+, O_2^+, CO_2^+, H^+, C^+, He^+, NO^+) up to SZA = 160°, using a rough empirical fit to the ion velocity field observed by Knudsen *et al.* (1981). The fit was designed such that the largest velocities occurred near the equator and then decreased sinusoidally for larger SZA; the simulated velocity profiles increased linearly with increasing altitude. Although a major advance in predicting nocturnal ion densities (the computed number densities were in very good agreement with observed values), the Cravens model was not self-consistent in the sense that it did not provide for feedback of the computed ion densities into the velocity field.

Theis *et al.* (1984) and Elphic *et al.* (1984b) took a different tack, constructing empirical models of the plasma flow dynamics using measured electron densities and temperatures from the electron temperature probe (OETP) on board Pioneer Venus Orbiter (PVO). Their models consisted of simplified (one-dimensional) solutions to Equation (1); ion temperatures needed to compute sonic velocities were assumed to be equal to the means of the electron and neutral atmosphere temperatures. They also included viscosity in their calculations of the horizontal velocity fields. Their results, shown in Figure 8, indicate the occurrence of high velocities at high altitude near the

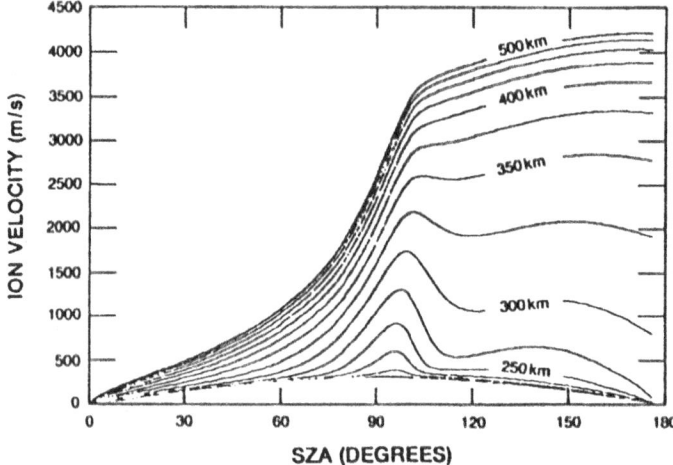

Fig. 8.   Solution to the momentum equation, neglecting plasma viscosity effects, but including ion neutral drag (after Theis *et al.*, 1984).

anti-solar point, an impossible condition. This behavior led them to conclude that viscosity, due either to ion collisions, magnetic fields or both, is essential to slow the plasma flow on the nightside.

In a later modification to those models, Elphic *et al.* (1984b) concluded that viscosity had not been properly included in the calculations. As a result, they concluded that viscosity is not very important to the ion dynamics and suggested that downward advection on the nightside is the dominant process for slowing the plasma. They also suggested the possibility of the occurrence of turbulence as a result of the large velocity shears. Some recent unpublished work by Whitten and Singhal supports this conjecture. Figure 9 shows vertical profiles of he horizontal ion velocity calculated by Elphic *et al.* (1984b). They agree reasonably well with the velocity profiles calculated by Whitten *et al.* (1984) using a different approach.

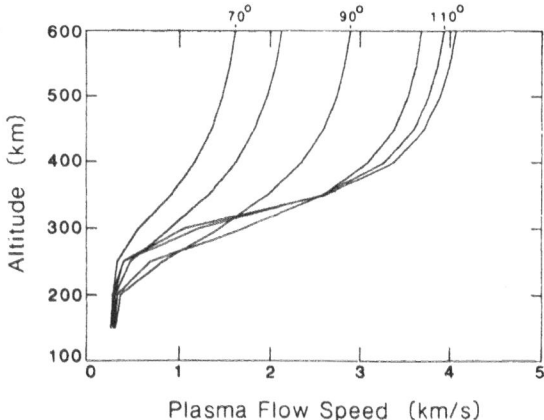

Fig. 9.   Flow speeds for the inviscid solution versus altitude for six solar zenith angle locations near the terminator. Note the strong velocity shear that develops at 300 km altitude just beyond about 90° (after Elphic *et al.*, 1984b).

Whitten and coworkers (Whitten *et al.*, 1984; McCormick *et al.*, 1987) developed a two-dimensional model in which ion number densities (O^+, O_2^+, CO_2^+, CO^+, He^+) and velocities (both horizontal and vertical) were computed. The horizontal velocity component and ion number densities are shown in Figures 6 and 10, respectively. It is apparent from the solid curve in Figure 6 that the computed horizontal component of the ion velocity is in very good agreement with observations; admittedly, the error bars on the nightside data are quite large, accommodating a broad range of allowed velocity predictions.

In the course of varying the parameters which influence the ion density, it was discovered that the computed nocturnal ion densities are very sensitive to the structure of the neutral atmosphere and to the presence of magnetic fields (McCormick *et al.*, 1987). In addition, the ion velocities are sensitive to the ratio between dayside and

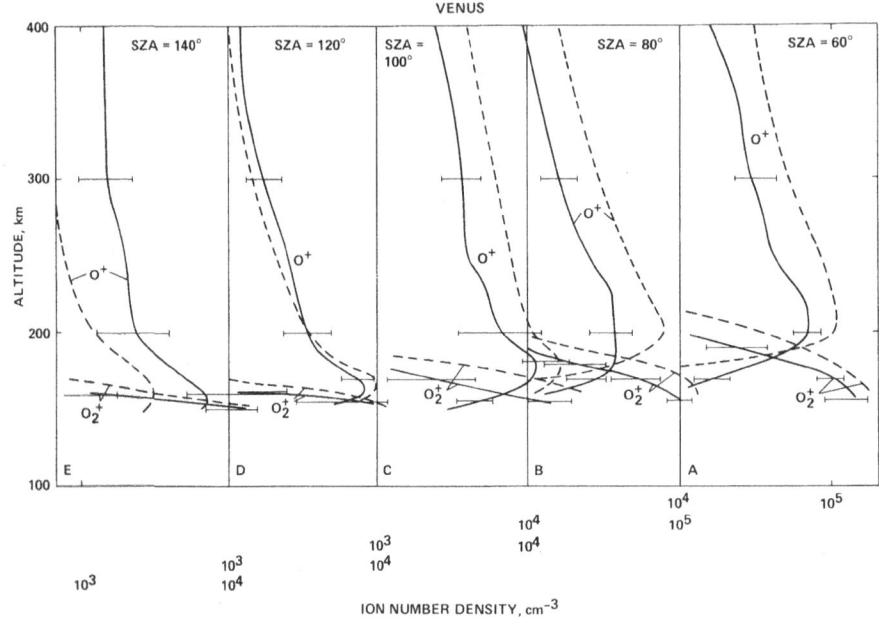

Fig. 10. Observed (solid lines) and calculated (broken lines) O^+ and O_2^+ densities at various solar zenith angles. The observed data (obtained from the ORPA experiment on board Pioneer Venus orbiter) are the median (50% quartile) values while the error bars are illustrative of the 25 and 75% quartiles.

nightside major ion number densities at high altitude. We obtain from Equation (9) (approximately)

$$v^2 = c^2 \ln[N_d/N_n], \qquad (12)$$

where $c = [2kT_p/m]^{1/2}$ is the plasma sonic velocity, and N_d and N_n are the dayside and nightside high altitude major ion number densities, respectively.

As an alternative to the pressure gradient-induced plasma flow, Perez-de-Tejada (1982) suggested that the plasma is dragged along by the solar wind which is coupled to the ionosphere through viscous forces. Unfortunately, for Venus it is not possible to distinguish between flow induced by a pressure gradient and flow induced by the solar wind interaction. It may be possible to do so for Mars (Singhal and Whitten, 1988), which may in turn shed more light on the corresponding situation with Venus. Some unpublished calculations have indicated that the shapes of the vertical profiles of velocity are not very different for the two mechanisms, certainly not within the measurement uncertainties.

4.2. TRANS-SONIC FLOW

It is useful to describe here the mechanism for the transition from subsonic to supersonic flow. In order that the flow become supersonic, it is essential that a converging-diverging nozzle be simulated in some way (e.g., Banks and Kockarts, 1973). In the Venus ionosphere the 'convergence' of the nozzle is supplied by the steady addition of mass (inertial drag) to the flowing ion gas by photoionization of atomic oxygen and by resonant charge exchange of O^+ with the hot atomic oxygen component at high altitude. The divergence on the nightside results from mass removal by vertical transport and ion chemical reactions low in the ionosphere. The nozzle 'throat' is located at or just behind the terminator. Solution of the steady-state momentum equation (1) requires special techniques to integrate through the 'critical points' at which the plasma velocity is equal to the sonic velocity. However, a time-dependent approach like that used by Whitten *et al.* (1984) avoids this problem by allowing the solutions to 'relax' to the steady-state.

4.3. VELOCITY CONVERGENCE ON THE NIGHTSIDE

As discussed previously, Singhal and Whitten (1987) constructed a spectral model of the dynamics of the Venus ionosphere in which they reduced the ion momentum equation to one dimension. Attenuation of the plasma velocity on the nightside was effected by multiplying the computed velocity by a factor $\sin \Theta$. This was in contrast to the suggestion of Theis *et al.* (1984) that ion viscosity is necessary to account for the slowing of the plasma on the nightside. In their calculations, Singhal and Whitten (1987) investigated the effects of viscosity on plasma velocities and found that for reasonable values, the velocity would be reduced by about 10% or less. The plasma velocities obtained by Singhal and Whitten without introducing viscosity are represented by the broken curve in Figure 6. Since it converges toward the anti-solar point, the velocity of the plasma must eventually vanish unless the electron-ion gas is convected out through the wake and lost to the planet. As the plasma velocity decreases through the transonic range, a shock wave should form. There is some evidence for shock wave formation at an SZA of about 140°; specifically, the sudden fluctuations observed in the velocity field. Knudsen *et al.* (1980a) suggested that the rapidly increasing ion temperatures occurring in the same region indicates that the cessation of mean nightward flow is due to a recompression shock in the plasma. However, Whitten *et al.* (1986) showed that the same amount of kinetic energy would be converted into heat from an adiabatic compression as from a recompression shock. Neither the abrupt decrease in velocity on the nightside nor the high ion temperatures at the anti-solar point have been successfully modeled.

4.4. PLASMA TEMPERATURE

Calculations of dayside temperatures of the ion and electron gases on Venus were done initially by Whitten (1969), who used a very simple model of electron heating efficiency and various assumptions about the heat flux from the solar wind. McCormick *et al.* (1976) used greatly improved calculations of the electron heating efficiency to arrive at high altitude dayside electron temperatures (4000 K) that proved to be quite close to those observed by Pioneer Venus Orbiter (PVO). Chen (1977) and Chen and Nagy (1978) extended the one-dimensional model computations as far as the terminator, the results of which were also quite close to the values observed by PVO (e.g., Miller *et al.*, 1980).

The next step was the development of two-dimensional models which could simulate heat transport to the nightside by means of the bulk flow of the plasma. Following a study of the energetics of the nightside ionosphere (Knudsen *et al.*, 1980b), Bougher and Cravens (1984) constructed a model which solved the ion energy equation (Equation (2)) by finite differencing in the vertical and horizontal directions. They used the plasma bulk velocity fields employed previously by Cravens *et al.* (1983) to compute ion number densities on both the day and the nightsides. Results obtained with a solar zenith angle-dependent horizontal velocity patterned after the ion velocities observed by Knudsen *et al.* (1981) are shown in Figure 11. It should be noted that Bougher and Cravens (1984) assumed the absence of nocturnal heat sources for the ion and electron gases. As a consequence, their predicted ion temperatures on the nightside (2000 K) were well below the observed temperatures in the central nightside ionosphere of 5000 K or larger (Miller *et al.*, 1980) but consistent with results of Singhal and Whitten (1986) for the case in which no external source was supplied to the nightside ionosphere.

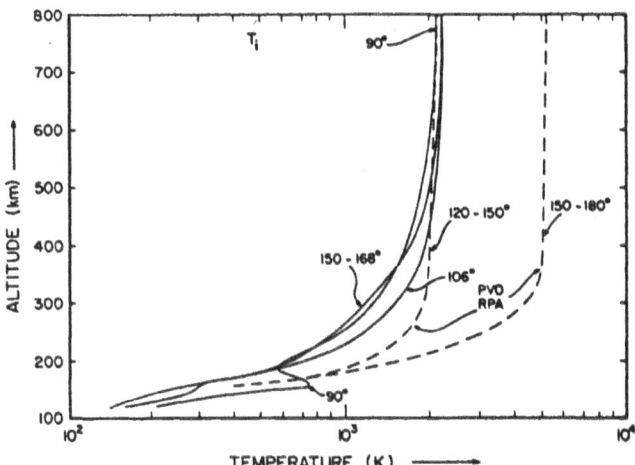

Fig. 11.   Calculated (solid lines) and observed (broken lines) ion temperatures for several solar zenith angles. The measured velocities are from the observational data of Knudsen *et al.* (1981) (after Bougher and Cravens, 1984).

Whitten and coworkers (Whitten *et al.*, 1986; Singhal and Whitten, 1986) developed spectral models which solved the ion and electron energy equations. That is, the solar zenith angle-dependence of the various parameters including the temperature were expanded in Legendre polynomials (axial symmetry was assumed), the argument of which was the cosine of the solar zenith angle. Vertical derivatives were simulated by finite differencing and horizontal plasma velocities were simulated by using the predictions of Whitten *et al.* (1984). Like Bougher and Cravens (1984), Whitten and coworkers found that the predicted nighttime temperatures were much too low (1500 K) and that a nighttime heat source of $1-2 \times 10^{-4}$ ergs cm^{-2} s^{-1} was necessary to satisfy observations. They also found that as the plasma flowed toward the nightside, the ion temperature first fell (to 1000 K with no nocturnal heat source) as a result of adiabatic expansion and then rose due to compression. This effect did not appear in the results of Bougher and Cravens (1984). The computed high altitude ion temperatures obtained by Singhal and Whitten are shown in Figure 12.

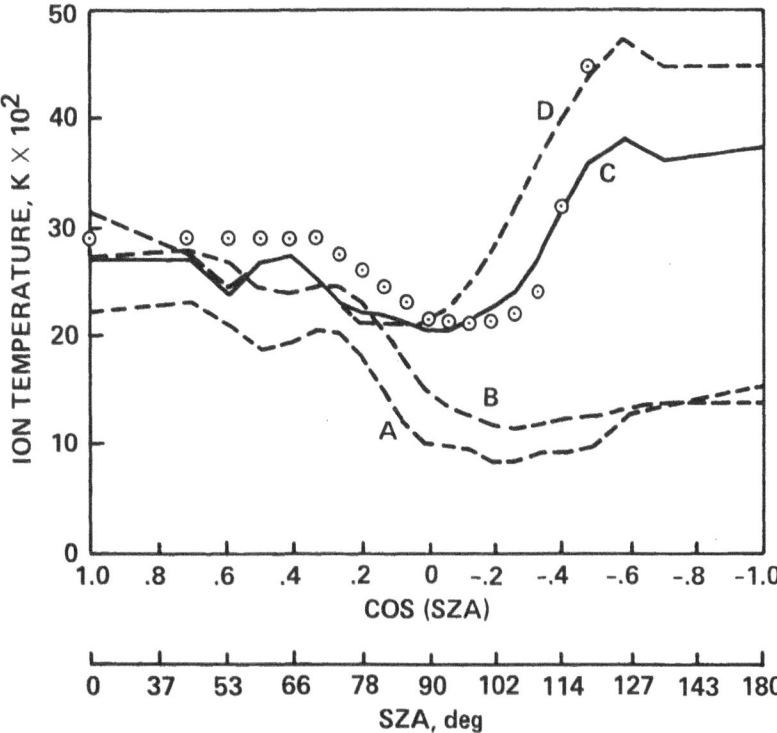

Fig. 12. Computed ion temperatures at high altitudes. Curve A: no external heat source to ion gas; curve B: heat to ions 1.4 [cos (SZA)]$^{1/4}$ on the dayside, 0 on the nightside; curve C: 1.4 uniform on dayside, 1.4/2 uniform on nightside; curve D: 1.4 uniform on dayside, 1.4/1.2 uniform on nightside. Circles denote the observed high-altitude ion temperatures. Heating rate units are eV cm^{-3} s^{-1} (after Singhal and Whitten, 1986).

5. Temporal and Spatial Variability

5.1. DAWN-DUSK ASYMMETRY

Miller and Knudsen (1987) have presented evidence of a dawn-dusk asymmetry in the ion flow. An example of this asymmetry is the observation that the divergence and convergence points are not coincident with the sub-solar and anti-solar ponts and are not opposite each other on the day and night sides. Figure 13 shows horizontal O$^+$ velocities in the ionosphere of Venus projected onto a latitude–longitude grid. The origin

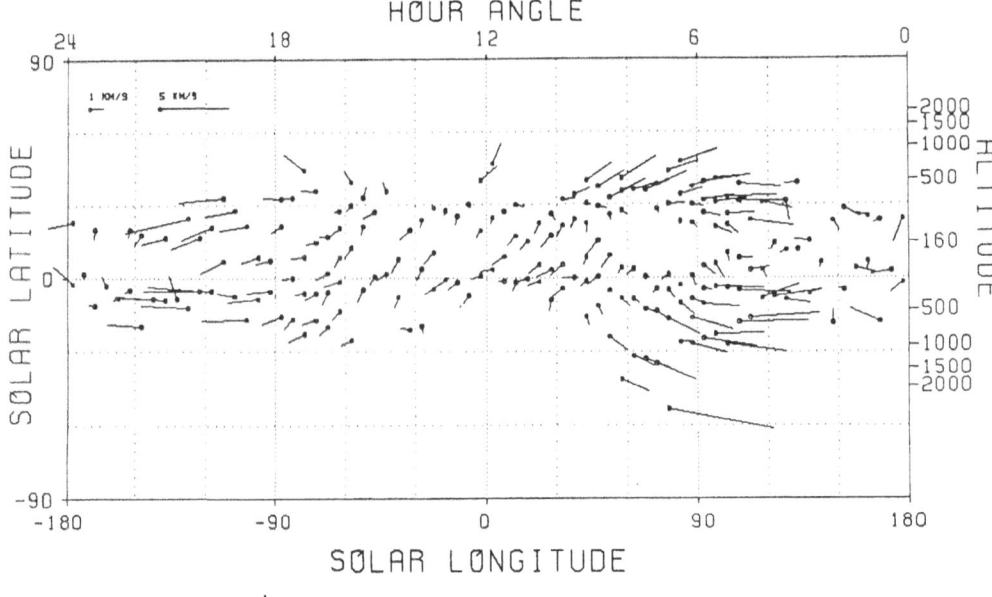

Fig. 13.   Horizontal O⁺ velocities measured by the ORPA. Periapsis of the PVO orbit is at about 15° N latitude.

is defined as the sub-solar point. Longitude is measured to the east of the sub-solar meridian. Because of the retrograde planetary and atmospheric rotation, the dusk terminator is at $-90°$ and the dawn terminator at $90°$. Periapsis occurs at approximately 15° north on this plot, with the inbound leg to the north and the outbound leg to the south. The flow pattern is seen to diverge from the dayside and converge into the nightside. However, the points of symmetry are displaced from the sub-solar and anti-solar points. The center of divergence is east of the sub-solar point, and the center of convergence is west of the anti-solar point. The directions of the displacements of symmetry points is consistent with the addition of a superrotation (or a constant westward velocity) to the axially-symmetric flow.

Figure 14 shows the average zonal component of the horizontal ion velocity at 250 km altitude as a function of the longitudinal angle measured eastward from the sub-solar point. Positive is toward the east. The flow is generally symmetric about the sub-solar meridian, but it can be seen to be offset by a few hundred meters per second toward the west. This dawn/dusk asymmetry can be interpreted as a superrotation of the ionosphere superimposed on the axially-symmetric flow (Miller and Knudsen, 1987). The superrotation is in the same sense as the superrotation of the neutral atmosphere, but is much larger than model calculations of neutral superrotation predict. An explanation of these data based on instrumental or geometric effects that would be interpreted as a 400 m s⁻¹ superrotation has been sought, but unsuccessfully.

The neutral atmosphere is observed to superrotate with a zonal wind speed of about

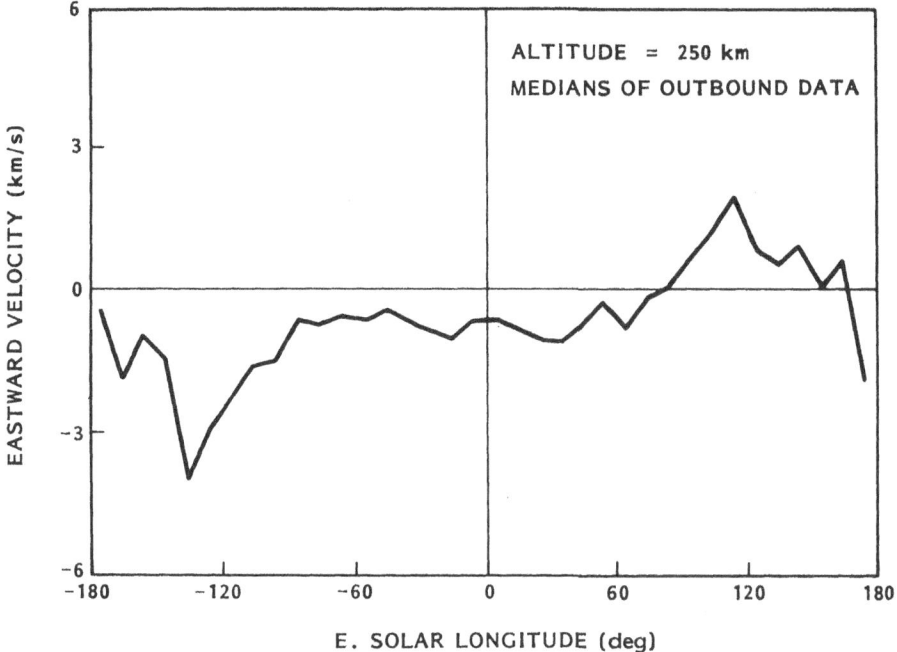

Fig. 14.   Average eastward component of the O^+ velocity at 250 km altitude (after Miller and Knudsen, 1987).

100 m s^{-1} or less at the cloud tops (Schubert, 1983). Although this is much less than the apparent superrotation velocity of the ionosphere, it may be that the motion of the neutral atmosphere serves to give a zonal bias to the ionospheric flow and creates a negative pressure gradient from the equator to the poles, resulting in the conditions necessary for the generation of superrotation. The requirements of cyclostrophic balance under such conditions can be estimated, and gives a superrotation velocity of about 1 km s^{-1} (Elphic *et al.*, 1984b).

Figure 15 shows zonal flow of the ions near the equator. These are average velocities determined from the outbound segments of the orbit only. The altitude scale in Figure 15 is expanded by a factor of four to emphasize the flow regions. The darker area is the region where eastward flow is observed; the lighter area, where westward flow is observed. On the dayside, the regions are seen to overlap in the dawn sector, with the boundary between the flow regimes increasing in altitude toward the sub-solar point. On the nightside, the flow appears to undergo a shock that randomized the flow field. The center of the region between the apparent shocks is seen to be displaced toward dawn by about $10°$.

The higher density and velocity on the dusk side indicate a greater flux into the nightside. This suggests that a higher ion density might be expected in the post-midnight sector. However, this is not observed. In fact, the only obvious asymmetry in the nightside that can be attributed to superrotation of the ionosphere is the offset of the

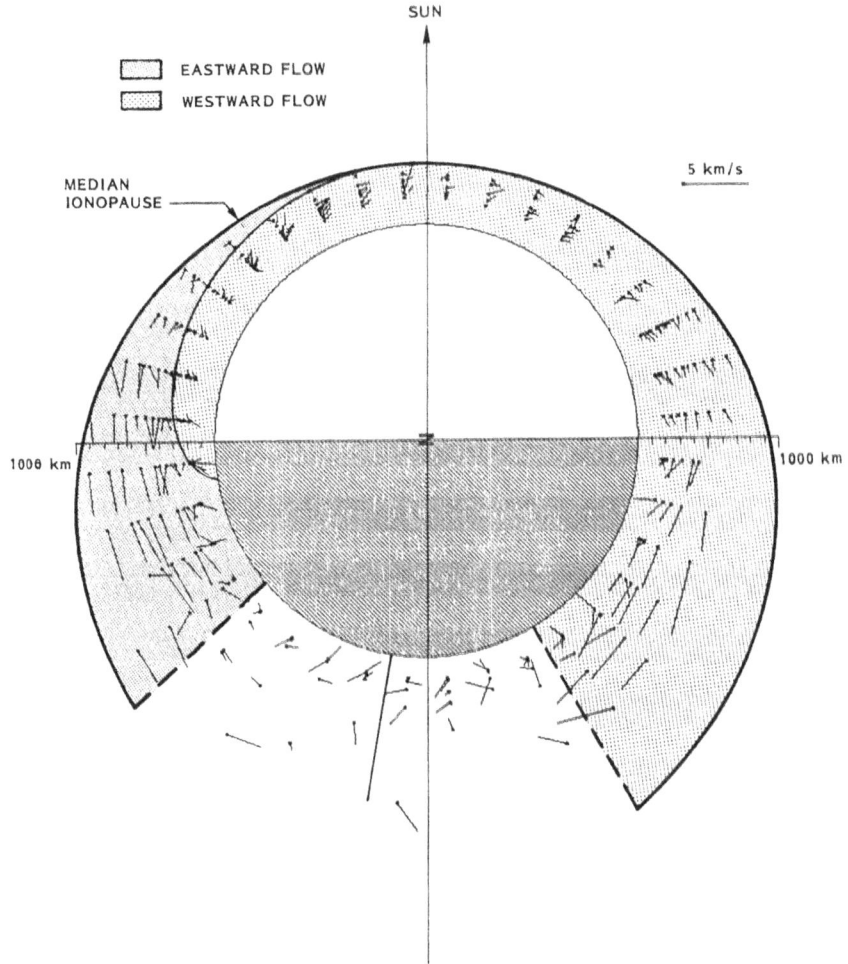

Fig. 15. Anti-Sunward O^+ velocity averages at $10°$ intervals in longitude. The altitude scale has been exaggerated by a factor of four relative to the planetary radius. The shaded area is approximately defined by the median altitude of the ionopause, with the nightward boundary near the point where the velocities become chaotic. The light and dark shading denote regions where the velocity is predominantly westward and eastward, respectively.

apparent recompression shock, as shown in Figure 15. The other major nightside asymmetry, the relatively high H^+ concentration in the dawn sector, was shown by Mayr *et al.* (1980, 1985) to be caused by changes in the neutral composition resulting from superrotation of the thermosphere.

The measured superrotation component generally decreases with increasing altitude. Figure 16 shows the difference between average nightward velocities compared with calculations based on the momentum equation (Equation (1)). The heavy curve in Figure 16 shows the difference between measured horizontal nightward velocities in the

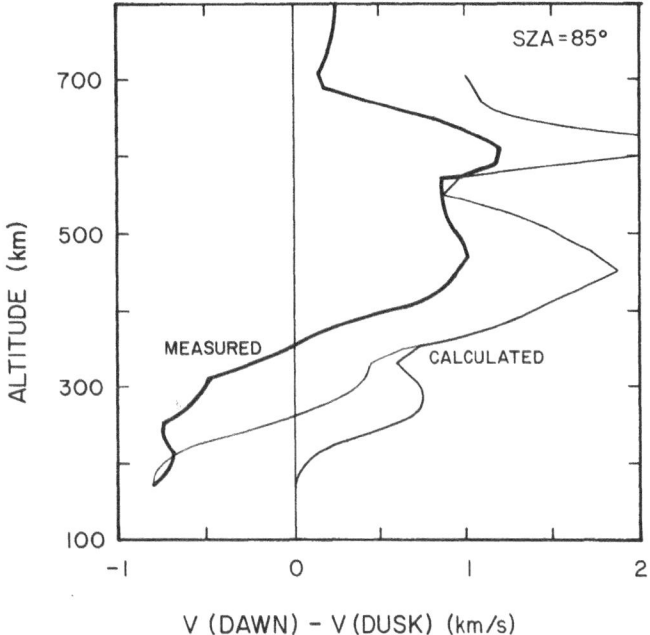

Fig. 16. Differences between average nightward velocity at the dawn and dust terminators. The heavy line is the measured difference at 85° SZA. The light line is the calculated difference based on the momentum equation. The line joining the two curves below 400 km altitude is the calculated difference assuming a 400 m s^{-1} superrotation of the neutral atmosphere (after Miller and Knudsen, 1987).

dawn and dusk hemispheres at 85° SZA. The difference is negative below about 400 km, indicating superrotation, but positive above 400 km, which indicates a prograde rotation component. This is not actually a superrotation, but a dawn-dusk asymmetry in the in velocity.

It must be said here that the possibility that the superrotation component is a systematic measurement error has not been entirely ruled out. The ram direction of the spacecraft is toward the southwest during periapsis, and thus has a component in the direction of the measured superrotation. At the time of this writing, nothing has been found in the ORPA measurement that would result in an incorrect superrotation velocity component.

Since the internal magnetic field of Venus is negligible or non-existent, and induced fields are generally small, plasma flow is primarily in response to a gradient in the plasma pressure. As a first approximation, Elphic et al. (1984b) described the velocity using a simplified one-dimensional steady-state momentum equation coupled with measurements of ionospheric densities. Miller and Knudsen (1987) used ion concentration and temperature measurements by the ORPA (Miller et al., 1984), together with the neutral model of Hedin et al. (1983) to evaluate this equation. The cumulative contribution of advection was approximated by beginning the calculation at the location of the median

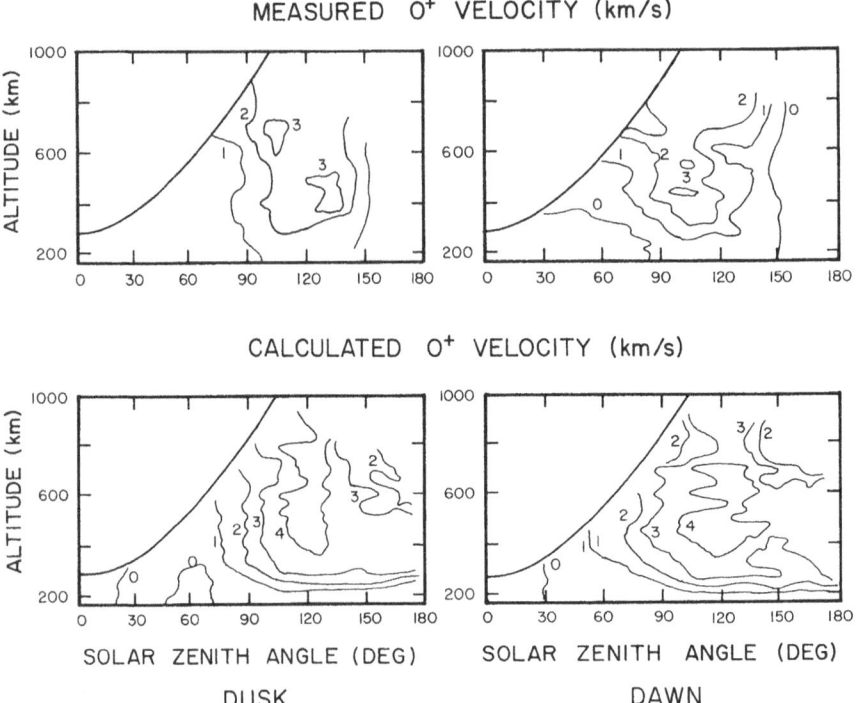

Fig. 17.   Measured and calculated O$^+$ velocity in the dawn and dusk hemispheres. Calculations were based
on a simplified momentum equation (after Miller and Knudsen, 1987).

ionopause (Knudsen *et al.*, 1982b). The velocity field based on the simplified momentum
equation is compared with the measured average velocity in Figures 16 and 17.

The dayside results are consistent with measurements, both qualitatively and quanti-
tatively. Contours in Figure 17 show a much greater vertical gradient in the velocity on
the dawn hemisphere than on the dusk hemisphere. The calculated results for the
nightside are probably inaccurate, because of, among other things, a strong downward
flow behind the terminator and the measured end of ordered flow at about 140° SZA
(Knudsen *et al.*, 1980a).

The difference in the computed velocities of the two hemispheres shows that the result
of the momentum equation gives a prograde asymmetry that disappears at low altitude.
The retrograde superrotation component at low altitude can be simulated by introducing
a high-speed superrotation in the neutral atmosphere. The effect of collisions with the
neutral atmosphere is dominant below 250 km, but decreases rapidly with altitude and
becomes negligible above about 350 km. Figure 16 compares the measured dawn/dusk
velocity difference at 85° SZA with the difference calculated with and without a neutral
atmosphere that superrotates at a speed of 400 m s^{-1}.

The velocity differences at all SZA are compared with the calculated differences in
Figure 18. It can be seen that the calculations agree quite well with the measurements.

DAWN – DUSK O⁺ VELOCITY DIFFERENCE (km/s)

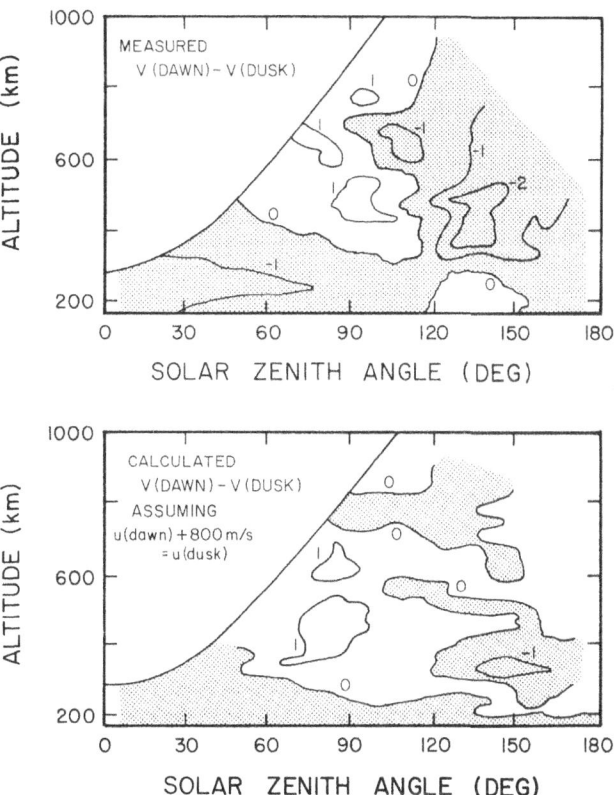

Fig. 18. Measured and calculated velocity differences between dawn and dusk hemispheres. A super-rotation of the neutral atmosphere of 400 m s⁻¹ has been included in the calculation (after Miller and Knudsen, 1987).

As before, calculated velocities on the nightside are probably in error because of influences on the velocity that are not directly related to the pressure gradient and not included in the model. For SZA less than approximately 110°, the calculation reproduces the general features shown in the measurement.

No one has suggested why the high-altitude ion velocities are greater at the dawn terminator than at the dusk terminator, although Miller and Knudsen (1987) showed that the difference is consistent with measured plasma pressure gradients. The asymmetry may be related to a similar dawn-dusk asymmetry in the ionopause height. Phillips *et al.* (1988) showed that the apparent asymmetry in ionopause height is a result of the combination of the aberration of the solar wind velocity due to the orbital motion of Venus and a measurement bias that is the result of the solar wind dynamic pressure being greater on the average during the time when the dusk ionosphere was sampled.

5.2. VARIABILITY IMPOSED BY FLUCTUATIONS IN THE SOLAR WIND

Experiments on Pioneer–Venus have shown large excursions in the height of the ionopause and in the local density, temperature, and velocity on a time-scale comparable to the 24-hour period of the Pioneer–Venus orbit. Ionospheric changes due to solar wind variations must certainly have a much shorter response time (i.e., Wolff et al., 1982). Statistical studies, however, have shown the median ionosphere to be well-behaved, and to vary smoothly with solar zenith angle (Knudsen et al., 1980a, 1981, 1982a, Miller et al., 1980, 1984; Theis et al., 1984). The range of variability in the dayside ionosphere is presented in the statistical treatment of the ionospheric parameters by Miller et al. (1984). Over the three Venus years while the Pioneer Venus periapsis was maintained in the ionosphere, about two-thirds of the measurements of ion density on the dayside were within a factor of two of the median. A similar variability is observed in the dayside ion temperature.

The variability in the nightside ionosphere is more than a factor of three greater than that on the dayside (Knudsen et al., 1986). The change of variability from the dayside to the nightside of Venus is apparently caused by the differing ionization sources on the two hemispheres (Theis et al., 1984). The dayside ionospheric variability is produced primarily by changes in solar EUV radiation (Taylor et al., 1981a, b; Elphic et al., 1984a), while changes in the nightside ionosphere are apparently the result of changes in the ion flux across the terminator (Knudsen et al., 1980a; Cravens et al., 1982).

Trans-terminator plasma flow can only be a significant source for the nocturnal ionosphere if the ionopause height is at least 300 km (Knudsen et al., 1987). Should that height be lowered, as during periods of high solar wind dynamic pressure or low solar activity, the flow will shut off and another source such as suprathermal electrons must be responsible for any observed nighttime ionosphere (Spenner et al., 1981; Cravens et al., 1982; Knudsen et al., 1987).

A simultaneous measurement of the nightside ion density and the flux of ionization across the terminator is not possible with the Pioneer–Venus orbit. Miller and Knudsen (1987) investigated a causal relationship between the ion density on the nightside and the ion flow past the terminator by correlating the variability of the ion density within the Venus nightside ionosphere with the variability of the height of the ionopause.

The relationship between nightside ion density and ionopause height was derived by normalizing the ion density measured on each specific orbit to the empirical model of Theis et al. (1984), and the corresponding ionopause height to the model of Knudsen et al. (1982b). There is essentialy no correlation with ionopause height in the dayside ionosphere between 170 and 200 km altitude. On the nightside, however, the ion density and ionopause height are correlated. Figure 19 is a scatter plot of normalized ionopause heights vs normalized nightside ion density, illustrating this direct relationship.

Cravens et al. (1982) suggested that the flux of ions into the nightside is directly related to the height of the ionopause. The variability in the ionospheric density is thus directly correlated to variations in ion flux into the nightside. Changes in ionospheric pressure

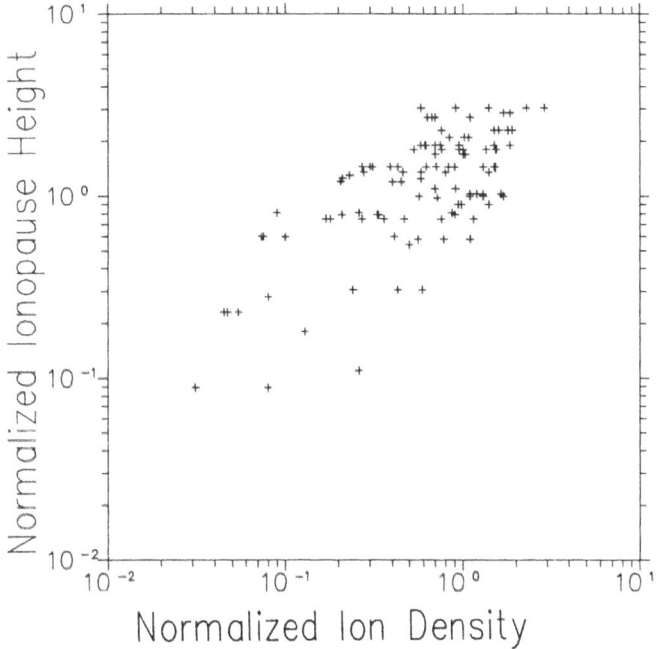

Fig. 19. Ionopause height plotted as a function of nightside ion density. Both quantities have been normalized by model calculations to allow for differences in location of the measurements (see text) (after Miller and Knudsen, 1987).

caused by variations in the solar EUV flux are not expected to be large from one orbit to the next, and changes in the solar wind pressure and the resulting changes in the height of the ionopause were suggested to cause the modulation of the nightside ionization by opening and closing the ion 'nozzle' at the terminator (Knudsen *et al.*, 1980a, 1981; Cravens *et al.*, 1982; Miller and Knudsen, 1987).

5.3. SOLAR CYCLE EFFECTS

Using six-month averages of solar wind data from Feldman *et al.* (1979), Knudsen *et al.* (1987) showed that the solar wind dynamic pressure does not vary significantly over a solar cycle, and, in fact, may be greater at solar minimum than at solar maximum. However, the solar EUV radiation has been observed to undergo very large variations, specifically by as much as a factor of about four (Oppenheimer *et al.*, 1981). As Knudsen *et al.* (1987, 1988) have shown, this variation in EUV irradiance very substantially reduces the ionospheric static pressure, mainly by decreasing the ionization rate of neutral atomic oxygen. As a result, the ionopause height is lowered until the solar wind dynamic pressure is balanced by the static pressure of the ionosphere or the solar wind is directly absorbed by the atmosphere of the planet (Luhmann *et al.*, 1987; Mahajan *et al.*, 1989). Because the dynamic pressure of the solar wind varies approximately as $\cos^2(SZA)$, one can expect to find a balance between dynamic and static pressures as the terminator is approached from the dayside.

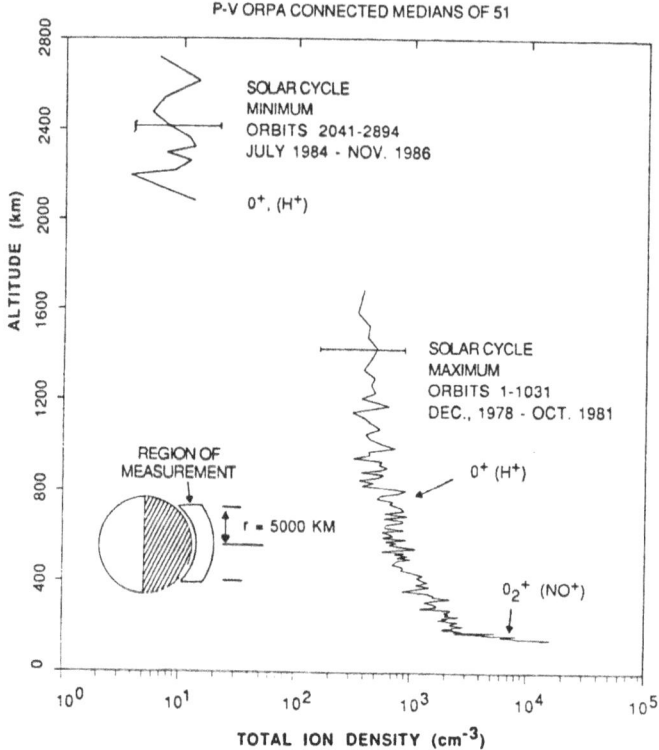

Fig. 20.   Median ion densities on the nightside at solar maximum and at solar minimum. The discontinuity is interpreted as indicating the lack of O^+ transport from the dayside at solar minimum (after Knudsen, 1988).

Figure 20 shows median ion densities on the nightside of Venus measured by the ORPA. The data shown below 1600 km altitude are from the period between December 1978 and October 1981, while periapsis was being maintained in the lower ionosphere. This was also during the maximum of the solar cycle. The data shown at altitudes higher than 2000 km in Figure 20 are from the period between July 1984 and November 1986, near solar minimum. The periapsis altitude had been allowed to rise after 1981, accounting for the difference in altitude ranges of the two sets of data.

Knudsen et al. (1987) interpreted the difference between the nighttime ion densities of solar maximum and solar minimum as being a result of the lowering ionopause affecting the source of O^+ ions. The effect of the lowering of the ionopause at solar minimum is to effectively cut-off the ion flow from the dayside to the nightside. With the source of O^+ ions reduced, the dominant ionization source is impact ionization by low-energy suprathermal electrons (Gringauz et al., 1979, 1983; Knudsen et al., 1985). Thus the low-altitude molecular layer remains about the same, but there is a substantial reduction in the density of the high-altitude O^+ layer (Knudsen, 1988).

6. Summary

Ion velocities measured in the Venus ionosphere show the predominant flow to be away from the sub-solar point, and reaching speeds than 3 km s^{-1} in the terminator region. The axially-symmetric flow has been modelled using the laws of momentum, energy, and mass conservation. The modelling has generally been successful in reproducing the ion velocities at solar zenith angles less than about 140°. The chaotic flow and high ion temperature near the anti-solar point of the planet have not been successfully modelled. It has been suggested that the change from nightward flow to a chaotic flow near 140° SZA is the result of a recompression shock wave at the point where the ion vleocity drops below the sonic velocity. The presence of a shock wave does not, however, explain the high ion temperatures that are observed.

Ion velocities are not strictly axially-symmetric, but, in a statistical sense, show a difference between the nightward flow in the dawn and dusk hemispheres. At altitudes below about 400 km, this asymmetry has the appearance of an ionospheric superrotation of about 400 m s^{-1}. There is also evidence of an ionospheric superrotation in the dawnward offset of the region of ion convergence on the nightside.

Above 400 km, the dawn-dusk asymmetry is in the opposite sense to the low-altitude superrotation component. Although no cause has been suggested for the greater nightward velocity at the dawn terminator, it has been shown to be consistent with a similar asymmetry in the plasma pressure gradients of the two hemispheres. The high-altitude asymmetry may, as is the case with a similar asymmetry in the ionopause height, be the result of differences in the solar wind and solar radiation at the times of the measurements.

Changes in the dynamic pressure of the solar wind have been shown to produce variability in the nightside ionosphere. During solar maximum, this variability is the result of transient increases in the solar wind pressure lowering the ionopause height and restricting the flow of O$^+$ ions to the nightside. The O$^+$ flux is restricted in a similar way during most of the solar minimum when the ionospheric pressure is reduced and allows the ionopause to remain near its lower limiting altitude.

Acknowledgements

This work was supported by NASA contracts NAGW1556 and NAGW869. The initial development of some sections of this manuscript was supported by Lockheed independent research funds.

References

Banks, P. M. and Kockarts, G.: 1973, *Aeronomy*, Academic Press, New York.
Bauer, S. J.: 1973, *Physics of Planetary Ionospheres*, Springer-Verlag, Berlin.
Bougher, S. W. and Cravens, T. E.: 1984, 'Two-Dimensional Model of the Nightside Ionosphere of Venus: Ion Energetics', *J. Geophys. Res.* **89**, 3837.
Brace, L. H. and Kliore, A. J.: 1991, 'The Structure of the Venus Ionosphere', *Space Sci. Rev.* **55**, 81.

Braginskii, S.: 1958, 'Transport Phenomena in a Completely Ionized Two-Temperature Plasma', *Soviet Phys. JETP* **6**, 358.

Chen, R. H.: 1977, 'The Venus Ionosphere', Ph.D. Thesis, University of Michigan, Ann Arbor.

Chen, R. H. and Nagy, A. F.: 1978, 'A Comprehensive Model of the Venus Ionosphere', *J. Geophys. Res.* **83**, 1133.

Cloutier, P. A.: 1984, 'Formation and Dynamics of Large-Scale Magnetic Structures in the Ionosphere of Venus', *J. Geophys. Res.* **89**, 2401.

Cloutier, P. A., Tascione, T. F., and Daniell, R. E., Jr.: 1981, 'An Electrodynamic Model of Electric Currents and Magnetic Fields in the Dayside Ionosphere of Venus', *Planetary Space Sci.* **29**, 635.

Cloutier, P. A., Tascione, T. F., Daniell, R. E., Jr., Taylor, H. A., Jr., and Wolff, R. S.: 1983, in D. M. Hunten, L. Colin, T. M. Donahue, and V. I. Moroz (eds.), 'Physics of the Interaction of the Solar Wind with the Ionosphere of Venus: Flow/Field Models', *Venus*, Univ. Arizona Press, Tucson, pp. 941–979.

Cloutier, P. A., Taylor, H. A., Jr., and McGary, J. A.: 1987, 'Steady State Flow/Field Model of Solar Wind Interaction with Venus: Global Implications of Local Effects', *J. Geophys. Res.* **92**, 7289.

Colin, L.: 1983, in D. M. Hunten, L. Colin, T. M. Donahue, and V. I. Moroz (eds.), 'Basic Facts about Venus', *Venus*, Univ. Arizona Press, Tucson, pp. 941–979.

Cravens, T. E., Kliore, A. J., Kozyra, J. U., and Nagy, A. F.: 1981, 'The Ionospheric Peak on the Venus Dayside', *J. Geophys. Res.* **86**, 11323.

Cravens, T. E., Brace, L. H., Taylor, H. A., Jr., Russell, C. T., Knudsen, W. C., Miller, K. L., Barns, A., Mihalof, J. D., Scarf, F. L., Quenon, S. J., and Nagy, A. F.: 1982, 'Disappearing Ionospheres on the Nightside of Venus', *Icarus* **51**, 271.

Cravens, T. E., Crawford, S. L., Nagy, A. F., and Gombosi, T. I.: 1983, 'A Two-Dimensional Model of the Ionosphere of Venus', *J. Geophys. Res.* **88**, 5595.

Cravens, T. E., Shinagawa, H., and Nagy, A. F.: 1984, 'The Evolution of Large-Scale Magnetic Fields in the Ionosphere of Venus', *Geophys. Res. Letters* **11**, 267.

Elphic, R. C., Brace, L. H., Theis, R. F., and Russell, C. T.: 1984a, 'Venus Dayside Ionospheric Conditions: Effects of Ionospheric Magnetic Field and Solar EUV Flux', *Geophys. Res. Letters* **11**, 124.

Elphic, R. C., Mayr, H. G., Theis, R. F., Brace, L. H., Miller, K. L., and Knudsen, W. C.: 1984b, 'Nightward Ion Flow in the Venus Ionosphere: Implications of Momentum Balance', *Geophys. Res. Letters* **11**, 1007.

Feldman, W. C., Asbridge, J. R., Bame, S. J., and Gosling, J. T.: 1979, 'Long-Term Solar Wind Electron Variations between 1971 and 1978', *J. Geophys. Res.* **84**, 7371.

Gombosi, T. I., Cravens, T. E., Nagy, A. F., Elphic, R. C., and Russell, C. T.: 1980, 'Solar Wind Absorption by Venus', *J. Geophys. Res.* **85**, 7747.

Gringauz, K. I., Verigin, M. I., Breus, T. K., and Gombosi, T.: 1979, 'The Interaction of Electrons in the Optical Umbra of Venus with the Planetary Atmosphere – The Origin of the Nighttime Ionosphere', *J. Geophys. Res.* **84**, 2123.

Gringauz, K. I., Verigin, M. I., Breus, T. K., and Shvachunova, L. A.: 1983, 'On the Prevailing Ionization Source in the Main Ionization Peak of the Nightside Ionosphere of Venus', *Kosmich Issled.* **21**, 746.

Hedin, A. E., Niemann, H. B., Kasprzak, W. T., and Seiff, A.: 1983, 'Global Empirical Model of the Venus Thermosphere', *J. Geophys. Res.* **88**, 73.

Heroux, L. and Hinteregger, H. E.: 1978, 'Aeronomical Reference Spectrum for Solar UV below 2000 Å', *J. Geophys. Res.* **83**, 5305.

Hinteregger, H.: 1979, 'Development of Solar Cycle 21 Observed in EUV Spectrum and Atmospheric Absorption', *J. Geophys. Res.* **84**, 1933.

Hinteregger, H.: 1981, 'Representations of Solar EUV Fluxes for Aeronomical Applications', *Adv. Space Res.* **1**, 39.

Kar, J. and Mahajan, K. K.: 1987, 'On the Response of Ionospheric Magnetization to Solar Wind Dynamic Pressure from Pioneer Venus Measurements', *Geophys. Res. Letters* **14**, 507.

Keating, G. M., Nicholson, J. Y., and Lake, L. R.: 1980, 'Venus Upper Atmosphere Structure', *J. Geophys. Res.* **85**, 7941.

Knudsen, W. C.: 1988, 'Solar Cycle Changes in the Morphology of the Venus Ionosphere', *J. Geophys. Res.* **93**, 8756.

Knudsen, W. C.: 1990, 'Role of Hot Oxygen in Venusian Ionospheric Ion Energetics and Supersonic Antisunward Flow', *J. Geophys. Res.* **95**, 1097.

Knudsen, W. C., Miller, K. L., and Spenner, K.: 1985, 'Improved Venus Ionopause Altitude Calculation and Comparison with Measurement', *J. Geophys. Res.* **87**, 2246.

Knudsen, W. C., Bakke, J. C., Spenner, K., and Novak, V.: 1979, 'Retarding Potential Analyzer for the Pioneer–Venus Orbiter Mission', *Space Sci. Instr.* **4**, 351.

Knudsen, W. C., Spenner, K., Miller, K. L., and Novak, V.: 1980a, 'Transport of Ionospheric O^+ Ions Across the Venus Terminator and Implications', *J. Geophys. Res.* **85**, 7803.

Knudsen, W. C., Spenner, K., Whitten, R. C., and Miller, K. L.: 1980b, 'Ion Energetics in the Venus Nightside Ionosphere', *Geophys. Res. Letters* **7**, 1045.

Knudsen, W. C., Spenner, K., Bakke, J., and Novak, V.: 1980c, 'Pioneer Venus Orbiter Palanar Retarding Potential Analyzer Plasma Experiment', *IEEE Transactions on Geoscience and Remote Sensing* **GE-18**, 54.

Knudsen, W. C., Spenner, K., and Miller, K. L.: 1981, 'Anti-Solar Acceleration of Ionospheric Plasma Across the Venus Terminator', *Geophys. Res. Letters* **8**, 241.

Knudsen, W. C., Banks, P. M., and Miller, K. L.: 1982a, 'A Model of Plasma Motion and Planetary Magnetic Fields for Venus', *Geophys. Res. Letters* **9**, 765.

Knudsen, W. C., Miller, K. L., and Spenner, K.: 1982b, 'Improved Venus Ionopause Altitude Calculation and Comparison with Measurement', *J. Geophys. Res.* **87**, 2246.

Knudsen, W. C., Miller, K. L., and Spenner, K.: 1986, 'Median Density Altitude Profiles of the Major Ions in the Central Nightside Venus Ionosphere', *J. Geophys. Res.* **91**, 11936.

Knudsen, W. C., Kliore, A. J., and Whitten, R. C.: 1987, 'Solar Cycle Changes in the Ionization Sources of the Nightside Venus Ionosphere', *J. Geophys. Res.* **92**, 13391.

Luhmann, J. G. and Cravens, T. E.: 1991, 'Magnetic Fields in the Ionosphere of Venus', *Space Sci. Rev.* **55**, 201.

Luhmann, J. G., Russell, C. T., and Elphic, R. C.: 1984, 'Time-Scales for the Decay of Induced Large-Scale Magnetic Fields in the Venus Ionosphere', *J. Geophys. Res.* **89**, 362.

Luhmann, J. G., Russell, C. T., Scarf, F. L., Brace, L. H., and Knudsen, W. C.: 1987, 'Characteristics of the Marslike Limit of the Venus–Solar Wind Interaction', *J. Geophys. Res.* **92**, 8545.

Mahajan, K. K. and Kar, J.: 1988, 'Planetary Ionospheres', *Space Sci. Rev.* **47**, 303.

Mahajan, K. K., Mayr, H. G., Brace, L. H., and Cloutier, P. A.: 1989, 'On the Lower Altitude Limit of the Venusian Ionopause', *Geophys. Res. Letters* **16**, 759.

Mayr, H. G., Harris, I., Niemann, H. B., Brinton, H. C., Spencer, N. W., Taylor, H. A., Hartle, R. E., Hoegy, W. R., and Hunten, D. M.: 'Dynamic Properties of the Thermosphere Inferred from Pioneer Venus Mass Spectrometer Measurements', *J. Geophys. Res.* **85**, 7841.

Mayr, H. G., Harris, I., Stevens-Rayburn, D. R., Niemann, H. B., Taylor, H. A., Jr., and Hartle, R. E.: 1985, 'On the Diurnal Variations in the Temperature and Composition: A Three-Dimensional Model with Superrotation', *Adv. Space Res.* **5**, 109.

McCormick, P. T., Whitten, R. C., and Knudsen, W. C.: 1987, 'Dynamics of the Venus Ionosphere Revisited', *Icarus* **70**, 469.

Merritt, D. and Thompson, K.: 1980, 'Thermal Energy Transport in the Venus Ionosphere: Classical and Saturated Electron Temperature Profiles', *J. Geophys. Res.* **85**, 6778.

Miller, K. L. and Knudsen, W. C.: 1987, 'Spatial and Temporal Variations of the Ion Velocity Measured in the Venus Ionosphere', *Adv. Space Res.* **7**(12), 107.

Miller, K. L., Knudsen, W. C., Spenner, K., Whitten, R. C., and Novak, V.: 1980, 'Solar Zenith Angle Dependence of Ionospheric Ion and Electron Temperatures and Density on Venus', *J. Geophys. Res.* **85**, 7759.

Miller, K. L., Knudsen, W. C., and Spenner, K.: 1984, 'The Dayside Venus Ionosphere. I. Pioneer–Venus Retarding Potential Analyzer Experimental Observations', *Icarus* **57**, 386.

Nagy, A. F. and Cravens, T. E.: 1988, 'Hot Oxygen Atoms in the Atmospheres of Venus and Mars', *Geophys. Res. Letters* **15**, 433.

Nagy, A. F., Cravens, T. E., Smith, S. G., Taylor, H. A., Jr., and Brinton, H. C.: 1980, 'Model Calculations of the Dayside Ionosphere of Venus: Ionic Composition', *J. Geophys. Res.* **85**, 7795.

Nagy, A. F., Cravens, T. E., Lee, J. H., and Stewart, A. E. F.: 1981, 'Hot Oxygen Atoms in the Upper Atmosphere of Venus', *Geophys. Res. Letters* **8**, 629.

Nakada, M. P. and Sullivan, E. C.: 1980, 'Thermal Diffusion Calculations for the Ionosphere of Venus', *J. Geophys. Res.* **85**, 171.

Niemann, H. B., Kasprzak, W. T., Hedin, A. E., Hunten, D. M., and Spencer, N. W.: 1980, 'Mass Spectrometric Measurements of the Neutral Gas Composition of the Thermosphere and Exosphere of Venus', *J. Geophys. Res.* **85**, 7817.

Oppenheimer, M., Babeu, S., and Brinton, H. C.: 1981, 'EUV Flux Variations During Solar Cycle 21 from AE–E He$^+$ Abundances', *J. Geophys. Res.* **86**, 825.

Paxton, L. J. and Stewart, A. E.: 1988, 'The Hot Oxygen Corona of Venus', *J. Geophys. Res.* (submitted).

Perez-de-Tejada, H.: 1982, 'Viscous Dissipation at the Venus Ionopause', *J. Geophys. Res.* **87**, 7405.

Phillips, J. L., Luhmann, J. G., Knudsen, W. C., and Brace, L. H.: 1988, 'Asymmetries in the Location of the Venus Ionopause', *J. Geophys. Res.* **93**, 3927.

Rohrbaugh, R. P., Nisbet, J. S., Bleuler, E., and Herman, J. R.: 1979, 'The Effect of Energetically Produced O_2^+ on the Ion Temperatures of the Martian Thermosphere', *J. Geophys. Res.* **84**, 3327.

Russell, C. T., Elphic, R. C., and Slavin, J. A.: 1980, 'Limits on the Possible Intrinsic Magnetic Field of Venus', *J. Geophys. Res.* **85**, 8319.

Russell, C. T., Luhmann, J. G., and Elphic, R. C.: 1983, 'The Properties of the Low Altitude Magnetic Belt in the Venus Ionosphere', *Adv. Space Res.* **2**, 13.

Schubert, G.: 1983, in D. M. Hunten, L. Colin, T. M. Donahue, and V. I. Moroz (eds.), 'General Circulation and the Dynamical State of the Venus Atmosphere', *Venus*, Univ. Arizona Press, Tucson, pp. 681–765.

Schunk, R. and Walker, J. C. G.: 1969, 'Thermal Diffusion in the Topside Ionosphere for Mixtures which Include Multiply-Charged Ions', *Planetary Space Sci.* **17**, 853.

Shinagawa, H. and Cravens, T. E.: 1988, 'A One-Dimensional Multispecies Magnetohydrodynamic Model of the Dayside Ionosphere of Venus', *J. Geophys. Res.* **93**, 11263.

Shinagawa, H., Cravens, T. E., and Nagy, A. F.: 1987, 'A One-Dimensional Time-Dependent Model of the Magnetized Ionosphere of Venus', *J. Geophys. Res.* **92**, 7317.

Singhal, R. P. and Whitten, R. C.: 1986, 'A Two-Dimensional Model of the Ionosphere of Venus: Thermal Structure', *Icarus* **67**, 325.

Singhal, R. P. and Whitten, R. C.: 1987, 'A Simple Spectral Model of the Dynamics of the Venus Ionosphere', *J. Geophys. Res.* **92**, 5735.

Singhal, R. P. and Whitten, R. C.: 1988, 'Horizontal Plasma Flow Velocities in the Ionosphere of Mars: A Test Case for the Solar Wind Interaction', *Ind. J. Radio Space Res.* **17**, 42.

Spenner, K., Knudsen, W. C., Whitten, R. C., Michelson, P. F., Miller, K. L., and Novak, V.: 1981, 'On the Maintenance of the Venus Nighttime Ionosphere: Electron Precipitation and Plasma Transport', *J. Geophys. Res.* **86**, 9170.

Taylor, H. A., Jr., Brinton, H. C., Bauer, S. J., Hartle, R. E., Cloutier, P. A., Michel, F. C., Daniell, R. E., Donahue, T. M., and Maehl, R. C.: 1979, 'Ionosphere of Venus: First Observations of the Effects of Dynamics on the Dayside Ion Composition', *Science* **203**, 755.

Taylor, H. A., Jr., Brinton, H. C., Wagner, T. C. G., Blackwell, B. H., and Cordier, G. R.: 1980a, 'Bennett Ion Mass Spectrometers on the Pioneer Venus Bus and Orbiter', *IEEE Transactions on Goescience and Remote Sensing* **GE-18**, 44.

Taylor, H. A., Jr., Brinton, H. C., Bauer, S. J., Hartle, R. E., Cloutier, P. A., and Daniell, R. E.: 1980b, 'Global Observations of the Composition and Dynamics of the Venus: Implications for the Solar Wind Interaction', *J. Geophys. Res.* **85**, 7765.

Taylor, H. A., Jr., Bauer, S. J., Daniell, R. E., Brinton, H. C., Mayr, H. E., and Hartle, R. E.: 1981a, 'Temporal and Spatial Variations Observed in the Ionospheric Composition of Venus – Implications for Empirical Modelling', *Adv. Space Res.* **1**, 37.

Taylor, H. A., Jr., Daniell, R. E., Hartle, R. E., Brinton, H. C., Bauer, S. J., and Scarf, F. L.: 'Dynamic Variations Observed in Thermal and Superthermal Ion Distributions in the Dayside Ionosphere of Venus', *Adv. Space Res.* **1**, 247.

Theis, R. F., Brace, L. H., Elphic, R. C., and Mayr, H. G.: 1984, 'New Empirical Models of the Electron Temperature and Density in the Venus Ionosphere with Application to Transterminator Flow', *J. Geophys. Res.* **89**, 1477.

Von Zahn, U., Fricke, K. H., Hunten, D. M., Krankowsky, D., Mauersberger, K., and Nier, A. O.: 1980, 'The Upper Atmosphere of Venus During Morning Conditions', *J. Geophys. Res.* **85**, 7829.

Whitten, R. C.: 1969, 'Thermal Structure of the Ionosphere of Venus', *J. Geophys. Res.* **74**, 5623.

Whitten, R. C. and Knudsen, W. C.: 1980, 'Simple Models of the Thermal Structure of the Venusian Ionosphere', *Icarus* **44**, 85.

Whitten, R. C., Baldwin, B., Knudsen, W. C., Miller, K. L., and Spenner, K.: 1982, 'The Venus Ionosphere at Grazing Incidence of Solar Radiation: Transport of Plasma to and in the Nightside', *Icarus* **51**, 261.

Whitten, R. C., McCormick, P. T., Merritt, D., Thompson, K. W., Brynsvold, R. R., Eich, C. J., Knudsen,

W. C., and Miller, K. L.: 1984, 'Dynamics of the Venus Ionosphere: A Two-Dimensional Model Study',
   Icarus **60**, 317.
Whitten, R. C., Singhal, R. P., and Knudsen, W. C.: 1986, 'Thermal Structure of the Venus Ionosphere: A
   Two-Dimensional Model Study', *Geophys. Res. Letters* **13**, 10.
Wolff, R. S., Stein, R. F., and Taylor, H. A., Jr.: 1982, 'The Dynamics of the Venus Ionosphere. 1. A
   Simulation of the Solar Wind Compressin of the Upper Dayside Ionosphere', *J. Geophys. Res.* **87**, 8118.

MAGNETIC FIELDS IN THE IONOSPHERE OF VENUS

J. G. LUHMANN

Institute of Geophysics and Planetary Physics, University of California, Los Angeles, CA 90024, U.S.A.

and

T. E. CRAVENS

Department of Physics and Astronomy, University of Kansas, Lawrence, KS 66045–2151, U.S.A.

Abstract. This review surveys the observations of the ionospheric magnetic fields of Venus as observed on the Pioneer Venus Orbiter and the models that have been developed to describe them over the last decade. The models for the 'large-scale' ionospheric field have developed to the advanced stage of one-dimensional, self-consistent, multi-fluid MHD models which provide a detailed picture of the field in the subsolar region for specific upper boundary conditions. In contrast, the models for the small-scale fields and the nightside fields have only reached a rudimentary stage. Much challenging work remains to be done on the origin of the ionospheric flux ropes and nightside ionospheric 'hole' fields. On the whole, the subject of the ionospheric fields would greatly benefit from 3-dimensional global MHD models with self-consistent treatments of the ionosphere.

Table of Contents

1. Introduction

The first *in situ* observations of magnetic fields in the ionosphere of Venus were made by the magnetometer carried aboard the Pioneer Venus Orbiter (PVO) in December

Space Science Reviews **55**: 201–274, 1991.
© 1991 *Kluwer Academic Publishers.*

1978 (cf. Russell and Elphic, 1979a, b). Prior to this discovery, various authors had predicted the existence of magnetic fields in the ionospheres of planets with negligible intrinsic magnetic fields which would be induced by the solar wind interaction (cf. Johnson and Midgely, 1969; Cloutier and Daniell, 1977), but only after the Pioneer Venus Orbiter actually entered the ionosphere of Venus in 1978 were we able to assert that we had indeed found an example of such a planet. Many of the details which had been left open in the early heuristic models of these fields could at last be filled in and tested against the empirical results. However, the road to increased understanding has not been smooth. As is well known, data from spacecraft are generally less than complete in their scope of measurements and in their spatial and temporal coverage of the planet under investigation. This leaves room for divergent interpretations, a situation which has proven both stimulating and a source of contention. Nevertheless, we feel that enough progress has been made over the past decade to warrant a review of the subject. The natural place to begin is with a consideration of our observational knowledge as it now stands.

2. Observations

2.1. General characteristics of the magnetic fields around Venus

The basic phenomenology of the interaction of the solar wind with Venus and the related magnetic topology in the equatorial plane as determined by observations on PVO are illustrated by Figure 1. The solar wind plasma with its frozen-in interplanetary magnetic field (IMF) perceives the dayside ionosphere as an obstacle to its supermagnetosonic flow and is shocked and diverted around it in the magnetosheath. The interplanetary magnetic field, which is effectively frozen into the magnetosheath flow, piles up on the front of the obstacle forming a 'magnetic barrier' as it is carried around it. The solar wind plasma is largely excluded from this barrier region wherein magnetic pressure seems to take up the bulk of the incident solar wind pressure. The geometry of the magnetic field on the nightside is related to the draping of the interplanetary field over the obstacle on the dayside. Some of the draped magnetic field apparently sinks into the wake of the planet to create a comet-like 'magnetotail' consisting of lobes of sunward and anti-sunward directed field whose polarities are governed by the cross-flow IMF orientation.

The location of the obstacle surface depends on the relative magnitudes of the incident solar wind dynamic pressure and the ionospheric plasma pressure (gas or thermal pressure + field pressure). If the ionospheric thermal pressure alone is equal to or greater than the incident solar wind pressure, the obstacle forms near the surface where these two pressures become equal. (The ionospheric plasma pressure thus replaces the dipole field pressure that determines a magnetospheric obstacle or magnetopause.) Figure 1(a) depicts this purely thermal pressure-determined 'ionopause'. In this case a clearly defined boundary exists between the magnetic field and plasma of the magneto-sheath and the ionospheric plasma. The magnetosheath field is for the most part excluded from the ionosphere by ionopause currents. If the ionospheric thermal pressure

a) Low Dynamic Pressure

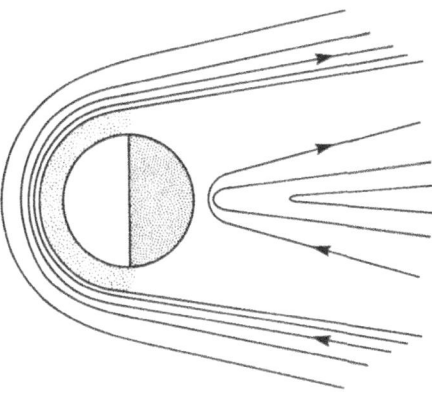

b) High Dynamic Pressure

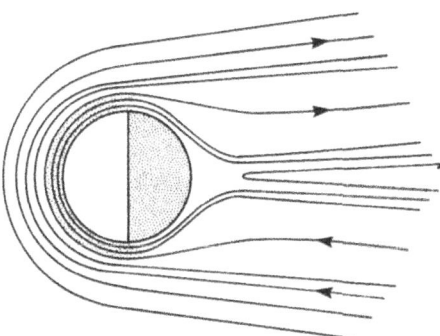

Fig. 1. Magnetic field configurations (not to scale) consistent with observations near Venus for the conditions of low and high solar wind dynamic pressure (compared to ionosphere thermal pressure). (From Luhmann *et al.*, 1980b.) The subsolar obstacle boundary, or ionopause, is typically located at 250–300 km altitude at solar maximum, compared to the planetary radius of 6053 km. The tail current sheet is normal to the plane shown.

is by itself insufficient to provide the obstacle, the interplanetary magnetic field, as we shall later discuss, is taken into the ionosphere (see Figure 1(b)) where it supplements the total ionospheric pressure with magnetic pressure. In this case the difference between the magnetic field in the magnetosheath and that in the ionosphere is less distinct.

Magnetic fields also appear in the ionosphere in a transitional range where the thermal pressure is still high enough to stand off the solar wind, but the ionopause is pushed below ~ 240 km so that collisional diffusion broadens the ionopause current layer, but for the most part we can think of the ionosphere as having two basic states: 'magnetized' and 'unmagnetized'.

'Unmagnetized' states are observed when the solar wind dynamic pressure is significantly less than the maximum ionospheric thermal pressure in the dayside ionosphere. They have the following characteristics:
– The ionopause is located at high altitudes (altitude $z > 300$ km) at all solar zenith angles.
– The ionopause layer, in which the electron density drops by over a factor of 100 from ionospheric values and in which the magnetic field strength dramatically increases with altitude is narrow ($\Delta z \approx 20$ km).
– Large-scale magnetic fields are not present in the dayside ionosphere; rather, small-scale (~ 10 km diameter) magnetic structures (flux ropes) are often present.
– A substantial ionosphere exists throughout most of the nightside (electron densities are typically $\approx 10^4$ cm^{-3} at the peak).
– Localized ionospheric 'holes' exist deep in the nightside. The ionospheric magnetic field in these holes is well-organized and roughly vertical.

'Magnetized' states are observed whenever the solar wind dynamic pressure has a value approaching or greater than the maximum subsolar ionospheric thermal pressure. Ionospheres in the magnetized state have the following characteristics:
– The dayside ionopause is located at low altitudes ($z < 300$ km).
– The ionopause layer is broad ($\Delta z \approx 80$ km).
– Large-scale, nearly horizontal magnetic fields of magnitudes of up to ~ 150 nT are present.
– The ionospheric magnetic field has a characteristic vertical structure with a minimum near 190 km and a maximum near 170 km.
– The nightside ionosphere is weak or absent.

Details of the observations that lead to these descriptions follow, but first it is important for the reader to consider that the PVO magnetometer has been the only magnetometer to have made *in situ* measurements of the magnetic fields in the ionosphere of Venus. Since data from this instrument provided the basis for essentially all of the results described here, one must appreciate the limitations of these observations.

Although PVO still orbits Venus, the ionospheric observations took place only during the approach to the maximum of solar cycle 21. The *in situ* measurements were made along orbital trajectories such as those illustrated in Figure 2. The periapsis of the orbit was maintained near ~ 150 km altitude for the main phase of the PVO mission which lasted about 2 Venus years (1 Venus year ≈ 224 Earth days), but was then allowed to rise above the dayside ionopause. Since the orbital period is ~ 24 hours, this means that the prime ionospheric data were obtained during the first ~ 450 orbits. The small ($15°$) inclination of the periapsis point north of the equator slightly breaks the symmetry of the inbound and outbound observations with respect to the subsolar point. The planet rotates under the orbit so that the dayside ionosphere was probed near periapsis for two ~ 112 day periods during the course of orbits 130–242 and 354–466, while the nightside ionosphere was probed in between.

The observations of the magnetic field vectors are usually cast in the usual Sun–Earth-planet coordinate system where x points toward the Sun, z is northward

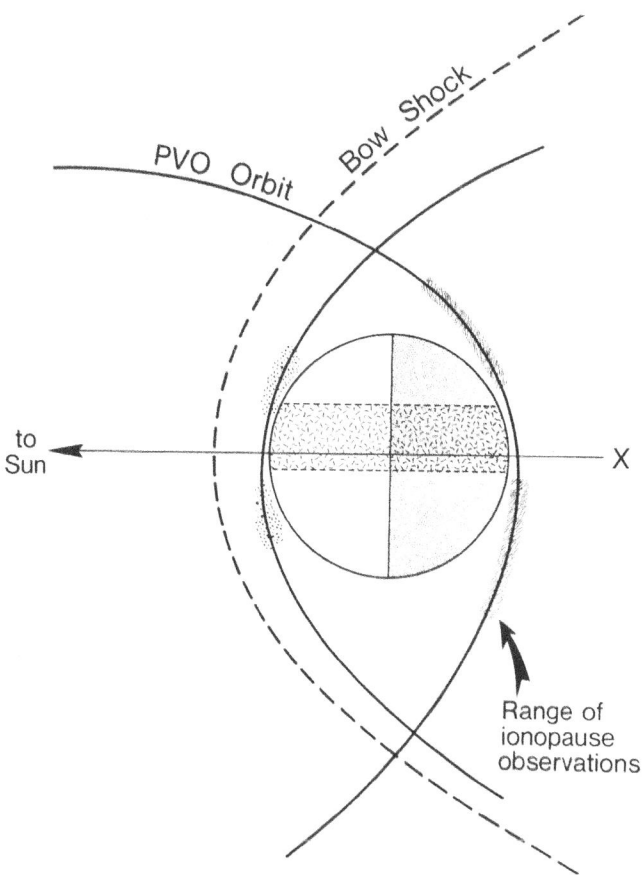

Fig. 2.   Illustration of the geometry of the PVO orbit in the noon-midnight plane for dayside and nightside periapsis. As the planet moves around the Sun, periapsis moves so that the ionosphere in the shaded region is covered every ~ 224 days.

along the perpendicular to the plane of the planet's orbit, and y is in the plane of the orbit pointing opposite to the direction of planetary motion. The maximum temporal resolution of the magnetic measurements is $\frac{1}{4}$ s, although 12 s overlapping averages are often used for analyses which do not require this level of detail. Motion of the spacecraft at a speed of ~ 10 km s^{-1} near periapsis thus implies a low altitude spatial resolution of a few kilometers.

2.2. THE DAYSIDE IONOSPHERE

The early observations (cf. Russell *et al.*, 1979; Elphic *et al.*, 1980) in the dayside ionosphere by the Pioneer Venus Orbiter experiments showed that there was some dichotomy in the appearance of the ionospheric field. On many occasions, the average magnetic field rapidly dropped in magnitude from its value of ~ 50–100 nT at the magnetosheath lower boundary to less than ~ 10 nT within the ionosphere. Moreover,

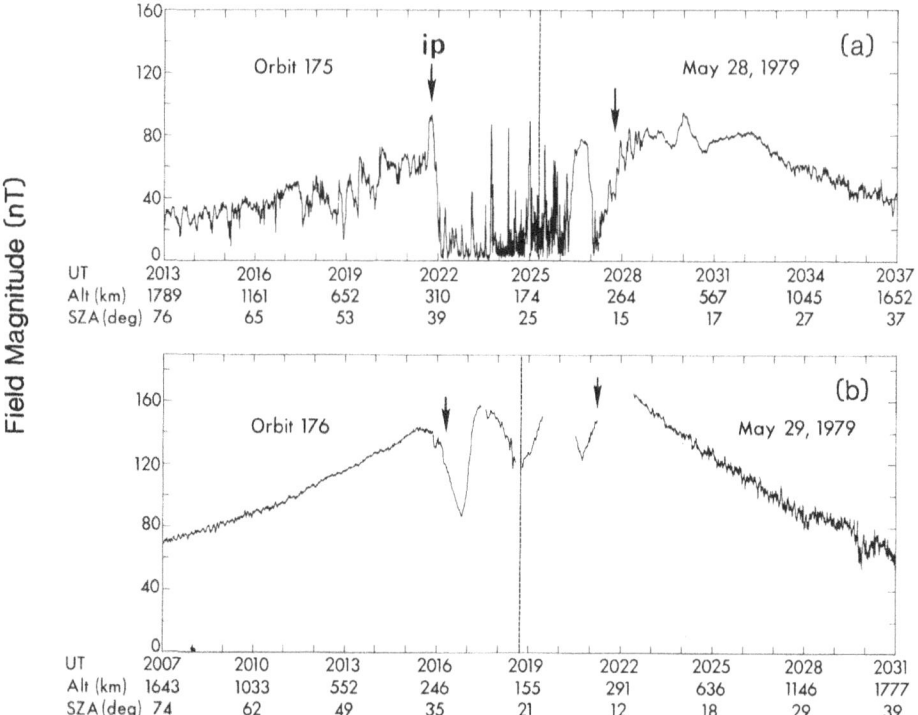

Fig. 3. Examples of time series of magnetic field data obtained along the trajectory of PVO when it flew across the dayside ionosphere of Venus. (From Elphic *et al.*, 1980.) Arrows marked IP show the approximate location of the ionopause, within which the cold, dense ionospheric plasma was found.

as illustrated by the time series of PVO data in Figure 3(a), this low background ionospheric field was punctuated with either rapidly varying or small scale structures. On other occasions, a larger scale field structure appeared which seemed to preclude the formation of the small-scale structures. This large scale structure, which is illustrated in Figure 3(b), consisted of a practically horizontal field with a magnitude of up to ~ 150 nT.

2.2.1. *Large-Scale Fields*

The large-scale ionospheric field was generally found to lie in a direction close to that of the overlying magnetosheath field, in accord with the schematic presented in Figure 1(b) (cf. Luhmann *et al.*, 1980). The statistics of its occurrence, reproduced here in Figure 4, indicated a preference for small solar zenith angles which must be explained by any model of this phenomenon. The maximum strength of the field has a dependence on solar zenith angle (SZA) with an upper envelope roughly consistent with a cosine (SZA) function as shown in Figure 5. Since the magnetosheath field also decreases with increasing SZA, this observation suggested that the ionospheric field is related to the

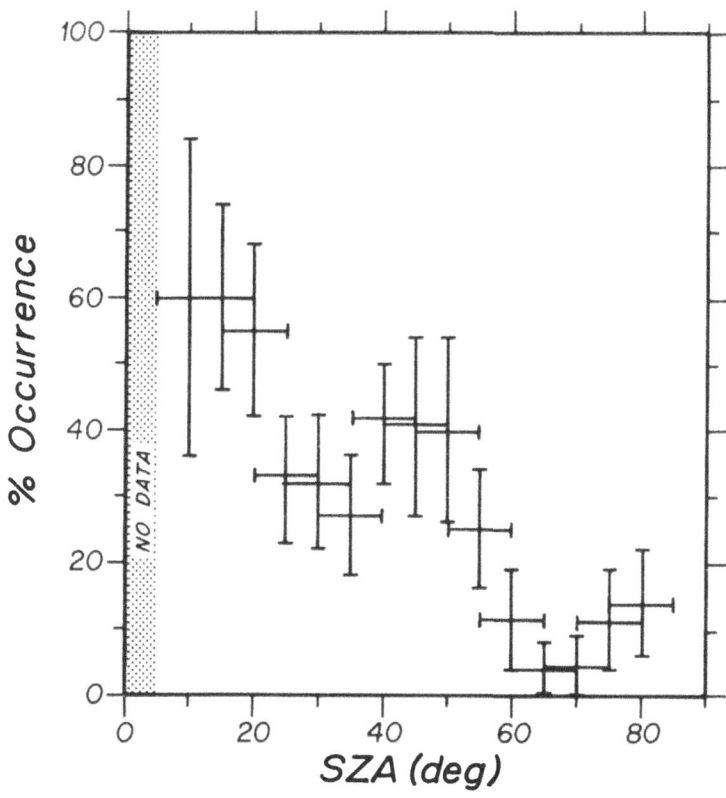

Fig. 4.   Occurrence rate of large-scale fields in the dayside ionosphere as a function of solar zenith angle.
(From Luhmann *et al.*, 1980a.)

magnitude of the overlying magnetosheath field magnitude (cf. Russell *et al.*, 1983; see
also Figure 4 of Luhmann *et al.*, 1980).

Perhaps the most striking feature of the dayside large scale ionospheric field was its
reproducible altitude profile (cf. Russell *et al.*, 1983). When plotted versus altitude, the
field magnitude almost always displayed the characteristics illustrated by the examples
in Figure 6. The key features were the local minimum near 190 km and the local
maximum near 170 km, which were observed irrespective of SZA (see also Russell *et al.*,
1983, Figures 2 and 3). This observation told us that there was a strictly horizontal
current layer associated with the large-scale field. A cautionary note is that the altitude
profiles are aliased by the orbit in the sense that the spacecraft trajectory traverses a
substantial horizontal distance from ionopause entry to exit. Thus, the higher altitudes
are sampled at higher latitudes. Nevertheless, the observed persistence of the
characteristic 'altitude' profile to large solar zenith angles suggested that the gradients
responsible for the structure were indeed vertical and not horizontal.

The dependence of the large-scale field structure on external conditions was
demonstrated by considering the relationship between the observed solar wind pressure

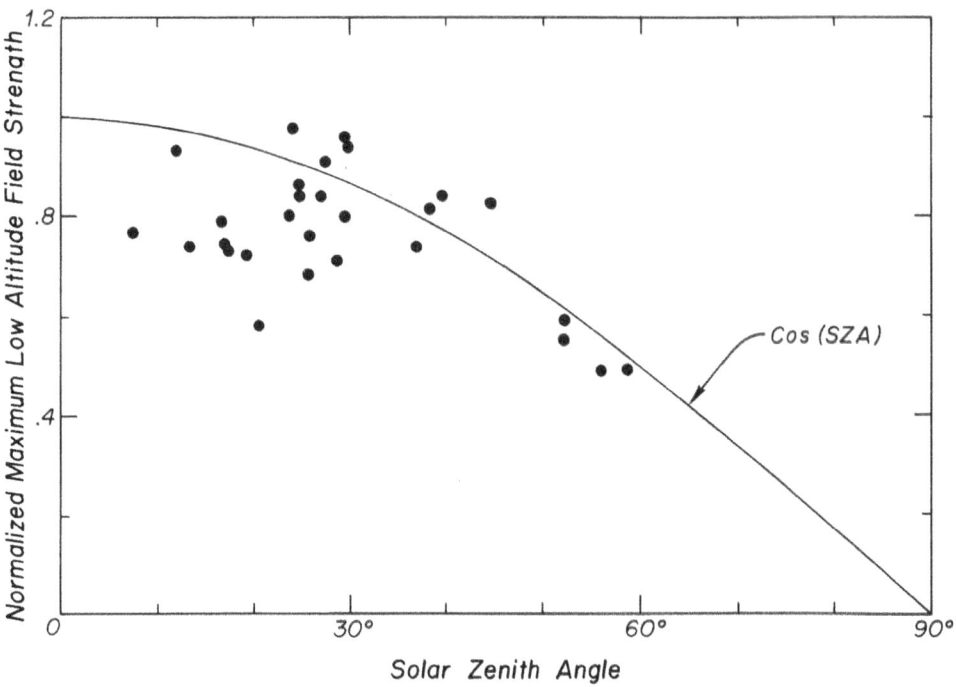

Fig. 5.   The maximum strength of the ionospheric magnetic field normalized to the extrapolated subsolar
field above the ionopause a function of solar zenith angle. (From Russell *et al.*, 1983.)

and the ionospheric field magnitude near periapsis as shown in Figure 7 (see also
Luhmann *et al.*, 1980). Such studies verified the earlier preliminary conclusions that the
large scale ionospheric fields were coupled to the presence of high upstream solar wind
dynamic pressures (cf. Elphic *et al.*, 1980). A later analysis (cf. Luhmann *et al.*, 1987),
from which Figure 8 is taken, showed that the field magnitude could increase enough
to *at least* balance the solar wind pressure when combined with the ionospheric plasma
thermal pressure. In fact, Figure 8 indicates that on occasion the field pressure in the
upper ionosphere exceeded the relatively constant thermal pressure by a factor of 3.
Thus β (the ratio of thermal to magnetic pressure) in the ionosphere can be much less
than 1. The large-scale fields also seemed to occur only when the ionopause was low,
as illustrated by Figure 9. They first became significant for ~ 240 km ionopause height
and became ubiquitous for the lowest ionopause heights of ~ 220 km (cf. Luhmann
et al., 1987). An ionopause altitude lower limit of ~ 220 km is reached as the incident
solar wind pressure approaches and then exceeds the maximum ionospheric thermal
pressure (cf. Brace *et al.*, 1983a, b; Phillips *et al.*, 1984). Since low ionopauses first occur
near the subsolar point where the solar wind incidence is nearly normal, the association
of the large-scale fields with both small solar zenith angles and high solar wind pressures
could be understood. As in Figure 10, one could use the observed solar wind pressure
and a model ionosphere to successfully predict where across the dayside hemisphere

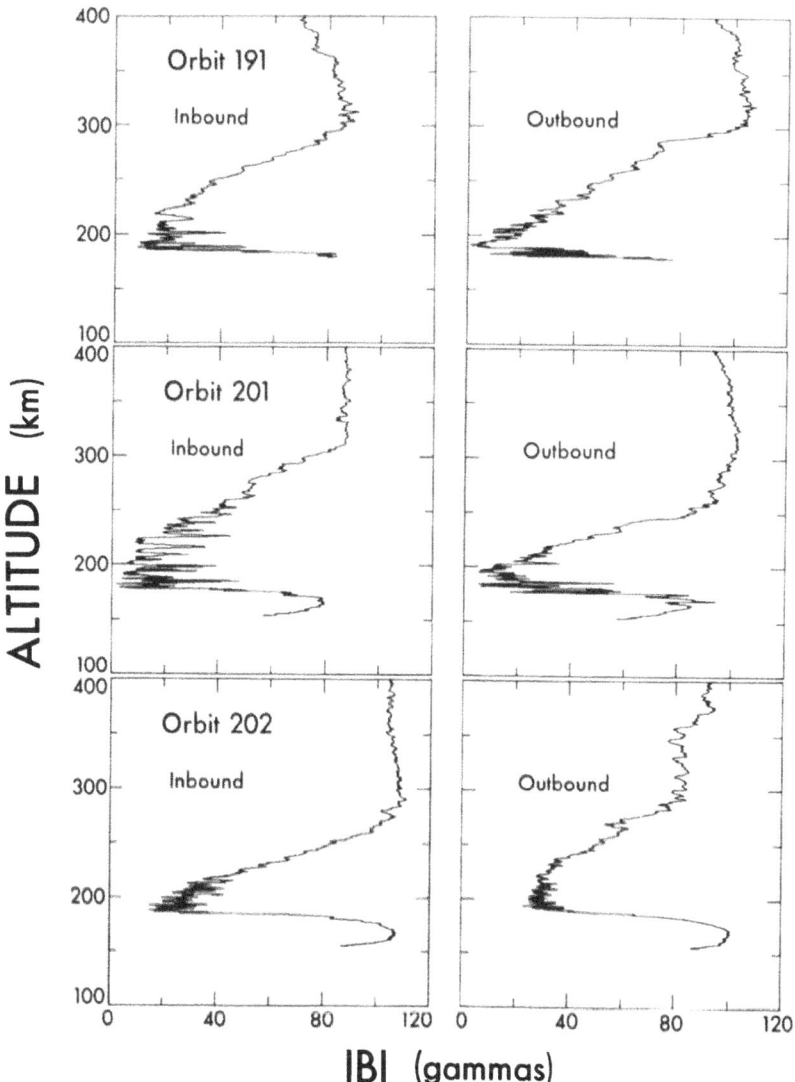

Fig. 6.   Altitude profile of magnetic field strength on three inbound and outbound passes through the Venus
ionosphere when a large-scale field was present. (From Russell *et al.*, 1983.)

the ionopause would be low enough (≤ 240 km) to cause large-scale fields to appear in
the ionosphere.

The large-scale horizontal fields affect the ionospheric plasma by virtue of their
control of both vertical heat conduction and vertical diffusion (cf. Cravens *et al.*, 1980;
Luhmann *et al.*, 1983). When large-scale fields of the greatest magnitudes are present,
the average observed electron temperature and density gradients are largest as illustrated
in Figure 11 (from Luhmann *et al.*, 1987). The ion temperature gradients appear relatively insensitive to the presence of the large scale field.

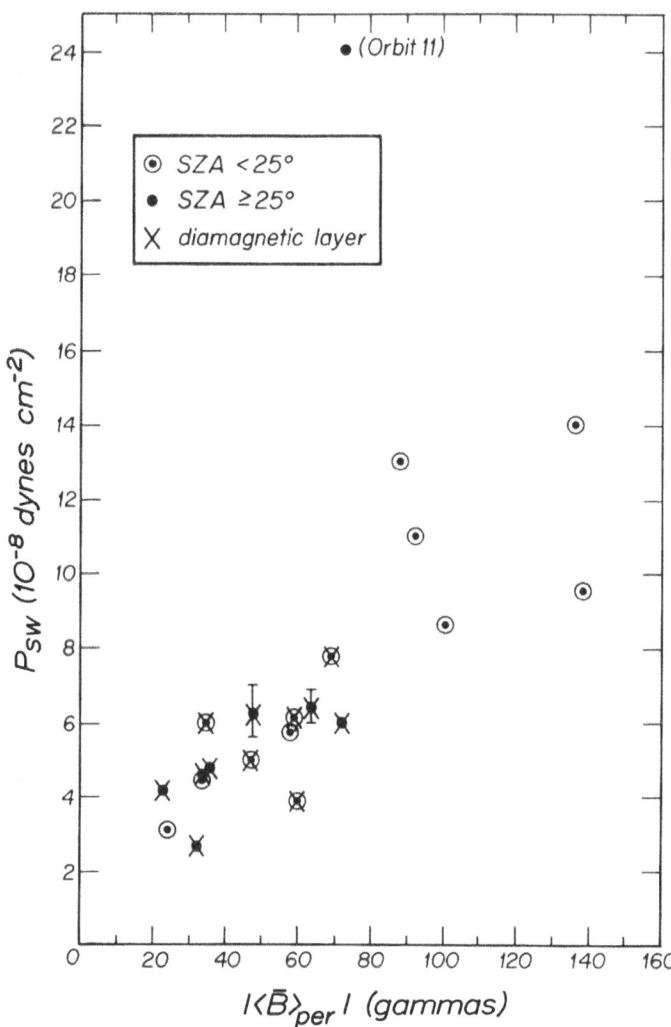

Fig. 7. Dependence of the average field strength at periapsis on solar wind dynamic pressure for a collection of orbits with large-scale ionospheric fields. The × indicates which orbits had a pronounced minimum in the altitude profile of the field near ~190 km. (From Luhmann *et al.*, 1980a.)

2.2.2. *Small-Scale Fields (Flux Ropes)*

Examination of a closeup of the small scale ionospheric field structures, akin to those apparent in Figure 3(a), immediately reveals several characteristic features. As illustrated by Figure 12, the orientation of the magnetic field from structure to structure appears quite random and bears no apparent relationship to that in the overlying magnetosheath. The structures are also largest and most numerous between ~200 km and 160 km altitudes.

The randomness of the field orientations is illustrated more graphically in Figure 13,

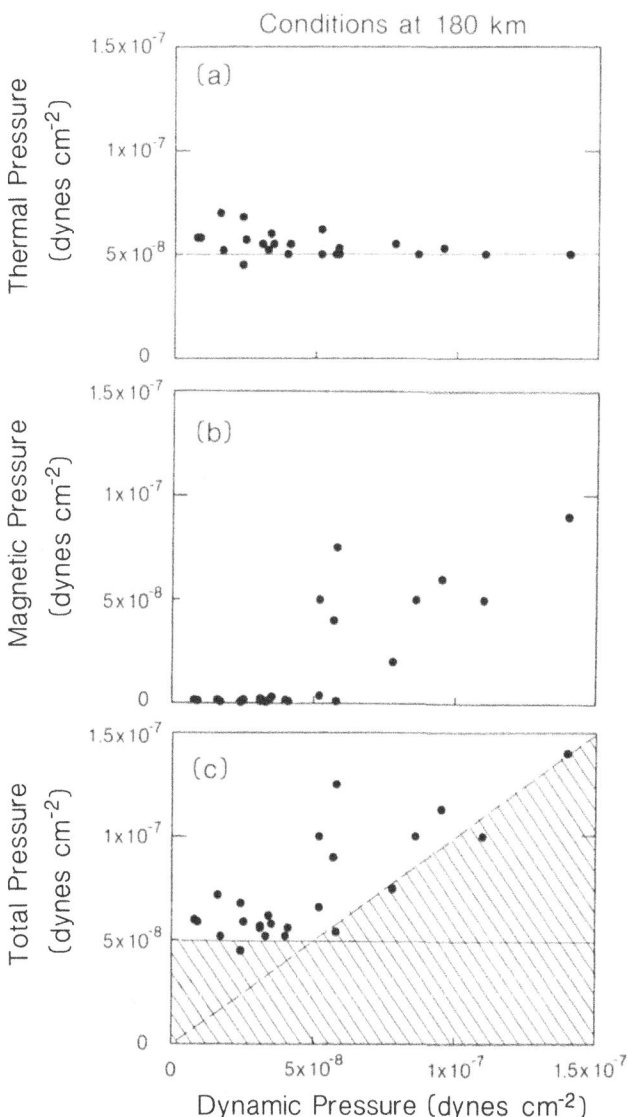

Fig. 8.   Illustration of the division between thermal and magnetic pressure in the ionosphere as a function of solar wind pressure. Note that at 180 km altitude the thermal pressure (a) is insensitive to solar wind pressure but the magnetic pressure (b) increases to keep the solar wind pressure balanced (c). (From Luhmann *et al.*, 1987.)

where the average horizontal components of the low altitude ionospheric fields are compared with the corresponding components of the prevailing low altitude magneto-sheath field. Only when the average low altitude field becomes larger than ~ 10 nT is there a hint of a correlation between the two (cf. Luhmann and Elphic, 1985). This can be understood as the signature of a weak large-scale field structure since the latter is

J. G. LUHMANN AND T. E. CRAVENS

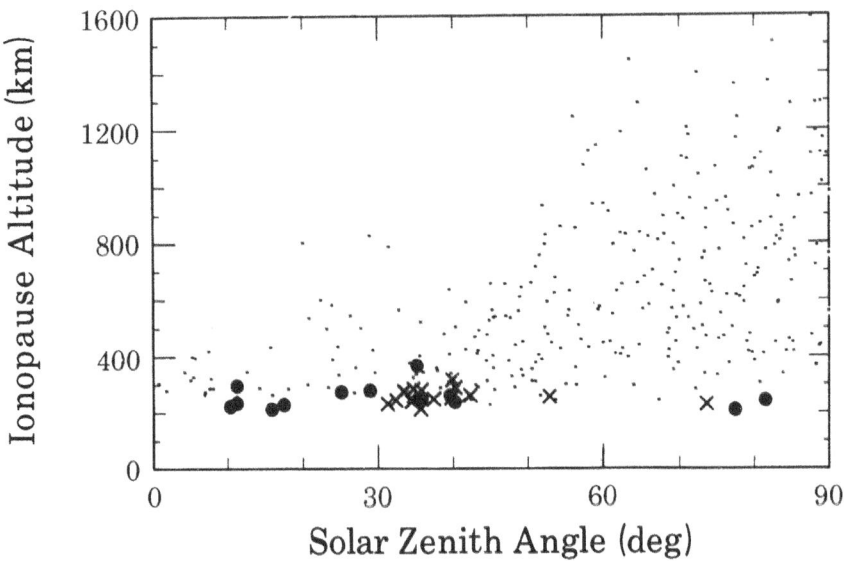

Fig. 9.   Ionopause altitudes observed at various solar zenith angles on the dayside. The larger points identify those orbits on which large scale ionospheric fields were observed on inbound (●) or outbound (×) legs. (From Luhmann *et al.*, 1986.)

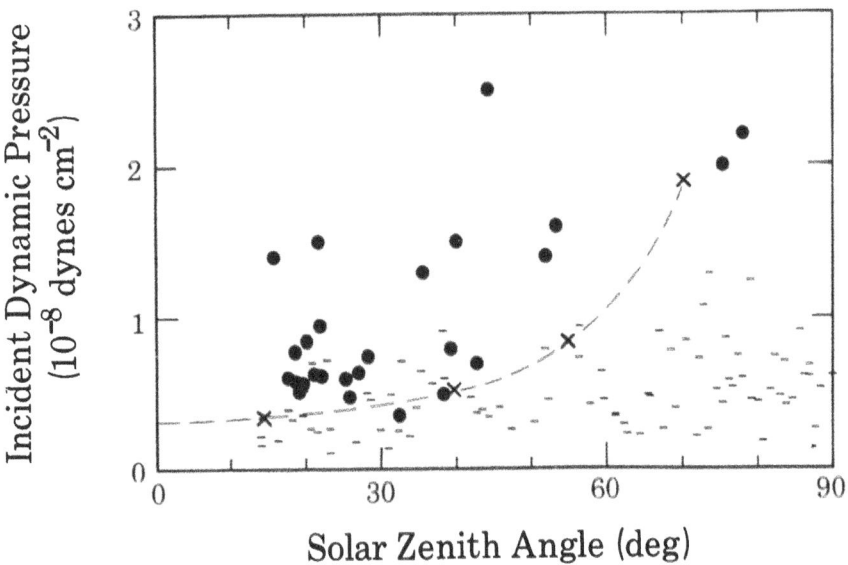

Fig. 10.   Locations of large-scale field occurrences are shown as dots and other orbits shown as dashes in this illustration that both solar wind pressure and SZA determine large scale field occurrence. (From Luhmann *et al.*, 1986.) The dashed line shows the locus of points where the solar wind pressure is equal to the (undisturbed) ionospheric thermal pressure at 240 km altitude. One expects magnetized orbits to fall above this line if the requirement for magnetization is that the ionopause is locally forced below 240 km.

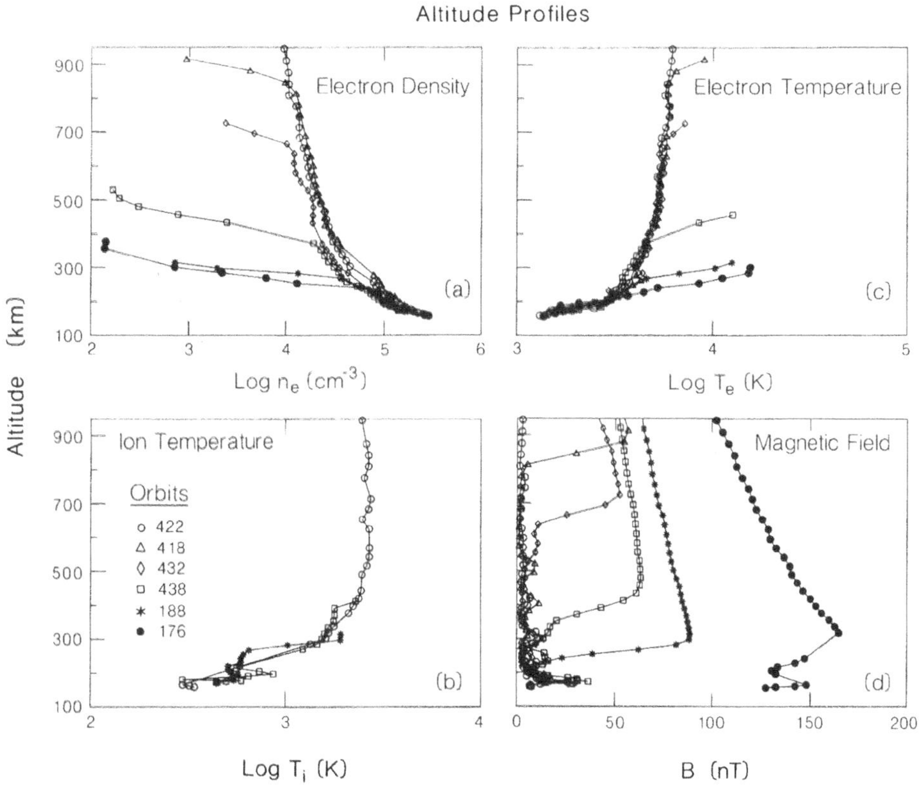

Altitude Profiles

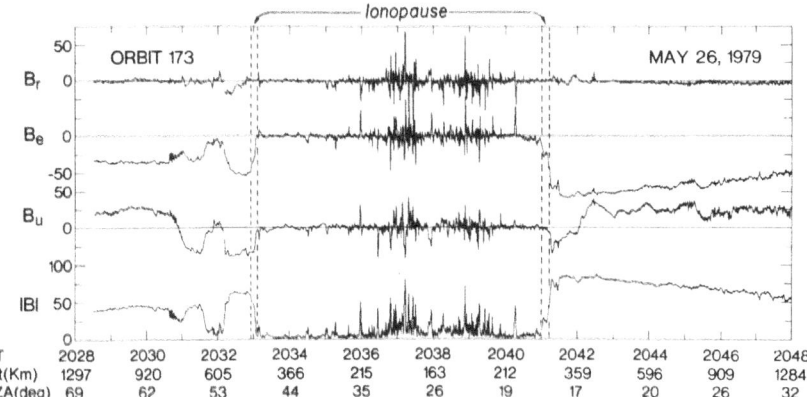

Fig. 11.   Examples of altitude profiles of various quantities measured along the Pioneer Venus trajectory. These were selected from the subsolar data to illustrate the response of the ionosphere to increasing solar wind dynamic pressures. (Similar variations with solar zenith angle also occur.) The different symbols identify the different orbits. Only the outbound leg of each orbit is shown because it is closer to a true altitude profile sample than the inbound leg. (a) Ionosphere densities measured by the Langmuir probe. (b) Electron temperatures from the Langmuir probe. (c) Ion temperatures measured by the retarding potential analyzer. (d) Magnetic field magnitude.

Fig. 12.   High time resolution time series of magnetic field data during an orbit when small-scale structures were present in the dayside ionosphere. The occurrence of some structure of larger scale within the ionopause current layer and the tendency for the low altitude structures to appear in double clumps with the intervening minimum at periapsis are typical features of this type of orbit.

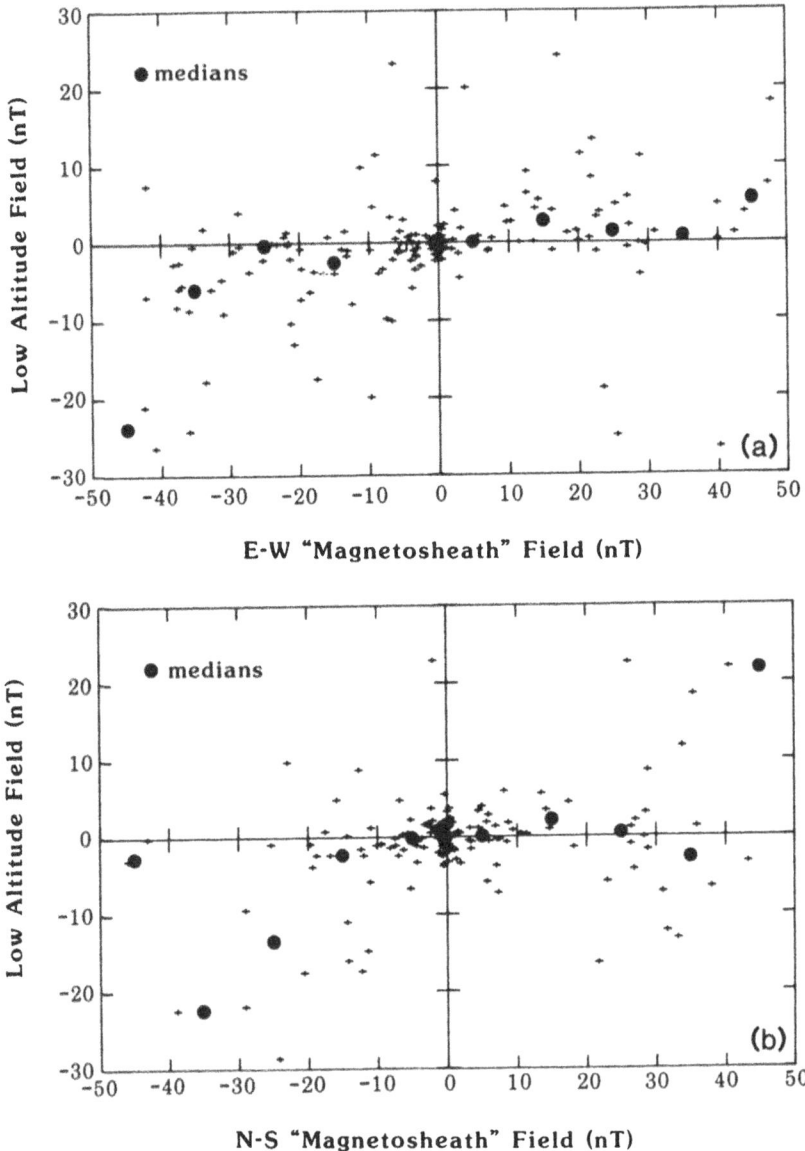

Fig. 13. Comparison of average low altitude (< 200 km) ionospheric field in east–west and north–south directions with field direction in the overlying ionopause and magnetosheath (270–600 km) for orbits with flux ropes. These data indicate a relationship between the field orientation in the ionosphere and the low altitude magnetosheath.

closely tied to the magnetosheath field orientation. Moreover, since the small-scale structures are observed only where the large-scale field is absent, or equivalently, where the local ionopause is high, one also expects transient intermediate or transitional states as the ionopause moves up and down in response to solar wind pressure changes.

The altitude distribution of the small-scale structures, which is described in detail by Figures 14–17, in fact demonstrates a similarity and perhaps a connection between the two types of ionospheric field structures. The altitude profile in Figure 14 from one leg of an orbit during which the small scale structures were present reinforces the observation made earlier that there is a minimum in the number and strength of the structures between the ionopause and ~ 200 km. This observation is made more quantitative by Figure 15, which shows the maximum field in the structures and their numbers over the range of ionospheric heights (cf. Elphic and Russell, 1983a, b). The peak occurrence rate, illustrated by the detailed lowest altitude statistics in Figure 16, occurs near 170 km. The large-scale field structures similarly exhibited a minimum field strength at ~ 190 km and a peak field strength at ~ 170 km.

Another statistic of interest concerns the size distribution of the individual structures. Figure 17 shows that they are smallest (~ 1 km) at the location of their peak occurrence and largest at the highest altitudes where the associated magnetic field magnitude is also weakest (cf. Elphic and Russell, 1983).

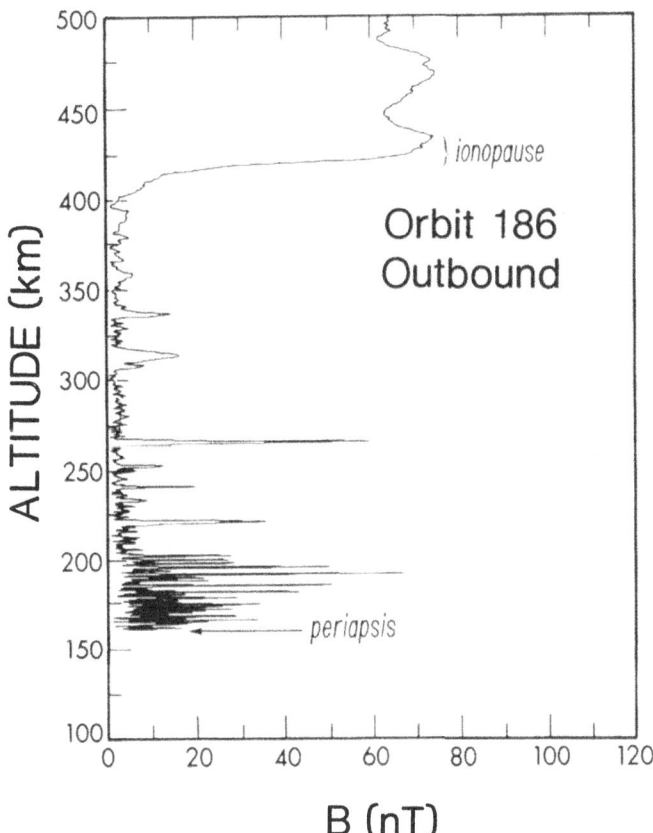

Fig. 14. 'Altitude' profile of magnetic field magnitude from a PVO orbit during which small-scale ionospheric field structures were observed. (From Elphic *et al.*, 1980.)

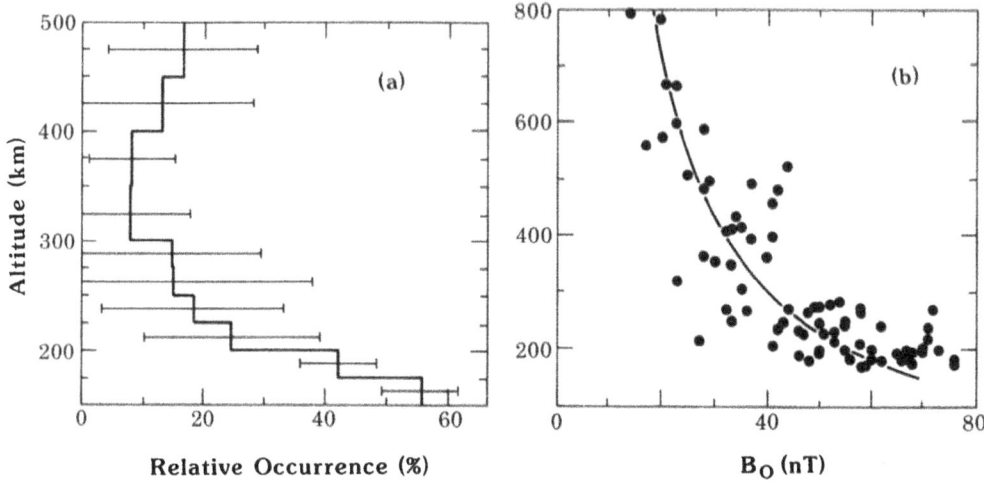

Fig. 15. (a) Altitude dependence of small-scale structure 'occurrence' and (b) altitude dependence of maximum field in flux ropes. (From Elphic and Russell, 1983b.) Notice that the rate of occurrence seems to maximize below ~ 200 km.

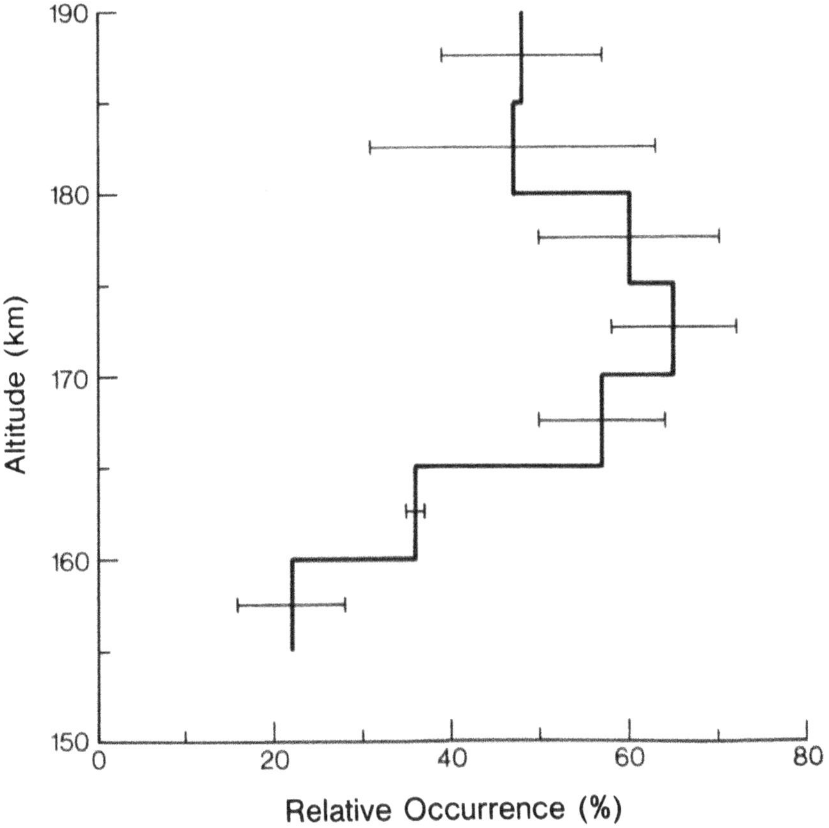

Fig. 16. Observed rate of occurrence of small-scale structures between 155 and 180 km altitude, in 5-km bins. Error bars are the standard deviations of the samples. (From Elphic *et al.*, 1983a.)

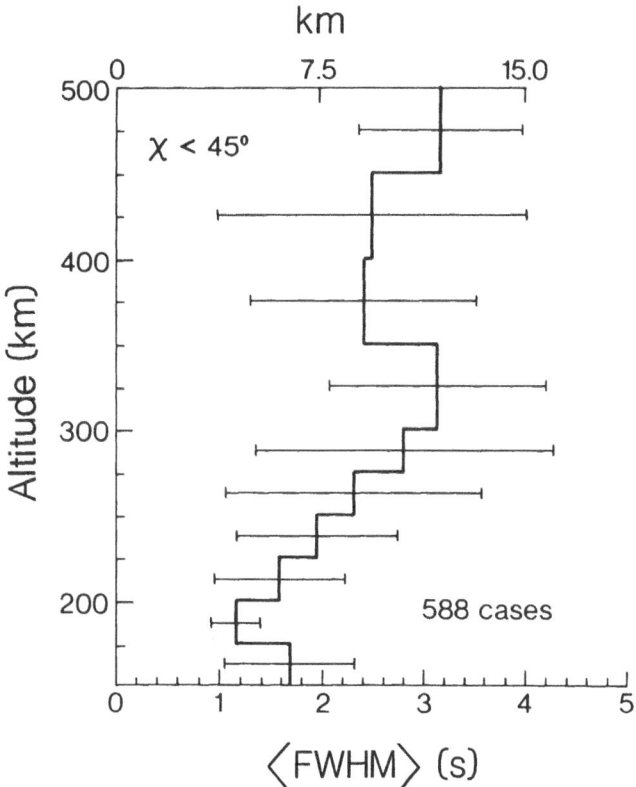

Fig. 17. Average full width at half-maximum (FWHM) of small-scale structures as a function of altitude for the subsolar region. Error bars denote standard deviations of the samples. Upper axis converts ⟨FWHM⟩ to an average scale in km. (From Elphic *et al.*, 1983a.)

The PVO included a Langmuir probe which made fairly high time resolution (~ 1 s) electron density measurements, but as the example in Figure 18 demonstrates, in most cases this may have been insufficient for correlating plasma and field structures. Localized density depressions might be expected if total pressure is a constant. Indeed, density depletions are apparent in conjunction with some of the larger structures, but others appear to be 'force-free' (cf. Elphic and Russell, 1983), showing no signatures in the plasma.

Perhaps one of the most intriguing characteristics of the small-scale structures was found soon after they were discovered when the magnetometer investigators applied the minimum variance analysis technique to the time series of the magnetic data (cf. Russell and Elphic, 1979). Figure 19 shows examples of the hodograms that resulted. For a particular subset of the structures defined as the 'small impact parameter' class (see Elphic and Russell, 1983a, b), clear elliptical traces were found in the display for the maximum variance coordinates, with relatively narrow linear or slightly curved traces in the orthogonal views. When less restrictive wave analysis techniques were applied to

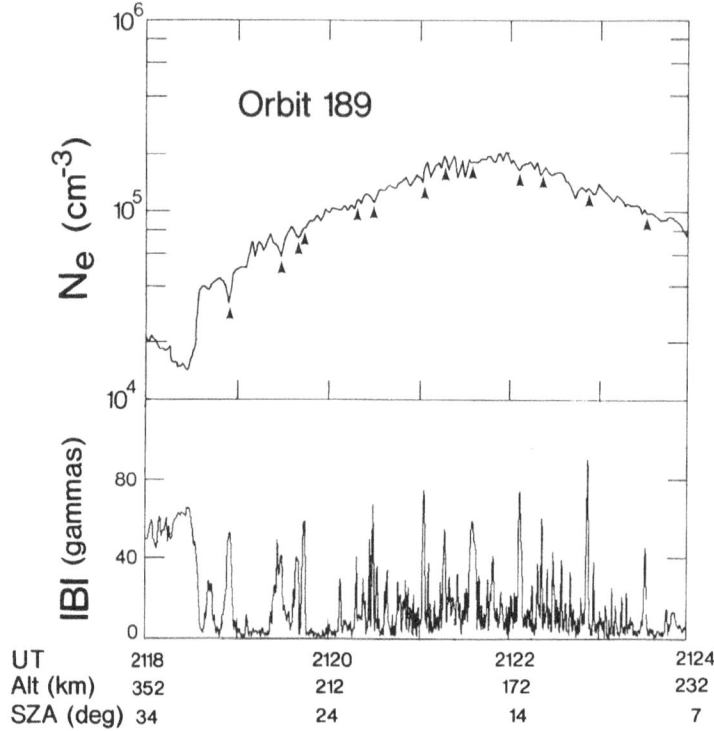

Fig. 18.   Illustration of the possible effect of small-scale magnetic structures on the ionospheric density measured by the PVO Langmuir probe. (From Elphic *et al.*, 1983a.)

the time series of the small-scale structures to determine gross properties such as 'polarization' and 'ellipticity', some organization by spatial coordinates such as solar zenith angle or local time was suggested by the results (cf. Luhmann, 1990). The analysis for the first dayside observing season showed the progression with orbit number (periapsis moves from the dawn to the dusk terminator) illustrated in Figures 20 and 21. The changes in behavior seen near the terminator reinforced what earlier investigators had already noted (cf. Elphic and Russell, 1983b). The small-scale field structures become more polarized and more 'linear' (i.e., more like monodirectional oscillations) at high solar zenith angles.

An example of a time series of ionospheric magnetic field data from the terminator region is shown in Figure 22 for contrast with Figure 12. The different, more wave-like character of the small-scale structures in the near-terminator ionosphere inspired a dedicated analysis on the part of some investigators (cf. Brace *et al.*, 1983a). They found that the 'terminator waves' appear to behave in a manner more or less consistent with a picture wherein horizontal magnetic fields lying across the terminator are alternately pointing toward and away from the Sun. These analyses also showed that there were accompanying wave-like changes in the ionospheric plasma density which were 90°

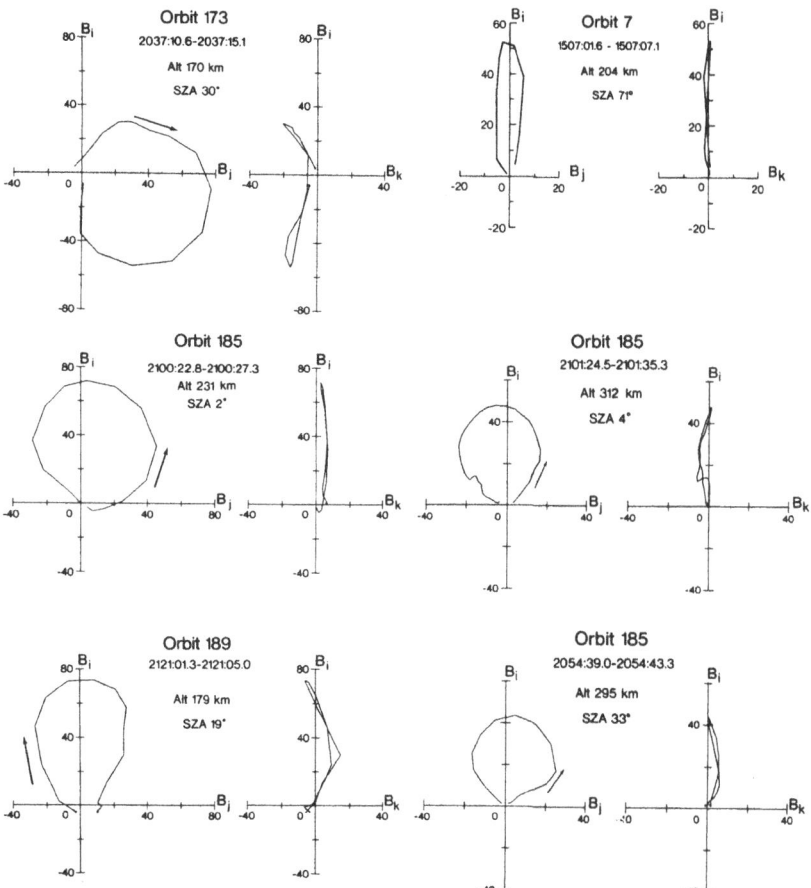

Fig. 19. Hodograms in principal axis coordinates of the field variation of small-scale structures observed on various PVO orbits. (From Elphic *et al.*, 1983a.)

out-of-phase with the changes in the magnetic field magnitude as also displayed in Figure 22. (Recall that plasma density effects were generally not detected in the subsolar region.) Similar structures continue beyond the terminator into the nightside for a $\sim 10°-20°$ solar zenith angle range. As indicated by Figures 20 and 21, they first appear $\sim 10°-20°$ in front of the terminator where there is a rather sudden transition from the characteristics that prevail over most of the dayside. A connection between these near-terminator ionospheric field structures and the twisted 'flux-rope'-like structures described above, if one exists, has not been established.

Other details of the observations of small-scale fields in the dayside ionosphere of Venus can be found in the recent review by Russell (1991).

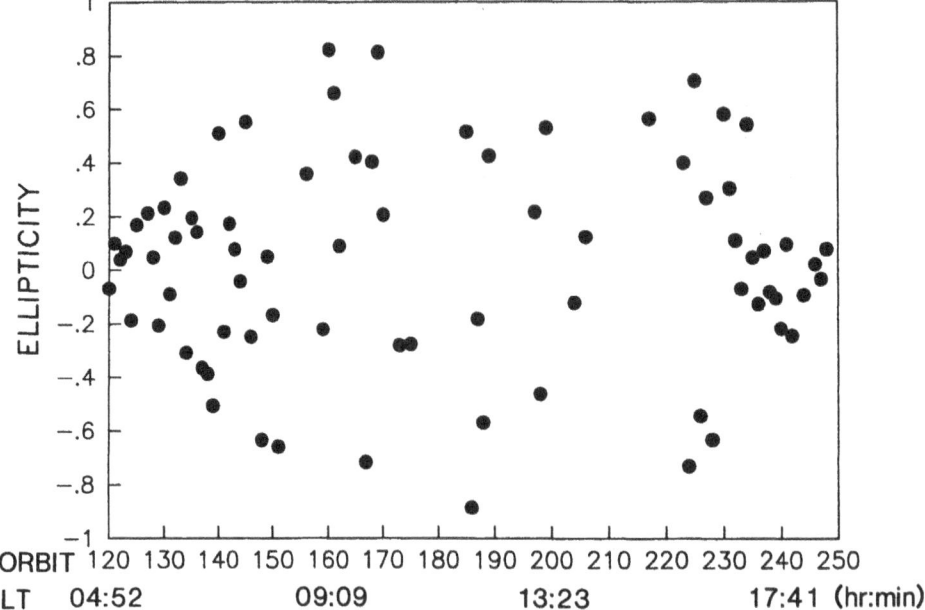

Fig. 20. Average ellipticities obtained from time series of the small-scale magnetic field structure data when power spectral analysis methods are employed. An ellipticity was computed separately for inbound and outbound legs of each orbit. A value of 1 denotes circular 'polarization' while 0 denotes linear polarization or a lack of helicity. Orbit 120 has its periapsis near the dawn terminator, orbit 185 is near-subsolar, and orbit 250 goes across the dusk terminator ionosphere.

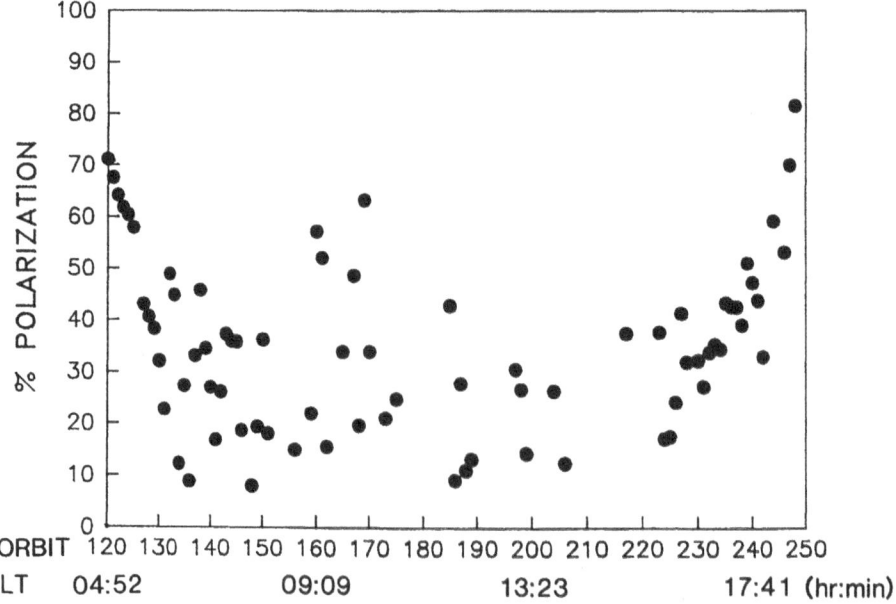

Fig. 21. Percent polarization for the data used in Figure 20. Zero denotes a lack of polarization while 100% denotes completely polarized 'waves'. (From Luhmann, 1990.)

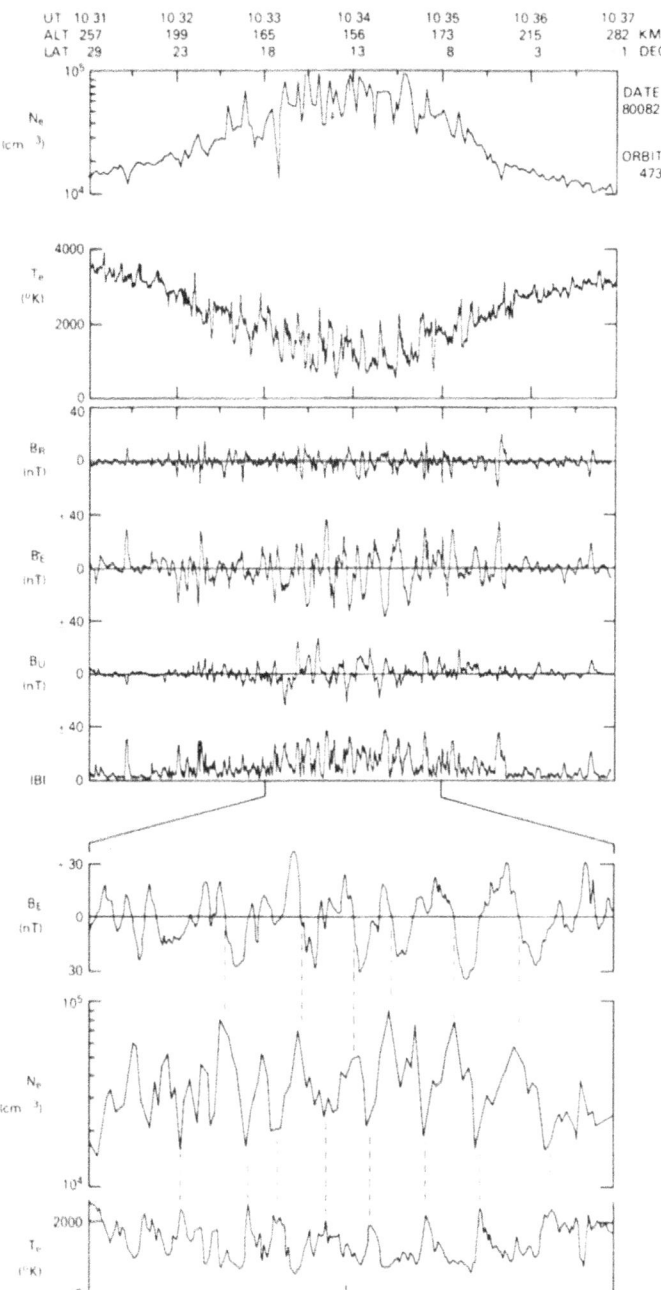

Fig. 22. The ionospheric and magnetic characteristics of terminator waves in orbit 473 (SZA = 100 degrees at periapsis). The upper five traces show a 6-minute segment of N_e, T_e, B_R, B_E, B_U, and **B**, while the lower panels display an expanded view of N_e, T_e, and B_E which show details of the wave patterns. Well ordered waves develop below about 175 km. The vertical dashed lines draw attention to persistent phase relationships between B_E, N_e, and T_e. (From Brace *et al.*, 1983a.)

2.3. The nightside ionosphere

2.3.1. *Large-Scale Fields*

Large-scale fields were also occasionally observed on PVO in the nightside ionosphere of Venus (cf. Russell *et al.*, 1979b; Luhmann *et al.*, 1981). As shown by the statistics of Figure 23, the maximum strength of these fields tended to be a few tens of nanoteslas in contrast to the ~ 100 nT dayside fields. Moreover, the large nightside fields occurred within a ~ 45° SZA range of the anti-solar point. The statistics of the large field observations displayed in Figure 24 show that there was a slight preference for early morning (west) which is opposite in definition to that of earth because of Venus' retrograde rotation. Like the large-scale dayside fields, the field strength of the nightside fields showed some relationship to the incident solar wind dynamic pressure. Figure 25

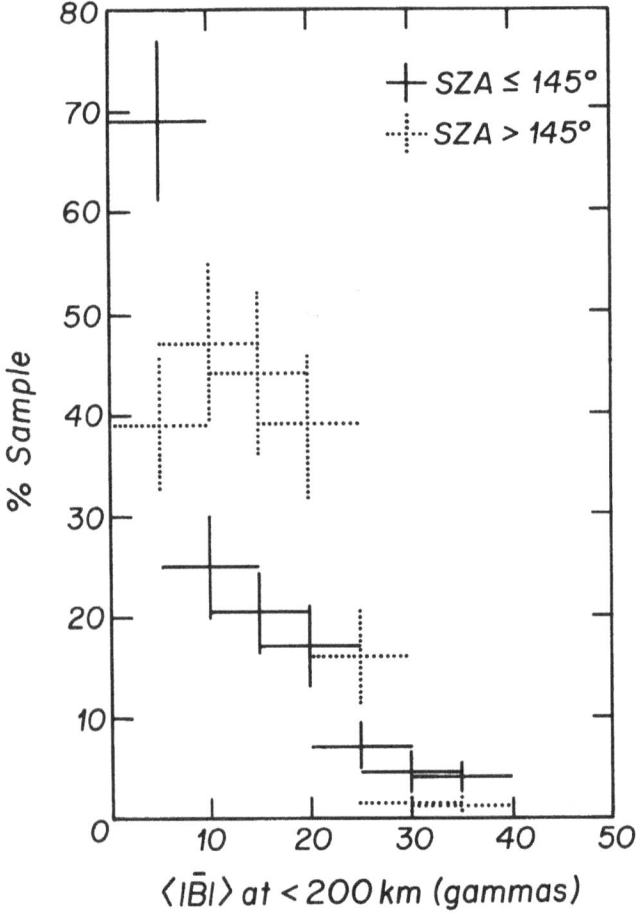

Fig. 23.   Distribution of nightside ionospheric field magnitude observed between ~ 150 and 200 km on the nightside of Venus. (From Luhmann *et al.*, 1980b.)

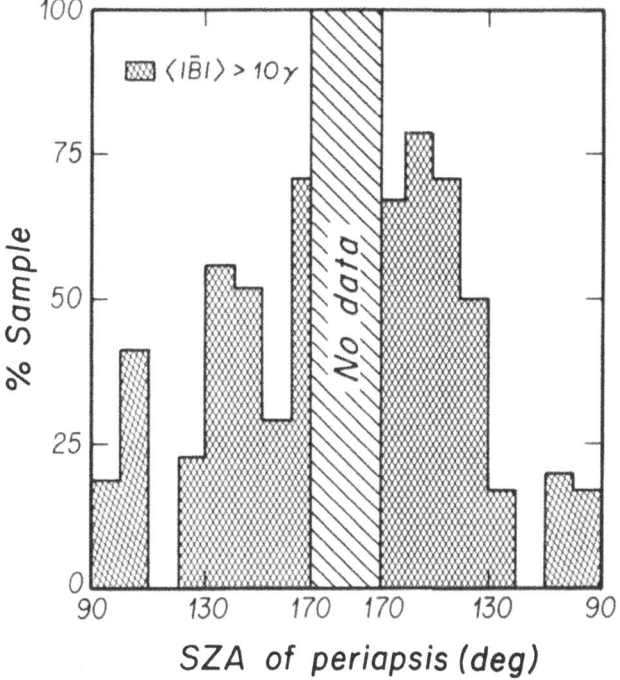

Fig. 24.   Solar zenith angle distribution of high ($> 10\gamma$) nightside field observations near periapsis. East and west are in the sense of Earth's, here measured with respect to the midnight meridian. (From Luhmann *et al.*, 1980b.)

illustrates that the field in the nightside flanks was low (< 10 nT) except when the solar wind pressure was elevated. In contrast, the field was as likely to be high (> 10 nT) as it was to be low in the anti-solar region for low solar wind pressure, although it was also high for high solar wind pressure as in the flanks. The nightside field was thus high near the anti-solar region even when solar wind pressure was low. It was high everywhere when solar wind pressure was high.

Also like the dayside high fields, the time series of the ionospheric magnetic field data from the nightside showed distinctive characteristics as illustrated in Figures 26 and 27. It appears that there are two basic types of nightside high fields during the period of PVO's reconnaissance. The type associated with low solar wind pressure appears as in Figure 26. Double or single intrusions of large steady field, oriented in the sunward or anti-sunward (x) direction and asymmetrically placed with respect to periapsis, are separated by low magnitude variable fields which have the characteristics of neither the terminator nor subsolar small-scale structures. The angle from the vertical, ψ, shows that in contrast to the dayside high fields, these nightside high fields are practically vertical (or equivalently, solar or anti-solar). As seen in the simultaneously obtained plasma data in Figures 26 and 27, the intrusions of large-scale field are accompanied by electron density and temperature depletions which have been called 'holes' in the literature (cf.

J. G. LUHMANN AND T. E. CRAVENS

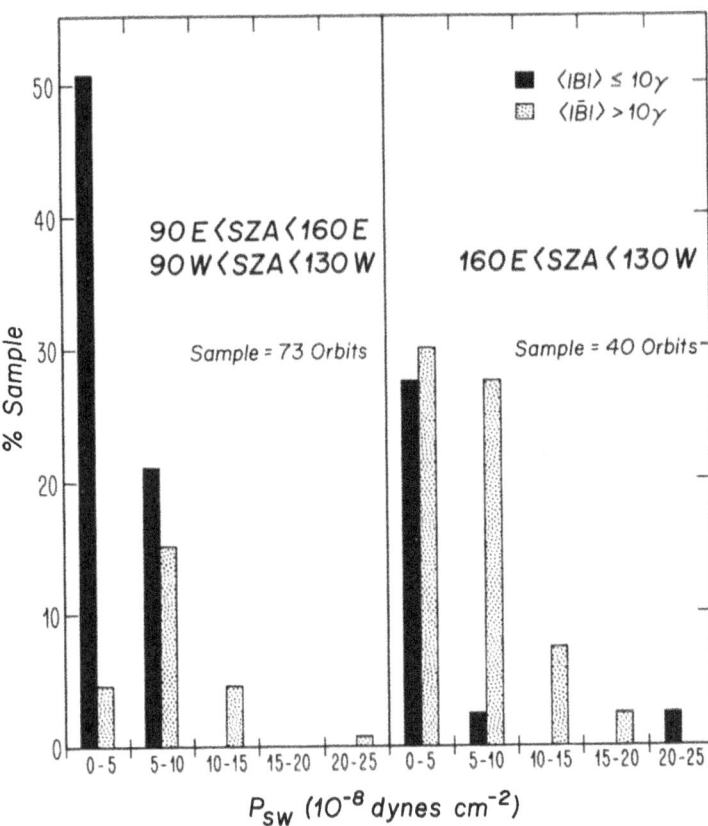

Fig. 25.   Occurrence frequencies of high and low average nightside field observations for various values of solar wind dynamic pressure.

Brace *et al.*, 1983a, b). Low-frequency (100 Hz) plasma waves have also been detected in these holes, together with a population of super-thermal electrons and ions, and a compositional change favoring light ions (cf. Luhmann *et al.*, 1983).

For the contrasting case of high solar wind pressure related high fields, the situation is generally as depicted in Figure 27. The angle ψ in this case shows that the orientation of the field is almost horizontal and slowly varies throughout the ionosphere (cf. Cravens *et al.*, 1982). At these times the electron density does not merely exhibit localized holes but is depleted everywhere in the observable ionosphere. The electron temperature is highest in the most depleted regions. Because of the general depletion of the density, these occurrences have been given the name of 'disappearing ionospheres'.

The example of the behavior of the plasma thermal and field pressures through a hole shown in Figure 28 indicates that pressure balance holds in these structures and therefore they are quasi-steady in nature (cf. Brace *et al.*, 1982). The distribution of hole magnetic 'polarities' (outward or inward) in longitude, reproduced in Figure 29, showed that the fields in the holes were probably not related to the presence of a large-scale

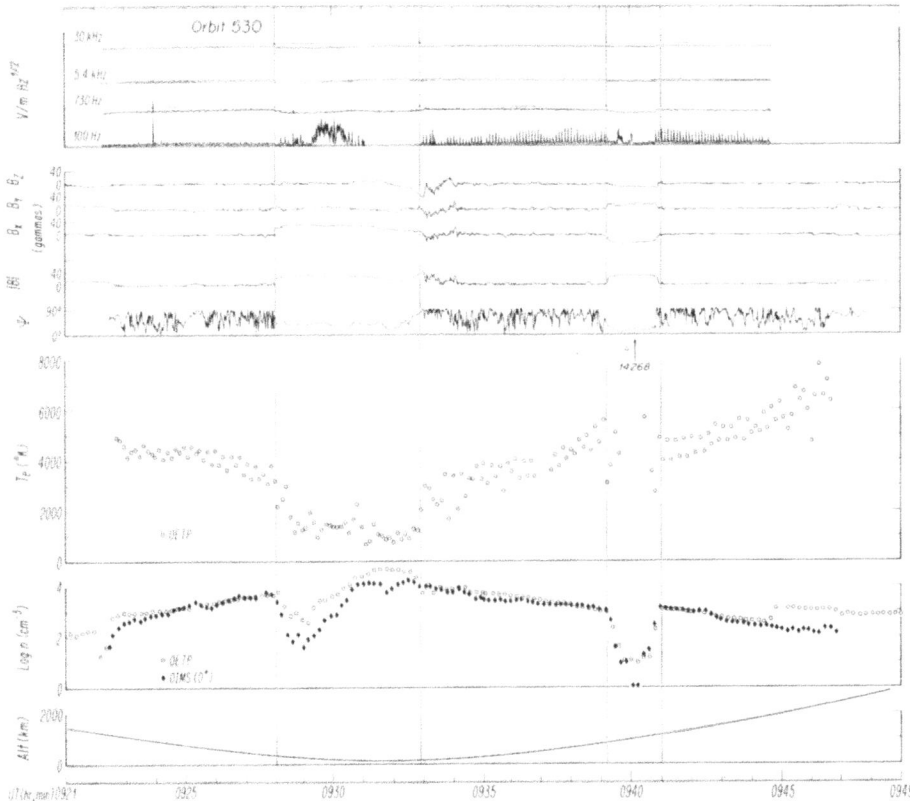

Fig. 26.   Data obtained near periapsis on orbit 530 from the PVO electric field, magnetometer, Langmuir probe (OETP), and ion mass spectrometer (OIMS) experiments in order from the top. The altitude above the surface is given at the bottom. The magnetic field components shown are in the solar ecliptic coordinate system; the angle ψ is between the radial direction and the local magnetic vector (e.g., $\psi = 0$ is radial). Note the location of the two 'square waves' of enhanced magnetic field (near periapsis) that are oriented at small angles to the radial, embedded in a typically weak, irregular nightside ionospheric field. The vertical lines delineate the approximate boundaries of the regions of enhanced magnetic field. (From Luhmann *et al.*, 1983.)

dipolar planetary field since they did not consistently reverse direction across the equator (cf. Phillips and Russell, 1987). On the other hand, organization by the upstream field direction produces the two-sided pattern shown in Figure 30. This pattern is akin to that found in the magnetotail at ~ 12 planetary radii downstream where draped interplanetary fields are considered responsible for the field structure (cf. Saunders and Russell, 1987).

No analyses have been carried out on the relatively indistinctive small scale, low magnitude fluctuating ionospheric fields that pervade the nightside ionosphere outside of holes on lower solar wind pressure orbits.

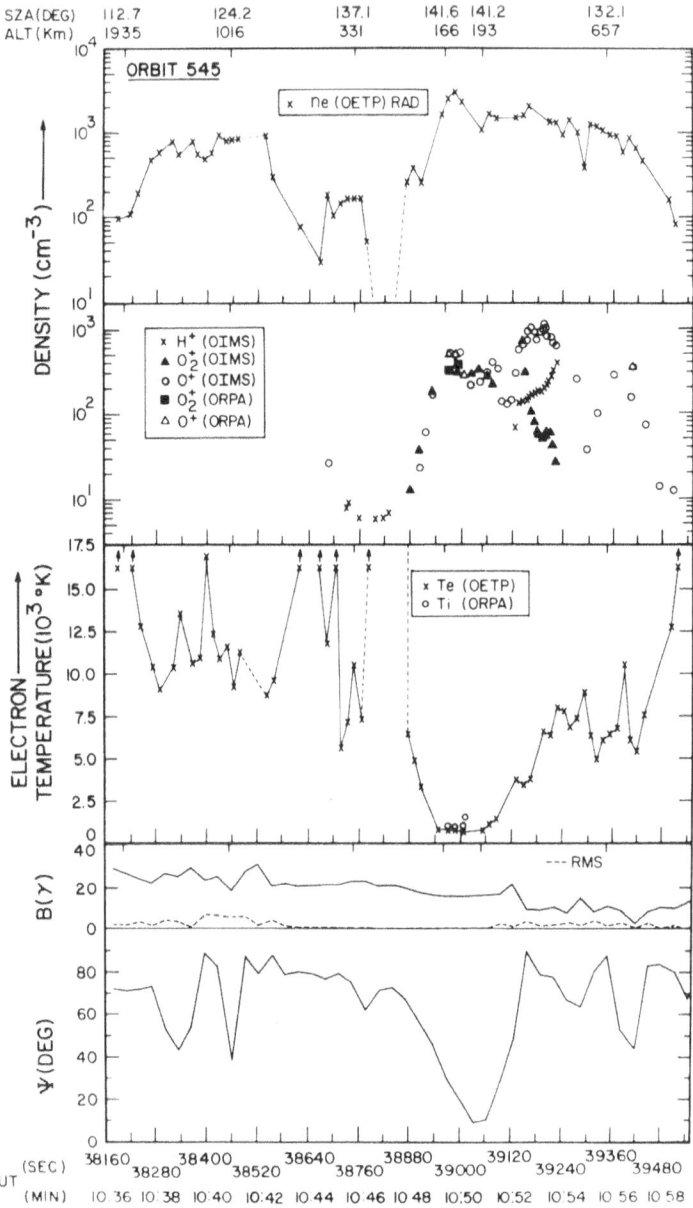

Fig. 27.   Electron density, ion density, electron and ion temperatures and magnetic field magnitude and orientation angle as observed on PVO orbit 545 during a 'disappearing' nightside ionosphere episode. Notice that ψ is usually large, denoting approximately horizontal fields. (From Cravens *et al.*, 1982.)

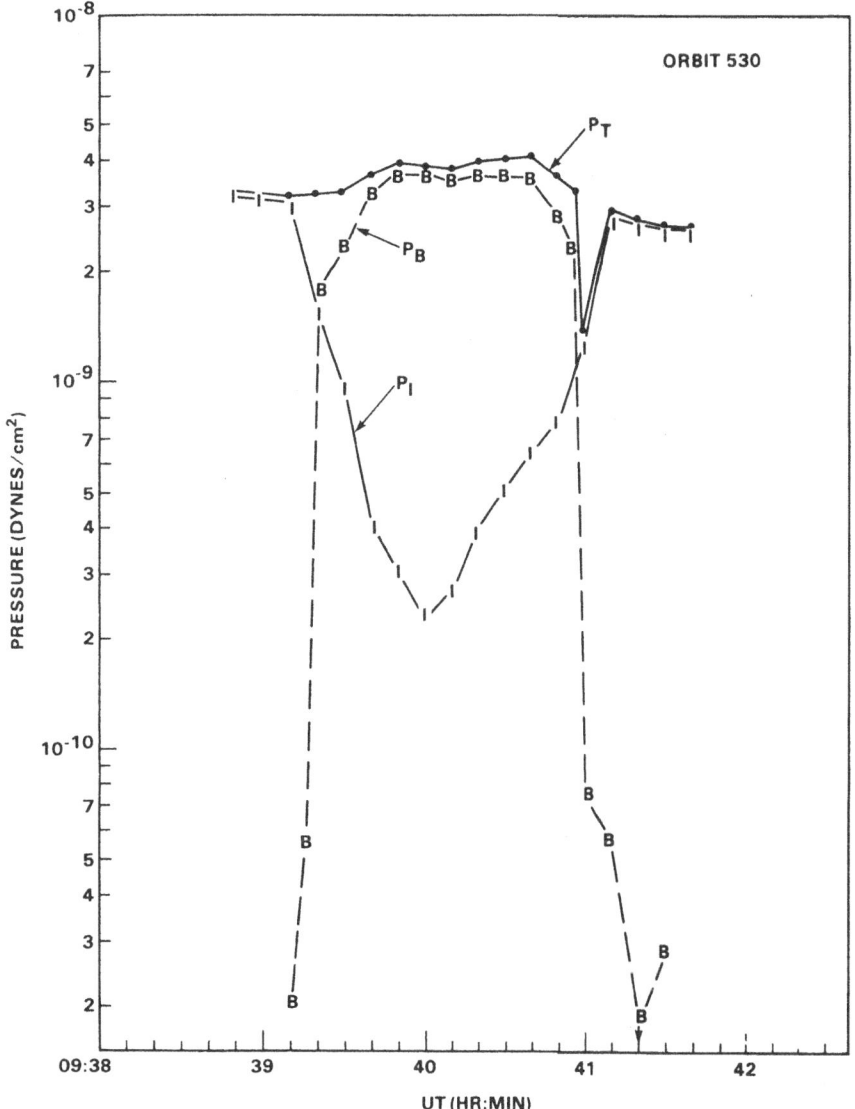

Fig. 28. The plasma pressure P_i and magnetic pressure P_B across the deep ionospheric hole of Figure 26. The total pressure P_T is approximately constant across the entire hole region, indicating that the magnetic field is adequate to exclude almost totally the ionospheric plasma. (From Brace *et al.*, 1982.)

3. Models

Three groups have carried out most of the theoretical investigations of the magnetic fields in the dayside ionosphere of Venus. The first group to attempt to make quantitative physical arguments relating to the generation of these fields was based at Rice University. Their published work on the subject commenced in 1969 (cf. Cloutier *et al.*,

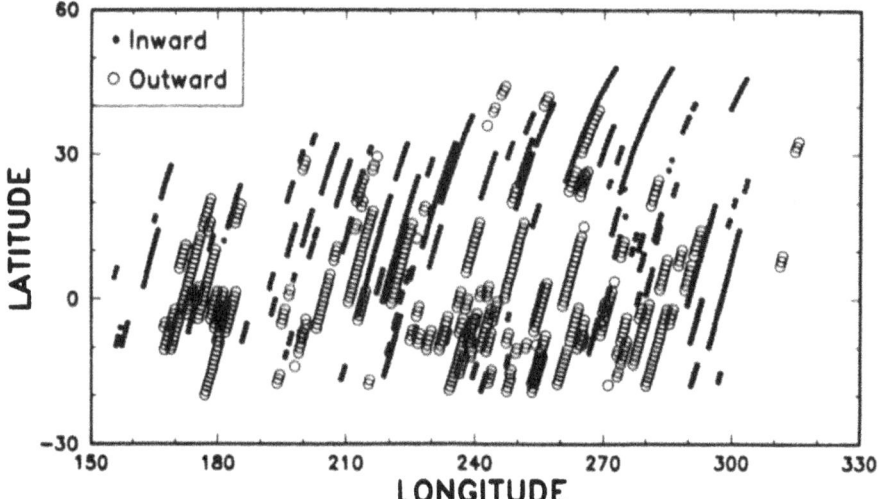

Fig. 29.   Mercator projection showing the geographic locations of the observed magnetic 'holes', defined as fields with greater than 10 nT radial component and inclination at least 45° from horizontal. Open and solid circles indicate outward and inward fields, respectively. (From Phillips *et al.*, 1987.)

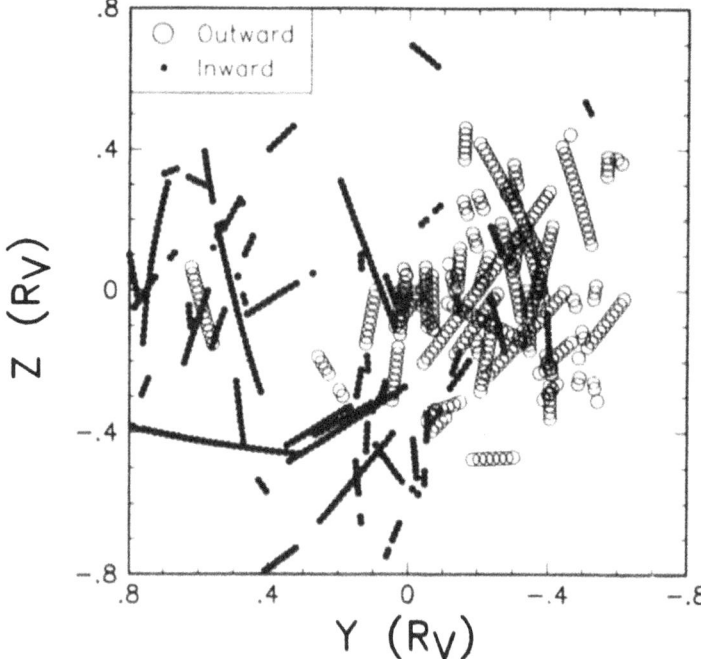

Fig. 30.   Same as Figure 29, except that the orbit segments have been rotated into a coordinate system in which the transverse interplanetary field is always directed to the right (+ y). Note the organization of the radial field polarity into two lobes consistent with the draping of the interplanetary field around the planetary obstacle. (From Phillips *et al.*, 1987.)

1969) in connection with the interpretation of some radio occultation profiles of the Martian ionosphere. (Mars was considered to represent another case of the weakly magnetized planet-solar wind interaction based on the small size of its bow shock as seen by Mariner 4 in 1965.) Another more comprehensive model was developed by the Rice group in the 1970s (cf. Cloutier and Daniell, 1973), but there were no new data to provide further guidance. In the early 1980s, many analyses of the ionospheric magnetic field data from the PVO magnetometer appeared in the literature (cf. references here and in Russell and Vaisberg, 1983). These observational studies at last provided constraints for the theoretical models. Shortly thereafter, the UCLA group also began working on models of the ionospheric field. These two groups were joined by a third group from the University of Michigan in 1984. A detailed review of the Rice models has been written by Cloutier et al. (1983).

The nightside fields have received considerably less attention, with the only existing published work being that by Grebowsky and coworkers on the formation of ionospheric holes. The most appropriate starting point for a discussion of models, however, is a review of the underlying physical equations.

3.1. THE APPLICATION OF MAGNETOHYDRODYNAMICS TO THE IONOSPHERE OF VENUS

The ionosphere of Venus is a partially-ionized plasma containing electrons and many ion species (e.g., CO_2^+, O_2^+, O^+, H^+). This plasma is produced by the photoionization of the neutral constituents of the upper atmosphere by solar extreme ultraviolet radiation. Ion-neutral chemical reactions and electron-ion recombination are both important processes in the lower ionosphere. The individual ion species and the electrons (with the important exception of superthermal electrons including photo-electrons) are adequately represented by drifting Maxwellian distributions, and thus a fluid treatment is generally valid (e.g., see Nagy et al., 1983).

Since the purely aeronomical attributes of the ionosphere of Venus have been reviewed before and are also discussed in other chapters in this volume, only those aspects relevant to the understanding of the ionospheric magnetic field will be reviewed here. A discourse on aeronomy in general can be found in the series of volumes by Banks and Kockarts (1973), and an appropriate introduction to magnetohydrodynamics can be found in a review by Siscoe (1982).

The set of equations which applies to the problem of the ionospheric magnetic field is that of multi-species magnetohydrodynamics (MHD) including aeronomical collision and production terms. These equations include Maxwell's equations and the fluid conservation equations which are the velocity moments of the Boltzmann equation (Nicholson, 1983). The latter include the continuity and momentum equations. In aeronomical applications, the set of moment equations is generally truncated at the energy equation which is required to find the temperature of a given species. However, a detailed discussion of the energy equation, or the energetics of the ionosphere of Venus, is beyond the scope of the present paper. Here we shall be dealing only with the continuity and momentum equations together with Maxwell's equations.

3.1.1. *Continuity Equation*

The continuity equation for ion species s is:

$$\frac{\partial n_s}{\partial t} + \nabla \cdot (n_s \mathbf{v}_s) = P_s - L_s,$$ (1)

where n_s is the number density of species s, $\mathbf{v}_s$ is the velocity vector of that species, and P_s and L_s are production and loss rates. P_s includes ion production due to photoionization and production due to ion-neutral chemistry. L_s includes losses due to both ion-neutral reactions and ion-electron recombination. CO_2^+ has the largest production rate of all ions at low altitudes, but O_2^+ is the major ion near the ionospheric peak due to the following fast chemical reaction:

$$CO_2^+ + O \rightarrow O_2^+ + CO.$$

The O_2^+ ions dissociatively recombine with electrons. A separate continuity equation for electrons is not used; the electron density, n_e, is taken to be the sum of all of the ion densities. The ionosphere is approximately in photochemical equilibrium below an altitude of about 200 km at Venus for all ions, which means that the left-hand side of Equation (1) can be neglected in that region ($P_s = L_s$). In the photochemical regime, the chemical lifetime of species s is much less than its transport time, and the scale height of the ionosphere is approximately twice that of the neutral gas if the recombination coefficient is roughly constant (e.g., if $L_s = \alpha n_e^2$ and P_s is proportional to neutral density). Above 200 km, the major ion species are controlled mainly by vertical (or horizontal) transport. O^+ becomes the major ion in this diffusion regime where the scale height is determined by $\mathbf{v}_s$ as well as P_s and L_s.

3.1.2. *Momentum Equation*

The momentum equation for species s is

$$n_s m_s \left[\frac{\partial \mathbf{v}_s}{\partial t} + \mathbf{v}_s \cdot \nabla \mathbf{v}_s \right] =$$

$$= -\nabla p_s + n_s e_s (\mathbf{E} + \mathbf{v}_s \times \mathbf{B}) + n_s m_s \mathbf{g} - n_s m_s \sum_j v_{sj}(\mathbf{v}_s - \mathbf{v}_j) + P_s m_s \mathbf{V}_{0s},$$ (2)

where m_s, p_s, and e_s are the mass, pressure, and charge, respectively, for species s. The neutral velocity, $\mathbf{v}_n$, is typically quite small compared with the plasma velocity. The vectors $\mathbf{E}$ and $\mathbf{B}$ are the ionospheric electric and magnetic fields, and $\mathbf{g} = -g\hat{z}$ is the acceleration due to gravity. The effective momentum transfer collision frequency v_{sj} between species s and j describes the collisional coupling between particles in the 'fluid' (Banks and Kockarts, 1973). The pressure is given by the equation of state: $p_s = n_s k_B T_s$, where k_B is Boltzmann's constant and T_s is the temperature of species s. Most of the terms in Equation (2) are self-explanatory: from the left, one finds the inertial and advection terms, the pressure gradient force term, the Lorentz force term, the gravitational term, the friction (or collision) term, and the 'mass-loading' term. The latter is added in the event that newly produced ions have a substantial initial velocity $\mathbf{V}_{0s}$. This would occur, for example, if the neutrals had a high velocity. There is no corresponding term related to L_s since that loss affects the momentum only through the change in n. Many useful approximations can be made to the ion-momentum equation, but first we consider those related to the electric field.

3.1.3. *Generalized Ohm's Law*

The most convenient way of finding the electric field in a quasi-neutral plasma is to use the electron momentum equation (cf., Siscoe, 1983). Solving the electron momentum equation for $\mathbf{E}$ gives:

$$\mathbf{E} = -\mathbf{v}_e \times \mathbf{B} - \frac{1}{n_e e} \nabla p_e - \frac{m_e}{e} \sum_j \nu_{ej}(\mathbf{v}_e - \mathbf{v}_j) +$$

$$+ \frac{m_e}{e} \left\{ -\left(\frac{\partial \mathbf{v}_e}{\partial t} + \mathbf{v}_e \cdot \nabla \mathbf{v}_e \right) + \mathbf{g} + P_e \mathbf{V}_{0e} \right\}. \tag{3}$$

This equation is the generalized Ohm's law (GOL). The first term on the right-hand side is the motional electric field and in much of space physics this is the only term retained. The next term is the polarization field, and the next is the Ohmic dissipation term (Ohmic dissipation is due to collisions of electrons with both neutrals and ions). The last term (in brackets) includes forces which are quite negligible in the ionosphere.

One can then use the definition of the current density $\mathbf{J}$:

$$\mathbf{J} = n_e e(\mathbf{v} - \mathbf{v}_e) \tag{4}$$

to write the GOL in terms of the average (all species) ion velocity $\mathbf{v}$ and the current density. With this substitution, and dropping the negligible terms, the generalized Ohm's law becomes

$$\mathbf{E} = -\mathbf{v} \times \mathbf{B} + \frac{1}{n_e e} \mathbf{J} \times \mathbf{B} - \frac{1}{n_e e} \nabla p_e + \eta \mathbf{J} - m_e \nu_{en}(\mathbf{v} - \mathbf{v}_n). \tag{5}$$

The second term on the right-hand side is the Hall term, which at least at Venus at solar maximum is usually much smaller than the $-\mathbf{v} \times \mathbf{B}$ term. The last term on the right-hand side is also quite small compared to the other terms. The remaining collision term is the Ohmic dissipation term, written as $\eta \mathbf{J}$, where $\eta = m_e \nu_{et}/n_e e^2$ is the resistivity and $\nu_{et} = \nu_{ei} + \nu_{en}$ is the total electron collision frequency. A standard expression for the current density in terms of the Hall and Pederson conductivities can be obtained by algebraic manipulation of Equation (5). One can see that for the situation where Ohmic currents prevail, one can use $\mathbf{J} = \sigma \mathbf{E}$ for the current.

3.1.4. *One-Fluid Momentum Equation*

It is often useful to consider all of the ion species together, use the GOL for the electric field, and thus obtain a single-fluid momentum equation from Equation (2). In particular, if one assumes that one ion species (with mass m_i) is dominant, and that the neutral parent atoms or molecules are stationary, this momentum equation takes the form

$$\rho\left(\frac{\partial \mathbf{v}}{\partial t} + \mathbf{v} \cdot \nabla \mathbf{v}\right) = \mathbf{J} \times \mathbf{B} - \nabla(p_e + p_i) - \rho\mathbf{g} - \rho v_{in}(\mathbf{v} - \mathbf{v}_n), \qquad (6)$$

where $\rho = m_i n_e$ is the mass density of the plasma and v_{in} is the ion-neutral momentum transfer collision frequency. The left-hand side of Equation (6) can usually be neglected for relatively slow changes in the flow speed v and if the plasma speed is much less than the thermal speed, in which case solving (6) for the flow speed $\mathbf{v}$ gives the major ion diffusion equation. Using Ampère's law, $\nabla \times \mathbf{B} = \mu_0 \mathbf{J}$, the $\mathbf{J} \times \mathbf{B}$ term can be expressed in terms of a magnetic curvature force, $\mathbf{B} \cdot \nabla \mathbf{B}/\mu_0$, and a gradient of magnetic pressure, $-\nabla(B^2/2\mu_0)$, where the magnetic pressure is $p_B = B^2/2\mu_0$ (mks units).

Under circumstances (i.e., in flux ropes) where both curvature and pressure forces contribute to $\mathbf{J} \times \mathbf{B}$, the static force balance equation (neglecting the gravitational and frictional forces) is simply

$$\mathbf{J} \times \mathbf{B} = \nabla(p_e + p_i). \qquad (7)$$

If straight field lines are assumed, the magnetic pressure gradient force is the only contribution from $\mathbf{J} \times \mathbf{B}$. In this case, the ionospheric plasma flows in response to gradients in the total pressure $p_{\text{total}} = p_e + p_i + p_B$, according to the force balance equation $\nabla p_{\text{total}} = 0$. This equation indicates that p_{total} is a constant for static conditions in the Venus ionosphere wherever collisions and gravity can be neglected. For example, constancy of total pressure has been used to explain the observed transition of magnetic pressure (on the magnetosheath side) to ionospheric thermal pressure across the (high altitude) ionopause for low solar wind dynamic pressure conditions.

Several other insights can also be gained from Equation (6). For example, if $\mathbf{B} = 0$ and static conditions ($\mathbf{v} = \mathbf{v}_n \approx 0$) prevail, the equation's solution is the hydrostatic (or diffusive equilibrium) solution, in which the plasma pressure (and the electron density approximately) falls off exponentially with the plasma scale height $H_p = k_B(T_e + T_i)/m_i g$. Another consideration is that if flux ropes are merely narrow filaments of straight field lines, the thermal plasma pressure should decrease through the structure in order to compensate for the increase in B and magnetic pressure, thus keeping the total pressure constant. However, if the structure is instead a force-free flux rope wherein the magnetic curvature force exactly balances the magnetic pressure gradient force (e.g., $\mathbf{J} \times \mathbf{B} = 0$), the plasma pressure should remain constant across the rope. In order to achieve the 'force-free' condition $\mathbf{J} \times \mathbf{B} = 0$, the current must flow only along the magnetic field, and hence the following condition is met:

$$\mathbf{J} = \nabla \times \mathbf{B} = \alpha \mathbf{B}, \qquad (8)$$

where α is a function only of the distance from the axis of the cylindrically symmetric magnetic structure.

3.1.5. *Magnetic Induction Equation*

The magnetic field is un unknown quantity in the momentum equation, and is found in MHD (e.g., Siscoe, 1983) from Faraday's law of induction:

$$\frac{\partial \mathbf{B}}{\partial t} = -\nabla \times \mathbf{E}. \tag{9}$$

Taking the electric field from the GOL (Equation (5)), one obtains the magnetic diffusion/convection (or induction) equation, also known as the dynamo equation

$$\frac{\partial \mathbf{B}}{\partial t} = \nabla \times (\mathbf{v}_e \times \mathbf{B}) - \nabla \times (D\nabla \times \mathbf{B}), \tag{10}$$

where $D = \eta/\mu_0$ is the magnetic diffusion coefficient. Some small terms were neglected in deriving Equation (10). For example, the curl of the pressure gradient term of Equation (5) is zero if gradients in the electron pressure are parallel to gradients in the electron density, as they almost always are. The first term on the right-hand side of Equation (10) describes the variation of the magnetic field due to the convection of magnetic flux with the plasma motion (or more precisely, electron bulk motion), and the second term describes the diffusion of the field associated with the Ohmic or collisional dissipation of the currents responsible for the magnetic field. The average ion velocity can be substituted for the electron velocity in Equation (10) if the Hall term is small, which is the case at Venus.

A useful parameter called the magnetic Reynolds number, $R_M = LV/D$, is found by taking the ratio of the order of magnitude of the diffusion term to the convection term in the induction equation, for magnetic structures with a scale size of L and for a plasma flow speed of V. If the plasma is collisionless and $D \approx 0$, then R_M is extremely small, and the magnetic field is 'frozen' into the plasma flow. But if L or V are very small, or D is large (large resistivity), then R_M is large and the magnetic field will rapidly diffuse away unless magnetic flux is resupplied to the system.

As we shall discuss below, the ionospheric plasma at Venus moves both horizontally in the anti-solar direction with speeds of the order of 1 km s^{-1} (Knudsen *et al.*, 1979) and also downward with speeds of the order of 50 m s^{-1} (Cravens *et al.*, 1984). However, since scale sizes in the vertical direction are much smaller than in the horizontal direction, and vertical plasma motion is more important in determining the magnetic field distribution over most of the dayside ionosphere. In this case, the one-dimensional induction equation:

$$\frac{\partial B}{\partial t} = -\frac{\partial}{\partial z}(WB) + \frac{\partial}{\partial z} D \frac{\partial B}{\partial z} \tag{11}$$

is sufficient to describe the ionospheric magnetic field. Here the field is assumed to be purely horizontal and the flow, at speed W, purely vertical; both depend only on the altitude z. Numerical solutions of this equation for the ionosphere of Venus are discussed below.

The magnetic Reynolds number for the ionosphere of Venus is

$$R_M \approx HW/D \approx 5 \times 10^9/D \, (cm^2 \, s^{-1}) \, ,$$

where values of W and scale height H typical for $z \approx 200$ km were used. The diffusion coefficient D decreases with increasing z due to the decreasing collision frequency. The interesting value $R_M \approx 1$ occurs at an altitude of $z \approx 150\text{--}180$ km where $D = 5 \times 10^9$. At altitudes greater than this altitude, the magnetic field is practically frozen into the plasma and magnetic flux is merely transported by the plasma convection. At lower altitudes, the field diffuses (or dissipates) and flux is removed. Many of these equations and concepts appear again below in the review of past modeling efforts.

3.2. THE DAYSIDE IONOSPHERE

3.2.1. *Large-Scale Fields*

Cloutier and a number of different coworkers at Rice University developed a progression of theoretical approaches to the problem of the solar wind interaction with a planetary ionosphere. The first work in this series (cf. Cloutier *et al.*, 1969) specifically addressed the observations by Mariner spacecraft that the topside scale height of the Martian ionosphere was about half of that expected from purely photochemical models of the Martian ionosphere (based on neutral atmosphere models), although one of the results of the model was an ionospheric magnetic field altitude profile. In this case, the authors considered one-dimensional flow of the solar wind plasma directly into the subsolar ionosphere using the mass-loaded gas dynamic formulation based on Biermann *et al.*'s (1967) treatment of the problem of the solar wind interaction with a cometary atmosphere. They argued that, as for a comet, a shock forms ahead of the ionosphere when the degree of solar wind mass-loading by the planetary plasma gets too great. This shock deflects the bulk of the solar wind plasma away from the stagnation point. According to their assumptions, a maximum downward velocity of ~ 1 km s^{-1} is possible in the ionosphere. They then integrated the fluid equations for the combined solar wind and ionospheric plasmas, together with Maxwell's equations, to obtain ionospheric altitude profiles of density, velocity, pressure and magnetic field. In the region above 200 km altitude, they assumed recombination can be neglected so that the continuity equation for a constant downward velocity gave an electron density (e.g., half that of a photochemical equilibrium ionosphere). Their corresponding altitude profile for B in the Martian ionosphere was constant down to ~ 300 km (assuming an ionopause at ~ 400 km) at a magnitude of 1.08 times the post-shock field (which may be up to 4 times the interplanetary field strength), and then rapidly dropped to a very low value by ~ 200 km. After this calculation was published, analyses by other investigators showed that the small plasma scale height could be explained by a colder-than-usual exosphere resulting from the low EUV flux at the time of the observations (e.g., Hogan *et al.*, 1972). This finding obviated the need for an explanation involving solar wind interaction effects; nevertheless, the authors of this model could arguably stand by it since we still do not know what the magnetic field looks like in the Martian ionosphere. Their later work, based on PVO *in situ* observations of ionospheric magnetic fields at Venus, did not use the same approach.

In the second series of papers from the same group, Ohm's law was invoked to solve the problem of the current distribution in planetary atmospheres. The authors in this case (cf. Cloutier and Daniell, 1973) initially used Spreiter *et al.*'s (1970) gas dynamic model of the magnetosheath and models of the planetary atmosphere and ionosphere to estimate the normal component of the current through the ionopause from the surface integral of the product of the Pederson conductivity, the tangential velocity and the magnetic field. This calculation was based on the premise that the currents are purely resistive, or given by

$$\mathbf{J} = \sigma \cdot \mathbf{E} ,$$

while E at the ionopause is given by the convection electric field $E = -V \times B$ where V is the sheath plasma velocity at the boundary and σ is the conductivity. The combination of a $-V \times B$ E field, which is usually associated with collisionless plasmas, and resistive current was deemed appropriate in their estimation. They then argued that these currents map through the ionopause from the magnetosheath into the ionosphere to cancel the compressed interplanetary magnetic field within the ionosphere. This requirement determined their ionopause height since the ionopause was assumed to be located at an altitude high enough to allow sufficient internal current density. In this first paper, no attempt was made to describe the configuration of the ionospheric currents except to say that they flow along lines of constant longitude in the noon-midnight plane (for perpendicular IMF).

The second paper in this series (Daniell and Cloutier, 1977) presented the calculation of these Ohmic currents assuming that they flow in a pattern that minimizes the Joule heating in the ionosphere. For this purpose, an iterative or variational technique was applied to find the electric field or potential consistent with a minimum in $\mathbf{J} \cdot \mathbf{E} = (\sigma \cdot \mathbf{E}) \cdot \mathbf{E}$ in a volume with the conductivity distribution of the ionosphere of Venus. The problem was reduced to two dimensions in the noon meridian plane, with the ionospheric magnetic field perpendicular to that plane and the ionopause altitude variation with solar zenith angle assumed to be that given by Spreiter *et al.* (1970) based on pressure balance between the incident solar wind and ionosphere. The authors had to adopt an initial ionospheric magnetic field distribution for the purpose of calculating the conductivity. This was given by the analytical expression

$$\mathbf{B}(h, \phi) = B_0 \cos\phi \; \frac{\exp((h - h_1)/H)}{1 + \exp((h - h_1)/H)} ,$$

where ϕ is solar zenith angle, h is altitude, h_1 is ionospheric density peak altitude, and H is scale height. Iteration was to lead to subsequently improved estimations of **B**. However, the final result given for the ionospheric field appears to be practically identical to their initial 'guess' which was based on intuitive grounds and a knowledge of the magnetic field distribution at the ionopause from the magnetosheath model. Aside from this, a more important criticism was brought to light by the authors themselves in a note added in proof which acknowledged the problem of the neglect of non-Ohmic currents in their analysis.

The third and final paper in this series attempted to add non-Ohmic or drift currents related to the ionospheric plasma pressure gradients (Cloutier and Daniell, 1979). Here the authors reasoned that as long as the collision frequency was much less than the gyro-frequency, one could simply add the diamagnetic drift current

$$J_D = \frac{\mathbf{B} \times \nabla \mathbf{p}}{B^2}$$

to the Ohmic $\mathbf{J}_E = \sigma \cdot \mathbf{E}$ which had been the focus of their earlier work. They assumed that if ∇p and B could initially be calculated from the external (magnetosheath) flow field model, the same iterative technique as used previously could be used to solve for $\mathbf{E}$. Further, assuming that the internal (ionospheric) plasma velocity is given by $\mathbf{V} = \mathbf{E} \times \mathbf{B}/B^2$ (in spite of the inclusion of collisional current dissipation), they added the continuity equation to the calculation:

$$\nabla \cdot (n\mathbf{V}) = P - L \,,$$

where n is density and P and L are the usual ion production and loss rates, respectively. A 'double iteration' was then used to solve for $\mathbf{E}$ (by the variational technique) and n sequentially. The authors remarked on the omission of any thermal diffusion process in their calculation of n, but they calculated a density profile for Mars that agreed with the profile observed on the Viking landers. They thus asserted that the topside ionosphere of Mars is in a 'convective equilibrium' where solar wind-driven convection establishes the topside density profile rather than photochemical equilibrium or dif- fusion. Since their calculated current density contours for the ionosphere appear much like those in the preceding calculations without drift currents, one can assume that their results on the ionospheric magnetic field were essentially unchanged by this addition. This observation aside, the initial assumption that drift and Ohmic currents can simply be added should be considered. Strictly speaking, the combined currents must satisfy the generalized Ohm's law (Equation (5)) which does not reduce to a simple addition of Ohmic and drift currents even when the collision frequency is much less than the gyro-frequency. The Ohmic current is obtained from the GOL when the only terms of significance are $\mathbf{E}$ and $\eta \mathbf{J}$. If pressure gradients, and hence drifts, are present in a resistive medium, the current must be described by the more complete generalized Ohm's law $\mathbf{E} + \mathbf{V} \times \mathbf{B} = \eta \mathbf{J}$. One can see that even for scalar η, the additional term $V \times B/\eta$ is not equal to the drift current $\mathbf{B} \times \nabla p/B^2$. The reason is that, when collisions are present, J is modified by drifts through their effects on $\mathbf{E}$ and $\mathbf{V} \times \mathbf{B}$. The 'collision- less' expression for J_D is no longer appropriate. Thus it seems that the basic premise of this model leaves some questions (see Vaisberg and Zelenyi, 1984).

The final phase in the progression of Rice models of currents and fields in planetary ionospheres was first described in a paper by Cloutier et al. (1981). Herein, the authors present three modifications to their earlier model including the effects of plasma depletion and magnetic field enhancement near the ionopause, velocity-shear induced Kelvin–Helmholtz (K–H) instabilities in the ionosphere, and the variability of the solar wind and IMF direction. It appears as if their two-dimensional ionospheric magnetic field configuration which we described earlier is adopted and varied according to an 'empirical latitude scaling relationship' to allow latitude variations in the current contours, which could also be arbitrarily varied with altitude in concert with the assumed ionopause height. The authors then considered the criterion for the K–H instability in a collisionless plasma for uniform **B**:

$$\rho(\mathbf{k} \cdot \mathbf{V})^2 > \frac{\mathbf{k} \cdot \mathbf{B}}{\mu_0} \, ,$$

where μ_0 is the permeability, ρ is the plasma mass density, **V** is the ionospheric convection velocity as before, and **k** is the wave number of a perturbation that can grow in the presence of a shear in the velocity perpendicular to **B**. They illustrated how waves developing at a shear interface can in principle twist up a magnetic field perpendicular to the flow, and argued that collisional diffusion will cause the magnetic field lines near the axis of a cylindrical flux rope to straighten out first because they are 'older'. Moreover, they considered that the altitude variation of the magnitude of the magnetic Reynolds number can explain why small flux ropes occur at low altitudes and large flux ropes occur at high altitudes. In their view, the condition $R_m = 1$ determines the likelihood of flux ropes forming, and σ varies with altitude to fulfill this condition for smaller structures at the lower altitudes where σ is larger. Applying an 'empirical' instability criterion

$$\mu_0 \rho V^2 / B^2 \cos^2 \theta > C \, ,$$

where θ is the angle between **V** and **B**, and C is some arbitrary constant, to their density, velocity and B field model, they determined where in the volume of the ionosphere flux ropes must form from the twisting of the larger scale field, and attempted to model a number of time series of magnetic field data obtained on PVO. These modeled time series presumably incorporated the adjustments of the current system altitude and strength mentioned above. The effects of an IMF that is not perpendicular to the noon–midnight meridian are qualitatively added by considering the relative geometry of a rotated draped field and anti-solar ionospheric convection. Thus there are many free parameters in this model. In brief, the results of this work suggested that flux ropes form locally, within the ionosphere, due to the presence of a large-scale magnetic field and a velocity shear in the ionospheric plasma flow. The large-scale field and flux rope structures coexist and are in fact intimately related. In the framework of this model, the observed large-scale ionospheric fields are located only in those subregions of the original structure where the criterion for K–H instability is not satisfied.

The authors' last elaboration on these ideas, in the form of a three-dimensional extension of this highly evolved model, is presented in Cloutier *et al.* (1983). In this paper a general set of steady-state MHD equations describing the ionosphere of Venus is solved under the assumption of purely horizontal magnetic fields and convection. The ionospheric velocity used here is self-consistent in that the B field configuration is affecting the velocity (a departure from the old frozen field assumption), but this part of the calculation is not included and not published. The authors argued that the velocity pattern that they calculated makes the magnetic field lines exhibit a belt-buckle-like configuration in the ionosphere. By applying the same K–H instability criterion as in their two-dimensional model, the authors modeled the three-dimensional topology of the region of stability, or large-scale field, in the ionosphere. They again compared their modeled locations of large scale fields to PVO observations, pointing out the differences that the relative orbit and large-scale field region geometries make in reproducing various details. They concluded that most of the structure of the large-scale field observed along the PVO orbit is latitudinal structure, although altitude structure plays some small role. The latitudinal boundaries are determined by the K–H instability. This is in contrast to the view that altitude structure is actually what determines the appearance of the large-scale field structures, and that the ionopause altitude determines the horizontal boundaries of those structures, which is the conclusion of the alternative models described below.

It should be mentioned that in all of the Rice group models, the absorption of solar wind plasma by the ionosphere is implicit. It is argued by the authors that the absorption imposes the electric field across the ionosphere that drives the internal currents associated with the field. Electric fields from other sources such as the temporal variability of B and the convection of magnetized ionospheric plasma are not explicitly considered. The idea that ionospheric currents will continuously flow simply in the presence of gradients in B if the magnetic field of the solar wind (devoid of plasma) diffuses into the ionosphere and is maintained in a state of non-uniformity by ionospheric convection and dissipation which is essential to the alternative models, is found only in one of the more recent works by Cloutier (1984). In this paper on the subject of the large-scale ionospheric fields, the author uses an approach that bears a closer resemblance to that of Luhmann (1984) and Cravens et al. (1984).

In this latest model (Cloutier et al., 1987), the steady-state continuity and momentum equations are solved together with an energy equation. These equations are integrated over altitude. Instead of using the induction equation to solve for B, as described above in the tutorial section, Cloutier et al. integrate the energy equation to find B. In order to do this integration, they assume empirical temperatures and a heat source function which is varied to make the resulting altitude profile of the magnetic field resemble a typical observed profile. However, their corresponding downward plasma velocity obtained from these calculations has a peak value of almost 1 km s^{-1}, which is super-sonic or near-supersonic. This large velocity appears to be associated with the large pressure gradient force they impose near 160 km. The pressure gradient at 160 km is very large in this model because the O_2^+ density at that altitude is roughly 10 times less than the observed density (cf. Brace et al., 1983a, b). Since this problem of correct densities has yet to be resolved, we next consider the alternative approach to the Rice models mentioned above.

3.2.1.1. *Diffusion/Convection Models.* The forerunner of this series of models was a 1981 paper in which an attempt was made to empirically deduce the altitude dependence of the ionospheric current density. In this paper, Luhmann and coworkers used the observed electron density and temperature throughout the ionosphere, together with the observed magnetic field near the ionopause, to integrate Ampère's Law

$$\frac{\partial B_x}{\partial z} = \mu_0 J_y = \mu_0 ne(v_{iy} - v_{ey})$$

downward. The ion and electron velocities were obtained from the momentum equations for the two species. The ion temperature, electric field, and neutral wind in the momentum equations were varied as free parameters in order to determine what values were necessary to produce the observed altitude profile of B_x. The authors concluded that sometimes the assumptions that the ion temperature is equal to one-half the electron temperature, and that the electric field and neutral wind effects are negligible, produce a good approximation to the magnetic field profile, and hence, to the current density profile within the ionosphere. They also examined the force balance (assuming steady conditions) in the highly magnetized ionosphere and demonstrated that the vertical force balance in the region observed by PVO is predominantly between the thermal and magnetic pressure gradient forces $\nabla p = \nabla(nk(T_i + T_e))$ and $\nabla B^2/2\mu_0$. It was pointed out that time-dependence and 3-dimensional gradients not easily deduced from the data could play a role in producing the observed magnetic fields as well.

(i) Kinematic Dynamo Versions. The first diffusion/convection model of the iono- spheric fields was developed for the purpose of examining the question of steady versus non-steady current systems as an explanation for the observations of the large scale field. The Rice models considered the distinctive altitude profiles of the dayside fields as evidence for spacecraft sampling of a 'permanent' ionospheric magnetic field structure with latitudinal gradients (cf. Cloutier *et al.*, 1983), whereas the UCLA view, based on data analyses, was that they could instead be characteristic of the 'decay phase' of an initially more uniform large-scale field. In particular, Russell *et al.* (1983) had argued that the large horizontal anti-solar plasma velocities observed in the upper ionosphere could produce the observed ~ 190 km altitude minimum in the field profiles by convecting the high altitude field away at a rapid rate compared to that at lower altitudes. It now appears that both of these conceptual views were accurate in some respects but not in others. The most recent quantitative models described below indicate that steady large-scale field structures can indeed exist when the ionopause is low, but they also evolve with time after the conditions at the ionopause (e.g., in the solar wind) change. However, these models also indicate that the observed structure can be largely explained by vertical gradients, and that vertical (not horizontal) velocities probably play the major role in determining the appearance of the characteristic altitude profiles.

   Luhmann *et al.* (1984) first attempted to resolve the issue of temporal evolution by considering the evolution of horizontal B, in a one-dimensional (slab) model of the ionosphere, according to the induction equation (Equation (9)) which, together with the ion and electron momentum equations

$$-\nabla p_i + nm_i \mathbf{g} + \tfrac{1}{2} nm_i \nu_{in}(\mathbf{v}_n - \mathbf{v}_i) - nm_i \nu_{ie}(\mathbf{v}_i - \mathbf{v}_e) + en(\mathbf{E} + \mathbf{v}_i \times \mathbf{B}) = 0 \,,$$

$$-\nabla p_e + nm_e \mathbf{g} + nm_e \nu_{en}(\mathbf{v}_n - \mathbf{v}_e) - nm_e \nu_{ei}(\mathbf{v}_e - \mathbf{v}_i) - en(\mathbf{E} + \mathbf{v}_e \times \mathbf{B}) = 0$$

and Ampère's Law

$$\mathbf{J} = (1/\mu_0)(\nabla \times \mathbf{B}) = ne(\mathbf{v}_i - \mathbf{v}_e)$$

gives the one-dimensional diffusion/convection or dynamo equation (Equation (11))

$$\frac{\partial B}{\partial t} = -\frac{\partial}{\partial z}(WB) + \frac{\partial}{\partial z} D \frac{\partial B}{\partial z}$$

as described above. The diffusion coefficient, which here takes the form

$$D = m_e(\nu_{en} + \nu_{ei})/(ne^2 \mu_0)$$

and vertical ionospheric drift velocity, $W = v_{iz} = v_{ez}$, are the key parameters in the field evolution as previously noted in the tutorial section.

The authors numerically solved the above equation using an observed profile of B as an initial condition, with empirical models for the densities and temperatures in the collision frequency-dependent diffusion coefficient, and assuming various magnitudes and altitude dependencies for the vertical drift velocity W. The altitude dependence of the collision frequencies used for evaluating the diffusion coefficient, and the diffusion coefficient are shown in Figures 31 and 32, respectively. These were assumed not to vary with time in the calculations. The altitude and field of the ionopause upper boundary were also held constant except for special cases where the effect of changing solar wind conditions was under investigation. This initial model demonstrated several key points about the dayside ionospheric magnetic field. First, it demonstrated that magnetic fields imposed in the ionosphere at the time of some magnetizing 'event' could persist for hours after that event. Secondly, it showed that both the altitude profile and rate of decay of the field at a particular altitude depended on the assumed vertical velocity magnitude and altitude profile. However, it did not address the nature of W.

Cravens *et al.* (1984) next applied an ionospheric model based on the coupled continuity and momentum equations for seven ions (CO_2^+, O_2^+, O^+, N^+, C^+, He^+, H^+), and an empirical model of the ion and electron temperatures, to calculate the actual value of W:

$$W = \frac{1}{nm_i \nu_{in}} \left[\frac{\partial p}{\partial z} + nm_i g \right] =$$

$$= -D_{in} \left[\frac{1}{n} \frac{\partial n}{\partial z} + \frac{T_e/T_i}{n_e} \frac{\partial n}{\partial z} + \frac{1}{T_i} \frac{\partial}{\partial z}(T_e + T_i) + \frac{m_i g}{kT_i} + \frac{1}{nkT_i} \frac{\partial}{\partial z} \left(\frac{B^2}{2\mu_0} \right) \right]$$

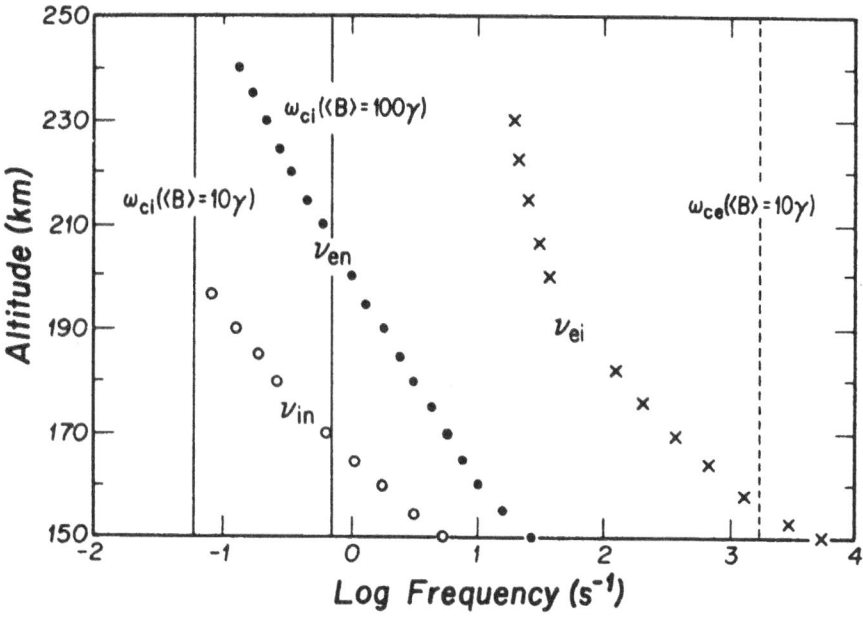

Fig. 31.   Collision frequencies versus altitude for the subsolar Venus ionosphere (v_{en}, v_{in}, and v_{ei} are electron-neutral, ion-neutral, and electron-ion collision frequencies). The dashed lines show the ion and electron gyro-frequencies ω_{ei} and ω_{ce} for different field strengths for comparison. The models of Theis *et al.* (1984) and Hedin *et al.* (1983) were used to compute the collision frequencies. (From Luhmann and Elphic, 1985.)

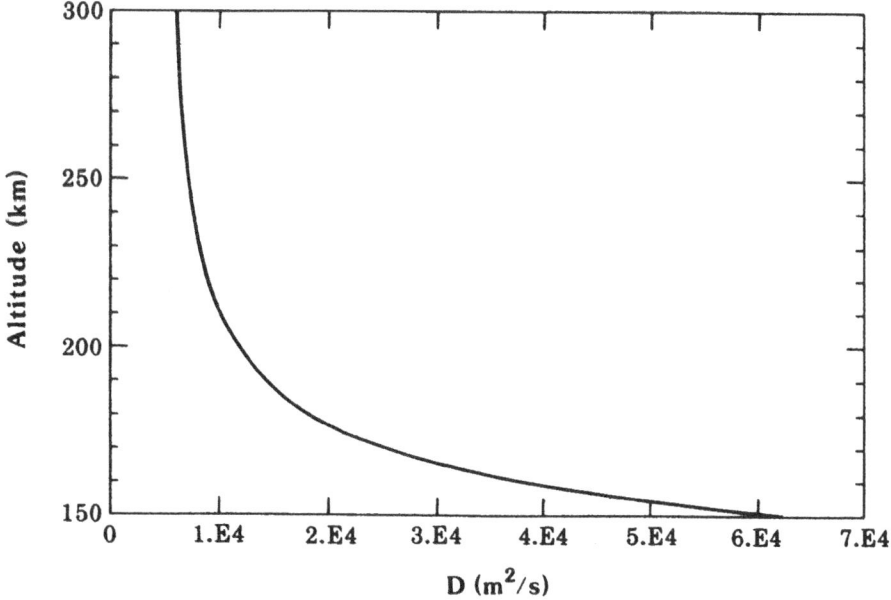

Fig. 32.   Diffusion coefficient ($D - m_e(v_{en} + v_{ei})/\mu_0 ne^2$) versus altitude for the subsolar Venus ionosphere. This profile is not very sensitive to solar zenith angle over the dayside. (From Luhmann and Elphic, 1985.)

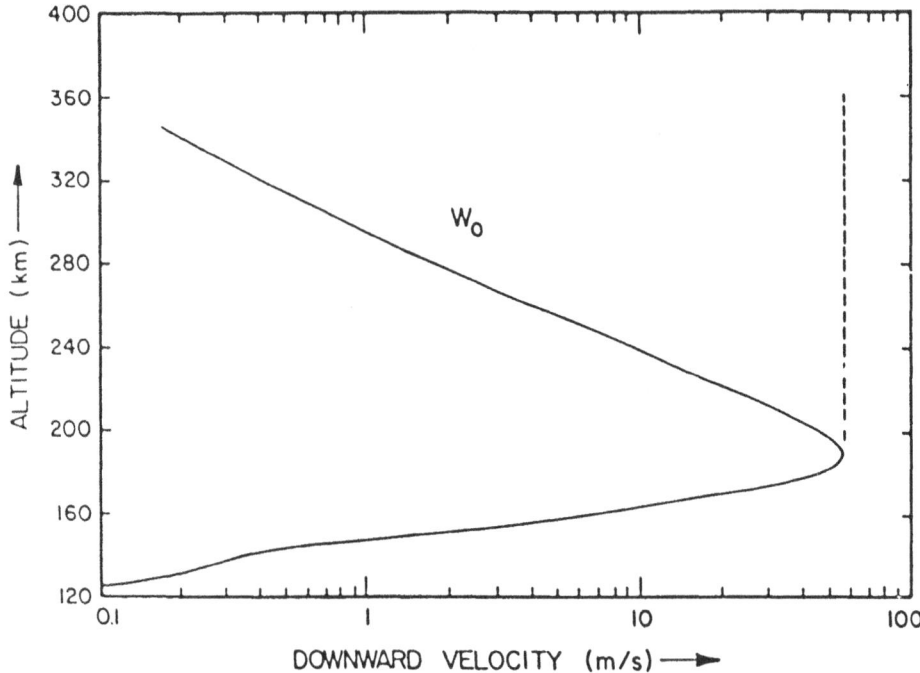

Fig. 33.   Calculated downward ionospheric plasma velocity, W, as a function of altitude (solid line). (From Cravens *et al.*, 1984.)

at first neglecting the contribution of the magnetic pressure term. This calculation gave a vertical velocity profile, shown here in Figure 33, which produced enduring magnetic fields with the distinctive observed altitude profile. It was now evident that the usual ionospheric vertical drift was downward, so that the downward velocity could maintain the magnetic field profile for long periods of time (hours) since flux can be constantly convected downward from the high altitude boundary. The characteristic 170 km altitude maximum and the minimum near 190 km appeared naturally in the solutions of the diffusion/convection equation as shown by the examples in Figure 34. The 190 km minimum is related to the velocity maximum near that altitude. At the bottom of the ionosphere, Ohmic dissipation destroys the magnetic field that was carried from above. This 'conveyor belt' picture of Venus' ionospheric field is illustrated schematically in Figure 35. Addition of the corresponding magnetic pressure term only slightly altered these profiles. Thus the distinctive magnetic field profiles appeared to be attributable to vertical, rather than horizontal, structure in the large-scale ionospheric field. These magnetic fields were also shown to persist for observationally significant periods of time.

We call these models kinematic dynamo models because they use prescribed diffusion coefficients and velocities in the diffusion/convection or dynamo equation. In the more correct, fully self-consistent MHD models, the velocities, densities, and temperatures in the plasma are also calculated because they are affected by the magnetic field.

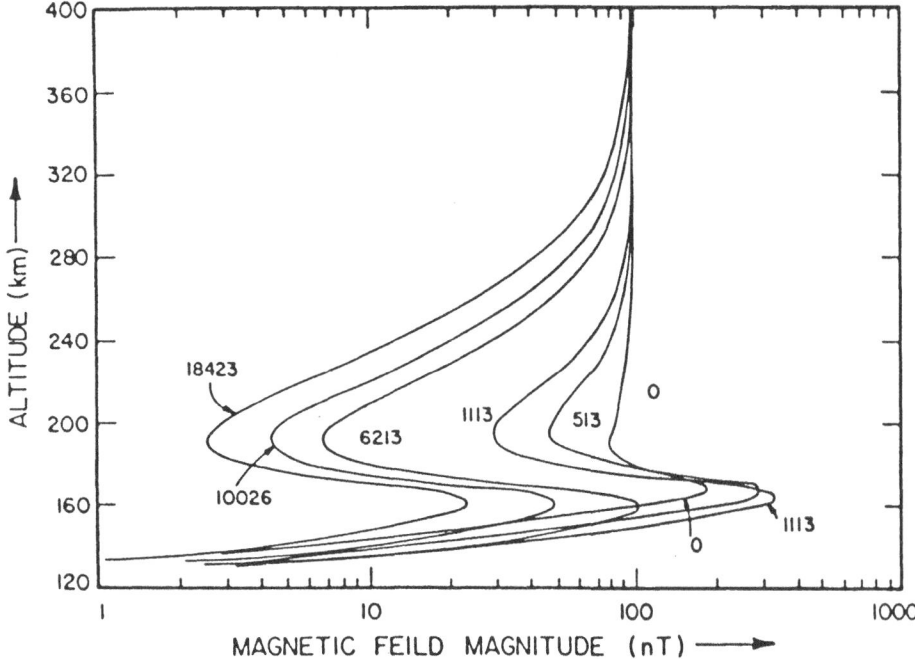

Fig. 34. Altitude profiles of calculated magnetic field strength labeled with time in seconds. The calculations assumed that the ionospheric plasma velocity was always equal to W. (From Cravens *et al.*, 1984.)

However, the apparent success of the kinematic dynamo approach encouraged several further applications which should be mentioned here. In particular, a study by Phillips *et al.* (1984) approximately incorporated solar wind conditions into the model, while Luhmann *et al.* (1988) attempted to extend it to the global picture of the large-scale fields.

Phillips *et al.* (1984) used the velocity profile of Cravens *et al.* (1984) in an investigation of the effect of ionopause height on the ionospheric field. Since the solar wind pressure controls both the ionopause height and the magnitude of B in the overlying magnetic barrier, it was assumed that the magnetic field magnitude at the upper boundary of the model (the ionopause) was that which would give a magnetic pressure equal to the local thermal pressure. The altitude of the upper boundary in the calculation was varied to see what effect these different boundary values of B, together with the fixed altitude dependence of D and W, would have on the solutions. In this case, the field was allowed to grow from a small initial value instead of decaying from a large value. The results are illustrated with some corresponding data in Figure 36. Although the calculations did not consider modifications to D and W near the ionopause, these calculations demonstrated how ionopause altitude and field strength might affect the ionospheric field profile. The calculated low altitude field maximum had magnitudes that depended on ionopause altitude, as observed, except for very low ionopauses when the calculation did not converge to a steady profile for the fixed values of D and W. This result indicated

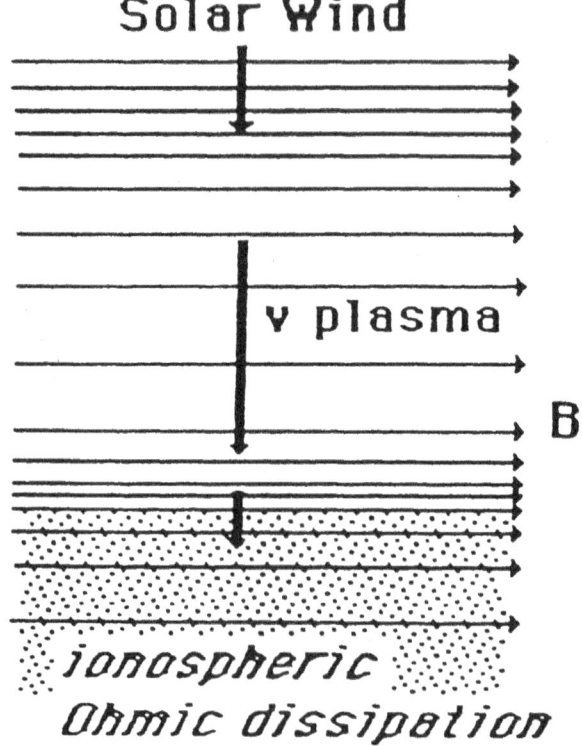

Fig. 35.   Schematic illustration of the vertical transport of magnetic flux from the magnetic barrier into the ionosphere, where it is redistributed by altitude-dependent vertical convection and diffusion. The field is ultimately destroyed at the bottom by Ohmic dissipation.

that the values of D and W must be significantly modified by the magnetic field at these times. A velocity profile inferred from plasma and field observations for a highly magnetized, low ionopause orbit indicated that the downward velocity for this example was indeed lower than the values in Figure 33. This revised profile provided a convergent solution to the dynamo equation in this attempt at 'self-consistency'.

The work of Phillips *et al.* also suggested a global picture for the diffusion/convection model where local vertical profiles were determined by the local height of the ionopause, which varies systematically across the dayside ionosphere. Lower ionopauses (altitudes < 300 km) tend to occur at the subsolar point, while high ionopauses are generally found near the terminator. Thus, as in the Rice models, a 'steady' ionospheric field structure can exist for a fixed ionopause position in this model. This field structure also has horizontal dependence in that the strongest fields are found in the subsolar region, but to a first approximation, the controlling coordinate is solar zenith angle and not latitude, and the observed altitude profile is produced by vertical structure and not horizontal structure. As in the Rice concept, an ionopause position that is lowered as a whole by increased solar wind pressure causes the region permeated by strong

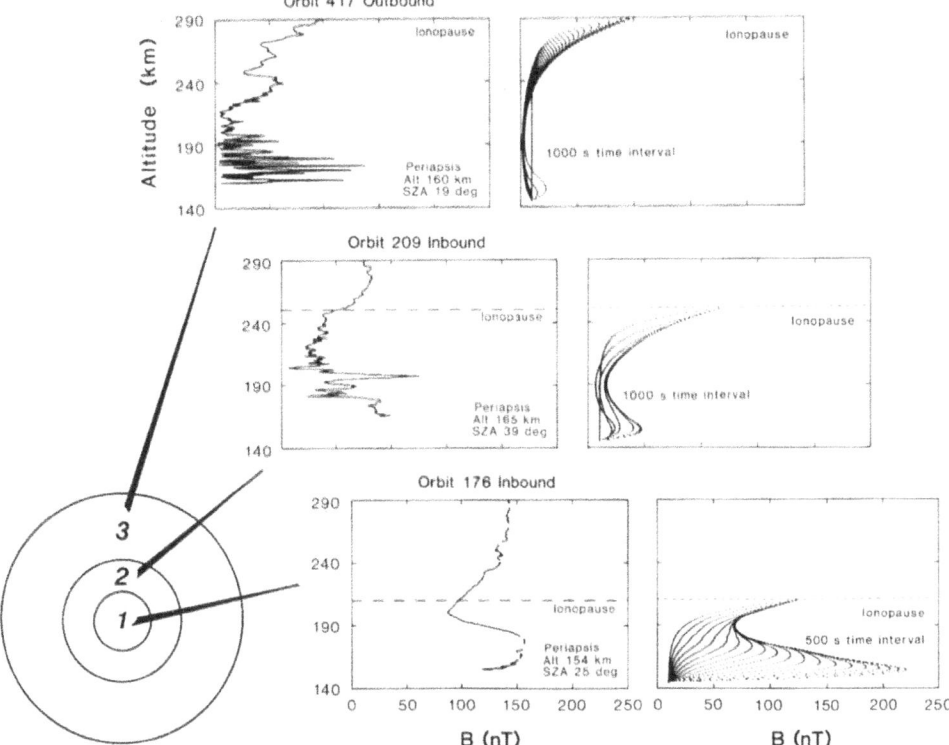

Fig. 36.   Schematic illustration of the general features of the ionospheric magnetic field observed on PVO compared to the results of the one-dimensional calculations from Phillips *et al.* (1984). The regions on the Venus dayside where the different types of profiles prevail because of the ionopause altitude variation with SZA are indicated by the 'bull's eye' pattern. (From Luhmann *et al.*, 1986.)

ionospheric field to grow. Conversely, an inflated ionopause can cause the large-scale field to practically disappear everywhere in the ionosphere. The nominal or average ionopause position at Venus during the PVO solar maximum measurements should maintain a small region of magnetized ionosphere near the subsolar point.

One problem with these global extrapolations is that they do not include the effects of the horizontal convection of the ionosphere that was once thought responsible for the field profiles. Although the one-dimensional models appeared quite successful, it was recognized that the anti-solar horizontal plasma drift speeds can greatly exceed the vertical velocity as indicated by the empirically modeled horizontal velocities in Figure 37 (Theis *et al.*, 1984). These large velocities are to a first approximation driven by day-to-night thermal pressure gradients in the plasma. Luhmann (1988) attempted to incorporate these drifts into a global model wherein the three-dimensional version of the diffusion/convection equation

$$\frac{\partial \mathbf{B}}{\partial t} = \nabla \times (\mathbf{V} \times \mathbf{B}) - \nabla \times D(\nabla \times \mathbf{B})$$

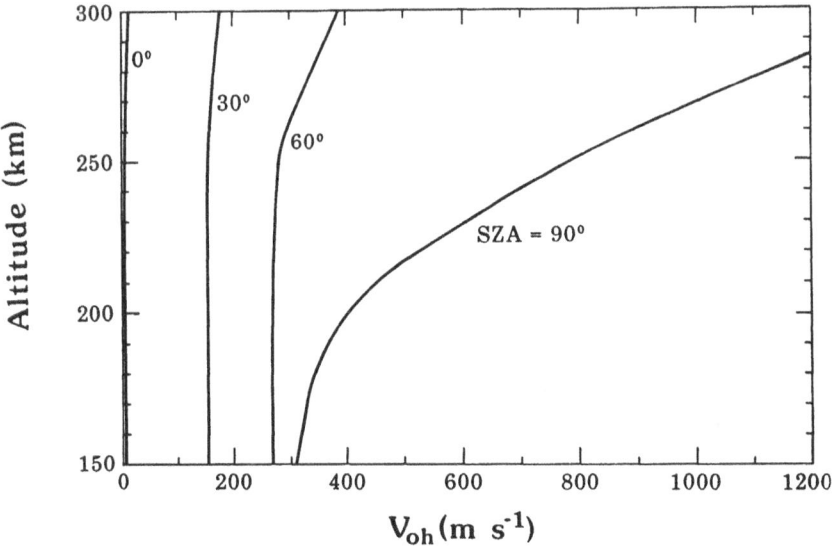

Fig. 37. Altitude profiles of the antisolar horizontal ionosphere drift velocities at various solar zenith
angles according to the model of Theis *et al.* (1984). (From Luhmann and Elphic, 1985.)

is solved in the kinematic dynamo approximation with the total ionospheric velocity field
described by the sum of the velocities illustrated in Figures 33 and 37. Their result, which
is illustrated by Figure 38, unfortunately did not have sufficient spatial resolution to
resolve the three-dimensional ionospheric field structure near its 170 km altitude peak.
However, the results did indicate that the ionosphere distorts the draped magnetosheath
field as it penetrates the ionosphere. This kind of distortion is indeed suggested by some
of the observations of adjacent magnetosheath and ionospheric field vectors as shown
in Figure 39. Any future efforts along these lines will have to deal effectively with the
problem of high spatial resolution in a curved layer of ionosphere that is locally thin but
horizontally extensive.

The distribution of magnetic fields in the ionosphere is so dependent on the
ionospheric plasma velocity in the magnetic diffusion/convection model, that it is
worthwhile to consider some additional details of that velocity. The plasma moves in
response to the forces on it, including those associated with gradients in both the
magnetic and thermal pressures. Thermal pressure gradients result from local imbalances
in heating or in the photochemical production and loss rates of ions. For altitudes
greater than 200 km, more O^+ ions are produced than are removed by chemistry. As
a consequence, the density of O^+ at high altitudes somewhat exceeds diffusive
equilibrium values and the O^+ plasma diffuses downward towards lower altitudes and
also flows horizontally from day to night. The downward flux of O^+ is removed by
ion-neutral recombination at lower altitudes in the photochemical region. Since the
plasma motion has both horizontal and vertical components, the magnetic field is also
convected in both dimensions, but there are some simplifying approximations that are

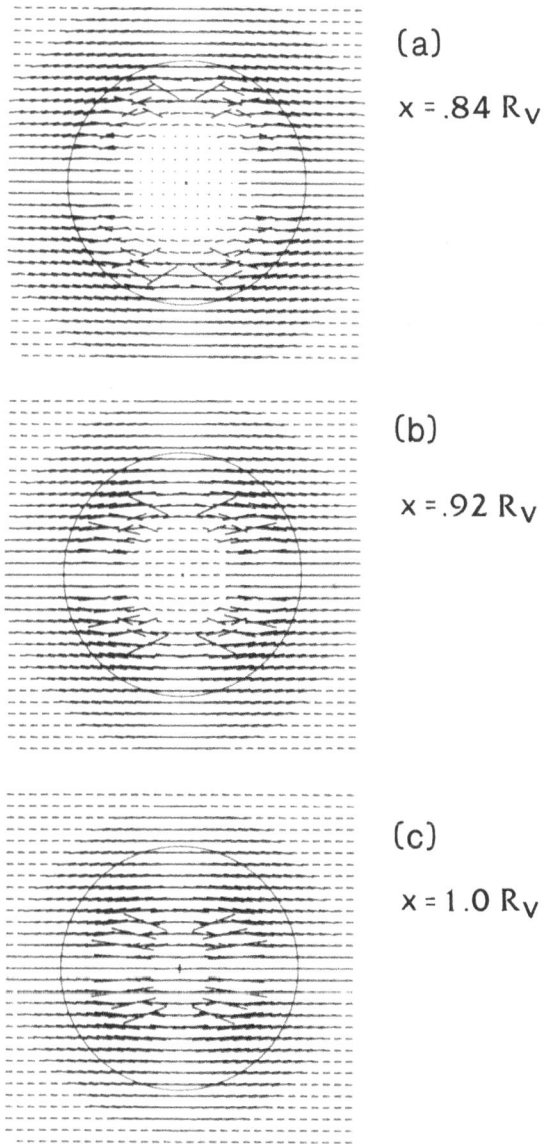

Fig. 38.   (a) View from the Sun of magnetic field vectors projected on the plane perpendicular to the x (upstream flow) axis located at $x = 0.84\ R_v$ from the center of the planet in the direction of the Sun. The vectors are centered at computational grid points. The vectors in the upstream solar wind at Venus typically have magnitudes ~ 10 nT. (b) Same as (a) but for $x = 0.92\ R_v$. (c) Same as (a) but for $x = 1.0\ R_v$. (From Luhmann, 1988.)

appropriate under different conditions. How the plasma moves depends on the altitude and solar zenith angle, as well as on the state of magnetization of the ionosphere. Krymskii and Breus (1988) discussed the large-scale convection pattern from the point

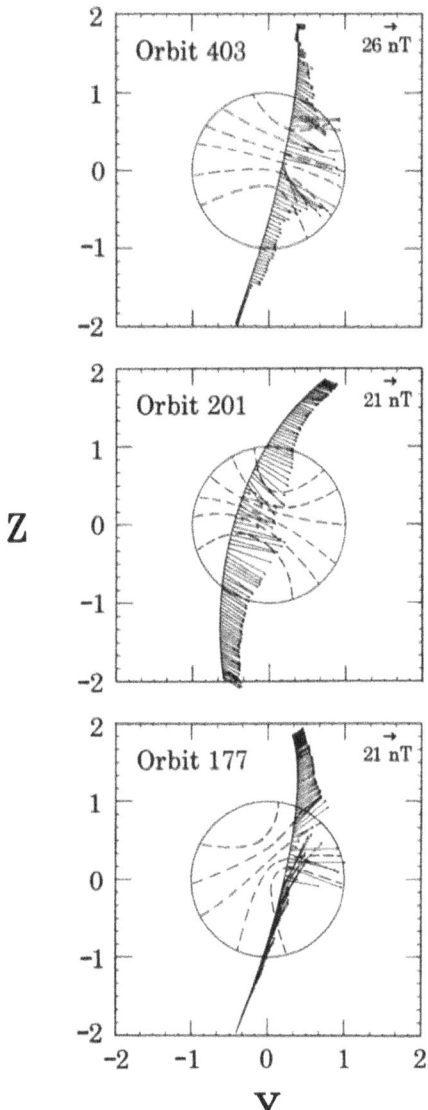

Fig. 39.   Examples of projected magnetic field vectors, as viewed from the Sun, observed along the PVO trajectory (from Luhmann *et al.*, 1987). As suggested by the dashed lines, the 'focusing' of the draped magnetic field toward the subsolar point which is seen in the model is also evident in these data.

of view of what should happen for different ionopause heights. For highly magnetized conditions, when the ionopause is low, the plasma flow is probably largely downward over most of the dayside. When the ionosphere is unmagnetized and the ionopause is high, the plasma also flows from day to night, especially at high altitudes and large solar zenith angles. For very low solar wind dynamic pressures, when the solar wind has very little effect on the ionospheric dynamics and large scale ionospheric fields are precluded, the plasma flow can actually be upward at high altitudes ($z > 250$ km), although it remains downward at lower altitudes. Excess O^+ production near 200–250 km is removed both via downward flow and by flow up and then horizontally towards the nightside (Knudsen *et al.*, 1980; Cravens *et al.*, 1984). The region of upward vertical flow would be confined to higher solar zenith angles where the ionosphere has a relatively high ionopause compared with the subsolar region. A schematic of this flow scenario is given in Figure 40. However, one can argue that for the purpose of understanding the large scale field, the downward transport of ions is the most important consideration. Even though the horizontal flow speeds can be much larger, the relevant vertical scale length (10's of km) is much less than the horizontal scale length ($L \approx R_v = 6053$ km) (Shinagawa and Cravens, 1988).

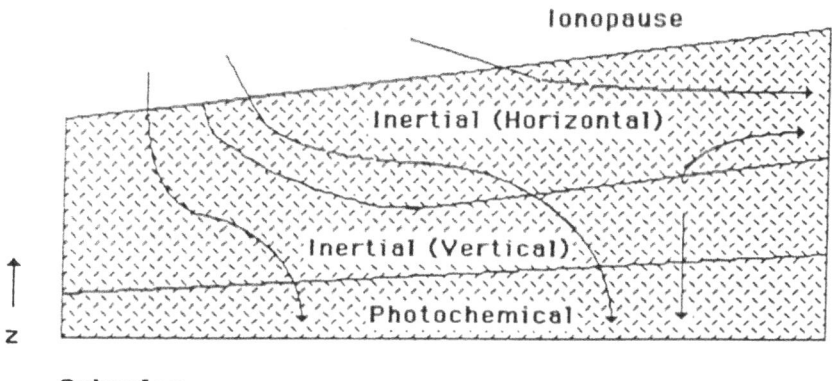

Fig. 40. Schematic illustration of the different regions of the ionosphere showing mainly vertical convection in the subsolar region and the increasing importance of horizontal and (possibly) upward convection at high altitudes and larger solar zenith angles.

Another important factor in the production of the large-scale field is the upper boundary condition. At this time we have only made simplifying assumptions about the vertical velocity and diffusion coefficient at the ionopause. Phillips *et al.* (1984) simply used the nominal models of vertical velocity and diffusion coefficient without considering the possibility of anomalous diffusion or altered (reduced?) velocity in the ionopause gradient. The Rice models described above invoked other boundary conditions as do the most recent MHD models described below. Observationally, all we know is that the solar wind appears to flow around an obstacle that is larger than the measured ionopause surface. This implies that solar wind plasma is deflected around a 'magnetic barrier' so that one should be able to treat the upper boundary in the one-dimensional models as a layer of horizontal magnetic field. We also know that the altitude of the boundary, like the field strength in the magnetic barrier, is determined by the incident solar wind pressure, and that there is an approximate pressure balance of external (field) and internal pressures through the ionopause. But we must assume some upper boundary value for the velocity, and we must assume that the diffusion coefficient is known. We also do not know whether the planetary ions or photoelectrons produced above the ionopause, which are apparently accelerated and removed by the solar wind, are important in the context of the ionospheric field problem. Similarly, we do not consider what happens to the ionospheric field during sudden compressions and expansions of the ionosphere in response to rapid changes in solar wind pressure at the upper boundary, although Wolff et al. (1982) suggest that ionospheric shocks and rarefaction waves occur regularly. Changes in the magnetic field at the upper boundary may also be expected to propagate through the ionosphere. These limitations of the present models should be kept in mind.

(ii) MHD Models.   As previously mentioned, plasma densities and velocities are assumed to remain unaltered in kinematic models of the magnetic field. However, the distribution of the magnetic field affects the motion of the plasma via the $\mathbf{J} \times \mathbf{B}$ force, and any alterations in the ion and electron velocity profiles will then affect the ion density profiles. Ideally, one should solve all of the time-dependent, three-dimensional, coupled MHD equations for all relevant ion species, with sufficiently high spatial resolution to resolve all of the relevant ionospheric structures. This has not yet been done; however, a time-dependent, one-dimensional, multi-species MHD model with high spatial resolution ($\Delta z \approx 2$ km) has been recently developed (Shinagawa et al., 1987; Shinagawa and Cravens, 1988).

Shinagawa et al. (1987, referred to as SCN) and Shinagawa and Cravens (1988, referred to as SC) solved the time-dependent, one-dimensional, continuity and momentum equations for the three ions O^+, O_2^+, and H^+, and a photochemical continuity equation for CO_2^+, together with the magnetic convection/diffusion equation. The background neutral atmospheric densities were taken from empirical models based on Pioneer Venus data (e.g., Hedin et al., 1983). Photochemistry appropriate for the above 4 ion species was used to calculate production and loss terms in the continuity equations. The aeronomy in the SCN and SC models is basically that found in Nagy et al. (1980, 1983) (also see other articles in this issue).

The SCN and SC models are very similar; they differ mainly in how the momentum equation is handled. In SCN, the diffusion velocity approach was taken and the time and convective derivatives of the ion velocity were neglected. In order to achieve this, it was necessary to designate one ion species as the major ion species (O^+). The advantage of this method is that the ion velocities are eliminated as dependent variables once the diffusion velocities are substituted into the respective continuity equations. In SC, each momentum equation was solved and the ion velocities (including horizontal and vertical components) were assumed to be dependent variables. This allows for more accurate treatment of rapid ionospheric variations, and requires no assumptions concerning the identification of a major ion species. Differences between the results of the two versions of the MHD model were not very significant with one exception. For a few of the cases, SC included an additional term in the continuity equation in order to approximate the loss of ions due to horizontal transport. This term was found using the horizontal momentum equation and assuming a Newtonian variation of the magnetic pressure with solar zenith angle. This loss term was not included in the magnetic induction equation and is thus not truly self-consistent, however, it adds another dimension to the model. Hence, the SC rather than the SCN results will be emphasized in this review.

Both the SCN and SC models did not include a solution of the electron and ion energy equations. Instead, temperature profiles based on measurements of T_e and T_i made by the PVO Langmuir probe and retarding potential analyzer were adopted (Brace *et al.*, 1979; Knudsen *et al.*, 1979a, b; Theis *et al.*, 1980; Miller *et al.*, 1980), and extended to lower altitudes using the theoretical results of Cravens *et al.* (1980). Omission of the explicit solutions of the energy equations was thought to be justified since the energetics of the Venus ionosphere are still not well understood, especially for highly magnetized conditions.

The lower boundary condition for the magnetic field in both SC and SCN models is $B = 0$, which is appropriate for a negligible intrinsic magnetic field at Venus. The upper boundary condition is $\partial B / \partial z = 0$ at 500 km altitude. The upper boundary condition on the vertical velocity is given by $E_y = -WB = $ constant, where E_y is the horizontal component of the electric field, which assumes that the other terms in the generalized Ohm's law are small at 500 km. It is assumed that large values of E_y are associated with large, or increasing, solar wind dynamic pressures (and, hence, large or increasing magnetic field strength in the magnetic barrier outside the ionopause).

Figures 41, 42, and 43 show comparisons of magnetic field, electron density, and

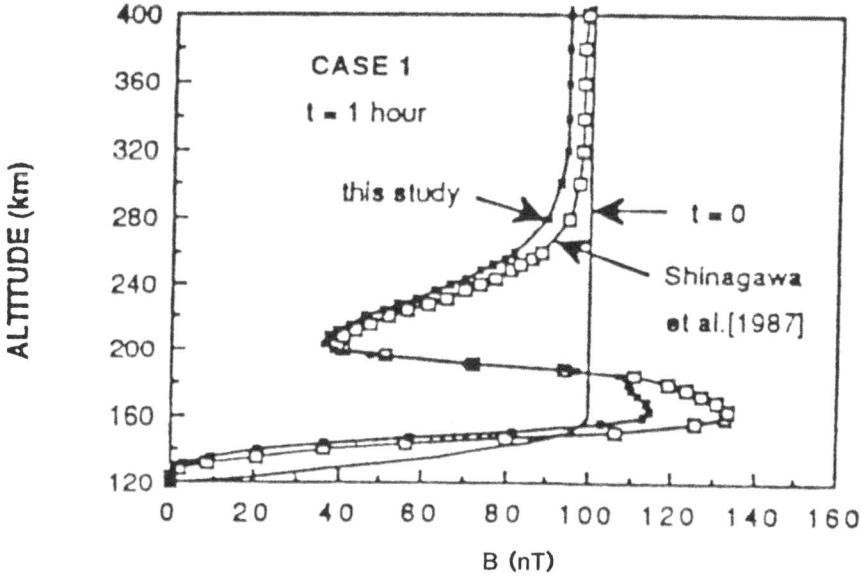

Fig. 41.   Comparison of calculated magnetic field profiles at $t = 1$ hour between the model with one major ion and the multi-species MHD model.

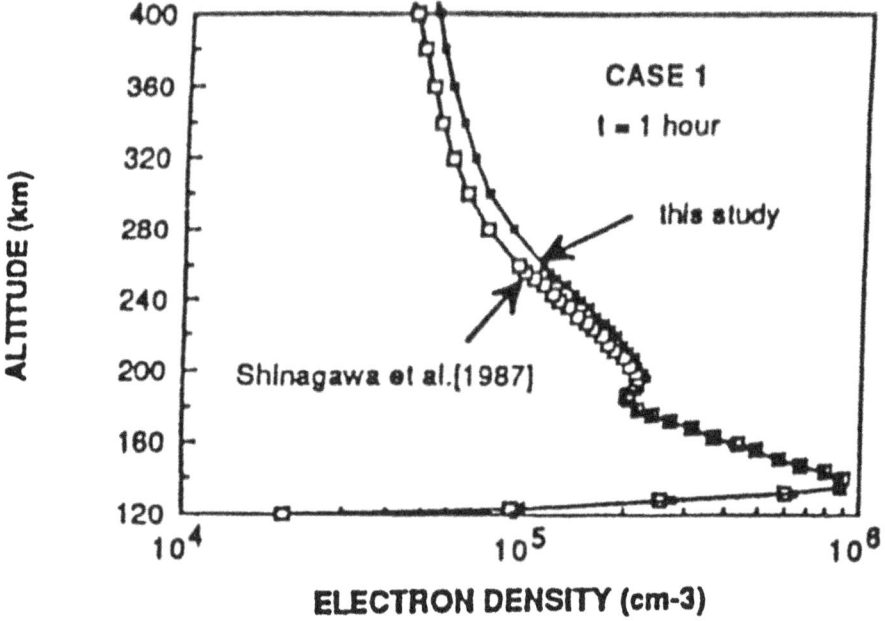

Fig. 42.   Comparison of calculated electron density profiles at t = 1 hour between the model with one major ion and the multi-species MHD model.

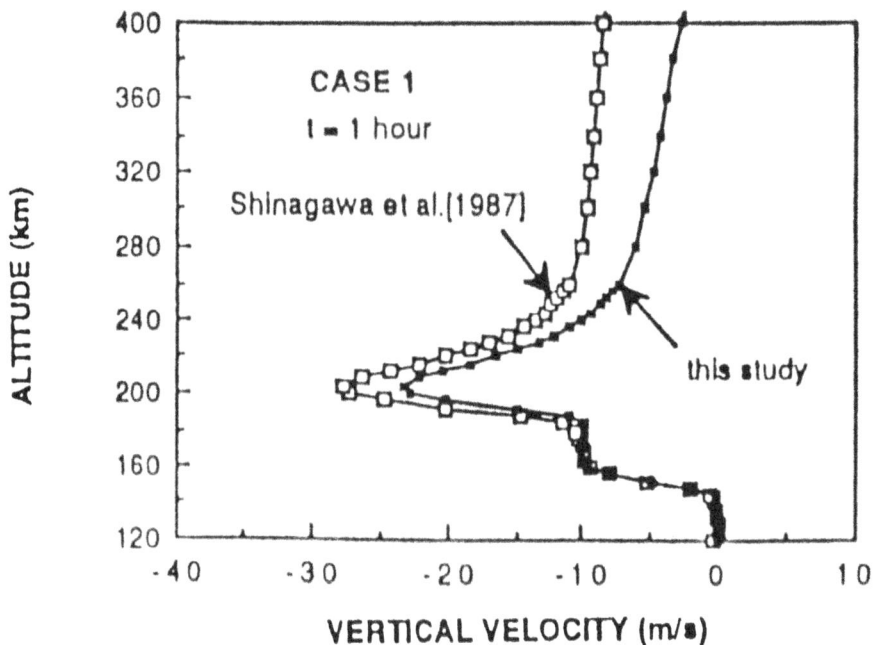

Fig. 43.   Comparison of calculated vertical plasma velocity profiles at t = 1 hour between the model with one major ion and the multi-species MHD model.

average ion velocity profiles for the two versions of the model. For this case, the upper boundary condition for the SC model was $E_y = 0$ or zero downward velocity. The ionospheric field gradually decays under these conditions since there is no resupply of magnetic flux from above. The results of both models are quite similar. The initial magnetic field profile is labelled with $t = 0$, and is independent of altitude above 160 km. The field profile (Figure 41) is well-evolved after an elapsed time of 1 hour and displays the characteristic minimum at 200 km and maximum at 170 km from the vertical velocity profile. The density profile (Figure 42) has a ledge near 200 km due to the perturbation of the downward velocity by the magnetic field gradient. The average plasma velocity profile (Figure 43) at $t = 1$ hour has a maximum downward velocity of 25 m s^{-1} near 200 km. Note that the maximum downward velocity for the field-free ionosphere (Figure 33) was almost 60 m s^{-1}. The magnetic field gradients in the upper half of the low-altitude magnetic layer retard the downward flow of ions, thus reducing this velocity. Figure 44 shows that each individual ion species has a somewhat different velocity profile. The velocities are generally downward, except for O$_2^+$ which flows upward at high altitudes. For times greater than 1 hour (not shown) the magnetic field continues to decay, reaching a maximum value of 30 nT in the low-altitude layer by $t = 6$ hours.

The steady-state ionosphere is magnetized if E_y at the upper boundary is sufficiently large that the resupply of magnetic flux from the magnetic barrier exceeds the rate of Ohmic dissipation of magnetic flux in the lower ionosphere. Figure 45 shows three

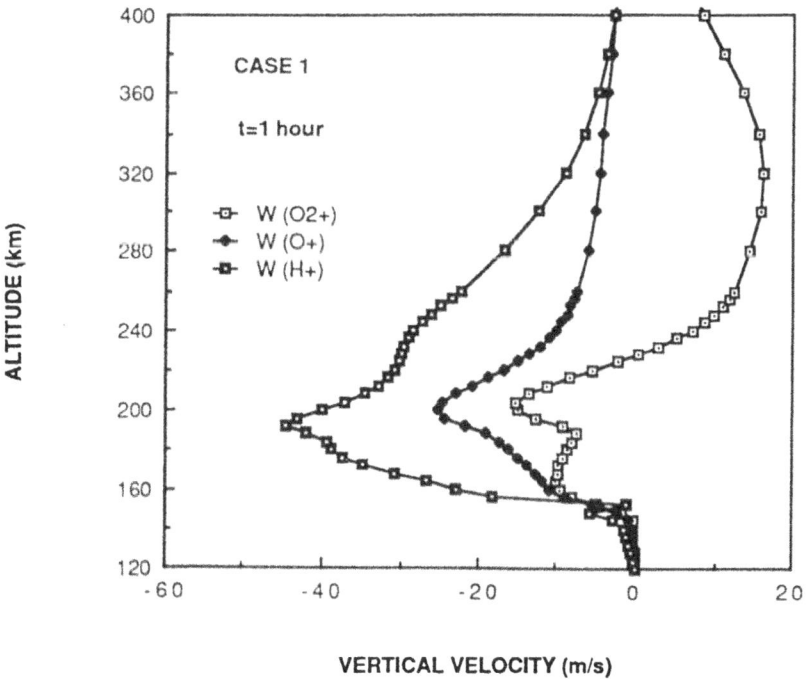

VERTICAL VELOCITY (m/s)

Fig. 44.   Calculated vertical velocity profiles for individual ions at $t = 1$ hour.

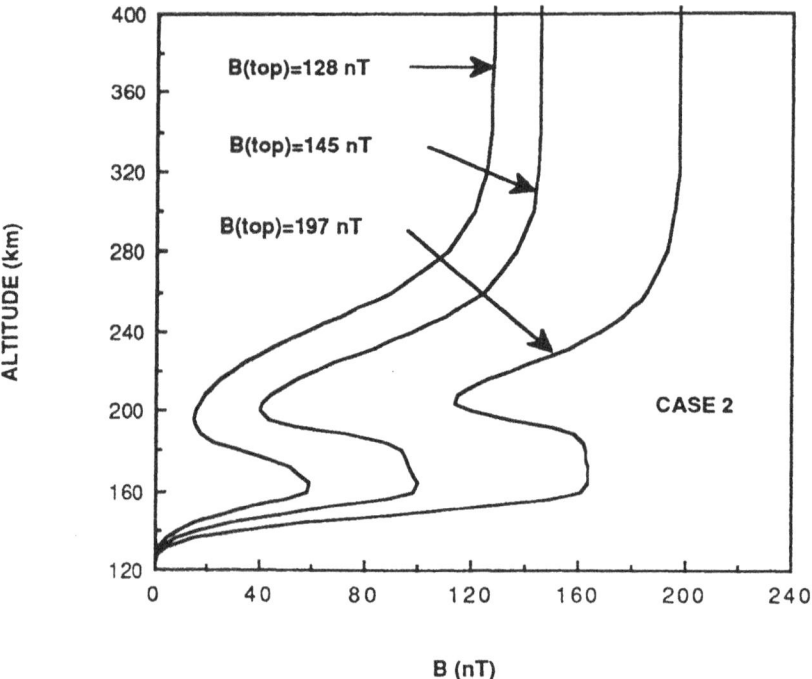

Fig. 45. Calculated magnetic field profiles for steady-state conditions for three different degrees of magnetization.

steady-state magnetic field profiles for upper boundary conditions of E_y = 0.5, 1.0, and 2.0×10^{-6} V m^{-1}, corresponding to downward velocities at the upper boundary of 4, 7, and 10 m s^{-1}. The peak downward plasma velocities for the low, medium, and high B cases are 35, 25, and 17 m s^{-1}, respectively. The horizontal electric field, E_y, for the high B case is shown in Figure 46. The electric field is independent of altitude ($\partial E_y/\partial z = 0$) as it should be for steady-state conditions, since Faraday's law in 1-D states that $\partial B/\partial t = \partial E_y/\partial z$. Notice that above 160 km, the field almost entirely consists of the motional electric field (see the generalized Ohm's law discussion in the tutorial), whereas below 150 km, the electric field is almost entirely due to the Ohmic dissipation term.

When E_y at the upper boundary is either too large or too small, or it is rapidly changing due to changing solar wind conditions, steady-state solutions are not appropriate. The ionospheric field will either decay or grow with time. The SC model considered both growing and decaying conditions, but only the latter is reviewed here. The assumed initial state of the ionosphere was magnetized. This initial 'steady' state was achieved by another model run in which ion loss due to an assumed horizontal transport was included (this loss term clears out ions in the magnetic barrier region above the ionopause). The loss term was also included in the case described here. To produce a decaying field, the vertical velocity at the upper boundary was gradually lowered from

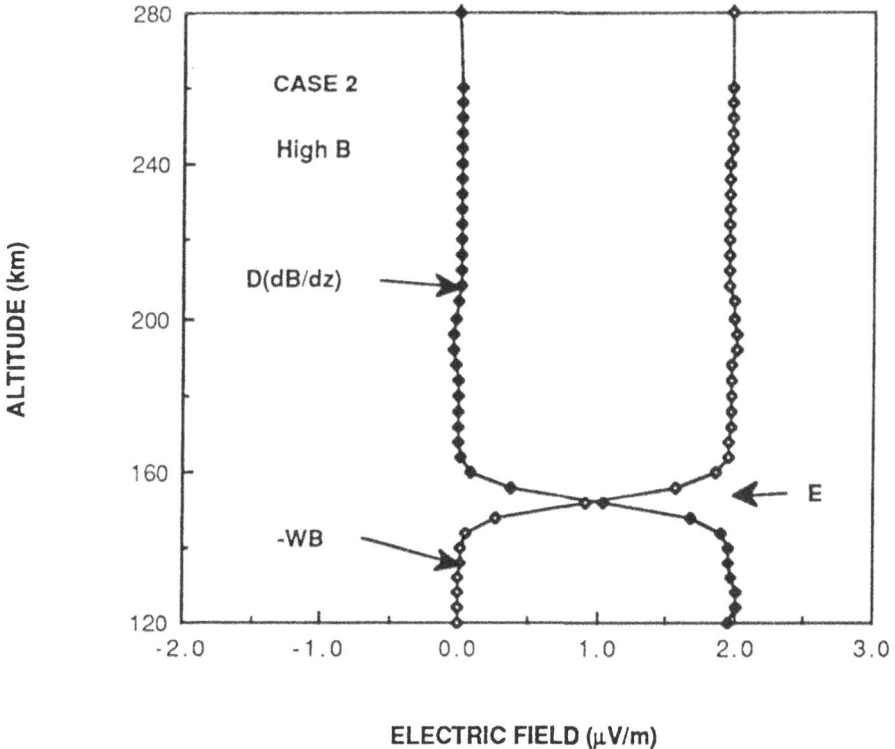

ELECTRIC FIELD (μV/m)

Fig. 46.   Calculated electric field in the highly magnetized ionosphere. The electric field profile for a moderately magnetized ionosphere (not shown) is almost the same but with a total electric field strength of 1 μV m^{-1}.

-20 m s^{-1} to 0 m s^{-1}. Figures 47 and 48 show magnetic field and electron density profiles, respectively, for several values of the elapsed time. During the first 2 hours, the field profile almost uniformly decreases in magnitude. As time elapses further, the ionopause becomes thinner. However, at about 10 hours, an important change occurs. The low altitude magnetic layer essentially disappears as a consequence of the barrier field moving out of reach of the large downward plasma flow associated with the ionospheric photochemistry (see the discussion of this flow earlier in this review). As the magnetic layer disappears, the electron density ledge near 200 km disappears. The growth process is similar to the decay process but in reverse, although the magnetic layer at 170 km grows more rapidly than it decays (in ~ 3 hours).

The MHD model of Shinagawa and Cravens (1988) which included a high altitude loss term is able to explain many features of the magnetized ionosphere of Venus, including the overall development and structure of both high and low ionopause cases. However, this model is most relevant to regions where horizontal transport is not dominant within the ionosphere proper. The next logical step in modeling both the large-scale magnetic fields in the ionosphere of Venus and the ionosphere itself is to

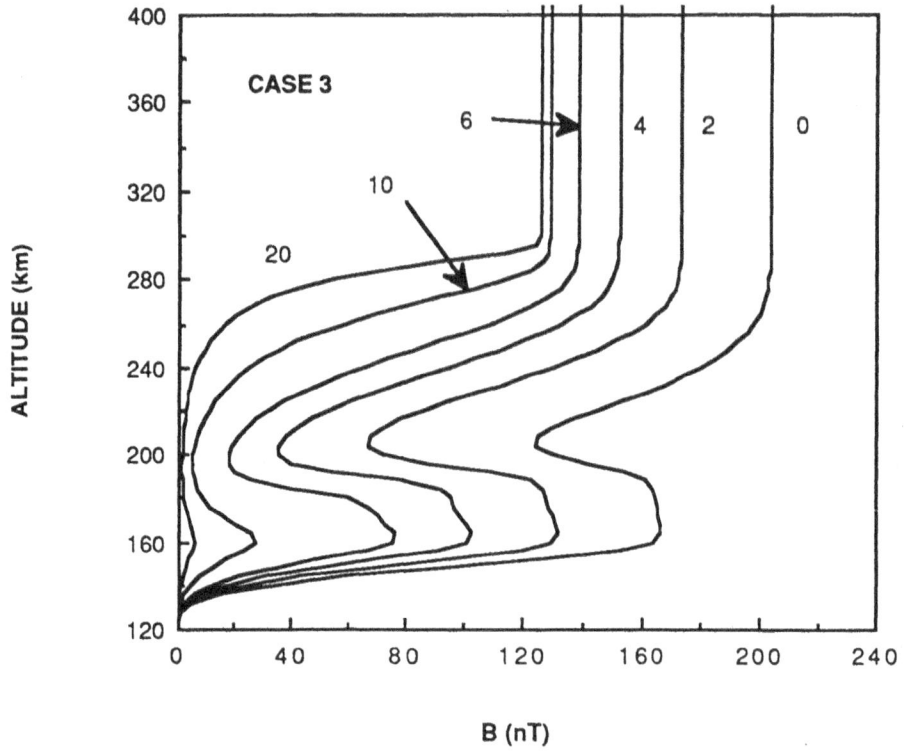

Fig. 47. Altitude profiles of the calculated magnetic field magnitude for $0 \le t \le 20$ hours. The upper boundary conditions are such that the field strength at the upper boundary (i.e., the solar wind dynamic pressure) is decreasing with time. Horizontal ion loss was included in this calculation.

develop a two- or three-dimensional self-consistent MHD model, which includes both vertical and horizontal transport processes.

3.2.2. *Small-Scale Fields*

Soon after the small-scale field structures in the dayside ionosphere were first detected, Russell and Elphic (1979) noticed that a selected subset of them could be modeled by a formula describing a cylindrical 'rope' of twisted field lines or 'flux rope':

$$\mathbf{B}(\rho) = B_\phi(\rho)\boldsymbol{\phi} + B_z(\rho)\mathbf{z}$$

where ρ, ϕ, and z correspond to the usual cylindrical coordinates. Such a structure automatically satisfies $\nabla \cdot \mathbf{B} = 0$. The field could be further described as

$$B_\phi(\rho) = B(\rho) \sin \alpha(\rho),$$

$$B_z(\rho) = B(\rho) \cos \alpha(\rho),$$

where $\alpha(\rho)$ is the helical pitch and $B(\rho)$ has the general form

$$B(\rho) = B_0 \exp(-\rho^2/a^2).$$

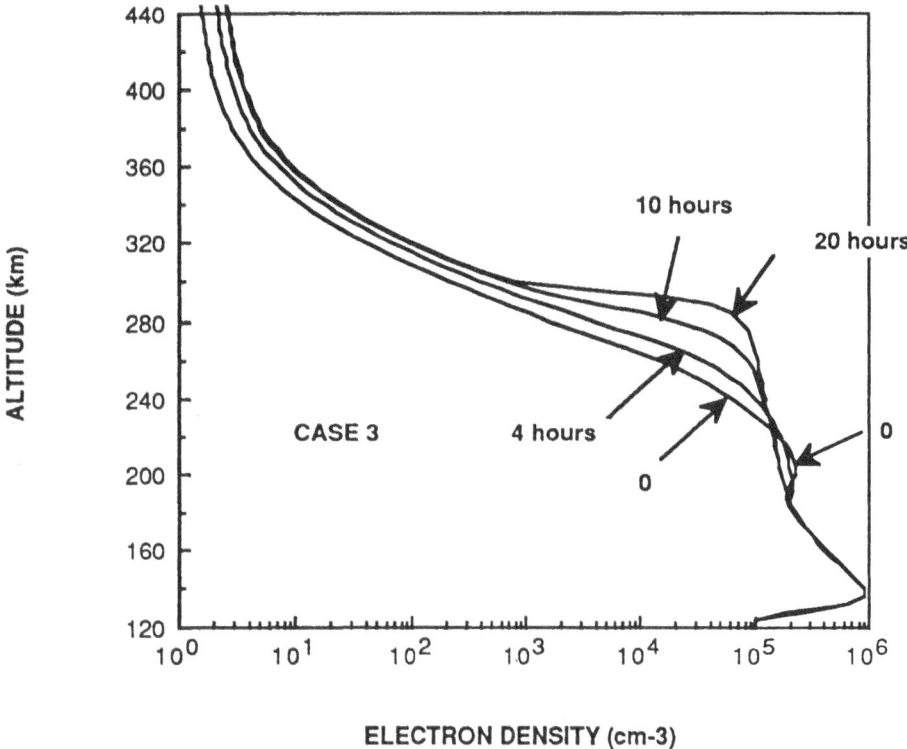

Fig. 48.   Altitude profiles of calculated electron densities for $0 \leq t \leq 20$ hours. The secondary peak at 200 km disappears and an ionopause-like structure develops at 300 km.

Here B_0 and a are constants describing the axial field strength and a determines the rate at which the field magnitude drops off with distance from the axis. With this description they were able to reproduce the distinctive hodograms of the structures as illustrated by Figure 49. Moreover, by fitting the data they were able to deduce the form of $\alpha(\rho)$ in the above formula. The typical example shown in Figure 50 shows that the flux ropes were best described by a structure with straight, strong axial field surrounded by ever more inclined and weakening helical field until the field was completely azimuthal and near zero magnitude at its outer boundary. This configuration is shown schematically in Figure 51.

Elphic and Russell (1983a, b) then used this analytical description to analyze the small-scale fields in ways that might elucidate their origins. Because only a subset of the small field structures fit the model well enough to determine such quantities as $\alpha(\rho)$, B_0 and the inclination i of the flux rope axes with respect to the horizontal, these investigations focused on the properties of up to several hundred examples selected from a variety of orbits from the two low altitude dayside passes of PVO. These examples were considered usable because of their small 'impact parameters' with respect to the flux rope axes (identified by the s in Figure 49). The other structures were considered

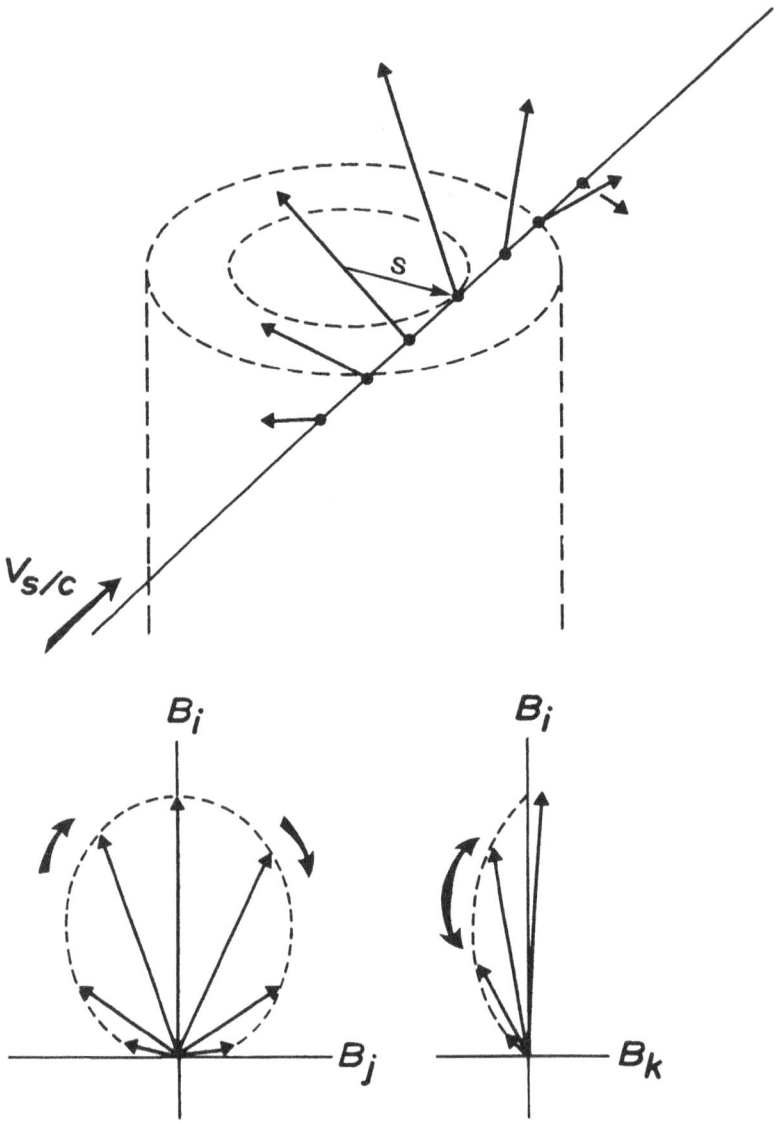

Fig. 49.   Schematic of a traversal of a small scale field structure. The axis is vertical, the spacecraft trajectory horizontal and slightly off the axis. The arrows at points along the trajectory represent sampled field vectors. The lower panel shows how these vectors, when cast into principal axis coordinates, trace out hodograms like those observed. B_i, B_j, and B_k refer to field components in the directions of maximum, intermediate, and minimum magnetic variation. (From Elphic and Russell, 1983.)

to be flux ropes which were sampled at too great an axial distance to be modeled. Of particular importance to the understanding of how these structures might form was the derived distribution of flux rope inclinations. These distributions, shown in Figure 52, indicated that the flux ropes are almost randomly inclined except at high altitudes near

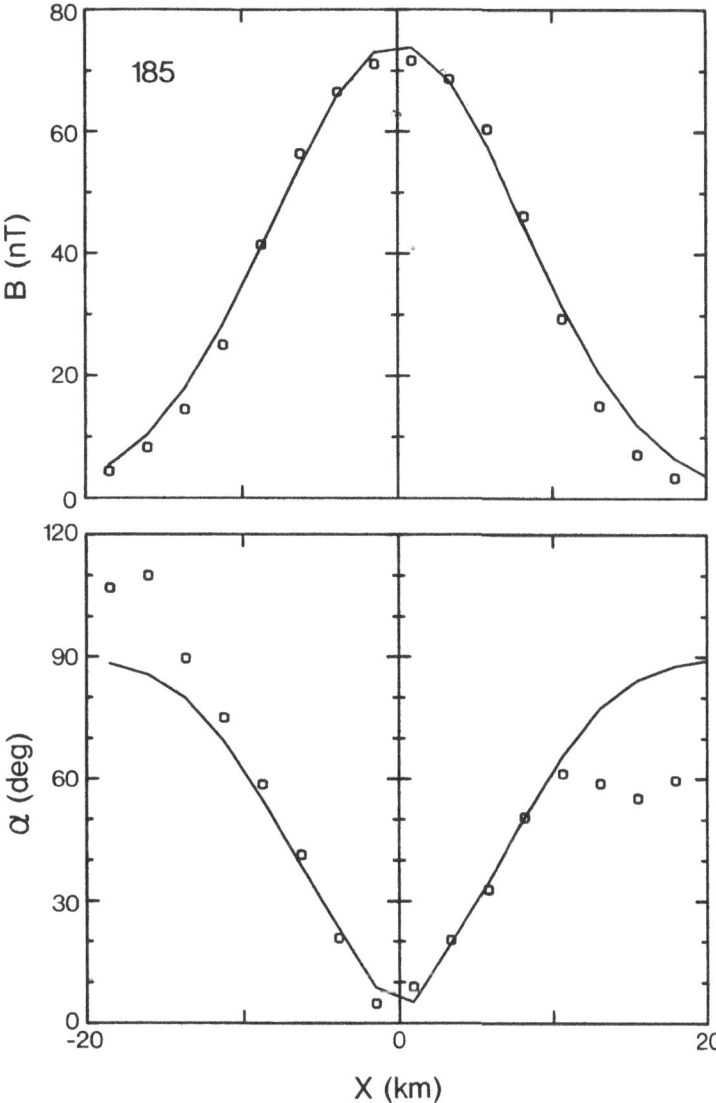

Fig. 50. Sample of a small scale field structure that has been converted to $B(\rho)$ and $\alpha(\rho)$ and plotted as a function of $X = v_s \Delta t$ where v_s is spacecraft velocity. Circles are data, and lines are the best fit model corresponding to the data points. (From Elphic and Russell, 1985.)

the terminator where they are horizontal. It was also found that the horizontal flux ropes are loosely 'wrapped' in the sense that their helical structure does not become fully azimuthal as in the classic flux rope picture. Elphic and Russell (1983c) then suggested that a helical kink instability which coiled the flux ropes themselves could be responsible for their apparently random orientations. Figure 53 is a schematic illustration of the scenario that this implies for the ionosphere of Venus. Although it is not clear that a

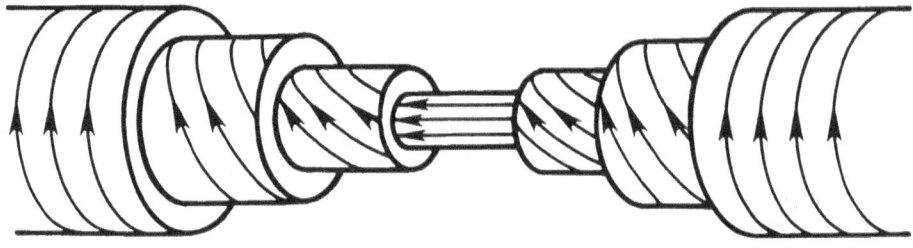

INTERIOR STRUCTURE OF FLUX ROPE

Fig. 51. Cartoon showing the inferred progression of helicity of magnetic field within a flux rope. (From Elphic and Russell, 1983.)

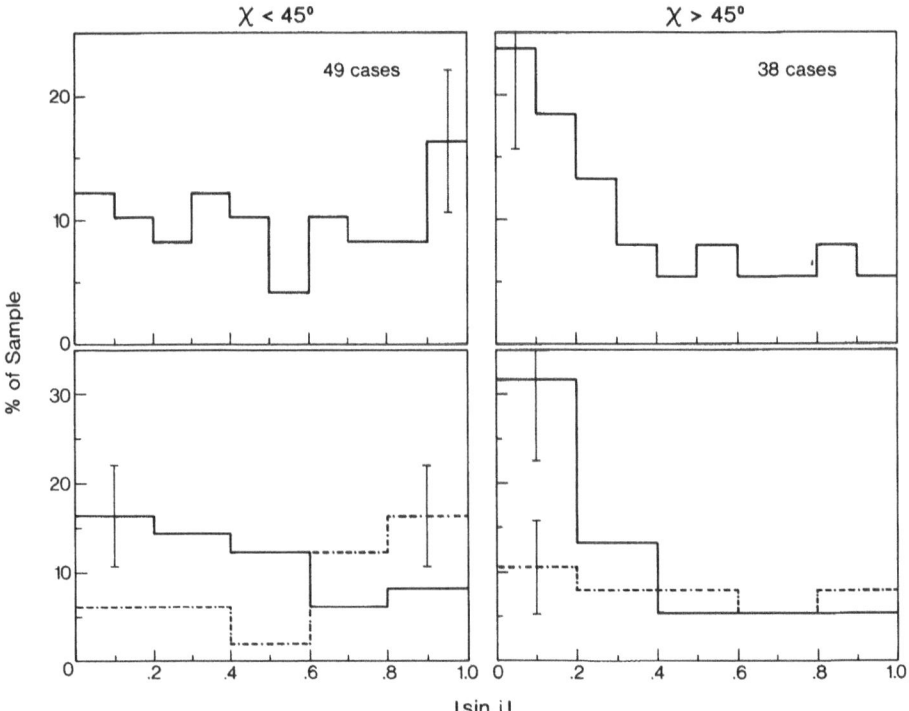

Fig. 52. (*Upper panels*) Flux rope inclination *i* with respect to local horizontal plotted as a distribution of $|\sin i|$ for all subsolar cases (left panel) and all terminator cases (right panel) in the small impact parameter subset. Plotted this way, a purely random distribution of flux rope orientations would yield a flat histogram in $|\sin i|$. (*Lower panels*) Flux rope inclinations for subsolar (left) and terminator (right) data sets when divided into cases above (solid line) and below (broken line) a selected altitude. For the subsolar cases, the dividing altitude is 200 km; for the terminator cases, it is 300 km. Error bars denote $\sqrt{N}$ uncertainty, where *N* is the total number of cases in the bin. (From Elphic and Russell, 1983.)

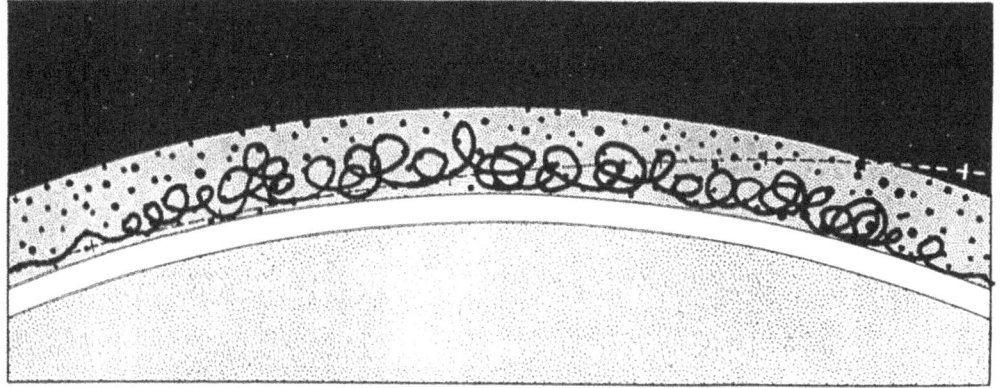

Distribution of Flux Ropes

Fig. 53.   Schematic of the distribution of ionospheric flux ropes suggested by the PVO data. The flux ropes appear to be horizontal at high solar zenith angles and randomly oriented in the subsolar region.

helical kink instability will work in a situation where collisions with neutrals are causing Ohmic dissipation of the field-producing currents, and where other external forces like gravity may be acting on the plasma in and around the flux ropes, this coiled flux rope picture seems to at least describe the observations of the small-scale fields. Determining their origin is another matter.

Figure 54 illustrates the most generally held view of the formation of flux ropes. Flux tubes from the magnetosheath overlying the ionosphere somehow get into the ionosphere where they are twisted and distorted as they sink. In this picture, they remain connected to the magnetosheath (and thus to the solar wind) in the flanks. Several ideas about the penetration of flux tubes through the ionopause have been advanced. The most developed, in a quantitative sense, is that which invokes a Kelvin–Helmholtz or shear instability at the ionopause (cf. Wolff *et al.*, 1980). In this model, shear in the flow boundary between the solar wind and ionosphere causes the development of perturbations of the interface. When these are perpendicular to the magnetosheath field orientation, they can grow and entrain flux tubes as illustrated in Figure 55. The flux tubes are subsequently pulled downward by virtue of the connection they maintain to the flowing solar wind. Because there is little or no shear in the velocity near the subsolar ionopause, this process works best near the flanks. An alternative suggestion has recently been made which invokes an instability of the subsolar ionopause related to local production of heavy ionospheric plasma on flux tubes at the stagnation point, but this idea has not advanced past he stage of qualitative discussion (cf. Russell *et al.*, 1987). The cartoons in Figure 56 compare a global picture of this process to that of the Kelvin–Helmholtz scenario. Data analysis may ultimately determine whether flux tubes from the magnetosheath are entrained by the ionosphere in the flanks or near the subsolar point or, equally likely, at both sites.

The possibility that the flux tubes are created locally from larger scale field structures

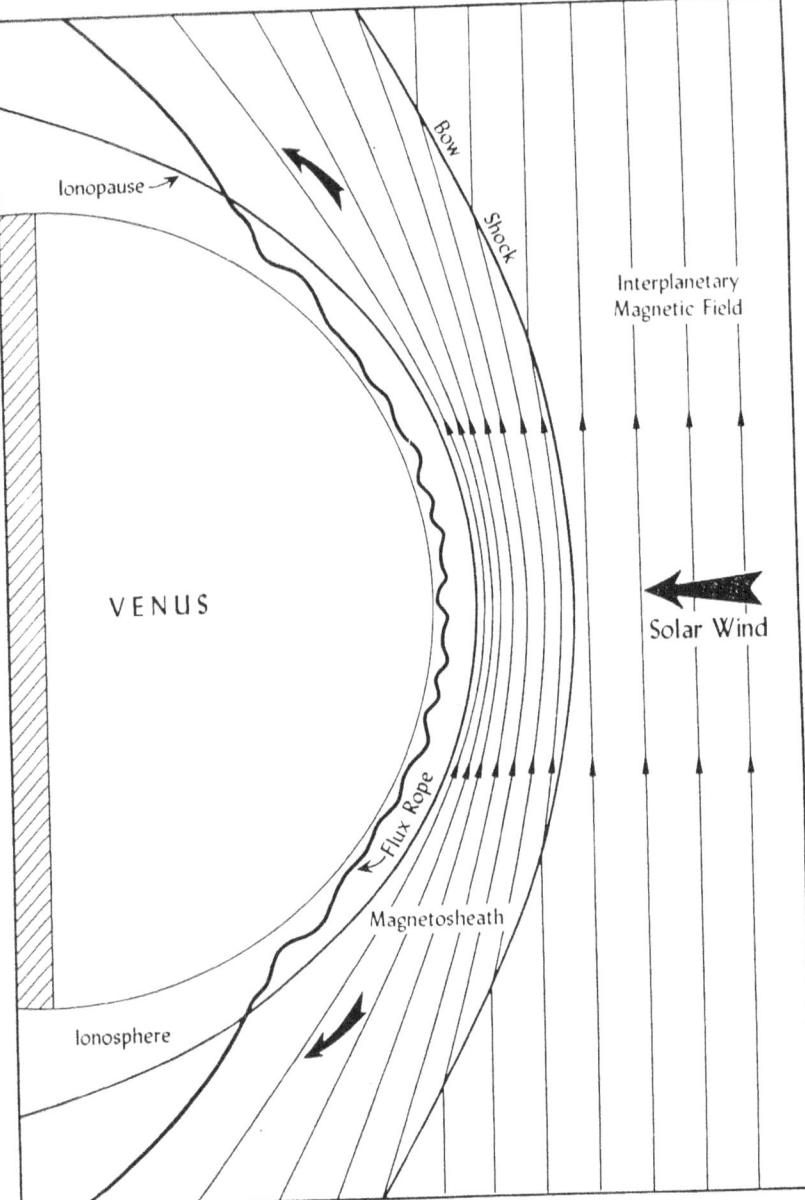

Fig. 54.   Schematic illustration of the possible manner in which flux ropes derive from and connect to the
magnetosheath. (From Elphic, 1980.)

also cannot be ignored. As discussed earlier, Cloutier *et al.* (1981) proposed that flux
ropes were due to the breakup of the large scale ionospheric field by shear instability
within the ionosphere rather than at its boundary. This poses some difficulty for the
production of flux ropes observed in the subsolar region where the shear is small, but

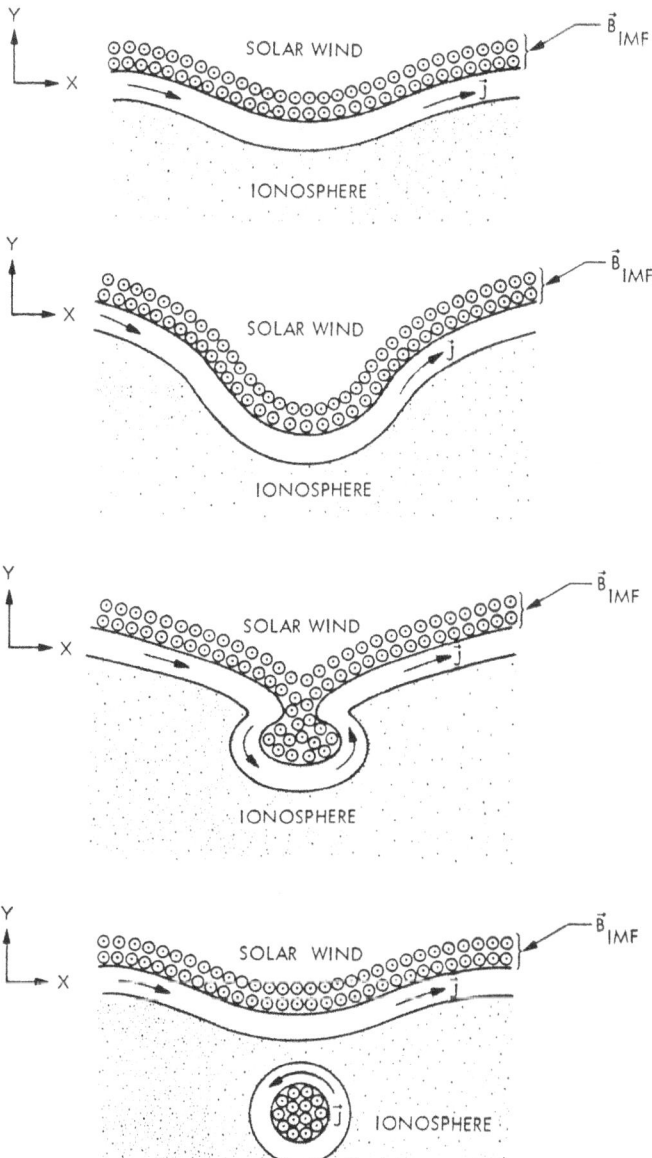

Fig. 55. Cartoon illustrating how the formation of a ripple on the ionopause by the Kelvin–Helmholtz instability can lead to the entrainment of a magnetosheath flux tube in the ionosphere. (From Wolff *et al.*, 1980.)

as pointed out in a different *in situ* model of flux rope origin by Luhmann and Elphic (1985), one certainly cannot ignore the possibility that the same ionospheric processes that create the large scale fields play a role in producing the small-scale fields. In the view of these latter authors, flux-rope like signatures can arise from redistribution of the

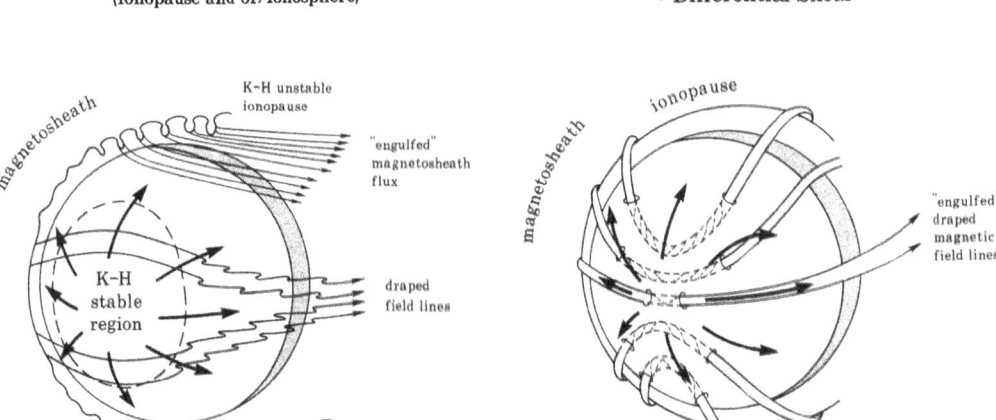

Kelvin–Helmholtz Instability   Subsolar Ionopause Instability
(Ionopause and or/Ionosphere)   + Differential Shear

Fig. 56.   Schematic comparison of two scenarios that have been suggested for flux rope generation.

weak large-scale field by small-scale wave-like or turbulent ionospheric motions. (The large-scale counterpart of this dynamo process is what produces the distinctive large-scale field profile.) This model also allows the construction of model hodograms, as shown in Figure 57, which resemble the observations. The small-scale field structure in this case is not so much a collection of classic discrete flux ropes as a widespread distribution of twisted and inhomogeneously spaced field lines with an occasional helix produced by vorticity. For example, one of the model hodograms results from a simulated spacecraft pass through the field structure illustrated in Figure 58. Thus both the magnetosheath field and the large-scale ionospheric field (which also comes from the magnetosheath) have been proposed as the source of 'flux ropes'.

In general, in the models where the flux ropes are formed at the ionopause, the structures must be transported by the background ionospheric drift down to the lower ionosphere where they were most frequently observed by the PVO magnetometer (see Figure 15). If a flux rope is strictly force-free, the plasma pressure (and plasma density to a lesser degree) inside the rope is the same as outside, and there is no buoyancy force on the rope. In this case, the rope will merely be convected with the background plasma flow (neglecting the magnetic tension force on the rope as a whole). Because of the nature of the plasma flow, flux ropes created anywhere will be convected downward and toward the nightside. Magnetic tension along the rope would tend to contribute to this downward motion. However, as Krymskii and Breus (1988) point out, if flux ropes are created in the high, thin ionopause current layer found in conjunction with unmagnetized ionospheres, there are two main problems with their transport down to the lower ionosphere. First, ropes at high altitudes are not perfectly force-free (see Figure 18), but have at least some thermal pressure and density depletions associated with them so that the $\mathbf{J} \times \mathbf{B}$ force is balanced by $\nabla(p_e + p_i)$ (see the tutorial). In fact, a rope that has just

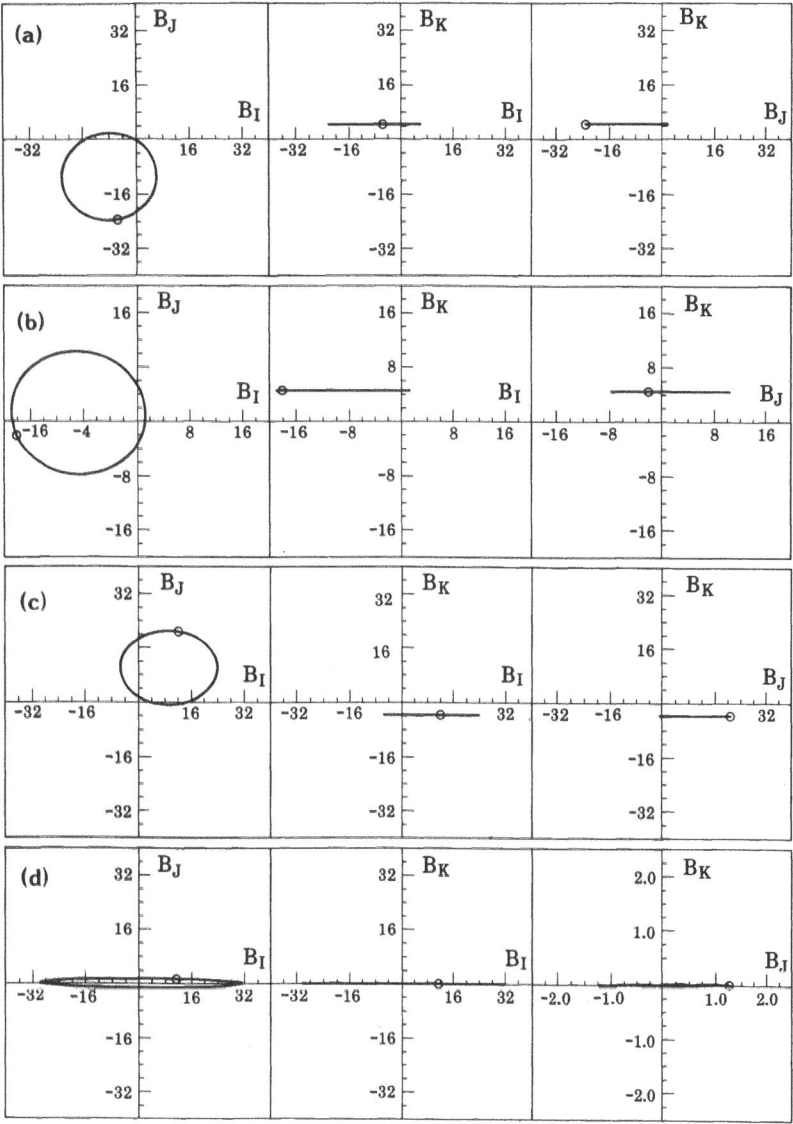

Fig. 57.   Hodograms for various wave-like field structures generated by wave-like velocity perturbations.
(From Luhmann and Elphic, 1985.)

been formed at the ionopause by the Kelvin–Helmholtz instability should be very
depleted of plasma. The consequence is that the flux rope will experience an upward
buoyancy force which counteracts the downward background plasma convection.
Second, and more serious, the plasma convection at high altitudes near the flanks where
the instability is likely to occur is in some locations upward and not downward (see the
schematic in Figure 40). Thus the downward tension force on a newly formed flux rope

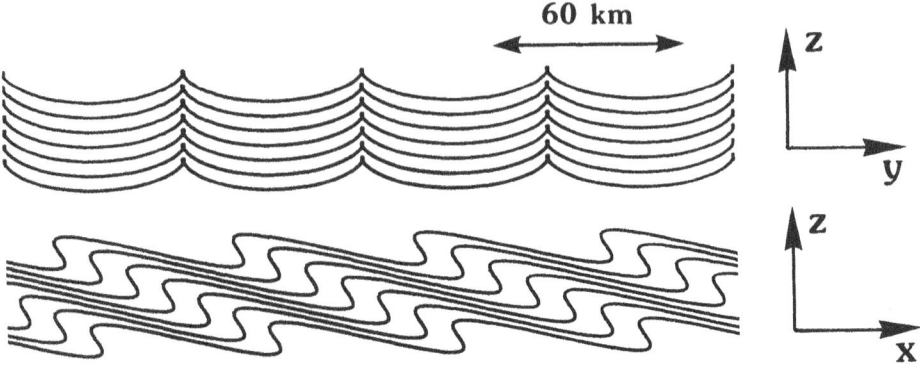

Fig. 58.　Examples of orthogonal views of field lines that produce hodograms like those shown in Figure 49. This type of field structure differs from the classical picture of flux ropes. (From Luhmann and Elphic, 1985.)

at the ionopause would have to overcome both a significant upward buoyancy force and upward background plasma convection. For these reasons, Krymskii and Breus concluded that flux ropes are created in the ionosphere following conditions of high solar wind dynamic pressure or in the subsolar ionosphere, which is often magnetized because of the low subsolar ionopause. However, their argument does not explain why flux ropes are seen long after a magnetizing 'event' and even when the ionopause is extremely high for several orbits.

As a whole, the current flux rope models are incomplete in the sense that most, like the ionopause Kelvin–Helmholtz instability model, describe the formation of flux ropes but not their subsequent evolution. One exception which was mentioned earlier is described in the paper by Elphic and Russell (1983c) which suggests that the apparently random orientation of the structures at low altitude is due to a 'helical kink' instability, wherein the flux ropes are themselves twisted into larger scale helices. No other model has addressed this key observation, or others such as the lack of organization of the 'polarity' or handedness of flux rope twists across the dayside ionosphere.

Models have similarly not dealt with the question of whether the terminator ionosphere wave-like field structures are related to flux ropes. One configuration of draped ionospheric fields that could in principle produce the terminator small-scale field structures is suggested by Figure 59. In this conceptual model, a wavy horizontal curent sheet is responsible for the alternating sunward–anti-sunward fields. However, this picture requires an interplanetary sector boundary passage or another substantial rotation of the interplanetary field to produce the overlying, oppositely directed ionospheric fields. Nevertheless, since such rotations can occur several times a day at Venus because of its proximity to the heliospheric current sheet, this situation is not out of the question. The connection of these fields to flux ropes can then be made either by invoking the subsolar turbulent 'dynamo' process to redistribute the related large scale ionospheric field, or even by invoking the more exotic scenario, depicted in Figure 60, where non-explosive magnetic field merging across the subsolar current sheet occurs

Terminator Waves

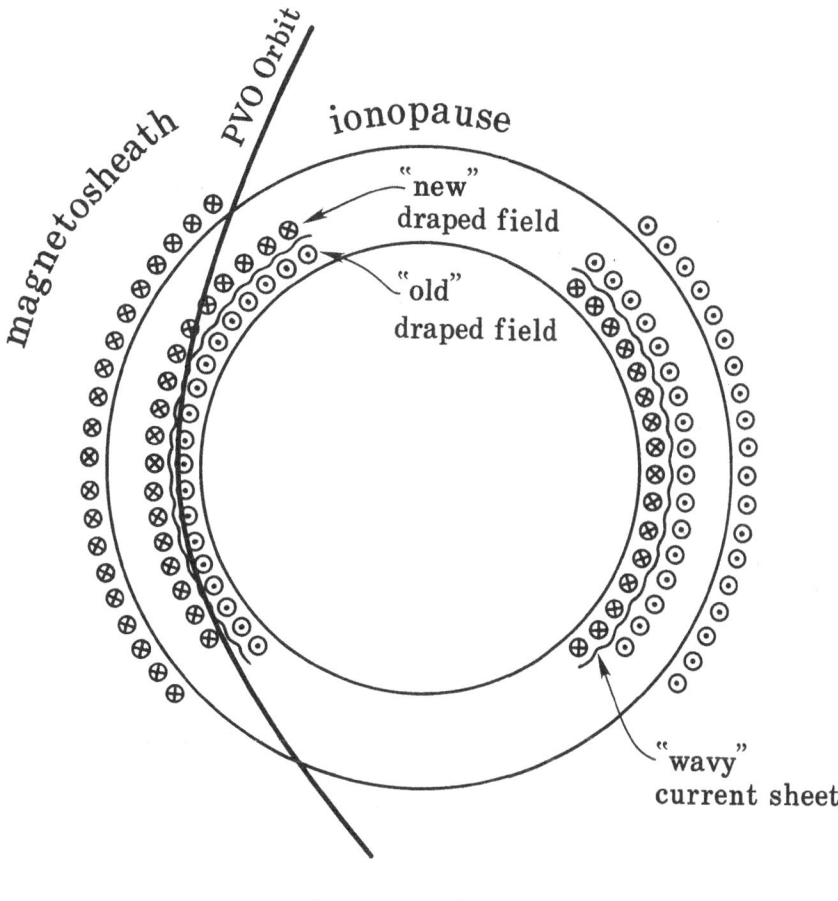

terminator plane view

Fig. 59.   A possible explanation for the observed nature of terminator waves. (From Luhmann, 1990.)

and collisional dissipation henceforth determines the field structure. But in the end, it is probably safe to say that flux rope modeling lags behind the large-scale field modeling in that it is both less quantitative and less conclusive.

3.3. THE NIGHTSIDE IONOSPHERE

3.3.1. *Large-Scale Fields*

In contrast to the dayside fields, there have been few efforts to model the large-scale nightside ionospheric fields which are connected with ionospheric holes and dis-

Dayside Merging/Dissipation

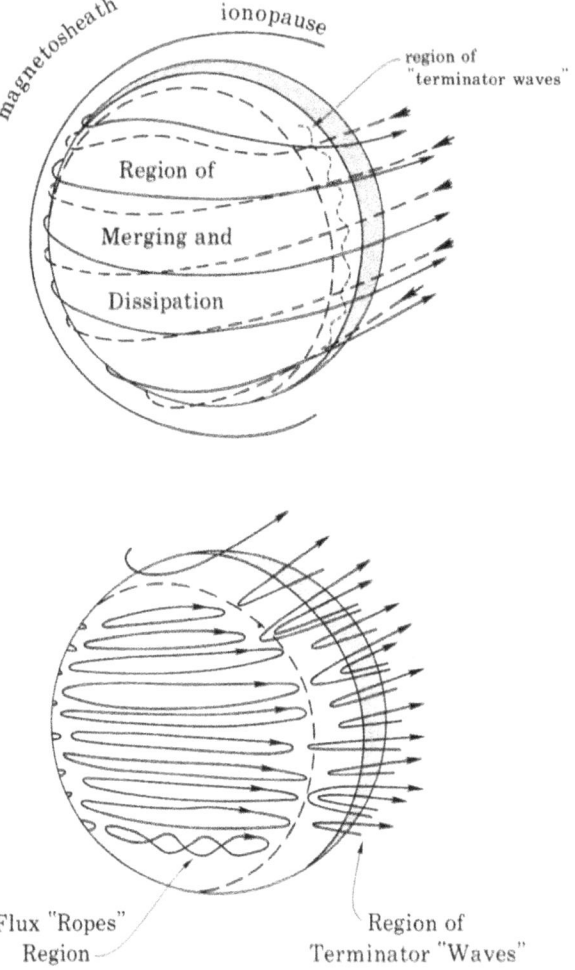

Fig. 60.   A possible scenario involving magnetic merging whereby both terminator waves and flux ropes form. (From Luhmann, 1990.)

appearing ionospheres. The reasons are multiple, ranging from the fact that the interplanetary field interacts with the dayside ionosphere first, to the fact that the related effects on the plasma density in the local (nightside) ionosphere demand a global and self-consistent MHD approach. Nevertheless, there have been several qualitative 'models' of nightside magnetic field line configurations that have been proposed.

Brace *et al.* (1982) sketched one of the first conceptual models of the magnetic field in holes illustrating how the field might first drape over the dayside ionosphere wherein

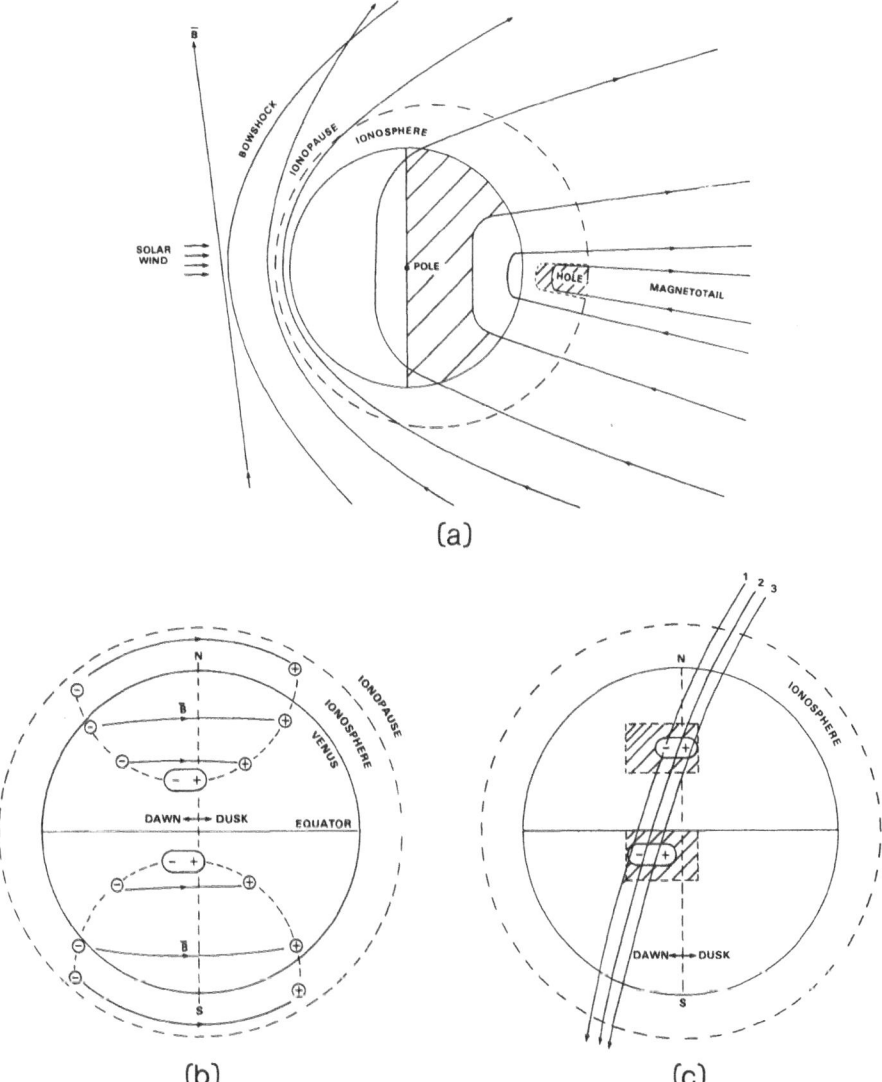

Fig. 61a.   A conceptual view of the evolution of an interplanetary field line entering the northern dayside ionosphere, convecting over the northern polar cap, and pulling out of the ionosphere to form a hole on the nightside somewhere north of the equator. Those field lines which convect over the southern pole would form a similar hole south of the equator. (From Brace *et al.*, 1982.)

Fig. 61b.   A view of the same scenario from the Venus tail. Interplanetary field lines, after being dragged through the polar ionosphere (top and bottom of figure) are convected equatorward by field line tension and ionospheric flow toward the antisolar point where they emerge from the ionopause at the encircled pluses and minuses. (From Brace *et al.*, 1982.)

Fig. 61c.   An illustration of the polarities of the *B* field expected in the holes. On orbit 1, one would see a downward magnetic field in both holes. On orbit 2, one might expect to see the field reversed in the two holes, while the field in orbit 3 would appear outward in both holes. (From Brace *et al.*, 1982.)

it gets entrained at low altitude. As shown in Figure 61, the low altitude segment of the entrained flux tube remains at low altitudes until it has been pulled or convected well past the terminator toward the anti-solar point. No attempt was made to explain either the flux tube motion in terms of forces or physical processes, or the mechanism by which the flux tubes that make up the holes become devoid of ionospheric plasma. Grebowsky and Curtis (1981) suggested that the picture proposed by Brace might involve kilovolt parallel electric fields along the holes analogous to those in high latitude ion troughs on Earth. They also elaborated further on the magnetic field geometry by proposing that

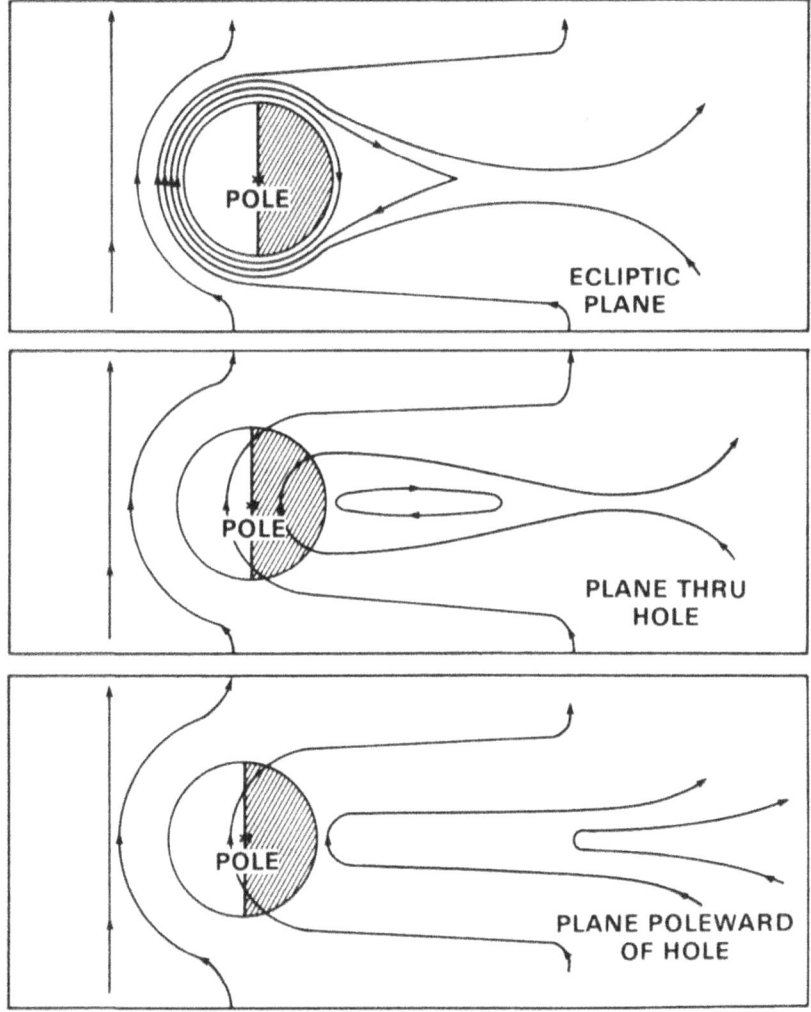

Fig. 62.   Schematic illustration of draped magnetic field lines, projected onto planes at various levels above the equator, showing how holes might be related to merging in the tail current sheet. (From Grebowsky and Curtis, 1981.)

reconnection takes place on the field lines of the holes as depicted in Figure 62, which shows their hypothetical field lines projected on three planes at different distances above the equatorial plane. An alternative suggestion by Knudsen *et al.* (1982), that the fields of the holes were the tail lobes of a weak intrinsic dipole, met with some resistance (cf. Luhmann and Russell, 1983) because the field polarity in the holes was consistent with that of the draped interplanetary field. It seems that a true understanding of the holes may only come about as a result of global modeling of the solar wind interaction with a planetary ionosphere. Needless to say, no attention has been paid to the modeling of the low level 'turbulent' field that pervades the ionosphere outside of the holes.

The large-scale horizontal field associated with the disappearing ionosphere phenomenon, like the holes, has received only qualitative treatment. Cravens and coworkers (1983) had shown that the field was associated with high solar wind dynamic pressures and thus low terminator ionopauses, which prohibit the normal nightward transport of ionization across the terminator. One possible picture of the field lines to invoke in this instance, since we know that high horizontal dayside fields are also present when the solar wind pressure is high, is that in Figure 1(b). Statistical studies of field draping by Marubashi *et al.* (1985) and Phillips *et al.* (1984) have suggested that there is indeed some closure of the magnetosheath field in the wake which is consistent with this picture. However, there is currently no quantitative theoretical argument that supports this conceptual model. This problem may also be solvable only by global modeling.

4. Conclusions

Here we have covered much of the history of the work which has been done on the ionospheric magnetic fields of Venus. Given the currently renewed interest in the exploration of Mars, it is perhaps appropriate to consider how the models that were developed for Venus might apply to our other neighbor in the solar system. Previous observations of the solar wind interaction with Mars indicate that the Martian intrinsic magnetic field must be weak (cf. Slavin and Holzer, 1982; Luhmann *et al.*, 1990). We also know that the subsolar ionosphere peak thermal pressure at Mars is usually lower than the incident solar wind pressure. If Mars behaves like Venus, the Martian magnetosheath field should pile up to form a magnetic barrier from which magnetic flux diffuses and convects into the ionosphere below to create a 'permanent' large-scale ionospheric field (cf. Luhmann *et al.*, 1987; Shinagawa and Cravens, 1989). Indeed, the few *in situ* measurements of the Martian ionosphere that we have from the Viking landers include electron and ion temperature altitude profiles that suggest the presence of a horizontal ionospheric field of ~ 40 nT magnitude (cf. Hanson and Mantas, 1988). Only *in situ* magnetic measurements in the ionosphere of Mars can unequivocally establish whether that field is intrinsic or of interplanetary origin, but thus far, no unambiguous evidence exists in favor of the former explanation over the latter. It is also worth noting that the Venus ionosphere at solar minimum, for which we have no *in situ* data, is similarly expected to be in a 'permanently' magnetized state since its ionospheric thermal pressure is typically exceeded by the solar wind pressure.

In closing, it is perhaps sufficient to say that our experience at Venus has given us a first detailed look at a planetary ionosphere that has a direct interaction with the solar wind. The knowledge gained from that experience has already been applied to comets (cf. Russell *et al.*, 1982; Breus *et al.*, 1986) (where the analogy is more limited than at Mars) and is likely to be applied to future studies of the interaction of flowing plasmas with unmagnetized or weakly magnetized obstacles such as planetary satellites and outgassing asteroids. The *in situ* observations of flux ropes have provided intellectual fodder for the solar physics community (cf. Russell, 1985) as well. As a whole, the topic of ionospheric magnetic fields at Venus has already received considerable attention and it has given us considerable scientific rewards in return. Future developments will no doubt include refinements of many of the aforementioned ideas (e.g., three-dimensional global MHD models of the large-scale fields) and resolution of the remaining problems (e.g., flux rope origin and evolution). With good fortune, we will also be able to carry out comparative studies with the ionospheric fields observed at Mars.

Acknowledgements

The authors are supported by NASA Grants NAG2–501 (JGL) and NAGW-1588 (TEC) from the Solar System Exploration Division.

References

Banks, P. M. and Kockarts, G.: 1973, *Aeronomy*, Academic Press, New York.

Biermann, L., Brosowski, B., and Schmidt, H.: 1967, *Solar Phys.* **1**, 245.

Brace, L. H., Elphic, R. C., Curtis, S. A., and Russell, C. T.: 1983a, *Geophys. Res. Letters* **10**, 1116.

Brace, L. H., Taylor, H. A., Jr., Gombosi, T. I., Kliore, A. J., Knudsen, W. C., and Nagy, A. F.: 1983b, in D. Hunten, L. Colin, T. Donahue, and V. Moroz (eds.), *Venus*, Univ. of Arizona Press, Tucson, p. 779.

Brace, L. H., Theis, R. F., Krehbiel, J. P., Nagy, A. F., Donahue, T. M., McElroy, M. B., and Pedersen, A.: 1979, *Science* **203**, 763.

Brace, L. H., Theis, R. F., Mayr, H. G., Curtis, S. A., and Luhmann, J. G.: 1982, *J. Geophys. Res.* **87**, 199.

Breus, T. K. Krymskii, A. M., and Luhmann, J. G.: 1987, *Geophys. Res. Letters* **14**, 499.

Cloutier, P. A.: 1984, *J. Geophys. Res.* **89**, 2401.

Cloutier, P. A. and Daniell, R. E., Jr.: 1973, *Planetary Space Sci.* **21**, 463.

Cloutier, P. A. and Daniell, R. E., Jr.: 1977, *Planetary Space Sci.* **25**, 621.

Cloutier, P. A. and Daniell, R. E., Jr.: 1979, *Planetary Space Sci.* **27**, 1111.

Cloutier, P. A., McElroy, M. B., and Michel, F. C.: 1969, *J. Geophys. Res.* **74**, 6215.

Cloutier, P. A., Tascione, T. F., and Daniell, R. E., Jr.: 1981, *Planetary Space Sci.* **29**, 635.

Cloutier, P. A., Tascione, T. F., Daniell, R. E., Jr., Taylor, H. A., and Wolff, R. S.: 1983, in D. Hunten, L. Colin, T. Donahue, and V. Moroz (eds.), *Venus*, Univ. of Arizona Press, Tucson, p. 941.

Cloutier, P. A., Taylor, H. A., Jr., and McGary, J. E.: 1987, *J. Geophys. Res.* **92**, 7289.

Cravens, T. E., Brace, L. H., Taylor, H. A., Jr., Russell, C. T., Knudsen, W. L., Miller, K. L., Barnes, A., Mihalov, J. D., Scarf, F. L., Quenon, S. J., and Nagy, A. F.: 1982, *Icarus* **51**, 271.

Cravens, T. E., Gombosi, T. I., Kozyra, J., Nagy, A. F., Brace, L. H., and Knudsen, W. C.: 1980, *J. Geophys. Res.* **85**, 7778.

Cravens, T. E., Nagy, A. F., Brace, L. H., Chen, R. H., and Knudsen, W. C.: 1979, *Geophys. Res. Letters* **6**, 341.

Daniell, R. E., Jr. and Cloutier, P. A., 1977, *Planetary Space Sci.* **25**, 621.

Elphic, R. C. and Ershkovich, A. I.: 1984, *J. Geophys. Res.* **89**, 997.

Elphic, R. C. and Russell, C. T.: 1983a, *J. Geophys. Res.* **88**, 58.

Elphic, R. C. and Russell, C. T.: 1983b, *J. Geophys. Res.* **88**, 2993.

Elphic, R. C. and Russell, C. T.: 1983c, *Geophys. Res. Letters* **10**, 459.
Elphic, R. C., Russell, C. T., Slavin, J. A., and Brace, L. H., 1980, *J. Geophys. Res.* **85**, 7679.
Grebowsky, J. M. and Curtis, S. A.: 1981, *Geophys. Res. Letters* **12**, 1273.
Hanson, W. B. and Mantas, G. P.: 1988, *J. Geophys. Res.* **93**, 7538.
Hedin, A. E., Niemann, H. B., Kasprzak, W. T., and Sieff, A.: 1983, *J. Geophys. Res.* **88**, 73.
Hogan, J. S., Stewart, R. W., and Rasool, S. I.: 1972, *Radio Sci.* **7**, 525.
Johnsen, F. S. and Midgely, J. E.: 1969, *Space Res.* **IX**, 760.
Knudsen, W. C., Spenner, K., Whitten, R. C., Spreiter, J. R., Miller, K. L., and Novak, V.: 1979, *Science* **205**, 105.
Krymskii, A. M. and Breus, T. K.: 1988, *J. Geophys. Res.* **93**, 8459.
Luhmann, J. G.: 1986, *Space Sci. Rev.* **44**, 241.
Luhmann, J. G.: 1988, *J. Geophys. Res.* **93**, 5909.
Luhmann, J. G.: 1990, in *Physics of Magnetic Flux Ropes*, AGU Monograph, AGU, Washington, DC.
Luhmann, J. G. and Elphic, R. C.: 1985, *J. Geophys. Res.* **90**, 12047.
Luhmann, J. G. and Russell, C. T.: 1983, *Geophys. Res. Letters* **10**, 409.
Luhmann, J. G., Elphic, R. C., and Brace, L. H.: 1981, *J. Geophys. Res.* **86**, 3509.
Luhmann, J. G., Elphic, R. C., Russell, C. T., Brace, L. H., and Hartle, R. E.: 1983, *Adv. Space Res.* **2**, 17.
Luhmann, J. G., Elphic, R. C., Russell, C. T., Mihalov, J. D., and Wolfe, J. H.: 1980, *Geophys. Res. Letters* **7**, 917.
Luhmann, J. G., Elphic, R. C., Russell, C. T., Slavin, J. A., and Mihalov, J. D.: 1981, *Geophys. Res. Letters* **8**, 517.
Luhmann, J. G., Phillips, J. L., and Russell, C. T.: 1987, *Adv. Space Res.* **7**, 101.
Luhmann, J. G., Russell, C. T., Brace, L. H., Taylor, H. A., Knudsen, W. C., Scarf, F. L., Colburn, D. S., and Barnes, A.: 1982, *J. Geophys. Res.* **87**, 9205.
Luhmann, J. G., Russell, C. T., and Elphic, R. C.: 1984, *J. Geophys. Res.* **89**, 362.
Luhmann, J. G., Russell, C. T., Scarf, F. L., Brace, L. H., and Knudsen, W. C.: 1987, *J. Geophys. Res.* **92**, 8545.
Luhmann, J. G., Russell, C. T., Brace, L. H., and Vaisberg, O. L.: 1990, in H. Kieffer and C. Snyder (eds.), *Mars*, Univ. of Arizona Press, Tucson.
Marubashi, K., Grebowsky, J. M., Taylor, H. A., Jr., Luhmann, J. G., Russell, C. T., and Barnes, A.: 1985, *J. Geophys. Res.* **90**, 1385.
Miller, K. L., Knudsen, W. C., Spenner, K., Whitten, R. C., and Novak, V.: 1980, *J. Geophys. Res.* **85**, 7759.
Nagy, A. F., Cravens, T. E., Smith, S. G., Taylor, H. A., Jr., and Brinton, H. C.: 1980, *J. Geophys. Res.* **85**, 7795.
Nagy, A. F., Cravens, T. E., and Gombosi, T. I.: 1983, in D. M. Hunten, L. Colin, T. Donahue, and V. Moroz (eds.), *Venus*, Univ. of Arizona Press, Tucson, p. 841.
Nicholson, D. R.: 1983, *Introduction to Plasma Theory*, John Wiley and Sons, New York.
Phillips, J. L. and Russell, C. T.: 1987, *J. Geophys. Res.* **92**, 2253.
Phillips, J. L., Luhmann, J. G., and Russell, C. T.: 1984, *J. Geophys. Res.* **89**, 10676.
Russell, C. T.: 1985, in M. Kundu and G. Holman (eds.), *Unstable Current Systems and Plasma Instabilities in Astrophysics*, IAU Publications, p. 25.
Russell, C. T.: 1990, in *Physics of Magnetic Flux Ropes*, AGU Monograph, AGU, Washington, DC.
Russell, C. T. and Elphic, R. C.: 1979, *Nature* **279**, 616.
Russell, C. T. and Vaisberg, O.: 1983, in D. Hunten, L. Colin, T. Donahue, and V. Moroz (eds.), *Venus*, Univ. of Arizona Press, Tucson, p. 873.
Russell, C. T., Elphic, R. C., and Slavin, J. A.: 1979a, *Science* **203**, 745.
Russell, C. T., Elphic, R. C., and Slavin, J. A.: 1979b, *Science* **205**, 114.
Russell, C. T., Luhmann, J. G., Elphic, R. C., and Neugebauer, M.: 1982, in L. Wilkening (ed.), *Comets*, Univ. of Arizona Press, Tucson, p. 561.
Russell, C. T., Luhmann, J. G., and Elphic, R. C.: 1983, *Adv. Space Res.* **10**, 13.
Russell, C. T., Singh, R. N., Luhmann, J. G., Elphic, R. E., and Brace, L. H.: 1987, *Adv. Space Res.* **7**, 115.
Shinagawa, H. and Cravens, T. E.: 1988, *J. Geophys. Res.* **93**, 11263.
Shinagawa, H. and Cravens, T. E.: 1989, *J. Geophys. Res.* **94**, 6506.
Shinagawa, H., Cravens, T. E., and Nagy, A. F.: 1987, *J. Geophys. Res.* **92**, 7317.
Siscoe, G. L., 1983, in R. L. Carovillano and J. M. Forbes (eds.), *Solar Terrestrial Physics*, D. Reidel Publ. Co., Dordrecht, Holland, p. 11.

Slavin, J. A. and Holzer, R. E.: 1982, *J. Geophys. Res.* **87**, 10285.

Spreiter, J. R., Summers, A. L., and Rizzi, A. W.: 1970, *Planetary Space Sci.* **18**, 2181.

Theis, R. F., Brace, L. H., Elphic, R. C., and Mayr, H. G.: 1984, *J. Geophys. Res.* **89**, 1477.

Vaisberg, O. L. and Zelenyi, L. M.: 1984, *Icarus* **58**, 412.

Wolff, R. S., Goldstein, B. E., and Yeates, C. M.: 1980, *J. Geophys. Res.* **85**, 7697.

Wolff, R. S., Stein, R. F., and Taylor, H. A., Jr.: 1982, *J. Geophys. Res.* **87**, 8118.

PLASMA WAVES AT VENUS

Institute of Geophysics and Planetary Physics, University of California at Los Angeles,
Los Angeles, CA 90024, U.S.A.

Abstract. Many significant wave phenomena have been discovered at Venus with the plasma wave instrument flow on the Pioneer Venus Orbiter. It has been shown that whistler-mode waves in the magnetosheath of the planet may be an important source of energy for the topside ionosphere. Plasma waves are also associated with thickening of the ionopause current layer. Current-generated waves in plasma clouds may provide anomalous resistance resulting in electron acceleration, possibly producing aurora. Ion-acoustic waves are observed in the bow shock, and appear to be a feature of the magnetotail boundary. Lastly plasma waves have been cited as evidence for lightning on Venus.

Table of Contents

1. Introduction

The Pioneer Venus mission, which recently celebrated ten years of on-orbit operations, has provided significant insights into the interaction of the solar wind with an unmagnetized body. Luhman (1986) has presented a thorough review of the interaction between the solar wind and the Venus ionosphere. To emphasize the difference between a magnetized and unmagnetized body we note that at the Earth the intrinsic field is so large that there is no direct contact between the solar wind and the planetary ionosphere. In fact the primary mechanism for coupling energy from the solar wind to the terrestrial magnetosphere is thought to be reconnection at the magnetopause. Energy is subsequently dissipated in the ionosphere through substorms. At Venus, on the other hand, there is little or no intrinsic field, and the solar wind interacts directly with the dense

Space Science Reviews **55**: 275–316, 1991.
© 1991 *Kluwer Academic Publishers.*

ionosphere. The interaction region is consequently at lower altitudes than at the Earth, and other processes, such as ion pick up, are expected to play a role in the coupling between the solar wind and the planet.

In this review we will touch on just one aspect of the solar wind – Venus interaction, that is plasma waves. As we will show later, plasma waves can be significant in the dynamics of the solar wind coupling to the ionosphere. For example, whistler-mode waves generated at the bow shock can propagate through the magnetosheath and supply energy to the ionosphere. Waves may also play a role in the generation of the ionopause current layer, affecting the thickness of this current sheet and providing dissipation. As another example, current sheets are observed in 'plasma clouds' within the magnetosheath, and it has been suggested that the waves observed within these clouds produce anomalous resistivity, resulting in particle acceleration which in turn may result in aurora, as at the Earth.

In addition to acting as means for energy transport and dissipation, waves can also be useful diagnostics of the plasma. We will present examples of waves as indicators of bow shock processes, a transition layer in the magnetosheath, and as supplemental diagnostics in determining the magnetotail boundary.

As will become apparent later on, any review of plasma waves at Venus will by necessity be a review of just one aspect of the work of the late Frederick L. Scarf. He was the Prncipal Investigator on the Pioneer Venus wave instrument; the only instrument thus far to sample plasma waves near Venus, and his interests were so diverse that he contributed significantly to every topic we review here. One of Scarf's interests was, of course, the question of lightning on Venus. That topic is reviewed elsewhere (Russell, 1991), and we will restrict the discussion here to the interpretation of the plasma wave data. In this review we shall consider the plasma wave evidence for lightning as just one aspect of the many plasma wave phenomena observed with the Pioneer Venus instrument.

The structure of this review is as follows. In the next section we will briefly summarize some of the properties of the plasma wave instrument flown on the Pioneer Venus Orbiter. We will further describe how the characteristic frequencies of a plasma relate to the actual frequency bands of the wave instrument. In subsequent sections we will discuss the wave observations from different regions around the planet. We have attempted to discuss the observations in a sequence starting at low altitudes through the ionopause, to the magnetotail and out to the bow shock and solar wind. These results are presented in Sections 3, 4, and 5. In Section 6 we will discuss some aspects of the plasma wave evidence for lightning on Venus. Lastly we will conclude with a few statements about future directions for research with the Pioneer Venus plasma wave instrument.

2. The Pioneer Venus Wave Instrument

The goals of the Pioneer Venus mission are described by Colin (1980). We refer the reader to this paper and other papers in the same volume (*J. Geophys. Res.*, December

1980) for a full description of the Pioneer Venus Orbiter, orbit parameters, and payload, together with a comprehensive collection of papers on results from the first six months of on-orbit operations.

The Pioneer Venus Orbiter Electric Field Detector (OEFD) was designed and built by Frederick L. Scarf and colleagues at TRW Defense and Space Systems (Scarf *et al.*, 1980c). The instrument and antenna as originally proposed were based on wave instruments flown on Pioneer 8 and 9, with a spacecraft element (e.g. boom) being used as an unbalanced dipole. However, because of mission constraints, the experiment was redesigned to use a small *Y*-shaped pair of antenna elements that were passively deployed when the launch vehicle fairing was ejected. One of the antenna elements is shown schematically in Figure 1, both before and after deployment. The deployed antenna configuration on the orbiter, together with a block diagram of the electronics circuit is shown in Figure 2.

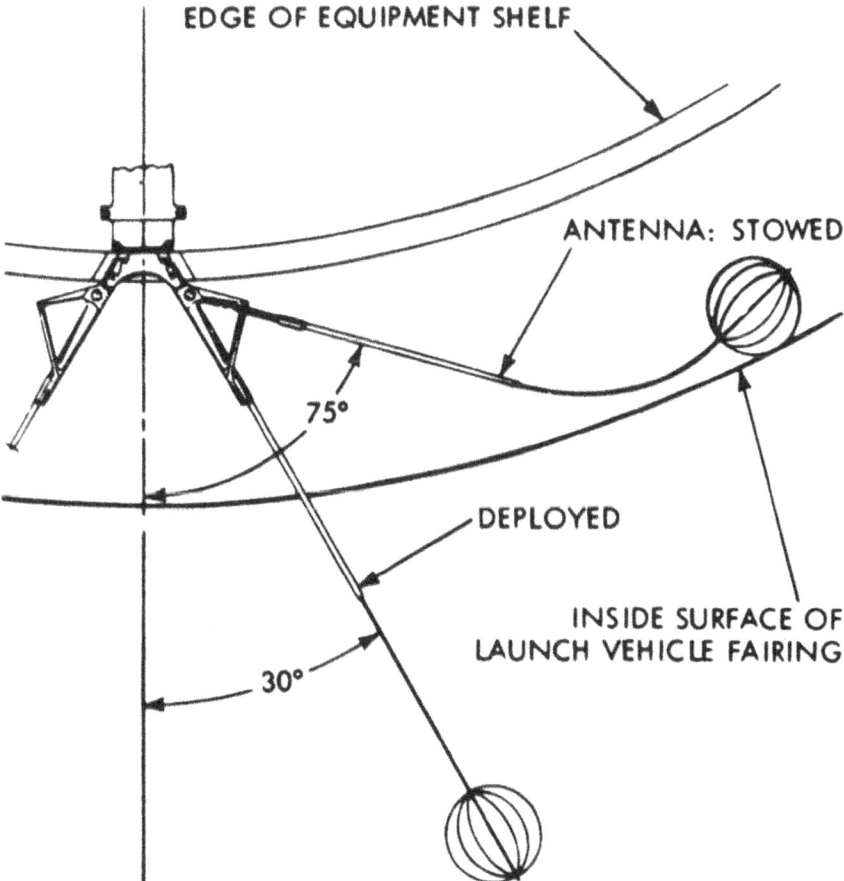

Fig. 1. The Pioneer Venus Orbiter Electric Field Detector (OEFD) antenna configuration (from Scarf *et al.*, 1980c). The figure shows one of the antenna elements, which is 0.76 m long, both stowed and deployed.

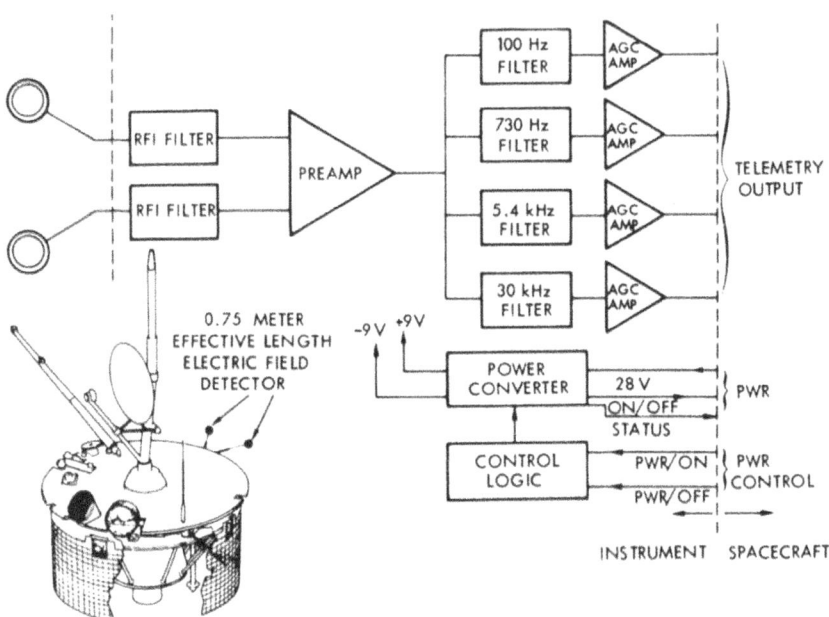

Fig. 2. Sketch of the Pioneer Venus Orbiter with all booms deployed, and a block diagram of the instrument electronics (from Scarf *et al.*, 1980c). The effective separation of the wire spheres on the OEFD antenna elements is 0.75 m.

The OEFD has the following characteristics: total mass including electronics 0.5 kg; power consumption around 0.5 W; no commanding, other than power on/off; 4 analog frequency channels sampled. The effective antenna separation is 0.75 m tip-to-tip. The four filters have 30% bandwidth with center frequencies of 100 Hz, 730 Hz, 5.4 kHz, and 30 kHz. The analog signal sent to the orbiter telemetry system is the voltage from Automatic Gain Control amplifiers, which have rise times of the order 50 ms, and decay times around 500 ms. Output from the four channels is read once per data minor frame. The time resolution of the data depends on the spacecraft telemetry rate, with the highest rate being four minor frames s^{-1}. The wave instrument can detect signals some 90 dB above threshold before saturation. The thresholds (in V^2 m^{-2} Hz^{-1}) are as follows: 3.242×10^{-11} (100 Hz); 4.834×10^{-12} (730 Hz); 4.891×10^{-13} (5.4 kHz); 7.321×10^{-14} (30 kHz). On-orbit background noise levels are typically some 7 dB higher than the instrument threshold. The background noise depends on frequency and whether or not the orbiter is in sunlight, with the 100 Hz channel showing the greatest variability.

Because of the limited frequency coverage of the OEFD, interpretation of the wave data requires knowledge of the plasma parameters to determine the characteristic frequencies of the plasma. We will cite some of the expressions for determining the characteristic frequencies here. First the electron plasma frequency is given by $f_{pe} = 9 \sqrt{n}$ kHz, where n is the electron density in cm^{-3}. A typical solar wind density of 10 cm^{-3} corresponds to $f_{pe} \sim 30$ kHz. The plasma frequency is 5.4 kHz in regions where the density is around 0.4 cm^{-3}, as may occur in the tail. The electron gyro-frequency, $f_{ce} = 28B$ Hz, where B is the ambient magnetic field in nT. In general the magnetic field in the solar wind is around 10 nT, while the field can be as large as 100 nT just above the Venus ionopause. Typical electron gyro-frequencies, hence, lie in the range 280 Hz to 2.8 kHz, although at times the gyro-frequency may be less than 100 Hz in the ionosphere where the magnetic field is shielded out by ionopause currents. In terms of plasma waves then, plasma oscillations will usually be observed in the 30 kHz channel, while whistler-mode waves will often be detected in the 100 Hz and 730 Hz channels.

Although ion waves have lower frequencies they need not be undetectable by the OEFD. The ion-plasma frequency and lower hybrid frequency are both lower than the electron plasma frequency and gyro-frequency, respectively, by the square root of the electron/ion mass ratio. Lower hybrid waves are, hence, probably undetectable, unless significantly Doppler-shifted through spacecraft motion. Ion-acoustic waves, on the other hand, have an upper cut-off at the ion-plasma frequency which can be some 100's

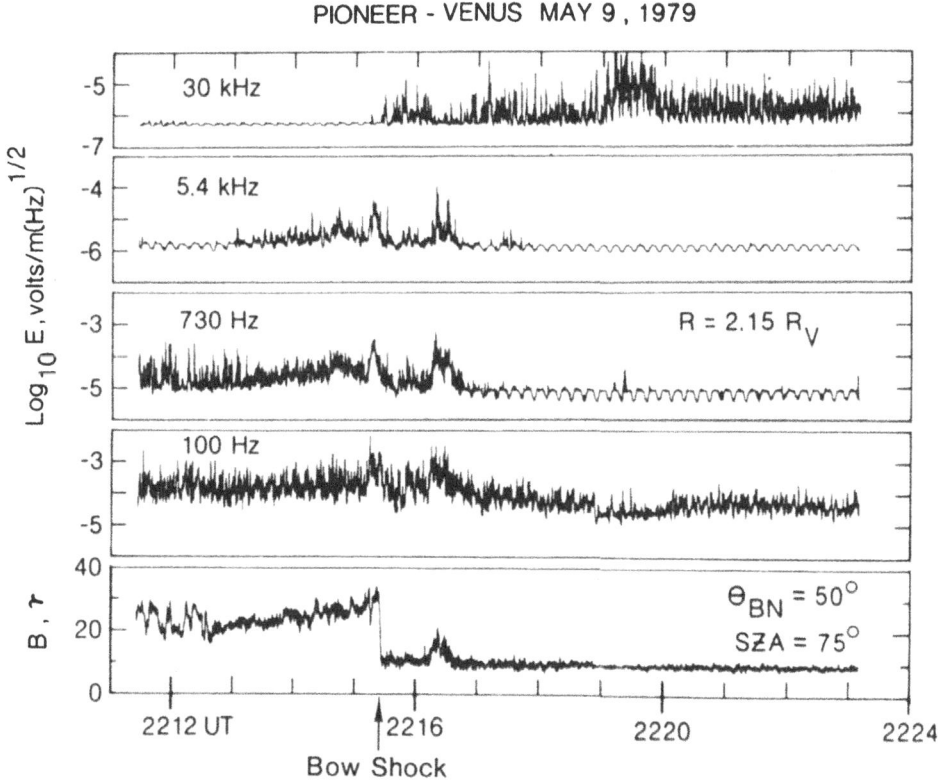

Fig. 3. Example of OEFD wave data associated with a bow shock crossing on May 9, 1979 (from Scarf *et al.*, 1980b).

of Hz, or even higher in the ionosphere. Doppler shift produced by the spacecraft motion can also be significant for ion-acoustic waves. The ion sound speed $\sim 10(T_e/\mu)^{1/2}$ km s^{-1}, where T_e is the electron temperature in eV and μ is ion mass with respect to proton mass. Near periapsis the orbiter velocity is around 10 km s^{-1}, which can be larger than the ion sound speed for a sufficiently low temperature, or heavy ion species. Ion-acoustic waves can, therefore, be Doppler-shifted to quite high frequencies.

As an example of the interpretation of data from the OEFD, Figure 3 shows waves observed in association with a bow shock crossing at 22:15:20 UT on May 9, 1979. The wave activity in the 30 kHz channel is interpreted as plasma oscillations driven by suprathermal electrons reflected at the shock, assuming density ~ 10 cm^{-3}. The wave noise in the 730 Hz and 5.4 kHz channels around the shock crossing is thought to be Doppler-shifted ion-acoustic noise. The modulation seen in these two channels after 22:17 UT is at the spin frequency, and may be associated with sheath effects, or shadowing of the antenna elements, which are close to the spacecraft body, as shown in Figure 2.

3. Waves in the Ionosphere and at the Ionopause

In this section we will review observations of waves in the ionosphere of Venus, and also at the ionopause. We will also address some results which might be included in the next section on the magnetosheath, that is current-driven instabilities. However, since these waves are sometimes associated with regions of high-density plasma above the ionosphere known as plasma clouds, we will include the discussion here. One iono-spheric phenomenon which we will defer to a subsequent section is the plasma wave evidence for lightning on Venus.

3.1. IMPACT IONIZATION NOISE

Curtis *et al.* (1985) investigated the change in noise level observed with the OEFD near periapsis in the nightside ionosphere. Figure 4, from Curtis *et al.* (1985), shows wave data from two orbits in the upper half of the figure plotted as log of amplitude, with the baseline for each channel being 10^{-6} V m^{-1} Hz$^{-1/2}$. The noise level increases in all the frequency channels around periapsis. Curtis *et al.* showed that the noise amplitude was well-correlated with the density of CO_2, as shown in the lower half of the figure where the wave electric field amplitude is divided by the square root of the neutral CO_2 density, plotted logarithmically. This ratio is constant around periapsis.

Curtis *et al.* showed that the enhanced wave noise around periapsis was due to the interaction between the spacecraft and the neutral atmosphere, causing ionization through impact. The orbiter velocity (10 km s^{-1}) is high enough to ionize CO_2 molecules through impact. The newly created electrons drift with respect to the CO_2^+ ions in the vicinity of the spacecraft since the electrons have a higher thermal velocity. This relative drift would generate an ion-acoustic instability, at or near the CO_2^+ plasma frequency, which for typical impact ion densities around 10^4 cm^{-3} gives an ion-plasma frequency ~ 3 kHz.

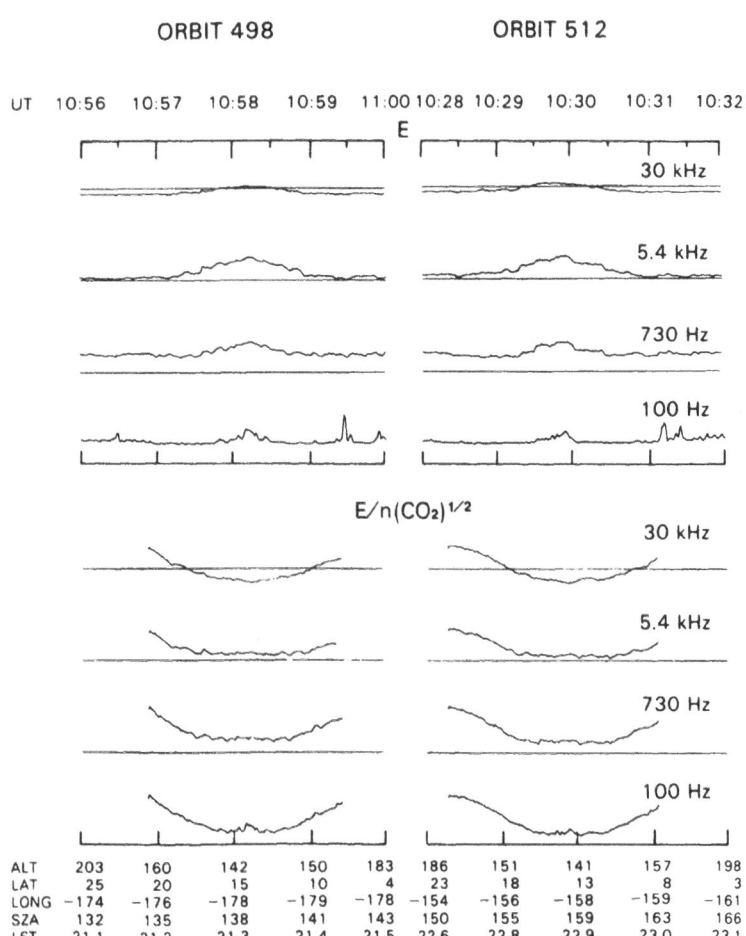

Fig. 4.   Comparison of wave electric field amplitudes and CO_2 density around periapsis (from Curtis *et al.*, 1985). The upper panels show log wave amplitude, while the lower panels show the wave amplitude normalized to the square root of the CO_2 density, log scale.

When discussing the generation mechanism for the waves it is important to note that they are generated in the spacecraft frame, as pointed out by Curtis *et al.* Consequently, even though the waves can have short wavelengths, their frequency is not Doppler-shifted due to spacecraft motion. Taking this into account the authors argued that the wave could not be due to a modified two-stream instability generating waves at the lower hybrid frequency, which for typical magnetic field strengths around 5 nT is ~ 0.7 Hz.

In addition the authors discounted an ion/ion drift instability associated with the relative drift of the newly created ions and the ambient ionosphere. Their arguments in this case were less straightforward than for the lower hybrid instability, being based on the correlation between the wave amplitude and the neutral density. Curtis *et al.* argued that the wave saturation amplitude will be inversely dependent on the neutral density (which determines the amount of impact ionization) for the ion/ion instability. The opposite is observed, and Curtis *et al.* present arguments as to why this would be the case for the ion-acoustic instability generated by the impact ion/electron drift.

As a last comment on the impact ion noise, we note that Curtis *et al.* restricted their analysis to the nightside of Venus. The impact noise is more readily observed on the nightside since the ambient noise level of the wave instrument is lower when the spacecraft is in shadow. Furthermore, the instrument noise is not strongly spin modulated, as occurs when the spacecraft is in sunlight. However, a brief survey of dayside passes carried out for the purposes of this review show some evidence for a similar phenomenon on those orbits which have a low (< 160 km) periapsis altitude on the dayside.

3.2. WHISTLER-MODE WAVES AT THE IONOPAUSE

One of the first results from the Pioneer Venus OEFD instrument concerned the observation of whistler-mode waves in the magnetosheath which appeared to be absorbed at the ionopause (Scarf *et al.*, 1979, 1980b; Taylor *et al.*, 1979). In the report on initial observations, Scarf *et al.* (1979) presented an example of the wave absorption at the ionopause, shown here in Figure 5. There is a sudden increase in the noise level of the 100 Hz waves just prior to 15:22 UT, when the orbiter leaves the ionosphere.

In their initial report, Scarf *et al.* estimated the total energy flux carried by the 100 Hz waves into the inosphere was around 0.05 erg cm^{-2} s^{-1}. This estimate assumed that the waves were electromagnetic whistler-mode waves. The OEFD instrument has relatively poor frequency coverage, and no wave magnetic field sensor was included in the spacecraft instrument complement. Scarf *et al.*, therefore, initially identified the waves as whistler-mode by analogy with observations in the Earth's magnetosheath.

In a subsequent paper, Taylor *et al.* (1979) presented additional arguments for interpreting the 100 Hz waves in the magnetosheath as whistler-mode waves. They expanded on the comparison with the terrestrial magnetosheath, and also pointed out that the waves did not display the polarization characteristics of electrostatic waves.

An important result presented by Taylor *et al.* as evidence for wave absorption at the ionopause is shown here in Figure 6. It can be shown that the energy of electrons resonant with low-frequency whistler-mode waves is given by

$$E_l = 0.5 m_e c^2 f f_{ce} / f_{pe}^2 \cos \theta \qquad (1)$$

for Landau resonance, where m_e is the electron mass, c is the speed of light, f is the wave frequency, f_{ce} is the electron gyro-frequency, f_{pe} is the electron plasma frequency, and θ is the angle of propagation for the waves with respect to the ambient magnetic field. For gyro-resonance, the energy is given by $E_c = (f_{ce}/f)^2 E_l$. These are the two energies shown in Figure 6(b), converted to temperature. The measured magnetic field strength (from the magnetometer) and electron density (from the Langmuir probe) were used to calculate the gyro- and plasma frequencies (Figures 6(c) and 6(d), respectively).

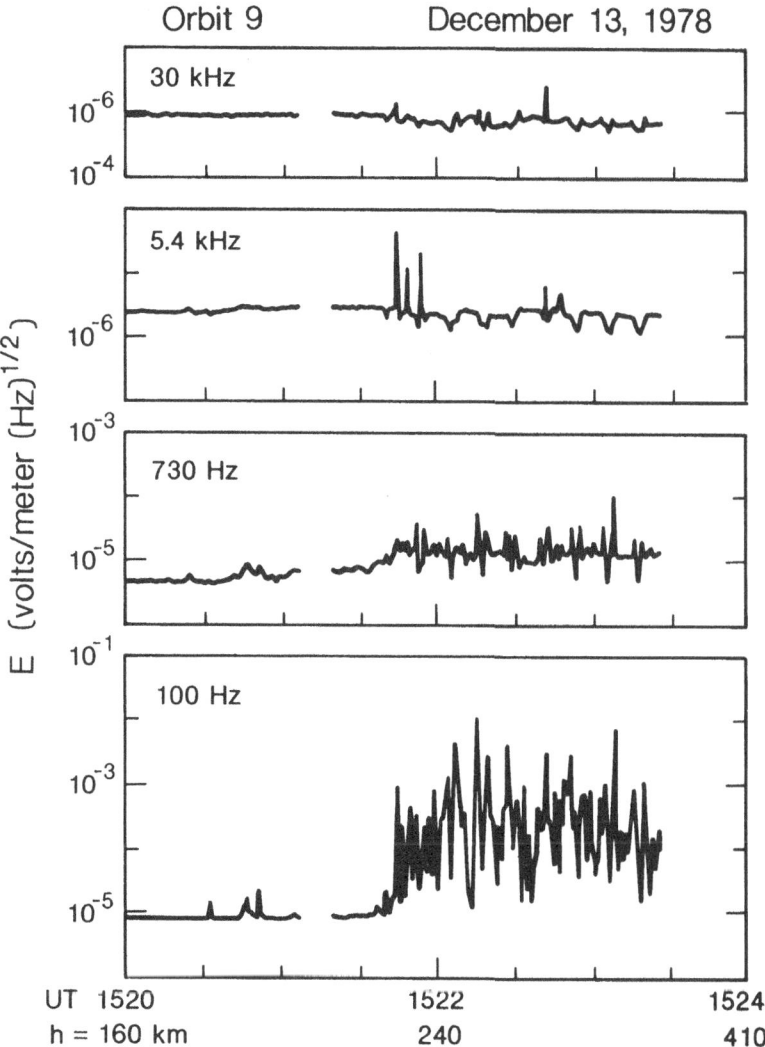

Fig. 5. Example of wave absorption at the ionopause (from Scarf *et al.*, 1979). The 100 Hz wave level decreases from an average level of 10^{-4} to 10^{-5} V m^{-1} Hz$^{-1/2}$ on entering the ionosphere, on the left side of this figure.

The electron temperature as measured by the Langmuir probe is also shown in Figure 6(b). The decrease in 100 Hz wave intensity (Figure 6(a)) occurs at the location where the Landau damping energy and the electron temperature are equal.

In their review of Pioneer Venus plasma wave observations, Scarf *et al.* (1980b) presented additional examples of the equality between electron temperature and Landau resonance energy where the 100 Hz wave amplitude decreased. They also pointed out that the estimated energy flux was comparable with that reported by Brace *et al.* (1979) and Knudsen *et al.* (1979) as a necessary electron heat input at the top of the ionosphere

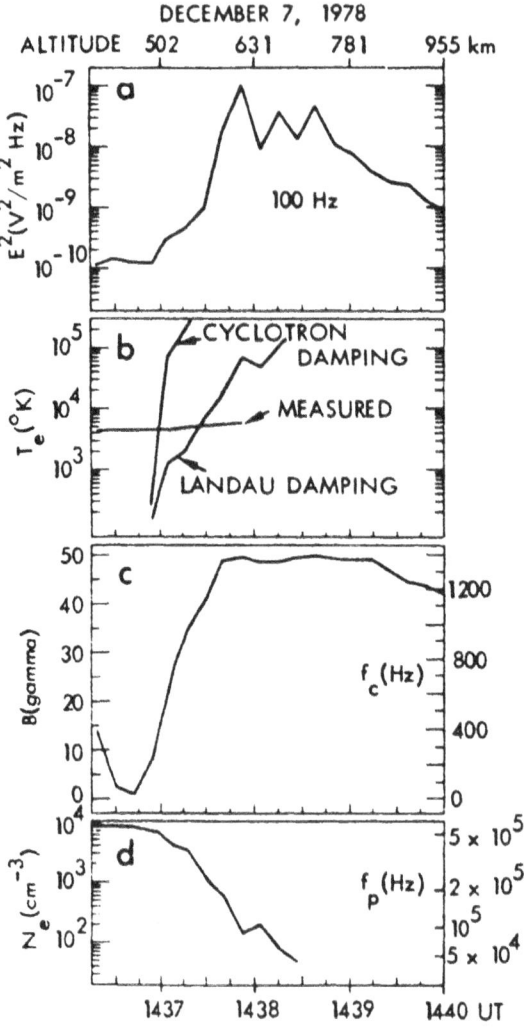

Fig. 6.   Evidence for Landau damping of whistler mode waves at the ionopause (from Taylor *et al.*, 1979). The upper panel shows wave intensity, the second panel shows the electron temperature and resonance energies, while the lower two panels show magnetic field and density.

for the observed electron temperature profiles. Models of the Venus ionosphere required similar energy flux inputs (Cravens *et al.*, 1979).

   Scarf *et al.* (1980b) also discuss the requirement that the waves must propagate across the ambient magnetic field if they are to heat the electrons. The Landau resonance involves the component of wave electric field along the ambient field, but whistler-mode waves propagating parallel to the ambient field are transverse polarized. Whistler-mode waves tend to propagate along the field due to refraction in a medium where the magnetic field and plasma density are smoothly varying. However, Scarf *et al.* showed that the ambient magnetic field can rotate significantly in very short spatial scales at the ionopause, and this may allow the waves to propagate across the magnetic field and so Landau resonate with electrons.

There is, therefore, strong evidence for whistler-mode waves at the dayside ionopause, and these waves may be the source of the energy required to be input into the ionosphere at the ionopause. However, there are some questions that still require answering. First, the energy flux estimate assumed fully electromagnetic (i.e., transverse) whistler-mode waves. If the waves are propagating obliquely to the ambient magnetic field, then they may have an electrostatic component, and the energy flux will be less than that estimated for purely transverse waves. A second effect that has not been included is the change in the instrument response as the spacecraft enters the ionosphere. We note that in Figure 5 the noise level changes for all the frequency channels at the ionopause. In particular, the spin ripple clearly observed in the 730 Hz and 5.4 kHz channel after 15 : 22 UT is absent in the ionosphere. While the cause of the interference is not known at this time, it appears likely that changes in the Debye length are responsible for the change in noise level. A change in the Debye length could affect the interference through changes in plasma sheath scale size, which in turn may change how the antenna couples to the plasma. For reference the Debye length for the ionospheric plasma parameters shown in Figure 6 is about 15 cm (at 14 : 36 : 30 UT), increasing to at least 1.5 m (at 14 : 38 UT). It is, therefore, possible that at least some of the change in the 100 Hz level is due to changes in the interference level. However, the 100 Hz amplitude can be as high at 10^{-3} V m^{-1} Hz$^{-1/2}$, sufficiently far above the normal noise level that a significant part of the signal must be due to plasma waves.

We now come to the other basic question concerning the identification of the waves as whistler-mode. As discussed above the identification is primarily by inference, and there is some evidence for suggesting that at least some of 100 Hz wave activity is actually generated at the ionopause itself. As can be seen in Figure 6, the wave intensity reaches a peak at the ionopause, which could be due to refraction effects, or local wave generation. Elphic *et al.* (1981) carried out a study of the variation of the ionopause thickness as a function of several parameters, including the wave noise level inside and outside the ionopause current layer. There was some evidence for suggesting that the waves affected the ionopause thickness, as shown in Figure 7. Wave levels were generally higher for the lower altitude ionopauses, which were thicker, and the thickness appeared to increase with wave level at the higher altitudes.

Elphic *et al.* made no attempt to identify the wave modes, but if the waves do affect the thickness, which is several ion gyroradii, there is the implication that the waves are interacting with the ions since the electron gyro-radius is much smaller than the ion gyro-radius. Figure 6 shows the energy for electrons in resonance with the whistler-mode. The ion-resonance energy will be larger than this by a factor of the ion/electron mass ratio. Figure 6 implies that the ions do not resonate with these waves. The results of Elphic *et al.*, therefore, suggests that the waves may be electrostatic ion-acoustic waves or lower hybrid waves at the ionopause. Although lower hybrid waves are actually on the same branch of the cold plasma dispersion relation as the electromagnetic

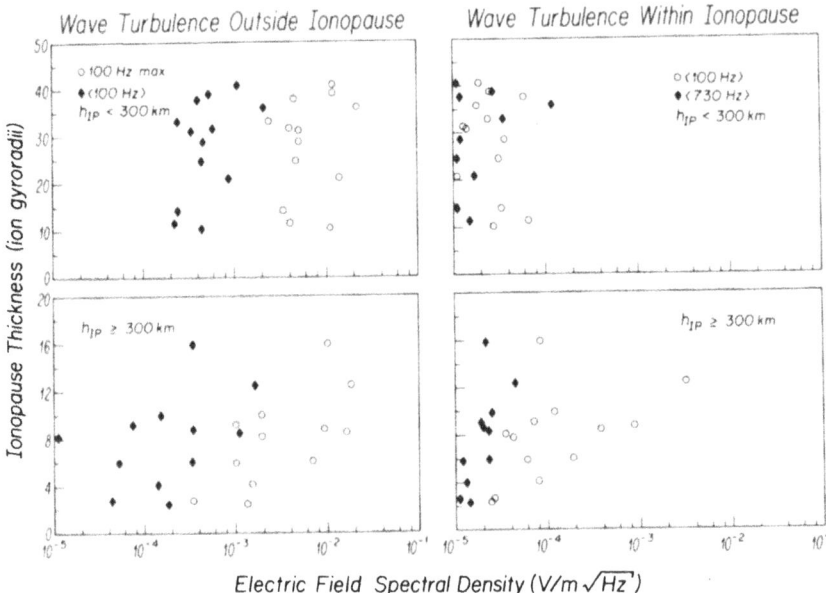

Fig. 7.   Variation of ionopause current sheet thickness with wave levels (from Elphic *et al.*, 1981). The left panels show wave levels outside of the ionopause at low and high altitudes. The right column shows wave levels in the ionopause current layer.

whistler-mode, the lower hybrid waves are strongly electrostatic and, therefore, likely to be generated locally because of their low-phase velocity. However, some Doppler shift would be required to bring the lower hybrid waves ($f_{LHR} \sim 30$ Hz for protons) into the 100 Hz channel range.

The generation of electrostatic waves at the ionopause was also considered by Taylor *et al.* (1981), who showed that at least some of the wave activity in the 100 Hz channel was associated with superthermal ions. As discussed by Taylor *et al.*, the Orbiter Ion Mass Spectrometer (OIMS) was designed to measure low-energy (< 1 eV) ions, but can also detect 'superthermal' ions, with energies in the range 10–90 eV. One such event is shown in Figure 8. The ionopause, as determined by the rapid change in the O^+ density, is encountered just after 21 : 10 UT. Prior to this time superthermal ions are detected, as is enhanced 100 Hz wave activity. Taylor *et al.* suggested that these waves are generated locally by recently photo-ionized ions 'picked up' by the flowing magneto-sheath plasma. A similar effect has been reported by Kasprzak *et al.* (1982), using additional ion data from the Orbiter Neutral Mass Spectrometer (ONMS), as shown in Figure 9.

It appears then that there are still some questions concerning waves at the ionopause that warrant further study. Are the waves generated locally at the ionopause or are they generated elsewhere, and propagate to the ionopause? The latter possibility allows the waves to mediate in the transport of energy into the top of the ionosphere, while the former allows the waves to play a role in the structure of the ionopause itself. A

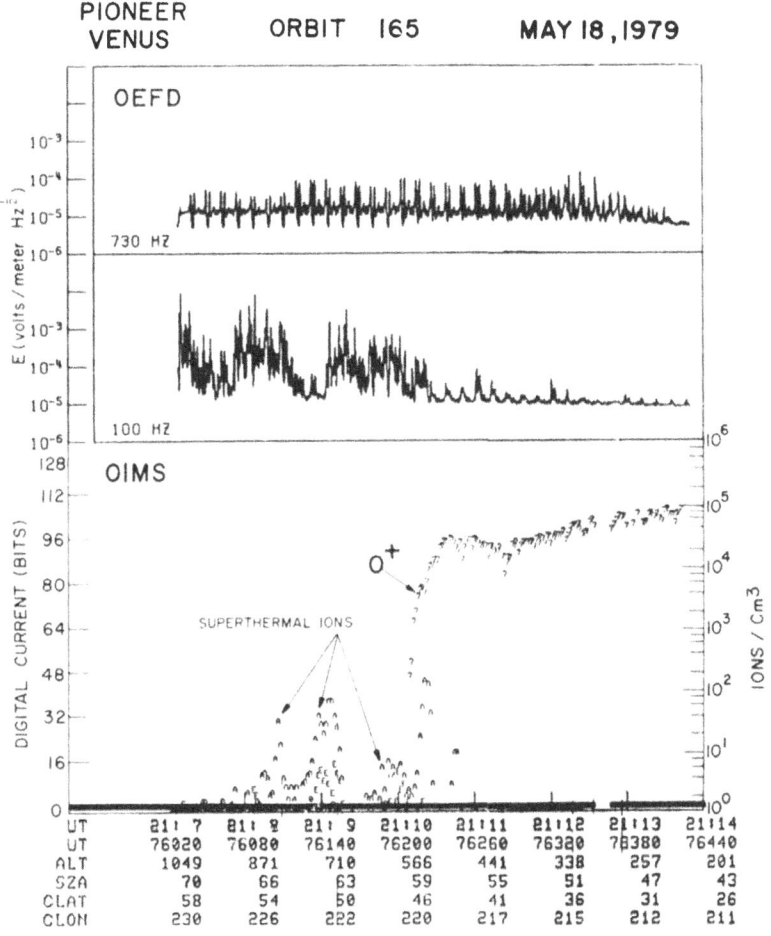

Fig. 8. Superthermal ions above the ionosphere, and the associated plasma wave activity (from Taylor *et al.*, 1981).

systematic study of the wave, field, and plasma parameters would help address these questions.

3.3. CURRENT-DRIVEN INSTABILITIES

Russell *et al.* (1982) reported on observations of the magnetic field and plasma wave structure associated with a plasma cloud above the Venus ionosphere. The main magnetic field signature of the cloud was a reversal in the B_x component, where the *x*-direction is along the Venus–Sun line. The B_x reversal was associated with a decrease in total field. Data from the Langmuir probe showed a plasma density enhancement at the same time as the field reversal. The presence of this diamagnetic cavity was interpreted by Russell *et al.* as indicative of a cloud of ionospheric plasma being 'gathered' by the magnetic field and accelerated due to the sling shot associated with the field reversal in the *x*-component. With regard to the associated plasma wave

ORBIT 403

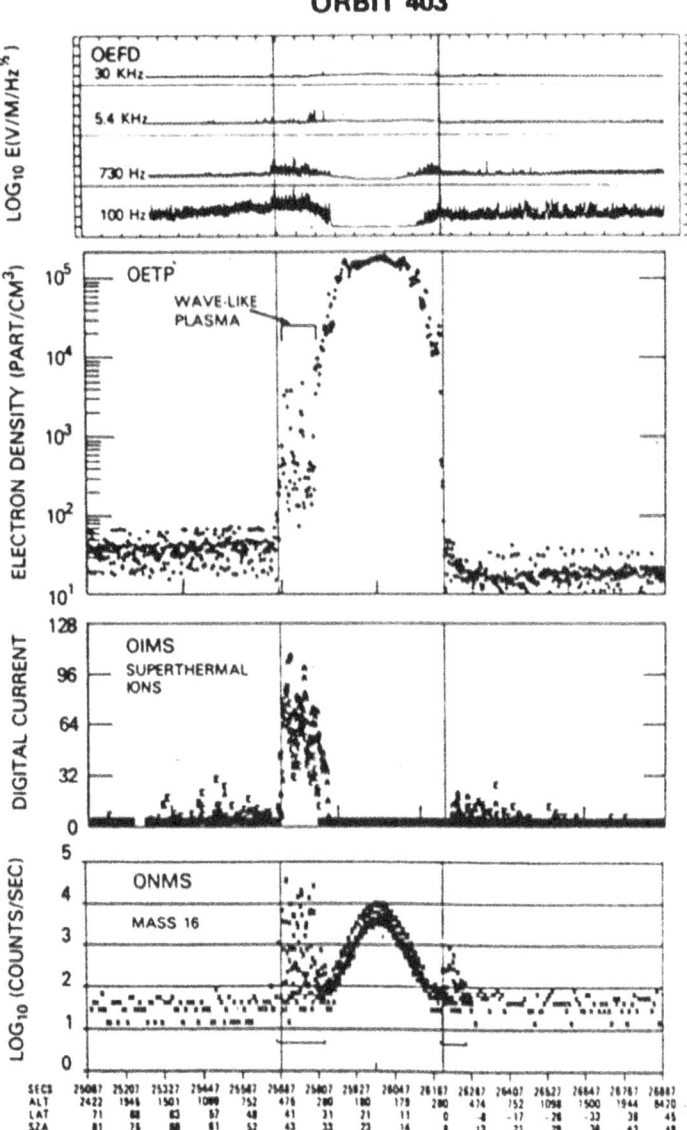

Fig. 9.   Another example of a superthermal ion layer outside of the ionosphere (from Kasprzak *et al.*, 1982).
A superthermal ion layer is detected by both OIMS and ONMS just prior to the inbound ionopause
crossing.

activity, only 100 Hz data were presented by Russell *et al.* Higher frequency wave
activity observed in conjunction with clouds was shown in a subsequent paper by Scarf
et al. (1985), examples of which are given in Figure 10.

Scarf *et al.* (1985) interpreted the broad-band wave noise observed within the plasma
cloud as being ion-acoustic waves driven unstable by the presence of field-aligned
currents. To support their interpretation they presented an instability analysis where
they discussed the required electron drift velocity for ion-acoustic instability. They found

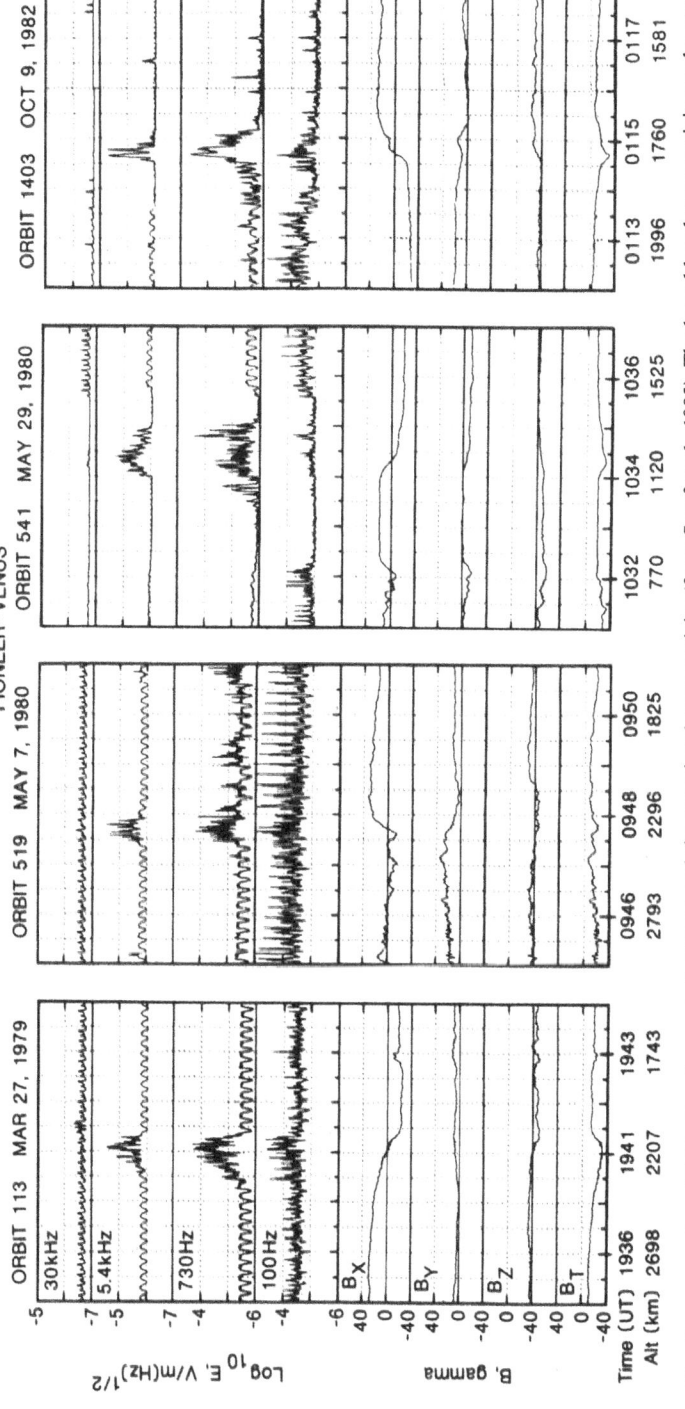

Fig. 10. Four examples of field-aligned current structures and the associated wave activity (from Scarf *et al.*, 1985). The broad-band wave activity peaks at the same time as the B_x field reverses and the total field decreases.

that the current density associated with the critical electron drift would be equivalent to the current density inferred from the observed change in magnetic field provided that the current channel size was of order the ion gyro-radius.

Additional evidence for the presence of drifting electrons was presented by Scarf *et al.*, using current-voltage ($I - V$) curves from the Langmuir probe on the Pioneer Venus Orbiter (OETP). In particular they found that for orbit 992 a large amount of small-scale structure was observed in the Langmuir probe $I - V$ curve in association with the current sheet filament, as shown in Figure 11. The enhanced electron density

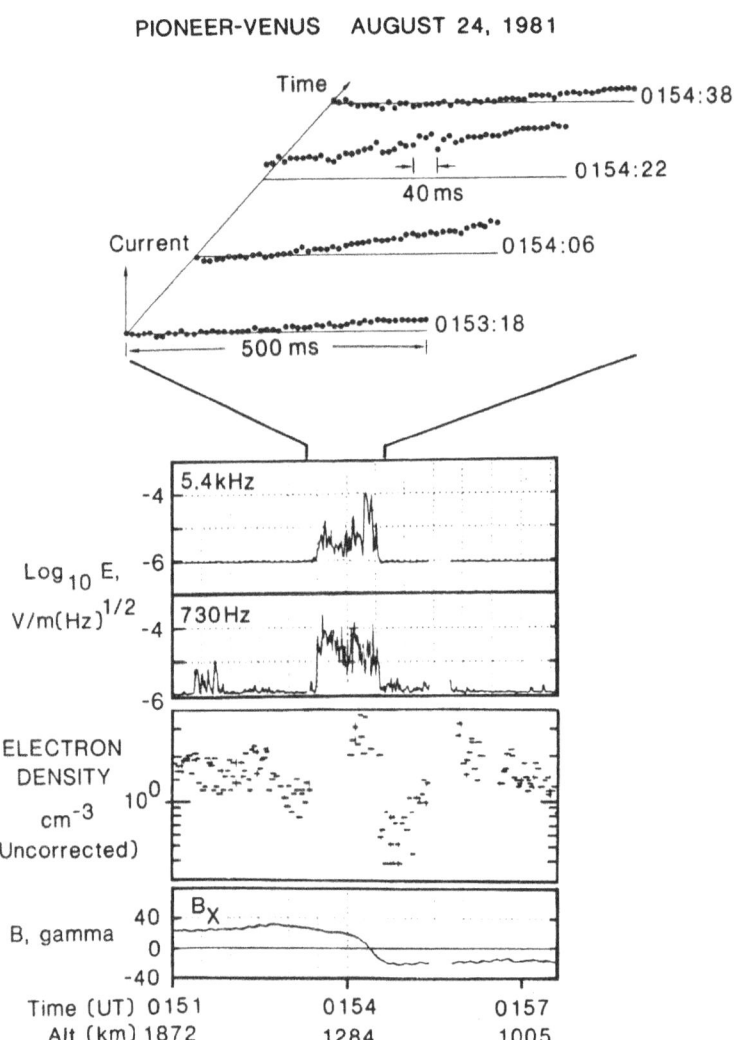

Fig. 11.   Example of Langmuir probe (OETP) anomalies associated with a current layer (from Scarf *et al.*, 1985). In particular note the fluctuations in the $I - V$ curve around 01 : 54 : 22.

shown in Figure 11 was interpreted as being due to secondaries produced by energetic (suprathermal) electron impacts on the spacecraft body. The suprathermal electrons were assumed to be the current carriers responsible for the magnetic field structure, and the generation of the observed waves. The suprathermal electrons also possibly resulted in the fine scale structure of the $I - V$ curves. Furthermore, anomalous resistivity associated with the waves could result in electron acceleration which could then produce detectable aurora at Venus.

The occurrence rate for the current-driven instability events was quite low, Scarf et al. reported the presence of 33 well-defined events out of 1500 orbits. However, because of the short amount of time per orbit spent at low altitude the authors considered this rate to be significant. Most of the 33 events were observed in the 90° to 140° solar zenith angle range and at altitudes below 3000 km. The range in solar zenith angle and altitude may reflect the actual spatial distribution of the current sheets, although restricted coverage due to orbital bias could affect the observed distribution.

The broad noise bursts are associated with current regions, and these currents may be carried by energetic electrons streaming along the magnetic field, although it is not clear that the field lines map to the Venus ionosphere as needed to produce aurora. In an attempt to test this hypothesis we have determined the orientation of the magnetic field and spacecraft location at the center of the diamagnetic cavity for the six events presented in detail by Scarf et al. (1985). Since the B_x component of the ambient field is near zero at this time, the $y - z$ projection of the field will determine whether or not the current will intercept the ionosphere (coordinates are defined in the Venus Solar Orbital system, which is analogous to GSE, with the Venus–Sun line defining x_{VSO} and z_{VSO} being perpendicular to the Venus orbital plane). We are assuming that the current responsible for the reversal in B_x is indeed field-aligned in the diagmagnetic cavity, and so only in the $y - z$ plane. Our analysis is further compromised by the use of the 12 s resolution Unified Abstract Data System (UADS) data. Somewhat surprisingly we find that for the six orbits the angle between the magnetic field and the radius vector perpendicular to the x-axis is 97° $\perp$ 14°. These results suggests that if the currents are field-aligned then they do not intercept the ionosphere, but are draped over the ionosphere. It is, therefore, not obvious that these currents will produce aurora.

Alternatively, it may be that the currents are perpendicular to the field in the cavity, in which case they will intercept the ionosphere. Perpendicular currents could be expected, since they will result in plasma acceleration due to $\mathbf{j} \times \mathbf{B}$ forces. However, perpendicular currents do not involve freely streaming electrons and, hence, are unlikely to generate ion-acoustic waves or auroral signatures. It is apparent that a more rigorous analysis than that presented here should be carried out to determine the nature of the current in the plasma clouds, the source of the plasma waves and their consequence.

4. Waves in the Magnetosheath and Tail

The region of shocked solar wind plasma flowing around the planetary obstacle has been referred to as either the magnetosheath (e.g., Mihalov and Barnes, 1981) or as the ionosheath (e.g., Perez-de-Tejada *et al.*, 1984). Near the planet, especially on the dayside, ionosheath may be a more descriptive name since the underlying obstacle is the ionosphere itself. On the other hand, downstream of the planet the low-density magnetotail is the obstacle to the solar wind flow, and magnetosheath is probably better. The two names are to some extent interchangeable, and both are extant in the Pioneer Venus literature. Since both names refer to the same region of shocked solar wind, and many features are reminiscent of the terrestrial magnetosheath, we have used magneto-sheath in this review, although the reader should be aware of the term ionosheath.

4.1. MAGNETOSHEATH BOUNDARY LAYER

Moving from the dayside to the nightside we will first discuss the plasma wave evidence for a 'viscous' interaction in the inner magnetosheath. Perez-de-Tejada (1982, and references therein) has argued that plasma waves may contribute to the viscosity of the magnetosheath plasma, resulting in a viscous boundary layer within the magnetosheath. Perez-de-Tejada *et al.* (1984, 1985) have presented plasma wave and particle observations as support for this hypothesis. Data from the Orbiter Plasma Analyzer (OPA) suggest that the inner magnetosheath is characterized by a less dense, but hotter plasma than that observed closer to the shock. Consequently, there does appear to be an inner boundary layer in the magnetosheath, although whether or not this layer is 'viscous' is still an open issue.

The OEFD wave data were used to support the inferred decrease in density; a change in the amplitude of plasma oscillations was cited as evidence for a boundary in the magnetosheath. A recently discovered interference problem may affect the interpretation of wave data within the magnetosheath, which we discuss here. This does not affect the analysis and conclusions of Perez-de-Tejada *et al.*

Figure 12, from Perez-de-Tejada *et al.* (1985), shows magnetosheath data obtained

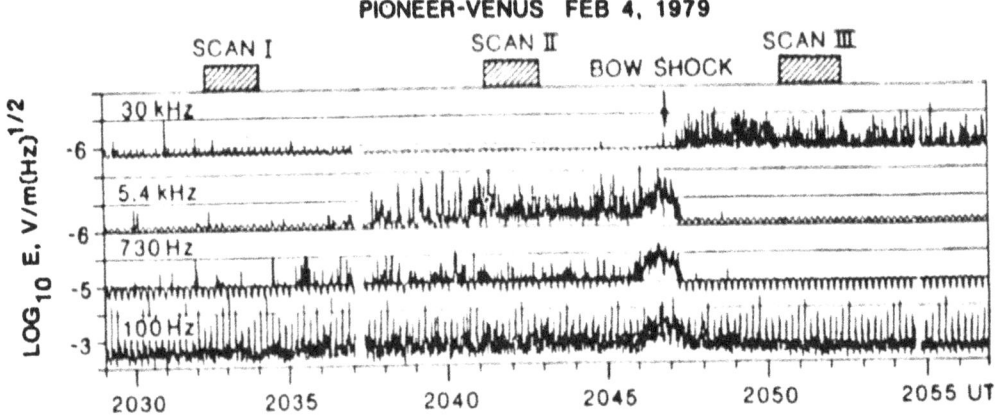

Fig. 12.   OEFD data during an outbound pass of the Venus magnetosheath (from Perez-de-Tejada *et al.*, 1985). The cross-hatched regins label those times that the Orbiter Plasma Analyzer performed angular scans.

post-periapsis. The wave data were acquired when the spacecraft was in sunlight and all the frequency channels show the characteristic spin modulation observed in sunlight (Scarf *et al.*, 1979). After 20:37 UT the 730 Hz and 5.4 kHz channels show enhanced wave activity. The change in noise level at these two frequencies was cited by Perez-de-Tejada *et al.* as evidence for a boundary in the magnetosheath. They further argued that the disappearance of the spike-like signals observed at 30 kHz prior to 20:35:30 UT indicated that the plasma density was enhanced after this time. Although not used as part of the discussion by Perez-de-Tejada *et al.*, the reduced signal level in 30 kHz after 20:37 UT might also be thought to be due to a transition in the magnetosheath. We want to caution the reader here about such an interpretation of the 30 kHz data.

Figure 13 shows a two-hour summary plot of OEFD 12 s peaks and averages for orbit 455. Periapsis on this day is around 09:54 UT. It will be noted that at 09:14 UT the 30 kHz peak level suddenly increases, while there is a subsequent decrease at 10:24 UT. For this orbit the change in noise level coincides with commands to turn the Orbiter Neutral Mass Spectrometer (ONMS) on and off, respectively (W. Kasprzak, personal communication). These times are roughly 40 min before and 30 min after

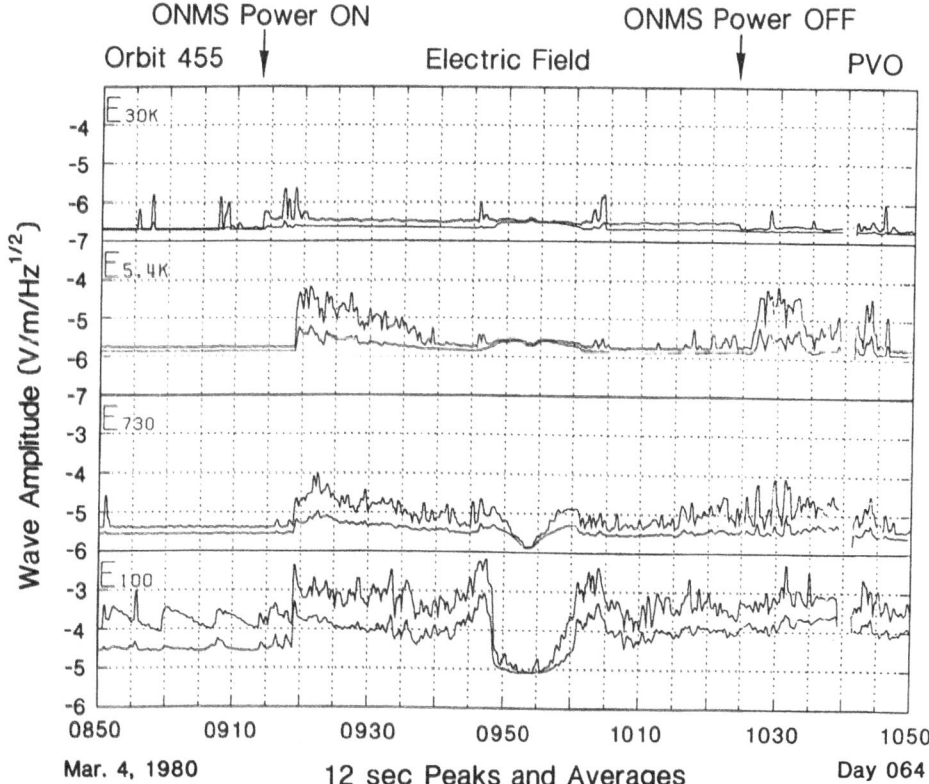

Fig. 13.   Two hours of OEFD data from orbit 455. The traces show 12 s peaks and averages for the four channels. The times of ONMS power on/off are indicated.

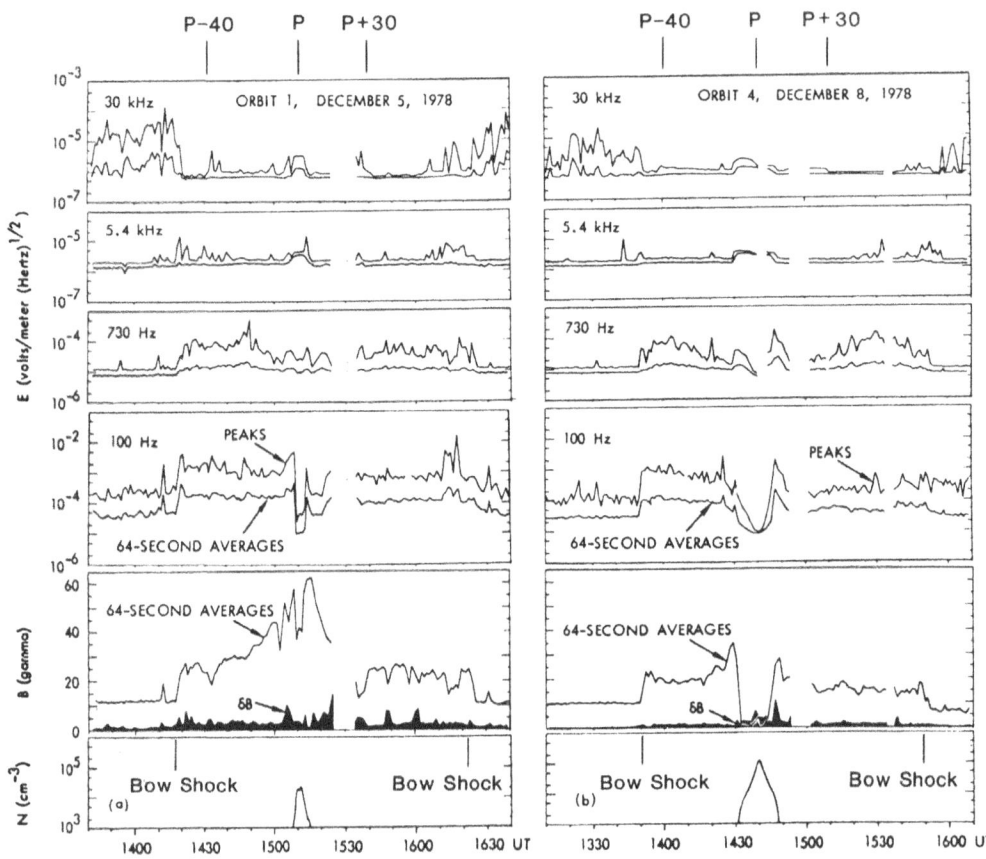

Fig. 14.   Two summary plots of OEFD data (after Scarf *et al.*, 1979). Note the change in the 30 kHz peak level 40 min prior to and 30 min after periapsis. Periapsis occurs around the electron density peak, shown in the bottom panel.

periapsis. The interference is seen in data from other orbits. Figure 14, taken from the preliminary results paper by Scarf *et al.* (1979), shows that the 30 kHz noise level changes at the same times with respect to periapsis as for orbit 455. To further emphasize the persistence of this feature, we have surveyed orbits 40–100, and a change in noise level 40 min before and 30 min after periapsis is observed on 83% of those orbits where the transition is not masked by other 30 kHz activity, such as plasma oscillations upstream of the shock. Although we have not verified on an orbit by orbit basis that the change in 30 kHz noise levels is associated with the powering on or off of the ONMS, interference from the ONMS is a likely cause since the signature occurs at the same time with respect to periapsis time on each orbit. In addition, there is a peak in the radiated emission from the ONMS in the 35 kHz to 40 kHz range (W. Kasprzak, personal communication), and it is possible that the OEFD 30 kHz channel (30% bandwidth) is sensitive to this interference.

On day 62 (Figure 12), periapsis occurs at 20:07 UT, and it will be noted that the decrease in 30 kHz activity occurs at 20:37 UT. We, therefore, conclude that the decrease in 30 kHz wave amplitude is not additional evidence for an increase in plasma density, although the spike-like noise prior to 20:35:30 UT may correspond to plasma oscillations.

A clear example of plasma oscillations observed in the magnetosheath in association with a transition in the ion energy flux spectra, as reported by Perez-de-Tejada *et al.* (1984), is presented in Figure 15. Around 19:53 UT there is an enhancement in the 100 Hz wave level, a rotation in the ambient magnetic field and a sharp reduction in the ion energy flux. While the 30 kHz waves marked as plasma oscillations are weak, they do not display the spin modulation seen in the interference signal prior to this time. Moreover, periapsis on this day is at 20:07 UT, and the change in 30 kHz noise is not associated with ONMS power on. The conclusion of Perez-de-Tejada *et al.* that supra-thermal electrons are present at this time is, therefore, supported by the OEFD data.

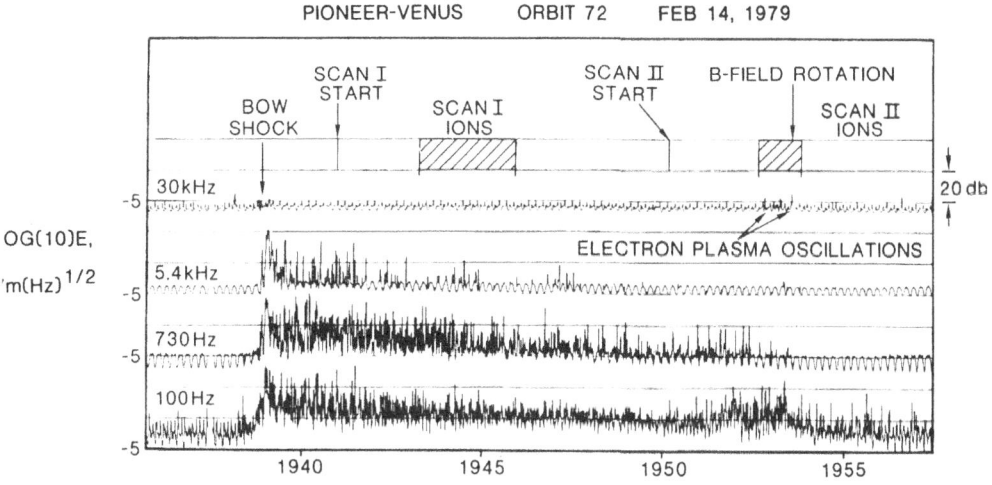

Fig. 15. Electric field data observed during an inbound crossing of the magnetosheath (from Perez-de-Tejada *et al.*, 1984). A transition region appears to be observed around 19:53 UT.

4.2. WAVES IN THE MAGNETOSHEATH

Further down tail, Intriligator and Scarf (1982) presented data from the Plasma Analyzer and the OEFD observed near apoapsis in the magnetosheath. They presented examples of correlations between the wave activity and the ion distributions observed, as shown in Figure 16. The small squares show the ion flux as a function of energy/charge. The positioning of the squares, together with their abscissa length are chosen to reflect the actual times that a particular energy/charge measurement was made. The lower two panels show the peak and average wave amplitudes for the 730 Hz

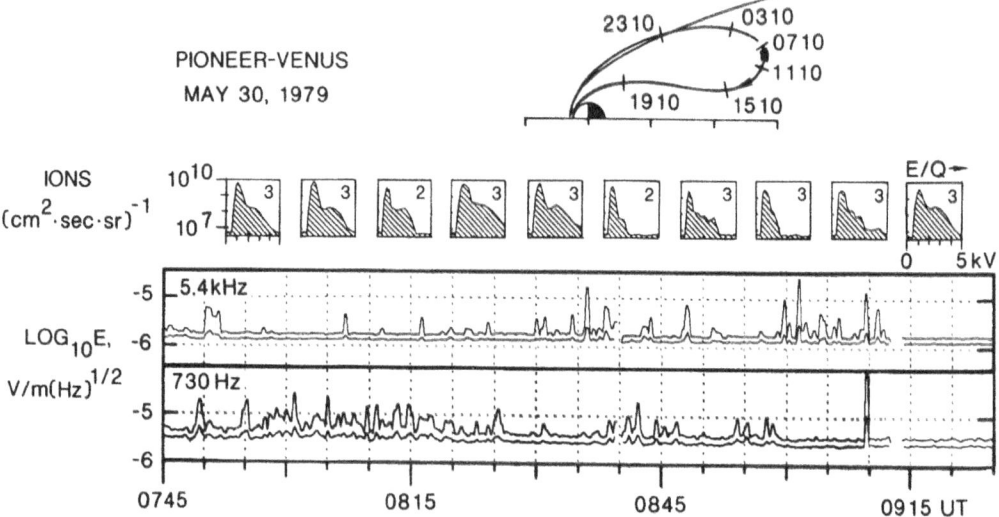

Fig. 16.   Plasma and wave data obtained in the magnetosheath near apoapsis (from Intriligator and Scarf, 1982). There is some degree of correlation between the wave activity in the different channels and the ion spectra.

and 5.4 kHz channels. The authors note that there is some correlation between the spectral shape and the wave activity. In general, 5.4 kHz wave activity appears to be associated with the more sharply-peaked distributions, after 09 : 00 UT, where there is a somewhat reduced flux of particles around 2.5 keV. On the other hand the broader distributions seem to be associated with 730 Hz activity. The second peak in the ion distributions is around 2.5 keV, while the primary peak is around 1 keV (e.g., at 08 : 30 UT). The second peak cannot be obviously interpreted as either a He^+ peak, or an O^+ peak, flowing at the same velocity as the primary (presumably) H^+ peak. If the secondary peak corresponds to pick up ions they have not yet attained the shocked solar wind flow speed.

Intriligator and Scarf interpret the waves as ion-acoustic waves Doppler-shifted by the high plasma flow with respect to the spacecraft, typically around 400 km s^{-1}. Doppler-shifting an ion wave to higher frequencies appears to be necessary since the plasma density is so high in these regions, and there should be no electron modes in the 730 Hz, 5.4 kHz range. However, it is not clear that Doppler-shifting alone can explain the signatures. In particular the Doppler shift might be expected to produce a broad-banded wave spectrum, since the underlying ion-acoustic wave exists from extremely low frequencies up to the ion plasma frequency. Other wave modes in addition to those present in a warm plasma may have to be considered to explain the observations, but such analysis is beyond the scope of this review.

4.3. Waves at the Magnetotail Boundary

Plasma waves at the tail boundary and in the magnetotail were investigated by Russell *et al.* (1981). In general, the tail boundary was readily observed in both the magnetic field data, and in the plasma waves. Figure 17 shows one tail pass from orbit 188. Because of the orbit characteristics, tail passes are usually obtained in the range 7–12 R_V downtail. It will be noted that within the magnetotail the magnetic field is highly structured, with large changes in the total field and many reversals in the x-component. The 5.4 kHz data show large fluctuations just outside the tail, with quieter levels within the tail itself. The magnetotail boundary is well defined in both the magnetic field and plasma wave data. The large depression in the wave level around 15 : 00 UT is due to orbiter entry into the optical shadow of the planet. The decrease in noise level in shadow may be due to reduced photo-electron emissions, and also due to reduced interference from the solar cells.

The tail region was also studied by Intriligator and Scarf (1984). They also presented data from orbit 188, which is shown in Figure 18. It will be noted that the ion density

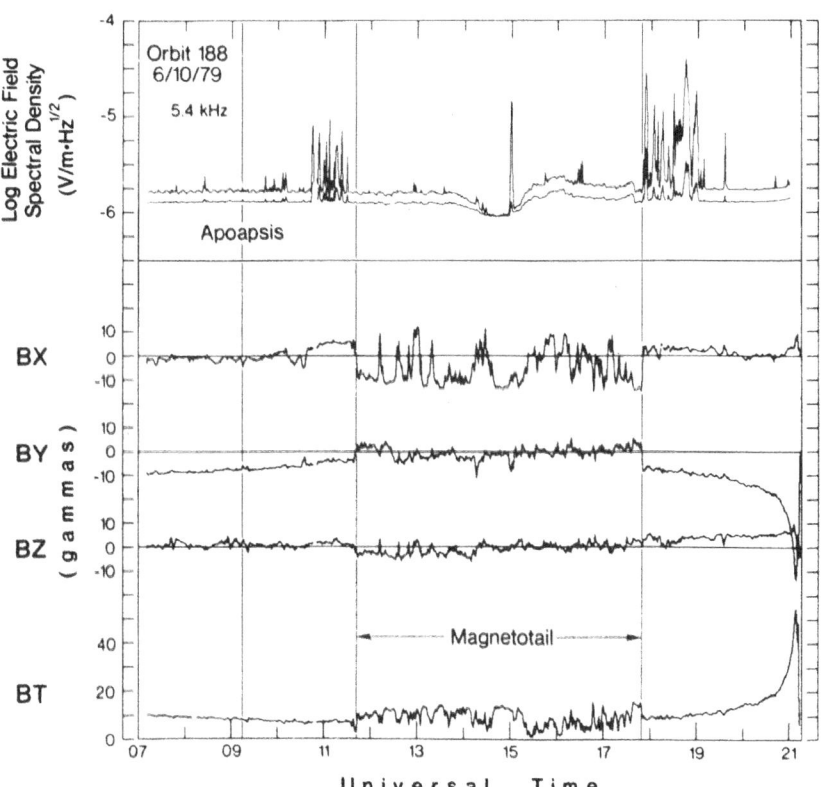

Fig. 17.   Magnetic field and plasma wave data for a pass through the magnetotail (from Russell *et al.*, 1981). The tail boundary is clearly observed in both the magnetic field and the plasma wave signatures.

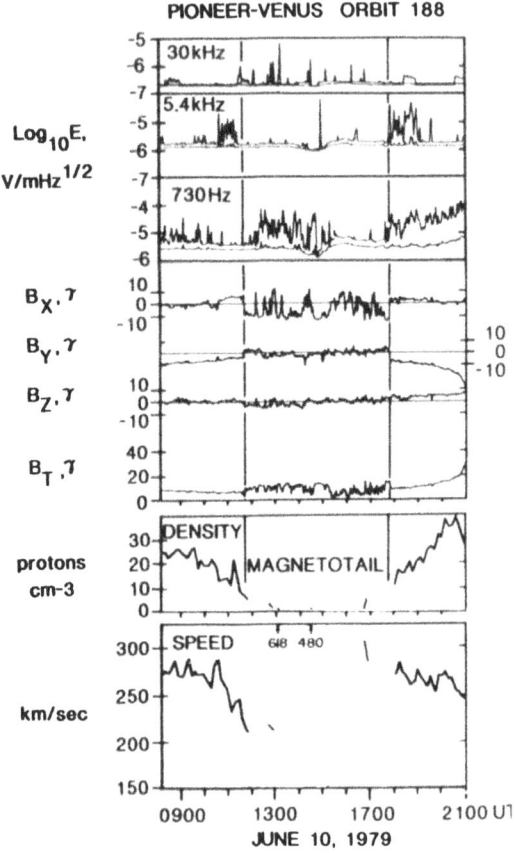

Fig. 18.   The same tail pass as shown in Figure 17, but with additional wave channels and ion densities
included (from Intriligator and Scarf, 1984).

and flow measurements are consistent with the tail boundary determination presented
in the previous figure. Figure 19, on the other hand, shows a case where the ion and wave
data do not give the same boundary as the magnetic field data. The ions suggest that
the magnetotail boundary is observed around 14:00 UT, while the structure in the
magnetic field suggests 13:00 UT might mark the tail boundary. The plasma wave data
are consistent with the ion data. It appears then that magnetometer data alone are
insufficient to determine the tail boundary; while particle data are preferable, in their
absence plasma wave data provide a useful diagnostic for verifying the magnetometer
data.

We are struck again by the need to clarify the nature of the wave modes observed by
the OEFD. Invoking Doppler shift of an ion-acoustic wave does not appear to fully
explain the observations. In Figures 18 and 19 there are regions where the 5.4 kHz
signal is extremely large, but the 730 Hz signal is near background. Our perhaps
simplistic expectation that the Doppler shift will produce a broad-banded spectrum may
be incorrect, or other wave modes may have to be invoked to explain the signatures.

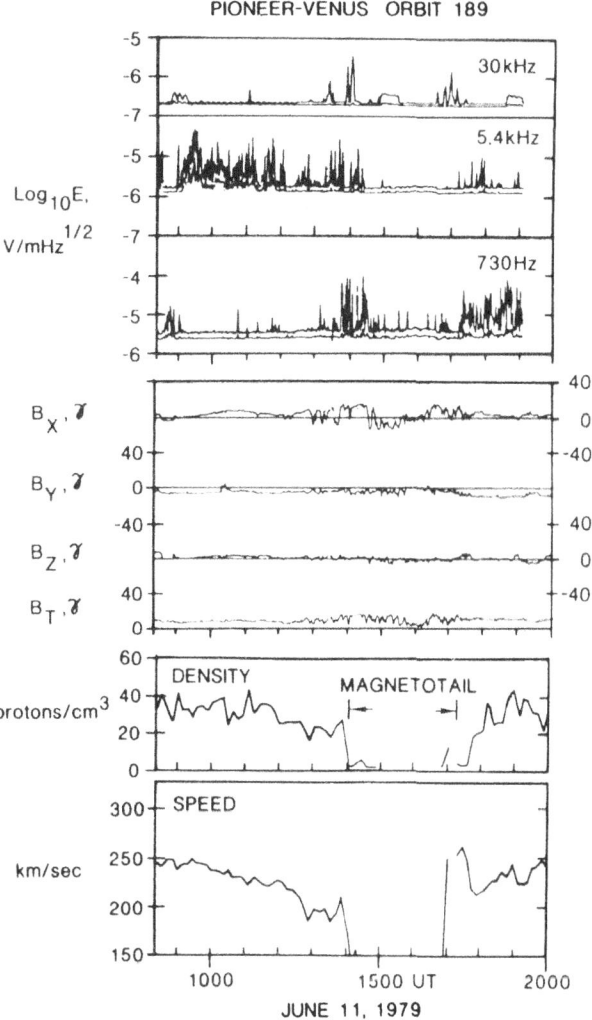

Fig. 19.   A tail pass (from Intriligator and Scarf, 1984) which shows some differences between the magnetotail cavity as determined from the magnetic field or from the plasma data.

5. Waves at the Bow Shock

Surprisingly little attention has been drawn to the wave phenomena associated with the bow shock at Venus. Early in the Pioneer Venus mission Scarf and colleagues presented many examples of wave signatures at the bow shock, but only recently has the topic of waves at the shock been re-addressed.

5.1. PLASMA WAVES IN THE SHOCK

Figure 14 contains some examples of shock crossings, presented by Scarf *et al.* (1979). Upstream plasma oscillations (30 kHz) are observed in the solar wind, with lower-frequency wave turbulence observed in the magnetosheath. Some of this turbulence is the whistler-mode wave activity discussed previously as a possible energy source to the ionosphere.

Scarf *et al.* (1980b) discussed shock observations in more detail, and presented a comparison between wave spectra at the terrestrial shock and at the Venus shock, shown here in Figure 20. The mean value of the average and peak spectra from 48 crossings of the Venus shock are shown in the left and right panels, respectively, together with corresponding average and peak spectra from the terrestrial shock, from Rodriguez and Gurnett (1975). Scarf *et al.* considered that the difference in wave levels between the Venus and terrestrial shock spectra at the low and high frequencies was significant and indicated that different instabilities were operating at the Venus shock. As can be seen in Figure 20, the terrestrial shock spectra appear to be dominated by waves with a broad peak around 1 kHz, identified as ion-acoustic waves. The enhanced level of 30 kHz signals at Venus was attributed to plasma oscillations being generated in the shock itself, not just upstream. The 100 Hz enhancement was attributed to generation of whistler-mode turbulence at the shock.

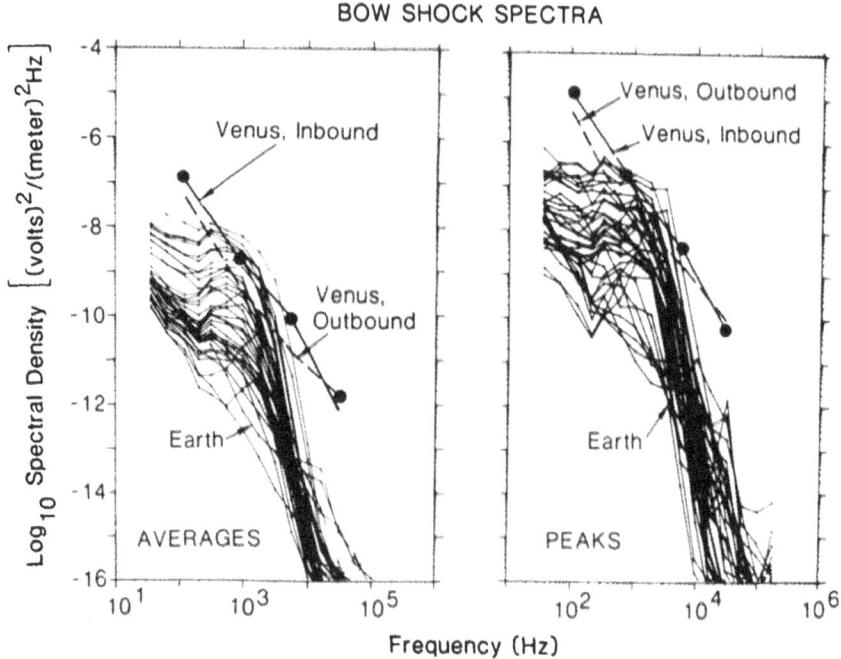

Fig. 20.   Comparison of bow shock spectra observed at Earth and at Venus (from Scarf *et al.*, 1980b). The terrestrial spectra are taken from the study of Rodriguez and Gurnett (1975).

Because the Pioneer Venus antenna (0.75 m) is so short with respect to the Imp-6 antenna (100 m tip-to-tip, Rodriguez and Gurnett, 1975), a comparison such as shown in Figure 20 does not reflect any differences in how the antenna couples to the plasma. In the solar wind typical Debye lengths are of the order 10 m, and as a consequence long antennas are resistively coupled to the plasma, while short antennas are capacitively coupled, as discussed, for example, by Scarf *et al.* (1968). A more direct comparison is shown in Figure 21. The data at left are from the Pioneer Venus Orbiter, while the data at right are from the wave instrument flown on board the AMPTE/CCE spacecraft. The AMPTE/CCE wave instrument is similar in design to the Pioneer Venus instrument (Scarf, 1985). Before comparing the data it is important to note that the AMPTE/CCE spacecraft apogee is near $9 R_e$, and so the solar wind conditions must be somewhat unusual for the shock to be observed so close to the Earth. The data for November 1, 1984 are discussed in more detail by Strangeway *et al.* (1988), who point out that the solar wind density was at least 40 cm^{-3} on this day. With this caveat in mind, it appears that the 100 Hz and 30 kHz signals are more nearly in agreement (note the different scales used), but the 730 Hz and 5.4 kHz signals are much more intense at the Earth. This contradicts some of the comparison shown in Figure 20, and it is apparent that a better knowledge of the intercalibration of the various wave instruments is required. This intercalibration is further complicated since the noise level of the AMPTE/CCE instrument is known to depend on the ambient plasma parameters (Strangeway *et al.*, 1988), and similar effects may be expected for the PVO instrument. As a last remark on the data in Figure 21, the upstream plasma oscillations observed at AMPTE are relatively weak, an order of magnitude less in amplitude than observed at Venus, because of the high solar wind density.

Another study which presented examples of wave observations at the shock was carried out by Slavin *et al.* (1980). Figure 22 shows one such shock crossing. This is the same shock shown in Figure 14, for orbit 1 outbound. We have included this figure because it is somewhat unusual, with almost no wave activity at the time of the shock crossing. On comparison with Figure 14 it is apparent that some upstream plasma oscillations are observed after the crossing. Slavin *et al.* attributed the lack of wave noise at the shock as being due to the transitional nature of the shock, the angle between the shock normal and the ambient magnetic field, θ_{bn}, ~51° and variable about this value.

5.2. WAVES AND THE SHOCK PRECURSOR

Recently, Knudsen *et al.* (1989) have used wave data at shock crossings to support their contention that a shock precursor (Brace *et al.*, 1985) is not really a precursor but is actually a tangential encounter with the shock itself, and Langmuir probe signatures attributed to mass loading of the solar wind were in fact due to an enhancement in a low temperature electron population trapped in the spacecraft potential. Mass loading may be expected to occur upstream of the bow shock at Venus because of the low altitude at which the shock is encountered. This mass loading could produce a precursor to the actual shock transition. Knudsen *et al.* showed that the density enhancement

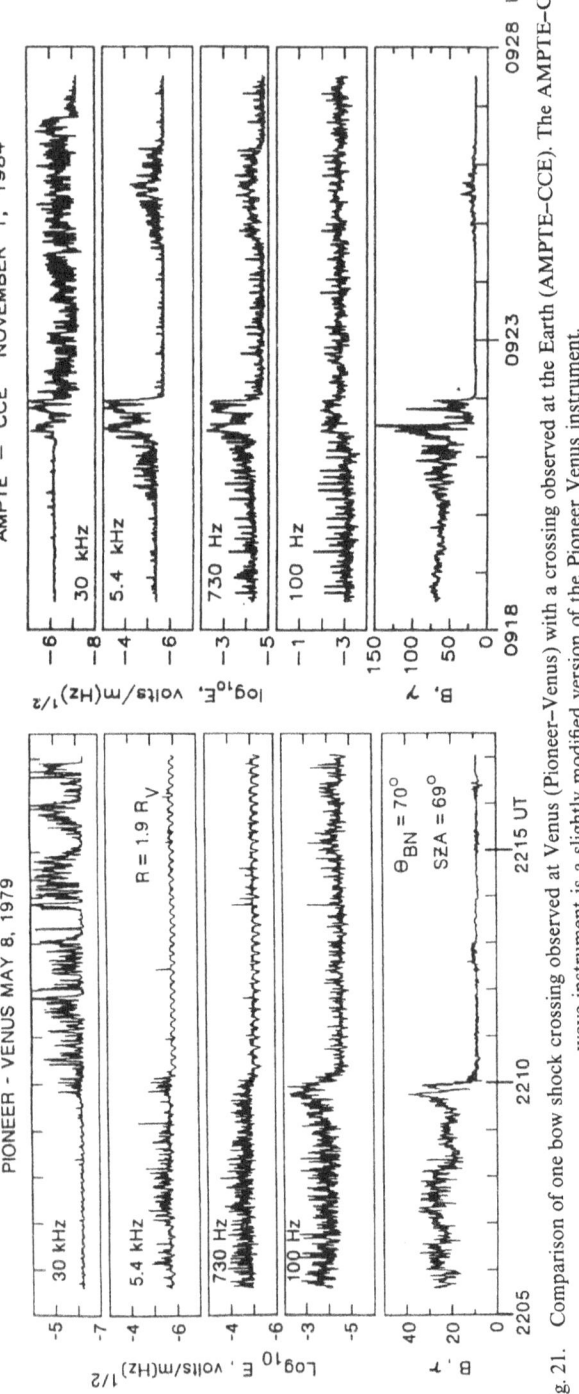

Fig. 21. Comparison of one bow shock crossing observed at Venus (Pioneer–Venus) with a crossing observed at the Earth (AMPTE–CCE). The AMPTE–CCE wave instrument is a slightly modified version of the Pioneer Venus instrument.

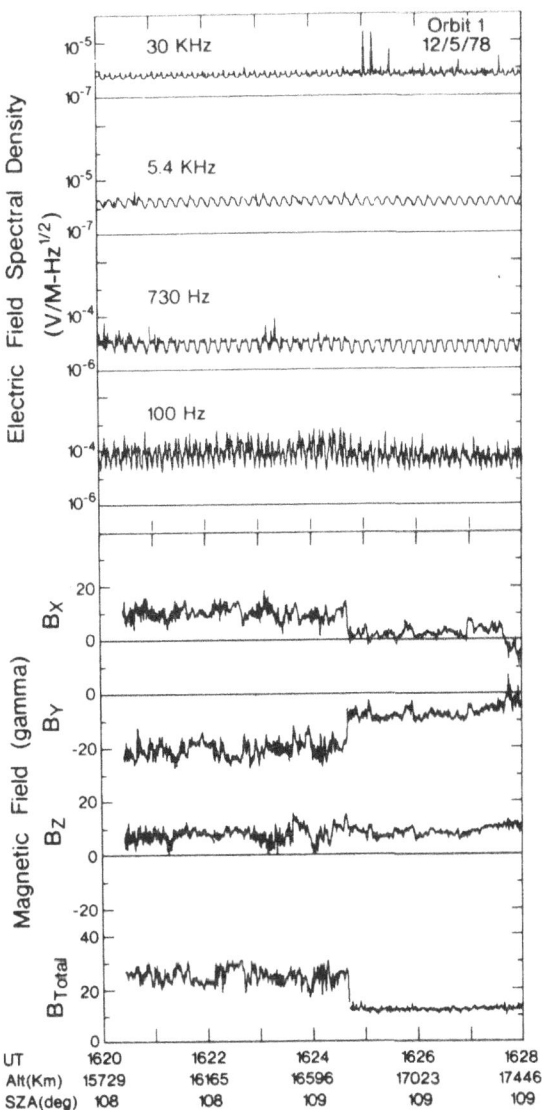

Fig. 22.   A quiet shock crossing (from Slavin *et al.*, 1980). Unlike many other shocks, there is almost no signature in the wave data at the time of the crossing (16 : 24 : 45 UT).

observed in the precursor region was associated with a low temperature (~ 1 eV) population, and this enhancement was independent of position relative to the shock.

Data from one grazing encounter is shown in Figure 23. There is enhanced 730 Hz wave activity (presumably ion-acoustic waves) in the region labelled as the shock transition region in Figure 23, together with some modulation of the ambient field and the solar wind density and temperature. In this figure the contribution to the temperature and density by the trapped population has been removed.

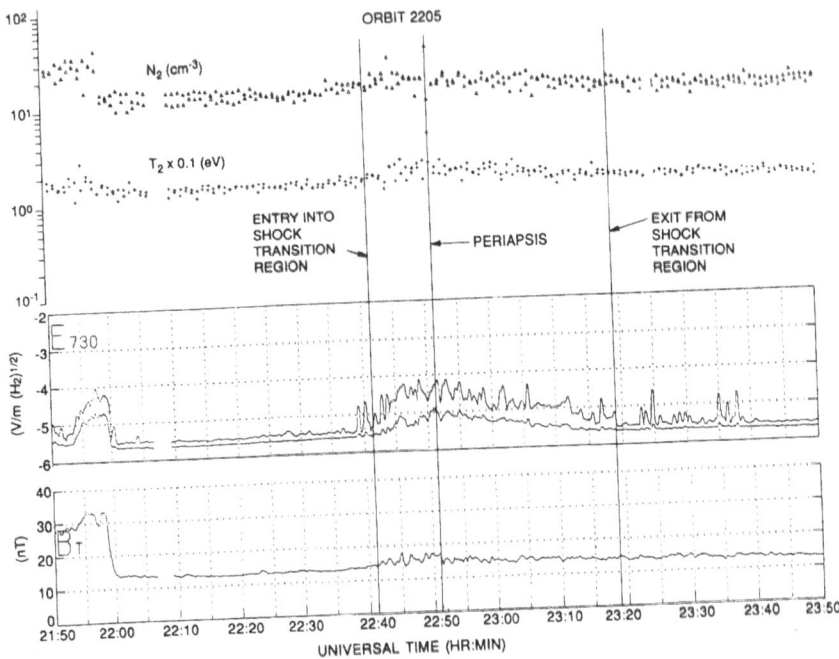

Fig. 23.   Example of a grazing shock encounter, as determined by Knudsen *et al.* (1989). The upper traces show the density and temperature of the solar wind plasma measured by the retarding potential analyzer (ORPA).

While it is reasonable to interpret the signatures as indicating a grazing shock encounter, we would like to point out that similar signatures are observed in the ion foreshock at the Earth (see the interval around 09 : 24 : 30 UT in Figure 21). In the ion foreshock, reflected ions play a similar role to that expected for ions introduced into to the solar wind through mass loading, giving signatures in the ambient field, plasma, and plasma waves upstream of the shock. Wave and field data alone do not appear to be sufficient to determine if structure such as that shown in Figure 23 is in fact a grazing shock encounter, the ion foreshock, or due to mass loading. Additional information, such as the presence of an electron population trapped in the spacecraft potential (as cited by Knudsen *et al.*), is required to help determine whether or not mass loading must be invoked to explain the observed signatures.

6. Plasma Waves as Evidence for Lightning on Venus

As mentioned in the Introduction, we will not present a full review of the plasma wave evidence or lightning here. We refer the reader to Russell (1991) for such a review. Instead we will address some remarks to the basic question of whether or not lightning needs to be invoked to explain the wave signatures observed in the nightside Venus ionosphere. When discussing the observations we will consider waves with frequencies below and above the gyro-frequency separately. We do this since different issues arise depending on the wave frequency with respect to the electron gyro-frequency. In addition the statistical properties of the bursts are different.

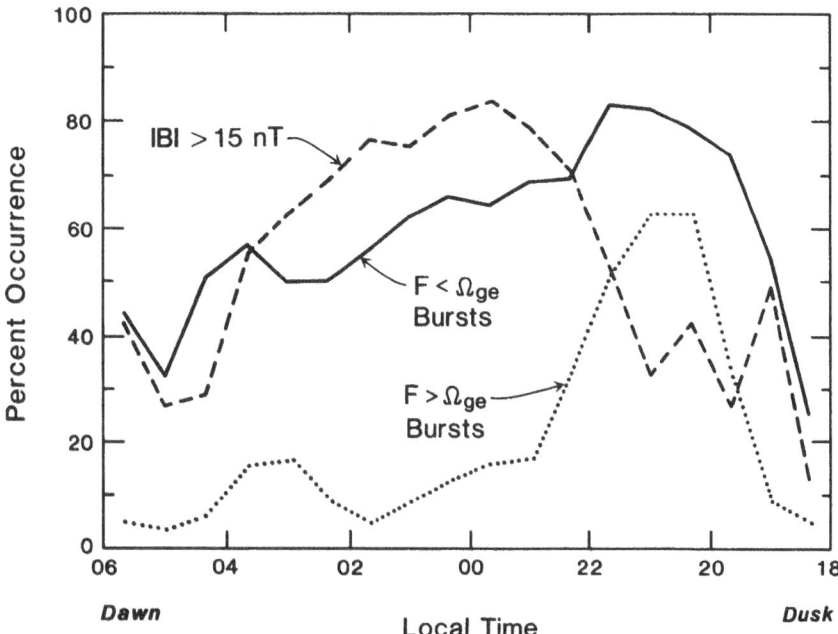

Fig. 24. Percent occurrence rate as a function of local time (from Russell *et al.*, 1989). Signals below the local gyro-frequency are observed throughout the nightside ionosphere, while higher frequency bursts are confined to the post-dusk local time sector.

Figure 24, from Russell *et al.* (1989), shows that the bursts with frequency less than the electron gyro-frequency are observed over most of the nightside, while the higher frequency bursts are confined to the dusk-midnight local time sector. The altitude distribution is also different for the different frequencies. Figure 25, also from Russell *et al.* (1989), shows that while the percent occurrence rate for the higher frequencies decrease rapidly with altitude, the 100 Hz channel only decreases slowly.

The dependence on local time and altitude shown in Figures 24 and 25 is consistent with an atmospheric, or lightning, source for the bursts detected with the OEFD. The 100 Hz signals propagate through the ionosphere as a whistler-mode wave, having gained access to a wave duct after propagating in the surface-ionosphere waveguide in an analogous manner to terrestrial lightning whistlers. In the Venus ionosphere the density holes may act as ducts since the ambient magnetic field is also enhanced (see Brace and Kliore (1991) for a description of ionospheric holes). The higher frequency signals, on the other hand, may represent a 'prompt' response to a lightning event, and suffer strong attenuation away from the source region.

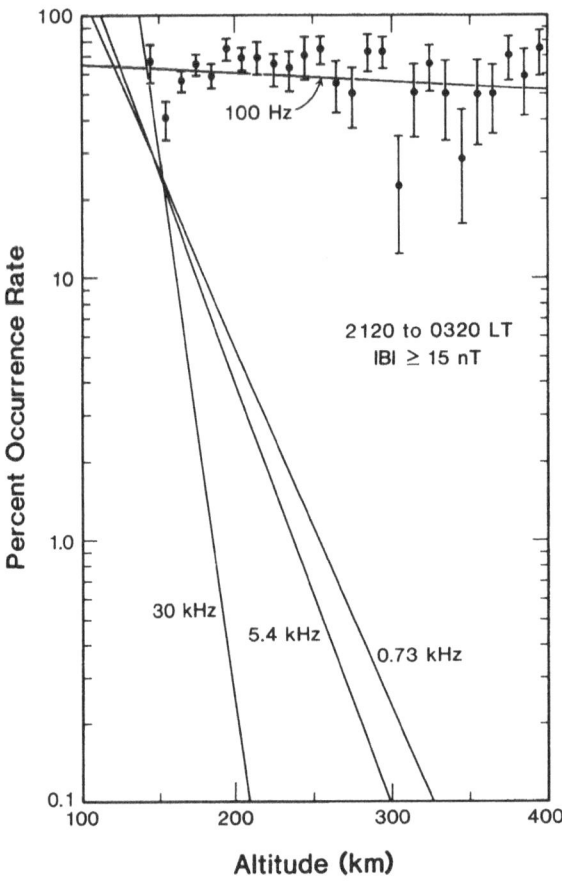

Fig. 25.   Percent occurrence rate as a function of altitude (from Russell *et al.*, 1989). The 100 Hz signals show little attenuation as a function of altitude.

While it is our opinion that lightning is a reasonable explanation for the nightside emissions, we shall discuss alternative explanations for the wave observations here in an effort to determine more rigorous tests of the lightning hypothesis. Although we shall separate the discussion by gyro-frequency, the wave phenomena are linked if we postulate a common source for them. For example, if the bursts with $f > f_{ce}$ can be explained without invoking lightning, then there is a strong implication that the $f < f_{ce}$ waves may not be due to lightning either, since the bursts in both frequency ranges are characterized by their impulsive nature.

6.1. IMPULSIVE SIGNALS BELOW THE GYRO-FREQUENCY

The analysis and interpretation of the impulsive signals observed in the nightside ionosphere of Venus as carried out by F. L. Scarf and co-workers was restricted to signals in the 100 Hz channel only. Scarf *et al.* (1980a) argued that since impulsive signals should in most cases be strongly attenuated for $f > f_{ce}$, only the 100 Hz channel data should be used in studying possible lightning events. Scarf and Russell (1983), for example, specifically rejected any events from their analysis if impulsive signals with $f > f_{ce}$ were detected simultaneously with bursts in the 100 Hz channel. This increases the likelihood that the 100 Hz waves are electromagnetic whistler-mode waves, not ion-acoustic waves which would be subject to Doppler shift. However, as noted above, the higher frequency impulsive signals also have to be explained by the lightning hypothesis.

Scarf and Russell (1988) used the lack of Doppler shift to counter a suggestion by Taylor *et al.* (1985) that the 100 Hz signals are generated by an *in situ* plasma instability. Taylor *et al.* noted that superthermal ions are observed in association with the holes, and so by analogy with the results of Taylor *et al.* (1981), it is possible that plasma instabilities are present within the holes. Although the association with holes does not preclude a lightning source for the waves, since the enhanced magnetic field and reduced density may allow the hole to act as a duct for whistler-mode waves, the possibility of an *in situ* instability should be investigated.

As discussed by Scarf and Russell (1988), an ion-acoustic instability appears unlikely as a source of the 100 Hz signals. First, there is some evidence that the wave are perpendicularly polarized with respect to the ambient magnetic field, as shown in Figure 26. Ion-acoustic waves, being electrostatic, are expected to be polarized along the ambient field. Second, the waves do not show any Doppler shift. Scarf and Russell pointed out that for typical nightside parameters the phase speed of an ion-acoustic wave is around 1 km s^{-1}. The orbiter velocity is ~ 10 km s^{-1}, giving a factor of ~ 10 Doppler shift. The upper cut-off for ion-acoustic waves is the ion-plasma frequency, which is ~ 500 Hz for O$^+$ ions with a density 100 cm^{-3}. Impulsive signals should, hence, be observed in the 100 Hz, 730 Hz, and 5.4 kHz channels. This is not the case for the events selected by Scarf, and these events cannot be explained in terms of ion-acoustic waves, unless the spacecraft velocity vector is perpendicular to the ion-acoustic wave vector (the Doppler shift will be less than a factor of two for angles within 6° of perpendicular).

More recently, Maeda and Grebowsky (1989) have investigated the possibility of *in situ* instabilities generating whistler-mode waves. In particular, they drew a comparison between the impulsive events observed with the OEFD and VLF saucers observed on low altitude spacecraft at the Earth. Maeda and Grebowsky argued that the narrow frequency bandwidth of the OEFD would result in spatial structure, such as the saucer emission cone, being aliased as temporal structure. Exactly how the OEFD would respond to a saucer-like emission depends on many factors such as the gyro-frequency, distance from the source, and source-size. However, if the source is sufficiently small (< 250 m), then the wave signature would be detected as an impulse with duration less than the sampling time of the instrument if the orbiter flew near the source region.

Leaving aside the question of spatial scales, there is another problem with the VLF saucer explanation. This concerns the differences between the terrestrial and Venus ionospheres. At the Earth there is a large intrinsic dipole, while the regions of enhanced magnetic field at Venus are induced fields, with strengths of a few tens of nT. As a

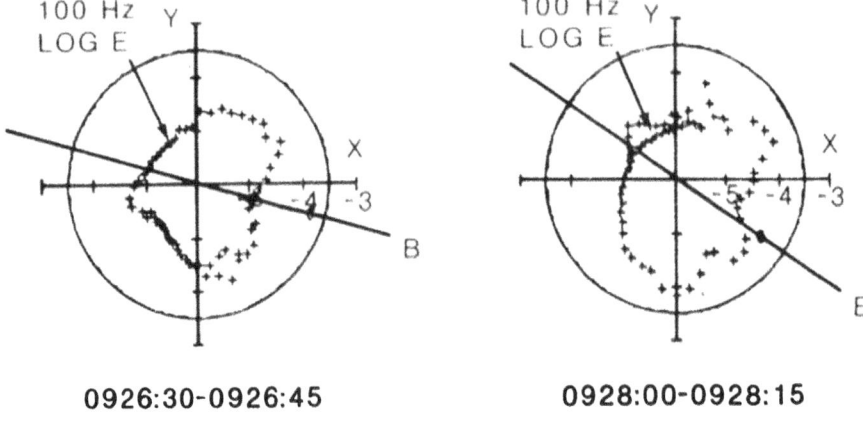

0926:30-0926:45   0928:00-0928:15

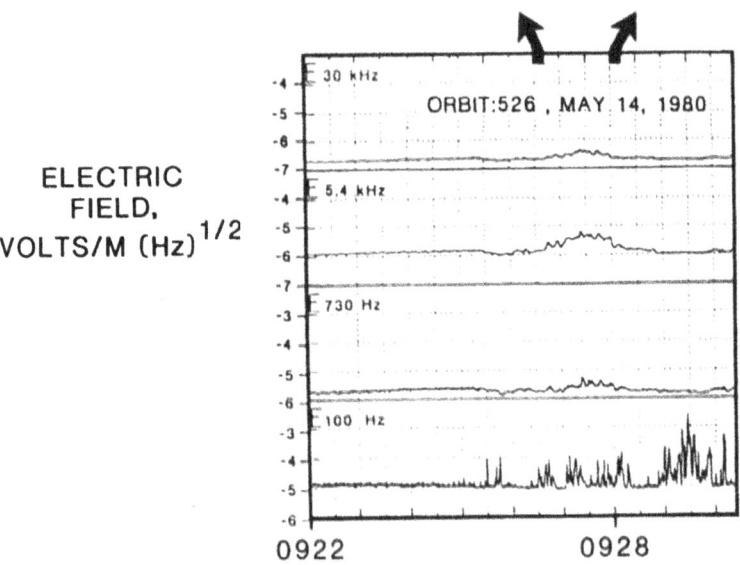

Fig. 26.   Polarization diagrams (from Scarf and Russell, 1988), together with the associated time series. For the two 15 s intervals shown, the wave electric field is maximum when the antenna direction is perpendicular to the average magnetic field direction.

consequence the plasma frequency is much higher than the electron gyro-frequency throughout the Venus ionosphere. This means that whistler-mode waves have extremely low-phase velocities. In Section 3.2 we discussed the resonant absorption of whistler-mode waves at the dayside ionopause, and it was shown that whistler-mode waves could be absorbed through Landau damping with ionospheric electrons. The same resonance conditions apply when considering wave growth, and Equation (1) gives the Landau resonance energy for electrons when $f \ll f_{ce} \cos \theta$, where θ is the propagation angle with respect to the ambient field. For electron densities around 10^3 cm^{-3}, and magnetic field strengths of ~ 20 nT, $f_{pe} \sim 300$ kHz and $f_{ce} \sim 500$ Hz. On substitution in Equation (1) the Landau resonance energy for 100 Hz waves ~ 0.15 eV. Most theories for saucer generation (e.g., James, 1976) require the whistler-mode waves to be generated on or near the resonance cone via a Landau resonance. Given the low-resonant energy it appears unlikely that VLF saucers can be generated in the Venus nightside ionosphere.

Maeda and Grebowsky also postulate that gyro-resonance may generate whistler-mode waves. Since the gyro-resonance energy is larger by a factor of $(f_{ce}/f)^2$, typical resonance energies will be about 4 eV for the plasma parameters cited above, and gyro-resonance is consequently a more likely instability mechanism. However, it is not clear that gyro-resonance will result in waves propagating only near the resonance cone, as required for the saucer emission cone. Inspection of the general plasma conductivity tensor (e.g., Clemmow and Dougherty, 1969) shows that the Landau resonance $\omega = k_{\parallel} v_{\parallel}$ is associated with the parallel wave electric fields, while the gyro-resonant terms $\omega - k_{\parallel} v_{\parallel} = \pm \Omega_e$ are associated with the perpendicular electric field terms. Parallel propagating whistler-mode waves are polarized perpendicular to the ambient field, and so are stable to Landau resonance. On the other hand, parallel propagating whistler-mode waves will be unstable to gyro-resonance. This implies that any emission cone associated with a gyro-resonant source will be filled, since waves will be driven unstable at angles other than the resonance cone. Indeed, this is why Landau resonance has been invoked for the saucer mechanism, ensuring that waves are generated for a narrow range of propagation angles.

Although the properties of the Venus nightside ionosphere make it unlikely that a VLF saucer-like mechanism exists, this does not exclude other whistler-mode instabilities. We have argued against the saucer mechanism on the grounds that the conditions required to make narrow-spatial structures, such as the saucer emission cone, do not exist in the Venus ionosphere. If, however, the wave signatures do not consist of isolated bursts then the argument against an *in situ* instability is weakened.

As a counter example to impulsive whistler-mode waves, we include data from a nightside hole region in Figure 27. In this figure the 100 Hz emissions are confined to the region of enhanced magnetic field, the impulsive signals outside of the hole region are interference and occur at twice the spin frequency. Since the ambient field is large and mainly in the anti-solar direction, and the orbiter is near 22 : 20 LT at 600 km altitude, the hole could duct whistler-mode waves generated in the atmosphere by lightning. However, we cannot discount alternative explanations for this event. First, although the 100 Hz waves show some variability, the signals do not consist of isolated impulses. This may reflect the presence of an active lightning source in the atmosphere, or it may be due to *in situ* instabilities, cf. the plasma cloud events in Figure 10. Second, while there are no impulsive signals at higher frequency, implying whistler-mode waves, we note that for this particular event the spacecraft is moving almost perpendicular to the Venus–Sun line, and Doppler shifts due to spacecraft motion would be small for waves propagating along the ambient field.

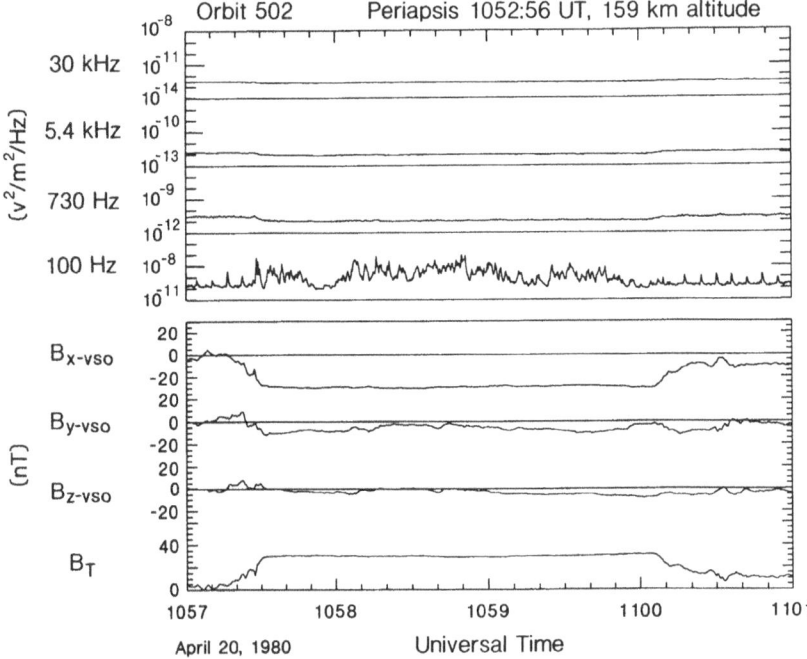

Fig. 27.   Wave and magnetic field data observed within an ionospheric hole. The hole was observed near
600 km altitude, and 100 Hz waves are detected throughout the enhanced field region.

To show examples of more impulsive 100 Hz emissions, data from the previous orbit
to that in Figure 27 are shown in Figure 28. Qualitatively, these data could be associated
with lightning, due to their transient nature. However, as pointed out by D. L. Carpenter
(personal communication), it is not clear that for this event whistler-mode waves could
propagate to the spacecraft from the atmosphere. In a horizontally stratified medium
in which the refractive index increases as a function of altitude, such as the bottomside
of the ionosphere, the wave vector will become nearly vertical through Snell's law.
Whistler-mode waves cannot propagate at angles greater than the resonance cone angle
with respect to the ambient field. For frequencies much larger than the lower hybrid
resonance the condition $f = f_{ce} \cos \theta$ gives the resonance cone angle. Using data
acquired at 10 : 54 UT in Figure 28 as an example, $f_{ce} \sim 600$ Hz, and the resonance cone
angle $\sim 80°$. At this time the spacecraft is near 21 : 00 LT, and the magnetic field is
almost parallel to the Venus orbital plane ($B_z \sim 0$). While B_x is large, so is B_y, and B_y
is of roughly the same magnitude as B_x. At this time, then, the magnetic field points
towards the dusk terminator, at nearly 90° to the radius vector. Whistler-mode waves
would consequently be required to propagate outside of the resonance cone.
As an argument against this point we note that the events around periapsis in
Figure 28 (from 10 : 53 to 10 : 55 UT) were not counted by Scarf and Russell (1983) as
lightning whistlers because of the presence of higher frequency emissions detected
simultaneously with the 100 Hz emissions. Nevertheless, the discussion presented

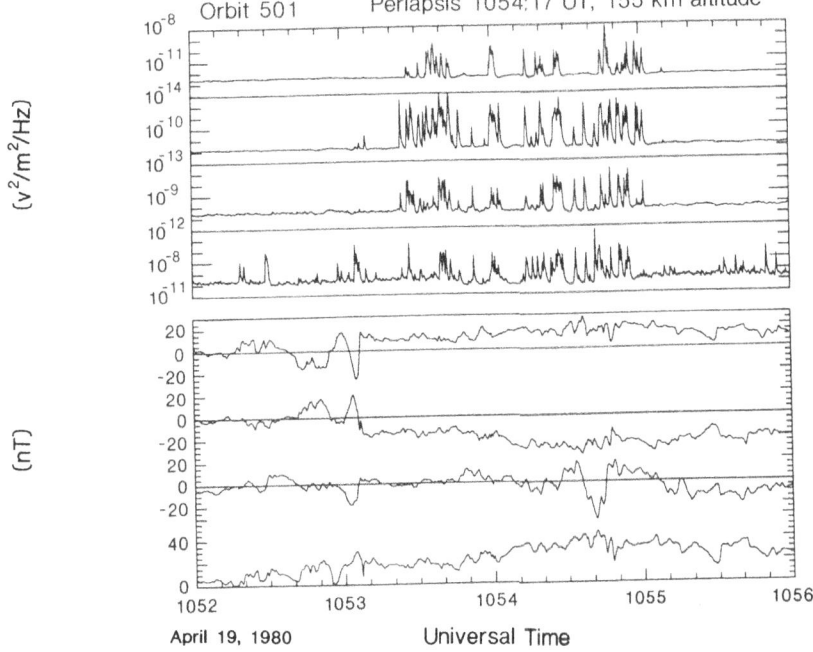

Fig. 28. Wave and field data observed near periapsis. In this case impulsive signals are observed in all channels. Periapsis altitude is 155 km.

above does indicate that the lightning whistler model for 100 Hz impulsive events could be subjected to more rigorous testing as follows. If we restrict our events to those in which 100 Hz signals are present with no higher frequency signals then we can ask:

(1) Do the events appear to be impulsive?

(2) Are the ambient field and spacecraft velocity vector sufficiently aligned to allow significant Doppler shift for low-phase velocity ion-acoustic waves?

(3) Is the magnetic field sufficiently far from horizontal that whistler-mode waves whose wave vector is near vertical can propagate inside the resonance cone?

These three tests do not preclude other events from being lightning-generated, but they do ensure that any event which satisfies them is most likely a lightning event. The first two tests imply that the signals are impulsive whistler-mode waves. The last test determines if whistler-mode waves can propagate from an atmospheric source.

6.2. IMPULSIVE SIGNALS ABOVE THE GYRO-FREQUENCY

While Scarf and co-workers restricted their analysis to 100 Hz signals only, Russell and colleagues considered impulsive events at any frequency as candidate lightning events. As an example of analysis of these higher frequency impulsive signals see Russell *et al.* (1988). As shown in Figures 24 and 25 the higher frequency events tend to be confined to lower altitudes and more restricted in local time. The higher frequency signals are consequently consistent with a strongly attenuated signal propagating away from the source region in the post-dusk local time sector.

In his review, Russell (1991) points out that from a theoretical point of view, no signal should propagate to the spacecraft from the atmosphere for frequencies above the gyro-frequency. Indeed, such high-frequency events could be interpreted as Doppler-shifted ion-acoustic waves, using the arguments of Scarf and Russell (1988), although it is not clear why the events would depend so strongly on local time and altitude. As a mitigating factor Russell notes that Kelley *et al.* (1985) have reported observations of a prompt transient signal observed in association with lightning events. This signal is polarized along the ambient magnetic field, even though the electric field should be shorted out by the high mobility of electrons along the field. From a phenomenological point of view perhaps the high-frequency events observed at Venus correspond to a similar process allowing a transient signal to propagate some distance through the ionosphere.

Russell speculates that electron collisions provide the mechanism for propagation of high-frequency signals to the Pioneer Venus orbiter. As we shall show here, it is unlikely that collisions alone can explain the observations. First, to emphasize how difficult it is for a signal to propagate through the ionosphere, the attenuation scale is given by the plasma skin depth c/ω_{pe} for frequencies $f_{ce} < f < f_{pe}$. Using 300 kHz as a representative plasma frequency, the skin depth is 0.16 km. In other words the wave intensity will decrease by 100 dB in about 2 km.

Collisions may be expected to provide anomalous wave dispersion if the collision frequency, ν, is comparable to the angular wave frequency, ω. As an estimate for the collision frequency we include the results of Luhmann *et al.* (1984) in Figure 29. This figure shows the electron-ion and electron-neutral collision frequencies, together with ion-collision frequencies for the dayside ionosphere. At 150 km altitude the two electron-collision frequencies are less than 10^3 s^{-1}. In the nightside we expect the collision frequencies to be lower.

First, as given by Luhmann *et al.*, the electron-ion collision frequency is approximately proportional to $n_e/T_e^{3/2}$, where n_e and T_e are the electron density and temperature, respectively. From Theis *et al.* (1980), the electron density decreases by nearly two orders of magnitude at 160 km in going from 0° to 180° solar zenith angle. Even at 105° SZA, the density has decreased by an order of magnitude. We, therefore, expect the electron-ion collision frequency in the nightside to be between 1 and 2 orders of magnitude smaller than shown in Figure 29.

For electron-neutral collisions the collision frequency is proportional to $NT_e^{1/2}$, where N is the neutral density. From Niemann *et al.* (1980), the neutral density typically decreases by an order of magnitude at 160 km altitude on going from dayside to nightside. The decrease is even more marked at higher altitude because of the smaller scale height on the nightside. Consequently, the electron-neutral collision frequency is also expected to decrease in the nightside ionosphere.

Although Niemann *et al.* and Theis *et al.* do not present data below 160 km, and perhaps the day–night asymmetry is less at lower altitudes, we do not expect the collision

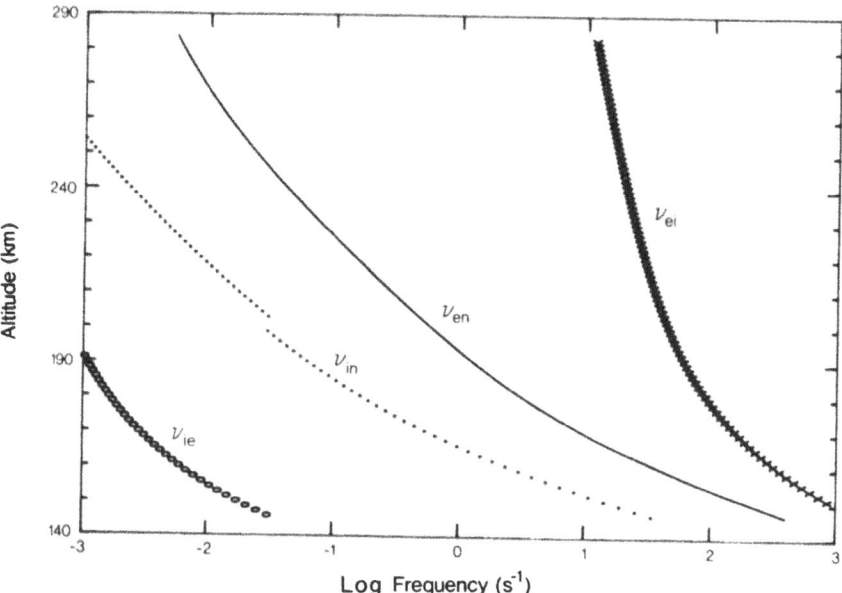

Fig. 29.   Collision frequencies as a function of altitude for the dayside ionosphere (from Luhmann *et al.*, 1984).

frequencies to be higher than shown in Figure 29. Collisions may affect the 730 Hz channel to some extent, but only weakly given the extremely short attenuation scale. Higher frequencies are mostly unaffected and it does not appear that collisions play a major role in allowing high-frequency emissions to propagate through the ionosphere in the nightside of Venus. This then is the outstanding question in the interpretation of the $f > f_{ce}$ emissions. How do these emissions propagate to the spacecraft if generated by a lightning discharge in the atmosphere? Terrestrial data (Kelley *et al.*, 1985) show evidence for an anomalous transient associated with lightning, and perhaps the high-frequency emissions at Venus are a similar phenomenon. However, some understanding of a mechanism for producing the observed signals would provide the lightning inter-pretation with a stronger foundation.

7. Conclusions

We will briefly summarize the major results presented in this review, and present some questions still outstanding concerning plasma waves at Venus.

IONOSPHERE AND IONOPAUSE

The noise peak at periapsis observed most clearly on the nightside is well explained by the impact ionization model of Curtis *et al.* (1985).

Whistler-mode waves are observed at the ionopause, and may be a significant source of energy for the topside ionosphere. However, evidence exists for the generation of instabilities at the ionopause itself, and more detailed studies of the wave properties at the ionopause should be carried out to determine the nature of the waves both within and above the current layer. Changes in instrument response as a function of the plasma parameters should also be investigated as possible contributors to changes in the wave signals.

Plasma waves are observed in current sheets, which are associated with plasma clouds. These currents may be carried by relatively energetic electrons, which in turn may be accelerated in regions of anomalous resistivity caused by the waves, and so produce aurora. A systematic study of the morphology of the plasma clouds and the associated currents and waves would be useful in determining if the currents are field-aligned, or transverse, and if the currents map to the ionosphere.

MAGNETOSHEATH AND MAGNETOTAIL

There is evidence for a boundary layer in the inner magnetosheath, with different levels of wave activity, and a hotter less dense plasma. Some features of the transition are reminiscent of the plasma clouds (field rotation and enhanced wave activity in the transition region). A comparison of the two would be useful.

Further behind the planet plasma waves are observed at the magnetotail boundary, and are a useful diagnostic for determining that boundary. These waves have been previously identified as ion-acoustic waves. It is our impression, however, that the nature of the waves should be clarified, as there are some inconsistencies with the mode identification.

BOW SHOCK

There are differences between the wave spectra observed at Venus and at the Earth. However, differences in antenna lengths make such comparisons difficult; we can carry out a comparison with the AMPTE/CCE spacecraft, but again better knowledge of the dependence of instrument response on plasma parameters is required.

The wave data have been used as an aid in identifying grazing encounters with the shock. A systematic study of waves at the shock would strengthen the use of wave data for such a purpose.

LIGHTNING

The question of lightning on Venus has been a particularly controversial topic. As noted by Russell (1991), impulsive wave events are often present in the nightside ionosphere, but the evidence for geographic clustering is not strong, and it appears that local time may better order the data. A large body of statistical data now exists, but questions such as how would impulsive electromagnetic waves generated in the Venus atmosphere propagate to the spacecraft have not as yet been adequately addressed. Alternative mechanisms for generating the signals have been suggested, and while *in situ* instabilities might not be expected to depend on altitude and local time in the observed manner, we can use the expected properties of locally generated waves to determine on an event-by-event basis those signals which are most probably lightning-related. It is also possible that studies of the plasma cloud phenomena, and the associated wave generation mechanisms would shed light on the feasibility of alternative generation mechanisms.

In closing, many different wave phenomena have been discovered with the Pioneer Venus wave instrument. Significant insights into various aspects of the solar wind–Venus interaction have come about through study of the waves. Nevertheless, many questions still remain and answering these questions should be as exciting as many of the results presented here.

Acknowledgements

I would like to thank C. T. Russell and J. G. Luhmann for many helpful discussions on Pioneer Venus. I would also like to acknowledge Frederick L. Scarf, who was the Principal Investigator for the OEFD until his untimely death on July 17, 1988. His knowledge, energy, and enthusiasm were boundless, and the invaluable advice and comments he would have given me on this review are sorely missed. This work was supported by NASA grant NAG 2–485.

References

Brace, L. H. and Kliore, A. J.: 1991, *Space Sci. Rev.* **55**, 81.

Brace, L. H., Curtis, S. A., Russell, C. T., and Scarf, F. L.: 1985, *Eos, Trans. AGU (Abstract)* **66**, 294.

Brace, L. H., Theis, R. F., Krehbiel, J. P., Nagy, A. F., Donahue, T. M., McElroy, M. B., and Pederson, A.: 1979, *Science* **203**, 763.

Clemmow, P. C. and Dougherty, J. P.: 1969, *Electrodynamics of Particles and Plasmas,* Addison-Wesley, London.

Colin, L.: 1980, *J. Geophys. Res.* **85**, 757.

Cravens, T. E., Nagy, A. F., Brace, L. H., Chen, R. H., and Knudsen, W. C.: 1979, *Geophys. Res. Letters* **6**, 341.

Curtis, S. A., Brace, L. H., Niemann, H. B., and Scarf, F. L.: 1985, *J. Geophys. Res.* **90**, 6631.

Elphic, R. C., Russell, C. T., Luhmann, J. G., Scarf, F. L., and Brace, L. H.: 1981, *J. Geophys. Res.* **86**, 11430.

Intriligator, D. S. and Scarf, F. L.: 1982, *Geophys. Res. Letters* **9**, 1325.

Intriligator, D. S. and Scarf, F. L.: 1984, *J. Geophys. Res.* **89**, 47.

James, H. G.: 1976, *J. Geophys. Res.* **81**, 501.

Kasprzak, W. T., Taylor, H. A., Brace, L. H., Niemann, H. B., and Scarf, F. L.: 1982, *Planetary Space Sci.* **30**, 1107.

Kelley, M. C., Siefring, C. L., Pfaff, R. F., Kintner, P. M., Larsen, M., Green, R., Holzworth, R. H., Hale, L. C., Mitchell, J. D., and Le Vine, D.: 1985, *J. Geophys. Res.* **90**, 9815.

Knudsen, W. C., Luhmann, J. G., Russell, C. T., and Scarf, F. L.: 1989, *J. Geophys. Res.* **94**, 197.

Knudsen, W. C., Spenner, K., Whitten, R. C., Spreiter, J. R., Miller, K. L., and Novak, V.: 1979, *Science* **203**, 757.

Luhmann, J. G.: 1986, *Space Sci. Rev.* **44**, 241.

Luhmann, J. G., Russell, C. T., and Elphic, R. C.: 1984, *J. Geophys. Res.* **89**, 362.

Maeda, K. and Grebowsky, J. M.: 1989, *Nature* **341**, 219.

Mihalov, J. D. and Barnes, A.: 1981, *Geophys. Res. Letters* **8**, 1277.

Niemann, H. B., Kasprzak, W. T., Hedin, A. E., Hunten, D. M., and Spencer, N. W.: 1980, *J. Geophys. Res.* **85**, 7817.

Perez-de-Tejada, H.: 1982, *J. Geophys. Res.* **87**, 7405.

Perez-de-Tejada, H., Intriligator, D. S., and Scarf, F. L.: 1984, *Geophys. Res. Letters* **11**, 31.

Perez-de-Tejada, H., Intriligator, D. S., and Scarf, F. L.: 1985, *J. Geophys. Res.* **90**, 1759.

Rodriguez, P. and Gurnett, D. A.: 1975, *J. Geophys. Res.* **80**, 19.

Russell, C. T.: 1991, *Space Sci. Rev.* **55**, 317.

Russell, C. T., Luhmann, J. G., Elphic, R. C., and Scarf, F. L.: 1981, *Geophys. Res. Letters* **8**, 843.

Russell, C. T., Luhmann, J. G., Elphic, R. C., Scarf, F. L., and Brace, L. H.: 1982, *Geophys. Res. Letters* **9**, 45.

Russell, C. T., von Dornum, M., and Scarf, F. L.: 1988, *J. Geophys. Res.* **93**, 5915.

Russell, C. T., von Dornum, M., and Scarf, F. L.: 1989, *Adv. Space Res.* **10**(5), 37.

Scarf, F. L.: 1985, *IEEE Trans. Geosc. Remote Sensing* **GE-23**, 250.

Scarf, F. L. and Russell, C. T.: 1983, *Geophys. Res. Letters* **10**, 1192.

Scarf, F. L. and Russell, C. T.: 1988, *Science* **240**, 222.

Scarf, F. L., Fredricks, R. W., and Crook, G. M.: 1968, *J. Geophys. Res.* **73**, 1723.

Scarf, F. L., Neumann, S., Brace, L. H., Russell, C. T., Luhmann, J. G., and Stewart, A. I. F.: 1985, *Adv. Space Res.* **5**, 185.

Scarf, F. L., Taylor, W. W. L., and Green, I. M.: 1979, *Science* **203**, 748.

Scarf, F. L., Taylor, W. W. L., Russell, C. T., and Brace, L. H.: 1980a, *J. Geophys. Res.* **85**, 8158.

Scarf, F. L., Taylor, W. W. L., Russell, C. T., and Elphic, R. C.: 1980b, *J. Geophys. Res.* **85**, 7599.

Scarf, F. L., Taylor, W. W. L., and Virobik, P. V.: 1980c, *IEEE Trans. Geosc. Remote Sensing* **GE-18**, 36.

Slavin, J. A., Elphic, R. C., Russell, C. T., Scarf, F. L., Wolfe, J. H., Mihalov, J. D., Intriligator, D. S., Brace, L. H., Taylor, H. A., Jr., and Daniell, R. E., Jr.: 1980, *J. Geophys. Res.* **85**, 7625.

Strangeway, R. J., Scarf, F. L., Zanetti, L. J., and Klumpar, D. M.: 1988, *J. Geophys. Res.* **93**, 14357.

Taylor, H. A., Jr., Daniell, R. E., Hartle, R. E., Brinton, H. C., Bauer, S. J., and Scarf, F. L.: 1981, *Adv. Space Res.* **1**, 247.

Taylor, H. A., Jr., Grebowsky, J. M., and Cloutier, P. A.: 1985, *J. Geophys. Res.* **90**, 7415.

Taylor, W. W. L., Scarf, F. L., Russell, C. T., and Brace, L. H.: 1979, *Science* **205**, 112.

Theis, R. F., Brace, L. H., and Mayr, H. G.: 1980, *J. Geophys. Res.* **85**, 7787.

VENUS LIGHTNING

C. T. RUSSELL

*Institute of Geophysics and Planetary Physics and Department of Earth and Space Sciences,
University of California, Los Angeles, CA 90024-1567, U.S.A.*

Abstract. Although it is not unanimously accepted, many independent observations lead to the conclusion that lightning is prevalent on Venus. The electromagnetic signals detected by all 4 Venera landers are most readily explained as generation by lightning. The Venera 9 spectrometer appears to have observed a lightning storm on one occasion. The Pioneer Venus plasma wave instrument detects waves both below the electron gyrofrequency that may be due to lightning and signals above the electron gyrofrequency but at very low altitudes that may be due to the near field of the lightning. The VLF observations suggest that Venus lightning must be an intra-cloud phenomenon which is most frequent in the afternoon and evening sector. The occurrence rate is likely to be greater than on Earth.

1. Introduction

Lightning occurs on Earth when charge accumulated in a portion of a cloud creates an electric field so strong that the neutral gas becomes ionized and an electrically conducting channel forms which allows the accumulated charge to be discharged or neutralized. About one-third of the time on Earth the ionized path leads to the ground, and most frequently a negative charge flows from the cloud into the Earth. About two-thirds of the time the discharge takes place between portions of the clouds which have become oppositely charged. The current surge through the ionized channel heats the air until it glows over much (but not necessarily all) of the channel. The current surge from each lightning flash also generates electromagnetic waves which propagate around the world. Finally, the heated channel sends a shock wave into the air and it is this shock wave we hear as thunder. The layman knows that lightning discharges are present if he sees the flash or hears thunder. These signatures are both quite unambiguous. However, the researcher is more likely to use electromagnetic techniques to study lightning occurrence because the electromagnetic effects travel further.

Venus is Earth's sister planet. Like many sisters there are a few common family traits but also some remarkable differences. Venus is only slightly less massive and slightly smaller than the Earth. However, it rotates in a retrograde sense. It has a dense atmosphere and intense surface pressures and temperatures. Like the Earth, Venus has clouds but Venus is completely cloud covered and its clouds consist of sulfuric acid, not water.

Should Venus have lightning like the Earth or not? The answer to this question is difficult because of our imperfect understanding of the charging mechanism in clouds and our incomplete knowledge of the behavior of the Venus atmosphere. Fortunately, now several different observational datasets exist which can be studied.

Well before these data were obtained, the existence of lightning on Venus was a

Space Science Reviews **55**: 317–356, 1991.

subject of speculation (Meinel and Hoxie, 1962). Nightside optical emissions observed with ground-based telescopes, called Ashen Light, have been reported for over three centuries (cf. Moore, 1965). Some have speculated that the Ashen Light is caused by lightning (Meinel and Hoxie, 1962; Krasnopol'sky, 1980) while others have dismissed the Ashen Light as an artifact and have pointed out that Venus lightning would be far too weak to cause the reported effect (Borucki *et al.*, 1981; Williams *et al.*, 1982). In any event, there have been no ground-based observations of lightning flashes on Venus despite several serious attempts by professional observers.

The desire to confirm this speculation led to the installation on the Venera 11, 12, 13, and 14 landers of instrumentation capable of detecting lightning (Ksanfomaliti, 1979). The Venera investigations examined the magnetic signature and, in the case of the Venera 11 and 12 landers, acoustic data were also obtained. Lightning detection was also one of the objectives of the Pioneer Venus Orbiter (PVO) electric field detector (Colin and Hunten, 1977). None of the early missions were specifically instrumented to detect the optical signature of lightning. Nevertheless use was made of both the Venera 9 visible spectrometer (Krasnopol'sky, 1980), and the Pioneer Venus star sensor (Borucki *et al.*, 1981), to search for lightning. Eventually a simple photometer, sensitive to optical flashes, was included in the VEGA balloon mission (Sagdeev *et al.*, 1986).

The fact that four Soviet instruments were devoted exclusively to lightning, one Soviet and one American instrument were partially devoted to lightning, and two additional instruments were used in the search, attests to the effort devoted to the question of Venus lightning. But is lightning just an atmospheric curiosity or is it important in the processes that occur on Venus? The answer lies partially in the rate of occurrence of lightning. On Earth lightning is thought to play a role in the fixation of nitrogen but it is not the only process by which this occurs (Borucki and Chameides, 1984). On Venus, lightning could produce an amount of CO and O_2 that could play a key role in maintaining the clouds by fueling the conversion of S and SO_2 to H_2SO_4 and also H_2O and SO_2 to S if its occurrence rate is much, much greater than that on Earth (Chameides *et al.*, 1979). Even if lightning is not much more frequent on Venus than on the Earth, it is important to determine the electrical activity of the clouds because of possible hazards to our space probes. No probes have flown through the region in which the VLF data suggests that lightning appears to be most frequently generated. Moreover, studying a terrestrial process in a different planetary context is a proven means of better understanding the terrestrial process.

It is the purpose of this review to examine the results of these experimental investigations and their implications. It will be argued that lightning occurring in the clouds of Venus is the most obvious explanation of the observed signals. However, before proceeding, it is worthwhile to examine the properties of terrestrial lightning even if the terrestrial analog for Venus is imperfect so that we can at least learn what questions to ask.

2. Terrestrial Lightning and Implications for Venus

There have been many reviews of the lightning discharge (Rinnert, 1982; Vonnegut, 1982; Levin et al., 1983; Williams et al., 1983; Uman, 1987). Most of the lightning on Earth is produced in strongly convective cumulus clouds but lightning can also be produced in volcanic clouds, dust storms, and snowstorms. While lightning associated with terrestrial volcanoes is rare in comparison with weather-related lightning, volcanic eruptions are probably the most active of all lightning generators (Vonnegut, 1982). These clouds vary in altitude from half a km to over 30 km. While there was some early speculation that Venus lightning was associated with volcanism (Scarf et al., 1980b; Scarf and Russell, 1983), as will be discussed in a later section such an association is very unlikely (Russell et al., 1988c, 1989a).

Terrestrial lightning occurs in two quite distinct forms, cloud-to-ground discharge and intra-cloud discharges. Such discharges occur when the electric field in a cloud exceeds the breakdown value which ranges from about $10^6 \, V \, m^{-1}$ in wet air to about $3 \times 10^6 \, V \, m^{-1}$ in dry air. Usually cloud-to-ground discharges begin in the cloud with a stepped leader carrying about 200–300 A and moving about 50–100 m per step. The stepped leader travels with an average velocity of about $1.5 \times 10^5 \, m \, s^{-1}$. This is followed by a return stroke in the same channel at a velocity of about $0.6 \times 10^8 \, m \, s^{-1}$ carrying 10–20 kA which decays in about 20 to 50 μs. There are usually 3 or more of these strokes in any one flash separated by about 40 ms and lasting about 0.2 s. Figure 1 shows the electric field during 6 different ground-to-cloud return strokes (Uman, 1987). The bottom two strokes show the first discharge of a series of strokes. The letter 'L' indicates the various steps of the stepped-leader. The upper four panels show subsequent strokes both with and without stepped leaders.

Discharges within a cloud are 2 to 4 times more frequent than cloud-to-ground discharges. There is no return stroke, per se, in an intra-cloud discharge but there is a recoil streamer which propagates about $2 \times 10^6 \, m \, s^{-1}$ over channel lengths of 1 to 3 km with peak currents of 1 to 4 kA. Figure 2 shows the magnetic field pulses during five intra-cloud discharges. The pulses are much more narrow than those of the cloud-to-ground return strokes.

A moderate cloud-to-ground flash will generate about $4.5 \times 10^8 \, J$ and a large one about $2 \times 10^{10} \, J$ of which about 10^{-3} or 10^{-4} of the power goes into optical radiation. Most of the energy released in a lightning flash is expended in ionizing the air and heating it. Intra-cloud discharges generate less energy per stroke than cloud-to-ground discharges but there are more such discharges.

Terrestrial lightning has both geographic and local time correlations. Figure 3 shows the global distribution of lightning discharges (Bliokh et al., 1980). There are three major centers of activity as a function of longitude; the Americas, Africa, and Indonesia. At any one time there are about 2000 active storm cells, each producing an average flash rate of about 1 every 20 s. Figure 4 shows the Universal Time distribution of lightning in each of these regions and the corresponding local time distribution. Each region has a similar local time profile maximizing about 16:00 LT when the surface temperature

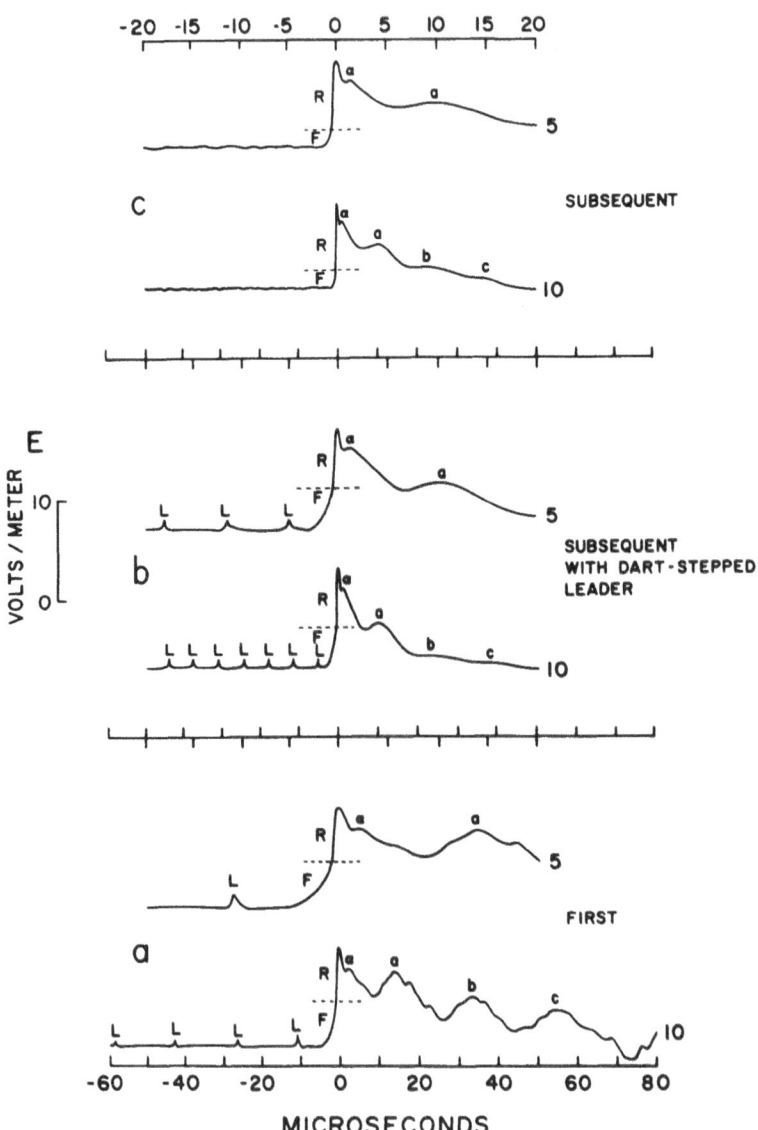

Fig. 1.   Radiation electric field due to cloud-to-ground discharges normalized to 100 km (a) first stroke with stepped leader; (b) subsequent stroke with leader; (c) subsequent strokes. Time-scale in microseconds with units of scale indicated beside each trace. Closest scale applies (Uman, 1987).

due to solar heating and convection is at a maximum. However, in some localized areas the maximum frequency of thunderstorms and precipitation occurs late at night (Vonnegut, 1982).

In order to create lightning there must be an abundance of a substance that can be readily electrified, a process to electrify these particles and a large-scale charge-

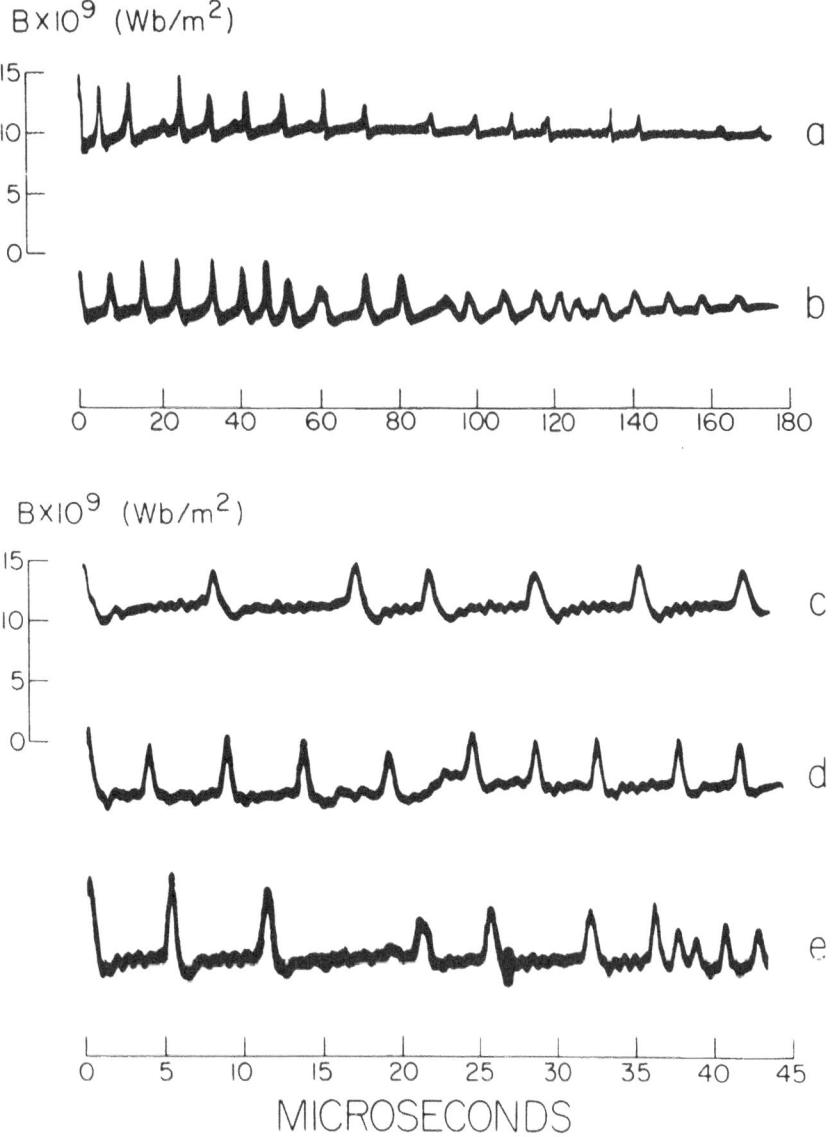

Fig. 2.   Magnetic field due to intra-cloud discharges within 50 km (Uman, 1987).

separation process. The electrification is proportional to the polarizability of a molecule which in turn is proportional to its dielectric constant. The dielectric constant for water is 80; and for H_2SO_4, which is present in the clouds of Venus, 110. The largest charges in terrestrial clouds occur at altitudes where the water has become supercooled to temperatures from -10 to $-40\,°C$ and where there are ice crystals present. Most of the charge resides in a layer 100's of m thick centered around the altitude at which the

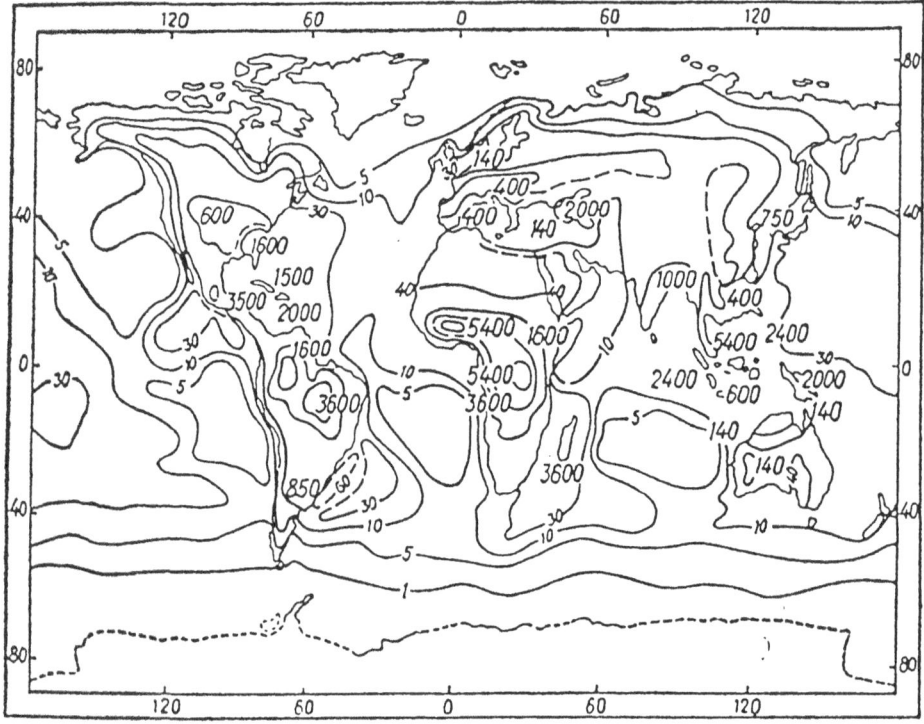

Fig. 3.   Global distribution of the annual number of lightning flashes per 100 km² (Bliokh *et al.*, 1980).

temperature is approximately – 15 °C. H_2SO_4, the main constituent of the Venus clouds, freezes at temperatures similar to that of water. The terrestrial clouds in which lightning discharges arise vary in height from 4 to 20 km. The distance between the effective centers of charge in a cloud is similar to the distance of the lower charge center to ground. On Venus the clouds in contrast occur at about 55 km which is much greater than the cloud layer thickness of about 10 km. Thus, it is much more difficult for a discharge to occur from the Venus cloud layers to ground. However, if a Venus discharge were due to vertical charge separation within a cloud one might expect it to have a discharge length comparable to that on Earth.

    The charge separation process is not well understood. One possible mechanism is that particles become charged differently according to size and then separate according to size when large particles fall more rapidly in updrafts under the action of gravity. One mechanism that has been proposed for the production of the charge gathered by these falling particles is ion production by cosmic rays in the atmosphere (Wilson, 1929). While this process may be too weak on the Earth, it may be much more productive on Venus which has no shielding magnetic field (Russell *et al.*, 1980; Phillips and Russell, 1987) and which is much closer to the Sun. It is also difficult to compare the expected magnitude of updrafts on Earth and Venus. While strong updrafts were observed on the

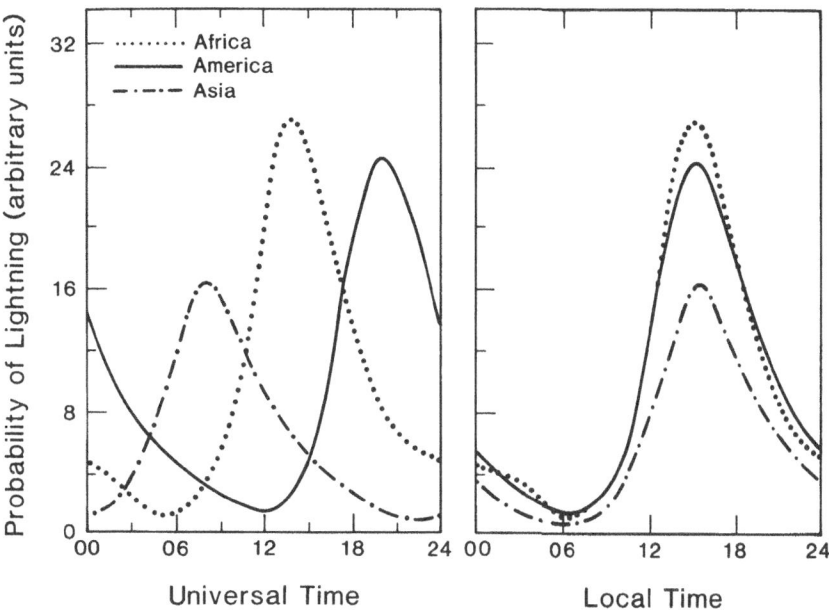

Fig. 4.   Universal time distribution and local time distribution in the three principal zones of activity (after Bliokh *et al.*, 1980).

VEGA balloons (Sagdeev *et al.*, 1986), no balloon or probe data are available at the local times where we would expect from our terrestrial experience that lightning would be present. Another possible electrification process would be charge exchange due to the collision of different sized ice particles. This explains several of the observed properties of the charge distribution in terrestrial clouds and could also work at Venus if Venusian clouds contained particles with properties similar to ice. In fact the average particle number density in the Venus clouds is comparable to that in the terrestrial clouds, as is the mean mass (Levin *et al.*, 1983). Nevertheless, the existence of particles large enough to fall in strong electric fields (~ 400 μm) has been questioned (Knollenberg and Hunten, 1980).

The power expended in a terrestrial storm increases as the fifth power of its size. Thus, the size of Venus clouds is an important factor in determining the possible energy released in lightning. We do not have any information at present on the sizes of these clouds. However, since the clouds on Venus occur at much higher altitudes than on Earth we might expect larger scale sizes, perhaps governed by the vertical scale height of the atmosphere.

Lightning also plays a role in the terrestrial global electric circuit. The positively charged upper layers of the cloud provide a conduction current of about 1 ampere to charge the ionosphere positively as illustrated in Figure 5. The clouds charge the Earth negatively. The ionosphere reaches a potential difference of 3×10^5 V relative to the Earth. This potential difference in turn generates a fair weather return current to the

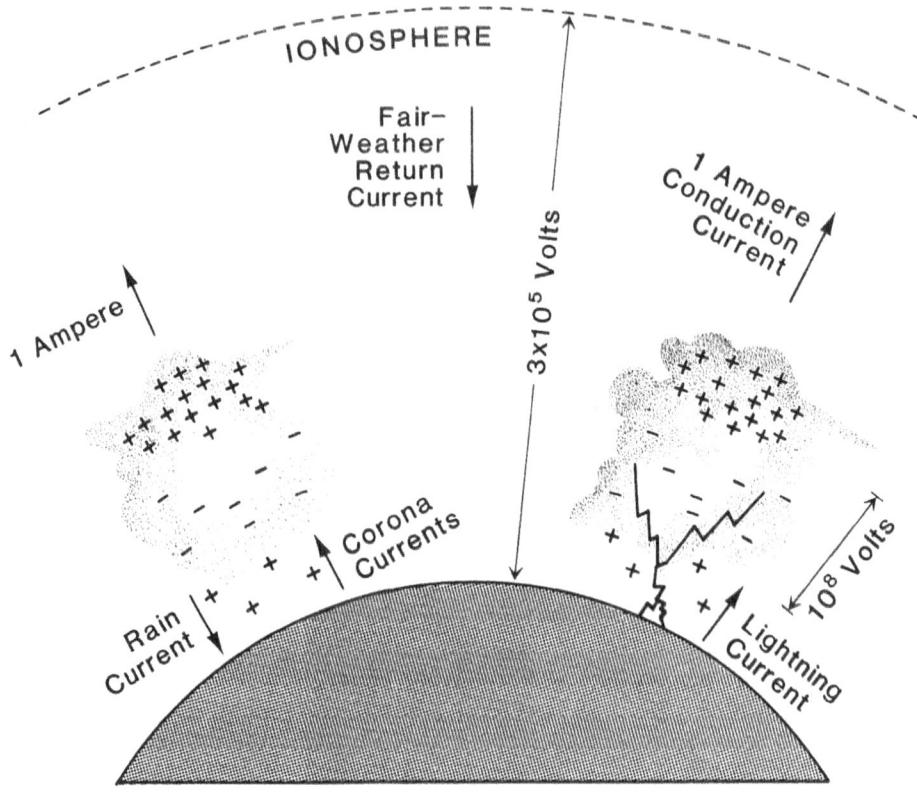

Fig. 5.   The atmospheric electric circuit.

Earth through the atmosphere whose conductivity is maintained by the cosmic-ray flux. The remaining part of the circuit back to the cloud consists of lightning discharges and coronal discharge from the surface. The strength of the global potential difference varies during the day as the lightning rate increases and decreases as shown in Figure 4. It also varies as the atmospheric conductivity changes. Similarly we would expect the potential difference on Venus to depend on the actual rate of lightning activity integrated over the planet, and the variations in conductivity in the Venus atmosphere.

In short, our terrestrial experience suggests that, because of the height of the cloud layer, any lightning on Venus would be intra-cloud lightning. On Earth, intra-cloud lightning occurs more frequently and consists of more and shorter individual pulses than cloud-to-ground discharges. The height of the Venus clouds should also weaken any geographic correlations but should not affect local time correlations. On Earth these local time correlations are strong and we might expect the same on Venus where the '4-day' winds transport the particulate matter in the clouds heated on the dayside into cooler regions beyond the terminator. Since the thickness of the clouds and breakdown voltage are expected to be similar in the terrestrial and the Venus atmospheres, the power and duration of an individual flash of terrestrial and Venus might be the same. However, the rate at which such flashes may occur should depend on the charging rate which might be controlled by the ion production rate and the velocity of updrafts which are unknown. Possibly the charging rate depends on the collision rate between large and small particles in the cloud. Thus we cannot predict *a priori* how often Venus lightning might occur.

Since in the later sections of this review we will be examining signals in the Venus ionosphere that appear to be generated by lightning, it is appropriate to review at this point the nature of VLF signals in the terrestrial ionosphere associated with terrestrial lightning. Figure 6 shows 5 s of signals recorded by the ISIS 2 satellite at 1400 km altitude (D. L. Carpenter, 1988, personal communication). The data have been spec-

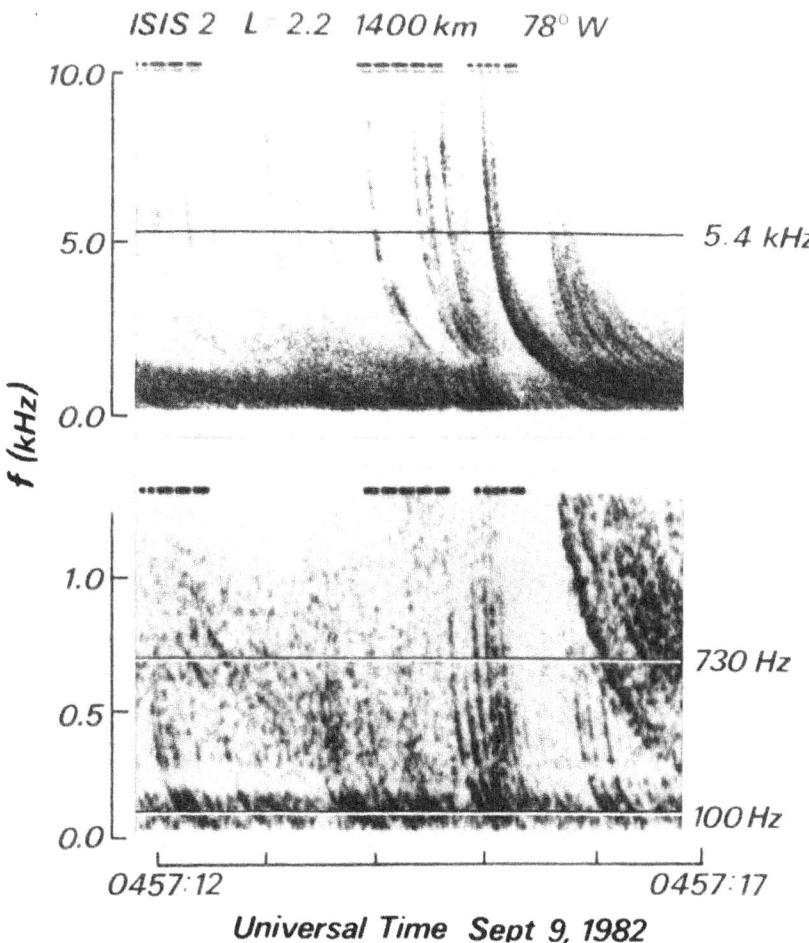

Fig. 6. Whistler mode waves in the terrestrial ionosphere as seen by ISIS-2 at 1400 km. The three horizontal lines mark the frequencies of the lowest three channels of the Pioneer Venus plasma wave instrument (Carpenter, personal communication).

trally analyzed and displayed versus time so that darkness represents the intensity of the signal. The top panel shows the band from 0 to 10 kHz and the lower panel shows an expansion of the top panel from 0 to 1.2 kHz. Also indicated for reference are the center frequencies of 3 of the narrow-band channels of the Pioneer Venus Orbiter plasma wave experiment. The signals are seen to cover a wide frequency range and are dispersed so that higher frequencies arrive first. This dispersion occurs because of the dependence of the velocity of the waves on their frequency with respect to the local electron gyrofrequency. Waves that have travelled the furthest have the greatest dispersion.

In this figure there are two groups of signals with quite different dispersion. Those with the least dispersion have propagated up from below the satellite. Those with the most dispersion have travelled along the magnetic field from the opposite hemisphere. The former signals are called fractional hop whistlers; the latter are called one-hop whistlers. The name whistler derives from the audio tone heard when these signals are played on a speaker. The plasma wave mode by which these waves propagate has been termed the whistler mode.

Figure 6 illustrates that terrestrial lightning generates frequent impulsive signals that can be observed in the ionosphere. The number of signals seen at 100 Hz seems

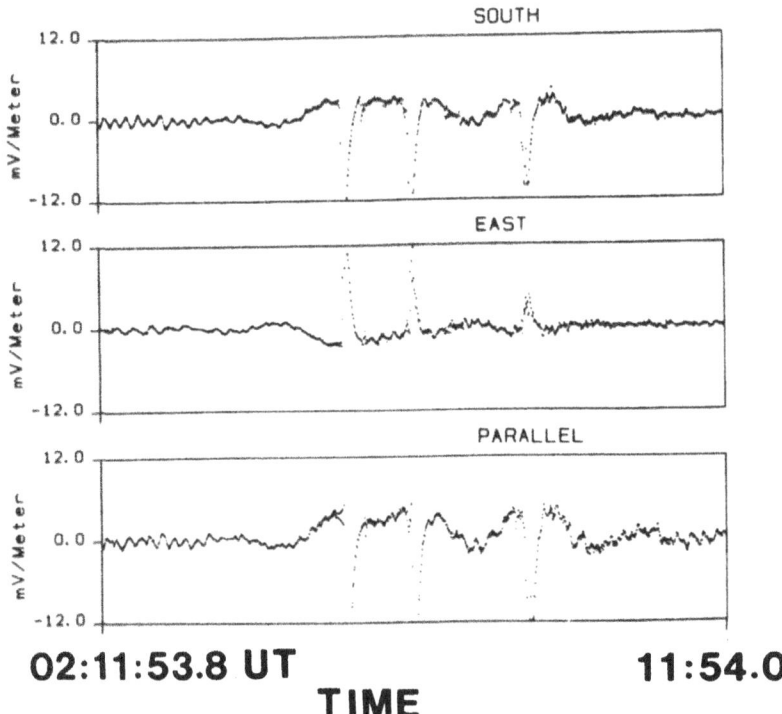

Fig. 7. Ionospheric electric field data from the rocket flight Thunder Hi transformed to geomagnetic coordinates. There are significant electric fields parallel to the magnetic field (Kelley *et al.*, 1985).

particularly frequent. We note, however, that the electron gyrofrequency plays an important role in the propagation of these signals. Whistler mode signals do not propagate above the electron gyrofrequency which on Venus is seldom above 500 Hz. Thus, if signals at higher frequencies are observed to penetrate into the Venus ionosphere, one must find a separate mechanism to explain that observation. That such a mechanism exists has been demonstrated by the rocket experiments of Kelley *et al.* (1985).

Figure 7 shows the electric field measured at 140 km by the instruments on the rocket 'Thunder Hi' above a thunderstorm. The data have been corrected for filter constants and transformed to geophysical coordinates. It is important to note that the electric field is strong, over 10 mV m^{-1}, and has a significant component parallel to the magnetic field. Presumably the collisions of electrons with ions and the neutral atmosphere has prevented them from immediately neutralizing this electric field parallel to the magnetic field. In the Venus ionosphere we might also expect the penetration of electric fields into the ionosphere for the same reason.

3. Venera Landers

The Venera 11 and 12 landers carried both a high sensitivity loop antenna and an acoustic sensor. While the acoustic sensor was saturated during descent, the VLF measurements of the loop antenna provided the most unambiguous evidence of lightning of all the Venus lightning investigations (Ksanfomaliti, 1979). The instrument returned amplitudes in 4 narrow-band channels centered at 10, 18, 36, and 80 kHz and in a wide-band channel. Figures 8 and 9 show the amplitude of the VLF signals as a function of altitude. The fact that the amplitude profiles look very different even though the descent trajectories were very similar suggests that the observed waves are not due to the interaction of the probe with the Venus atmosphere. Rather the signal variation is consistent with a temporally varying source like lightning. Figure 10 shows high-resolution VLF data during the weak activity at high altitudes on the Venera 12 landing. Here the pulses occur infrequently enough that they can be counted by eye. In this stretch there is a burst about every 12 s on the average. Later during the descent the rate becomes too great to count. Venera 13 and 14 carried electromagnetic sensors similar to those on Venera 11 and 12 but with no acoustic sensors. Instead they monitored the coronal discharge current from the spacecraft. No discharge currents were detected (Ksanfomaliti, 1983). The activity on Venera 13 and 14 was similar to that on Venera 12 and significantly less than on Venera 11.

The landing area for the probes was close to the subsolar point and far away ($\sim$ 8000 km) from the equivalent region in which terrestrial lightning is most frequent. This may be the reason why in Figures 8 and 9 the signal strength falls at low altitude. Simply, when the spacecraft reached the surface they could have been shadowed from the distant source of the VLF signals as illustrated by Figure 11. On the surface the instruments continued to operate but only Venera 12 saw a burst of VLF noise and only for a brief period as shown in Figure 12. Lightning discharges close enough not to be

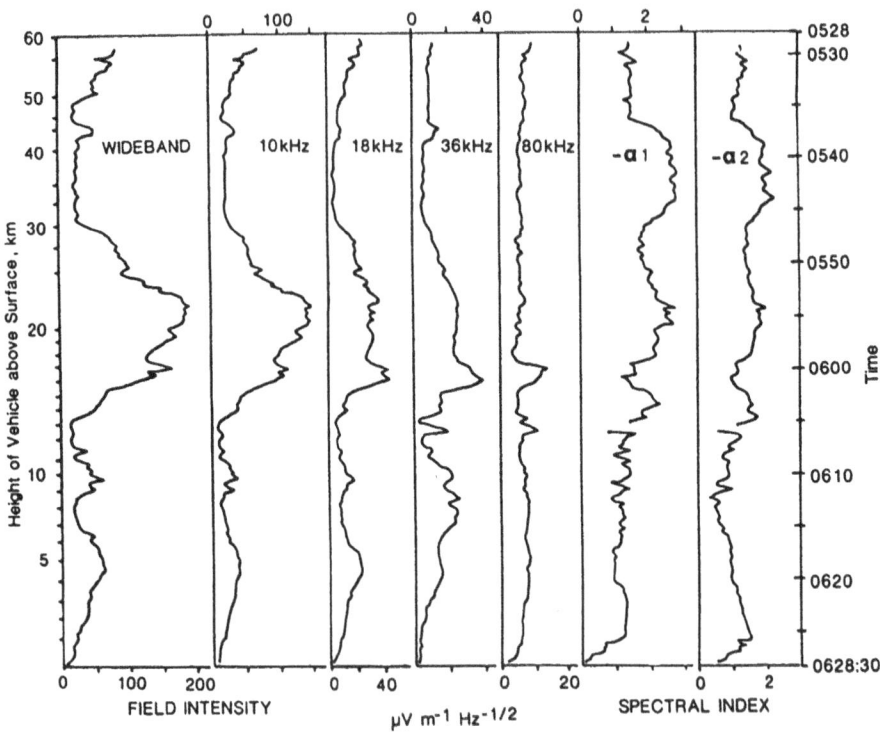

Fig. 8. The altitude distribution of VLF signals observed by Venera 11. The column on the left gives the wideband intensity. The next four columns give the intensity at 10, 18, 30, and 80 kHz. The last two columns give the spectral indices of the signals (Ksanfomaliti *et al.*, 1983).

shadowed appeared to be rare. No acoustic signals other than those associated with spacecraft operations were observed after landing.

The Venera VLF instruments included an impulse counter. The mean rate was 16.5 s⁻¹ overall. Below 20 km altitude the rate dropped to 13 s⁻¹ and below 5 km it dropped to 10 s⁻¹. These rates are much greater than terrestrial rates. The rates obtained during a test run on a clear day in the terrestrial atmosphere were similar to the lowest rates obtained on the Venera 12 entry from 25 to 12 km altitude and while the instrument was on the surface of the planet. On the Venera 11 landing the rates were an order of magnitude higher than the terrestrial rate until the vehicle landed.

Estimates of the source location are difficult. The detection of spin modulation during descent (Ksanfomaliti *et al.*, 1983) indicates a small angular source as opposed to an extended source and perhaps therefore a distant one. Ksanfomaliti *et al.* (1983) have estimated that Venera 11 and 12 were able to monitor lightning discharges from 7.5% of the planet and thus by extrapolating the number of discrete burst sites, they suggested that there were about 50 sites over the entire planet compared to the 2000 that one would detect on Earth. However, their extrapolation depends on the assumption that storms

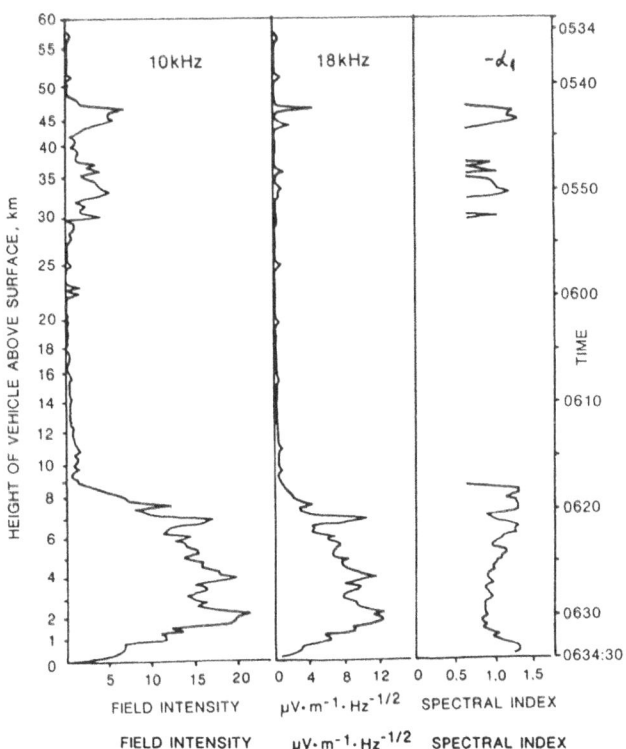

Fig. 9.   The altitude distribution of signals seen by Venera 12. The left-most column gives the amplitude at 10 kHz, the middle column gives the amplitude at 18 kHz and the right-most column gives the spectral index (Ksanfomaliti *et al.*, 1983).

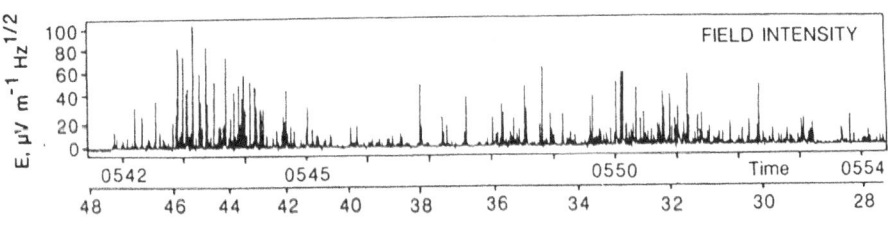

Fig. 10.   High-resolution measurements of the wideband field intensity on Venera 12 (Ksanfomaliti *et al.*, 1983).

are uniformly distributed, and thus ignores the possibility that they could be more frequent on Venus beyond the radio horizon of the Venera landers.

Ksanfomaliti (1979) assumed a cloud discharge that generated an energy of 10^9 J for 10^{-4} s at a range of about 1800 km. However, if instead the duration of an individual stroke was 10^{-6} s as suggested by Figure 2 at a range of 7000 km as appropriate for a late afternoon source, then the energy of a single stroke would be only 3.8×10^7 J

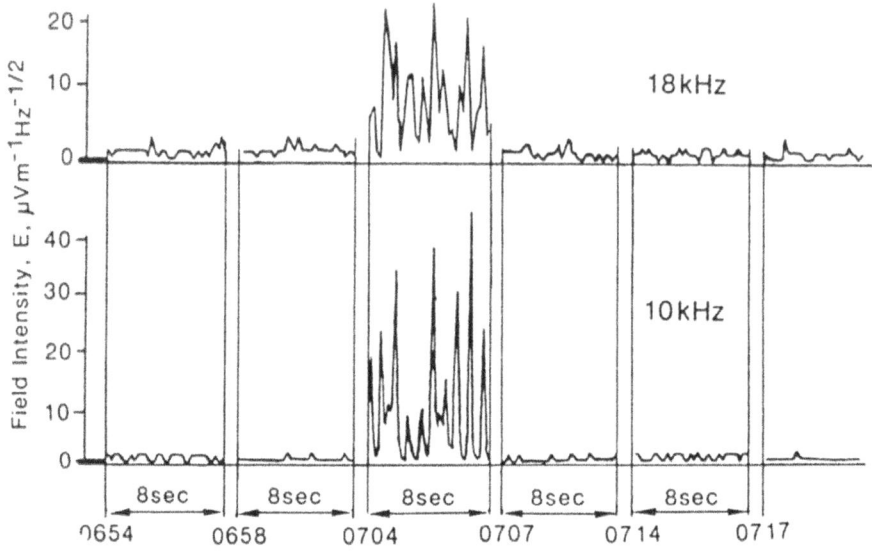

Fig. 11.   Interpretive sketch explaining the disappearance of signals at low altitudes on the Venera descents (Ksanfomaliti *et al.*, 1983).

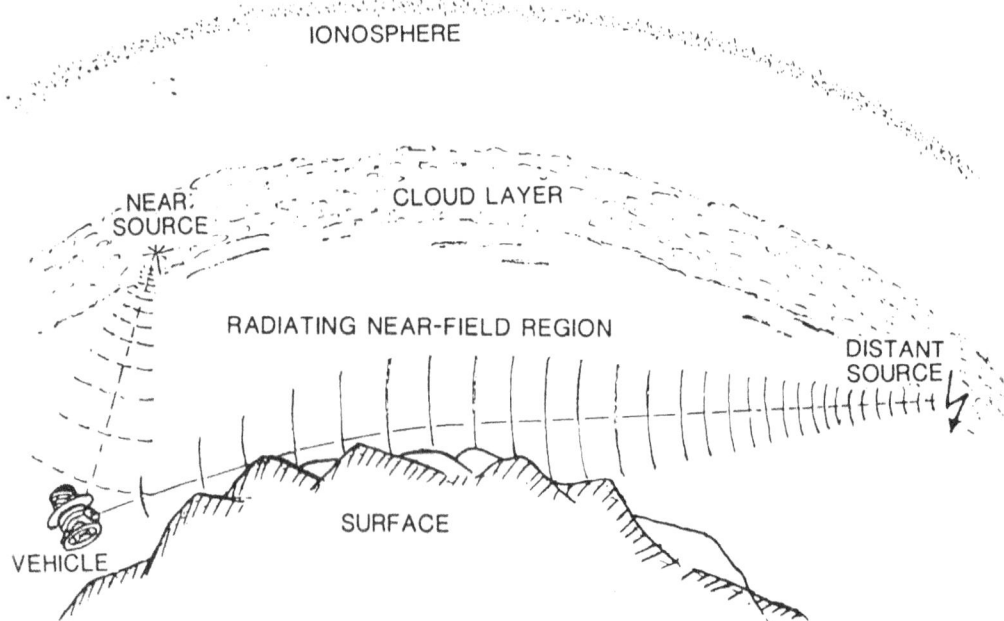

Fig. 12.   A burst of VLF noise while Venera 12 was sitting on the surface (Ksanfomaliti *et al.*, 1983).

assuming an inverse radial distance dependence in the surface-ionosphere wave guide. The total energy for a flash then depends on the number of these 'K' strokes in a flash. Figure 2 suggests that 20 is not an unreasonable number. If so, then the energy in a flash might be 8×10^8 J. However, this value depends linearly on the number of strokes per flash which actually occur on Venus, a number on which we have no information. Thus, the Venera VLF data provide almost no constraints on the energetics of Venus lightning.

It is very difficult to determine the global lightning rate on Earth from observations at a single site even if the waveform of the signals were measured over a very broad bandwidth. At Venus, the Venera landers monitored amplitudes only in narrow bandwidths with modest temporal resolution. Moreover, all of the landers landed near noon local time, far away from the late afternoon sector where our terrestrial experience suggests lightning may occur most frequently. Thus we should be cautious in our interpretation of rates of occurrence based on the data from the Venera landers. They clearly show the presence of electric discharges. The amplitude and occurrence rate of these signals is highly variable as one might expect from a weather related phenomenon, and the observations suggest that the source of these discharges is far away from local noon.

4. Optical Studies

On October 26, 1975 the Venera 9 visible spectrometer detected a period of apparent optical flashes on the nightside of Venus at 19:30 LT and 9° S latitude (Krasnopol'sky, 1983a). The flashes were seen over a period of 70 s corresponding to 10 sweeps of the instrument. The field of view of the instrument moved 450 km during this period so that the region containing flashes was much greater than a typical terrestrial storm. Such flashes were never seen when the instrument pointed away from the planet. Also the measured intensity of the flashes diminished at longer and shorter wavelengths where the instrument sensitivity diminished. Thus these flashes appeared to be real.

The fall off at longer wavelengths was not so steep as to indicate strong atmospheric absorption so Krasnopol'sky (1983b) deduced that the emission must be occurring in the cloud layer or above. Moreover, Borucki (1982) points out that the Venus atmosphere is strongly absorbing at wavelengths shorter than 500 nm and that the fact that as much energy was received per unit wavelength below this wavelength as above indicates even more strongly that the source is at high altitudes. The instrument did not see well-separated individual strokes in each flash as might be expected in terrestrial cloud-to-ground discharges. Rather the flashes appeared to be somewhat continuous for about 0.25 s consisting of about 10–20 strokes. Thus, they more closely resemble terrestrial intra-cloud discharges.

The optical energy in each flash was about 2.5×10^7 J corresponding to a total energy of about 7×10^9 J (Borucki and McKay, 1987). Although the Venusian strokes dissipate more energy than terrestrial strokes, their longer pulse duration causes the peak optical power to be much less (Levin et al., 1983). The region surveyed by Venera 9 over its lifetime was 10^7 km². Since only this one storm was found the average rate was 45 flashes $km^{-2} yr^{-1}$. While this rate is about 6 times that on Earth, the energy released is much greater because the flash energy estimated above is also greater than on Earth by a factor of about 20. Thus, if the observations of this one storm by Venera 9 is typical of Venus conditions, the energy released by lightning in the Venus atmosphere is about 100 times that on Earth. While this number seems very large, we note that no *in situ* data have been obtained in the region of the atmosphere over which the Venera data were obtained so we cannot judge the reasonableness of these estimates.

The Pioneer Venus star sensor has also been used in an attempt to detect lightning (Borucki *et al.*, 1981). Observations were initially made for a period of 20 min centered at periapsis for each of 36 orbits on the dawnside of midnight. The star sensor could not be directed directly at Venus because the night airglow in the visible is sufficiently intense to saturate the sensor. Thus, the star sensor was pointed off the limb of the planet and scattered light in the optics of the instrument was used to search for lightning bursts. The total observing time amounted to the equivalent of 10 s over the whole nightside. The event rate detected in this search was similar to the event rate for 'false alarms' seen when the sensor was pointed well away from the planet. What the instrument should have seen for any given occurrence rate depends on several factors. The star sensor has a frequency response so that signal duration is important. Also, the detectability of signals drops off with increasing range. For a terrestrial distribution of intensities, the PVO star sensor would theoretically detect all flashes at a periapsis of 150 km, 50% at a range of 1500 km and only 1% at a range of 5000 km. For dim strokes (2.8×10^8 W) Levin *et al.* (1983) estimate 85 flashes $km^{-2} yr^{-1}$ could occur on Venus and still be consistent with the star sensor data. No optical measurements were made of the evening region over which strong VLF signals are seen. Most recently, the Pioneer Venus star sensor has begun to be used again to look for the scattered light arising from lightning discharges. However, the analysis of these data is not yet complete.

Photometers were installed in the VEGA balloons which could look for lightning (Sagdeev *et al.*, 1986). These detectors failed to detect an unambiguous signature of lightning. However, these balloons flew over the dawn sector, where VLF occurrences are weak.

5. Orbital VLF Studies

As discussed above, terrestrial lightning generates VLF signals which may be detected to great altitudes in the ionosphere of the Earth. This is to be expected because electromagnetic VLF waves can travel with little attenuation in the ionosphere of the Earth with its strong magnetic field. A surprise is that electric fields parallel to the magnetic field can also appear in the ionosphere. In the simplest models of a magnetized plasma, these electric fields are neutralized by rapid electron motion along the magnetic field. At Venus the magnetic field in the ionosphere is much weaker than on Earth. For typical magnetic field strengths in the night ionosphere of about 15 nT, electromagnetic VLF waves from lightning could reach the Pioneer Venus spacecraft by whistler mode propagation just like on Earth. However, signals above the electron gyrofrequency should not propagate to the spacecraft. As we will argue below, waves from lightning appear to be able to propagate into the ionosphere a short distance even when they are above the electron gyrofrequency. It is possible that the same mechanism (collisions?) that allows the parallel electric fields to enter the terrestrial ionosphere also allows the penetration of the high-frequency signals into the night ionosphere of Venus.

5.1. INITIAL RESULTS

The Pioneer Venus spacecraft includes a simple plasma wave instrument which monitors the wave power in 4 narrow ($\pm 15\%$) frequency bands centered at 100 Hz, 730 Hz, 5.4 kHz, and 30 kHz (Scarf et al., 1980a). The antenna consists of two wire cages about 12 cm across deployed on 1 m booms about 0.8 m apart. The instrument has a fast response with a rise time of several tens of ms but a much slower decay time of about $\frac{1}{2}$ s. In sunlight the noise level of the instrument exceeds all but the strongest natural emissions. However, in the shadow of the planet the instrument appears to be quite sensitive. At orbit injection Pioneer Venus periapsis was in sunlight and not until a month later did the spacecraft begin to be in shadow as it passed through closest approach. During this period the spacecraft observed the first electric field signals that appeared to be due to lightning (Taylor et al., 1979). The signals had all the characteristics expected for lightning-generated signals. They were intense, impulsive and broadband. They occurred on all 4 frequency channels and lasted less than 0.5 s because the signals were shorter than the decay constant of the instrument. This type of signal is illustrated in Figure 13 for the 30 kHz channel. The signals appeared to be more frequent at lower altitudes suggesting a low altitude or atmospheric source.

After analyzing a larger sample of data, Scarf et al. (1980b) decided to use a more conservative definition of a possible lightning-generated signal. Since signals above the electron gyrofrequency should not propagate to the spacecraft in a collisionless plasma, Scarf et al. (1980b) defined lightning events as impulsive signals at 100 Hz which were unacccompanied by signals at other frequencies. Furthermore, the magnetic field had to be strong enough that the electron gyrofrequency well exceeded 100 Hz. Since Venus clouds were 50 km or above the Venus surface, it is unlikely that cloud-to-ground strokes occur. However, since volcanoes also generate clouds and lightning on the Earth, Scarf et al. (1980b) looked for a possible correlation of the emissions with volcanoes seen in the radar-derived topography.

In this study only the times of 'events' were noted. No records were kept of times when signal propagation conditions appeared to be favorable for propagation to the satellite but no signal was observed. Nevertheless, Ksanfomaliti et al. (1983) stated that lightning occurred only 2% of the time that PVO was in darkness. To convert this to a lightning flash rate one would need to know both the number of flashes in an event as defined by Scarf et al. (1980b) and the area of the ionosphere 'illuminated' by one flash. The events as identified by Scarf et al. in fact could last many seconds and, hence, contain many flashes, and there were no studies conducted that could determine the area of the planet monitored by the VLF receivers. Thus it was not possible to convert this rate to a flash rate per unit area for comparison with the terrestrial rate.

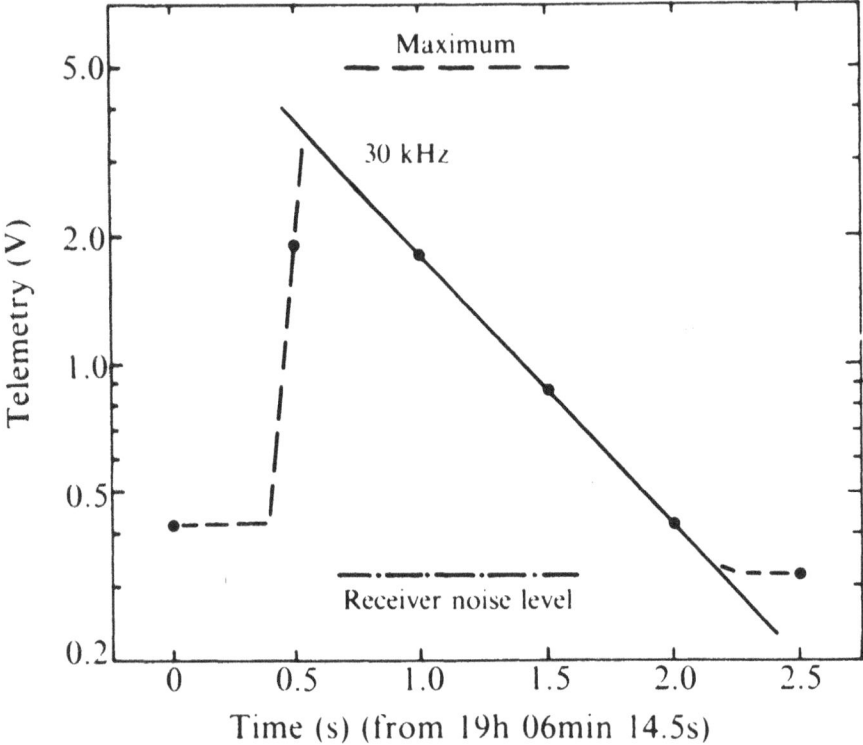

Fig. 13. Decay of 30 kHz signal strength in Pioneer Venus plasma wave instrument due to naturally occurring signals in the night ionosphere. Note the logarithmic voltage scale (Taylor *et al.*, 1979).

A larger sample of data was analyzed by Scarf and Russell (1983) who examined records for the first 1185 orbits of Pioneer Venus. These orbits covered 5 complete traversals of the night ionosphere (or observing seasons) plus part of a sixth. Again, possible lightning events were defined as impulsive signals which occurred when the local electron gyrofrequency was greater than 100 Hz, and when there were no accompanying signals at higher frequencies. No criteria were placed on the signals as to whether the magnetic field line connected the satellite to the atmosphere but a later check showed that this was true in every case. This survey covered 14% of the Venus surface. In this study 65% of the observed signals came from regions near Phoebe, Beta, and Atla which were topographically high regions where volcanism was suspected. The data seemed to be consistent with a small number of surface sources, strengthening in the minds of the authors a possible link between volcanism and the lightning events. This study reported a rate of 2.4 bursts km^2 yr^{-1}. However, this burst rate cannot be considered to be a flash rate both because there are many flashes in a burst and because many bursts were not counted because they failed one of the conservative selection criteria. Furthermore, the area of surface monitored at one time is not known, nor is known the size of the discharge which could cause a burst at Pioneer Venus. As before there was no attempt to normalize the distribution for observing time.

Scarf and coworkers later attempted to revise the definition of an event to make it more nearly equal to a flash by counting each impulse in a group of impulses. However, the time resolution of the instrument still limited this approach so that the flash rate was still under-estimated. In this study they also chose a threshold just above that of the instrument of 20 μV m^{-1} Hz$^{1/2}$ at 100 Hz. They required that the electron gyrofrequency be greater than 100 Hz and that the magnetic field line through the spacecraft intersect the planet. Examining data from the first 2124 orbits (9.5 observing seasons), Scarf *et al.* (1987) found 4240 burst covering the altitude range from 150–2900 km. After the first three seasons, the periapsis altitude of Pioneer Venus was allowed to rise because there was not enough propellant on board to counteract the orbital evolution caused by solar gravitational perturbations. As periapsis rose, the seasonal total number of bursts fell, indicating that the event rate was greatest at low altitudes. An attempt to show this altitude dependence using the rate versus altitude independent of season was flawed because of the non-uniform sampling of altitudes within any season. As before no attempt was made to normalize the occurrence rates.

While to Scarf and coworkers the possible lightning 'sources' seemed to cluster geographically on Venus the association with topography was not as evident to other researchers. Taylor *et al.* (1985) noted that the lightning events often occurred at the time of a depression in the measured ion densities which they called an ion trough. Taylor *et al.* suggested that it was likely that the plasma wave disturbances resulted from 'energetic and dynamic processes to be expected in the vicinity of the magnetic field and plasma configurations involved' rather than due to lightning. They attributed the weak geographic association of the lightning source regions as due to the fact that Pioneer Venus had merely passed over mountainous regions, rather than plains, during its mission to date.

In reply, Scarf (1986) pointed out that there is no one-to-one correlation between ion troughs and the identified lightning events. Events occur without troughs and troughs without lightning events. No attention was given to the issue of the statistical significance of the correlation with surface features. Taylor *et al.* (1986) responded to Scarf by showing even more associations of lightning events and troughs. Their conclusion is expressed in their statement: 'In view of the obvious high degree of association between the plasma depletions and the 100 Hz noise, we have no reasonable doubt that many of the signals attributed to Lightning are in fact stimulated near the satellite and thus are unrelated to lightning in the ower atmosphere'. The criticism of the purported geographic correlation was quite valid but neither Scarf, nor Taylor and coworkers, made an attempt to normalize the observations for observing time. The association with ion troughs, on the other hand, could be explained by processes consistent with either source mechanism. The ion density and electron density are anti-correlated with the magnetic field strength (Luhmann *et al.*, 1982). Since strong magnetic fields were necessary to identify an event, all events were associated necessarily with strong fields and statistically these had ion troughs associated with them. Moreover, since electromagnetic waves travel faster in lower density regions, waves with the same energy flux will have stronger associated electric fields. Thus, electromagnetic waves will be more readily detected by the electric field antennas of the Pioneer Venus instrument in low density regions. In short, the association between ion troughs and the possible 'lightning events' did not help in distinguishing possible source mechanisms.

As noted above, Scarf and coworkers changed their definition of an event after their 1983 study although they did not publish any new results until 1987. During this interval, which is described here for historical perspective, Taylor solicited and analyzed an interim listing of 'lightning events' identified by Scarf *et al.* This analysis formed the basis of further publications arguing against the lightning and volcanism hypotheses (Taylor and Cloutier, 1986; Taylor *et al.*, 1987). In the former study, Taylor and Cloutier repeat the claim that the geographic association is weak based on an analysis like that carried out by Scarf and Russell (1983), but with a larger data set. They continued to stress as they did in their earlier work the assertion that the association of the plasma waves with the density depletions is an argument against a possible lightning source. In addition, they further argued that the waves could not be electromagnetic because the wavelength of the waves is much larger than that of any ion trough. The wavelength they calculate, 3000 km, is the wavelength of 100 Hz waves in a vacuum. In fact the correct wavelength of these waves in the locally measured medium is less than a km. Thus the waves could easily propagate in a duct.

In a second paper, Taylor *et al.* (1987) made additional arguments against the whistler wave and lightning source hypothesis. First, they claimed that the number of events drops off at the very lowest altitudes and hence the waves could not be propagating up from below. Second, they claimed that the waves appear more often than expected when the satellite is moving perpendicular to the magnetic field and hence might be Doppler-shifted electrostatic waves. The first argument assumes that the waves are as easily detectable at lower altitudes as at high. However, the electric field amplitude is lower in the denser regions of the ionosphere for the same energy flux. Moreover, the criteria used by Scarf and coworkers discriminate against low altitude events where the events usually occur over a broad frequency band than a narrow one. Finally, the second argument is based on the assumption that the magnetic field and velocity vector are random with respect to each other in 3-D space. But if the magnetic field were radial, as is often the case where the events are observed, the spacecraft velocity would be nearly perpendicular to the magnetic field at all times.

If the lightning interpretation for the VLF waves is correct, then the waves should be electromagnetic below the electron gyrofrequency. If the waves were electrostatic, as Taylor and Cloutier (1986) and Taylor *et al.* (1987) suggest, the lightning hypothesis for the source of these waves would have to be abandoned. Scarf and Russell (1988), therefore, performed two tests to determine whether the waves were electromagnetic or electrostatic. The first test was to calculate the Doppler shifting expected for electrostatic ion acoustic waves and to see if the signals appear to have such Doppler broadenings. As illustrated in Figure 14, Doppler-shifts are expected to occur which would spread the observed spectrum over a bandwidth of several kHz depending on the direction of

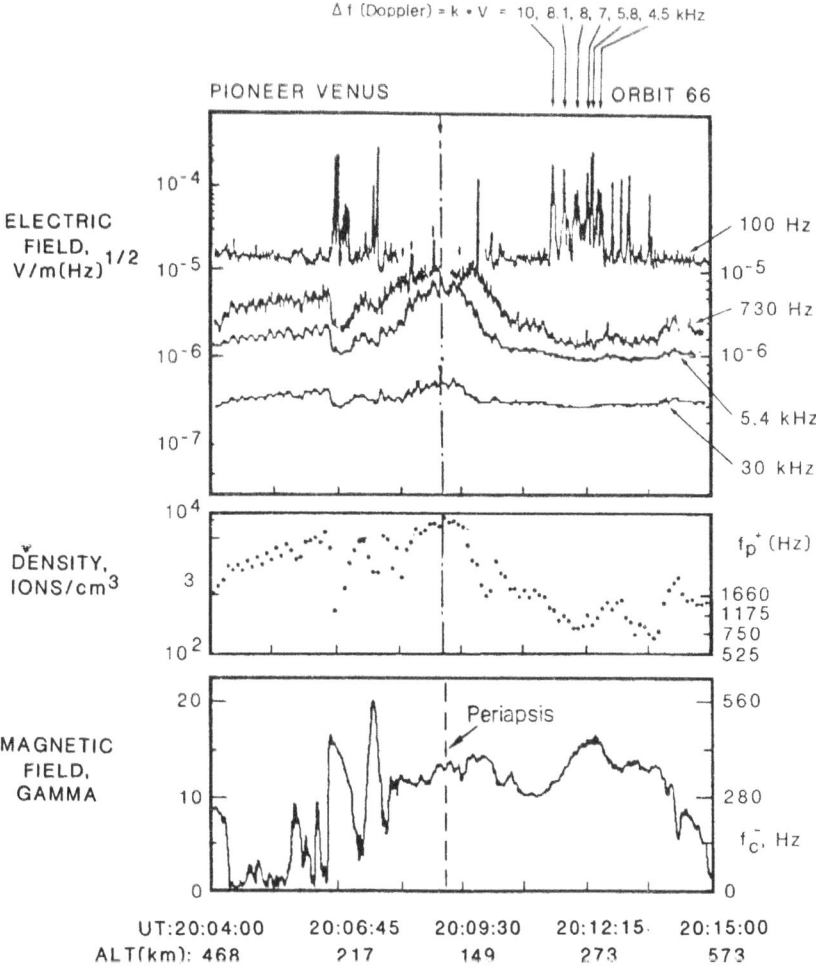

Fig. 14.   The electric field amplitude, the plasma density and magnetic field strength near periapsis on orbit 66. The Doppler shifts shown at the top of the figure expected if these waves were electrostatic are not observed (Scarf and Russell, 1988).

the velocity vector of the spacecraft to the wave vector. No such broadening is seen. In this example no waves are seen except in the 100 Hz channel. Such examples are common in the data.

The second test was to calculate the direction of maximum electric field perturbation with respect to the background magnetic field as the satellite spun about its axis. The direction of maximum electric field should be perpendicular to the magnetic field for an electromagnetic wave and parallel for an electrostatic wave. Since the spin rate of Pioneer Venus is such that 'lightning events' rise and fall in much less than a spin period, individual bursts of noise may occur at any orientation of the electric antenna relative to the magnetic field. However, as Figure 15 illustrates, when the magnetic field is

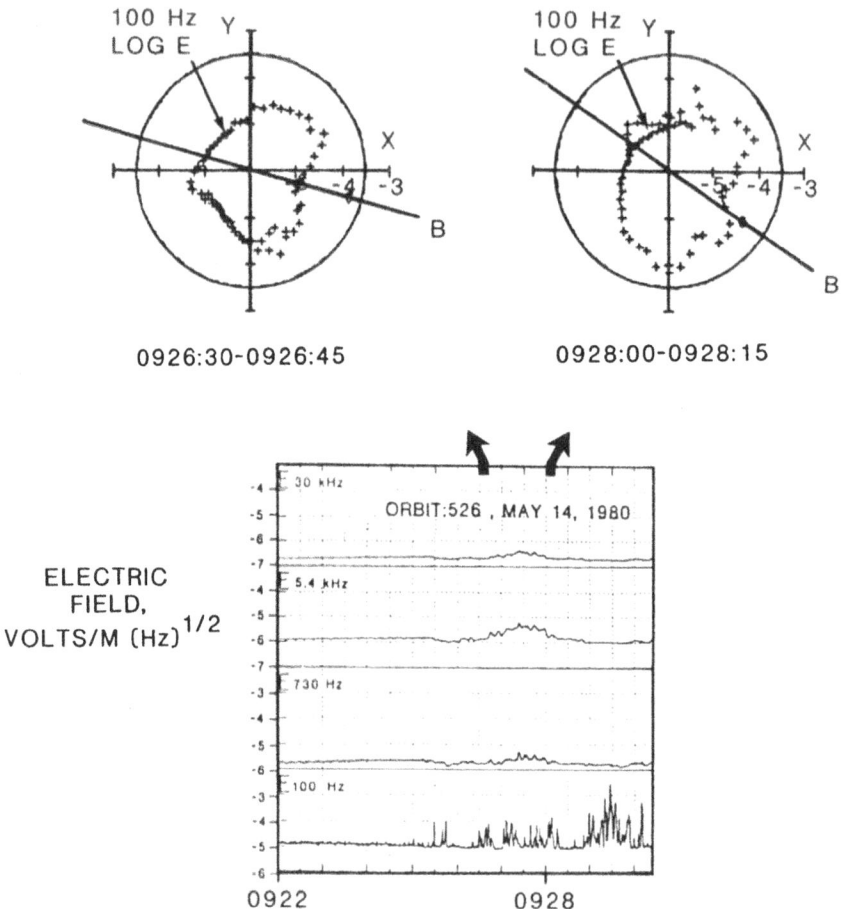

Fig. 15.   Amplitude of 0.1 kHz signals at low altitude in night ionosphere plotted versus the angle in the spin plane. The amplitudes are larger in the directions perpendicular to the magnetic field than parallel to it as expected for an electromagnetic wave (Scarf and Russell, 1988).

predominantly in the spin plane, the signal strength perpendicular to the magnetic field is greater than that along the magnetic field.

The net result of these interchanges could be summarized as follows: the geographic associations reported by Scarf and coworkers were justifiably questioned, but neither side performed a definitive test of their claim. Thus the possibility of a volcanic source for the lightning was certainly unproven and possibly suspect. On the other hand, the suggestion that the lightning events were electrostatic waves of some other origin related to the ion troughs remained unsubstantiated. The waves appear to be electromagnetic as originally proposed (see also Section 4.5), and the association with ion troughs can be due to several causes one of which is propagation from the atmosphere below. Thus, the initial hypothesis that lightning occurs on Venus remained intact.

5.2. ABOVE THE ELECTRON GYROFREQUENCY

The early observations of impulsive nighttime VLF bursts had been interpreted as possibly due to lightning at all frequencies both above and below the local electron gyrofrequency (Taylor *et al.*, 1979). However, since only whistler mode waves below the electron gyrofrequency would have access to the spacecraft in a homogeneous collisionless plasma, Scarf *et al.* (1980) adopted the above mentioned more conservative definition so that only impulsive emissions occurring on the 100 Hz channel were considered as possible candidates for lightning induced events. Singh and Russell (1986) took issue with this restriction. They pointed out that bursty signals like the 100 Hz signals occurred in the night ionosphere at all frequencies. They showed that there were many of these signals at low altitudes in the night ionosphere. Furthermore, as periapsis altitude rose, the signal occurrence at the higher frequencies also dropped. They postulated 2 mechanisms whereby propagation of signals above the electron gyrofrequency might occur. These included wave scattering on electron density irregularities (Singh and Russell, 1986) and 'radiation' from electron density holes in the ionosphere (Singh *et al.*, 1987). Taylor *et al.* (1987) took issue with the Singh and Russell work on two grounds. First, the new definition was inconsistent with that earlier used by Scarf *et al.* (1980) and second, the high-frequency noise occurred at both low and high altitude. Singh and Russell (1987) responded that the Scarf *et al.* (1980) definition had been overly conservative and excluded many other possible lightning signals and pointed out that as periapsis rose the signal occurrence decreased so there must be an altitude dependence to the signal occurrence. Taylor and Cloutier (1988) then criticized the Singh and Russell (1986) interpretation because their own analyses showed that telemetry noise was present in the data from the Scarf electric field detector. However, the major difference in the studies of Taylor and Cloutier (1988) and Singh and Russell (1986) besides the separation into spike and non-spike noise was the fact that Taylor and Cloutier included data obtained in sunlight in their analysis of at least non-spike noise while Singh and Russell did not (Russell and Singh, 1989). No tally of telemetry errors was kept by Singh and Russell (1986) so it is impossible to tell how many such errors could have been mistaken for real signals but perhaps half of the signals counted in the 30 kHz channel which has the lowest natural impulse rate were due to unrecognized telemetry noise Russell and Singh, 1989). Even if the telemetry noise contribution were less than this the distributions plotted by Singh and Russell (1986) were merely qualitative. Since they did not normalize their event occurrences by observing time, nor did they present thresholds or event selection criteria, they could not obtain a quantitative picture of the dependence of the signal occurrence rate on altitude. A truly quantitative study required a different experimental protocol.

5.3. QUANTITATIVE OCCURRENCE RATES: HIGH FREQUENCIES

The most recent study attempted to derive quantitative occurrence rates by noting both where signals are detected and where they are not detected (Russell *et al.*, 1988a). All data were examined when Pioneer Venus was in the unilluminated ionosphere at a solar zenith angle greater than 90° and within 1.05 Venus radii of the extended Sun–Venus line. To be characterized as impulsive or bursty the magnitude of the signal had to vary in 30 s by an amount comparable to the mean of the signal and to exceed thresholds of $1 \times 10^{-5}\,\mathrm{V\,m^{-1}\,Hz^{1/2}}$ at 0.73 kHz, $3 \times 10^{-6}\,\mathrm{V\,m^{-1}\,Hz^{-1/2}}$ at 5.4 kHz and $9 \times 10^{-7}\,\mathrm{V\,m^{-1}\,Hz^{1/2}}$ at 30 kHz. The plots used were those available from the National Space Science Data Center. Each 30-s interval was classified. In addition to the 4 electric field measurements, the magnetometer data were examined to eliminate periods in which telemetry errors occurred. Figures 16(a) and 16(b) show typical signals classified in this study. The signals labelled 'b' are interference associated with the motion of the antenna into the spacecraft wake. The signals labelled 'a' and 'c' are typical of those thought to be associated with lightning. As we discuss below, the broadband nature of the 'a' signals implies they are probably from a source immediately below the satellite. Such signals are seen at lowest altitudes. Since the 'c' signals occur only on the 100 Hz channel, the spacecraft does not appear to be in the near field of the lightning stroke. Probably these signals have propagated a large distance to the spacecraft and come from an extended region. The signals labelled 'e' are probably associated with the solar wind interaction, since they are broadband and occur near the edge of the wake at high altitudes. All impulsive signals of types 'a', 'c', and 'e' were included in the study of Russell *et al.* (1988a, b, c; 1989a, b, c).

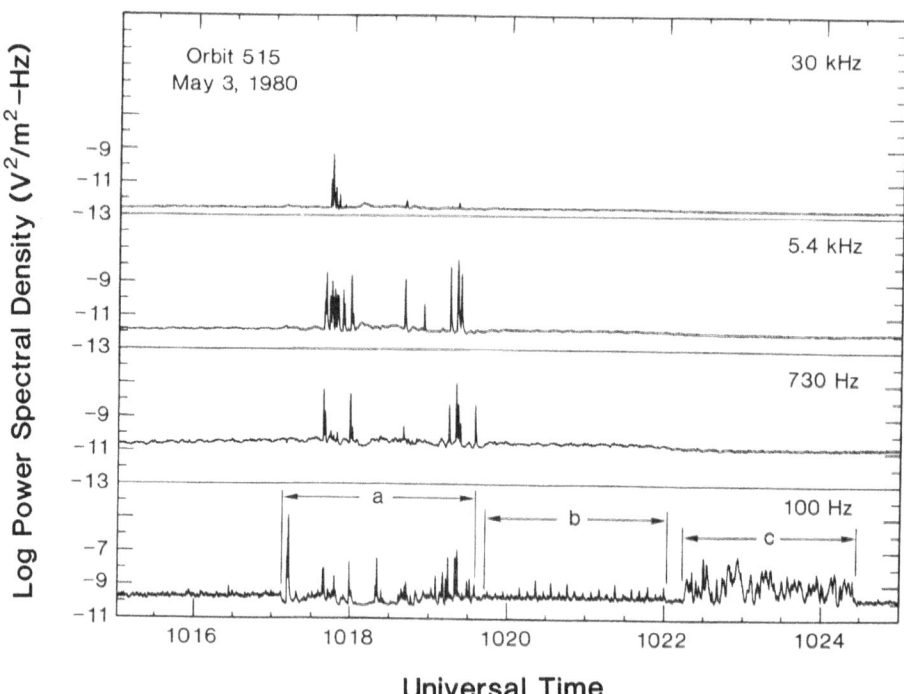

Fig. 16a.   Plasma wave amplitudes versus time on orbit 515 showing types of signals observed. Type 'b' is due to the electric field antenna entering the wake of the spacecraft (Russell and Scarf, 1990).

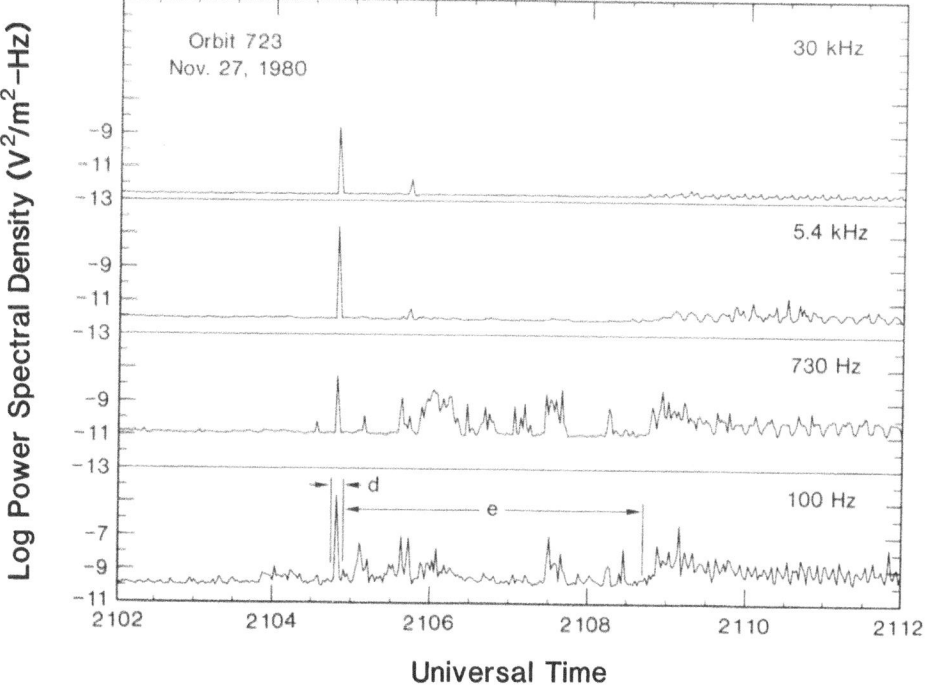

Fig. 16b.   Further amplitudes on orbit 723 showing additional wave types. Type 'd' is a telemetry dropout
(Russell and Scarf, 1990).

The initial examination of these data showed that the occurrence rate decreased with
increasing altitude at all frequencies, suggesting that at least some of the signals present
were generated below the ionosphere. It also showed that the occurrence rates varied
from year to year (Russell *et al.*, 1988a). Figure 17 shows the altitude dependence of the
signals at the 3 highest frequencies (Russell *et al.*, 1989a).

The magnetic field strength affects the occurrence rate of these signals but the
direction of the magnetic field has only a small or negligible effect (Russell *et al.*, 1988b).
Figure 18 shows the occurrence rate versus the center frequency of each channel as
normalized by the local gyrofrequency. Strongest fields are to the left-hand side of each
trace. At lowest frequencies, below $\frac{1}{4}$ of the electron gyrofrequency, the occurrence rate
becomes less dependent on magnetic field strength. At higher frequencies the occurrence
rate decreases rapidly as the frequency increases relative to the electron gyrofrequency.

The signals above the electron gyrofrequency are useful for mapping the source
locations precisely because they attenuate rapidly. Figure 19 shows the occurrence rate
at 730 Hz as a function of latitude and local time for each of the first three seasons
(Russell *et al.*, 1989a). During the first half of the second season Venus passed behind
the Sun preventing telemetry transmission to Earth. Thus only $2\frac{1}{2}$ seasons of low altitude
data are available. In the middle panel the region covered by Borucki's star sensor

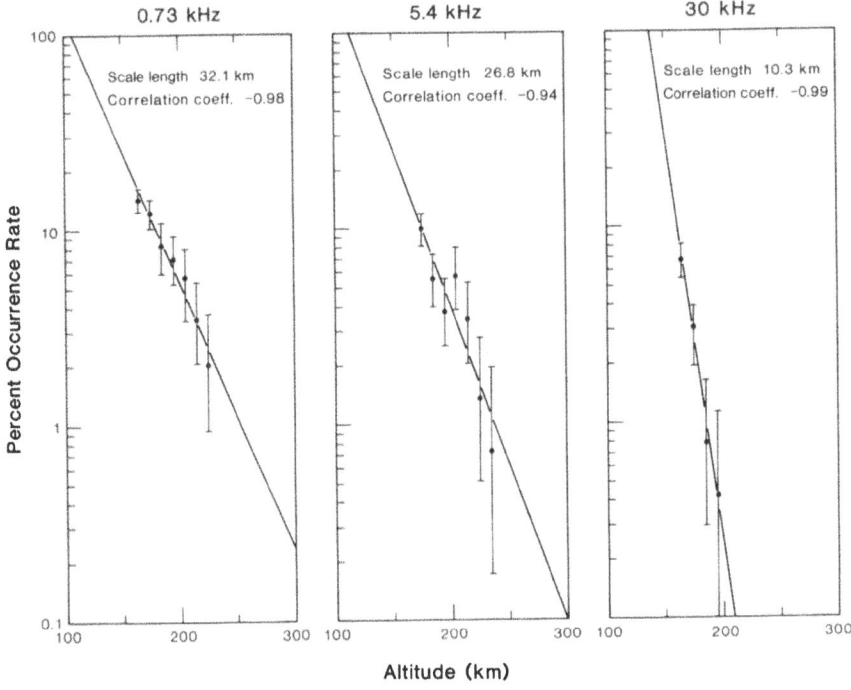

Fig. 17.   Altitude dependence of occurrence rate of impulsive bursts in the Venus nightside ionosphere above the electron gyrofrequency. An occurrence is defined as at least one impulsive signal rising above a threshold value in a 30 s period (Russell *et al.*, 1989a).

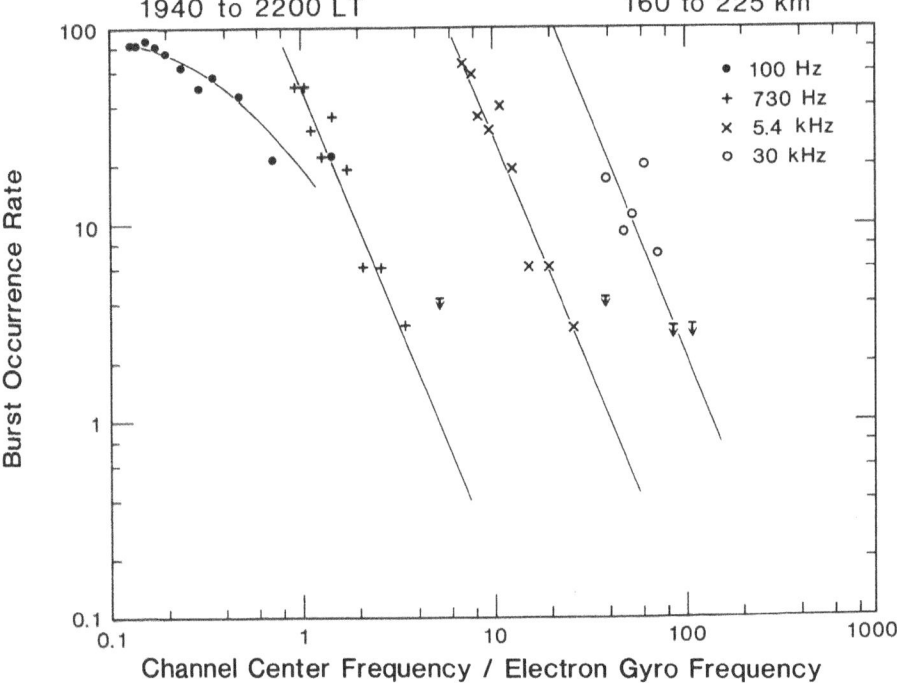

Fig. 18.   Dependence of signal occurrence rate versus frequency normalized by the local electron gyrofrequency. Strong fields are to the left-hand side of each trace (Russell *et al.*, 1988b).

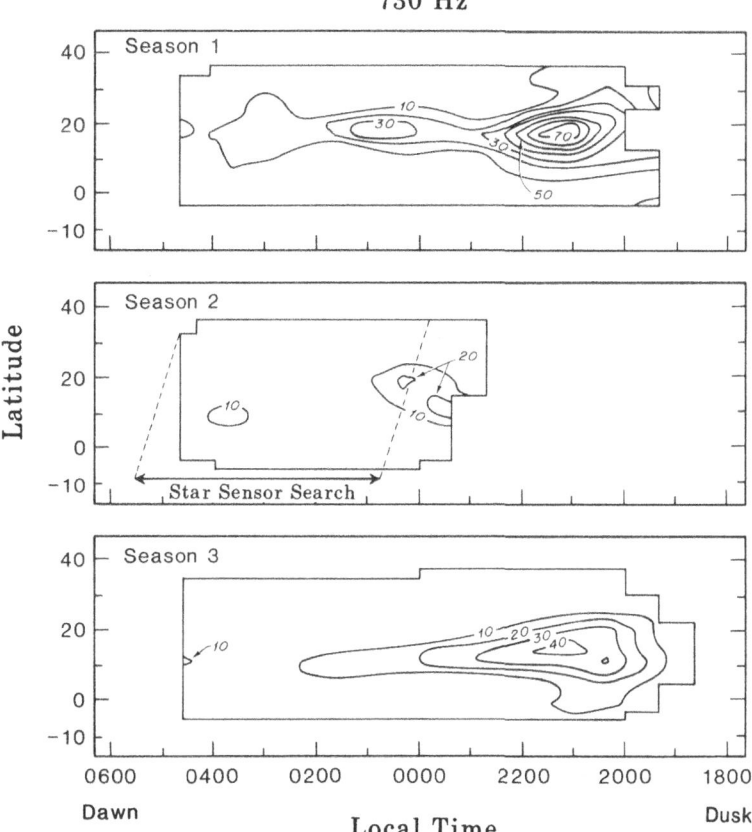

Fig. 19.   Contour maps of the occurrence rate of bursts at 730 Hz in local time and latitude for each of the three observing seasons. The interval during the second observing season over which the star sensor search for lightning was conducted is also shown (Russell *et al.*, 1989a).

survey (Borucki *et al.*, 1981) is also indicated. The first and third seasons show a very strong local time dependence with a region of highest occurrence rate on the dusk side which is confirmed during the second season by the relative absence of signals on the dawn side. The star sensor search was clearly performed over a region of infrequent plasma wave activity.

We can correct for the altitude dependence of these signals as shown in Figure 17 at each frequency and combine them to get the corrected map of occurrence shown in Figure 20. This figure also shows the path of the VEGA balloons and the location of the reported sighting of lightning by Venera 9. At the peak of the disturbances at about 21:00 LT, impulsive signals occur within over 70% of the 30 s observing periods, but on the dawn side impulsive signals occur about 10% of the time. Thus it is not surprising that the VEGA balloons mission and the Pioneer Venus star sensor search saw no significant activity while the Venera 9 spacecraft detected lightning. Both optical and

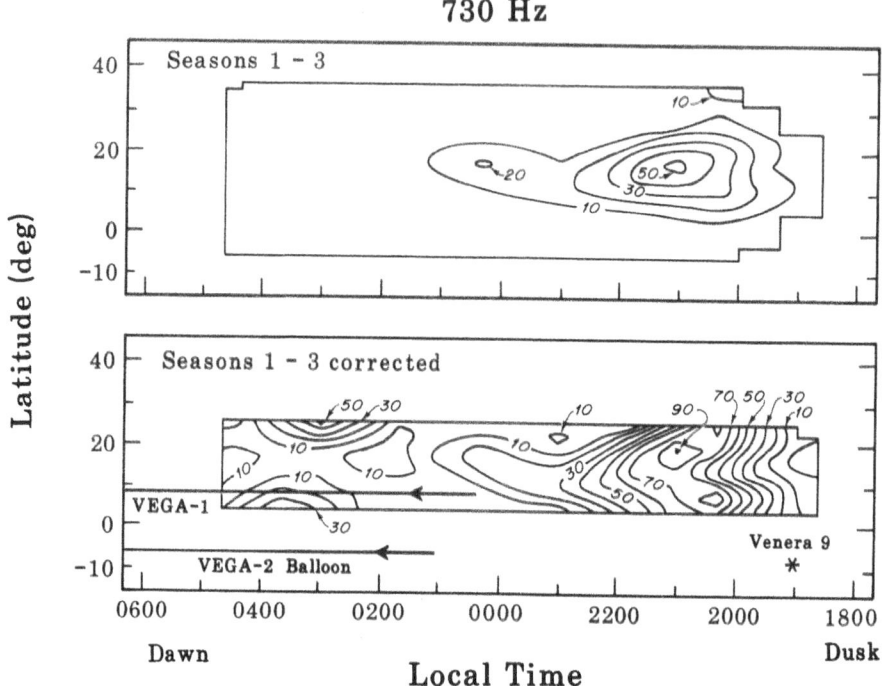

Fig. 20.   The occurrence rate at 730 Hz averaged over the three seasons as a function of local time and latitude (*top panel*). The same data corrected for the altitude dependence of the occurrence rate (*bottom panel*). The track of the VEGA balloons and the location of the Venera 9 lightning storm are also shown.

plasma wave data are consistent with an evening maximum in lightning activity. Lightning activity is also more frequent in Figure 20 toward the equator (Russell *et al.*, 1989a).

We believe that the local time variation seen on the dawn side of the peak activity in Figure 20 is due to variations in the source because the ionosphere does not vary much across midnight to dawn but the decrease in occurrence rate from 21:00 LT to 18:00 LT appears to be due to the increasing difficulty of the signals to propagate through an increasingly dense ionosphere. This is illustrated in Figure 21 which shows the occurrence rate from 22:00 LT to 18:00 LT plotted versus the density at 150 km of the VIRA ionosphere. There is a very strong anti-correlation of -0.99. It appears that the source of these waves extends well into the dayside but that the dense dayside ionosphere prevents them from reaching the spacecraft.

The studies above all examined occurrence rates. We expect and observe that signals occur with varying amplitudes. Since a fixed threshold was used in the above studies, a decrease in the average amplitude will lead to a decrease in the occurrence rate. The decreasing occurrence rate with increasing altitude implies a decrease in amplitude with altitude. In Figure 22 we confirm this expectation by plotting the upper quartile ampli-

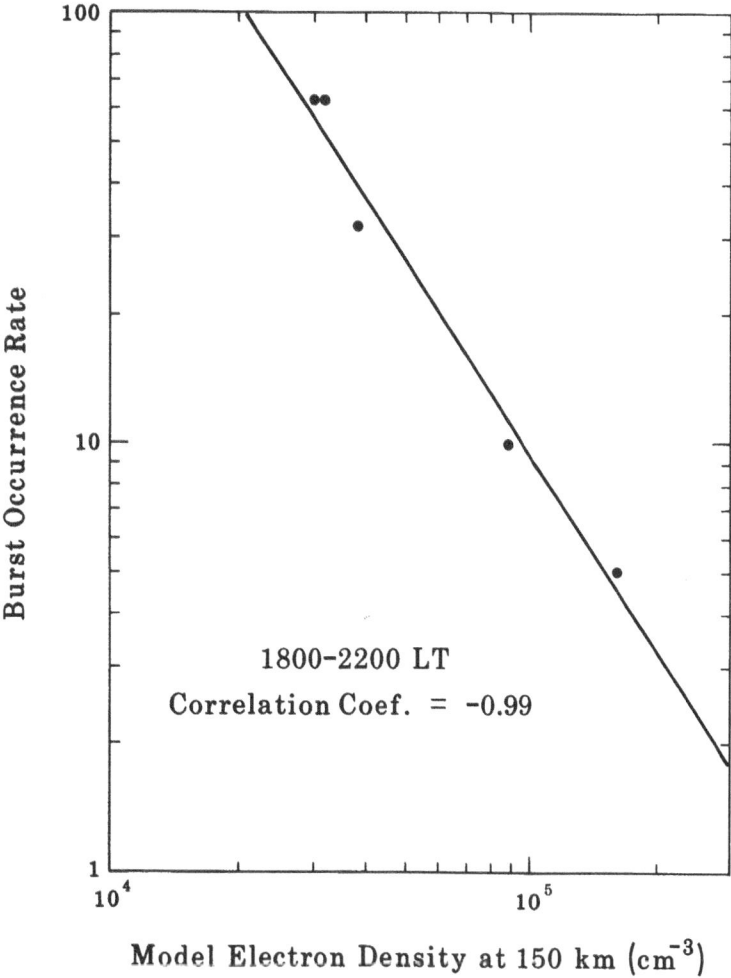

Fig. 21.   Occurrence rate versus ionospheric density from 22:00 to 18:00 LT (Russell *et al.*, 1989a).

tude of the peak power of bursts in the 730 Hz channel and the 5.4 kHz channels as a function of altitude (Russell *et al.*, 1989b). Over most of the altitude range the power decreases with increasing altitudes. In fact it decreases 1 to 2 orders of magnitude in 30 km. At the lowest altitudes the signal decreases also. This could be due to several factors. First, there is little data below 160 km so the decrease is not as statistically significant as at higher altitudes. Second, the amplitude fall off could be due to a propagation effect. The index of refraction varies rapidly with altitude in this region. The energy flux of the wave could be decreasing with increasing altitude while the electric power increased. Further it could be due to mode conversion from electromagnetic waves to electrostatic waves which might not create the electrostatic waves at lowest altitudes. While it is conceivable that there is some unknown plasma instability at about

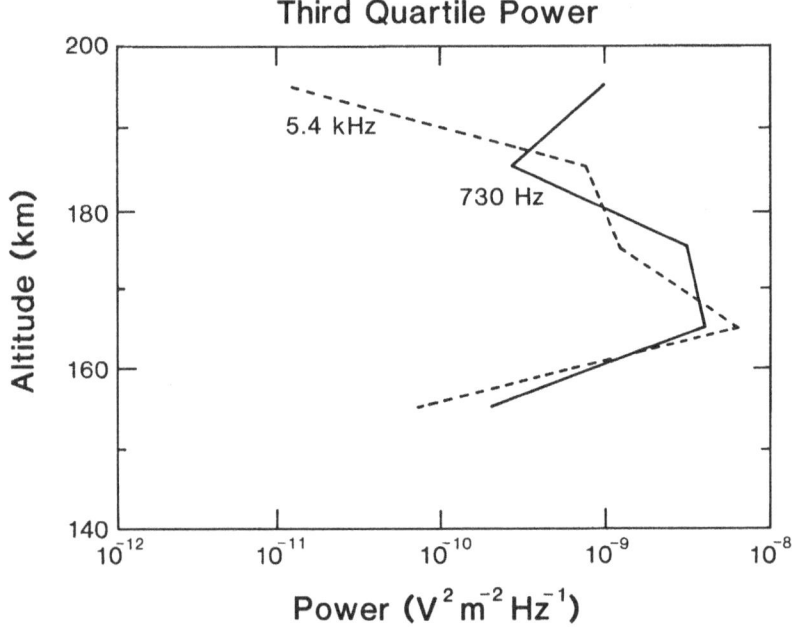

Fig. 22.   Wave power at 730 Hz and 5.4 kHz as a function of altitude in 10 km bins for local times from 19:40 to 22:00 and $B \geq 15$ nT. The power shown is the level exceeded by 25% of the intervals, i.e., the third of upper quartile. Medians are not shown because too many intervals would have been at or close to the instrument threshold (Russell *et al.*, 1989b).

165 km in altitude, we know of no property of the ionosphere or solar wind interaction which has a local time distribution event remotely similar to that of these signals. Lightning generation in the Venus clouds is the most likely candidate.

There is still some mystery about how these signals would reach the PVO spacecraft. In the Earth's ionosphere the electric impulse associated with lightning discharges is clearly seen parallel to the magnetic field where it should be shorted out (Kelley *et al.*, 1985). Perhaps collisions cause this. On Venus, a similar effect may occur. It is unlikely that the effect postulated by Singh and Russell (1986) is operative because the signals can be seen when the local magnetic field is horizontal. The Singh and Russell (1986) mechanism requires a vertical field.

5.4. QUANTITATIVE OCCURRENCE RATES: BELOW THE ELECTRON GYROFREQUENCY

If we treat the 100 Hz signals which generally occur below the electron gyrofrequency in the same manner as those above it we get some similar results and some different results. Figure 23 shows the local time of occurrence of the 100 Hz signals when B is greater than 15 nT, together with the occurrence rate of fields greater than 15 nT and with the occurrence rate of emissions above the electron gyrofrequency. This latter quantity was inferred earlier to point to the source region of the lightning emissions

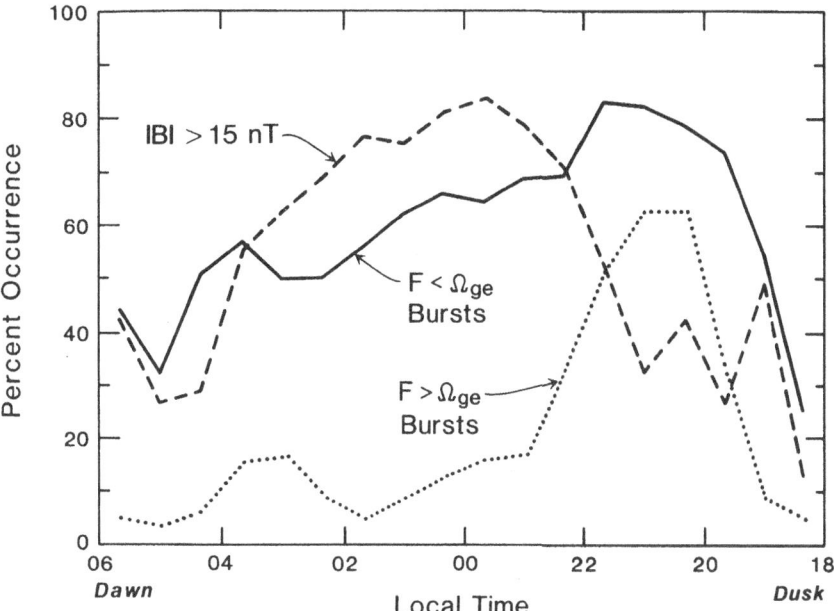

Fig. 23.   Occurrence versus local time rate of strong magnetic fields ($B \geq 15$ nT) of 100 Hz impulsive signals with frequency, F, below the local electron gyrofrequency (Ω_{ge}) and impulsive signals above the local electron gyrofrequency (Russell *et al.*, 1990).

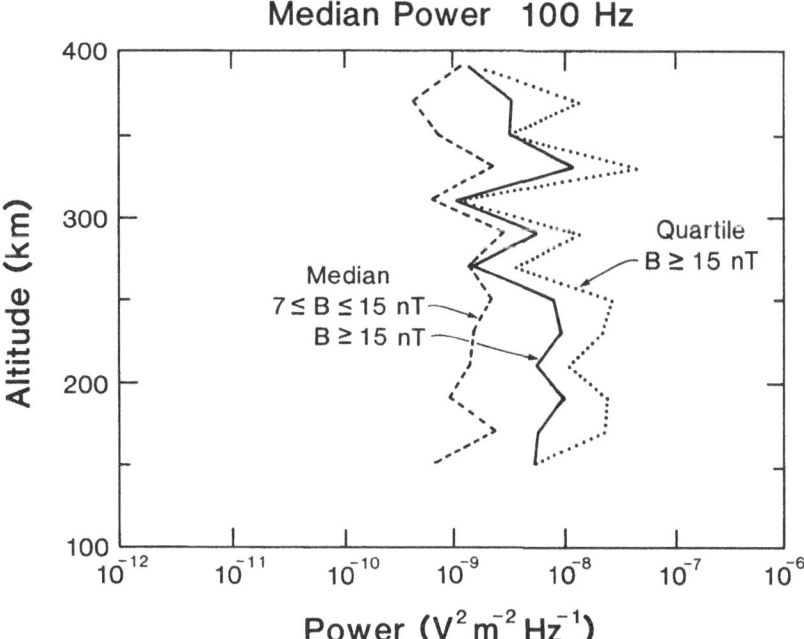

Fig. 24.   Median wave power at 100 Hz as a function of altitude in 20-km bins for local times from 19:40 to 03:20 for two different magnetic field strengths. Dotted line gives upper quartile for $B \geq 15$ nT (Russell *et al.*, 1989b).

because of the rapid decrease in occurrence (and amplitude) with altitude. The 100 Hz occurrence rate does not resemble this curve nor does it resemble the occurrence of strong fields. If it did we would infer that the occurrence of 100 Hz waves was controlled solely by the magnetic field. Rather the 100 Hz waves peak over the so-called 'source region' but extend far across the nightside gradually decreasing in occurrence. We deduce from this that the 100 Hz signals may be generated in the 'source region' but that they can propagate a long way in the Venus-ionosphere wave guide before they enter the ionosphere. We recall that we can say nothing about the possible extension of the source region into the afternoon sector because the ionospheric density appears to control the access of signals to the spacecraft.

The altitude dependence of the 100 Hz signals is weak. Figure 24 shows the altitude dependence of the occurrence rate for two ranges of magnetic field, 7 to 15 nT and greater than 15 nT. The two rates differ as will be discussed in the following section but there is little change with altitude for either curve.

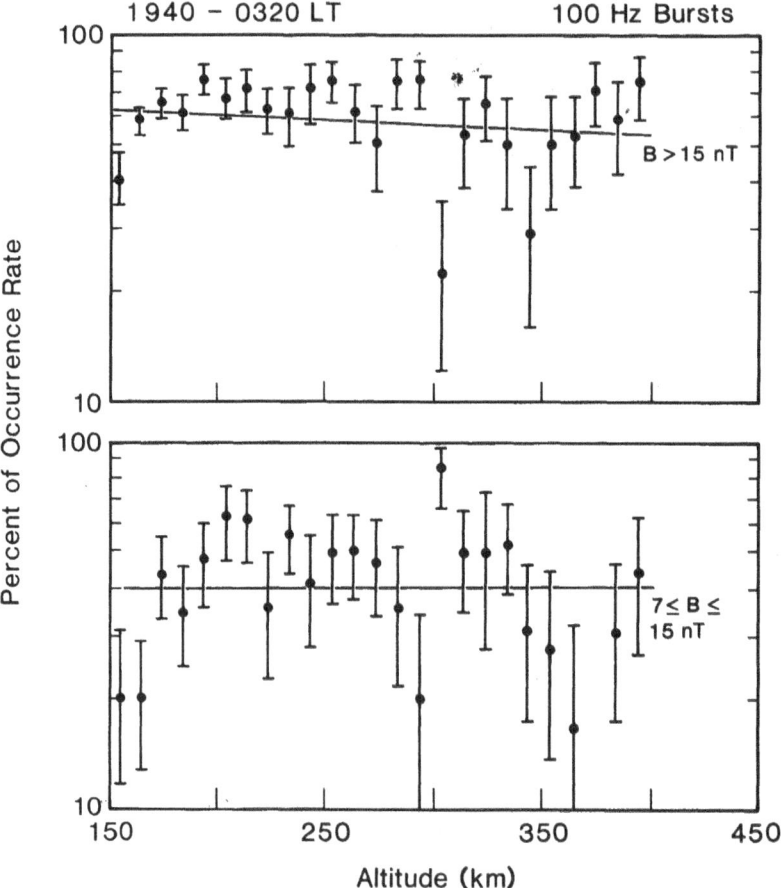

Fig. 25. Occurrence rate of 100 Hz bursts as a function of altitude for two different ranges of field strength (Russell *et al.*, 1990).

Another way to examine the fall off with altitude is to examine the average power in each 30-s interval as a function of altitude as is illustrated in Figure 25. The median power is shown for two ranges of field strength. The upper quartile for field strengths greater than 15 nT is also shown. The quartile illustrates that the variability of wave power is not large at 100 Hz. Also the altitude variation in power at 100 Hz is weak as implied by the slight change in occurrence rate with altitude. It appears that the 100 Hz signals can propagate long distances with little damping either in the Venus-ionosphere wave guide or in the Venus ionosphere. This is in great contrast to the signals above the electron gyrofrequency which seem to reach barely 200 km.

5.5. ESTIMATES OF THE POYNTING FLUX

To calculate the Poynting flux or energy flux of the wave ideally we need to know all six magnetic and electric field components. However, we only have one component, an electric field component measured in an arbitrary plane but averaged over that plane by the rotation of the spacecraft. We also can calculate the index of refraction of the medium because we know the wave frequency, the plasma density, and the magnetic field strength, except that we do not know the direction of propagation of the wave. In the absence of any other information we will assume the wave is a whistler-mode, electromagnetic wave propagating along the magnetic field and the wave component measured is that perpendicular to the magnetic field. This assumption may not be correct because of refraction effects in the high density regions near periapsis. Thus, the treatment of the Poynting flux is at best approximate.

Under these assumptions the Poynting flux in W m^{-2} becomes

$$0.0796NE^2 , \tag{1}$$

where N is the index of refraction, E^2 is the measured wave power in V^2 m^{-2} Hz^{-1} and the coefficient accounts for the 30 Hz bandwidth of the electric field instrument. We note that because of the 0.5 s decay of the detectors, this estimate of the Poynting flux will be high by an amount that depends on the occurrence rate of the signals. Since the occurrence rate of the 100 Hz signals varies little with altitude or local time over the region of our study (Russell et al., 1988a) this effect should not significantly affect the calculated altitude or local time variations.

Making the conversion to Poynting flux we obtain the altitude profile in Figure 26 (Russell et al., 1989b). There is little variation with altitude and little dependence on field strength. It appears as if the Poynting flux on Venus is quite constant. The slight decrease is consistent with a smal amount of damping as the waves propagate. This behavior in turn is consistent with the assumption of the waves being electromagnetic.

This calculation can be performed as a function of local time also. In Figure 27 we show the index of refraction which alters the electric amplitude of the wave as it propagates. In the local time range from 21:00 to 03:00 the index of refraction stays about constant but the wave energy fluctuates. The bottom panel shows the 100 Hz median wave power. The median power is at instrument threshold after about 01:20 LT and before about 19:40 LT. Thus it seems improper to calculate the median Poynting

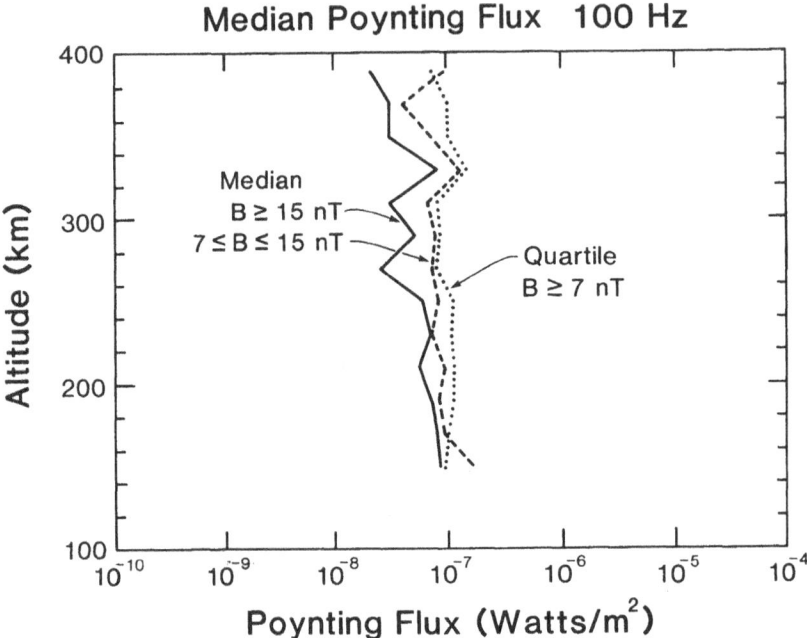

Fig. 26.   Median Poynting flux of 100 Hz waves as a function of altitude in 20-bins for local times 19:40 to 03:20 for two different magnetic field strengths. Poynting flux has been calculated assuming that waves are electromagnetic waves propagating parallel to the magnetic field and that the electric amplitude measured is the component perpendicular to the magnetic field. Dotted line shows the upper quartile for magnetic fields greater than 7 nT (Russell *et al.*, 1989b).

flux outside of these limits. The middle 2 panels show the median and upper quartile Poynting fluxes. The wave energy flux is greatest above the region hypothesized to be the source region based on the higher frequency data (Russell *et al.*, 1988b, 1989a, b, 1990).

5.6. GEOGRAPHIC CORRELATIONS

The altitude distributions and local time distributions of signals above the electron gyrofrequency indicate that these signals do not propagate far. Thus they provide our best tracers of the source region of the bursts, and we can use them to test the hypothesis that there is topographic control of the sources. Figure 28 shows the occurrence rate of emissions at 730 Hz as a function of planetary longitude and latitude for each of 3 observing seasons (Russell *et al.*, 1988c). Similar occurrence rates are seen at 5.4 and 30 kHz. The limited north–south extent of the emissions is due to the rapid decrease in occurrence with increasing altitude of these emissions combined with the variation of altitude with latitude of the Pioneer Venus orbit. Because Venus rotates slowly in a retrograde sense, these maps are essentially the mirror images of the local time maps. Some differences in values of the contour levels are present because of these maps were not restricted to the same altitude ranges as the local time maps.

The occurrence rates show some structure in longitude. There is minimal overlap from

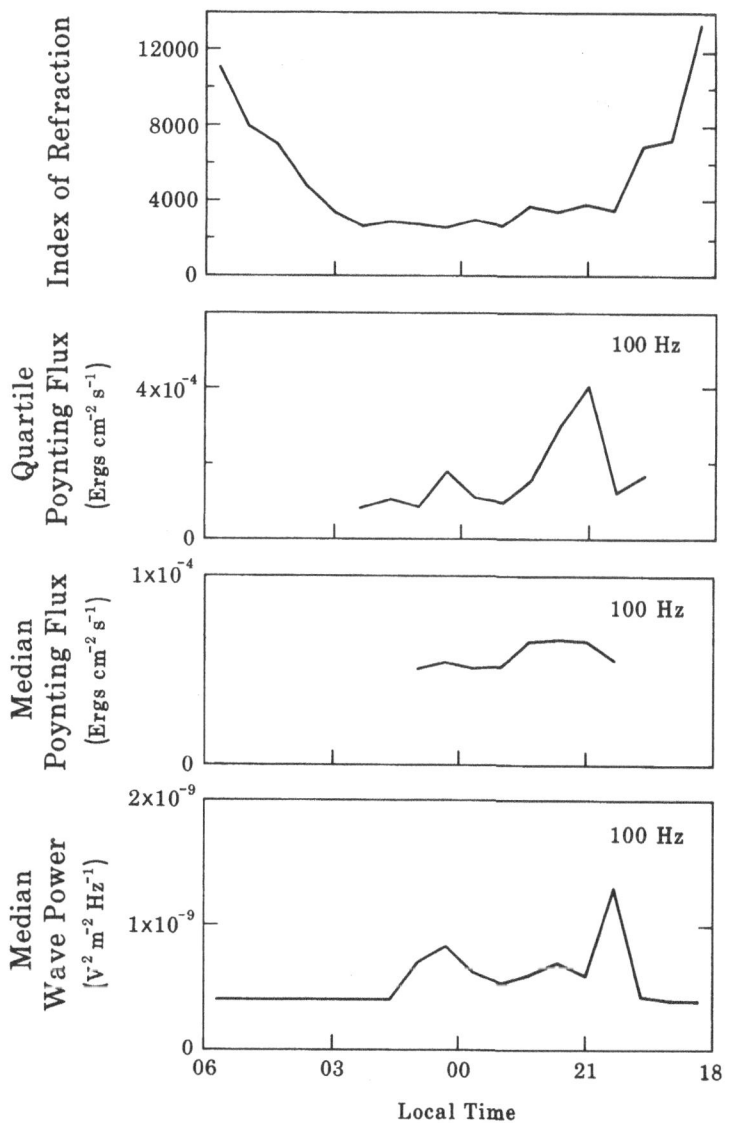

Fig. 27. Local time variation of median index of refraction (*top*), wave power (*bottom*), and quartile and median Poynting flux (*center*) in 10° bins at all altitudes 160 km to 400 km and for all magnetic field strengths. The median Poynting flux is shown only when the median wave power exceeds the instrument threshold (Russell *et al.*, 1989b).

observing season to observing season, partly because of a loss of some of the second observing season as Venus passed behind the Sun as seen from the Earth. However, in the region of overlap there seems to be some coherence from season-to-season. While at first glance this coherence may seem to be evidence for geographic or topographic control, the regions of enhanced occurrence rates are not solely over mountainous

730 Hz Burst Occurrence

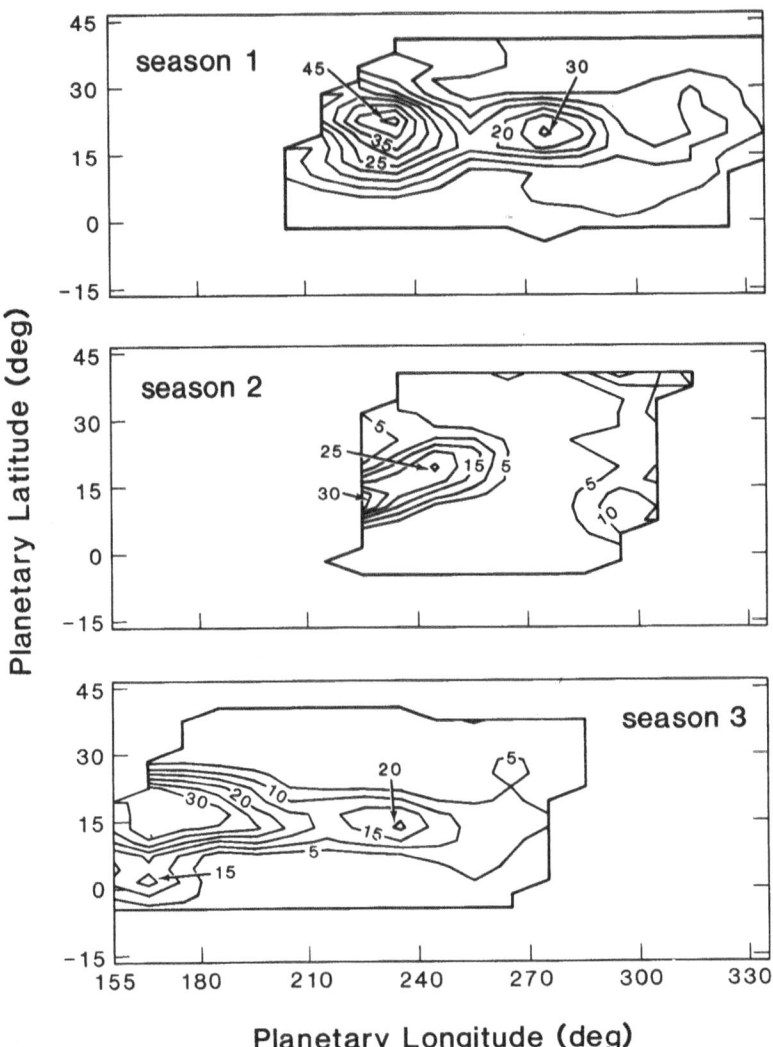

Fig. 28.   Planetary longitude-latitude map of signal occurrence rate at 0.73 kHz for each of the first 3 seasons of Pioneer Venus data (Russell *et al.*, 1988c).

terrain. Thus, it is most likely that the regions of enhanced occurrence represent the times during which the spacecraft was at local times that placed it over the 'source' regions identified earlier. Thus, the primary correlation seems to be with local time and not geography.

Below the electron gyrofrequency, i.e., at 100 Hz, the signals appear to propagate much farther in longitude and hence are not as narrowly confined in either plots of local time or geographic longitude. Thus, a misidentification of local time correlations as geographic correlations seems not to have been the cause for the weak associations between Scarf's 'lightning wave' events and mountainous terrain. The answer seems most probably to be due to the fact that coverage was not uniform planet-wide and that more data were collected over the mountains. There is no convincing evidence for volcanoes as the source of the Venus lightning bursts.

6. Discussion and Conclusions

The mapping of the occurrence rate of bursty VLF emissions in the dark ionosphere of Venus as a function of local time allows us to reconcile some of the often times apparently conflicting evidence concerning Venus lightning. The source region clearly extends into the evening hours to about 22:00 LT. We do not know where in local time the source region starts because this is screened from the sensors on Pioneer Venus by the dayside ionosphere. It is quite possible that the source region like on Earth begins in early afternoon. However, this region has never been probed by either US or USSR spacecraft or balloons. These measurements were all made in the post-midnight and morning through noon sectors. Figure 20 shows the tracks of the VEGA 1 and 2 balloons plotted on an occurrence rate map. The balloon trajectories took them over the region of lowest occurrence rate. Thus it is not surprising that the VEGA optical sensors may have seen no lightning emissions. The original lightning search by Borucki *et al.* (1981) using the Pioneer Venus star sensor was also over this region as indicated on Figure 19. The Venera-9 optical detection, on the other hand, was over the region in which VLF signals occur frequently. It is also now clear that lightning is an intra-cloud phenomenon on Venus and not associated with high mountains or volcanoes. This interpretation is illustrated in Figure 29.

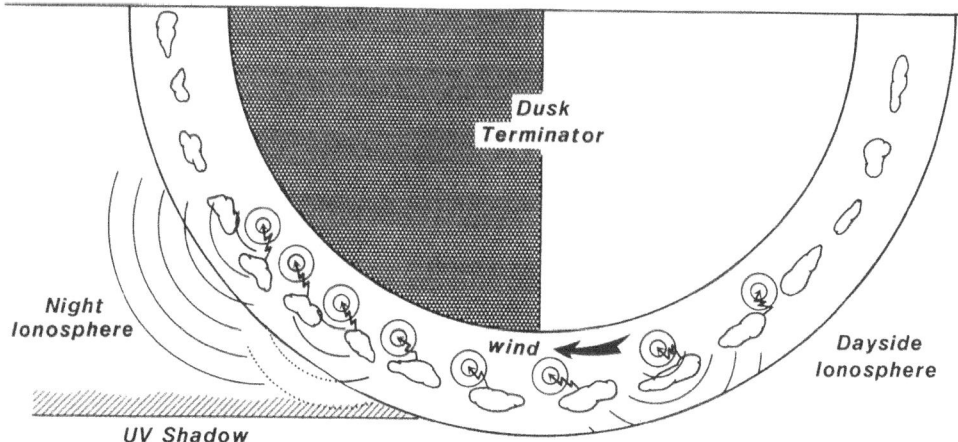

Fig. 29.   Interpretive sketch of the region of occurrence of lightning on Venus (Russell *et al.*, 1990).

It is difficult to compare the Venus lightning rate with that on Earth in part because of the lack of comparable data sets. One controlled experiment was run on the Venera landers in which the lightning instrument was flown in the Earth's atmosphere on a clear day. The rate of detection of signals was similar to the lowest rates seen in the Venus atmosphere. The highest rates observed (i.e., during the Venera-11 landing) were an order of magnitude higher.

A powerful technique to assess the total energy associated with the source of the bursts would be the measurement of the Poynting flux of the waves. This has been attempted with the Pioneer Venus wave instrument but the instrument is limited to the measurement of a single component of the 6 necessary, and many assumptions have to be made about factors such as the direction of propagation which is unmeasured and may be variable. Moreover, the electric field measurement, from which the Poynting flux is deduced, covers a very narrow frequency band. Without spectral information across the full bandwidth over which lightning signals are generated we cannot determine an accurate energy flux for the waves. Nevertheless, we can determine that the observed energy flux at 100 Hz is consistent with the hypothesis that the Venus lightning rate is similar to or greater than that on Earth (Russell *et al.*, 1989b).

We have not addressed the question of stroke or flash rate on Venus. Since terrestrial strokes last from microseconds to milliseconds we probably cannot hope to deduce a stroke duration or rate with the Pioneer Venus plasma wave detector whose rise time is 10's of milliseconds. However, it may be possible to deduce a flash rate where a flash consists of many strokes along the same channel since a flash can last a major fraction of a second. This problem is being addressed but no results are as yet available.

More work needs to be done with the available data on the polarization of the waves. Thus far there is only one published example of the amplitude of the waves as a function of direction to the magnetic field. This study needs to be repeated on many more signals. Furthermore, ray-tracing calculations need to be performed to see how electromagnetic energy will propagate through the rapidly varying density of the lower ionosphere.

While many tests can still be done with the existing data, the data set is limited. The Pioneer Venus sensor is electric whereas a VLF magnetic sensor would have been more appropriate. Also, only one component of the six wave components (electric and magnetic) were measured. A simple wave experiment in low altitude orbit around Venus could provide maps of the wave energy flux over the night ionosphere. However, this would provide only information below the local electron gyrofrequency and only on the nightside. To probe the dayside and to extend our frequency range above the electron gyrofrequency we need measurements in the atmosphere. A balloon mission could give adequate spatial coverage and temporal coverage.

In conclusion, there still seems to be much to do before all the questions concerning Venus lightning are answered. However, we can answer with some assurance the most basic question regarding Venus lightning. It is clear that there is evidence for lightning on Venus which suggests that in many ways it is similar to that on Earth. This evidence points to a source in the Venus clouds. It also seems that the rate of occurrence is comparable to, or possibly greater than, that on the Earth.

Acknowledgements

The author thanks both F. L. Scarf and R. N. Singh for getting him involved in this research. He is grateful to R. J. Strangeway for many helpful discussions regarding the Pioneer Venus instrument, the data and lightning in general. Discussions with W. J. Borucki, D. D. Sentman, and L. V. Ksanfomaliti have also been quite helpful as has correspondence with many, many colleagues. A special thanks goes to M. von Dornum who performed all of the UCLA statistical studies of the VLF signals. This work was supported by the National Aeronautics and Space Administration by research grant NAG2-501.

References

Bliokh, P. V., Nicholaenko, and Fillippov, Yu. F.: 1980, *Schumann Resonances in the Earth-Ionosphere Wave Guide*, P. Peregrinns Ltd.

Borucki, W. J.: 1982, *Icarus* **52**, 354.

Borucki, W. J. and Chameides, W. L.: 1984, *Rev. Geophys. Space Phys.* **22**, 363.

Borucki, W. J. and McKay, C. P.: 1987, *Nature* **328**, 509.

Borucki, W. J., Dyer, J. W., Thomas, G. Z., Jordan, J. C., and Comstock, D. A.: 1981, *Geophys. Res. Letters* **8**, 233.

Chameides, W. L., Walker, J. C. G., and Nagy, A. F.: 1979, *Nature* **280**, 820.

Colin, L. and Hunten, D. M.: 1977, *Space Sci. Rev.* **20**, 451.

Kelley, M. C. *et al.*: 1985, *J. Geophys. Res.* **90**, 9815.

Knollenberg, R. G. and Hunten, D. M.: 1980, *J. Geophys. Res.* **85**, 8039.

Krasnopol'sky, V. A.: 1980, *Kosmich. Issled.* **18**, 429.

Krasnopol'sky, V. A.: 1983a, *Venus*, Univ. of Arizona Press, Tucson, pp. 459–483.

Krasnopol'sky, V. A.: 1983b, *Planetary Space Sci.* **31**, 1363.

Ksanfomaliti, L. V.: 1979, *Kosmich. Issled.* **17**, 747.

Ksanfomaliti, L. V.: 1983, *Kosmich. Issled.* **21**, 279.

Ksanfomaliti, L. V., Scarf, F. L., and Taylor, W. W. L.: 1983, in D. M. Hunten, L. Colin, T. M. Donahue, and V. I. Moroz (eds.), *Venus*, Univ. of Arizona Press, Tucson, pp. 565–603.

Levin, Z., Borucki, W. J., and Toon, O. B.: 1983, *Icarus* **56**, 80.

Luhmann, J. G. *et al.*: 1982, *J. Geophys. Res.* **87**, 9205.

Meinel, A. B. and Hoxie, D. T.: 1962, *Comm. Lunar Planetary Lab.* **1**, 35.

Moore, P.: 1965, *The Planet Venus*, Macmillan, London.

Phillips, J. L. and Russell, C. T.: 1987, *J. Geophys. Res.* **92**, 2253.

Rinnert, R.: 1982, in H. Vollaad (ed.), *Handbook of Atmospherics,* Vol. 2, CRC Press, Bsca Raton, pp. 100–133.

Russell, C. T. and Scarf, F. L.: 1990, *Adv. Space Res.* **10**(5), 125.

Russell, C. T. and Singh, R. N.: 1989, *Geophys. Res. Letters* **16**, 1481.

Russell, C. T., Elphic, R. C., Luhmann, J. G., and Slavin, J. A.: 1980, *11th Proc. Lunar Planetary Sci*, 1897–1906.

Russell, C. T., von Dornum, M., and Scarf, F. L.: 1988a, *J. Geophys. Res.* **93**, 5915–5921.

Russell, C. T., von Dornum, M., and Scarf, F. L.: 1988b, *Planetary Space Sci.* **36**, 1211.

Russell, C. T., von Dornum, M., and Scarf, F. L.: 1988c, *Nature* **331**, 591.

Russell, C. T., von Dornum, M., and Scarf, F. L.: 1989a, *Icarus* **80**, 390.

Russell, C. T., von Dornum, M., and Strangeway, R. J.: 1989b, *Geophys. Res. Letters* **16**, 579.

Russell, C. T., von Dornum, M., and Scarf, F. L.: 1990, *Adv. Space Res.* **10**(5), 37.

Sagdeev, R. Z. *et al.*: 1986, *Science* **231**, 1411.

Scarf, F. L.: 1986, *J. Geophys. Res.* **91**, 4594.

Scarf, F. L. and Russell, C. T.: 1983, *Geophys. Res. Letters* **10**, 1192.

Scarf, F. L. and Russell, C. T.: 1988, *Science* **240**, 222.

Scarf, F. L., Taylor, W. W. L., and Virobik, P. F.: 1980a, *IEEE Trans. Geosci. Remote Sensing* **GE18**, 36.

Scarf, F. L., Taylor, W. W. L., Russell, C. T., and Brace, L. H.: 1980b, *J. Geophys. Res.* **85**, 8158.

Scarf, F. L., Jordan, K. F., and Russell, C. T.: 1987, *J. Geophys. Res.* **92**, 12407.

Signh, R. N. and Russell, C. T.: 1986, *Geophys. Res. Letters* **13**, 1071.

Singh, R. N. and Russell, C. T.: 1987, *Geophys. Res. Letters* **14**, 571.

Singh, R. N., Russell, C. T., and Scarf, F. L.: 1987, *Adv. Space Res.* **7**(12), 285.

Taylor, H. A., Jr. and Cloutier, P. A.: 1986, *Science* **234**, 1087.

Taylor, H. A., Jr. and Cloutier, P. A.: 1987, *Geophys. Res. Letters* **14**, 568.

Taylor, H. A., Jr. and Cloutier, P. A.: 1988, *Geophys. Res. Letters* **15**, 729.

Taylor, H. A., Jr., Cloutier, P. A., and Zheng, Z.: 1987, *J. Geophys. Res.* **92**, 9907.

Taylor, H. A., Jr., Grebowsky, J. M., and Cloutier, P. A.: 1985, *J. Geophys. Res.* **90**, 7415.

Taylor, H. A., Jr., Grebowsky, J. M., and Cloutier, P. A.: 1986, *J. Geophys. Res.* **91**, 4599.

Taylor, W. W. L., Scarf, F. L., Russell, C. T., and Brace, L. H.: 1979, *Nature* **282**, 614.

Uman, M. A.: 1987, *The Lightning Discharge*, Academic Press, Orlando.

Vonnegut, B.: 1982, *Handbook of Atmospherics*, Vol. 1, CRC Press, Boca Raton.

Williams, M. A., Thomason, L. W., and Hunten, D. M.: 1982, *Icarus* **52**, 166.

Williams, M. A., Krider, E. P., and Hunten, D. M.: 1983, *Rev. Geophys. Space Phys.* **21**, 892.

Wilson, C. T. R.: 1929, *J. Franklin Inst.* **208**, 1.

STRUCTURE, LUMINOSITY, AND DYNAMICS OF THE VENUS THERMOSPHERE

J. L. FOX

Institute for Terrestrial and Planetary Atmospheres and Department of Mechanical Engineering, State University of New York at Stony Brook, Stony Brook, NY 11974, U.S.A.

and

S. W. BOUGHER

Lunar and Planetary Laboratory, University of Arizona, Tucson, AZ 85721, U.S.A.

Abstract. We review here observations and models related to the chemical and thermal structures, airglow and auroral emissions and dynamics of the Venus thermosphere, and compare empirical models of the neutral densities based in large part on *in situ* measurements obtained by the Pioneer Venus spacecraft. Observations of the intensities of emissions are important as a diagnostic tool for understanding the chemical and physical processes taking place in the Venus thermosphere. Measurements, ground-based and from rockets, satellites, and spacecraft, and model predictions of atomic, molecular and ionic emissions, are presented and the most important sources are elucidated. Coronas of hot hydrogen and hot oxygen have been observed to surround the terrestrial planets. We discuss the observations of and production mechanisms for the extended exospheres and models for the escape of lighter species from the atmosphere. Over the last decade and a half, models have attempted to explain the unexpectedly cold temperatures in the Venus thermosphere; recently considerable progress has been made, although some controversies remain. We review the history of these models and discuss the heating and cooling mechanisms that are presently considered to be the most important in determining the thermal structure. Finally, we discuss major aspects of the circulation and dynamics of the thermosphere: the sub-solar to anti-solar circulation, superrotation, and turbulent processes.

1. Introduction

CO_2 was first identified as a constituent of the Venus atmosphere by Adams and Dunham (1932), who, in their search for evidence of O_2 and H_2O in ground-based infrared spectra observed these CO_2 bands at 7820, 7883, and 8689 Å. Over the following forty years, spectral resolution in the infrared improved and allowed the abundances of HCl, HF, and CO in the atmosphere to be determined (e.g., Connes *et al.*, 1967, 1968, 1969). A review of this early period in Venus spectroscopy has been presented by Young (1972) and more recently by von Zahn *et al.* (1983). That the atmosphere is composed mostly of CO_2 was suggested by Connes *et al.* (1967), Belton *et al.* (1968), and by Moroz (1968), but confirmation was not obtained until the first *in situ* measurements were performed in 1967 by instruments on the Soviet Venera 4 entry probe (Vinogradov *et al.*, 1968). The experiments were simple: the pressure of a sample of atmospheric gas was determined before and after adsorption by a chemical absorber. A description of the instruments on Venera 4, and on two similar spacecraft that were launched two years later, Veneras 5 and 6, have been presented by Vinogradov (1971). The major constituents and their mixing ratios (used hereafter to mean fraction

Space Science Reviews **55**: 357–489, 1991.

of particles by number) in the lower atmosphere were reported by Kuz'min (1970, 1971) as $95 \pm 2\%$ CO_2, $2-5\%$ N_2, $0.1-1\%$ H_2O and less than 0.4% O_2.

Up to the homopause, the composition of the thermosphere of a planet is nearly the same as that of the bulk atmosphere, at least with respect to the major constituents that are not affected by photochemistry; above the homopause photolysis of the molecular constituents and diffusive separation in the gravitational field cause lighter constituents to become increasingly abundant. Exploration of the thermosphere of Venus began in October 1967 when two spacecraft, the U.S. flyby Mariner 5 and the Soviet Venera 4 carried ultraviolet photometers to Venus. Both instruments were designed to detect resonance emission features of H at 1216 Å and of O at 1304 Å, from which column densities and altitude profiles of the H and O densities could be derived. No signal was reported at 1304 Å by either the Mariner 5 photometer (Barth *et al.*, 1967) or the Venera 4 instrument, which observed the atmosphere in darkness below about 350 km (Kurt *et al.*, 1968). Both detected strong Lα emission. The Venera 4 intensities were used to derive a nightside density profile of H atoms from 1 to about 4 Venus radii. From the scale height of the H densities and the upper limit on the O densities Kurt *et al.* (1968) deduced, more or less correctly, that 'the Venus nightside atmosphere is cold . . . and undergoes a sharp transition to interplanetary space'. The Mariner 5 Lα channel detected emission that was characterized by an abrupt increase in scale height as the minimum distance of the photometer line-of-sight to the center of the planet exceeded about 9000 km (Barth *et al.*, 1967; Barth, 1968). From the scale height, the ratio of the temperature to the mass of the species responsible for the of the inner component was derived to be about 350 K amu^{-1}. Exospheric temperatures as low as 350 K were considered unlikely, based partly on model calculations (McElroy, 1969) that were reinforced by prejudices arising from the analogy with Earth, where the solar fluxes are a factor of two lower and the average exospheric temperature is about 1000 K. Therefore, species heavier than H, including D and H_2, were considered also as the source of the low altitude component, the latter assumed to produce emission from photo-dissociative excitation rather than resonance scattering (e.g., Barth, 1968; Donahue, 1968; Wallace, 1969; McElroy and Hunten, 1969). Later work showed that the actual exospheric temperature was even lower than the 'unlikely' value of 350 K (Anderson, 1975). The measurement of the outer, larger scale height component was the first detection of the hot hydrogen corona, although it was not immediately identified as such.

That the thermospheric abundance of O was low was suggested by the non-detection of 1304 Å emission and by profiles of the electron density in the daytime ionosphere that were obtained from the radio occultation experiment on Mariner 5 (Kliore *et al.*, 1969). The electron density profile, shown in Figure 1, exhibited a single maximum that was interpreted as an F_1 peak, a Chapman layer consisting of a molecular ion (initially believed to be CO_2^+) produced by photoionization and photoelectron impact ionization and destroyed by dissociative recombination. The absence of a distinguishable O^+ (F_2) peak above the F_1 layer was interpreted as implying that the mixing ratio of atomic O in the Venus thermosphere was small (Stewart, 1968; McElroy, 1969; Kumar and Hunten, 1974).

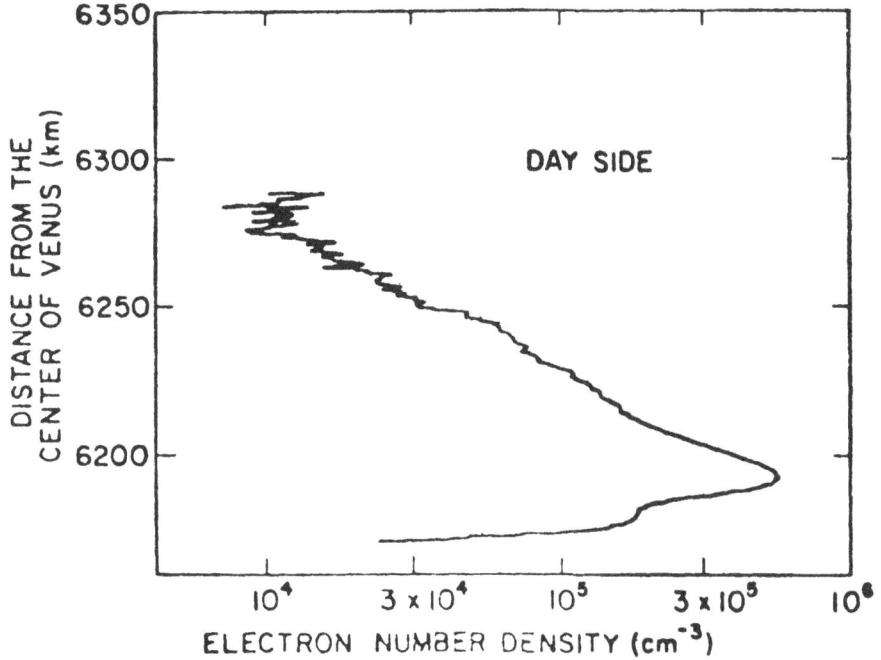

Fig. 1.   Electron density profile as measured by the radio occultation experiment on Mariner 5. Taken from Kliore *et al.* (1969).

Mariner 10 flew by Venus in February 1974, carrying an objective grating spectrometer with channel electron multipliers at the wavelengths of 9 atomic and molecular emission features between 200 and 1700 Å, two zero-order channels and a background channel at 430 Å (Broadfoot *et al.*, 1977). The spectral features that the Mariner 10 instrument was designed to detect are listed in Table I. Strong emissions were detected in the positions of the helium, hydrogen, oxygen, CO, and carbon features in the spectrum (Broadfoot *et al.*, 1974).

In October 1975, the Soviet spacecraft Veneras 9 and 10 were inserted into orbit about Venus. The spacecraft were equipped with hydrogen and deuterium absorption cells in front of an NO counter to study the hydrogen corona, and visible spectrometers to record nightglow emission features in the 3000 to 8000 Å wavelength range (Keldysh, 1977; Krasnopol'sky, 1983). Four band systems of O_2 were detected in the Venus nightglow with maximum volume emission rates appearing near 100 km. Upper limits were placed on other species and processes from the absence of emission features at 5577 Å (the atomic oxygen green line), 5893 Å (the sodium D-line doublet) and 3914 Å (the (0, 0) N_2^+ first negative band) (Krasnopol'sky, 1978, 1979). Unfortunately, the sensitivity of the detectors decreased by a factor of 5 after the first month of operations, so the number of orbits yielding useful data was limited.

In December 1978 ten separate space probes arrived at Venus, including the Pioneer Venus (PV) Orbiter and Multiprobe, which included the Bus and four probes that were

TABLE I

Features detected by the photometers on Mariner 10 and Veneras 11 and 12

Species	Wavelength (Å)	Mariner 10	Venera 11	Venera 12
He^+	304	no	yes	yes
He	584	yes	yes	yes
Ne	740	no	no	no
O^+	834	n/a[a]	yes	n/a
Ar	869	no	n/a	no
Ar	1048	no	no	no
H	1216	yes	yes	yes
O	1304	yes	yes	yes
C	1657	yes	yes	yes
CO	1480–1500	yes	yes	yes

[a] No detector at that wavelength.

released prior to entry into the Venus atmosphere, and the Soviet Veneras 11 and 12 flyby spacecraft and landers. The composition of the atmosphere below the clouds was measured by a mass spectrometer (LNMS) (e.g., Hoffman *et al.*, 1980a) and gas chromatograph (LGC) (e.g., Oyama *et al.*, 1980) on the Pioneer Venus sounder probe, neutral mass spectrometers on the Venera 11 and 12 landers (e.g., Istomin *et al.*, 1979; Grechnev *et al.*, 1980) and a gas chromatograph on the Venera 12 lander (Gel'man *et al.*, 1979). A comparison of the results of these measurements has been presented by Hoffman *et al.* (1980b). CO_2 and N_2 are found to make up 99.9% of the atmosphere. The range of values of the N_2 mixing ratio was 2.5 to 4.5%. Von Zahn *et al.* (1983) have reviewed the available measurements of N_2 in the bulk atmosphere and recommend an abundance of $3.5 \pm 0.8\%$; that value has also been adopted by the Venus International Reference Atmosphere (von Zahn and Moroz, 1986) (see Section 2.2).

The Venera 11 and 12 flybys carried UV spectrophotometers with detectors at 9 wavelengths with bandpasses about 12 Å wide and a zero-order channel that provided an integral of the intensity over the wavelength region from 300 to 1700 Å. The wavelengths of and species responsible for the emission features that the instruments were designed to detect are also listed in Table I. The Venera instruments were similar to, but more sensitive than, the Mariner 10 instrument (Bertaux *et al.*, 1981; Krasnopol'sky, 1986a). Unlike the Mariner 10 instrument, however, during the flyby the slits were not oriented parallel to the limb and consequently the spatial resolution was insufficient to derive altitude profiles of the intensities; except for Lα and the He 584 Å emission, only surface brightnesses were reported (Kurt *et al.*, 1979, 1983; Bertaux *et al.*, 1981). The resonance emissions of Ar and Ne were not detected, but the first observations of the O^+ and He^+ emission features at 834 and 304 Å, respectively, were reported.

The Pioneer Venus bus and orbiter carried a number of instruments that have provided information about the structure and composition of the thermosphere. A magnetic sector neutral mass spectrometer (BNMS) aboard the bus measured neutral densities from about 650 km to below the homopause, near 128 km (e.g., von Zahn,

1977; von Zahn *et al.*, 1980). The orbiter payload included a quadrupole neutral mass spectrometer (ONMS) to measure thermospheric densities of CO_2, O, CO, N_2, N, and He (e.g., Niemann *et al.*, 1979, 1980a, b), and an atmospheric drag experiment (OAD) to measure the total mass density profile (e.g., Keating, 1977; Keating *et al.*, 1980). The orbiter ultraviolet spectrometer (OUVS) was designed to measure intensities of airglow emission features of neutral atmospheric atoms, molecules, and ions. A complete description of the OUVS has been presented by Stewart (1980). Briefly, the instrument consists of 250-mm $f/5$ Cassegrainian telescope, an Ebert–Fastie monochromator of 125 mm focal length, and two photomultipliers, sensitive in the ranges 1100 to 1800 Å and 1600 to 3300 Å. The OUVS instrument can be commanded to scan the spectrum, or the wavelength may be fixed. Images of the planet at a fixed wavelength may be obtained from successive scans of the disk that are made as the spinning spacecraft approaches the planet from apoapsis. When the spacecraft is near periapsis altitude profiles of the emissions may be obtained. Among the features that the OUVS has detected are the γ and δ bands in the NO nightglow, the 1304, 1356, and 2972 Å features of O, H Lα, the 1561 and 1657 Å features of C, and the CO fourth positive bands in the dayglow, and the 1304 and 1356 Å auroral emission features.

We present here a brief history of investigations into and our current understanding of the structure, emission features, and dynamics of the Venus thermosphere. In Section 2, we present a description of the basic structure of the thermosphere, including the exospheric temperature and chemical composition, mostly as determined by Pioneer Venus *in situ* measurements. We compare empirical models of the major thermospheric constituents. In Section 3, we review the knowledge that has been gained by remote sensing of airglow and aurora, both Earth-based and from American and Soviet spacecraft. Measurements and models of emission features of the species, O, CO, N_2, O_2, C, NO, and He and their ions are described. Discussion of emission features of atomic hydrogen is deferred until Section 4, where we describe the hot atom coronas and atmospheric escape. Section 5 contains a detailed discussion of the thermal structure and heating and cooling mechanisms, mostly as given by models. The last part of the paper, Section 6, describes the circulation and dynamics of the thermosphere.

2. Basic Thermospheric Structure

2.1. EXOSPHERIC TEMPERATURE

The prejudice that Venus exospheric temperatures, T_∞, should not be too much smaller than terrestrial values, persisted until about 15 years ago, in spite of early indications that the temperatures were lower. The scale height of the H profile derived from Venera 4 Lα intensities close to the planet implied temperatures on the Venus nightside near 100 K, but a specific value was not presented by Kurt *et al.* (1968), who merely remarked that the nightside atmosphere was cold. Stewart (1968) first suggested the interpretation that the two-scale height Lα profile measured by Mariner 5 was H atoms at two different temperatures, which would imply that the temperature of the inner component on the dayside was about 350 K. Over the next few years, further measurements and models ruled out H_2 and D at higher temperatures as sources of the emissions (see Section 4 for further details). The higher temperatures, near 700 K, computed by McElroy (1969) could be accommodated only if CO_2^+ was considered to be the major ion. Fehsenfeld *et al.* (1970) showed that in the presence of a small admixture of O, CO_2^+ is converted rapidly to O_2^+ through the reaction

$$CO_2^+ + O \rightarrow O_2^+ + CO . \tag{1}$$

Kumar and Hunten (1974) showed that a model ionosphere with an exospheric temperature of 350 K and O_2^+ as the major ion fit the Mariner 5 electron density profiles. Mariner 10 measurements of the He corona gave a T_∞ of about 375 $\pm$ 105 K (Kumar and Broadfoot, 1975). Anderson (1976) reanalyzed the Mariner 5 Lα data and reported for the inner component a dayside temperature of 275 K and a nightside temperature of 150 K. These low values were later confirmed by the Pioneer Venus *in situ* measurements. A value of T_∞ of 275 $\pm$ 15 K was derived from the He densities measured by the Bus Neutral Mass Spectrometer at 08 : 30 local time, 37.9° S latitude (von Zahn *et al.*, 1980). Keating *et al.* (1980) constructed an empirical model from the OAD data obtained in the early PV orbits; dayside values of T_∞ were about 280–300 K. Airglow emission profiles measured by the OUVS were consistent with an exospheric temperature of about 275 K (Stewart *et al.*, 1979). Hedin *et al.* (1983) derived a global average temperature of 228 K from ONMS data taken between December 1978 and August 1980 for a narrow latitude range of a few degrees around 16° N. Variations with solar activity during this limited time period were observed to be only 10% of the change seen in the terrestrial thermosphere. The variation in the exospheric temperatures with solar flux in the model of Hedin *et al.* (1983) is 0.14 K per unit $F_{10.7}$, the 10.7 cm solar flux in units of 10^{-22} W m^{-2} Hz^{-1} at 1 AU.

The temperature at the lower boundary of the thermosphere, near 100 km, takes on values near 175 K and, unlike the exospheric temperature, varies little with solar zenith angle (Seiff *et al.*, 1980). The thermospheric temperature structure will be discussed in detail in Section 5.

2.2. MODELS OF THE VENUS THERMOSPHERE

Information about the composition and structure of the thermosphere of Venus, obtained largely from *in situ* measurements, has been summarized in several empirical models. Von Zahn *et al.* (1980) presented a model of the morningside atmosphere (hereafter the BNMS model) that is based mostly on measurements of densities of CO_2, He, N_2, and CO made by the BNMS. The 10.7 cm solar flux index ($F_{10.7}$) was about 189 at the time of the bus entry. The densities of O in the model were derived from the ratio O/CO_2 of 3 reported at 167 km by Niemann *et al.* (1980b) from ONMS measurements. The mass densities measured by the BNMS agree fairly well with those measured by the OAD and merge smoothly near 130 km with densities measured by the accelerometers on the entry probes (Seiff *et al.*, 1980). Altitude profiles of the major constituents in the model are shown in Figure 2. The exospheric temperature is 275 K and the temperature at 100 km, the lower boundary of the model, is 178 K.

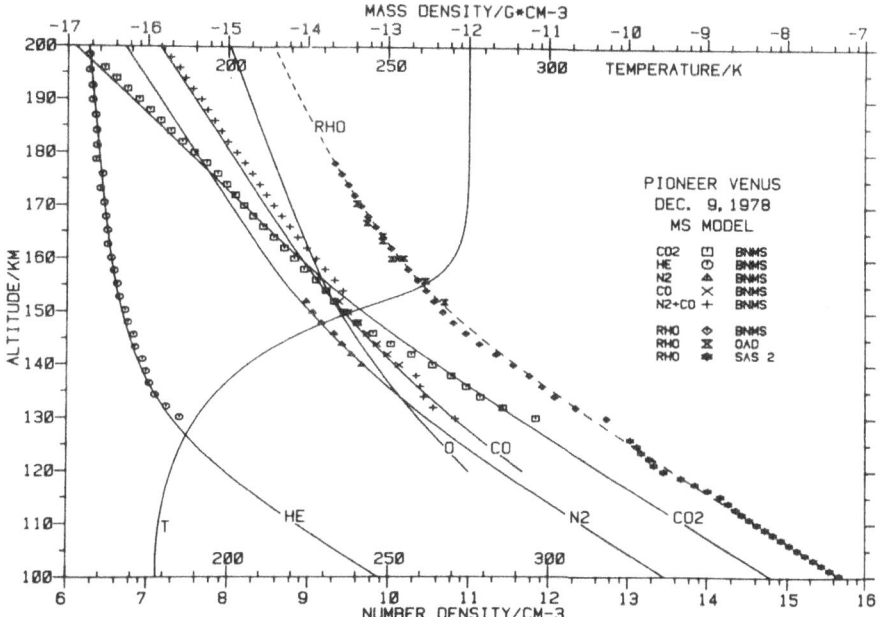

Fig. 2.   Number densities of He, N_2, CO, N_2 + CO, and CO_2 measured by the BNMS versus altitude, interpolated to integer altitudes. Also shown are mass densities measured by the large probe accelerometer from 100 to 126 km, those derived from the BNMS between 130 and 180 km, and those from orbiter drag data (OAD) between 150 and 170 km. The lines represent number densities, mass density and temperature from the BNMS model. Taken from von Zahn *et al.* (1980).

The density profiles of He, CO, and N_2 were used by von Zahn *et al.* (1980) to derive an expression for the eddy diffusion coefficient K (in units of $cm^2\, s^{-1}$):

$$K = \frac{A}{n^{1/2}},\qquad\qquad (2)$$

where n is the total number density in cm^{-3}. The constant A was determined to be within a factor of 2 of 1.4×10^{13}. The molecular diffusion coefficient D has a steeper altitude dependence, since it is proportional to n^{-1}. K and D are equal at the homopause, which for N_2 is near 136 km. The molecular diffusion coefficient for He is larger than that of N_2 and the He homopause is located lower, near 130 km. Extrapolating the mixing ratios to the lower atmosphere, the composition of the lower atmosphere is derived to be about 95.5% CO_2, 4.5% N_2, and 12 ppm He. The mixing ratio of He is very sensitive to the uncertainty in the eddy diffusion coefficient and is good to only about a factor of three. The probably uncertainty in the N_2 mixing ratio is about 25%.

Hedin *et al.* (1983) presented a global empirical model of the Venus thermosphere (hereafter referred to as the VTS3 model) based on measurements of CO_2, O, CO, N_2, He, and N made by the Pioneer Venus ONMS. The VTS3 model incorporates variations in the densities with latitude, local time and solar activity based on three diurnal cycles of observations. The altitude range of the orbiter measurements is from about 142 to 250 km, but data from the entry probe accelerometers (Seiff *et al.*, 1980), the BNMS (von Zahn *et al.*, 1980), and theoretical considerations were used to extend the model to 100 km. Dissociation of CO_2 is the primary source of O and CO in the lower thermosphere; the production is balanced by three-body recombination

$$O + O + CO_2 \rightarrow O_2 + CO_2 \tag{3}$$

below 110 km (Massie *et al.*, 1983). Thus the mixing ratio of CO should be equal, to a first approximation, to the sum of the mixing ratio of O plus twice that of O_2. In the VTS3 model, the mixing ratios of O and CO are assumed to be equal at 100 km, since photochemical considerations indicate that the mixing ratio of O_2 is small (Yung and DeMore, 1982). Data from the PV sounder probes, shown in Figure 3 compared to the ONMS model densities, indicate that latitude and local time variations of the mass densities near 100 km are small (Seiff *et al.*, 1980), so pressure gradients at the lower boundary in the model are constrained to be minimal. The mass densities measured by the ONMS are about 60% less than those given by the simpler PV OAD, the accelerometers on the probes and the BNMS. Ionospheric models also indicate that the ONMS densities are low (e.g., Cravens *et al.*, 1981). It has been suggested that the gas flow into the ONMS ion source at satellite velocities may deviate from that assumed in the data reduction, but efforts to quantify the effect experimentally have been unsuccessful (Hedin *et al.*, 1983). The model densities presented by Hedin *et al.* (1983) have, however, been normalized by a factor of 1.63 to bring them into agreement with the drag data. Since the nightside temperatures are much lower than the dayside, the midnight densities above 100 km are less than the noon densities, and above the homopause, diffusive separation is more effective, causing the nighttime mixing ratios of lighter constituents at a given altitude to be larger. For example, in the model for $F_{10.7} = 200$ at the equator, the mixing ratio of O at 135 km is 7.3% at noon and 17% at midnight; atomic oxygen becomes the dominant component above 155 km at noon and above 140 km at midnight. The noon and midnight models are shown in Figure 4. The variability of the neutral densities is found to be larger on the nightside than on the dayside: the standard deviations are 15–23% on the dayside and 50–60% on the nightside (Hedin *et al.*, 1983; Keating *et al.*, 1986). Although the VTS3 model is based on observations near high solar activity, a full range of solar activity variations has been incorporated into the model using the 10.7 cm flux at 1 AU ($F_{10.7}$) as a parameter. Thus, the low solar activity models should be considered somewhat speculative. It is interesting to note, however, that Kim *et al.* (1989) found that the solar cycle minimum electron density profiles measured by the PVO radio occultation experiment (ORO) are well reproduced by photochemical equilibrium models that employ the VTS3 model, but efforts to reproduce the solar cycle maximum ORO profiles suggest that the VTS3 densities are too low below 135–140 km.

The BNMS model can be compared to the model for conditions appropriate to the PV Bus entry: 39°, $F_{10.7} = 189$ and 08:30 local time. The CO_2 densities at 100 km in

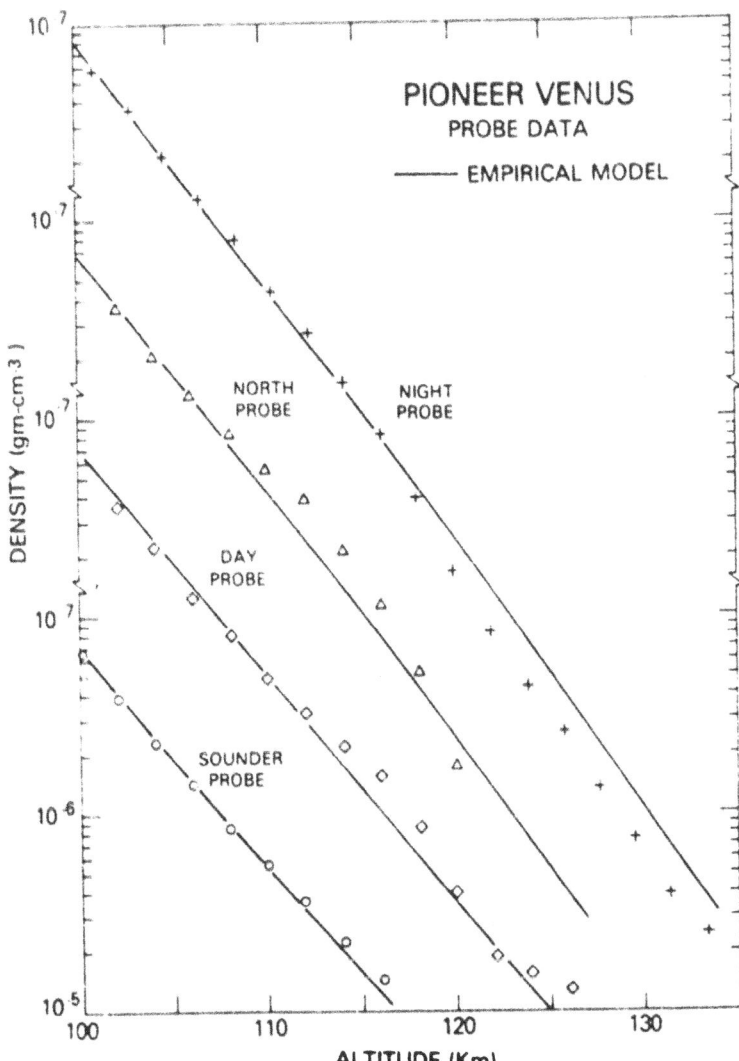

Fig. 3.   Total mass densities from the sounder probes as a function of altitude. Corresponding density profiles from the VTS3 empirical model are shown as solid lines. Taken from Hedin *et al.* (1983).

the BNMS model are smaller by about 15% compared to the VTS3 model, but the thermospheric temperature are higher below and lower above 145 km. The CO_2 densities at about 160 km are 15% larger in the BNMS model. The mixing ratio of O at 135 km is 7.9% in the VTS3 model and 7.0% in the BNMS model, the near equality indicating only that the BNMS values are based on ONMS data. The mixing ratios of N_2 at 135 km are similar: 6.3 and 8.3% for the BNMS and VTS3 models, respectively, but the CO mixing ratios differ by nearly a factor of three: 14.7% (BNMS) and 5.7% (VTS3). The He densities in the two models are equal to within 10%, but the mixing ratio is smaller by about 30% in the BNMS model.

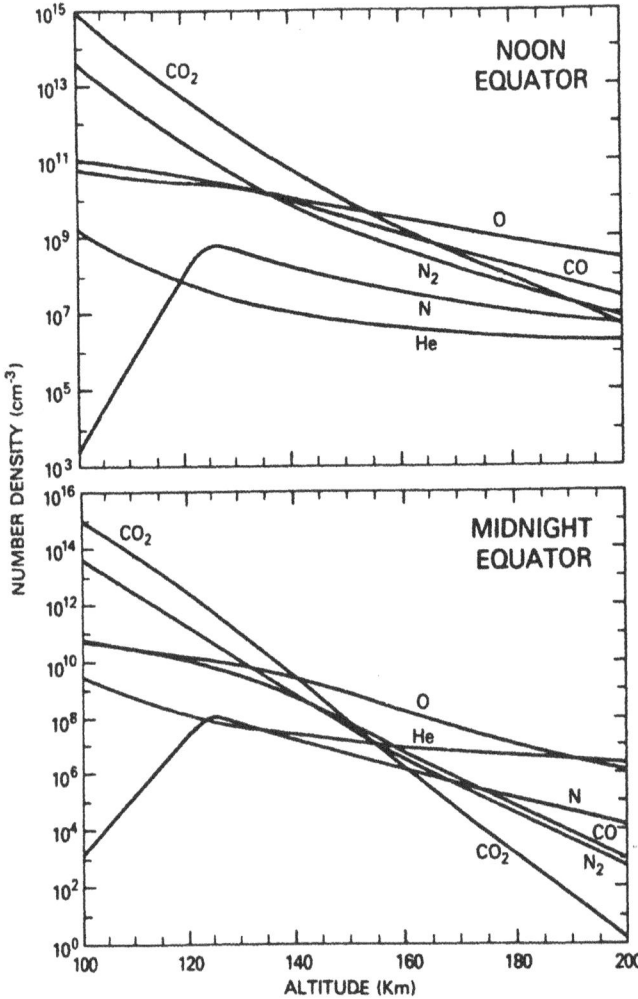

Fig. 4. VTS3 model densities as a function of altitude at noon and midnight at 0° latitude. Taken from
Hedin *et al.* (1983).

Massie *et al.* (1983) constructed daytime (08 : 00 and 16 : 00 hr local time) and midnight models of the densities of CO_2, O, CO, N_2, He, and O_2 in Venus thermosphere
over the altitude range 100 to 180 km by solving the one-dimensional continuity equation
constrained by the PV *in situ* measured neutral densities. No morning-evening asymmetry was included, although a bulge is observed in the light constituents near the
morning terminator (see Section 4). The downward fluxes of N and O near midnight
were inferred from the observed NO and O_2 nightglow intensities (see Section 3), and
the nightside density profiles of N and O were constrained to fit the measurements.
Massie *et al.* preferred the BNMS measurements of CO_2 and N_2, which were about 1.8

and 2.4 times the non-normalized ONMS values, and the ONMS measurements of O and CO. Alternative values of O and CO that were 1.6 times the ONMS values were also given. The densities of O_2 in their dayside model are computed from the continuity equation assuming that the major production reaction is three-body recombination (Equation (3)) and loss is by photodissociation and three-body recombination of O and O_2:

$$O + O_2 + M \rightarrow O_3 + M .$$ (4)

At the lower boundary, the mixing ratio of O_2 was set equal to 1×10^{-3}, that computed by Yung and DeMore (1982) from mesospheric chemistry. The dayside and nightside models of Massie et al. with non-normalized values for O and CO densities are shown in Figures 5(a) and 5(b).

A comprehensive model of thermospheric composition, part of the Venus International Reference Atmosphere (VIRA), has been constructed by Keating et al. (1986) based on information available through 1985. The thermosphere is divided into two regions, the lower region includes altitudes from 100 to 150 km and the upper region

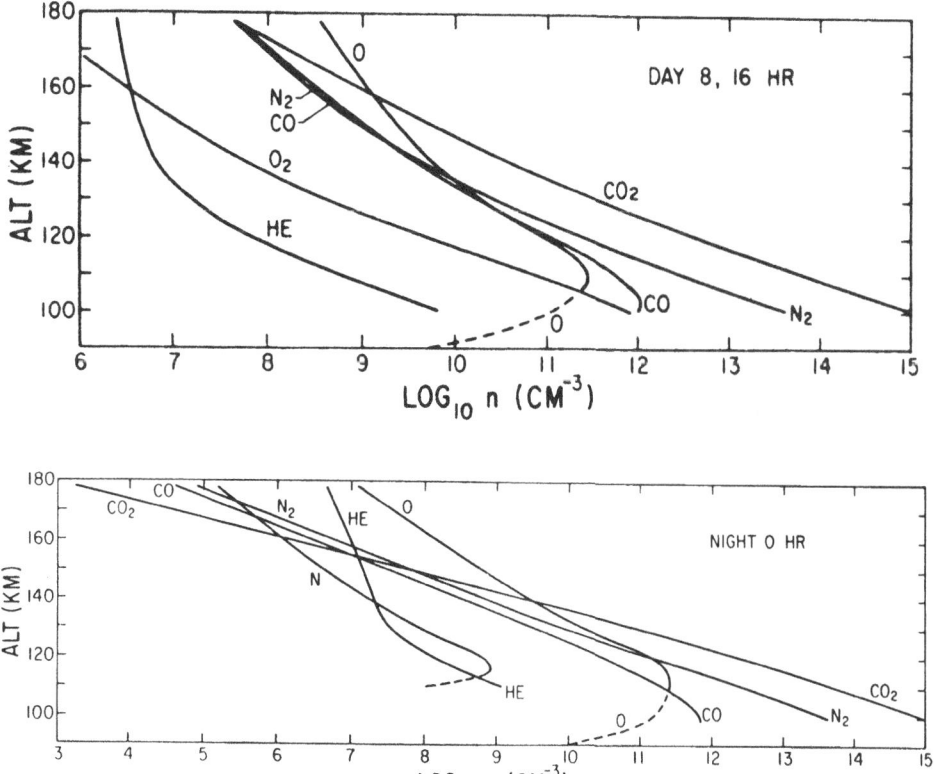

Fig. 5.   (a) Neutral number densities from the dayside (08:00 and 16:00 hours local time) model of Massie et al. (1983). (b) Neutral densities from the midnight model of Massie et al. (1983).

is that above 150 km. The models for noon and midnight are shown in Figures 6(a–d). Diffusive equilibrium is assumed for all species in the upper region; the composition is based on the measurements of the ONMS, normalized to the OAD data. The normalization factors used in the VIRA models, 1.83 for CO_2 and 1.58 for O, are slightly

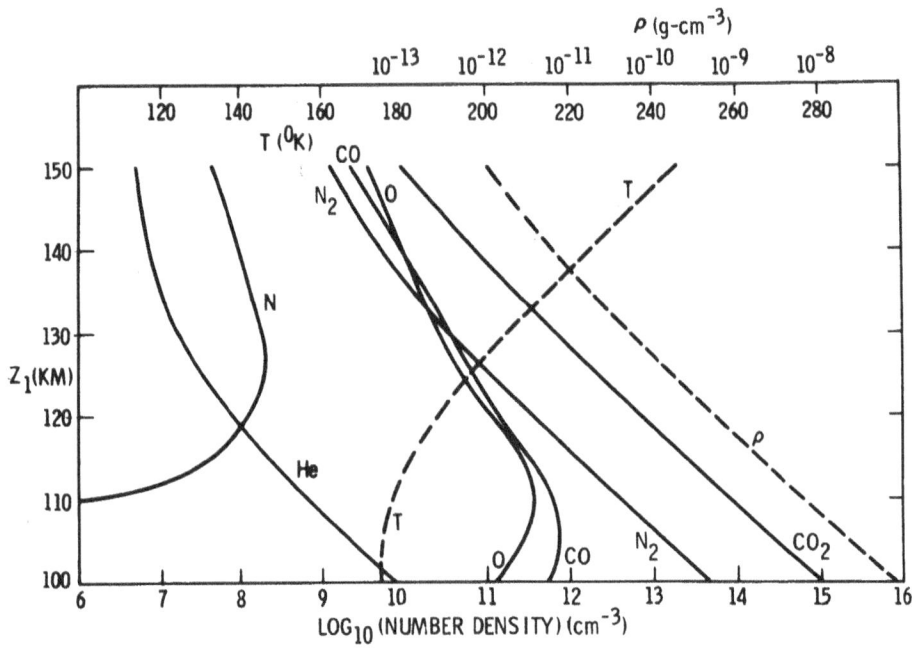

Fig. 6a.

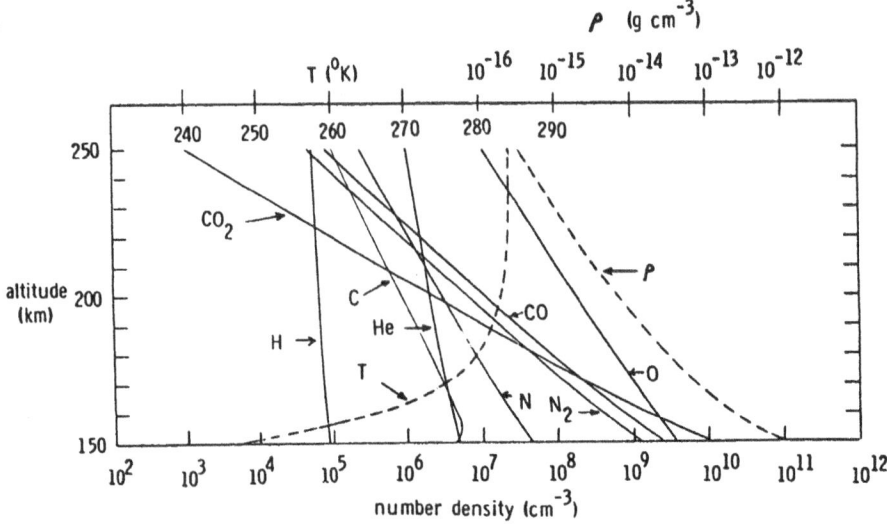

Fig. 6b.

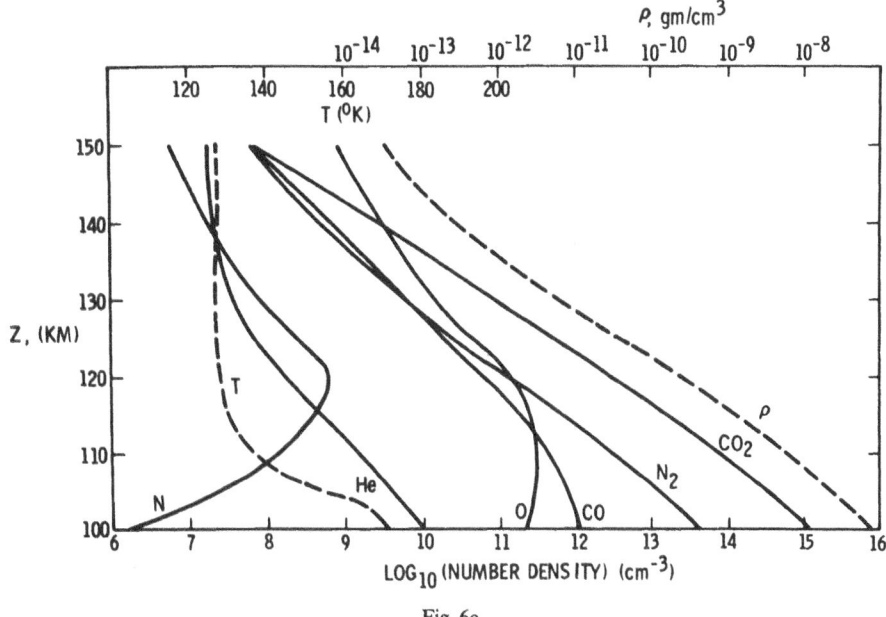

Fig. 6c.

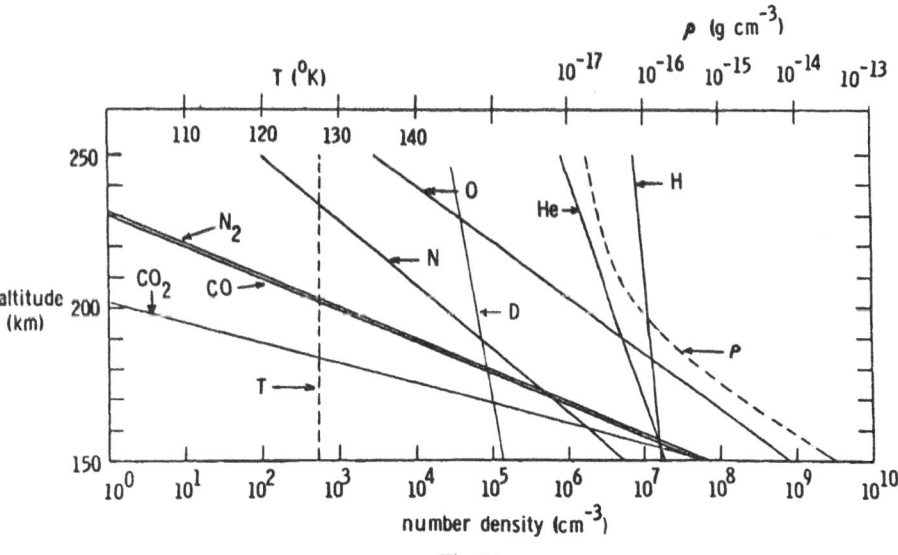

Fig. 6d.

Fig. 6. Neutral number densities as a function of altitude (z) from the VIRA model. (a) Dayside 100 to 150 km. (b) Dayside 150 to 250 km. (c) Nightside 100 to 150 km. (d) Nightside 150 to 250 km. Taken from Keating *et al.* (1986).

different from the factor 1.63 used for all species in the VTS3 model. Normalization factor for CO and N_2 could not be derived and were arbitrarily assumed to be the same as for O.

The VIRA models nominally apply to moderately high solar activity $F_{10.7} = 150$, but Keating *et al.* (1986) give an empirical expression for the change in the temperature and in the log of the number densities of CO_2, N_2, CO, O, He, and N as a function of $F_{10.7}$ and the 81 day mean value of $F_{10.7}$. The expressions are based on the VTS3 model and on the drag data, but they are simpler than those used in the VTS3 model. The use of the 10.7 cm solar flux as an index of ultraviolet fluxes and thermospheric densities and emissions is widespread, but not firmly grounded. For example, Tobiska *et al.* (1988) show that upper thermospheric heating in the terrestrial atmosphere is dominated by absorption of chromospheric emissions, whereas $F_{10.7}$ correlates better with transition region and cool coronal emissions.

The VIRA model can be compared to the VTS3 model appropriate to $16°$ N latitude and $F_{10.7} = 150$. The total number densities at noon and 135 km are comparable, but the O and CO mixing ratios are about 5% in the VTS3 model and somewhat larger, about 6% in the VIRA model. The VTS3 model for high solar activity ($F_{10.7} = 200$), however, has larger mixing ratios of about 7% for O and CO at 135 km. Possibly the solar activity variations in the O abundance assumed in the VTS3 model are larger than those implied by the VIRA model, although there have been no *in situ* measurements at low solar activity. In the VIRA model, the O density profile exhibits a maximum near 110 km, as the photochemical model of Massie *et al.* (1983) shows, whereas the O densities continue to increase with decreasing altitude in the VTS3 model. *In situ* measurements of O are not available below the altitude of minimum periapsis of the PV Orbiter, near 140 km, and the theoretical profiles of O are preferred below 140 km. At midnight, the mixing ratios of O and CO at 135 km are 26 and 8%, respectively, in the VIRA model compared to 17 and 5%, respectively, in the VTS3 model. Both models give N_2 mixing ratios near 100 km of about 4%, but because of different assumptions about the temperatures and eddy diffusion coefficients, at 135 km the noon mixing ratio of N_2 is 7.3% in the VTS3 model and 5.3% in the VIRA model; at midnight, the N_2 mixing ratio is 9.8% in the VTS3 model and 8.6% in the VIRA model.

3. Emission Features

3.1. INTRODUCTION

Atmospheric emission features are usually categorized as dayglow, nightglow, or aurora. Dayglow emissions are produced by the direct interaction of solar radiation or photo-electrons with atmospheric gases, and by prompt chemiluminescent reactions. Night-glow is luminosity produced by chemiluminescent reactions of species that are produced during the day or transported from the dayside. Auroral emissions are usually considered to be those that are produced by impact of particles other than photoelectrons. Although Venus does not have an intrinsic magnetic field, ultraviolet emission features attributed to particle impact have been detected on the nightside of the planet in images from the PV orbiter ultraviolet spectrometer (Phillips *et al.*, 1986; see Section 3.2).

The sources of dayglow emission features include photoionization and excitation

$$X + h\nu \rightarrow X^+{}^* + e , \tag{5}$$

electron impact excitation

$$X + e \rightarrow X^* + e , \tag{6}$$

and electron impact ionization and excitation

$$X + e \rightarrow X^+{}^* + 2e . \tag{7}$$

In the foregoing expressions, X is either an atomic or molecular species, and the asterisk (*) represents an excited state of the species that can decay to a lower state with the emission of radiation. In addition, dayglow emission features attributable to fragments or fragment ions of a molecular species AB may be produced by photodissociative excitation

$$AB + h\nu \rightarrow A + B^* , \tag{8}$$

electron impact dissociative excitation

$$AB + e \rightarrow A + B^* , \tag{9}$$

photodissociative ionization and excitation

$$AB + h\nu \rightarrow A + B^+{}^* + e \tag{10}$$

and electron impact dissociative ionization and excitation

$$AB + e \rightarrow A + B^+{}^* + 2e . \tag{11}$$

Excited species that may emit to the ground state by a dipole-allowed transition can be produced by resonance scattering of solar photons by an atomic species X

$$X + h\nu \rightarrow X^* \rightarrow X + h\nu \tag{12}$$

and fluorescent scattering by a molecular species AB

$$AB(v) + h\nu \rightarrow AB^*(v') \rightarrow AB(v'') + h\nu' . \tag{13}$$

The probability of an atom resonantly scattering a solar photon per unit time is called the g factor, which is related to the oscillator strength f by the expression

$$g = (\pi F) \, \frac{\pi e^2}{mc^2} \, \lambda^2 f, \tag{14}$$

where πF is the photon flux in units of photons $cm^{-2} s^{-1} \mathring{A}^{-1}$ and λ is the wavelength. In these units, the constant $\pi e^2/mc^2$ has the value 8.829×10^{-13} cm. For fluorescent scattering in a molecular band system, it is convenient to define an excitation rate $q_{v'}$ that is the probability per unit time of producing the excited state in the vibrational level

v' by absorption of a photon

$$q_{v'} = (\pi F)\lambda^2 \, \frac{\pi e^2}{mc^2} \, f_{0,\,v'} \tag{15a}$$

or

$$q_{v'} = (\pi F) \, \frac{\omega'}{\omega''} \, \frac{\lambda^4}{8\pi c} \, A_{v'0}\,, \tag{15b}$$

where ω' and ω'' are the statistical weights of the upper and lower states and $A_{v'0}$ is the absolute transition probability of the transition from vibrational level v' of the upper state to $v = 0$ of the ground state. The g factor for production of emission in the (v', v'') band is then given by

$$g_{v'v''} = q_{v'} \, \frac{A_{v'v''}}{\sum\limits_{v} A_{v'v}}\,. \tag{16}$$

A pedagogical discussion of atmospheric emissions and their measurement can be found in Barth (1969).

3.2. ATOMIC OXYGEN EMISSIONS

The atomic oxygen emissions at 1304 and 1356 Å arise from the transitions

$$O(^3S^0) \rightarrow O(^3P) + h\nu(1302, 1304, 1306 \text{ Å}) \tag{17}$$

and

$$O(^5S^0) \rightarrow O(^3P) + h\nu(1356, 1358 \text{ Å})\,, \tag{18}$$

respectively. The $O(^3S^0)$ state is connected to the ground state by a dipole-allowed transition with an oscillator strength of 0.048 (Doering *et al.*, 1985) so resonance scattering of solar radiation is an important source of the emission in the dayglow. Other sources include electron impact excitation of O, and electron-impact and photo-dissociative excitation of CO and CO_2. The sources of the 1356 Å emission are the same, except that, since the transition from the ground state of O to the $^5S^0$ state is spin forbidden, resonance scattering is not important.

Mariner 5 and Venera 4 searched for the 1304 Å resonance triplet of O, under the assumption that photodissociation of CO_2 and diffusive separation should cause O to become the major constituent of the atmosphere at high altitudes. The 1304 Å intensity is a measure of the integrated O density above the altitude of CO_2 absorption, although because the emission is optically thick, a radiative transfer model is necessary to interpret the measured brightness. No emission was initially reported for the Mariner 5 spectrophotometer (Barth *et al.*, 1967) or the Venera 4 instrument (Kurt *et al.*, 1968). Venera 4 observed the atmosphere below 300 km in shadow. Strickland (1973) suggested that the thickness of the airglow layer was not within the resolution of the

Mariner 5 instrument. According to von Zahn *et al.* (1983), the three channels of the Mariner 5 spectrophotometer were allowed to go off scale as they swept across the bright disk of the planet, and the oxygen channels did not recover before the bright limb had been crossed. Several years later, a reanalysis of the Mariner 5 airglow data by Anderson (1975) revealed the presence of the 1304 Å triplet in the region of the terminator with an intensity of 670 R. By scaling to the subsolar point by analogy with terrestrial data, a nadir intensity of 11 kR was inferred and an O mixing ratio of 10% was derived.

In 1967 Moos *et al.* (1969) detected emission from Venus near 1300 Å using a low-resolution rocket-borne spectrometer. Moos and Rottman (1971) later measured the FUV spectrum (1200–1900 Å) of Venus using a moderate resolution spectrometer and reported preliminary values for the intensities of the O emissions. The spectrum is shown here in Figure 7. Rottman and Moos (1973) refined the analysis and reported

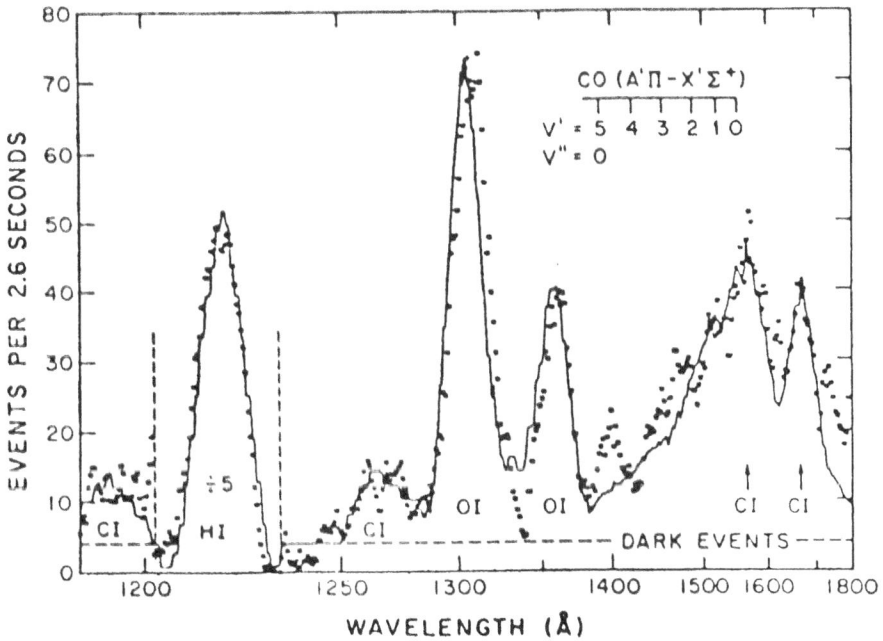

Fig. 7.   Rocket spectrum of Venus (points) compared with a best-fit synthetic spectrum. The 'mystery' feature at 1400 Å was identified by Durrance *et al.* (1980) as the (14, 5) CO fourth positive band. From Rottman and Moos (1973).

disk intensities of 5.5 ± 0.5 and 2.7 ± 0.5 kR for the 1304 and 1356 Å emissions, respectively. They estimated a mixing ratio for O of 10% at a CO_2 column density of 4×10^{16} cm^{-2}, the CO_2 column density above the altitude of absorption reported by Strickland *et al.* (1972) for Mars. Their production compares well to values found from Pioneer Venus *in situ* measurements. For example, in the VIRA dayside model, the O/CO_2 ratio is 13% at a column density of 4×10^{16} cm^{-2}. The agreement is, however, largely fortuitous. The analysis of Rottman and Moos was based on the Venus model of Strickland (1973), except for the cross sections for electron impact excitation of O, which were adopted from Zipf and Stone (1971). These cross sections have recently been shown to be too large, by a factor of about 7 for the 1304 Å emission and a factor of 2 for the 1356 Å emission (Stone and Zipf, 1974; Zipf and Erdman, 1985; Zipf, 1986). Rottman and Moos found electron impact excitation to be the major source of both of the emissions. Strickland (1973) computed intensities for the emissions using a model atmosphere of McElroy (1969) with an exospheric temperature of 700 K. He could not reconcile the 1304 Å intensity nor the ratio of 1304 to 1356 Å intensity of about 2 with the model even with several tens of percent O, based on excitation parameters (g values for resonance scattering and 'effective' g values for electron impact excitation) derived from analyses of terrestrial airglow and Mariner 6, 7, and 9 airglow from Mars (Strickland et al., 1972). Strickland's (1973) assumed impact excitation rate for the 1304 Å emission was a factor of 3 to 4 less than that derived from the cross section reported by Zipf and Stone (1971), but only about a factor of 2 larger than the currently accepted value. Strickland suggested, among other possibilities, that another source of emission might be present in the 1356 Å signal. Durrance et al. (1981) later showed that the (14, 4) CO fourth positive band contaminates the 1356 Å signal.

The Mariner 10 spectrophotometer also detected strong emission at 1304 Å (Broadfoot et al., 1974); the brightness was 17 kR, about 3 times the Moos and Rottman (1971) rocket values and an order of magnitude larger than the values reported from Mariners 6 and 7 for Mars (Strickland et al., 1972). Although the O abundance was difficult to model accurately, the conclusion that the atmosphere of Venus contained more O than that of Mars was inescapable.

In 1978, the Venera 11 and 12 spectrophotometers measured disk brightness at 1304 Å of 6.4 kR, from which Bertaux et al. (1981) inferred an atomic oxygen column density of 1.9×10^{12} cm^{-2}. They also concluded that electron impact on O must be an important source of $O(^3S^0)$, since the intensity was not flat across the disk. The importance of electron impact can be evaluated by reference to Table II, where we present computed excitation rates of various excited states of atmospheric species produced in electron-impact processes for models appropriate to high and low solar activity based on the neutral VTS3 models computed by Fox (unpublished calculations; cf. Fox, 1982). Table III shows the production rates due to photoprocesses for the same models. The integrated electron-impact excitation rate of $O(^3S^0)$ for the high solar activity model is only 3.4×10^8 cm^{-2} s^{-1}, but the apparent emission rate will be enhanced by radiative transfer.

Stewart et al. (1979) reported the first measurements of the UV airglow from the Pioneer Venus OUVS, including the oxygen emission lines; they noted that the 1356 Å emission was limb brightened and ascribed this effect to the production of energetic $O(^5S^0)$ atoms in dissociative excitation of CO. Cross sections are not available for production of $O(^5S^0)$ in electron impact or photodissociative excitation of CO. Table III shows that electron impact and photodissociative excitation of CO_2 are negligible sources of the 1304 Å emission. In their study of the CO fourth positive band system,

TABLE II

Integrated column production rates for excited electronic species due to electron impact for model Venus thermospheres based on Pioneer Venus data. The high solar activity (High SA) model is that of Hedin *et al.* (1983) for $F_{10.7} = 200$ and the low solar activity (Low SA) model for $F_{10.7} = 74$. The solar fluxes are the SC No. 21REFW and F79050N spectra of Hinteregger (private communication; see also Torr *et al.*, 1979) for the low and high solar activity models, respectively.

Species	Source	Column production rate $(10^6 \text{ cm}^{-2} \text{ s}^{-1})$	
		High SA	Low SA
$CO(a^3\Pi)$	CO_2	2.1(4)	7.0(3)
	CO	1.1(4)	2.1(3)
$CO(A^1\Pi)$	CO_2	5.5(2)	1.8(2)
	CO	4.0(3)	7.5(2)
$CO_2^+ (A^2\Pi_u)$	CO_2	3.7(3)	1.2(3)
$CO_2^+ (B^2\Sigma_u^+)$	CO_2	2.2(3)	7.2(2)
$CO^+ (A^2\Pi)$	CO	1.1(3)	2.0(2)
$CO^+ (B^2\Sigma^+)$	CO	3.5(2)	6.5(1)
$O(^1D)$	O	1.4(4)	2.1(3)
$O(^1S)$	CO_2	6.2(3)	2.1(3)
	O	1.1(3)	1.7(2)
$O(^3S^0)$	CO_2	4.0(1)	1.3(1)
	O	3.0(2)	4.6(1)
	CO	2.0(0)	4.0(−1)
$O(^5S^0)$	CO_2	4.0(1)	1.3(1)
	O	8.8(2)	1.4(2)
$N_2^+ (A^2\Pi_u)$	N_2	2.6(2)	7.9(1)
$N_2^+ (B^2\Sigma_u^+)$	N_2	8.8(1)	2.7(1)

Durrance *et al.* (1981) found that the limb brightening at 1356 Å is the result of contamination of the 1356 Å emission by the (14, 4) CO fourth positive band at 1352 Å.

Meier *et al.* (1983) modeled the atomic oxygen emissions using a Monte-Carlo technique for solution of the radiation transport equations. They found that the incorporation of partial frequency redistribution also enhanced the predicted limb brightening. Their model calculations reproduced the measurements if the O densities from the BNMS model were reduced by about 40%, but their analysis was based on cross sections for electron impact production of emission at 1304 and 1356 Å reported by Stone and Zipf (1974) that were later reduced by a factor of 2.8 (Zipf and Erdman, 1985; Zipf, 1986). A model of the terrestrial dayglow oxygen emissions (Meier *et al.*, 1985) suggested that the Stone and Zipf cross sections should be scaled down by 40%. A reanalysis of the Pioneer Venus data using the renormalized cross sections showed good agreement between the O densities necessary to explain the measured emission and the BNMS model densities (Paxton and Meier, 1986).

Remote sensing can be a reliable method of determining thermospheric densities, as

TABLE III

Column production rates of some excited electronic states of species in the
Venus thermosphere due to photodissociative excitation or photoionization
and excitation. The models used are as described in Table II.

Species	Source	Column production rate ($10^6 \, \mathrm{cm}^{-2} \, \mathrm{s}^{-1}$)	
		High SA	Low SA
$CO(a^3\Pi)$	CO_2	1.8(4)	7.9(3)
$CO(A^1\Pi)$	CO_2	7.8(2)	3.1(2)
$CO_2^+(A^2\Pi_u)$	CO_2	9.0(3)	3.6(3)
$CO_2^+(B^2\Sigma_u^+)$	CO_2	1.0(4)	4.1(3)
$CO^+(A^2\Pi)$	CO	3.4(3)	7.4(2)
$CO^+(B^2\Sigma^+)$	CO	1.3(3)	2.5(2)
$O(^1D)$	CO_2	1.6(6)	6.5(5)
$O(^1S)$	CO_2	2.1(5)	7.6(4)
$O(^3S^0)$	CO_2	1.2(1)	6.1(0)
$N_2^+(A^2\Pi_u)$	N_2	2.0(3)	7.8(2)
$N_2^+(B^2\Sigma_u^+)$	N_2	4.1(2)	1.5(2)

Meier and Anderson (1983) have suggested, but in practice, the use of 1304 and 1356 Å
intensities to infer atomic oxygen densities both in the terrestrial atmosphere and in
planetary atmospheres has been plagued by uncertainties. The major sources of error
in models are in the electron impact cross sections and in the solar fluxes, although,
if limb scans of an optically thick emission are available, the altitude distribution of the
scatterer can be obtained from the shape of the intensity profile independent of the solar
flux or the calibration of the instrument (e.g., Anderson et al., 1987). In this case, the
solar flux can be derived from the intensity of the emission. The accepted values for the
electron impact cross section have changed by nearly a factor of 3 in recent years,
although there is now substantial agreement between the excitation cross sections
measured using electron energy loss methods (Vaughan and Doering, 1986; Doering
and Vaughan, 1986; Gulcicek and Doering, 1988; Doering and Gulcicek, 1989) and the
renormalized emission cross sections. There is also a wide range in the values of the
solar photon fluxes used in or derived from resonance scattering calculations, even
allowing for solar activity variations. Both the rocket flights of Moos and Rottman
(1971) and the initial PV OUVS measurements were during periods of high solar
activity. Strickland (1973) required a photon flux integrated over the multiplet of
$5 \times 10^9 \, \mathrm{cm}^{-2} \, \mathrm{s}^{-1}$, whereas the calculations of Meier et al. (1983) were consistent with
a total flux of $1.4 \times 10^{10} \, \mathrm{cm}^{-2} \, \mathrm{s}^{-1}$. Link et al. (1988) required a 1304 Å irradiance of
$7 \times 10^9 \, \mathrm{cm}^{-2} \, \mathrm{s}^{-1}$ to explain the terrestial dayglow experiments at high solar activity.
Experimental measurements of the total photon flux in the 1300 to 1310 Å range, of
which the 1304 Å triplet accounts for about 85–90% (Mount and Rottman, 1981), also
vary widely. At high solar activity, reported values range from $9.5 \times 10^9 \, \mathrm{cm}^{-2} \, \mathrm{s}^{-1}$ from
the F79050N spectrum of Hinteregger (private communication; see also Torr et al.,

1979) to 1.54×10^{10} from rocket measurements (Mount and Rottman, 1981); at low solar activity from 4.8×10^9 from the SC No. 21REFW spectrum of Hinteregger to $7.1-7.3 \times 10^9$ (Mount and Rottman, 1983, 1985) to $1.21 \times 10^{10} \text{ cm}^{-2} \text{ s}^{-1}$ from the Solar UV Spectral Irradiance Monitor (SUSIM) data (Van Hoosier *et al.*, 1987). There are, however, independent reasons to believe that the Hinteregger fluxes are too low. Ogawa and Judge (1986) used a rocket-borne Ne ionization chamber to measure the integrated solar flux between 50 and 575 Å, with an estimated error of 7.3%. Their results indicate that the Hinteregger fluxes in this wavelength region are low by a factor of 2 at low solar activity and by 30% at high solar activity. Given the uncertainties in the parameters used in modeling the oxygen emissions, densities derived from these measurements should probably not be considered reliable to better than a factor of 2.

Emissions at 1304 and 1356 Å have been detected by the Pioneer Venus OUVS in images of the nightside of Venus. The emissions appear in bright patches that vary in size and intensity; the sheer spatial and temporal variability of the emissions suggests that they are produced by particle impact. The morphology of the aurora has been described by Phillips *et al.* (1986). Figure 8 hows a series of brightness images recorded

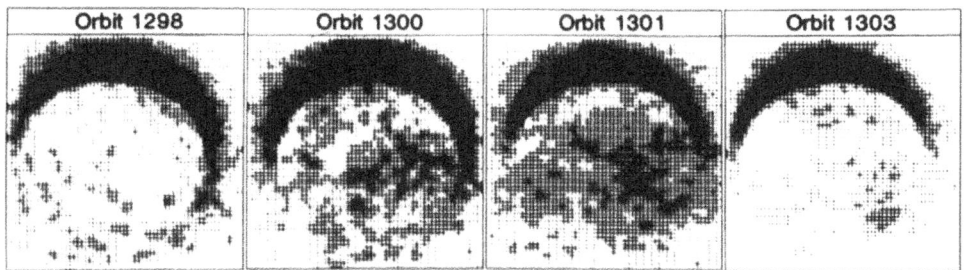

Fig. 8.   Series of brightness images from the PV OUVS for the orbital sequence of June 25–July 1, 1982 with apoapsis near midnight. Note expansion and intensification of emissions to fill the nightside in orbit 1301. Missing orbits were not imaged at 1304 Å. From Phillips *et al.* (1986).

by the OUVS for four nearly contiguous orbits. The intensity of the 1304 Å emission is usually about 10 R, but values near 100 R have been recorded; the 1304/1356 ratio is about 6. Fox and Stewart (1990) have presented evidence that the emissions are produced by impact of very soft ($\sim$ a few eV) electrons. More energetic electrons would penetrate too deeply and produce high intensities of the Cameron bands of CO, which are not observed, and ion densities that exceed observed values. Both radiative recombination of O^+ and precipitation of O^+, the sources in the terrestrial equatorial nightglow, are ruled out for Venus because the 1304 and 1356 Å emissions are produced with comparable intensities by these mechanisms and because the predicted intensities due to radiative recombination are less than 10^{-3} R (Julienne *et al.*, 1974; Abreu *et al.*, 1986). In addition, soft electrons have been observed in the Venus wake by both the PV orbiter retarding potential analyzer (ORPA) (e.g., Knudsen and Miller, 1985) and the plasma analyzers on board Veneras 9 and 10 (e.g., Gringauz *et al.*, 1979). Fox and

Stewart have modeled the intensities that would be produced by precipitation of electrons with the same energy spectrum and fluxes as the downward traveling portion of the ORPA spectrum. The total fluxes required were determined to be between 8 and 28% of the ORPA values.

The energies of the electrons producing the aurora are constrained to be low so that they lose their energy in the upper thermosphere where O is the dominant constituent of the atmosphere. If the electrons penetrate down to the part of the atmosphere where CO_2 is the most important constituent (below about 140 km) unobserved emissions, such as the CO Cameron bands, would be observed with large intensities and the O emissions would be excited less efficiently, producing a larger ratio of ionization to excitation for a given 1304 Å intensity. An alternative explanation has been proposed by J. Grebowsky (private communication). He has suggested that the same effect would be realized if the depth of penetration of the electrons were limited by the presence of a horizontal magnetic field in the nightside ionosphere, rather than by their energy.

The excitation of ground state oxygen atoms by electron impact, photodissociation and electron-impact dissociation of CO_2 and CO, and dissociative recombination of O_2^+ can produce O atoms in the ^1S and ^1D states, which can either be quenched or decay by the emission of photons. $O(^1D)$ produces the atomic oxygen 'red line' in the process

$$O(^1D) \to O(^3P) + h\nu(6300 \text{ Å}).$$ (19)

Ninety-five percent of $O(^1S)$ decays to $O(^1D)$ producing the atomic oxygen 'green line':

$$O(^1S) \to O(^1D) + h\nu(5577 \text{ Å}).$$ (20)

Only 5% decays to the ground state producing an ultraviolet photon of wavelength 2972 Å

$$O(^1S) \to O(^3P) + h\nu(2972 \text{ Å}).$$ (21)

Dayglow intensities of 730 R, 48 kR, and 2.4 kR were predicted for the 6300, 5577, and 2972 Å emissions, respectively by Fox (1978) and Fox and Dalgarno (1981) in a low solar activity pre-Pioneer Venus model. The major source of both $O(^1D)$ and $O(^1S)$ is photodissociative excitation of CO_2. Since pre-Pioneer Venus models differed from current models mostly in the mixing ratios of O and CO, the intensities should not be highly model-dependent. $O(^1D)$ is strongly quenched by CO_2, so most of the $O(^1D)$ produced does not radiate. The rate coefficient for quenching of $O(^1S)$ by CO_2

$$O(^1S) + CO_2 + O(^1D) + CO_2$$ (22)

is $3.3 \times 10^{-11} \exp(-1320/T)$ cm^3 s^{-1} (Atkinson and Welge, 1972) and the quenching altitude, where the lifetime against collisional deactivation is approximately equal to the radiative lifetime, is about 100 km. The altitude profile for production of $O(^1S)$, shown in Figure 9 is characterized by two peaks, the upper one due mainly to absorption of solar photons in the 1100–1140 Å range and the lower peak, near 115 km is due to photodissociation of CO_2 by Lα.

LeCompte et al. (1989) have reported intensities of the 2972 Å emission measured by

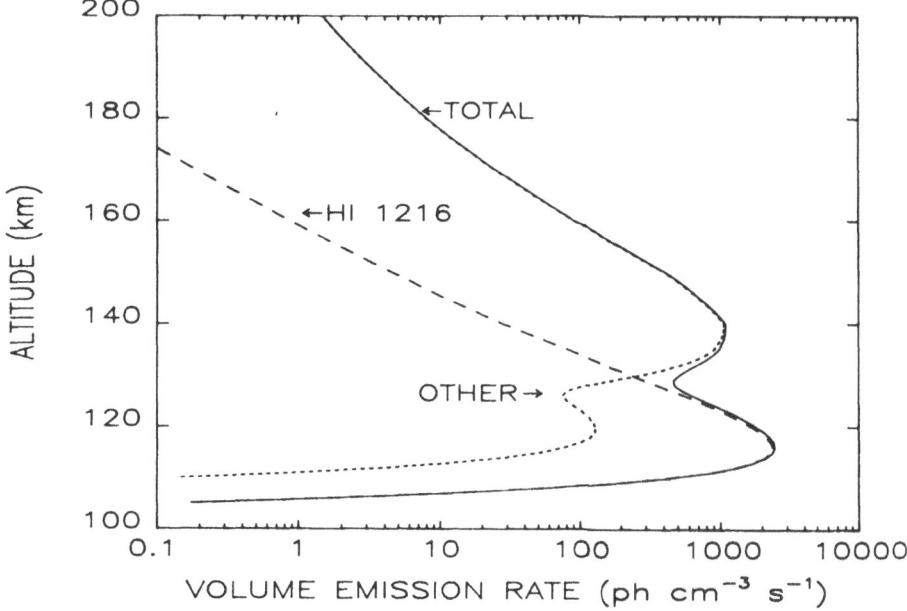

Fig. 9.   Volume production rate of 2972 Å emission from photodissociation of CO_2 by Lα (dashed line), from the rest of the solar spectrum (dotted line) and total (solid line). From LeCompte *et al.* (1989).

the Pioneer Venus OUVS. The zenith intensity reported for orbit 187, for which high solar activity was appropriate, is 7.0 kR, compared to 2.4 kR predicted by the low solar activity model of Fox and Dalgarno (1981). LeCompte *et al.* analyzed limb profiles from two orbits. They showed that photodissociation is the major source of $O(^1S)$ below 175 km. Above that altitude, dissociative recombination of O_2^+ and a contribution from the CI 2967 Å ($^5S \rightarrow {}^3P$) emission dominate the production profile. Fairly good agreement was obtained between the measurements and the model at high altitudes, although the model intensity deviates from the measurements at altitudes below 130 km. They suggested that the difference arises from oscillations in the temperature profile that result from the passage of gravity waves. Gravity waves have been shown to affect terrestrial airglow emissions (e.g., Porter *et al.*, 1974; Taylor *et al.*, 1987). Emission from $O(^1S)$ is sensitive to temperature because the rate coefficient for quenching of $O(^1S)$ by CO_2 (reaction 22) depends exponentially on temperature. Oscillatory structures appear in the temperature profiles inferred from neutral densities measured by several PV instruments, including the BNMS (von Zahn *et al.*, 1980), the ONMS (Kasprzak *et al.*, 1988), and by the accelerometers on the probes (Seiff *et al.*, 1980). The temperature profile that LeCompte *et al.* use for their standard model is the VTS3 model of Hedin *et al.* (1983), which is not based on ONMS measurements below 140 km. It has been suggested that the temperatures below 140 km in the VTS3 model are too low (Keating *et al.*, 1986; see also Section 2). LeCompte *et al.* tested a constant (positive) offset in the temperature profile from the VTS3 model, but did not consider a temperature profile in which the deviation increases as altitude decreases. Their assertion that the temperature profile must return to the VTS3 profile at the lower boundary is not entirely convincing. As they admit, LeCompte *et al.* test deviations of the VTS3 temperature profile without changing the neutral densities in a consistent way. More important, the model intensity profile is normalized to the measurements at low altitudes, where the source of the signal is mostly Rayleigh scattering. At 105 km the 2972 Å volume production rate vanishes, so the sensitivity of the model to the quenching coefficient and, hence, to the temperature near that altitude is small.

Visible spectrometers on Venera 9 and 10 searched for emissions on the nightside at 6300 and 5577 Å, but found none. Upper limits have been placed on the intensities of the green and red lines of 10 and 20–25 R, respectively (Krasnopol'sky, 1981, 1986, and private communication, 1988). In the nightglow, the emissions are produced mostly in dissociative recombination of O_2^+. Fox (1990a) has combined a model for the vibrational distribution of O_2^+ on the nightside with rate coefficients computed by Guberman (1987, 1988) for production of $O(^1D)$ and $O(^1S)$ in dissociative recombination of $O_2^+(v)$ from different vibrational levels, v. Altitude profiles of the volume emission rates are shown in Figure 10. The integrated overhead intensities are 1–2 R for the green line and about 46 R for the red line. The intensity of the green line is sensitive to the vibrational distribution of O_2^+, but the yield of $O(^1D)$ and, therefore, the emission rate of the red

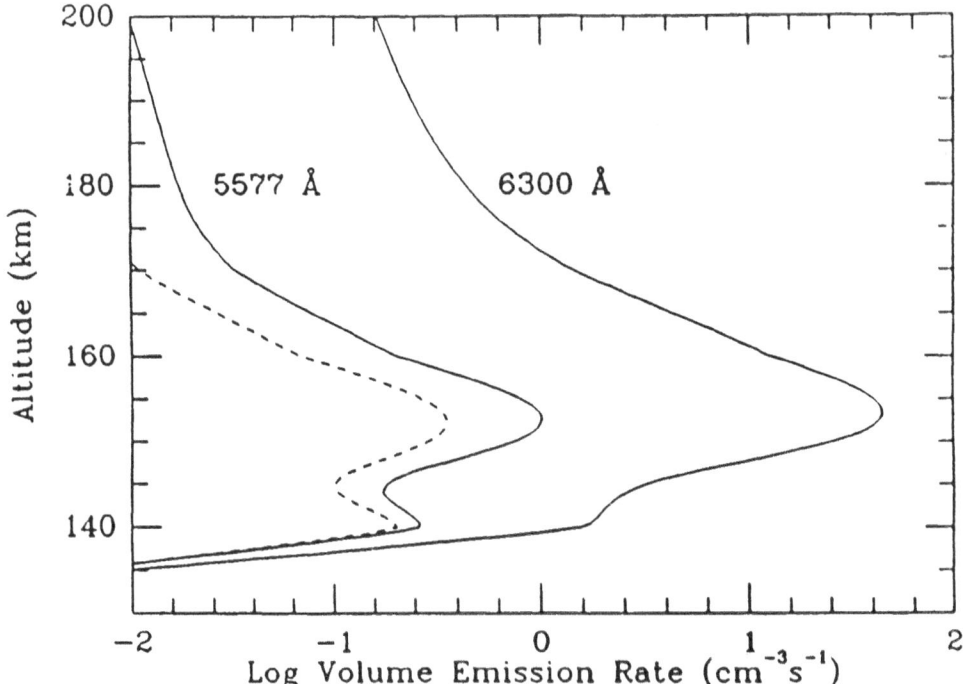

Fig. 10.   Altitude profiles of the volume emission rates of the red and green lines of atomic oxygen in the Venus nightglow. Solid lines are computed assuming $k_{43} = 1 \times 10^{-10}$ cm^3 s^{-1}. The dashed line is for $k_{43} = 6 \times 10^{-10}$ cm^3 s^{-1}. From Fox (1990a).

line is less sensitive to the vibrational distribution. The predicted intensity of the red line exceeds the Venera upper limit for the particular model ionosphere chosen. The night-side ionosphere, at least at times of high solar activity, is produced mainly by transport of O^+ from the dayside (Knudsen *et al.*, 1980; Spenner *et al.*, 1981; Cravens *et al.*, 1983), although it has been suggested that at times of low solar activity, electron impact is the most important ion source (Gringauz *et al.*, 1979; Breus *et al.*, 1985; Knudsen, 1988). The nightside ionosphere is highly variable, sometimes disappearing nearly completely (Cravens *et al.*, 1982). The model ionosphere of Fox (1990a) corresponds to undisturbed, relatively 'full-up' conditions for a moderate downward flux of O^+ of 1×10^8 cm^{-2} s^{-1} and may not be representative of any single measurement.

3.3. O^+ EMISSIONS

$O^+(^2D)$ and $O^+(^2P)$ are metastable states of O^+ that lie 3.32 and 5.02, respectively, above the ground state of O^+. Radiation of $O^+(^2D)$ produces a doublet at 3726 and 3728 Å. $O^+(^2P)$ may radiate to the $O^+(^2D)$ state producing a doublet at 7319 and 7329 Å, or to the ground $O^+(^2S)$ state producing an emission feature at 2470 Å. These states are produced in the atmosphere of Venus mainly in photoionization and electron impact ionization of O. The yields of $O^+(^2D)$ and $O^+(^2P)$ are about 37 and 31%, respectively, in photoionization and 42 and 22%, respectively, in electron impact ionization of O (see Kirby *et al.*, 1979; Burnett and Rountree, 1979). Emission from $O^+(^2D)$ in dipole forbidden and consequently its radiative lifetime is long, about 2.4×10^4 s (Seaton and Osterbrock, 1957). Most of the loss of $O^+(^2D)$ is via chemical reactions and the importance of this species is, therefore, in its effect on the ion chemistry. For example, charge transfer reactions of $O^+(^2D)$ have been shown to be important for the production of N_2^+, CO^+, and N^+ (Fox, 1982).

No measurements of intensities of the emission from $O^+(^2D)$ and $O^+(^2P)$ have been reported. Altitude profiles of the volume emission rates computed for a Pioneer Venus high solar activity model by Fox (1982) are shown Figure 11. The integrated overhead intensities are given here in Table IV.

The O^+ emission triplet at 834 Å arises from the dipole allowed transition $2s2p^4\,^4P \rightarrow 2s^2p^3\,^4S^0$, so resonance scattering of solar radiation by O^+ is a potential

TABLE IV

Integrated overhead intensities (volume production rates integrated over the layer) of emissions arising from metastable states of O^+ and N (adapted from Fox, 1982)

Transition	Wavelength (Å)	Intensity (R)
$O^+(^2P) \rightarrow O^+(^2D) + h\nu$	7319, 7329	720
$O^+(^2P) \rightarrow O^+(^4S) + h\nu$	2470	200
$O^+(^2D) \rightarrow O^+(^4S) + h\nu$	3726, 3728	4.3
$N(^2D) \rightarrow N(^4S) + h\nu$	5200	16
$N(^2P) \rightarrow N(^2D) + h\nu$	10,400	2.5
$N(^2P) \rightarrow N(^4S) + h\nu$	3466	160

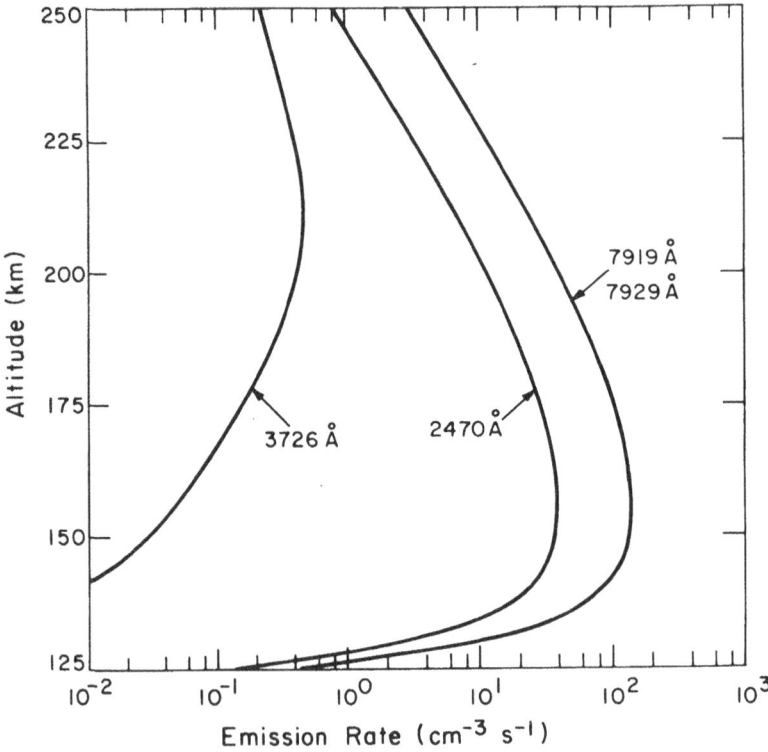

Fig. 11.   Volume emissions rates of the transitions $O^+(^2D) \to O^+(^4S)$, $O^+(^2P) \to O^+(^2D)$ and
$O^+(^2P) \to O^+(^4S)$ at 3726, 7319–7329, and 2470 Å. From Fox (1982a).

source, as are electron impact and photoionization of $O(^3P)$. The first measurements
of the intensity of the O^+ emission line at 834 Å were made by Venera 11 in December
1978. Bertaux *et al.* (1981) reported a maximum brightness on the disk of 156 R, with
fluctuations of less than 20% across the disk. They interpreted their measurements as
showing that resonance scattering of solar radiation is not important, as has also been
found for the Earth. In the terrestrial ionosphere photoionization of O produces a low
altitude source that is reonantly scattered by the F2 region O^+ ions at higher altitudes
(e.g., Carlson and Judge, 1973; Feldman *et al.*, 1981; Kumar *et al.*, 1983b). Presumably
the same is true for Venus also.

3.4. ATOMIC C AND C^+ EMISSIONS

From their low-resolution rocket spectrum of Venus, Moos *et al.* (1969) reported the
existence of a signal longward of 1500 Å that they attributed tentatively to broadband
fluorescence with an intensity of 10–100 kR. Marmo and Engleman (1970) proposed
that the signal could be explained by the presence of 14 kR of emission in the 1657 Å
resonance line of atomic carbon. Although the suggestion was influenced by the
erroneous belief that the thermosphere was largely decomposed under the influence of

solar ultraviolet radiation (see Section 3.11), the conclusion was at least partially correct. Rottman and Moos (1973) reported the first positive detection of emission lines of C in the Venus ultraviolet spectrum obtained by their rocket-borne FUV spectrometer (see Figure 7). The reported intensities were 2.4 $\pm$ 1.2 kR and 4.0 $\pm$ 1.5 kR for the 1561 and 1657 Å emissions, respectively. The Mariner 10 UV spectrophotometer also detected strong emission at 1657 Å, with a measured brightness of 30 kR (Broadfoot *et al.*, 1974). The Venera 11 and 12 EUV spectrophotometers recorded a signal at 1657 Å, but the maximum brightness on the disk (10–15 kR) was close to the background level (Kurt *et al.*, 1979; Bertaux *et al.*, 1981). The resolution of the Venera instrument was insufficient to exclude some emission in the fourth positive bands of CO, but Bertaux *et al.* (1981) estimated that 85% of the signal could be assigned to the CI line. Spectra recorded by the Pioneer Venus OUVS also showed features at 1561 and 1657 Å and limb profiles have been reported by Paxton (1985). A typical limb profile for the 1657 Å emission is shown in Figure 12.

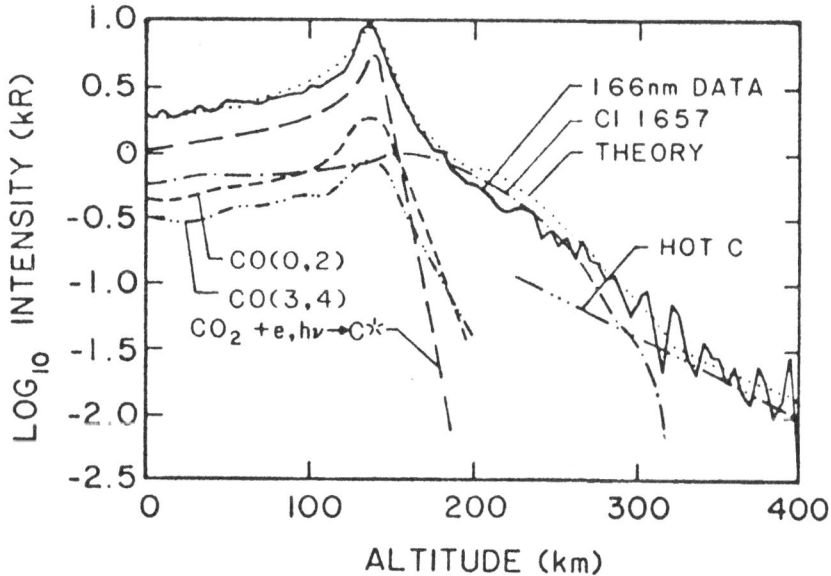

Fig. 12.   Comparison of calculated limb intensity and PV OUVS 1657 Å limb observations. The modeled sources include the resonant scattering of solar photons by atomic carbon, the CO fourth positive bands and electron and photon impact excitation of CO_2. Taken from Paxton (1985).

Possible sources for the CI emission include resonance scattering by atomic carbon, electron impact excitation of atomic carbon and photodissociative- and electron-impact dissociative excitation of CO_2 and CO. Fox and Dalgarno (1981) modeled the intensities due to electron impact dissociative excitation of CO_2 and showed that those sources could not explain the intensities. In order to determine the source due to resonance scattering, densities of C must be known, but because of the production of C from fractionation of CO_2 (Niemann *et al.*, 1980), the densities were below the detection threshold of the mass spectrometers on Pioneer Venus. McElroy and McConnell (1971) modeled the atomic carbon in the Venus atmosphere, but were unable to explain the large intensities using a model atmosphere with a thermospheric CO mixing ratio equal to the measured abundance above the clouds, 4.5×10^{-5} (Connes *et al.*, 1968). Pioneer Venus *in situ* measurement showed that thermospheric mixing ratios are much larger than those in the mesosphere (see Section 2). The first successful models of the atomic carbon densities based on Pioneer Venus neutral models were constructed by Krasnopol'sky (1982b) and by Fox (1982b). These models showed that the major source of atomic carbon in the Venus thermosphere is photodissociation of CO. The most important sink is reaction with O_2 (McElroy and McConnell, 1971):

$$C + O_2 \rightarrow CO + O, \tag{23}$$

which proceeds with a rate coefficient of $3.3 \times 10^{-11} \, cm^3 \, s^{-1}$ (Braun *et al.*, 1969). There are, however, no measurements of the densities of O_2 in the Venus thermosphere. Fox treated the O_2 mixing ratio, f_{O_2} as a free parameter and computed the atomic carbon density profiles shown in Figure 13. The profiles exhibit maximum C densities at 145–155 km of 5×10^6 and 3×10^7 for O_2 mixing ratios of 3×10^{-3} and 1×10^{-4}, respectively. Fox also showed that photoionization of C is the major source of C^+ and used the measured C^+ densities (Taylor *et al.*, 1980) to infer that the mixing ratios of

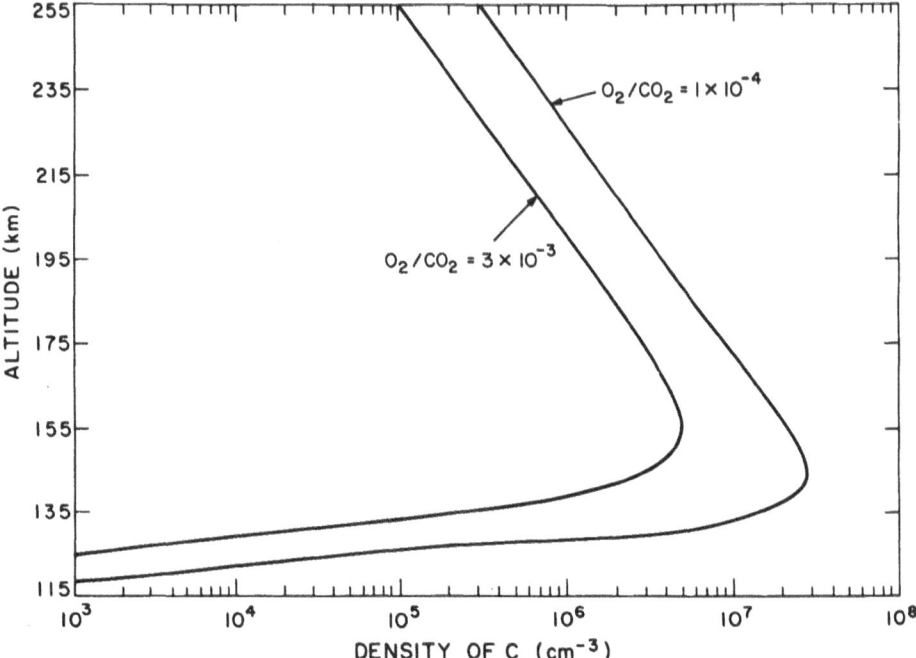

Fig. 13.   Computed number densities of neutral atomic carbon as a function of altitude for O_2/CO_2 mixing ratios of 3×10^{-3} and 1×10^{-4}. The profile for the larger O_2/CO_2 mixing ratio was adopted by the Venus International Reference Atmosphere (Keating *et al.*, 1986). Taken from Fox (1982b).

O_2 must be closer to the lower value. Krasnopol'sky (1982b) computed C densities using an O_2 mixing ratio of 2×10^{-3}, based on the mesospheric photochemical model of Krasnopol'sky and Parshev (1981) and obtained similar C densities to those in the larger f_{O_2} model of Fox.

Paxton (1985) modeled CI limb intensities at 1657 and 1561 Å made by the PV OUVS, including the contamination of the signal by several CO fourth positive bands. Figure 12 shows a comparison of the measured and computed limb intensities of the 1657 Å emission due to various sources. He reported CO_2 unattenuated photodissociative excitation frequencies of 2×10^{-10} and 4×10^{-10} s^{-1} for the 1561 and 1657 Å lines, respectively; for photodissociation excitation of CO the corresponding unattenuated frequencies were 1.8×10^{-9} and 2.7×10^{-9} s^{-1}. Resonance scattering of solar radiation was found to be the most important source of the emissions. Paxton used a radiation transport model to compute the resonance scattering intensities and to derive the atomic carbon densities. The best fit to the emission rates was obtained for a C density profile similar to the lower density profile of Fox (1982b), for which the O_2/C_2 mixing ratio was 3×10^{-3}. Thus a discrepancy exists between the C profiles derived from the C^+ densities and those derived from the CI airglow. Paxton has suggested that the problem is with the ion chemistry. It is also possible that the rate coefficients on which the atomic carbon models are based are inaccurate. In any case, the CO photodissociation rates computed in the models above were based, in part, on low-resolution photoabsorption cross sections in the wavelength region longward of the ionization threshold at 885 Å measured by Cook et al. (1965). Recent evidence has shown that the absorption in this wavelength region is due to discrete absorption into predissociating states rather than continuum absorption (Yoshino et al., 1988; Letzelter et al., 1987). Fox and Black (1989) constructed high-resolution photodissociation cross sections for rotational temperatures appropriate to the Venus thermosphere from a line list compiled by van Dishoeck and Black (1988) from information communicated privately to them by Stark, Smith, Yoshino, and Parkinson. They showed that the CO photodissociation rate, and the corresponding C production rate, are reduced by a factor of two over the values obtained using the low-resolution cross sections. This may have important consequences for the chemistry of atomic carbon in the Venus thermosphere that remain for future work to determine.

3.5. CO AND CO$^+$ EMISSIONS

CO was first identified in the atmosphere of Venus from ground-based infrared observations of the first overtone band at 2.35 μm by Sinton (1963) and by Moroz (1964). Connes et al. (1968) observed several rotational lines in the CO fundamental ($v = 1 \rightarrow 0$) vibration-rotation band and derived a mixing ratio above the cloud tops of 4.5×10^{-5}. Young (1972) applied new information about the line widths to revise that value to 5.5×10^{-5}. In situ measurements were carried out only below the clouds by the PV sounder probe gas chromatograph (LGC), which measured CO mixing ratios of 20–32 ppm from 52 to 22 km (Oyama et al., 1980) and by the gas chromatograph on Venera 12, which measured an abundance of 28 ppm (Gel'man et al., 1979).

The first microwave detection of CO in the Venus atmosphere was that of Kakar *et al.* (1976), who observed the 0.26 cm ($J = 0 \rightarrow 1$) pure rotational transition of CO. Microwave measurements are potentially more sensitive to lower pressures and therefore to higher altitudes than infrared measurements. Kakar *et al.*, as well as Schloerb *et al.* (1980) and Wilson *et al.* (1981) observed a strong diurnal variation in the intensities, and concluded that the CO mixing ratio decreases by a factor of 10 from day to night in the range of total pressure 2–100 mb (about 65–85 km). Clancy and Muhleman (1985a) found that the CO abundance between 80 and 90 km peaks near 08 : 30 local time, but above 95 km, the peak was found to occur near the anti-solar point. The magnitude of the diurnal variation is a factor of 2–4 for both altitude ranges. Between 100 and 105 km, the derived mixing ratios are in the range 4×10^{-4} to 1×10^{-3} at night and 2×10^{-4} to 5×10^{-4} during the daytime.

In situ measurements of thermospheric CO by the PV ONMS and BNMS show that the mixing ratio at 140 km is much larger than the microwave measurements indicate at 90 km (Niemann *et al.*, 1980b; von Zahn *et al.*, 1980; see Section 2). The thermospheric model of Massie *et al.* (1983) predicts larger CO densities at the lower boundary than the measurements of Wilson (1981) or Clancy and Muhleman (1985a) indicate, but given the uncertainties the agreement is adequate. That the model does not show the observed diurnal variation in the CO densities may be the results of its one-dimensional nature. Clancy and Muhleman (1985b) have suggested that the high altitude diurnal variation is a result of the sub-solar to anti-solar circulation, the return flow of which is in the upper mesosphere.

The fourth positive band system of CO is the most intense band system in the 1200 to 1800 Å region of the dayglow spectrum of Venus. It arises from the dipole allowed transition

$$CO(A^1\Pi; v') \rightarrow CO(X^1\Sigma; v'') + hv. \tag{24}$$

Potential sources include photodissociative excitation or electron impact dissociation of CO_2:

$$CO_2 + (e, hv; E > 13.5 \text{ eV}) \rightarrow CO(A^1\Pi; v') + O + (e, \quad), \tag{25}$$

electron impact excitation of CO:

$$CO(X^1\Sigma; v'') + e(E > 8.02 \text{ eV}) \rightarrow CO(A^1\Pi; , v') + e, \tag{26}$$

dissociative recombination of CO_2^+

$$CO_2^+ + e \rightarrow CO(A^1\Pi; v') + O, \tag{27}$$

and fluorescent scattering by ground state CO,

$$CO(X^1\Sigma; v) + hv \rightarrow CO(A^1\Pi; v') \rightarrow CO(X^1\Sigma, v'') + hv'. \tag{28}$$

The branching ratio for production of $CO(A^1\Pi)$ in dissociative recombination of CO_2^+ (reaction 27) is about 5% (Gutcheck and Zipf, 1973).

The observed signal longward of 1400 Å in the rocket spectrum reported by Rottman and Moos (1973) was interpreted as being due to the atomic carbon features and to the CO fourth positive band system. The total intensity of the band system was determined by fitting a synthetic spectrum to the measurement (see Figure 7). Except for an unidentified feature at 1400 Å (later identified by Durrance et al. (1980), see below) a good fit to the measured spectrum was obtained and a brightness of 25 ± 5 kR was reported. The process(es) responsible for the observed emission could not be unequivocably determined, but the source due to fluorescent scattering was judged to dominate the production, and a CO mixing ratio of 10% at a CO_2 column of 4×10^{16} cm^{-2} was inferred. The PV ONMS later confirmed the implication that the atmosphere of Venus contains larger amounts of CO than the atmosphere of Mars. The VTS3 model for $F_{10.7} = 200$ and 12 : 00 hr local time exhibits a CO mixing ratio of 12% at 137.5 km, the altitude at which the CO_2 column density is about 4×10^{16} cm^{-2}.

The Mariner 10 EUV spectrophotometer recorded a brightness at 1480 Å of 55 kR that was attributed to emission in the CO fourth positive bands (Broadfoot et al., 1974). The spectrophotometers on Venera 11 and 12 also recorded the intensity at 1500 Å (see Table I). Significantly smaller disk brightnesses of 2.1 and 2.7 kR for Venera 11 and 12, respectively, were reported (Bertaux et al., 1981). The disagreement with the Mariner 10 data was attributed by Bertaux et al. (1981) to contamination of the airglow signal by Rayleigh scattering.

Durrance et al. (1980) presented observations of the Venus dayglow from 1250 to 1430 Å made by the PV OUVS and Durrance et al. (1981) measured the Venus dayglow spectrum from 1280 to 1380 Å at high (0.4 Å) resolution with the International Ultraviolet Explorer (IUE) satellite. The strongest features in this region of the spectrum, other than the atomic oxygen emission at 1304 and 1356 Å, were shown to be due to the $(14, v'')$ progression of the fourth positive band system excited by fluorescent scattering of solar Lα. As pointed out by Kassal (1976), this source is larger than fluorescent scattering in all the rest of the fourth positive bands for CO column densities larger than 1×10^{17} cm^{-2}. The overlap of the rotational lines of the (14, 0) fourth positive band with the solar Lα line are shown in Figure 14. The unidentified feature at 1400 Å in the rocket spectrum of Rottman and Moos (1973) (see Figure 7) was identified by Durrance et al. (1980) as the (14, 5) fourth positive band.

The CO_2 absorption cross section is about 8×10^{-20} cm^2 at 1216 Å (Nakata et al., 1965); consequently solar Lα penetrates to a column density of about 1.25×10^{19}, or to about 110 km. The CO_2 cross sections in the region of the CO emissions, 1250 to 1400 Å, are in the range $2-9 \times 10^{-19}$ cm^2, so the emission bands will be absorbed if they are produced below about 120 km. Because the (14, 5) bands at 1400 Å is unblended and produced almost entirely by fluorescent scattering, measurement of the intensity of this feature could provide a valuable remote sensing technique for determining CO abundances in the Venus thermosphere (Durrance et al., 1980).

Durrance (1981) presented spectra with a resolution of 4 Å recorded by a sounding rocket telescope and spectrometer and constructed synthetic spectra using high-resolution solar fluxes. He showed that the (14, 3) and (14, 4) bands contaminate the O emission features at 1304 and 1356 Å; potentially 67% of the 1356 Å feature was

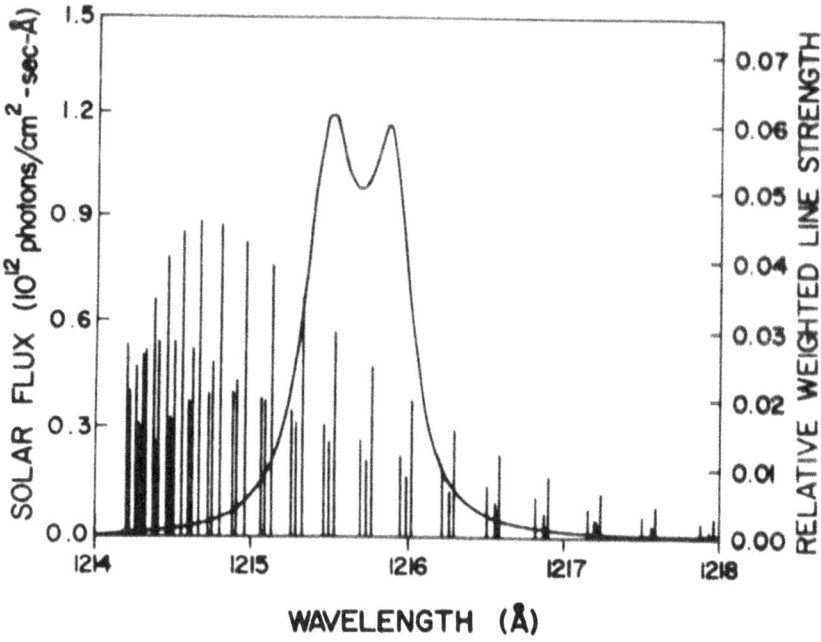

Fig. 14.   The rotational lines of the (14, 0) CO$^+$ fourth positive band showing the overlap with the solar
Lα line. Taken from Durrance *et al.* (1980).

estimated as arising from CO, thus bringing the ratio of 1304/1356 Å intensities to about
8, in agreement with the terrestrial value. Fox and Dalgarno (1981) presented altitude
profiles of the sources of the CO fourth positive band system from a pre-Pioneer Venus
model shown here as Figure 15. Ninety percent of the observed intensity of the band
system is due to fluorescent scattering, with the other four sources contributing about
equally to the observed intensity. The computed integrated overhead volume emission
rate is about 20 kR.

The Cameron band system of CO appears in the 1800 to 2600 Å region of the
spectrum. It arises from the dipole forbidden transition

$$CO(a^3 \Pi; v') \to CO(X^1 \Sigma; v'') + h\nu.   (29)$$

The CO($a^3 \Pi$) state is metastable with a relatively short lifetime of about 8 ms (Johnson,
1972; Lawrence, 1971). The potential sources are the same as those for the fourth
positive band system, except for fluorescent scattering. The Cameron band system is
the brightest feature in the Martian ultraviolet airglow measured by the UV spectrometers on Mariners 6, 7, and 9 (e.g., Stewart *et al.*, 1972; Barth *et al.*, 1971). The PV OUVS
longer wavelength channel measured spectra in the region of the Cameron bands, but
no observations or analyses have been reported. Table V gives intensities predicted by
a pre-Pioneer Venus model of Fox and Dalgarno (1981) and unpublished calculations
for a model based on Pioneer Venus data. The source due to dissociative recombination

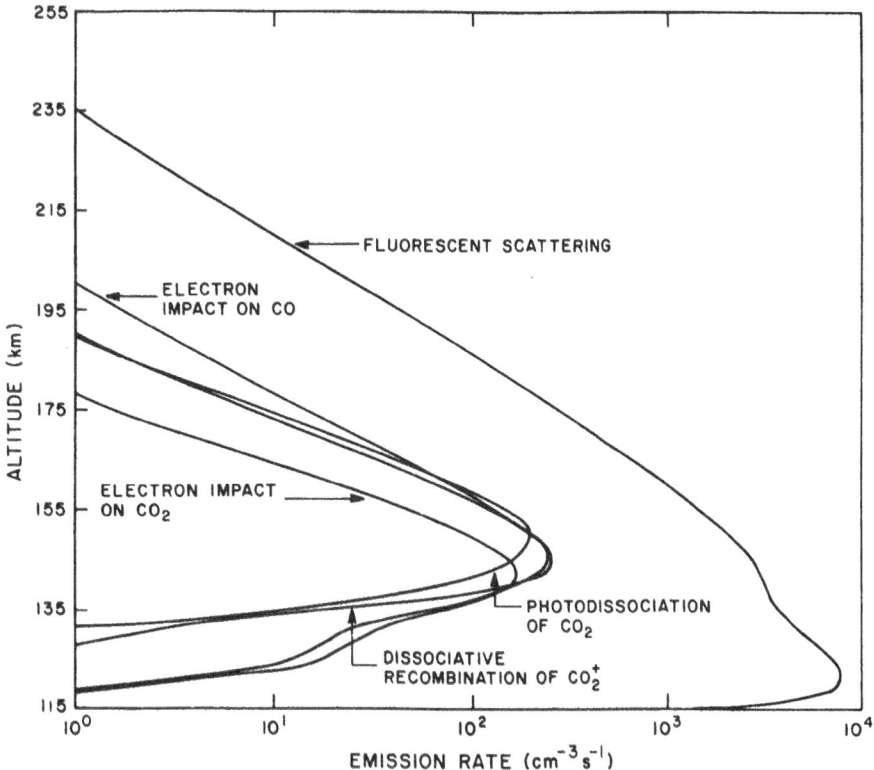

Fig. 15.   Altitude profiles of the sources of the CO fourth positive band system. Fluorescent scattering of Lα in the $v' = 14$ progression is responsible for 90% of the intensity. The shoulder above the peak in the fluorescent scattering source is due to absorption of shorter wavelength photons. Taken from Fox and Dalgarno (1981).

TABLE V

Computed Cameron band overhead integrated intensities due to different sources for high solar activity (high SA) and low solar activity (low SA) models

Source	Intensities (kR)	
	Low SA[a]	High SA[b]
$CO_2 + h\nu$	6.7	20.8
$e + CO_2$	6.7	20.9
$e + CO$	1.3	11.1
$CO_2^+ + e$	5.1	4.5
Total	20	57

[a] From Fox and Dalgarno (1981) (pre-Pioneer Venus model).
[b] Pioneer Venus model, based on neutral densities from Hedin *et al.* (1983) (cf. Fox, 1982, 1985).

of CO_2^+ is overestimated in the pre-Pioneer Venus model because the atomic oxygen densities were too low and consequently, the densities of CO_2^+ are too large.

The source due to electron impact excitation of CO_2 is uncertain, because the normalization of the cross section is a matter of some controversy. Excitation functions have been measured by several workers but reported values for the maximum value of the cross section range from 3×10^{-17} to 3×10^{-16} cm^2 (Freund, 1971; Ajello, 1971; Wells *et al.*, 1972; Erdman and Zipf, 1983). Conway (1981) constructed a synthetic spectrum of the Martian dayglow from 1800 to 2800 Å, and by comparing the model and measured intensities, found that a cross section with a maximum value of 7×10^{-17} cm^2 was consistent with the data. Taking into account new information about the lifetime of the $CO(a^3\Pi)$ state, Erdman and Zipf (1983), however, reported that laboratory data indicated a maximum value of the cross section closer to 2.4×10^{-16} cm^2, more than three times the Conway value. The electron-impact excitation rates in Tables II aand V are based on the cross sections of Conway. Analysis of the Pioneer Venus Cameron band data, if they are obtainable, might contribute to a resolution of this issue.

The CO^+ comet tail and first negative bands arise from the transitions: $A^2\Pi \to X^2\Sigma^+$ and $B^2\Sigma \to X^2\Sigma^+$, respectively. The thresholds for production of the A and B states are 2.53 and 5.65 eV, respectively (Herzberg, 1950). Therefore, the comet tail bands appear in the visible spectrum and the first negative bands in the ultraviolet. Potential production mechanisms include photoionization and electron impact ionization of CO, photodissociative and electron impact dissociative ionization of CO_2, and fluorescent scattering of solar radiation by ground state CO^+. No emissions have been observed that are attributable to either band system. Paxton (1988) has proposed that observations of these band systems by instruments on future spacecraft could help to determine the fraction of the mass-28 ion densities (which were measured by the PV OIMS (e.g., Taylor *et al.*, 1980)) that is due to CO^+, rather than to N_2^+. He has predicted a nadir intensity of the (0, 0) first negative band at 2191 Å of 615 R; for the (3, 0) comet tail band at 4011 Å, the predicted nadir intensity is 45 kR. Tables II and III show that for high solar activity the predicted integrated overhead intensities due to photoionization and electron impact ionization of CO alone are 4.5 kR and 1.6 kR for the comet tail and first negative band systems, respectively.

3.6. CO_2^+ EMISSIONS

The Fox–Duffendack–Barker $(A^2\Pi_u \to X^2\Pi_g)$ band system consists of several narrow bands between 2800 and 5000 Å. The ultraviolet doublet at 2883 and 2896 Å is produced by the (0, 0, 0)–(0, 0, 0) transition between the $B^2\Sigma_u^+$ state and the $X^2\Pi_g$ ground state. Both band systems are produced by photoionization and electron impact ionization of CO_2 and by fluorescent scattering of sunlight. They appear as prominent emissions in the Mariner 6, 7, and 9 UV spectra of Mars (Barth *et al.*, 1971, 1972) and should be present in Venus spectra as well. No intensities have, however, been reported from the Pioneer Venus OUVS.

Intensities of both CO_2^+ band systems were predicted by Dalgarno and Degges

(1971). They computed g values for fluorescent scattering at Venus of 4.9×10^{-2} and 5.2×10^{-3} for the $A \to X$ and $B \to X$ band systems, respectively. The total sub-solar zenith intensities were 107 kR and 31 kR, with 87 and 9 kR from fluorescent scattering for the Fox–Duffendack–Barker band system and the ultraviolet doublet, respectively. The contribution from fluorescent scattering was overestimated because at that time CO_2^+ was assumed to be the major ion.

The low solar activity model of Fox and Dalgarno (1981) predicted a total intensity of 15 kR for the Fox–Duffendack–Barker band system and 9.2 kR for the ultraviolet doublet. The most important source of both emissions is photoionization of CO_2. The branching ratios in photoionization are uncertain, since photoelectron spectroscopy and fluorescence measurements imply different values (Samson and Gardner, 1973; Wauchop and Broida, 1972; Lee and Judge, 1972; Leach et al., 1978a). It has been suggested that a mixing of the A and B states occurs before emission, so that about 50% of the ionization into the B state leads to emission in the A state (Samson and Gardner, 1973; Leach et al., 1978b). The predicted intensities based on the fluorescence branching ratios are 19 kR for the Fox–Duffendack–Barker bands and 5.4 kR for the ultraviolet doublet. Revised calculations based on a Pioneer Venus model for high solar activity are shown in Table VI. The integrated overhead intensity of the Fox–Duffendack–

TABLE VI

Computed integrated volume emission rates of CO_2^+ features in the Venus dayglow for a model (high solar activity) based on Pioneer Venus data

Source	Intensity (kR)	
	$A^2\Pi_u \to X^2\Pi_g$	$B^2\Sigma_u^+ \to X^2\Pi_g$
Photoionization of CO_2	9.0 (14.1[a])	10.2 (5.1[a])
Electron impact ionization of CO_2	3.7	2.2
Fluorescent scattering	4 7	0.25
Total	17.4 (23[a])	12.6 (7.5[a])

[a] Computed assuming a 50% crossover of B to A before radiating.

Barker band system is 17.4 kR, with 12.7 kR from electron impact ionization and photoionization and 4.7 kR from fluorescent scattering. For the ultraviolet doublet the predicted intensity from photoionization and electron impact ionization is 12.4 kR and that from fluorescent scattering is negligible, only 0.25 kR, for a total integrated over-head intensity of 12.6 kR. If a 50% crossover from B to A occurs before radiating, the total intensities of the $B \to X$ and $A \to X$ band systems would be 7.5 and 23 kR, respectively.

3.7. He AND He$^+$ EMISSIONS

The He abundance in the lower atmosphere has not been measured directly. The large probe mass spectrometer (LNMS) had the capability of measuring He densities, but the mixing ratio indicated (f_{He}) by the LNMS data, about 450 ppm, was considered unreliable because of contamination by the He–N$_2$ mixture used to fill the probe to detect leaks (Hoffman et al., 1980a; Donahue and Pollack, 1983). Thus bulk atmosphere mixing ratios have been obtained only by extrapolation from upper atmospheric values obtained from measurements by the PV BNMS and ONMS instruments and from airglow analyses. Because the mass of He is much less than the average molecular weight, this procedure is extremely sensitive to assumptions about the profiles of eddy diffusion coefficients and temperatures. The eddy diffusion coefficients presented by von Zahn et al. (1980) were based on a fit to the He profile measured by the BNMS above 130 km (see Section 2). The He homopause was located at 130 km, the lower boundary of the measurements. The morningside model gives a value for f_{He} at this altitude of about 50 ppm. An extrapolation to 100 km yields a mixing ratio of about 12 ppm. The VTS3 model is based largely on ONMS measurements, but the He profile below 130 km is based on BNMS values. Nonetheless, Hedin et al. (1983) found the altitude of the He homopause to be lower, about 125 km, and the lower atmosphere mixing ratio to be correspondingly lower; values of 1–2 ppm are obtained at the lower boundaries of the noon and midnight models. Massie et al. (1983) tested the sensitivity of their model to the value of the coefficient A in the numerator of the von Zahn et al. (1980) expression for the eddy diffusion coefficient (Equation (1)); they found that changing the value from 0.8 to 2.0 changed f_{He} for the lower atmosphere from 2 to 9 ppm. For the value $A = 1.4 \times 10^{13}$ preferred by von Zahn et al., they derived a helium mixing ratio of 5 ppm.

Thermospheric He densities above 140 km given by the VTS3, BNMS, and VIRA models are, however, in fairly good agreement. He densities at 150 km for solar zenith angles less than about 60° are 4–5 $\times$ 10^6 cm^{-3}. ONMS data show a strong diurnal variation in the He densities that is out of phase with the heavier constituents, with larger densities at night than during the day and a distinct bulge in the He densities in the dawn sector (Hedin et al., 1983). Figure 16 shows the densities of species included in the VTS3 model at 100 and 150 km as a function of hour angle. The VIRA model for 04:00 hr local time shows a density at 150 km of 1.2 $\times$ 10^8 cm^{-3}; closer to midnight the density is less, about 7 $\times$ 10^7 cm^{-3}.

He resonance radiation at 584 Å was detected by the both the Mariner 10 and Venera 11 and 12 spectrophotometers (Broadfoot et al., 1974; Bertaux et al., 1981). A maximum limb intensity of 600 R was measured by Mariner 10. The Mariner 10 bright limb profile from 200 to 800 km was analyzed by Kumar and Broadfoot (1975). They used a radiative transfer model that assumed spherical-symmetry and complete frequency redistribution to fit the shape of the limb profile. The exospheric temperature and the g factor were taken to be free parameters. The 'best fit' exospheric temperature was 375 $\pm$ 105 K. Takacs et al. (1980) claim that a reanalysis of the data suggests that a temperature of 270 K is a better fit to the data. A look at Figure 17, taken from Kumar and Broadfoot (1975), does seem to suggest, with the full benefit of hindsight, that a lower exospheric temperature would be an equally good fit. The line center solar flux derived from the best fit g factor was 2 $\times$ 10^{10} cm^{-2} s^{-1} Å^{-1} at 1 AU. A He density

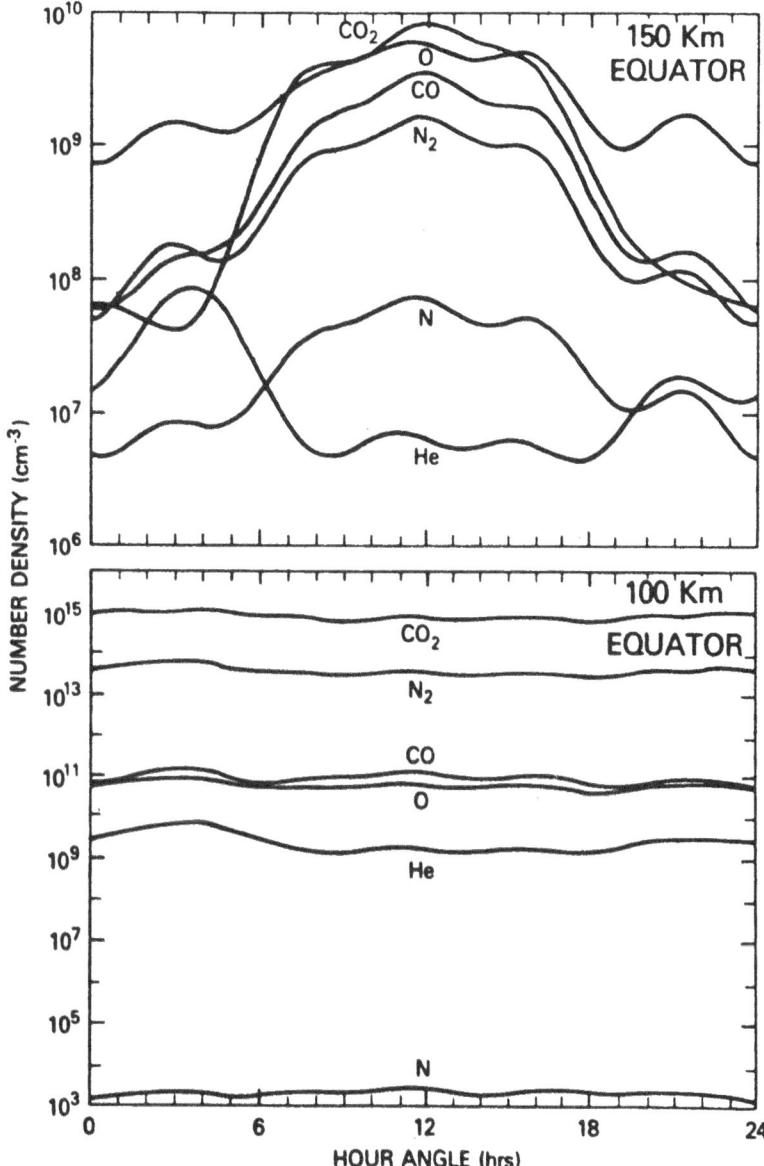

Fig. 16.   Neutral number densities as a function of local time at 100 and 150 km, 0° latitude from the VTS3 model. Note that the daytime/nighttime ratio increases with increasing mass of the species. For helium (and hydrogen) the densities are larger on the nightside and peak in the predawn sector, forming the 'helium (or hydrogen) bulge'. Taken from Hedin *et al.* (1983).

of 2×10^6 cm^{-3} was derived at the assumed homopause (145 km), where the density of CO_2 was assumed to be 2×10^{11} cm^{-3}. Von Zahn *et al.* (1983) comment that if the lower (and more accurate, according to our present knowledge) temperature is assumed, the He abundance is almost exactly that determined by PV. We find that the column

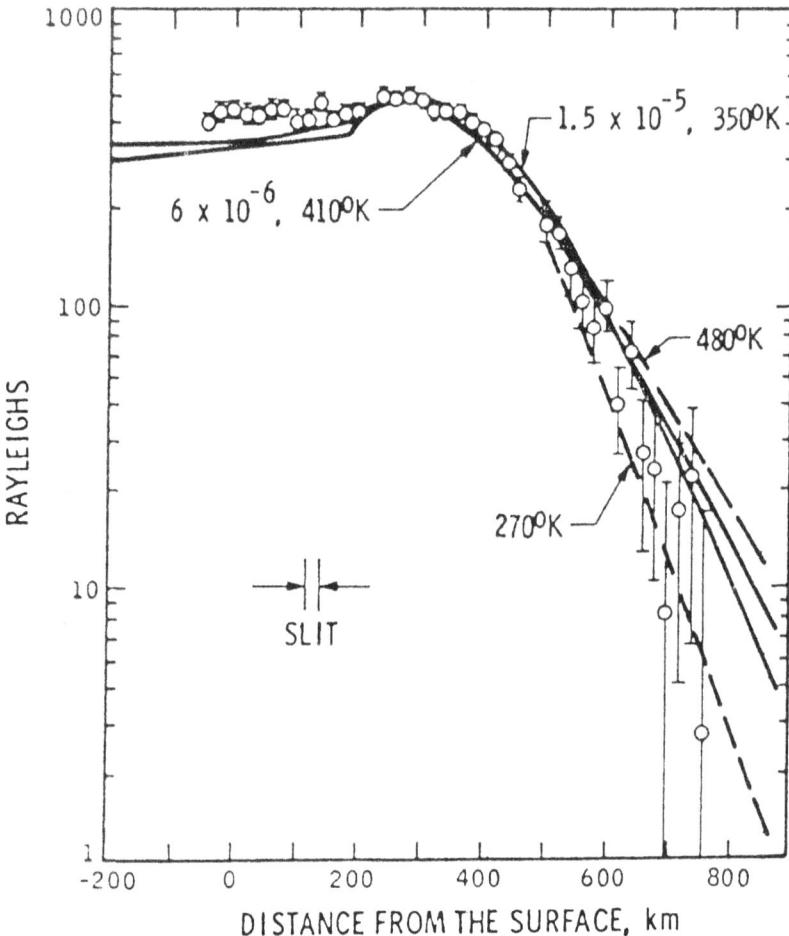

Fig. 17.   Mariner 10 584 Å brightness as a function of distance from the limb. The points were derived by subtracting the background noise level. The dashed lines represent altitude variations based on a single scattering model. The solid lines are the result of radiative transfer calculations including multiple scattering. The calculations are normalized to the peak observed brightness. From Kumar and Broadfoot (1975).

density above 145 km, the presumed CO_2 absorption altitude, deduced from the Mariner 10 measurements is about 2×10^{13} cm^{-2}, whereas the corresponding value for the BNMS model is about 2.9×10^{13} cm^{-2}. Because the exospheric temperature is lower than Kumar and Broadfoot assumed, the CO_2 absorption altitude is, however, lower in Pioneer Venus models. The CO_2 absorption cross section at 584 Å is about 2.3×10^{-17} cm^{-2}, so the absorption altitude is about 140 km in the VIRA dayside model and 136 km in the BNMS model. The He column density above 136 km in the BNMS model is about 3.7×10^{13} cm^{-2}.

The He 584 Å disk brightnesses measured by Venera 11 and 12 were presented by Bertaux *et al.* (1981). The maximum intensity on the disk was 270–280 R. Several analyses of the Venera data have appeared. Kurt *et al.* (1983) modeled the intensity across the disk, using the He optical depth as a free parameter. The best fit He optical depth, in the range 2–4, is independent of the solar flux and the instrument calibration. The derived He column density above the CO_2 absorption altitude, taken to be 140 km, is about $1.9–2.1 \times 10^{13}$ cm^{-3}, almost a factor of two less than the value from the Pioneer Venus models. Bertaux *et al.* (1985) presented a summary of similar results; a helium optical depth of 2.5, corresponding to a density of 1.6×10^6 cm^{-3} at 150 km, was required to fit the intensity profile across the disk. The implied column density is smaller than the PV value. Chassefiere *et al.* (1986) revised the required optical depth to 3.5 ± 1.5, for a number density at 150 km of 2.2×10^6 cm^{-3}. They suggest that the discrepancy with the Mariner 10 data may be due to poor absolute calibration of the Mariner instrument. Krasnopol'sky (1983a) has suggested that the Mariner 10 instrument had a more serious problem with stray light than the Venera instruments had. Although the hydrogen densities derived from Mariner 5 and 10 Lα data are a factor of 2 larger than the values derived from PV OUVS data (Paxton *et al.*, 1985, 1988) and from the *in situ* PV OIMS H$^+$ densities (Brinton *et al.*, 1980), Paxton *et al.* (1985) showed that the fault lay with the use of an isothermal spherically-symmetric radiative transport scheme. They also suggested that fly-by observations do not provide enough constraints to determine the distribution of scatterers. This may be the case for the He 584 Å data also. Indeed, given the large solar zenith angle variation of He densities, especially the existence of the pre-dawn bulge, one would expect a spherically-symmetric radiative transfer model, as all of the above models are including that of Kumar and Broadfoot (1975), to be a poor approximation. Furthermore, the analysis of Kurt *et al.* (1983) is isothermal and assumes a temperature in the scattering region of 300–400 K, whereas the dayside temperature varies from about 200 K near 140 km to about 275 above 180 km.

The solar fluxes at the 584 Å line center that are assumed or derived are different even for similar solar activities, illustrating the inadequacy of the analyses and/or a calibration problem with the instruments. Bertaux *et al.* (1981) determined a value from the Hinteregger 79050N solar fluxes, although there is some evidence that the Hinteregger fluxes are too low, by up to a factor of 30% at high solar activity (Ogawa and Judge, 1986; see Section 3.3). If so, the problem would be exacerbated, because a larger solar flux implies smaller He densities. Furthermore, Hinteregger presents the photon fluxes in the lines as integrated values and it is necessary to assume a lineshape to derive the line center fluxes; Bertaux *et al.* (1981) state merely that an appropriate line shape was assumed, but give no details. Doschek *et al.* (1974) found the shape of the 584 Å line to be approximately Gaussian with a slightly flattened core. Ogawa *et al.* (1984) report a value for the full width at half maximum of about 0.1 Å from a rocket measurement, in good agreement with previously measured values. For a Gaussian shape with a width of 0.1 Å, the line center solar flux implied by the Hinteregger integrated flux would be about 4.9×10^{10} cm^{-2} s^{-1} Å^{-1}. Other analyses fit the shape of the intensity profile and derive a value for the solar flux at line center that reflects the uncertain calibration of the instrument. Kurt *et al.* (1983) derive a value of 3.3×10^{10} cm^{-2} s^{-1} Å^{-1}, which

they state (incorrectly) agrees with the fluxes reported by Hinteregger. Bertaux *et al.*
(1985) give a line center photon flux of 8.6×10^9 cm^{-2} s^{-1} Å^{-1} for the best fit of the
BNMS model to their disk data, and report that a column density about a third of the
BNMS data gives a better fit to the shape of the intensity data. The value required by
Kumar and Broadfoot (1975), at a time of lower solar activity, was
2×10^{10} cm^{-2} s^{-1} Å^{-1}. The solar activity variation determined from the (high solar
activity) F79050N and the (low solar activity) SC No. 21REFW solar spectra of
Hinteregger is a factor of about 3, although if the Hinteregger fluxes are low by a factor
of 2 at low solar activity and by 30% at high solar activity (Ogawa and Judge, 1986),
the solar activity variation is reduced to a factor of 2.

The He$^+$ line at 304 Å was not detected by Mariner 10, although one of the channels
was designed to detect emissions at that wavelength. It was, however, measured by
Venera 11 and 12, due to the greater sensitivity of the spectrophotometer. Bertaux *et al.*
(1981) computed a column density of He$^+$ of 2×10^{11} cm^{-2}, which is larger by an order
of magnitude than the column density of about 2×10^{10} cm^{-2} implied by the PV OIMS
measurements.

3.8. N_2 AND N_2^+ EMISSIONS

No emission arising from the band systems of N_2 has been detected in the Venus
dayglow, even thought N_2 comprises 3–4% of the atmosphere. The intensities of the
most important bands of the major band systems, including the Vegard–Kaplan
$(A^3\Sigma_u^+ \rightarrow X^1\Sigma_g^+)$, the first positive $(B^3\Pi_g \rightarrow A^3\Sigma_u^+)$, the reverse first positive
$(A^3\Sigma_u^+ \rightarrow B^3\Pi_g)$, the $W^3\Delta_u \rightarrow B^3\Pi_g$, the second positive $(C^3\Pi_u \rightarrow B^3\Pi_g)$, and the
Lyman–Birge–Hopfield $(a^1\Pi_g \rightarrow X^1\Sigma_g^+)$ band systems were predicted by Fox and
Dalgarno (1981). These bands, the upper states of which are connected to the ground
state by dipole forbidden transitions, are produced only by photoelectron impact excitation of ground state N_2 in the dayglows of the terrestrial planets. The Vegard–Kaplan,
second positive and Lyman–Birge–Hopfield bands emit in the ultraviolet. The (1, 9)
Vegard–Kaplan band at 3199 Å, which falls between the (2, 0) and (3, 0) bands of the
Fox–Duffendack–Barker band system, and which may have been observed in the
Martian dayglow spectrum (Fox and Dalgarno, 1979), was predicted to have an intensity
of 77 R and should be detectable. The (3, 0) Lyman–Birge–Hopfield band at 1354 Å
may contaminate the signal at the O 1356 Å feature; the predicted intensity at 1354 Å
was 25 R for low solar activity model of Fox and Dalgarno (1981). Calculations based
on the O 1356 Å emission cross sections of Stone and Zipf (1974) indicated that the
1354 Å intensity would contribute little to the observed 1356 Å signal. Revised calculations, using the renormalized 1356 Å electron impact emission cross sections of Zipf
and Erdman (1985) give a lower integrated intensity for the 1356 Å emission of about
150 R at low solar activity (see Table II). Hence, the (3, 0) Lyman–Birge–Hopfield band
may contribute about 15% of the observed intensity.

The N_2^+ Meinel $(A^2\Pi_u \rightarrow X^2\Sigma_g^+)$ and first negative $(B^2\Sigma_u^+ \rightarrow X^2\Sigma_g^+)$ band systems can
be produced by photoionization and electron impact ionization of N_2, by fluorescent
scattering of solar radiation, and by electron induced fluorescence (Degen, 1981). The

latter source is negligible in the dayglow. The Meinel band system occurs in the infra-red; the most intense band in the first negative band system is the (0, 0) band at 3914 Å. No emission at that wavelength was recorded by the visible spectrometers on Venera 9 and 10. An upper limit of 45 R has been placed on the overhead integrated intensity at 3914 Å (Krasnopol'sky, 1978).

Fox and Dalgarno (1981) predicted the intensities of the Meinel and first negative bands due only to photoionization and electron impact ionization. Paxton (1988) has suggested that intensities of the N_2^+ first negative bands could be used to determine the fraction of the mass-28 ion densities due to N_2^+. He has predicted an overhead intensity for the 3914 Å band of 17 kR, using the neutral Hedin model appropriate to the conditions of Pioneer Venus orbit 185, but this prediction must be in error. Fox and Dalgarno (1983, 1985) have computed the vibrational distribution of N_2^+ in the terrestrial and Martian atmospheres, where vibrationally excited N_2^+ is produced mainly in fluorescent scattering of solar radiation in the first negative and Meinel band systems. The same model has been adapted for the ionosphere of Venus (cf. Fox, 1982a) and the emission rates in the N_2^+ band systems computed. The results are presented in Table VII. The integrated overhead intensity of the entire first negative band system is

TABLE VII

Computed overhead integrated intensities of the Meinel ($A \rightarrow X$) and first negative ($B \rightarrow X$) band systems of N_2^+ for a high solar activity ($F_{10.7} = 200$) VTS3 model

Source	Intensities (kR)	
	Meinel	First negative
Photoionization	2.0	0.41
Electron impact	0.26	0.00088
Fluorescent scattering	1.7	1.4
Total	4.0	1.8

only 1.8 kR. The integrated overhead intensity of the 3914 Å band is 790 R; for the (1, 0) Meinel band the predicted intensity is 910 R. Fluorescent scattering is the major source of both band systems; only about a quarter of the emission of the 3914 Å is from photoionization and electron-impact ionization.

3.9. N EMISSIONS

The $N(^2D)$ and $N(^2P)$ states of atomic nitrogen are metastable, with radiative lifetimes of about 10^5 and 12 s, respectively. They are produced in photodissociation and electron impact dissociation of N_2 and in dissociative recombination of N_2^+. The transition $N(^2D) \rightarrow N(^4S) + h\nu$ produces a photon of wavelength 5200 Å; the transition of $N(^2P)$ to the $N(^2D)$ state leads to emission at 10400 Å; the transition to the ground $N(^4S)$ state leads to emission at 3466 Å. Like $O^+(^2D)$ and $O^+(^2P)$, the importance of these metastable states lies in their effect on thermospheric and ionospheric chemistry, rather than in strong emission rates (Fox, 1982a). No measurements of the emissions have been reported, but the intensities have been modeled for a high solar activity Pioneer Venus model by Fox (1982a). The computed integrated overhead intensities are given in Table IV.

3.10. NO NIGHTGLOW

The ultraviolet night airglow was first mapped by the PV OUVS during the first diurnal period of the mission (Stewart and Barth, 1979; Stewart et al., 1980; Gerard et al., 1981). The airglow was detected and identified as the gamma and delta bands of nitric oxide (Feldman et al., 1979; see also Stewart and Barth, 1979) excited by the radiative recombination of N and O atoms. Since no emissions suggestive of a nightside source of N and O atoms were present, it was proposed that the airglow was the result of these atoms being transported from their source on the dayside to the nightside where recombination occurs. Thus the morphology and brightness of the airglow can provide a sensitive tracer of the thermospheric circulation. Other PV measurements, particularly the hydrogen and helium nightside bulges (Taylor et al., 1984; Hedin et al., 1983) (see Figure 16), substantiate this concept of a strong day-to-night Venus thermospheric circulation (Mayr et al., 1980). Modeling efforts have been very successful in validating the proposed transport mechanism yielding the NO nightglow distribution and intensity. However, chemical questions remain regarding the dayside N-atom production available for transport to the nightside (Bougher et al., 1990a).

Planet-wide observations by the OUVS at solar maximum initially showed that the NO(0, 1) δ-band airglow typicaly exhibits a bright spot reaching 2.8 kR near 02 : 00 local time just south of the equator (Stewart et al., 1980). This statisticaly averaged peak brightness is roughly four times the dark disk average vertical emission of 780 R as calculated for 25-orbits sampled early in the PV mission (Stewart et al., 1980). The intensities reported by Stewart et al. (1980) have been revised as described by Bougher et al. (1990a). Briefly, recent recalibration of the OUVS instrument has yielded an improved sensitivity for this early PV period, 52.5 counts $s^{-1} kR^{-1}$. An improved algorithm was also developed that conserved photon counts, revealing a factor of two error in previous NO nightglow data processing. Finally, the calculation of the Stewart et al. (1980) dark disk airglow intensity omitted a latitudinal weighting factor for area; its incorporation further reduces the observed dark disk intensity. The updated nightglow dark-disk average and peak intensities are now $400-460 \pm 120$ R and 1.9 ± 0.6 kR, respectively, for the 1978–1980 OUVS observational period. Individual orbit nightglow patches were seen to vary significantly in intensity and location on time-scales of an Earth day or less (Bougher, 1980; Stewart et al., 1980; Bougher et al., 1990a). The mean altitude distribution of the NO airglow was determined from limb profiles obtained near periapsis, with a 5 km vertical resolution. The altitude of the peak emission was found to be 115 ± 2 km (Gerard et al., 1981).

The OUVS sensitivity was shown to degrade substantialy after orbit insertion (December 1978). Subsequent monitoring of the nightglow beyond the first three diurnal periods (1979–1980) incorporated this effect. The PV spacecraft periapsis altitude was too high during 1984–1986 for the OUVS to obtain meaningful limb profiles of the solar minimum nightglow. However, several measurements during this period from apoapsis showed an average nightside bright patch intensity roughly a factor of three smaller than observed previously during 1979–1980. A substantial solar cycle variation in the night-glow emission is thus confirmed (Stewart, 1989, private communication).

The NO nightglow is particularly valuable in constraining the Venus thermospheric circulation in the absence of measured winds (Bougher, 1980; Stewart et al., 1980; Bougher et al., 1990a). The average brightness distribution map of Stewart et al. (1980) (revised in Bougher et al. (1990a)) can be viewed as an approximate map of the downward flux of N-atoms being supplied from the dayside. Correspondingly, the intensity of the resulting nightside emission is proportional to the dayside column production of N-atoms; i.e., dayside average net column production must balance the nightside average column destruction of ground state N-atoms. Since the NO emission occurs well below the nightside homopause, vertical transport other than molecular diffusion must be important, either due to eddy mixing and/or large-scale winds. In addition, the local time position of the average peak emission (and bright patch intensity) provides evidence for asymmetries in the otherwise mean sub-solar-to-anti-solar thermospheric circulation (Stewart et al., 1980; Bougher et al., 1990a). Superimposed retrograde zonal winds having a period of 4–6 days are also inferred from measured helium densities (Mayr et al., 1980; Mengel et al., 1989).

The basics of the Venus dayside odd-nitrogen chemistry are summarized in Figure 18. The major processes are little different from those operating on the Earth. CO_2, in fact, plays a similar role on Venus as O_2 does in the lower thermosphere of the Earth; i.e., both species react rapidly with $N(^2D)$ but not with $N(^4S)$. The initial steps leading to the production of odd-nitrogen through N_2 bond-breaking are accomplished through the same processes. The principal sources are (1) the dissociation of N_2 by electron impact, (2) the dissociative ionization of N_2 by electrons, and (3) the predissociation of N_2 by 80–100 nm photons. An effective branching ratio ($f = 0.5$) is typically chosen for the partitioning of $N(^2D)$ and $N(^4S)$ atoms that are produced (Cleary, 1986). The sequential reactions $N(^2D) + CO_2$ producing NO followed by NO + $N(^4S)$ yielding N_2 are by far the most important in reducing the total dayside production of excited and ground state N-atoms to the net available for nightside transport. The reaction producing NO^+ followed by its dissociative recombination serve to redistribute the forms of atomic-N. A standard value of $g = 0.75$ for branching to $N(^2D)$ is usually chosen (Kley et al., 1977). In addition, partial recovery of $N(^4S)$ atoms is obtained by quenching of $N(^2D)$ by O and CO above the $N(^4S)$ peak. The precise value of the $N(^2D) + O$ quenching coefficient is very important for understanding the terrestrial odd-nitrogen chemistry (cf. Fesen et al., 1989).

A preliminary study of the neutral and ion odd nitrogen chemistry was made by Rusch and Cravens (1979) soon after the first Pioneer Venus results were available. Their model predicted that ground state $N(^4S)$ atoms were the major neutral odd nitrogen species, with a peak concentration of about 2×10^7 cm^{-3} near 130 km. The integrated

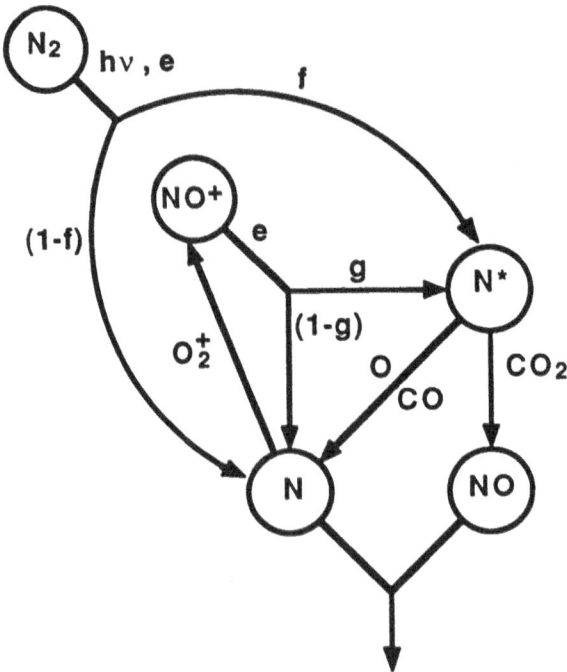

Fig. 18.   Simplified Venus dayside odd-nitrogen chemical scheme. Taken from Bougher *et al.* (1990a).

column production of atomic nitrogen produced in their model (for 70° SZA) was 4.5×10^9 cm^{-2} s^{-1}. Fox (1982) predicted a slightly larger maximum N(^4S) density of 4×10^7 cm^{-3}. Later mapping studies and the nightside chemical-diffusive modeling of Stewart *et al.* (1980) showed that the originally reported dark disk nightglow intensity (780 R) required a downward flux of N-atoms of 1×10^{10} cm^{-2} s^{-1}. This flux should be reduced by a factor of nearly two due to the recent downward revision of the intensities reported by Stewart *et al.* (1980). No correlation was found between the nightglow brightness and the 10.7 cm solar fluxes for the observational period. The nightside vertical diffusion model, and additional limb profile studies (Gerard *et al.*, 1981), furthermore, placed the peak volume emission rate altitude at 115 km using eddy diffusion of the same order as that used by von Zahn *et al.* (1980) for maintaining observed dayside densities. Thus, a determination of the altitude of the emission was found to be a sensitive indicator of the required nightside eddy diffusion profile (Stewart *et al.*, 1980). Either global dynamics and/or small-scale mixing could be the underlying mechanism responsible. An adequate supply of dayside N-atoms appeared to be available for sustaining the observed nightglow.

This conclusion was challenged by Krasnopol'sky (1983), who concluded, based on a simplified dayside atomic-N model, that virtually all the dayside produced N-atoms would be required on the nightside to account for the NO nightglow. His dayside total column production of both ground state (N(^4S)) and excited (N(^2D)) atoms was

1.1×10^{10} cm^{-2} s^{-1}. Virtually all of the N(^2D) would have to be quenched to N(^4S) rather than be converted to NO to sustain the required dayside to nightside flux. Presumably, this problem has been resolved with the reduction in the observed emission intensities.

Gerard *et al.* (1988) re-examined this problem using a 1-D photochemical-diffusive model that calculated N(^4S), N(^2D), and NO at selected SZA over the dayside. The daytime ionosphere and odd-nitrogen chemistry were coupled to obtain N(^4S) profiles and local time distributions for comparison with PV measurements obtained with the neutral mass spectrometer (Kasprzak *et al.*, 1980; Hedin *et al.*, 1983). The average amount of total N-atoms produced on the dayside was estimated at 1.3×10^{10} cm^{-2} s^{-1}. A maximum N(^4S) concentration of about 8×10^7 cm^{-2} was calculated at 124 km near LT = 17:00. Calculated N(^4S) concentrations near 150 km were substantially less than those given by the empirical model of Hedin *et al.* (1983) (see Figure 19). However, a reasonable agreement with the original PV densities (Kasprzak *et al.*, 1980) was obtained. Approximately 50% of the dayside source was required for nightward transport to maintain the originally reported NO nightglow

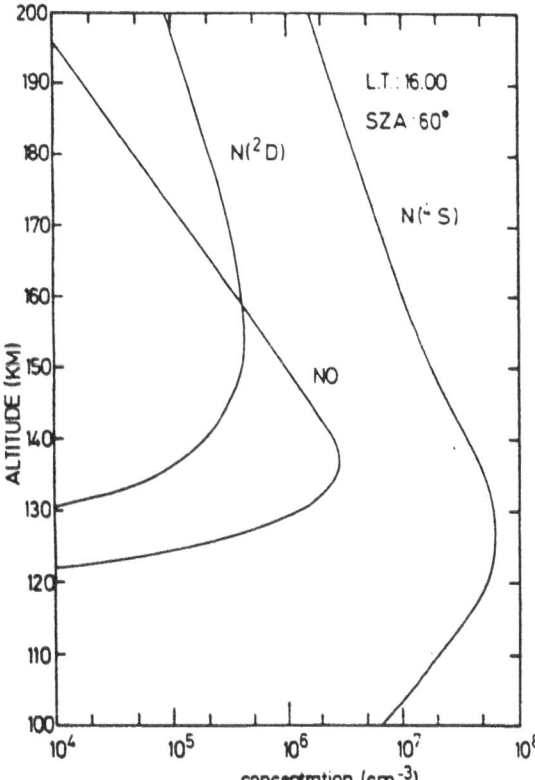

Fig. 19.   Vertical distribution of N(^4S), N(^2D), and NO number densities as a function of altitude calculated at the equator for a local time of 16:00. Taken from Gerard *et al.* (1988).

intensities; the fraction would be smaller for the revised intensities. This estimate reflects a higher branching ratio for the $(0, 1)$–δ band resulting from NO recombination (0.32, after McCoy (1983)), which produces a required N-flux of 7×10^9 cm^{-2} s^{-1}. The model showed little sensitivity to the value of the efficiency of the N(^2D) production by N$_2$ dissociation ($f = 0.5$ to 0.7); also, the calculated N(^4S) was found to be rather insensitive to the N(^2D)–O quenching coefficient for values less than 2×10^{-12} cm^3 s^{-1}. This is so because the reaction of N(^2D) with CO$_2$ yielding NO dominates all other N-atom destruction at the peak of the N(^4S) density.

A comprehensive treatment of this flux problem requires the use of a three-dimensional model coupling the hydrodynamical flow with photochemical production and loss of odd-nitrogen. The results of the Gerard *et al.* (1988) detailed photochemical model were subsequently used in a 3-D Venus thermospheric general circulation model (VTGCM) to parameterize the sources and sinks of atomic nitrogen in the Venus dayside thermosphere (Bougher *et al.*, 1990a). The VTGCM (see Section 6.2), an adaptation of the terrestrial NCAR TGCM, calculates global distributions of CO$_2$, CO, and O consistent with the 3-D model day–night temperature contrasts and corresponding large-scale winds. Calculations of the observed nightglow distribution and intensity provide a further means to validate the basic model circulation and thermospheric structure. The model N-atoms produced are transported to the nightside by the VTGCM circulation and destroyed primarily by radiative recombination with O atoms, and mutual destruction with NO. NO, N(^4S), and N(^2D) are incorporated as minor species having no impact on the VTGCM major species, temperatures, or winds. NO and N(^2D) are treated assuming photochemical equilibrium, while N(^4S) was subject to diffusion and large-scale transport, in addition to chemistry.

Results for solar maximum conditions indicate that the recently revised dark-disk average NO(0, 1) δ-band intensity at 198.0 nm (400–460 ± 120 R), based on statistically averaged PV OUVS measurements, can be reproduced with minor modifications in chemical rate coefficients and global mean eddy diffusion. The calculated average dayside production of total N-atoms is about 9.4×10^9 cm^{-2} s^{-1}; chemical losses resulting from NO production and subsequent N(^4S) destruction reduce this total production by a factor of 2 to 4. The nightward hemispheric flux of N-atoms required to yield 340 to 580 R of dark-disk average airglow is ~ 2.5–3×10^9 cm^{-2} s^{-1}, which is roughly 30% of the total dayside N-atoms produced. This is realized for VTGCM cases using standard rate coefficients and branching ratios (Gerard *et al.*, 1988), or slightly reduced NO production through a lower limit N(^2D) + CO$_2$ rate coefficient (3×10^{-13} cm^3 s^{-1}) (Piper *et al.*, 1987). A similar nightglow intensity could be simulated for somewhat weaker NO production coupled with enhanced global mean eddy diffusion ($K_t \leq 2 \times 10^7$ cm^2 s^{-1}). A large N(^2D) + O quenching coefficient, of the magnitude measured by Jusinski *et al.* (1988), is not required for adequate dayside N(^4S) production yielding observed nightglow intensities. The VTGCM peak intensity of 1.2 kR, calculated at LT = 01 : 00–03 : 00 near the equator, is in good agreement with PV OUVS observations (see Figure 20(a)). The altitude of the nightside peak emission layer (113 km) is also in agreement with PV OUVS data, and occurs where chemical

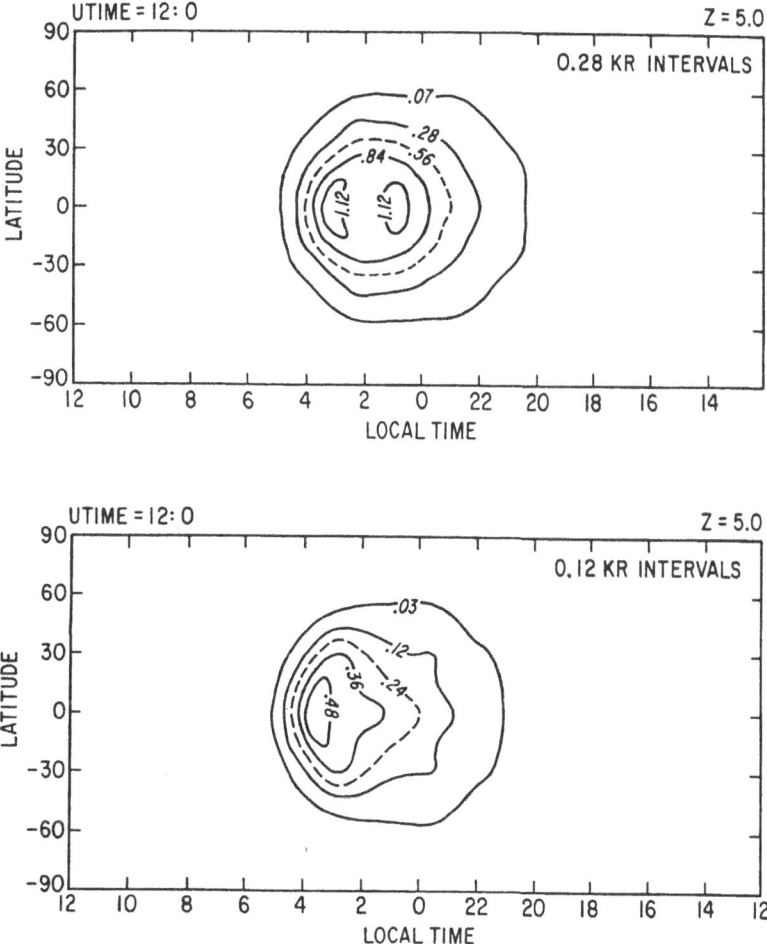

Fig. 20.   VTGCM NO(0, 1) δ-band vertical intensity distributions: (a) a solar maximum simulation, corresponding to OUVS data of 1978–1980, and (b) a solar minimum simulation, appropriate to the 1984–1986 period. The vertical intensity distribution corresponds to the vertically integrated volume emission rate. Taken from Bougher *et al.* (1990a).

and transport (global wind plus eddy diffusion) lifetimes are comparable. Use of enhanced global mean eddy diffusion ($K \sim 2 \times 10^7$ cm² s⁻¹) also yields atomic O and N densities that are consistent with observed PV values. Peak N densities of 6.5 to 9.2×10^7 cm⁻³ are obtained at LT = 17:00 at about 130 km. The simultaneous production of observed dayside N-densities, corresponding to Hedin *et al.* (1983) values, and observed NO nightglow intensities requires a slightly slower wind system (terminator horizontal winds ~ 150 m s⁻¹) than currently simulated by the VTGCM.

Corresponding solar minimum calculations predict that the NO nightglow intensities should be very sensitive to the net dayside production of N-atoms. The factor of two decrease in the N_2 bond-breaking rate combined with the reduced VTGCM solar-driven winds causes a reduction of nearly a factor of 3 in the dark-disk intensities. Figure 20(b) illustrates the VTGCM nightglow integrated vertical intensity distribution for solar minimum conditions; comparison can be made with Figure 20(a). This solar cycle variation is similar to that observed by the PV OUVS.

These VTGCM calculations confirm that the transport mechanism originally proposed is correct. Since this airglow is a good tracer of the thermospheric circulation, it provides a means to study remotely the variation of the Venus winds. The nightglow also provides a sensitive constraint on the magnitude and distribution of the global winds in the absence of available measurements, much like helium (Mengel et al., 1989). Furthermore, the Venus odd-nitrogen chemistry, since it is similar to that of the Earth, is also adversely impacted by the use of the large $N(^2D) + O$ quenching rate measured by Jusinski et al. (1988). It cannot be used for Venus and yield good dayside N-profiles plus NO nightglow intensities resembling Pioneer Venus observations. On Earth, present-day odd-nitrogen chemical schemes (e.g., Roble et al., 1987; Fesen et al., 1989) and observed $N(^2D)$ 520 nm airglow (Frederick and Rusch, 1977) cannot be reconciled with this large quenching rate. If $N(^2D) + O_2$ is the only important terrestrial NO source in the lower thermosphere, then the $N(^2D)$ production rate must be increased to compensate for the large quenching rate. Recent measurements of this rate coefficient, however, have yielded values near 1×10^{-12} cm^3 s^{-1}, a factor of 20 less than the Jusinski et al. (1988) rate (Piper, 1989; Miller et al., 1988). Further (Venus) progress requires future simultaneous measurements of $N(^2D)$, $N(^4S)$, and NO in the Venus lower thermosphere (Gerard et al., 1988). NO densities can also be derived from Venus nightside ion chemistry for comparison to VTGCM model values.

3.11. O_2 EMISSIONS

The first measurements of O_2 densities in the Venus atmosphere are those from the PV sounder probe gas chromatograph. Mixing ratios of 44 and 16 ppm were measured at 52 and 42 km; it appears that the mixing ratio decreases with altitude (Oyama et al., 1980). The gas chromatography experiments on Venera 13 and 14 measured an O_2 mixing ratio of 18 ppm for the altitude range 35 to 58 km (Mukhin et al., 1983). The PV sounder probe mass spectrometer determined an upper limit to the mixing ratio below the cloud tops of about 30 ppm (Hoffman et al., 1980a).

Above the cloud tops only upper limits have been measured. A ground-based search for absorption in the oxygen 'A band', the (0, 0) transition of the infrared atmospheric band system of $O_2(b^1\Sigma_g^+ \rightarrow X^3\Sigma_g^-)$, allowed Traub and Carleton (1974) to place an upper limit on the column abundance of O_2 of 1×10^{-6}; Trauger and Lunine (1983) reduced this number to 3×10^{-7}.

Explaining these small O_2 mixing ratios in the mesosphere has been a major goal of mesospheric photochemical modeling. Oxygen atoms produced in the photolysis of CO_2

$$CO_2 + h\nu \rightarrow CO + O \tag{30}$$

recombine very slowly by the process

$$CO + O + M \rightarrow CO_2 + M . \tag{31}$$

At 300 K, the rate coefficient for reaction (31) is about 4.5×10^{-36} cm^6 s^{-1}, but the rate coefficient for the recombination of O atoms via

$$O + O + M \rightarrow O_2 + M \tag{3}$$

is much larger, about 1×10^{-32} cm^6 s^{-1} (Baulch et al., 1980). Photolysis of CO_2 would cause the entire thermosphere to dissociate in a few weeks without effective recombination processes. The near-absence of O_2 in the mesosphere indicates that some probably catalytic process occurring in the Venus atmosphere acts to reform CO_2 from the O produced in reaction (30) or from the O_2 formed in reaction (2). On Mars, odd hydrogen compounds are implicated in the recombination of O and CO. A simple example of an odd-hydrogen catalytic cycle that effects recombination is

$$H + O_2 + M \rightarrow HO_2 + M, \tag{32a}$$

$$HO_2 + O \rightarrow OH + O_2, \tag{32b}$$

$$CO + OH \rightarrow CO_2 + H, \tag{32c}$$

with a net reaction of

$$CO + O \rightarrow CO_2 \tag{32d}$$

(Hunten and McElroy, 1970). On Venus, an extreme dearth of water vapor in the upper mesosphere precludes the presence of sufficiently large quantities of odd-hydrogen compounds. Compounds of chlorine have, however, been known to be present since HCl was identified in ground-based infrared spectra by Connes et al. (1967). The photochemical models of Sze and McElroy (1975), Krasnopol'sky and Parshev (1981, 1983), and Yung and DeMore (1982) rely largely on catalytic cycles involving Cl compounds, such as

$$ClCO + O_2 + M \rightarrow ClCO_3 + M, \tag{33a}$$

$$ClCO_3 + O \rightarrow Cl + CO_2 + O_2, \tag{33b}$$

$$Cl + CO + M \rightarrow ClCO + M, \tag{33c}$$

with a net reaction of

$$CO + O \rightarrow CO_2, \tag{33d}$$

to effectively recombine CO and O.

The model of Krasnopol'sky and Parshev (1981, 1983), which is presented here in Figure 21, shows that O_2 forms a layer 15 km thick near 90 km. The thermospheric density of O_2 is consistent with a mixing ratio of 2×10^{-3} at 135 km. The upper boundary of the model of Yung and DeMore (1982) is at 110 km, where the mixing ratio of O_2 is about 1×10^{-3}. The model of Krasnopol'sky and Parshev has been criticized by DeMore and Yung (1982) as including as a major step in destruction of O_2 the reaction:

$$ClCO + O_2 \rightarrow CO_2 + ClO. \tag{34}$$

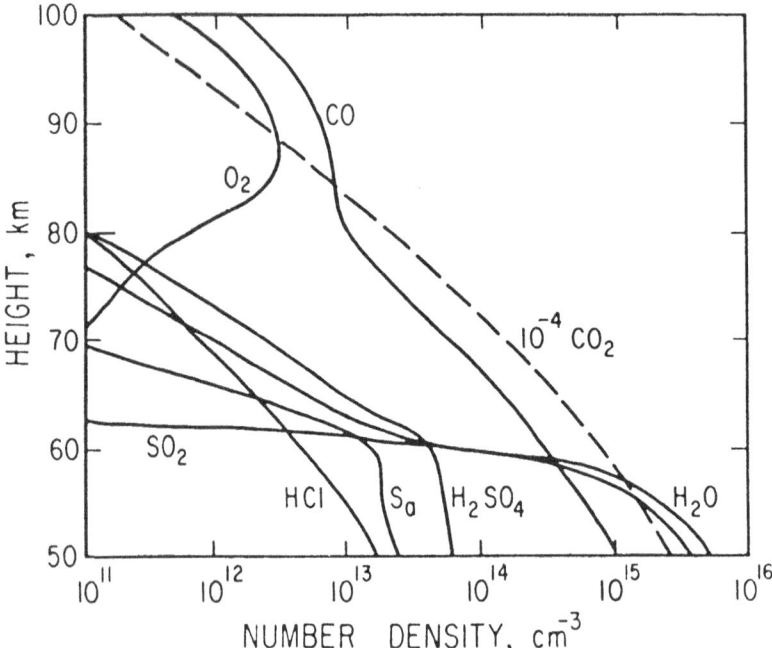

Fig. 21. Computed altitude profiles of the number densities of some neutral species in the Venus atmosphere. The CO_2 profile (dashed line) is a factor of 10^4 larger than shown. From Krasnopol'sky and Tomashova (1980).

Laboratory studies have shown the reaction above does not occur, the ClCO and O_2 combining preferentially in the three-body recombination reaction (33d). Krasnopol'sky and Parshev (1983) have criticized the model of Yung and DeMore for using a constant H_2O mixing ratio of 1×10^{-6} throughout the region above the clouds, when observations show that immediately above the clouds the mixing ratio is close to 200 times that value.

The chlorine and sulfur cycles have been found to be interconnected. DeMore et al. (1985) investigated the reaction of chlorine with SO_2 both with and without O_2 and they also discussed the implications for the Yung and DeMore (1982) model. A new reservoir species for chlorine, SO_2Cl_2 (sulfuryl chloride) was proposed, potentially increasing the total abundance of chlorine compounds in the mesosphere. Chlorine was found to catalyze the oxidation of SO_2 to H_2SO_4, thus providing another catalytic cycle for destruction of O_2.

All mesospheric models must explain the dayglow and nightglow intensities of 1.5 and 1.2 MR observed for the 1.27 μm infrared atmospheric band, from ground-based measurements of Connes et al. (1979). This band, which is the most intense feature in the Venus airglow, arises from the transition

$$O_2(a^1\Delta_g; v' = 0) \rightarrow O_2(^3\Sigma_g^-; v'' = 0) + h\nu(1.27 \ \mu m) \,. \tag{35}$$

The intensity of this feature is presumably not indicative of the ambient density of ground state O_2, but rather of its formation rate by recombination of oxygen atoms. In the dayglow, $O_2(a^1\Delta_g)$ can also be produced by photolysis of ozone, but the near equality of the intensities of the day and night airglows shows that this source is minor for Venus. The model of Yung and DeMore (1982) required that the yield of $O_2(a^1\Delta_g)$ be 30% in reaction (3) and 67% in the reactions:

$$HO_2 + O \rightarrow OH + O_2 , \tag{36}$$

$$Cl + O_3 \rightarrow ClO + O_2 , \tag{37}$$

$$ClO + O \rightarrow Cl + O_2 . \tag{38}$$

Subsequently, the yields of $O_2(a^1\Delta_g)$ and $O_2(b^1\Sigma_g^+)$ have been measured for the reactions above and several other postulated odd-chlorine reactions and the values have been found to be very small, generally less than 4% (Choo and Leu, 1985; Leu and Yung, 1987). Ali et al. (1986) have measured the rate for direct production of the $a^1\Delta_g$ state in reaction (3) and found that the yield is only about 7%, but that in the presence of O_2, the yield increases, indicating that the state is formed efficiently by energy transfer from a higher state. This conclusion is supported by an analysis of the oxygen bands in the terrestrial nightglow performed by Bates (1988b). Bates showed that the emission efficiency of the band systems originating in the $A^3\Sigma_g^+$, $A'^3\Delta_u$, and $c^1\Sigma_u^-$ states is small, indicating that the states are quenched efficiently, probably to the $a^1\Delta_g^+$ and $b^1\Sigma_g^+$ states, producing emission in the atmospheric and infrared atmospheric bands. In order to account for the emission rates at 1.27 μm in the terrestrial nightglow it is necessary to assume a large efficiency (~ 0.75) for the sum of direct recombination of O-atoms and energy transfer from higher states, and an additional, as yet unidentified source (e.g., McDade et al., 1987; Lopez-Moreno et al., 1988; Lopez-Gonzales et al., 1989).

Leu and Chung (1987) show that the total rate of CO_2 photolysis on Venus is about 8×10^{12} cm^{-2} s^{-1}, so the maximum emission rate of 1.27 μm emission would be 4 MR, if all the recombination reactions proceeded with a quantum yield of 1.0. The measured yields above imply an intensity of 0.05 MR of 1.27 μm emission from the chlorine reactions, whereas the observed value is 1.2–1.5 MR. Leu and Yung argue, therefore, that the 1.27 μ emission cannot arise from recombination of O atoms formed in photolysis of CO_2, and suggest an alternative source of O-atoms: photolysis of SO_2 introduced into the upper mesosphere by intermittent injections from the region near the cloud tops. The discrepancy for Venus is, however, rather more severe than the terrestrial nightglow problem, and seems to require a reconsideration of the basic theory of production of the 1.27 μm emission.

Massie et al. (1983) modeled the thermospheric densities of O_2, but at the lower boundary of the model, 100 km, the O_2 mixing ratio was constrained to the value indicated in the model of Yung and DeMore (1982). In the thermosphere, between 100 and 110 km photolysis of CO_2 is balanced by reaction (3). The computed mixing ratio of O_2 at 135 km is about 1×10^{-3}.

Thermospheric O_2 densities theoretically can be determined from the chemistry of atomic C. The major loss process for C in the thermospheres of Venus and Mars is the reaction

$$C + O_2 \rightarrow CO + O \tag{23}$$

(McElroy and McConnell, 1971; Krasnopol'sky, 1982; Fox, 1982b). Although no measurements of C densities are available, since the predicted densities were just below the sensitivity of the PV ONMS, the PV orbiter ion-mass spectrometer measured densities of C^+ (e.g., Taylor *et al.*, 1980). Fox (1982b) used the measured C^+ densities to constrain the O_2 mixing ratio; values of 10^{-4} or less were derived. Paxton (1983) used the intensities of atomic carbon emissions measured by the PV OUVS (see Section 3.4) to constrain the O_2 densities, and he derived an O_2 mixing ratio of 3×10^{-3}. As pointed out in Section 3.4, all of the models above are based on outdated photochemistry, and should be revised.

An outstanding success of the Soviet Venera 9 and 10 missions was the measurement of the visible nightglow of Venus in October 1975. The strongest features in the nightglow spectrum, shown in Figure 22, were identified by Lawrence *et al.* (1977) as the $v' = 0$

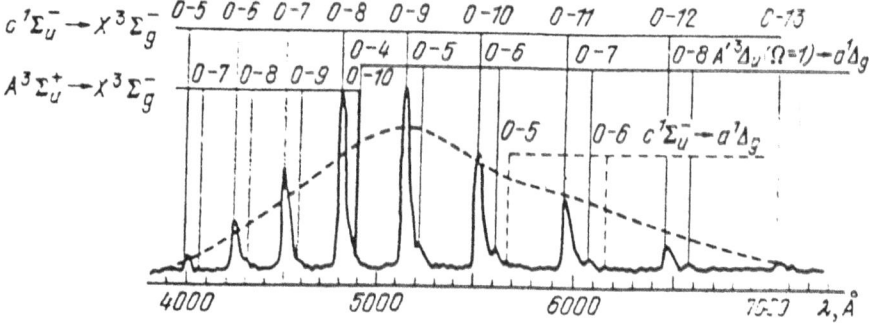

Fig. 22.   Visible night airglow spectrum measured by Venera 9 and 10. The dashed line is the instrument sensitivity. Taken from Krasnopol'sky and Tomashova (1980).

progression of the $O_2(c^1\Sigma_u^- \rightarrow X^3\Sigma_g^-)$ Herzberg II band system; an average intensity of 2.7 kR was assigned to the band system (Krasnopol'sky, 1981). Three more band systems of O_2 were later identified in the Venus spectrum: the Herzberg I $(A^3\Sigma_u^+ \rightarrow X^3\Sigma_g^-)$ bands, the Chamberlain $(A'^3\Delta_u \rightarrow a^1\Delta_g)$ bands and the Slanger $(c^1\Sigma_u^- \rightarrow a^1\Delta_g)$ bands. The measured intensities of the Herzberg I and the Chamberlain bands were 140 and 200 R, respectively. All of these band systems are formed in recombination of atomic oxygen produced by photolysis of CO_2 on the dayside of the planet and transported to the nightside (Lawrence *et al.*, 1977; Slanger, 1978). The maximum of the airglow layer is located near 100 km, the altitude of the maximum O density, and is about 15 km thick. Variations of about 10 km in the altitude of the maximum were observed. The emission shows a maximum at about 8–28° S latitude; diurnal variations of the nightglow intensities, shown in Figures 23, exhibit a maximum about 00 : 30 local time, consistent with transport of O produced on the dayside by CO_2

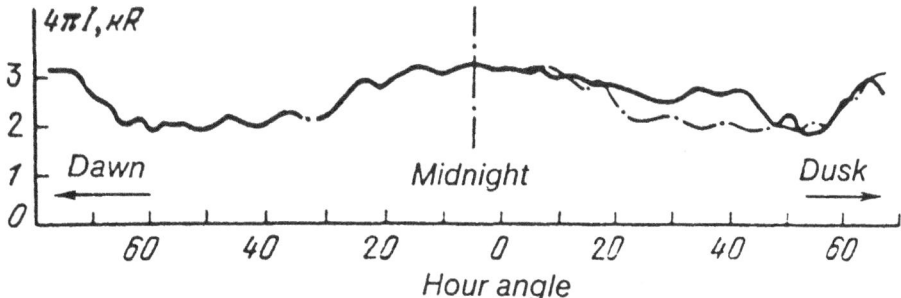

Fig. 23.   Diurnal variation of the visible night airglow as measured by Venera 9 and 10. The dashed line is the symmetric continuation of the left part of the curve. From Krasnopol'sky and Tomashova (1980).

photolysis across the terminator and subsidence at the anti-solar point. A constraint similar to that placed on the Venus solar minimum thermospheric structure and circulation by the measured NO nightglow emissions is given by using the Herzberg II nightglow observations of the Venera 9 and 10 instruments. The VTGCM could be used to examine this visible airglow, provided that total temperature is incorporated explicitly and that its model lower boundary is extended from 105 to ~ 90 km (Bouger *et al.*, 1988b, 1990a). Such a study could yield information about the thermospheric circulation in the vicinity of 100 km.

Subsequent to the identification of the Herzberg II bands in the Venus nightglow, the band system was identified in the terrestrial nightglow, where the major molecular features are the Herzberg I bands (Slanger and Huestis, 1981). The dominance of different band systems in the terrestrial and Venus nightglows has led to many laboratory and modeling studies in an attempt to explain the differences. Slanger (1978) and Kenner *et al.* (1979) showed that the presence of CO_2 enhances emission in the Herzberg II bands, possibly due to its higher efficiency as a third body. Krasnopol'sky *et al.* (1976) proposed that the non-appearance of the Herzberg II bands in the Martian nightglow, as measured by the Mars 5 space probe, suggests that they are suppressed by efficient quenching by O_2. Subsequently, the rate coefficient for quenching of the $O_2(c^1\Sigma_u^-; v = 0)$ state by O_2 has been measured by Kenner and Ogryzlo (1983); the reported value is 3×10^{-14} cm^3 s^{-1}, whereas that for quenching by O is a factor of 200 larger. From studies of the dependences of the band systems on [O], [O_2], and [M], Kenner *et al.* (1979) found that the ratio of the intensity of the Herzberg II bands to that of the Herzberg I bands is proportional to [M]/[O]. They suggested that the Herzberg II bands dominate on Venus because the maximum O recombination takes place in a region of higher total density and lower atomic oxygen density. Although this would imply an even greater dominance of the Herzberg II bands on Mars, where the total density in the region of maximum O recombination is about 10 times that in the Venus atmosphere and 10^3 times that in the terrestrial atmosphere (Krasnopol'sky, 1981), it is possible that quenching renders the total emission rate small. Indeed, from calculations of the transition probabilities of the O_2 band systems (Bates, 1988a; values corrected in Bates, 1989), Bates (1988b) has shown that the efficiency of emission relative to quenching in both the Herzberg I and II band systems in the terrestrial nightglow is on the order of 1–4%.

From an analysis of terrestrial and Venusian airglow profiles, Krasnopol'sky (1986b) proposed that the weakly bound $^5\Pi_g$ state, produced efficiently in three-body recombination of O atoms, is a common precursor for the five metastable states of $O_2(A, A', c, a, b)$, the lower states being produced by quenching or energy transfer to another O_2 molecule. This suggestion was based on calculations of the efficiency of production of various states of O_2 by Wraight (1982) and Smith (1984) that showed that the $^5\Pi_g$ state is produced with about a 66% efficiency. Bates (1988b) proposed that the fraction be reduced to 50% to account for the smaller potential well depth of the $^5\Pi_g$ state, about 0.14 eV as computed by Partridge *et al.* (1990, private communication to Bates), compared to the value 0.23 eV adopted by Wraight from the calculations of Saxon and Liu (1977). He proposed the fractions of associations into the seven lowest electronic states of O_2 shown in Table VIII. The efficiency of production of the A state

TABLE VIII

Fraction of direct associations (reaction 39) into various states of O_2. (From Bates, 1988b.)

State	Fraction
$X^3\Sigma_g^-$	0.12
$a^1\Delta_g$	0.07
$b^1\Sigma_g^+$	0.03
$c^1\Sigma_u^-$	0.04
$A'^3\Delta_u$	0.18
$A^3\Sigma_u^+$	0.06
$^5\Pi_g$	0.50

is only about 6%, 50% larger than that of the c state. Bates showed that at temperatures near 200 K redissociation of the $^5\Pi_g$ state is efficient and much larger than the potential rate for energy transfer to any of the lower states. Furthermore, because of this redissociation, the measured value of the three-body recombination rate coefficient underestimates the actual recombination rate. Temperatures are slightly lower in the Venus atmosphere near 100 km, however, so redissociation may not be as important there, and the dissociation energy itself is still somewhat uncertain.

The excited states of O_2 are relevant also for production of $O(^1S)$ and subsequent radiation in the atomic oxygen 'green line' at 5577 Å in the terrestrial nightglow. It is generally agreed that the production of $O(^1S)$ proceeds by the two-step mechanism proposed by Barth (1964):

$$O + O + M \rightarrow O_2^* + M ,$$ (39)

$$O_2^* + O \rightarrow O_2 + O(^1S) ,$$ (40)

rather than the single step mechanism proposed by Chapman (1931). Bates (1981) has reviewed the development of the theory for production of the green line in the terrestrial atmosphere. The identity of O_2^* in reactions (39) and (40), the 'Barth precursor', is uncertain. The A, A', and $^5\Pi_g$ states and the c state for $v > 0$ are sufficiently energetic. No green line emission was detected by the visible spectrometers on Veneras 9 and 10, and an upper limit of 10 R has been placed on its intensity (Krasnopol'sky, 1981, 1986b). This suggests that the precursor is not present in the Venus atmosphere. The A state was ruled out based on the large required rate coefficient for quenching by O and its altitude profile (e.g., Slanger, 1978; Thomas, 1981; Llewellyn et al., 1980). Krasnopol'sky (1981) showed that, if the A' state were the precursor, the green line emission would be two orders of magnitude larger than the observed value, but that the identification of the vibrationally excited c state as the precursor did not contradict any available data. Indeed, the Herzberg II system in the terrestrial nightglow consists of bands originating in vibrational levels with $v = 4$, 5, and 6, whereas only bands originating in $v = 0$ have been observed on Venus. With this assumption, Krasnopol'sky derived a rate coefficient of 2.5×10^{-10} cm^3 s^{-1} for the quenching of the c state by O:

$$O_2(c^1\Sigma_u^-) + O \rightarrow O_2 + O(^1S) ; \tag{41a}$$

a value of 5.9×10^{-10} cm^3 s^{-1} was measured by Kenner and Ogryzlo (1983) for the reaction

$$O_2(c^1\Sigma_u^-) + O \rightarrow \text{products} . \tag{41b}$$

It is probable, however, that the rate coefficients for energy transfer and quenching depend upon the vibrational level of the O_2, but the vibrational distributions in the atmosphere and sometimes in the laboratory, are uncertain. Moreover, Krasnopol'sky's 1981 analysis preceded the calculations of Wraight (1982) and Smith (1984) and was based on a statistical distribution of states of O_2 formed in the three-body recombination. He has recently reconsidered the problem and has concluded that the $^5\Pi_g$ state is a more likely precursor, but this analysis assumes a larger well depth for the state than current calculations indicate and no significant redissociation (Krasnopol'sky, 1986b).

Combining measured intensities of the band systems (Greer et al., 1986; Slanger and Huestis, 1981) with computed transition probabilities (Bates, 1988a, 1989), Bates (1988c) has estimated the average rate coefficients for energy transfer, reaction (40), that would be necessary for each of the potential states to be the precursor in the terrestrial atmosphere and compared the requirements to measurements made on oxygen-argon afterglows by Slanger and Black (1976). Only the $c^1\Sigma_u^-$ and $^5\Pi_g$ states are not eliminated by this procedure. Information on the reverse process

$$O(^1S) + O_2(X^3\Sigma_g^-) \rightarrow O(^3P) + O_2(c^1\Sigma_u^-) , \tag{42}$$

was used to further constrain the rate coefficient for energy transfer from the c state. The densities of $O_2(c; v = 0)$ present in the Venus atmosphere can be determined from the observed emission in the Herzberg II system, but no information is available about vibrationally excited states. From the Venus nightglow data, Bates also derived upper limits for the transfer rate coefficients for the A state, which is marginally consistent, and for the A' state, which is inconsistent, with the terrestrial required rate coefficients. He does note, however, that the rate coefficients may vary with vibrational quantum number, so the constraints are not rigid. The upper limit on the energy transfer rate coefficients for the $^5\Pi_g$ state from Venus data is consistent with terrestrial requirement, but there is no positive evidence in its favor.

3.12. O_2^+ EMISSIONS

The O_2^+ second negative band system arises from the transition $O_2^+(A^2\Pi_u \rightarrow X^2\Pi_g)$. The $O_2^+(A^2\Pi_u)$ state is produced in the Venus ionosphere mostly in fluorescent scattering of solar radiation by ground state O_2^+, since the ambient densities of O_2 in the Venus thermosphere are too low for photoionization and electron impact ionization to be significant sources. In order to determine the role of fluorescent scattering in producing vibrationally excited O_2^+, Fox (1985) computed the intensities of the bands in the second negative and first negative $(b^4\Sigma_g^- \rightarrow a^4\Pi_u)$ band systems. The second negative bands are weak and spread over a large wavelength range. The intensities summed in 50 Å intervals are shown in Figure 24 for two assumptions about the (unknown) rate coefficient for the vibrational quenching of $O_2^+(X^2\Pi_g; v)$ by O:

$$O_2^+(v) + O \rightarrow O_2^+(v-1) + O . \tag{43}$$

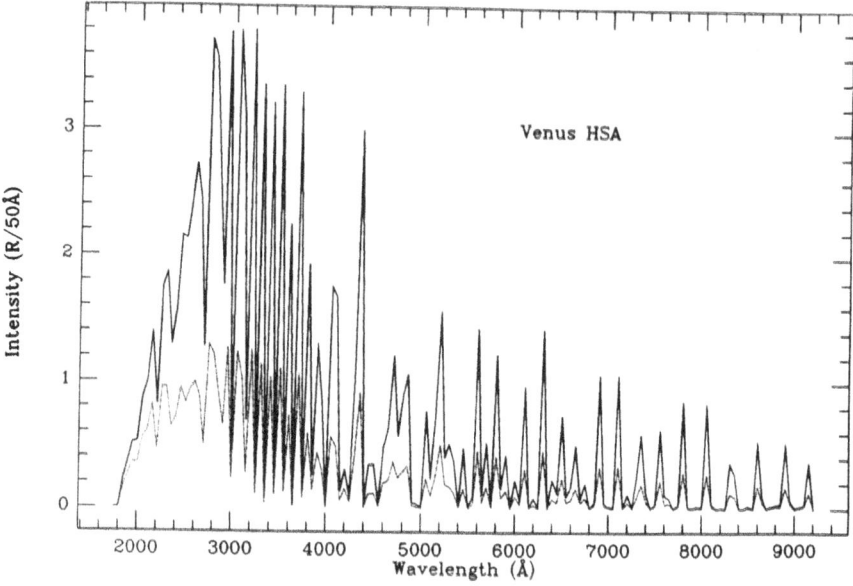

Fig. 24.   Computed intensities of the second negative band system, for a high solar activity (HSA) model, summed in 50 Å intervals. The solid line is for $k_{43} = 1 \times 10^{-10}$ cm^3 s^{-1}. The dotted line is for $k_{43} = 6 \times 10^{-10}$ cm^3 s^{-1}.

The integrated overhead intensity of the second negative band system for a rate coefficient of 1×10^{-10} cm^3 s^{-1} is 230 R. If the rate coefficient is near 6×10^{-10} cm^3 s^{-1}, the intensity is reduced to 43 R. The first negative band system is very weak on Venus due to the near absence of O_2; the computed intensities are 0.75–1.5 R.

4. Hot Atom Coronas and Escape

4.1. INTRODUCTION

This ingenious application of the properties of molecules to planetary despoiling of the weak, is due to an Irish gentleman whose name for the moment escapes me; and facts appear to support it.

PERCIVAL LOWELL (1894)

The idea of the escape of atoms from planetary atmospheres, the Moon and the Sun was first put forward in a paper read to the Royal Society by J. J. Waterston in 1845 (Chamberlain, 1963; Jeans, 1925). Because the paper contained 'certain inaccuracies' (Jeans, 1925), only an abstract was published (Waterston, 1846). The manuscript itself did not appear until 1892, when Lord Rayleigh had it published for historical interest (Waterston, 1892). The intervening development of the kinetic theory of gases and the Maxwell distribution law allowed the fundamental idea of evaporation of the energetic tail of the thermal distribution to be introduced (Stoney, 1898); the theory was later refined by Sir James Jeans (1925), whose name the thermal escape process now bears.

The idea of a planetary corona was first introduced by Stoney (1868). He remarked that at sufficient heights, atomic collisions would be so rare that most of the atoms traveling upward would follow ballistic orbits, eventually falling back to the denser parts of the atmosphere. The altitude above which collisions cease to be important is called the exobase or critical level and the region of the atmosphere above the exobase is the exosphere. A comprehensive theory of the exosphere and atmospheric evaporation has been presented by Chamberlain (1963) and a pedagogical treatment of the subject can be found in Chamberlain and Hunten (1987). Other non-thermal escape processes have subsequently been found to be more important than thermal (Jeans) escape for the terrestrial planets (e.g., Liu and Donahue, 1974a, b; Hunten, 1982). Excellent reviews of thermal and non-thermal escape processes have been presented by Hunten and Donahue (1976) for the escape of hydrogen from the terrestrial planets, and by Hunten (1982) for the coronas and escape of species from all the terrestrial bodies in the solar system.

The exobase is mathematically defined as the altitude where the mean free path l is equal to the atmospheric scale height. The mean free path is defined by the expression $l = (n\sigma)^{-1}$, where n is the total number density and σ is the collision cross section. The probability that a particle, moving upward from the exobase with sufficient velocity will actually escape without suffering another collision is e^{-1}. The condition $l = H$, therefore, reduces to $nH\sigma = 1$ or, equivalently, to $N = \sigma^{-1}$, where N is the column density. Since

a typical collision cross section is about 3×10^{-15} cm^{-2}, in practice the exobase is
located near the altitude above which the column density is about 3.3×10^{14} cm^{-3}. For
the high solar activity ($F_{10.7} = 200$) VTS3 model the exobase altitude is near 210 km at
noon and 154 km at midnight. The VIRA model for 16° N and $F_{10.7} = 150$ places the
exobase near 201 km at noon and 161 km at 22:00 hr local time. In the BNMS model,
which applies to 08:30 hr local time, the exobase is near 191 km. In the region of the
pre-dawn bulge in the densities of H and He, the column density of H$_2$ is important in
determining the altitude of the exobase (Kumar et al., 1983).

Whether the trajectory of a particle traveling upward from the exobase is ballistic
(bound) or escaping (free) is determined by the total energy E, which is the sum of its
kinetic and potential energies:

$$E = \frac{1}{2}mv^2 + \int_{\infty}^{r_c} \frac{mGM}{r^2} \, dr, \tag{44}$$

where m and v are the mass and velocity of the particle, G is the gravitational constant,
M is the mass of the planet, and r is the distance from the center of the planet; the
subscript c refers to the critical level or exobase. Expression (44) reduces to

$$E = \tfrac{1}{2}mv^2 - mg_c r_c, \tag{45}$$

where g_c is the gravitational acceleration at the exobase. If the total energy is negative,
the particle is bound; if the total energy is positive the particle may escape. The escape
velocity, v_{esc}, is then defined by the condition $v_{esc} = (2gr)^{1/2}$. For Venus, the escape
velocity is about 10.2 km s^{-1}.

In the Jeans escape process, particles with velocities greater than the escape velocity
in the high-energy tail of the Maxwellian distribution may escape if they are oriented in
the upward hemisphere (above the horizon). The escape flux, ϕ_J is given by

$$\phi_J = \frac{n_c u}{2\sqrt{\pi}} (1 + \lambda_c) \exp(-\lambda_c), \tag{46}$$

where

$$\lambda = \frac{GMm}{rkT} = \frac{mgr}{kT}. \tag{47}$$

λ is the gravitational potential energy in units of kT and u is the modal velocity of a gas
in thermal equilibrium at temperature T,

$$u = (2kT/m)^{1/2}. \tag{48}$$

Sometimes a correction factor is applied to the expression for the escape flux to account
for the suppression of the tail of the distribution due to the escape of the energetic
particles (Chamberlain, 1963; see also Hunten, 1982). Application of Equation (46) to

the escape of H from Venus, for the VIRA model with an exobase near 200 km, gives a value for λ of about 11 and a thermal escape flux of about 22 H atoms cm^{-2} s^{-1} at noon; at midnight the escape flux is several orders of magnitude smaller, due to the low nightside temperatures and the resulting large value of λ. These fluxes are insignificant compared to the actual escape fluxes that have been inferred from models of the hot H corona.

Hunten (1973a) was the first to recognize that the escape rate of a light species from a planetary atmosphere may be controlled by diffusion of the species to the exobase, rather than by the escape process itself. The escape flux of H from Venus is limited by the rate of transport of H through the middle atmosphere to the upper atmosphere; the limiting upward flux, ϕ_l of a species i with mixing ratio f_i can be estimated as

$$\phi_l \approx \frac{b_i f_i}{H_a} , \tag{49}$$

where H_a is the average scale height of the atmosphere and b_i is the binary collision parameter (Hunten, 1973a, b; Hunten and Donahue, 1976). b_i is defined by $b_i = D_i n_T$, where D_i is the molecular diffusion coefficient, evaluated as an appropriate average of the binary diffusion coefficients D_{ij} of the species i diffusing through the major atmospheric constituents j. D_i is usually computed as

$$\frac{1}{D_i} = \sum_{j \neq i} \frac{n_i/n_T}{D_{ij}} . \tag{50}$$

The expression (49) above is usually evaluated at the homopause, with the mixing ratio taken from a suitable altitude in the middle atmosphere, but above the cold trap. The limiting flux obtains if and only if the mixing ratio is constant with altitude. The effect of photochemistry can be eliminated if all chemical forms of the species are counted in the calculation of f_i. H is found in the mesosphere in the form of HCl, H_2O, and H_2 and in the thermosphere mostly as H. In order to evaluate the limiting flux of H we must first determine the mixing ratios of H-containing compounds in the middle atmosphere and at the homopause. The abundances of H, H_2, and H_2O will be discussed below.

4.2. H AND H_2 DENSITIES

No *in situ* measurements of H densities in the thermosphere are available, but H$^+$ densities were reported from PV ion mass spectrometer data (e.g., Taylor *et al.*, 1980). Brinton *et al.* (1980) showed that, in the altitude region where H$^+$ is in photochemical equilibrium, the major source of H$^+$ is the nearly thermoneutral charge transfer of O$^+$ to H

$$O^+ + H \rightarrow H^+ + O , \tag{51a}$$

while loss is via the reverse reaction

$$H^+ + O \rightarrow H + O^+ \tag{51b}$$

and by reaction with CO_2

$$H^+ + CO_2 \rightarrow HCO^+ + O \ . \tag{52}$$

Thus the steady-state density of H can be computed from ONMS and OIMS data using the expression

$$[H] = \frac{[H^+]}{[O^+]}\left([O]\ \frac{k_{51b}}{k_{51a}} + \frac{k_{52}}{k_{51a}}\ [CO_2]\right). \tag{53}$$

Brinton *et al.* found that photochemical equilibrium is a valid approximation below 200 km on the dayside and below 170 km on the nightside. A large variation with solar hour angle was found, with maximum densities occurring in the pre-dawn sector near 04 : 00 hr local time. Figure 25 shows the H densities derived by Brinton *et al.* at 165 km

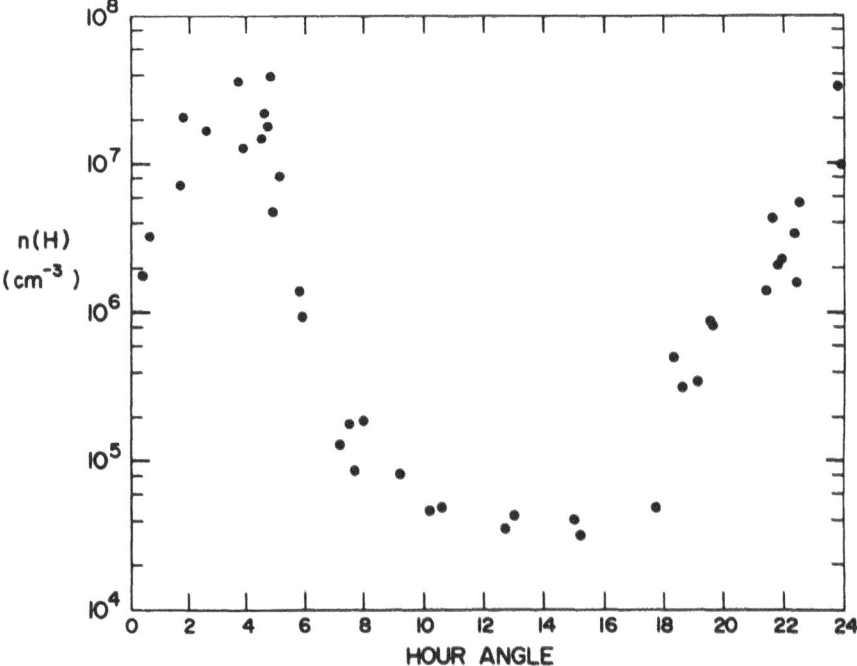

Fig. 25.   Diurnal variation of atomic hydrogen concentration in the Venus thermosphere, derived from *in situ* measurements of ion and neutral composition. Data were obtained near 165 km altitude on 25 orbits of the Pioneer Venus spacecraft between December 1978 and July 1979. These values are based on non-normalized ONMS data and should be increased by a factor of about 1.63 (see text). Taken from Brinton *et al.* (1980).

as a function of hour angle. Since this study was done, a renormalization of the ONMS data by a factor of 1.63 has been suggested (Hedin *et al.*, 1983) (see Section 2), so values of [H] presented by Brinton *et al.* should be multiplied by this factor also. Taylor *et al.* (1984, 1985), obtained similar results in a later analysis of three years of PV data. The peak densities at 165 km were in the range $2-5 \times 10^7$ cm^{-3}, with a night-day ratio of about 400. No evidence for a long-term variation in the H densities was observed, as would be expected for the time period of the observations, when solar fluxes declined by less than 10%. The large scale height of H near 200 km on Venus, about 270 km for a temperature near 275 K, suggests that the densities of H do not change much with altitude over the altitude range from 150 to 250 km. In actuality, H is not in diffusive equilibrium above the homopause. Its density distribution deviates slightly from diffusive equilibrium because of the large flux through the exobase; the apparent scale height will be somewhat smaller than the diffusive equilibrium scale height.

The strongest resonance feature of atomic hydrogen is the Lα line at 1216 Å; it arises from the $2p \rightarrow 1s$ electronic transition. Radiation from Venus at Lα was detected by the ultraviolet spectrometers on Mariner 5 (Barth *et al.*, 1967) and Venera 4 (Kurt *et al.*, 1968). The Mariner 5 intensities were later reduced by a factor of 0.7 to account for a post-flight recalibration of the instrument (Broadfoot *et al.*, 1974). A mean disk intensity of 40 kR has been derived for the Mariner 5 observations (Anderson, 1976). Rottman and Moos (1973) observed Lα with their moderate resolution rocket-borne spectrometer and reported an intensity of 27 kR. As discussed in the Introduction, Mariner 5 limb profiles showed a two-scale height distribution of H, the source of which was debated for several years (Barth *et al.*, 1967). Stewart (1968) first suggested that the two components were due to H at two different temperatures. Barth (1968) proposed that the inner component was a mass-2 species, either H$_2$ or D, and the outer component H. Barth favored the H$_2$ interpretation, while other workers preferred the idea that the inner component was due to deuterium (Donahue, 1968; Wallace, 1969; McElroy and Hunten, 1969). Wallace *et al.* (1971) struck a fatal blow to the deuterum hypothesis when they measured Venusian Lα with a rocket-borne UV spectrometer capable of resolving the H and D Lα lines and failed to detect the deuterium line. In addition, other evidence suggested that the thermospheric temperatures were lower than the 700 K or more required for the scale height to be that of a species with a mass of 2. Kumar and Hunten (1974) advanced the two-temperature hypothesis by showing that the ion densities derived from the radio-occultation data were consistent with a neutral atmosphere model with a low exospheric temperature of 350 K; they further suggested that the hot H component could be produced by ion-neutral reactions such as

$$O^+ + H_2 \rightarrow OH^+ + H , \tag{54}$$

is sufficient H$_2$ were available.

The two-temperature interpretation was established by a re-analysis of the Mariner 5 bright limb and dark disk Lα data that employed a spherical radiative transport code (Anderson, 1976). Dayside temperatures of 275 ± 50 K and 1020 ± 100 K and critical level densities of 2×10^5 cm^{-3} and 1.3×10^3 cm^{-3} were determined for the cold and hot H components, respectively. For the nightside, temperatures of 150 ± 50 K and 1500 ± 200 K, and critical level densities of 2×10^5 cm^{-3} and 1×10^3 cm^{-3} for the cold and hot components, respectively, were reported. The temperatures derived by Anderson are in very good agreement with those later reported from Pioneer Venus data.

The dayside densities are significantly larger and the nightside densities are smaller than the values derived by Brinton *et al.* (1980) from Pioneer Venus data. The large diurnal variation in densities shows that the spherically-symmetric radiative transfer model used by Anderson is not valid, especially for the nightside. In addition, Anderson placed the critical level at 250 km, significantly higher than Pioneer Venus models indicate, although, as we have seen, the H densities do not vary rapidly with altitude.

The Mariner 10 spectrophotometer detected a strong signal at Lα but the intensity was less than at the time of the Mariner 5 encounter (Broadfoot *et al.*, 1974). Takacs *et al.* (1980) used a spherically-symmetric radiative transfer model to analyze the emissions on the bright limb. The derived values for the temperatures and densities were similar to those of Anderson (1976): 275 ± 50 K and 1.5×10^5 cm^{-3} for the temperature and density of the dayside cold component, respectively; and 1250 ± 100 K and 500 ± 100 for the temperatures and density of the dayside hot component, respectively. Takacs *et al.* found that a spherically-symmetric model was inadequate to analyze the dark disk data.

In addition to the visible spectrometers, Venera 9 and 10 were equipped with Lα photometers to measure the intensities at 1216 Å. Bertaux *et al.* (1978) analyzed the data obtained using a spherically-symmetric, isothermal radiative transfer model. They found that the data could be fit by a two-temperature model, but the fit was not better than that of a one-temperature model at 500 K with an exobase density of 1.5×10^4 cm^{-3}, values which are bracketed by those of the hot and cold components from the Mariner 5 and 10 analyses. The spacecraft also carried resonance-absorption cells filled with hydrogen and deuterium that were heated by tungsten filaments to produce H and D atoms, respectively. The reduction in the intensity of the planetary Lα line by the absorption cells would provide an indication of the line width and therefore of the temperature. No absorption was observed for the D_2 cell. The H_2 cell data showed that the temperature increases abruptly above about 3000 km, near the altitude above which the Mariner analyses showed the hot component to dominate.

The Venera 11 and 12 ten-channel UV photometers were also sensitive at Lα (Kurt *et al.*, 1979). Bertaux *et al.* (1981) reported sub-solar zenith intensities of 38–42 kR on the disk, similar to the values reported by Anderson (1976) from Mariner 5 data. Anderson showed, however, that the bright disk intensities are due mostly to Rayleigh scattering by CO_2. Since the Venera photometer aperture was not parallel to the limb, the altitude resolution in the limb scan was limited. Analysis showed that the data could be fit by a two-temperature model with temperatures of 400 and 700 K.

Reports of Pioneer Venus OUVS Lα limb scan observations have been presented by Paxton *et al.* (1985, 1988). Paxton *et al.* (1988) used a radiative transfer model that allowed for variations of temperature and density with altitude, latitude, and local time to analyze 20 orbits covering the first three years of operation of the PV orbiter. The inferred density of H at the exobase averaged 6×10^4 cm^{-3}; this compares favorably with the daytime average value of about 6.5×10^4 cm^{-3} determined from the analysis of Brinton *et al.* (1980). They attributed the factor of 2–3 discrepancy with the Mariner 5 and 10 analyses to the use of a more sophisticated radiative transfer model. The column density of H above 110 km, the altitude below which CO_2 strongly absorbs Lα, averages $3.6 \pm 1 \times 10^{13}$ cm^{-2}.

4.3. D/H RATIO

Interest in determining the D/H ratio in the Venus atmosphere began when the 2-scale height distribution of H was first measured by Mariner 5 (Barth *et al.*, 1967; see Section 4.2). The interpretation that the inner component was due to D at the same temperature as the outer H component required a D/H ratio of about 10% (McElroy and Hunten, 1969; Wallace, 1969). Although the derived ratio, which was enhanced over the terrestrial value by a factor of about 10^3, became meaningless after this hypothesis was shown to be invalid, the idea of some enrichment of D due to differential escape of H and D remained a possibility.

The PV OIMS detected an ion with a mass of 2, which could be either H_2^+ of D^+ (e.g., Taylor *et al.*, 1980). Originally this ion was believed to be H_2^+, with the H_2^+ produced mainly by direct photo- and electron-impact ionization of H_2. H_2 would be destroyed by reactions with O^+ (reaction 54) and with CO_2^+:

$$H_2 + CO_2^+ \rightarrow CO_2H^+ + H. \tag{55}$$

Kumar *et al.* (1981) derived an abundance of 10 ppm H_2 at the homopause from a model of the chemistry of H_2^+. McElroy *et al.* (1982), Cravens *et al.* (1983), and Rodriguez *et al.* (1984) have emphasized that a large mixing ratio of H_2 provides a significant source of H through the reactions (54) and (55) above, which cannot be balanced by the known loss mechanisms and is difficult to reconcile with the measured abundances of H.

McElroy *et al.* (1982) suggested that the mass-2 ion measured by the PV OIMS was D^+ rather than H_2^+, the atmosphere having become enriched in D relative to H by preferential escape of H. Donahue *et al.* (1982) were able to measure the D/H ratio in the lower atmosphere with the PV LNMS, by comparing the H_2O and HDO peaks during the time that the inlet to the mass spectrometer was clogged by sulfuric acid from cloud droplets. A value of 1.6×10^{-2} was reported.

Hartle and Taylor (1983) identified the mass-2 ion as D^+ by comparing the scale height of the mass-2 ion to that of H^+ in the photochemical equilibrium region and the diurnal variation of the mass-2 ion to that of H^+. They reasoned that the altitude profile of the ratio D^+/H^+ should vary as $\exp(\Delta z/H(1))$, where $H(1)$ is the scale height of a mass-1 species, but the ratio of H_2^+/H^+ should vary as $[O^+]^{-1}\exp(\Delta z/H(1))$. The observed altitude variation of the ratio supported the identification of the species as D^+, as did the observation that both the mass-2 ion and H^+ exhibit bulges in the pre-dawn sector, with densities an order of magnitude larger near 04:00 hr local time than during the day. From the measured ratio D^+/H^+ Hartle and Taylor derived a D/H ratio of 1.7×10^{-2} between 155 and 160 km, which, when projected down to the presumed homopause at 132 km yields a value of $2.2 \pm 0.6 \times 10^{-2}$, in fairly good agreement with the measured value of Donahue *et al.* (1982).

Kumar and Taylor (1985) modeled the pre-dawn bulge ionosphere, where the maximum densities of H^+ and D^+ (as well as those of light neutral species) are found. By analyzing two orbits of PV data, they obtained additional strong evidence that the mass-2 ion was D^+. The reaction of O^+ with H_2 (reaction (54)) would strongly suppress the O^+ densities; a depletion in O^+ at times of enhanced mass-2 ion was not observed. Kumar and Taylor derived D/H ratios of 2.5 and 1.4% for PV orbits 117 and 120, respectively, and placed an upper limit on the H_2 abundance at the homopause of 0.1 ppm.

Using the high-resolution mode of the IUE satellite, Bertaux and Clarke (1989a) searched for the deuterium $L\alpha$ line, which is located 0.33 Å shortward of H $L\alpha$, but found no evidence for it. An upper limit was determined for its intensity of 300 R, compared to 21 kR measured for the H $L\alpha$ line. From the non-detection, they placed an upper limit on the D/H ratio at 100 km of $3.6 \pm 1.5 \times 10^{-3}$, a factor of more than 4 lower than the values estimated from the LNMS data by Donahue et al. (1982) and Hartle and Taylor (1983). Bertaux and Clarke suggested that the LNMS result for HDO may indicate a chemical fractionation process, and that the mass-2 ion may indeed be H_2^+, thus requiring the presence of 10 ppm H_2. They did not, however, account for the lack of suppression of O^+ in the bulge region, or for the large source of H atoms implied by such a large abundance of H_2. Donahue (1989) pointed out the inconsistency in their inference of 10 ppm H_2 at the homopause and the 0.7 ppm H required by the H $L\alpha$ data. He also challenged their conclusion that a D/H ratio of 1.6×10^{-2} would lead to 2.5 kR of D $L\alpha$, as ignoring the upward flux of D due to the global circulation, which will produce a different vertical distribution from diffusive equilibrium. Donahue found that a predicted intensity of 1.0 ± 0.4 kR is more appropriate and that the upper limit implied by 300 kR of D $L\alpha$ is about 5×10^{-3}. Bertaux and Clarke (1989b) suggest that the larger D/H ratio measured by Donahue et al. (1982) should produce observable limb brightening of the D $L\alpha$ emission and that a search of the data collected by the PV OUVS should reveal such a limb brightening.

According to Yung and DeMore (1982), the ratio of H_2 to H at the homopause is fixed by the photochemistry of H_2O in the mesosphere and should have a value of about 10%. For the VIRA standard noon model, an extrapolation of the H densities downward to 130 km, the approximate altitude of the H homopause, yields a mixing ratio of about 3 ppm, although this extrapolation, like that for He, is extremely sensitive to the assumed value of the eddy diffusion coefficient and should not be considered accurate to more than a factor of 2–3. The mixing ratio of H_2 at the homopause would, therefore, be about 0.3 ppm.

H_2 has been measured below the cloud tops by the PV large probe gas chromatograph (LGC). The mixing ratio appears to decrease toward the surface; the reported values at 52,42, and 22 km are 200, 70, and 10 ppm. The gas chromatographs on Veneras 13 and 14 had a threshold sensitivity for detection of H_2 of 0.8 ppm. A value of 25 ppm was measured within the clouds for the altitude range 49–58 km (Mukhin et al., 1983).

The mesospheric photochemical models of Yung and DeMore (1982) and Yatteau (1983) required much lower mixing ratios of H_2 above the cloud tops than were measured in or below the clouds or derived from the chemistry of H_2^+ at the homopause by Kumar *et al.* (1981). The catalytic processes that recombine CO and O in the lower atmosphere of Venus are assumed to involve odd-chlorine compounds, rather than odd hydrogen as in the Martian atmosphere, where the abundance of water vapor is larger (see Section 3.11). H_2 renders Cl ineffective as a catalyst because the reaction

$$H_2 + Cl \rightarrow HCl + H \tag{56}$$

produces HCl, which is a relatively unreactive 'reservoir' species for odd chlorine. Yung and DeMore presented three models that differed mainly in the mixing ratio of H_2; values of 20 ppm, 0.5 ppm, and 1×10^{-7} ppm were tested. Only the extremely hydrogen-deficient model could account for the observed nighttime depletion of CO near the cloud tops reported by Schloerb *et al.* (1980), Wilson *et al.* (1981), and Clancy *et al.* (1981) from microwave observations.

Other problems with these models have, however, become apparent. First of all, as Krasnopol'sky and Parshev (1983) have noted, a constant mixing ratio of 1 ppm H_2O was used over the entire mesosphere, in contrast to the PV orbiter infrared radiometer (OIR) measurements, which show that, near the cloud tops at least, the mesosphere is not as dry as assumed. Second, some of the O_2 production reactions involving odd chlorine that were proposed to lead to production of $O_2(a^1\Delta_g)$, explaining the observed strong 1.27 μ emission have been shown, by Yung and coworkers, to produce $O_2(a^1\Delta_g)$ very inefficiently (see Section 3.11). It also appears that the diurnal variation of CO in the Venus mesosphere is more complicated than early microwave observations indicated. Clancy and Muhleman (1985a) showed that the CO mixing ratio in the altitude range 80–90 km is depleted by a factor of 2–4, but at 95 km is *enhanced* by a factor of 2–4 on the nightside relative to the dayside. The maximum in the 80–90 km region occurs in the early morning rather than at mid-day. Clancy and Muhleman (1985b) have, however, proposed that the midnight bulge above 90 km is a result of the sub-solar to anti-solar circulation rather than a photochemical effect. Nonetheless, it appears that the photochemical processes that govern the recombination of O and CO and the destruction of O_2 are not yet well understood.

4.4. H_2O DENSITIES

The abundance of water is unknown in the thermosphere of Venus and controversial in the lower atmosphere. Below the cloud tops, the Venera 9 and 10 narrow-band photometry measurements imply an H_2O mixing ratio of 300 ppm, accurate to within a factor of 2 in the altitude range 20–40 km (Ustinov and Moroz, 1978). The Venera 11 and 12 descent probes carried scanning infrared spectrophotometers that recorded the spectrum of the Venus daytime sky. Synthetic spectra constructed by Moroz *et al.* (1980) indicated that the H_2O abundance is about 20 ppm near the surface increasing to 200 ppm in the clouds. From measurements made by the net flux radiometers on the four PV probes, Revercomb *et al.* (1985) have reported that the water abundance below the clouds varies with latitude, with mixing ratio of 20 to 50 ppm at 60° increasing to about 500 ppm at the equator. The gas analyzers aboard the Soviet Vega 1 and 2 spacecraft indicated that the water vapor is concentrated in a 30 km layer centered at

40 km, with a maximum mixing ratio on the order of 1000 ppm at about 50 km, decreasing to about 20 ppm near the surface (Surkov et al., 1987).

For the atmosphere above the cloud tops, ground-based infrared absorption measurements have indicated an H_2O mixing ratio of 1 ppm (Fink et al., 1972; Barker, 1975). Subsequently space probes have yielded conflicting results. In the cloud layer, from 49 to 58 km, the gas chromatography experiments on Veneras 13 and 14 yielded a value of 700 $\pm$ 300 ppm (Mukhin et al., 1983). The mixing ratios measured by the Vega gas analyzers appear to decrease over the altitude range 60 to 75 km, with values in the range 1–100 ppm (Surkov et al., 1987). The error bars, however, are more than a factor of 10 near 75 km. Above about 65 km the PV OIR showed a large diurnal variation in the water vapor abundance from about 100 ppm in the early afternoon to less than the detection threshold of 6 ppm at night (Schofield et al., 1982). The abundance near the cloud tops is probably determined by the saturated vapor pressure of H_2O over sulfuric acid and should decrease with increasing altitude. Krasnopol'sky (1985) has computed a sulfuric acid 'trap function', which relates the water vapor and H mixing ratios at 90 km to that in the lower atmosphere. For an assumed lower atmosphere mixing ratio of 200 ppm, the mixng ratio of H_2O is about 0.5 ppm at 90 km and that of all H-containing compounds is about 3 ppm.

4.5. DIFFUSION LIMIT FOR ESCAPE OF H

On Venus, the atmosphere above the cloud tops, the mesosphere, has generally been chosen as the region in which to evaluate the mixing ratio of H-containing compounds, in order to calculate the diffusion limiting upward flux of H (Hunten and Donahue, 1976; Walker, 1977). The mixing ratio of HCl has been determined from ground-based infrared absorption measurements to be about 0.6 ppm (Connes et al., 1967). In view of the influence of the clouds, care must be exercised in choosing the altitude range for determining the abundance of water vapor. Krasnopol'sky (1985) asserts that sulfuric acid affects the H_2O mixing ratio up to about 75 km, so the abundance at the cloud tops is not an appropriate value to use in computing the diffusion limited flux. The abundance of H_2 in the clouds has been reported as 25 ppm by Mukhin et al. (1983) from Venera 13 and 14 gas chromatography, but mesospheric models cannot accommodate such large mixing ratios (Yung and DeMore, 1982; Yatteau, 1983). If it is assumed that the abundance of H_2 above the cloud tops is negligible, and that the abundance of H_2O above 75 km is 1 ppm, then the total abundance of H is 2.6 ppm. This is good agreement with the H-mixing ratio extrapolated to the H homopause at about 130 km of about 3 ppm, but the latter value is accurate to not better than a factor of 2.

In order to compute the flux through the homopause, the binary collision parameter $(b = Dn_T)$ must be evaluated. We assume here that the major form of H at the homopause is atomic hydrogen. The diffusion coefficient of H through the atmosphere can be computed as a suitable average of the binary diffusion coefficients for H through each of the major atmospheric constituents (cf. Chamberlain and Hunten, 1987). The values of the diffusion coefficients assumed are shown in Table IX. The diffusion coefficients of H through N_2 is taken from Banks and Kockarts (1973), and is taken

TABLE IX

Parameters A and s used to compute diffusion coefficients for H through the major gases in the Venus thermosphere. The diffusion coefficient in $cm^2 \, s^{-1}$ is given by $D = AT^s/n_T$, if the temperature T is in K and the total density n_T is in cm^{-3}.

System	$A \; (10^{17} \, cm^{-1} \, s^{-1} \, K^{-s})$	s
H–O[a]	7.25	0.71
H–CO$_2$[b]	3.87	0.711
H–N$_2$[c]	4.87	0.698
H–CO[d]	4.87	0.698

[a] Cooper *et al.* (1984).
[b] Obtained by scaling the values for H–O$_2$ from Banks and Kockarts (1973) for the larger mass and radius of CO$_2$.
[c] Banks and Kockarts (1973).
[d] Assumed the same as for H–N$_2$.

to be the same as that for H–CO. The H–CO$_2$ diffusion coefficient was estimated by scaling D for H–O$_2$ for the larger mass and radius of CO$_2$ compared to O$_2$ (cf. Banks and Kockarts, 1973). The H–O diffusion coefficient is taken from the calculation of Cooper *et al.* (1984). The average value for b_H at 130 km in the VIRA dayside model is $1.8 \times 10^{19} \, cm^{-1} \, s^{-1}$, and the diffusion limited flux is $9.0 \times 10^7 \, cm^{-2} \, s^{-1}$. If the mixing ratio of H$_2$ is as large as 25 ppm in the mesosphere, and we consider this to be unlikely, the diffusion-limited flux could exceed $10^9 \, cm^{-2} \, s^{-1}$.

4.6. MODELS OF HOT AND ESCAPING HYDROGEN

McElroy and Hunten (1969) showed that the Mariner 5 Lα data implied a Jeans escape flux for H of only about $6 \times 10^5 \, cm^{-2} \, s^{-1}$, for an assumed exospheric temperature of about 700 K. The exospheric temperature is now known to be less than 300 K and the magnitude of the thermal escape flux has been shown to be negligible (see Section 4.1); subsequently the bulk of the escape has been assumed to arise from non-thermal mechanisms. Kumar and Hunten (1974) evaluated several sources of hot hydrogen, H*, including momentum transfer from the solar wind to the atmospheric gases, resonant charge transfer from H$^+$ to H

$$H^+ + H \rightarrow H^* + H^+ , \tag{57}$$

and the near-resonant process

$$H^+ + O \rightarrow H^* + O^+ . \tag{58}$$

Since ion temperatures are higher than neutral temperatures at high altitudes, the neutral H formed in these reactions will be hotter than thermal. For an H$_2$ abundance of 1–2 ppm at the homopause, the major source of energetic hydrogen atoms was reaction (54), which is followed by dissociative recombination

$$OH^+ + e \rightarrow O + H . \tag{59}$$

Chamberlain (1977) did the first detailed model calculations of the Venus exosphere, in which he attempted to reproduce the distribution of hot H inferred from Mariner 5 Lα measurements, based on charge exchange of H^{+*} with thermal H (reaction (57)) as the source of H*. Assuming a spherically-symmetric exosphere, he solved Liouville's equation (the collisionless Boltzmann equation) for the distribution of H and found that a higher ionopause than observed was necessary to reproduce the inferred densities. The ionopause has, however, been found from Pioneer Venus measurements to be highly variable (Taylor et al., 1979). An escape flux of 2×10^7 H atoms cm^{-2} s^{-1} was derived.

The Venera 9 and 10 Lα line shape measurements showed that the temperature increases abruptly near an altitude of 3000 km, from which Bertaux et al. (1978) inferred that the hot H is not generated near the exobase. They preferred charge exchange from solar wind protons as the source of hot H. Kumar et al. (1978), however, argued that the experiment could not distinguish an increase in temperature from a net Doppler shift.

Kumar et al. (1978) evaluated the proposed sources of non-thermal H atoms and concluded that reactions (54) and (59) were the most important sources. The production of CO$_2$H$^+$ in reaction (55) followed by dissociative recombination of the ion

$$CO_2H^+ + e \rightarrow CO_2 + H \tag{60}$$

would also provide a significant, but lower altitude source of H atoms. In order to reconcile these large sources with the low thermospheric densities of H inferred from Mariner 5 and 10 data, Mayr et al. (1978) proposed that the H atoms created in this process are swept to the nightside by thermospheric winds. Kumar et al. found that the dayside densities could be reduced by a factor of 1000 by this process. An evaluation of the escape rates due to the various processes led to the conclusion that the largest source of escaping H atoms was charge exchange with solar wind protons followed by reacceleration of the newly formed low-energy protons. An escape flux of 10^7 H atoms cm^{-2} s^{-1} was derived.

Cravens et al. (1980) presented the first model of the Mariner 5 hot hydrogen distribution based on atmospheric and ionospheric structure information from Pioneer Venus. Using a two-stream approach to evaluate the fluxes and densities of hot H at the exobase, they showed that the unexpectedly hot and dense nightside corona results from the high ion temperatures and densities observed by the PV orbiter (e.g., Taylor et al., 1980; Miller et al., 1980; Knudsen et al., 1979). They found that half the source of hot H was due to charge exchange from H$^+$ and O$^+$ (reactions (57) and (58)), and, for the H$_2$ densities adopted, half was due to the reaction of O$^+$ with H$_2$ (reaction (54)). Only preliminary data about the pre-dawn H bulge was available, so the importance of reaction (57) was underestimated for the nightside; the H$_2$ densities in their model were taken from Kumar and Hunten (1974), so the importance of reaction (45) was overestimated. Hodges and Tinsley (1981), however, found that charge exchange alone was sufficient to explain the Mariner 5 and 10 Lα profiles. They constructed a 3-D (non-spherically symmetric) Monte-Carlo model of the Venus densities and temperatures, and found that the most important source is the charge exchange from H$^+$, with charge exchange from O$^+$ contributing only near the exobase. They criticized the hot H source proposed by Bertaux *et al.* (1978), charge exchange from solar wind protons, as requiring the deceleration of solar wind neutralized protons from 400 km s^{-1} to sub-escape speeds and concluded that the magnitude of this source of hot H was negligible. The charge exchange sources adequately reproduced the day/night ratios of the hot H concentrations and scale heights. A factor of 2 uncertainty in the hot hydrogen measurements did not, however, exclude the presence of another source, such as reaction (54). The computed global average escape rate from charge exchange was about $1.8 \times 10^7 \text{ cm}^{-2} \text{ s}^{-1}$. Kumar *et al.* (1981) predicted an escape rate of $1 \times 10^8 \text{ cm}^{-2} \text{ s}^{-1}$ from reaction (54) based on the H_2 densities they derived from the assumption that the mass-2 ion was H_2^+.

The D/H ratio measured by Donahue *et al.* (1982), and inferred from the D^+/H^+ ratio, an enhancement of a factor of 100 over the terrestrial value, was interpreted as implying at least a hundred-fold depletion of the initial H inventory of the planet over the age of the solar system. Donahue *et al.* argued that most non-thermal escape mechanisms discriminate against D, but when the abundance of H_2 was enhanced by a factor of 100, hydrodynamic escape would take over, so the initial abundance of water vapor may be much larger than required by the D/H ratio. Hydrodynamic or transonic escape, also called *blow-off*, is a thermal loss process in which the atmosphere expands rapidly outward. The high frequency of collisions renders the process describable by the macroscopic theory of fluid dynamics, rather than by the microscopic kinetic theory. The dynamics of atmospheres undergoing hydrodynamic blow-off of hydrogen have been discussed by Watson *et al.* (1981) and the salient features of hydrodynamic escape have been reviewed by Hunten (1973b, 1982). Mass fractionation in hydrodynamic escape may be large, but the process is not as mass-selective as Jeans escape (Zahnle and Kasting, 1986; Hunten *et al.*, 1987). Blow-off becomes an appropriate description of the escape process above the altitude where λ, the reduced gravitational potential energy, is about 2. At this altitude, the rate of expansion of the atmosphere reaches the speed of sound; alternatively, the average thermal velocity of a molecule is approximately equal to the escape velocity. Hydrodynamic escape thus becomes the major escape mechanism if and when the $\lambda = 2$ level moves below the exobase (Walker, 1982).

McElroy *et al.* (1982) suggested another non-thermal mechanism for the escape of hydrogen that discriminates almost completely against deuterium: reaction with hot O produced in dissociative recombination of O_2^+ at the exobase. This process, which was first suggested for He and H escape from Mars by Knudsen (1973), can be represented by

$$O_2^+ + e \rightarrow O^* + O^* \tag{61}$$

followed by

$$O^* + H \rightarrow O + H^*. \tag{62}$$

Using a Monte-Carlo method and assuming isotropic scattering, McElroy *et al.* computed the fraction of collisions (reaction (62)) leading to H atoms with energies greater than the escape energy as 15% for an O^* with initial velocity of 5.6 km s^{-1} (2.6 eV) colliding with a 300 K H-atom. A planetary average H atom escape rate due to

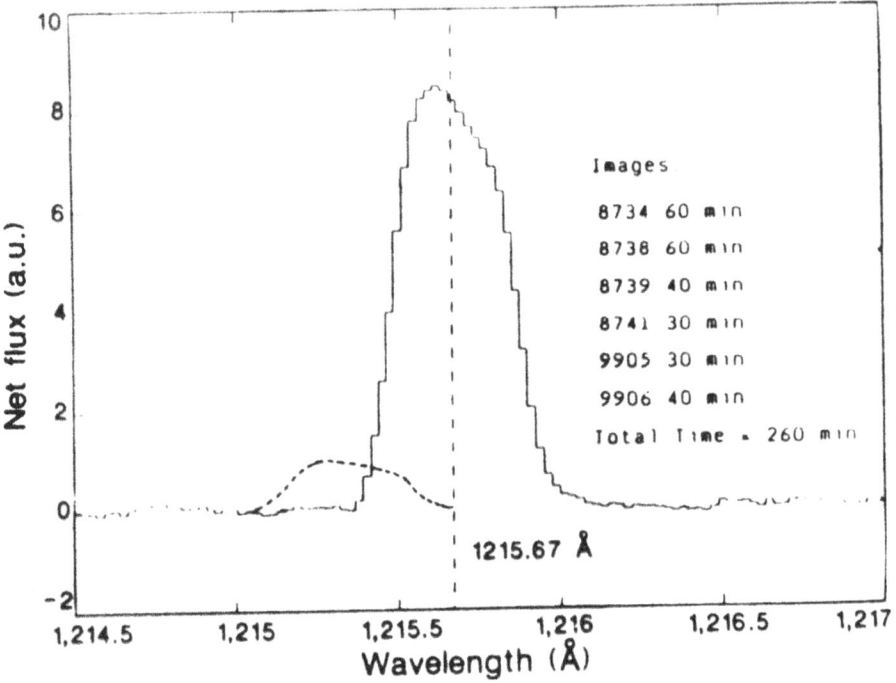

Fig. 26.   Hydrogen Lα spectra of bright disk emission from Venus obtained with the IUE large aperture and summing the six best images (solid line). The line intensity corresponds to $21 \times 10^3 R$; the long-dashed line shows the line center. The short-dashed line shows the profile calculated for deuterium Lyman alpha emissions of $2.5 \times 10^3 R$ (corresponding to a D/H ratio of 1.6×10^{-2}). The numbers refer to spectra reference numbers and exposure times. AU denotes 10^5 IUE flux units. Taken from Bertaux and Clarke (1989).

reaction (62) of 8×10^6 cm^{-2} s^{-1} was obtained. Much smaller escape rates due to charge transfer from H$^+$ (reactions (57) and (58)) were obtained, due partly to the assumption of a low (2000 K) ion temperature, compared to PV values, especially on the nightside, where ion temperatures have been shown to exceed 5000 K (Miller *et al.*, 1980).

The escape fraction in reaction (62) computed by McElroy *et al.* (1982) was shown to be overestimated by Cooper *et al.* (1984), who derived a general formula for the rate coefficient for the production of a particle with a specific energy colliding with the atoms of a thermal gas. The cross sections for O–H elastic scattering were computed using semi-empirical and theoretical interaction potentials, and the effect of the anisotropy of the scattering was included. Figure 27 shows their calculated frequency of production of a hydrogen atoms with various kinetic energies upon collision with an O-atom traveling at 5.6 km s^{-1}. The computed fractions of escaping H-atoms were 5.1, 6.9, and 8.5% for temperatures of 100, 200, and 300 K, respectively, about half the values computed by McElroy *et al.* (1982). In addition, the O-atom velocity of 5.6 km s^{-1} used in the calculations of McElroy *et al.* was computed assuming that the dissociative

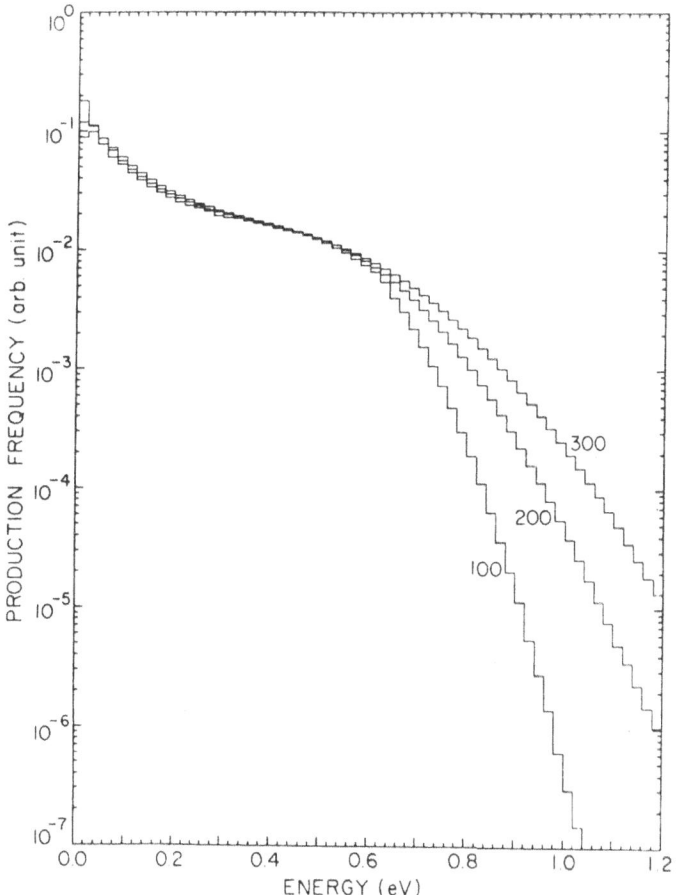

Fig. 27.   The frequency with which a hydrogen atom with energy E is produced in collision with a 2.5 eV oxygen atom at temperatures of 100, 200, and 300 K. Taken from Cooper *et al.* (1984).

recombination of O_2^+ produces one $O(^1D)$ and one $O(^3P)$ atom. As pointed out in Section 3.2, it has become clear from calculations and measurements that the yields of excited stated in DR of O_2^+ (v) depend on the vibrational level, v (e.g., Guberman, 1987, 1988; Bates and Zipf, 1980). Further discussion will be deferred to the discussion of the hot O corona in Section 4.7.

Rodriguez *et al.* (1984) computed the exospheric distribution and escape rates of H by an approximate numerical solution to the time-independent Boltzmann equation. The model parameters were based on PV data, including the pre-dawn bulge in the H densities inferred by Brinton *et al.* (1980), and the high ion temperatures measured by the ORPA (Miller *et al.*, 1980). Alternative O_2^+ densities were taken from the PV OIMS measurements (Taylor *et al.*, 1980) and the PV orbiter radio occultation (ORO) profiles (Kliore *et al.*, 1979). The H_2 abundance was assumed to be less than 0.5 ppm, in accord with the mesospheric photochemical models (Yatteau, 1983; Yung and DeMore, 1982).

A more realistic cross section for O*–H elastic scattering was employed, similar to that computed by Cooper *et al.* (1984). The assumed exobase altitudes, 190 km and 150 km on the day- and nightsides, respectively, are somewhat lower than those used by other workers and computed here (see Section 4.1). Rodriguez *et al.* found that collisions with hot O (reaction (62)) and H^+–H charge transfer (reaction (57)) were comparable sources of hot hydrogen, with the latter contributing more to escape. On the nightside, if the OIMS measurements were used for the densities of O_2^+, reaction (62) was found to produce adequate hot hydrogen; if the ORO densities, which are significantly smaller than the OIMS values, were used, it was necessary to include reaction (57). An average escape flux of about 10^7 cm^{-2} s^{-1} was computed.

Hodges and Tinsley (1986) investigated the effect of H–H^+ charge exchange on the velocity distribution and escape of exospheric H in an update of their previous three-dimensional Monte-Carlo model of the Venus exosphere (Hodges and Tinsley, 1981). Their computed exospheric H-atom densities, presented as a function of solar hour angle, show that, due to lateral transport, at increasing planeto-centric distances, the pre-dawn bulge becomes less pronounced; outward of 8000 km, the distribution becomes nearly spherically symmetric. Hodges and Tinsley (1986) suggest that a measurement of the line profiles of Lα due to resonance scattering of solar photons should carry a signature of their source; they have computed the Lα line profiles for the charge exchange source of hot H. The profiles are essentially a superposition of a narrow and a broad Gaussian, corresponding to the thermal and hot hydrogen components, with asymmetries due to escape and lateral flow. An average planetary escape rate due to charge exchange of 2.8×10^7 cm^{-2} s^{-1} was computed.

Bishop (1989) has discussed modifications to the exospheric distribution and escape that result from radiation pressure due to resonance scattering of solar Lα. The effects on the distribution of H are largely confined to distances from the center of Venus greater than two Venus radii, where densities are small. A small increase in the hydrogen escape flux of 2×10^6 cm^{-2} s^{-1} was computed.

The most comprehensive global model of non-thermal escape mechanisms and their evolution through the history of the planet is that of Kumar *et al.* (1983). The sources of hot H that they considered included charge-exchange of H^+ with H and impact of hot O with H. Their calculations for the production rates due to the reaction of O^+ with H_2 were contingent upon the identification of the mass-2 ion as H_2^+ rather than D^+, the currently accepted identification. The pre-dawn bulge region of the atmosphere was used as an indication of the structure of the thermosphere-ionosphere at earlier times, when the hydrogen abundance was larger. At present, the most important source of hot H is charge exchange of H^+–H (reaction (57)), with a global average escape rate of 1.2×10^7 cm^{-2} s^{-1}, and most of the flux occurs over the nightside hemisphere. Impact of hot O with H is slightly less important; the global average escape rate is 8×10^6 cm^{-2} s^{-1}. At earlier times, the importance of this source declines as the H abundances increases and the exobase rises above the O_2^+ peak. Figure 28 shows the evolution of the H escape flux due to various mechanisms as a function of the atomic hydrogen mixing ratio. The calculations of Kumar *et al.* show that when the H mixing

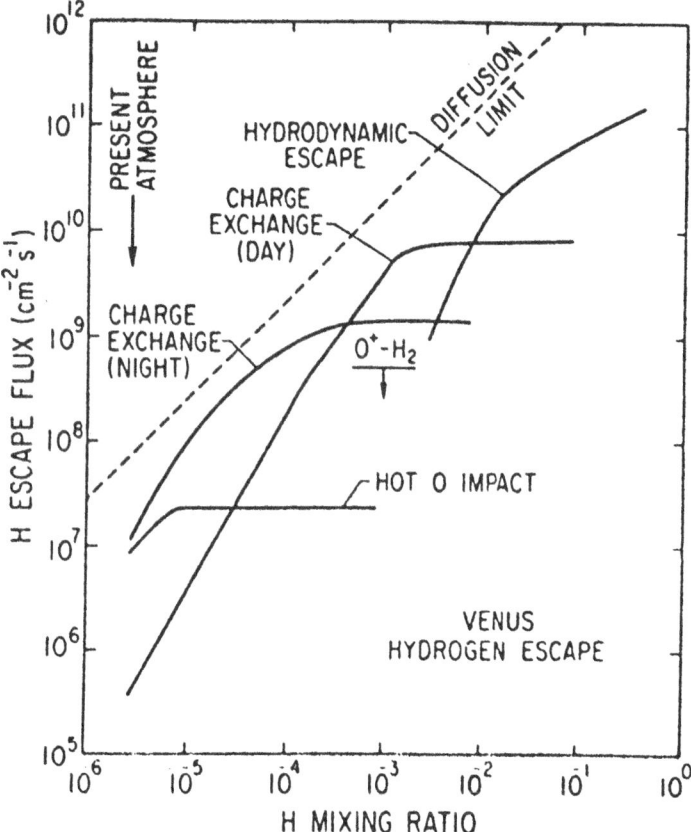

Fig. 28. Hydrogen atom escape flux on Venus as a function of H mixing ratio at the homopause is shown for various escape mechanisms. This H mixing ratio is equivalent to twice the H_2O vapor mixing ratio at the cold trap. The H mixing ratio at the homopause for the present atmosphere is 2.5×10^{-6}. Taken from Kumar *et al.* (1983).

ratio reaches 2×10^{-3}, the H density determines the exobase level and the escape rate of H levels off at a maximum value of 7.5×10^9 cm^{-2} s^{-1}. When the mixing ratio of H exceeds a value near 8×10^{-3}, a transition from non-thermal escape to hydrodynamic escape takes place.

Escape fluxes of H may be derived if the density profile of H below the exobase is known, because the actual scale height is slightly smaller than the diffusive equilibrium value. From $L\alpha$ profiles recorded by the PV OUVS, Paxton *et al.* (1988) deduced a value of $7.5 \pm 1.5 \times 10^7$ cm^{-2} s^{-1} for the flux of H through the exobase near the sub-solar point. The model calculations discussed above show that the escape flux is a quarter to a third of this value. The remainder of this flux could be accounted for by the sub-solar to anti-solar circulation, and it would be balanced by a downward flux on the nightside.

Detailed models for the evolution of the Venus atmosphere have been presented by Watson *et al.* (1981), Kasting and Pollack (1983), and Watson *et al.* (1984). These models assume that the ultimate source of the escaping hydrogen is water vapor, the predicted initial abundance of which is, therefore, much greater than at present. Toward earlier times, the larger amount of water vapor in the atmosphere increases the infrared opacity and causes the tropospheric temperature to rise. A critical value of the insolation exists for which liquid water cannot exist on the surface and the evaporation of the water produces a run-away greenhouse (Ingersoll, 1969). The insolation at Venus is proposed to be above the critical value. Kasting *et al.* (1984) showed that an increase in the water vapor mixing ratio also causes a decrease in the tropospheric lapse rate, causing the cold-trap to move to higher altitudes and lower pressures. The resulting increase in the mixing ratio of water vapor at high altitudes, the parameter that determines the atomic hydrogen escape rate, is dramatic. The predicted initial water abundance varies greatly from one model to another, depending on such factors as the mixing ratio of onset of hydrodynamic escape, the H/D fractionation factor during such escape, the magnitude of the hydrodynamic escape flux (which depends on assumptions made about the magnitude of the solar flux in the past), and the outgassing rate of volatiles from the interior (e.g., Watson *et al.*, 1981, 1984; Donahue *et al.*, 1982; Kasting and Pollack, 1983; Kumar *et al.*, 1983).

An alternative scenario has been proposed by Grinspoon and Lewis (1988). The current abundance of water vapor in the atmosphere could be in steady state as a result of a balance between loss by escape of H and a cometary source. As in the case of continuous slow outgassing, an enhanced D/H ratio could result from this exogenous source, but the value depends on the assumed rate of impact of large comets, which is somewhat stochastic and difficult to estimate. Nevertheless, a large initial abundance of water may not be necessary to explain the enhanced D/H ratio.

A detailed discussion of atmospheric evolution is beyond the scope of this review. The interested reader is referred to a review of the evolution of the atmosphere of Venus presented by Donahue and Pollack (1983) and to a book on the origin and evolution of planetary atmospheres (Atreya *et al.*, 1989), especially to the articles by Hunten *et al.* (1989) and Kasting and Toon (1989) contained therein.

4.7. HOT OXYGEN CORONA

Dissociative recombination of molecular ions produces energetic fragments that are important to the thermal structure, coronas and escape from planetary atmospheres. A short review of the importance of dissociative recombination in aeronomy has been presented by Fox (1989a). Dissociative recombination of O_2^+ can proceed according to a number of energetically allowed channels, with exothermicities as:

$$O_2^+ + e \rightarrow O(^3P) + O(^3P) + 6.98 \text{ eV} , \tag{63a}$$

$$\rightarrow O(^1D) + O(^3P) + 5.02 \text{ eV} , \tag{63b}$$

$$\rightarrow O(^1S) + O(^3P) + 2.79 \text{ eV} , \tag{63c}$$

$$\rightarrow O(^1D) + O(^1D) + 3.05 \text{ eV} , \tag{63d}$$

$$\rightarrow O(^1D) + O(^1S) + 0.83 \text{ eV} . \tag{63e}$$

Since the escape energy per unit mass for Venus is 0.54 eV amu^{-1}, none of the channels produce oxygen atoms with enough energy to escape the gravitational field of the planet. Oxygen atoms produced above the exobase will, however, travel along ballistic trajectories in the exosphere, producing a corona of hot oxygen atoms, similar to the hot hydrogen corona. A hot oxygen corona has been observed to surround the Earth (Yee *et al.*, 1980; Yee and Hays, 1980) and the density distribution has been modeled by Yee *et al.* (1980). The existence of hot oxygen coronas around Venus and Mars was predicted by Wallis (1978). The corona has a significant effect on the interaction of the planet with the solar wind (e.g., Biermann *et al.*, 1967; Wallis, 1972, 1978, 1982). A review of hydrogen and oxygen coronas of Venus and Mars has been presented by Nagy *et al.* (1990).

Bertaux *et al.* (1981) reported the existence of 'a strange feature' in the intensity profile recorded at 1304 Å by the spectrophotometer on Venera 11. Over a range of about 5000 km beyond the bright limb, a nearly constant intensity of about 500 R was measured. Because this feature was not observed by Venera 12, it was tentatively interpreted as being due to the presence of 'a sporadic hot component of exospheric oxygen', and an average density of 1.6×10^3 cm^{-3} was inferred.

Limb scans of the PV OUVS also showed the existence of an exospheric signal at 1304 Å (Nagy *et al.*, 1981). The oxygen atom density profiles derived from the intensities exhibit two scale heights, corresponding to an inner thermal component and an outer hot component. The hot component, shown in Figure 29, dominates beyond an altitude of about 350 km, where the O density is about 5×10^4 cm^{-3}. The scale height of the

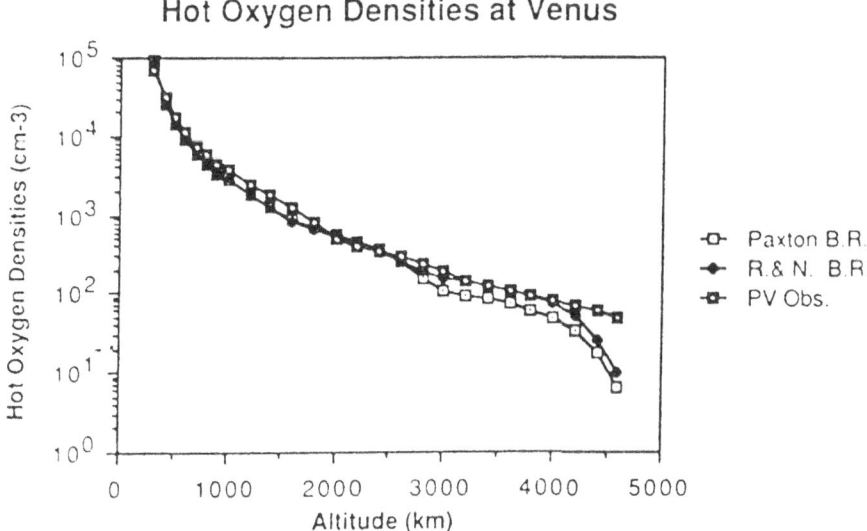

Fig. 29.   Measured and calculated hot oxygen densities at Venus for different branching ratios in the dissociative recombination of O_2^+ (reactions 63). 'Paxton B.R.' denotes branching ratios from Paxton (1983). 'R & N B.R.' denotes branching ratios from Rohrbaugh and Nisbet (1973). The experimental values are derived from PV OUVS 1304 Å data. From Nagy and Cravens (1988).

hot component is approximately 400 km, indicating a 'temperature' of 6300 K. Nagy
et al. (1981) constructed the first models of the hot O corona on both the dayside and
the nightside using PV data. In addition to dissociative recombination of O_2^+, they
considered the charge exchange processes

$$O^+ + O \rightarrow O^* + O^+ \tag{64}$$

and

$$O^+ + H \rightarrow O^* + H^+, \tag{65}$$

as sources for the energetic oxygen, but found these reactions to be relatively unimportant. A modified two-stream approach, similar to that used in the hot hydrogen model
(Cravens et al., 1980) was used to compute the flux of O at the exobase; the exospheric
distribution was then determined by solving Liouville's equation. Nagy et al. obtained
hot O densities a factor of 4–5 larger than the densities derived from the PV measurements. The discrepancy was later attributed to the use of preliminary ion densities that
were too high, and was resolved with the use of appropriate O_2^+ densities (Nagy and
Cravens, 1988). The corrected calculated hot O density profile is shown also in
Figure 29. In addition, the earlier paper (Nagy et al., 1981) assumed very low exobase
altitudes of 172 and 143 km on the day- and nightsides, respectively, compared to the
values computed here, which would also tend to exaggerate the magnitude of the O
source.

Reasonably successful models of the hot O corona were also presented by McElroy
et al. (1982b) and by Paxton (1983). McElroy et al. (1982b) presented the first calculations of the hot O column density as a function of solar zenith angle from 0 to 90°.
The results, presented here in Figure 30, show that the column density varies from about
3×10^{12} cm^{-2} at the subsolar point to 1×10^{12} cm^{-2} at the terminator.

Models of the hot oxygen corona depend on assumptions made about the relative
importance of the channels (63a–e) in dissociative recombination of O_2^+. No definitive
measurements of the yields of $O(^3P)$, $O(^1D)$, and $O(^1S)$ in dissociative recombination
of the lower vibrational levels of $O_2^+(v)$ are available. It has become clear that the
channel by which the dissociative recombination proceeds depends greatly on the
vibrational state of the ion (Guberman, 1983, 1987, 1988; Bates and Zipf, 1980;
Queffelec et al., 1989). Zipf (1970) determined yields of 0.1 and 0.9 for $O(^1S)$ and $O(^1D)$,
respectively, but these values were later withdrawn because the vibrational distribution
of O_2^+ in the experiment was unknown. Guberman (1987, 1988) carried out ab initio
calculations of the rate coefficients for production of $O(^1S)$ and $O(^1D)$ in dissociative
recombination of O_2^+ in various vibrational levels. Some of the rate coefficients he
obtained for a temperature of 300 K are shown in Table X. Specifically, he showed that
the rate coefficient for production of $O(^1S)$ increases nearly two orders of magnitude
from $v = 0$ to $v = 2$! Unfortunately, Guberman did not calculate the rate coefficient for
the channel that leads to two ground state atoms (reaction (63a)), so the yields of excited
states cannot be derived from his calculations alone. Recently, Queffelec et al. (1989)
have studied the yields of ground and excited O atoms for $O_2^+(v)$ in vibrational levels

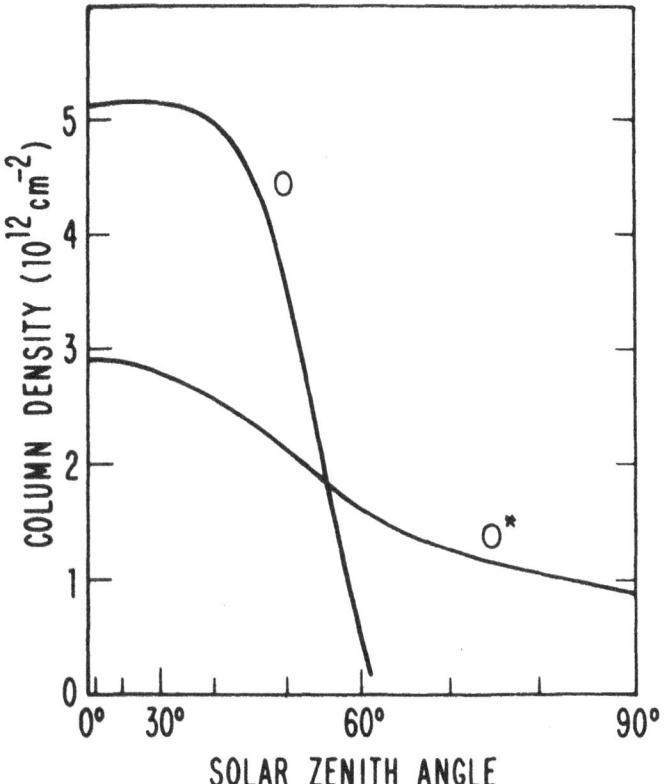

Fig. 30.   Column densities of neutral oxygen above Venus plasmapause as a function of solar zenith angle. The abscissa has been scaled in accord with fractional surface area. Thermal oxygen (O) and hot oxygen atoms (O*) contributed equally to the total amount of neutral O above the plasmapause, approximately 8×10^{30} atoms. Taken from McElroy *et al.* (1982b).

around $v = 9$ using a plasma flow tube experiment. The reported yields were 0.44, 1.0, and 0.56 for $O(^1S)$, $O(^1D)$ and $O(^3P)$, respectively. An average yield of about 0.1 was obtained for the yield of $O(^1S)$ in dissociative recombination of O_2^+ in an unknown distribution of vibrational levels 0, 1, and 2, in disagreement with Guberman's calcu-

TABLE X

Dissociative recombination coefficient ($cm^3 s^{-1}$) α for production of $O(^1D)$ and $O(^1S)$ from $O_2^+(v)$ in different vibrational levels at 300 K. From the *ab initio* calculations of Guberman (1987, 1988).

v	$\alpha(^1S)$	$\alpha(^1D)$
0	3.0×10^{-10}	2.2×10^{-7}
1	7.3×10^{-9}	1.8×10^{-7}
2	2.4×10^{-8}	1.2×10^{-7}

lations. The calculations of Guberman and the experimental study of Queffelec *et al.* (1989) do show, however, that the vibrational distribution of O_2^+ needs to be known in order to compute the production rates of excited states or the energy distribution of the O atoms.

Fox (1985) has modeled the vibrational distribution of O_2^+ in the thermosphere of Venus; at the exobase 55% of the O_2^+ ions are predicted to be vibrationally excited. Using the rate coefficients of Guberman and measured values for the total dissociative recombination rate, Fox (1990b) has computed the yields of the channels above and the fractions of O atoms with various energies produced at the exobase on Venus. The major uncertainty in this calculation is the yield of channel (63a), which is obtained as a difference between the measured total rate and the rates of the channels computed by Guberman. The total rate for dissociative recombination almost certainly depends on the vibrational level of O_2^+; that effect was accounted for only by the choice of the rate coefficient of Mul and McGowan (1978) over that of Alge *et al.* (1983), since the merged beam experiments are expected to contain a larger fraction of vibrationally excited ions, and are therefore more appropriate to exobase altitudes, than flowing afterglow values, which pertain to thermalized conditions for temperatures of 200 to 600 K. The branching ratios for the channels (63a–e) derived from the model of Fox (1990b) are shown in Table XI, where they are compared to those assumed in other models. The branching

TABLE XI

Yields of various channels in dissociative recombination of O_2^+ computed or assumed in calculations of the hot oxygen coronas

Channel	Energy (eV)	Yield			
		a	b	c	d
$^3P + {}^3P$	6.98	0.325	0.22	0.0	0.334
$^3P + {}^1D$	5.01	0.30	0.55	1.0	0.393
$^3P + {}^1S$	2.79	0.05	0.0	0.0	0.0
$^1D + {}^1D$	3.04	0.275	0.13	0.0	0.233
$^1D + {}^1S$	0.82	0.05	0.10	0.0	0.038

(a) Derived by Rohrbaugh and Nisbet (1973) from data of Zipf (1970). Used by Nagy *et al.* (1981) and Nagy and Cravens (1988).
(b) Determined by Paxton (1983) from a best fit to the PV OUVS hot oxygen data.
(c) Assumed by McElroy *et al.* (1982a).
(d) Derived by Fox (1990b) from calculations of Guberman (1987, 1988) and the measurements of Mul and McGowan (1979).

ratios agree surprisingly well with those derived by Rohrbaugh and Nisbet (1973) from the measurements of Zipf (1970), that were used in the models of Nagy *et al.* (1981) and Nagy and Cravens (1988). The agreement is poorer with the values obtained by Paxton (1983), who varied the branching ratios to obtain agreement with the PV hot oxygen data. McElroy *et al.* (1982) assumed that all the dissociative recombinations proceed by channel (63b). While this choice of branching ratios disagrees with those of other workers, the average energy of the O atoms is approximately correct, so fairly good agreement with experiment was obtained.

4.8. ESCAPE OF O

The large escape flux of H, the source of which is a putative large initial inventory of water or a continuous endogenous or exogenous source (see Section 4.6), raises the issue of the fate of the O atoms remaining behind, given the dearth of O_2 in the Venus atmosphere. The O atoms produced in dissociative recombination of O_2^+ are not energetic enough to escape on Venus, as they do above the Martian exobase. Most workers have assumed that the O atoms 'disappear' by oxidizing either crustal iron, Fe or Fe^{2+} to Fe^{3+}, or atomic carbon or CO from the planet's initial endowment of volatiles to produce some of the CO_2 that is in the atmosphere today (McElroy et al., 1982; Donahue et al., 1982; Lewis and Prinn, 1984). Prinn (1985) has described some of the problems with these scenarios, which include the requirement that a fair fraction (1–3%) of the interior of Venus be exposed to the atmosphere over the first 0.3 billion years or so of its existence. Models of the evolution of the atmosphere indicate that the weathering process may be accelerated by the presence of liquid water on the surface (Kasting et al., 1984) or by the presence of a molten surface due to a runaway greenhouse (Watson et al., 1984).

McElroy et al. (1982b) and Wallis (1982) proposed that O escapes from the Venus atmosphere by photoionization and electron impact ionization of thermal and hot O above the plasmapause, and subsequent pick-up of O^+ by the solar wind. McElroy et al. estimated the total quantity of O atoms as 8×10^{30} and the loss rate about 6×10^6 cm^{-2} s^{-1}. They asserted that this escape flux was about half the escape flux of H, and that the escape rates of H and O are regulated by the oxidation state of the lower atmosphere. This analysis was, however, based on the H escape flux of 1.2×10^7 cm^{-2} s^{-1} computed by McElroy et al. (1982a), which is smaller by a factor of two or more than subsequent models indicate (see Section 4.6).

Recently, Luhmann and Kozyra (1990) have shown that approximately 90% of the O^+ ions created by photoionization above the plasmapause and picked up by the solar wind reimpact the planet, rather than being swept away. Significant and perhaps even enhanced escape still occurs due to sputtering of O atoms by the precipitating O. They computed a global loss rate of 2×10^{25} O atoms s^{-1} or an escape flux of 6×10^6 cm^{-2} s^{-1}, which is comparable to the escape flux computed by McElroy et al. (1982a).

4.9. C AND N HOT ATOM CORONAS

Paxton (1983) modeled the production rate of hot carbon due to knock-on from the hot O produced in dissociative recombination of O_2^+ near the exobase (reactions 63(a–e)):

$$O^* + C \rightarrow C^* + O . \tag{66}$$

A smaller contribution, about 10% of that from reaction (66), was determined to arise from dissociative recombination of CO^+ :

$$CO^+ + e \rightarrow C^* + O^*.\tag{67}$$

The altitude distribution of C^* was derived from PV OUVS limb scans of the high altitude optically thin intensity at 1657 Å, assumed to arise from resonance scattering of solar radiation by atomic carbon. A C/O ratio of about 1% at the exobase was determined and the scale height of the altitude profile indicates a temperature that is approximately twice thermal.

Limb profiles of the signal at 1493 Å, a feature produced by resonance scattering of solar radiation by $N(^2D)$, were modeled by Paxton to determine the density distribution of hot $N(^2D)$ (Keating *et al.*, 1986). The major source was determined to be dissociative recombination of N_2^+, with lesser contributions from dissociative recombination of NO^+ and knock-on from hot O.

4.10. He ESCAPE

Thermal escape of He from the atmosphere of Venus is negligibly small, but escape is possible by photoionization and electron impact ionization of He atoms above the ionopause, followed by pick-up of the resulting He^+ ions by the solar wind. Kumar and Broadfoot (1975) used Mariner 10 data to estimate an escape rate of 2×10^5 He atoms $cm^{-2} s^{-1}$ due to this source. This estimate was based on limited observations of the ionopause altitude, which has been found to be quite variable by PV observations (Taylor *et al.*, 1979). Prather and McElroy (1983) have estimated the total escape rate as $1 \times 10^6 cm^{-2} s^{-1}$, with an accuracy of about $\pm 30\%$. It is interesting that they find that about 20% of the He above the ionopause is produced by knock-on from hot O, the process suggested by Knudsen (1973) for escape of He from Mars. Chassefiere *et al.* (1986) found, from Venera 11 and 12 observations of the He emission at 584 Å, that the abundance of He is a factor of two smaller than the value derived from PV data and, therefore, the estimated escape rate, $5 \times 10^5 cm^{-2} s^{-1}$, is also therefore smaller by a factor of 2.

In the terrestrial atmosphere, escape of ^4He is balanced by production in radioactive decay of uranium and thorium in the crust. The terrestrial escape flux is about $2 \times 10^6 cm^{-2} s^{-1}$ (MacDonald, 1963). Prather and McElroy conclude that the comparable magnitude of the sources of ^4He on Earth and Venus suggests comparable abundances of uranium and thorium and, along with independent data about other volatiles, indicates that Venus and Earth had similar origins.

5. Thermal Structure

5.1. GLOBAL MEAN HEAT BALANCE AND THERMAL STRUCTURE

The calculation of a reasonable dayside Venus mesosphere/thermosphere heat budget yielding observed temperatures has been an ongoing problem for nearly two decades. The difficulty is illustrated in Figure 31 (Keating and Bougher, 1987), which compares

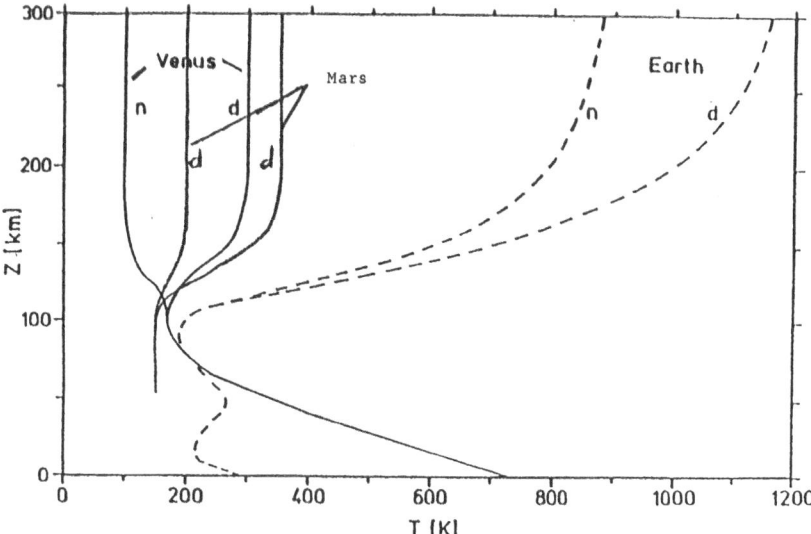

Fig. 31.   Temperatures of the neutral atmospheres of Earth, Venus, and Mars. Adapted from Schubert *et al.* (1983).

Venus, Earth, and Mars dayside thermospheric temperature profiles; Venus, although closest to the Sun, has the coldest temperatures above 100 km. This implies that, unlike the terrestrial thermosphere, a simple balance of EUV heat and molecular conduction may not be sufficient to maintain the observed relatively cold (≤ 300 K) temperatures. Several modeling efforts have attempted to determine what additional cooling mechanisms might be responsible. Uncertainties in the role of eddy conduction, the O–CO_2 collisional excitation rate, the EUV heating efficiencies, and the EUV solar fluxes themselves make the job very difficult. Recently additional constraints have been provided by the Pioneer Venus neutral thermospheric data, new laboratory measurements, and terrestrial modeling and observational efforts. Model simulations can be made in which EUV heating efficiency, CO_2 15-μm cooling, and eddy conduction are considered to be free model parameters. However, the range of realistic heating efficiencies reduces the possibilities considerably (see below).

5.2. EUV HEATING EFFICIENCIES

Heating due to the absorption of solar ultraviolet radiation occurs by photoionization and photodissociation. In photodissociation, the energy in excess of the dissociation threshold may be converted to internal or kinetic energy of the neutral fragments. Kinetic energy can be converted directly to heat through elastic collisions, or to internal energy through inelastic processes, such as collisional excitation of rotational or vibrational modes of molecules. Excited vibrational or electronic states may radiate, which results in cooling if the radiation escapes from the atmosphere, or they may be quenched, releasing their internal energies as heat; rotational energy is usually rapidly thermalized. In photoionization, the excess energy is carried away by the ejected electron, which can produce further ionization, dissociation or excitation. Some of the chemical energy of the ions is released as heat in exothermic charge transfer reactions and ultimately in neutralization through dissociative recombination.

It is traditional and convenient to define the heating efficiency, ε, as the fraction of solar energy absorbed at an altitude that is converted locally to heat. Heating efficiencies for Venus were first computed by Henry and McElroy (1968). Their computed mean heating efficiency, about 60%, was based on the (incorrect) assumption that CO_2^+ was the major ion in the ionosphere. The predicted global mean exospheric temperature was about 700 K (McElroy, 1969). Stewart and Hogan (1969) estimated heating efficiencies about half that value from studies of the Martian thermosphere in which the heating efficiency was varied to reproduce the available data. They reasoned that the long Venus day would allow radiative equilibrium to be established on the dayside, so the computed (dayside) exospheric temperature was the same as McElroy's mean value (Hogan and Stewart, 1969). This contention was criticized by McElroy (1970), who pointed out that the Venus atmosphere could not sustain the large day–night density and temperature gradients.

In the years that followed, the exospheric temperature was determined to be much lower (≤ 300 K) than these early models indicated. Dickinson (1976) and Dickinson and Ridley (1977) computed 1-D and 2-D models of the thermal structure in the Venus thermosphere; they found that the low exospheric temperatures required very small values of the heating efficiency. They proposed that much of the energy involved in solar energy absorption would be taken up as vibrational and rotational excitation of CO_2, which has a larger number of degrees of freedom than the diatomic molecules that dominate the terrestrial atmosphere. Fox and Dalgarno (1981) computed heating efficiencies in the thermosphere for three models with different assumptions about the fraction of the excess energy that is converted to vibrational excitation in photodissociation and chemical reactions. In the standard model, for the assumption that 50% of the excess energy was converted to vibrational excitation, the heating efficiencies varied from about 18 to 22% over the altitude range from 115 to 235 km. Studies of the thermal structure of the Venus thermosphere in light of the knowledge gained from Pioneer Venus indicated that even heating efficiencies of that magnitude could not be accommodated and suggested that the heating efficiencies were on the order of 10% (Dickinson and Bougher, 1986; Hollenbach et al., 1985).

Fox (1988) considered the likelihood that 10% heating efficiencies could be justified from a molecular viewpoint. Both Hollenbach et al. and Dickinson and Bougher partially justified the low heating efficiencies by the assumption that 90% or more of the exothermicity of the quenching reaction

$$O(^1D) + CO_2 \rightarrow O(^3P) + CO_2 \tag{68}$$

is converted to vibrational excitation of CO_2. Fox adduced evidence that reaction (68) proceeds via a collision complex, which should produce statistical energy partitioning if the lifetime of the complex is long enough. The vibrational distribution that would be produced for statistical energy partitioning was computed by applying information theory (e.g., Levine and Bernstein, 1974; Bernstein and Levine, 1975). It was found that only 55% of the excess energy appears as vibrational energy. An upper limit was determined from a survey of the literature, which indicated that a value of 75% would be high, but not unreasonable. After a consideration of the probable values and upper limits for energy deposited as vibrational excitation in photodissociation (about 25%, with an upper limit of about 35%) and chemical reactions (50–60%), and energy released in dissociative recombination of O_2^+, including a model of its vibrational distribution, altitude profiles of the heating rates, shown here in Figure 32, were

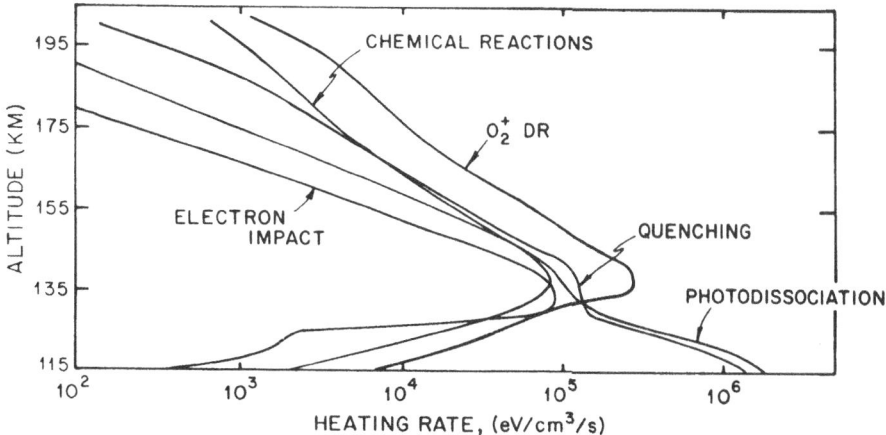

Fig. 32. Altitude profiles of heating rates due to the major source of neutral heating in the Venusian thermosphere from 115 to 200 km from the standard model, in which most probable values for the fraction of energy deposited as vibrational excitation in various elementary processes were employed (see text). Taken from Fox (1988).

computed. The major source of heating above about 130 km is dissociative recombination of O_2^+ (reaction (63)). Below that altitude, photodissociation and quenching of metastable electronic states are about equally important. Heating efficiencies for standard and lower limit models were computed and the results are shown in Figure 33. The standard model employs the probable values for the fraction of energy deposited as vibrational excitation in the processes discussed above and the lower limit model uses the upper limits for fraction of energy deposited as vibration excitation. Below 125 km, the heating efficiencies for the lower limit and standard models are about 16 and 22%, respectively, and above 130 km, they are 22 and 25%, respectively. The heating efficiencies are slightly higher and the difference between the standard and lower limit models is smaller than the calculations of Fox and Dalgarno (1981) indicate. Fox (1988) found that heating efficiencies on the order of 10% or less were difficult to justify with reasonable assumptions about molecular processes in the Venus thermosphere.

The F1 ion density peak at 140 km ($\tau = 1$ for EUV), composed mostly of O_2^+, should correspond closely to the maximum in per volume (eV cm^{-3} s^{-1}) heating if O_2^+ dissociative recombination and related ion-neutral chemical reactions dominate other heating

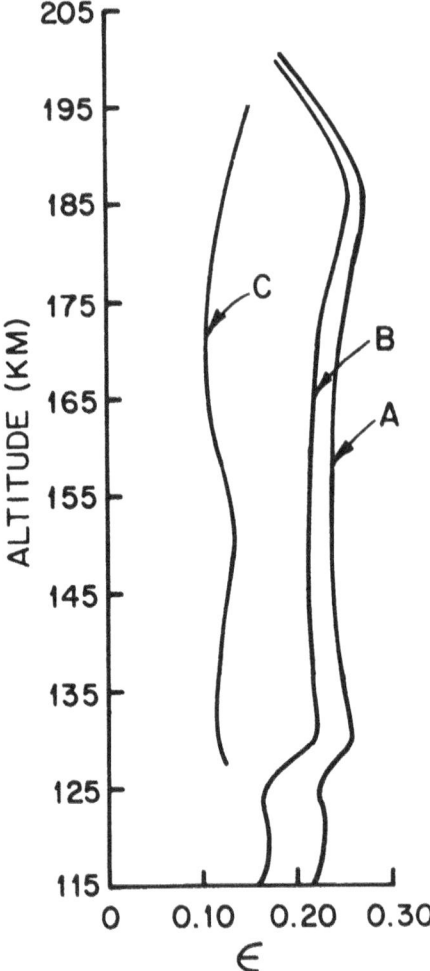

Fig. 33.   Computed heating efficiencies for the altitude range 115–200 km. The curve labeled A is the standard model and B is the lower limit model. The curve labeled C is taken from Hollenbach *et al.* (1985). Taken from Fox (1988).

mechanisms (see Figure 32). But the magnitude of the heating at $\tau = 1$ is a function of the specific EUV fluxes and the heating efficiency used. The height of this level depends on the column of absorber and thus on the neutral atmosphere chosen. Some of the discrepancies that arise in the temperature calculations and the heating efficiencies may be due to the use of different inputs. An effort by the authors is underway to compute heating efficiencies and model thermal structure using common background atmospheric models and solar fluxes.

5.3. THE DAYSIDE HEAT BUDGET

One-dimensional modeling efforts for global mean conditions date back to those of McElroy (1968, 1969) and Dickinson (1972), for a nearly pure CO_2 atmosphere, before the advent of Venus general circulation models. In retrospect, it appears that the approach of calculating temperatures based upon a balance of radiative sources and sinks and thermal conduction is adequate to address the Venus dayside heat balance problem. Large-scale upwelling winds appear to provide little adiabatic cooling affecting the dayside thermospheric heat budget (see Section 6.3). Therefore, circulation models are not required to explain the observed dayside temperatures. As a result, 1-D model simulations for various combinations of input parameters provide a useful and relatively efficient method for examining heating and cooling processes on the Venus dayside.

Dickinson (1972) chose to address both the Venus mesosphere and thermosphere (65 to 200 km), where, as for the Earth, a non-LTE (NLTE) treatment is required. In particular, his 1-D radiative transfer calculation utilized a two-level NLTE formulation for the CO_2 vibrationally excited states. The primary radiative transfer was assumed to occur only in the 15-μm and 4.3-μm band systems, for which relative populations of the v_2 bending mode and v_3 asymmetric stretch vibrational levels were painstakingly calculated. Near-IR heating was realized through the collisional relaxation of the 15-μm fundamental band. Several isotopic and hot bands yielding 15-μm emission were also considered. Dickinson (1972) calculated a global mean temperature profile assuming an altitude independent EUV efficiency of 30%, as suggested by early arguments of Stewart (1972). This large efficiency was determined with the assumption that all the $O(^1D)$ energy is transformed to heat. Below 115 km, IR processes by themselves are in balance, with heating and cooling rates decreasing from 300 K day^{-1} at 115 km to 1 K day^{-1} at 65 km. The EUV absorbed above 160 km (≤ 2500 K day^{-1}) was almost entirely offset by downward molecular conduction. This conducted heat was deposited between 120 and 140 km, and was balanced by 15-μm cooling. Temperatures ranged from a minimum of 180 K at 122 km (the mesopause) to 475 K at the top. Dickinson found that the mesopause occurs at essentially the lowest pressure at which CO_2 15-μm cooling is balanced by EUV heating.

The relative uncertainty in the EUV heating efficiency, as well as other parameterizations, lead Dickinson (1976) to conduct several sensitivity studies in which the EUV efficiency, eddy mixing, and CO_2 collisional excitation by atomic-O (enhancing 15-μm emission, see below) were varied. Studies of Dickinson (1976) and Fox and Dalgarno (1981) both indicated that 30% is probably an upper limit heating efficiency. Dickinson (1976) suggested that the very rapid collisional quenching of $O(^1D)$ by CO_2 is consistent with the concomitant excitation of upper CO_2 vibrational-rotational states. Vibrationally excited states at the low pressures of the Venus thermosphere lose their energy to IR airglow, rather than by being quenched. Assuming that a significant fraction of the kinetic energy of hot atoms, ions, and molecules is also transferred to CO_2 vibrations, increased cooling (or a reduced heating efficiency) is obtained. Efficiencies of 3 to 30% were examined, yielding global mean exospheric temperatures of 250 to 475 K (compared with later Pioneer Venus values of ~ 215 K). Thermospheric temperatures can be further reduced in the presence of O-atoms, for which the collisional excitation of the CO_2 v_2 bending mode:

$$CO_2(v_1, v_2, v_3) + O^* \rightarrow CO_2(v_1, v_2 + 1, v_3) + O \tag{69}$$

is more efficient. Dickinson argued that O excites the CO_2 15-μm level possibly ten times as rapidly per collision as do self-collisions (corresponding to a de-excitation rate of $k_{CO_2-O} = 5 \times 10^{-14}$ cm^3 s^{-1}) (see Section 5.4). This assumption resulted in a further reduction of model exospheric temperatures by 25–50 K. Dickinson considered the introduction of eddy cooling to be the least satisfactory method for modifying thermospheric temperatures (see Section 5.5). Calculations showed that eddy cooling using eddy diffusion coefficients $\leq 10^6$ cm^2 s^{-1} does not significantly alter thermospheric temperatures. However, values as large as 10^7 cm^2 s^{-1} have a substantial impact on the dayside heat budget. Temperatures near the mesopause are particularly sensitive to the amount of eddy heat conduction prescribed. Dickinson made the common assumption that the same eddy coefficient applies to eddy diffusion and eddy conduction; that is, mass and heat were assumed to be transported with the same efficiency.

Post-Pioneer Venus heat budget studies were first carried out by Hollenbach et al. (1985), who stressed the role of eddy cooling in reproducing the observed thermospheric temperature structure using a one-dimensional model. They suggested that the changing Venus circulation, as simulated by an effective eddy coefficient, adjusts to maintain a nearly constant thermospheric temperature, yielding exospheric values over the solar cycle that depart little from 300 K. Heating efficiencies of 12–15% were internally calculated and used for all simulations. The large $O + CO_2$ de-excitation rate (the inverse of reaction (69)) suggested by Sharma and Nadile (1980, 1981) and Gordiets et al. (1982) ($\sim 1 \times 10^{-12}$ cm^3 s^{-1}) was found to be incompatible with their model calculations. Later circulation model studies of Bougher et al. (1986) showed that Venus dynamics has little impact on the dayside thermospheric heat budget; i.e., the circulation is not an effective thermostat regulating dayside temperatures (see Section 6.3).

Venus heat budget studies were also carried out by Bougher (1985) and Dickinson and Bougher (1986) using an updated version of the previous NTLE mesosphere/thermosphere code (Dickinson, 1972, 1976). They stressed the role played by CO_2 15-μm cooling in maintaining observed temperatures. Several modifications were made that affected the calculated exospheric temperatures significantly. Solar EUV and FUV fluxes were taken from the Torr et al. (1979) and Torr et al. (1980) tabulations for day 78348, corresponding to near solar maximum conditions ($F_{10.7} = 200$) of the early in situ Pioneer Venus observations (1978–1980). These changes essentially doubled the UV heating rates from those previously calculated, making the problem of obtaining low exospheric temperatures that much more difficult. In addition, Pioneer Venus measurements indicated that dayside atomic-O densities are about 3 times larger than previously thought (von Zahn et al., 1983). This meant that the O/CO_2 ratio at the dayside ionospheric peak (~ 140 km) was 17–20% rather than 5.6%, as calculated earlier by Dickinson and Ridley (1977). This improvement increased the efficiency of excitation of the CO_2 vibrational $v_2 = 1$ state by O collisions, thereby enhancing 15-μm emission and cooling where NLTE conditions prevail. Finally, the O + CO_2 energy transfer rate itself was increased from that previously used, based in part on recent studies of the Earth's lower thermosphere as reviewed by Dickinson (1984), and observational estimates for this rate coefficient provided by terrestrial rocket measurements (Sharma and Nadile, 1981; Stair et al., 1985). A value of $5–8 \times 10^{-13}$ cm^{-3} s^{-1} at 300 K was adopted for this rate (Dickinson and Bougher, 1986; Bougher et al., 1986), in contrast to 2×10^{-13} used in Dickinson (1984) for Earth, and about 5×10^{-14} used in previous Venus studies (e.g., Dickinson, 1976). This amounts to a ~ 150-fold increase over CO_2 self-collisions.

It was found that the observed Pioneer Venus global mean temperatures above 140 km could be simulated using strong 15-μm cooling alone for balancing $\sim 10\%$-efficient EUV heating. The maximum EUV heating rate (~ 1100 K earth day^{-1}) was offset by a combination of molecular conduction and 15-μm cooling. The region above 160 km, where molecular conduction removes most of the heat, is nearly isothermal. Thus, the exospheric temperature of Venus was shown to be rather insensitive to the precise details of molecular thermal conduction, if CO_2 cooling is indeed so strong. The near IR heating rate was nearly doubled from that of Dickinson (1972), with its peak altitude rising from 115 to 130 km. The global mean model was also decomposed into a dayside and nightside model, with solar IR heating partitioned between the two in order to simulate hydrodynamical effects. An improved 15-μm cooling rate was obtained by averaging these day and night profiles, yielding a global mean exospheric temperature of 220 K. This is consistent with a dayside mean value of 300 K, and a corresponding nightside value of 130 K. Very efficient eddy cooling was shown to be required to yield observed temperatures in the face of larger EUV heating efficiencies (10–18%). However, such a possibility was generally ruled out since Venus circulation model studies (Bougher et al., 1986) require that eddy diffusion must be less than half that proposed by previous one-dimensional eddy diffusion models (e.g., von Zahn et al., 1980). Hydrodynamic model large-scale winds appear to be very important in maintaining the observed dayside density profiles (see Section 6.6). Therefore, the requirement that the eddy coefficient for diffusion and conduction be the same constrains the cooling to be rather small ($K \leq 2 \times 10^7$ cm^2 s^{-1}).

The low thermospheric heating efficiencies of 10–15% used by Hollenbach et al. (1985) as well as by Dickinson and Bougher (1986) have recently been shown to be improbable (Fox, 1988) (see Section 5.2). The narrow range of permissible EUV heating efficiencies (22–25% above 135 km) requires a re-examination of the Venus dayside heat budget. The NLTE model of Dickinson and Bougher (1986) has recently been updated to determine what reasonable combination of cooling parameters and EUV fluxes might enable observed temperatures to be simulated using EUV heating efficiencies consistent with the calculations of Fox (1988) (cf. Bougher et al., 1988a). The model EUV fluxes used have been revised according to Torr and Torr (1985), where solar maximum values are reduced as a result of a rocket calibration. The effect on the heating rates permits an increased EUV efficiency of 12% to be adopted. If dayside solar maximum temperatures as large as 310 K are permitted, in accord with the Hedin *et al.* (1983) empirical model, then an additional 1% increase of the EUV efficiency to 13% is possible. Finally, recent terrestrial low thermosphere heat balance studies of Roble *et al.* (1987) suggest that an $O + CO_2$ de-excitation rate similar to that initially proposed by Sharma and Nadile (1980) ($\sim 1 \times 10^{-12}$) is not unreasonable. Its incorporation into the Venus NLTE code enhances the CO_2 cooling further, so that an EUV efficiency of 14% can be used. Figure 34 illustrates the cooling requirements for observed dayside

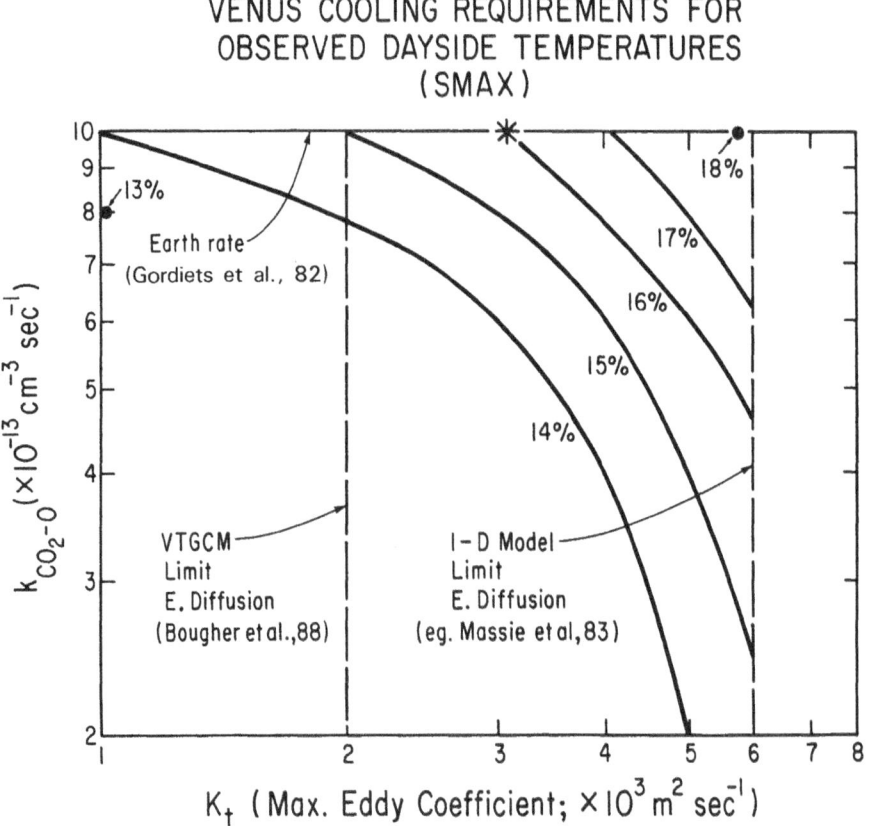

Fig. 34. Venus cooling requirements for observed dayside temperatures. Solar maximum conditions ($F10.7 = 200$). Based on the NCAR NLTE energy balance code.

temperatures; values along the ordinate correspond to the case of no eddy cooling. A further increase of EUV efficiencies seems to require cooling by eddy conduction. This, in fact, appears valid for moderate eddy coefficients with $K \leq 2 \times 10^7 \mathrm{cm}^2 \mathrm{s}^{-1}$, which is consistent with an upper limit for eddy diffusion incorporated in Venus odd-nitrogen studies (Bougher *et al.*, 1990a) (see Section 3.10). The use of EUV efficiencies larger than 15%, as Fox (1988) has suggested they are, seems impossible at the present time, unless the requirement that eddy diffusion and conduction coefficients be equal is relaxed. The low efficiencies required may, however, also be a due to the uncertainty of the solar EUV fluxes and the O–CO_2 energy exchange rate being used in model simulations.

5.4. IR HEATING AND 15-μm COOLING

Both IR heating and CO_2 15-μm cooling appear to be quite important in the Venus upper atmosphere. Dickinson and Bougher (1986) showed that IR heating and cooling offset one another below 130 km. Between 130 and 160 km, where the temperature increase is greatest, EUV heat is balanced by CO_2 15-μm cooling and molecular thermal conduction. Only above 160 km does molecular conduction balance EUV heating to maintain the observed thermospheric temperatures (see Figure 35).

The 15-μm cooling arises mostly from transfer of atomic-O kinetic energy to CO_2 vibrational energy; the magnitude of the cooling rate depends upon the O–CO_2 energy exchange rate and the relative amount of atomic-O in the Venus dayside thermosphere.

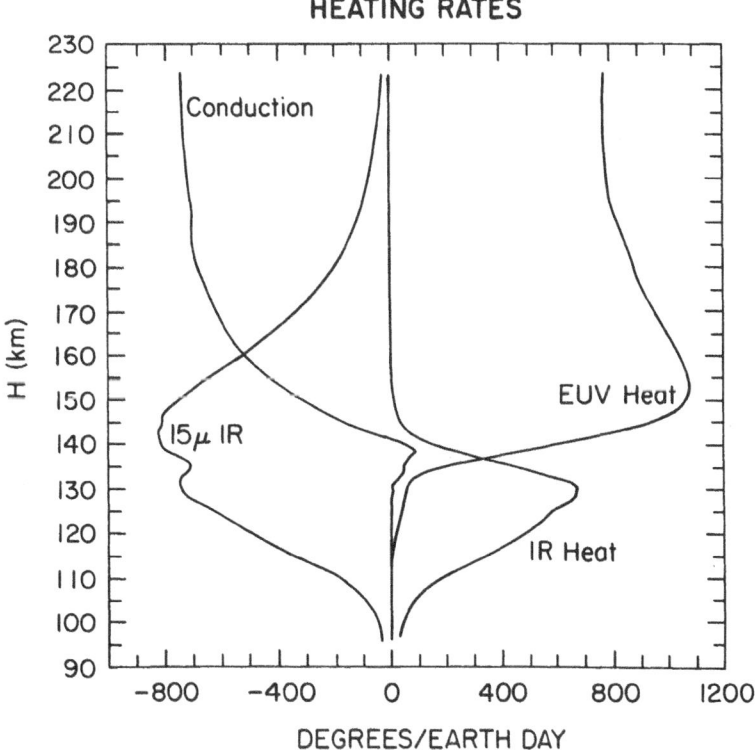

Fig. 35. Global average heating/cooling rates (K day^{-1}) for the Venus thermosphere. EUV heating efficiency = 10%. CO_2–O de-excitation rate = 8×10^{-13} cm^3 s^{-1} at 300 K. Taken from Dickinson and Bougher (1986). Later NCAR NLTE model calculations (Figure 34) make use of revised EUV fluxes, stronger CO_2 15-μm cooling, and larger heating efficiencies ($\sim 14\%$), yielding the same basic thermal balances shown here.

A pedagogical discussion of translation-vibration energy transfer can be found in Yardley (1980). The $O-CO_2$ excitation rate has not been measured at low temperatures, but may be large. In general the measured quantity that is commonly quoted is the rate coefficient for quenching or de-excitation (the reverse of reaction (69)). The rate coefficient for collisional excitation, k_{69}, is related to that for quenching, k_{-69}, by expression

$$k_{69} = k_{-69} \times g \exp(-\Delta E/kT),\qquad(70)$$

where ΔE is the vibrational spacing and g is the degeneracy of the upper state, equal to $v_2 + 1$ for the doubly-degenerate CO_2 bending mode. In the high temperature regime, the temperature dependence of the probability, P, of de-excitation in a collision, to which the rate coefficient is proportional, is given by the Landau–Teller expression (Landau and Teller, 1936)

$$P \sim \exp(-T^{-1/3}).\qquad(71)$$

In the Landau–Teller model and its later refinement and extension to polyatomics, the SSH model (Schwarz et al., 1952), energy transfer results from a 'direct' interaction involving the exponential repulsive potential between the species. This strong temperature dependence is characteristic of models involving 'impulsive' energy transfer of all types (e.g., McClelland et al., 1979). At low temperatures, the energy transfer probabilities are found to deviate substantially from the Landau–Teller values. The deviation is small for noble gases; the probabilities are found to decrease with decreasing temperature over a large range of temperature, but level off at very low (150 K) temperatures (Inoue and Tsuchiya, 1975; Simpson et al., 1977). For some species, usually efficient quenchers, the deviation from the Landau–Teller model can be several orders of magnitude at low temperatures and, below a certain temperature, the probability of quenching can increase with decreasing temperature. For example, the rate coefficient for quenching of the v_3 mode of CO_2 by N_2 is large and exhibits a minimum at about 1500 K (Moore et al., 1967; Rosser et al., 1969). This behavior for a near-resonant $V-V$ transfer was successfully explained by Sharma and Brau (1969), who incorporated the effect of long-range attractive multipole forces into the model for vibrational relaxation. Large rates and negative temperature dependence were also observed for de-activation of CO_2, H_2O, and D_2O at temperatures below 1000 K (Buchwald and Bauer, 1972). Significant departures from the Landau–Teller temperature dependence are also observed in de-activation of the asymmetric stretch mode of CO_2 by several molecular species at temperatures below about 400 K (Bauer et al., 1987). Quenching of the bend-stretch manifold of CO_2 by H_2 is rapid, even though the vibrational modes have dissimilar energies. Allen et al. (1980) have reported a rate coefficient of 5×10^{-12} cm^3 s^{-1} at 295 K; the rate coefficient increases with decreasing energy to 7.5×10^{-12} cm^3 s^{-1} at 170 K. The corresponding quenching probabilities are 7.4×10^{-3} and 1.4×10^{-2}. $V-T, R$ transfer in other systems such as NO–NO has been reported to be rapid, with quenching probabilities on the order of 10^{-3} and increasing with decreasing temperature below 400 K (Yardley, 1980).

Quenching of small molecules by atoms can be anomalously fast if the atom has an open shell (Chu et al., 1983). H, Cl, and F quench CO_2 rapidly compared to noble gases, such as He, Ne, and Xe (Chu et al., 1983; Flynn and Weston, 1986). The large cross sections and the negative temperature dependences at low collision energies for such systems have been ascribed to the formation of long-lived collision complexes in the presence of strong attractive chemical, hydrogen bonding or Van der Waals forces (Ewing, 1978; Lin et al., 1979; Parmenter and Seaver, 1979; Gordon, 1981; McClelland et al., 1979). The lifetime of a collision complex is longer at low collision energies, and the longer-lived the collision complex, the greater the probability that the energy in the vibrational mode may flow into the weak bond of the complex and cause its dissociation. The $V - T, R$ transfer probability for HF–HF, a system with strong hydrogen bonding, is about 1% at room temperature and increases with decreasing temperature (cf. Poulson et al., 1978). For CO_2–H, the large rates even at high energies have been ascribed to the presence of a chemically reactive channel, $CO + OH \rightarrow CO_2 + H$, and to the formation of a stable intermediate radical, HCO_2 (Flynn and Weston, 1986).

Atomic oxygen is a very efficient quencher of vibrational excitation, probably due to its reactivity, especially where de-activation can take place by atom exchange. The de-excitation rate coefficients for $CO_2^{\dagger}$ (the dagger denotes vibrational excitation) and $O_2^{\dagger}$ by O are about 9×10^{-13} cm^3 s^{-1}; these rates are about 25 times faster than that for quenching of $N_2^{\dagger}$, where atom exchange is not possible (McNeal et al., 1974; Fernando and Smith, 1969; Eckstrom, 1973). Isotope exchange studies of NO with ^{18}O showed that the rate coefficient for O atom exchange at 298 K is 3.7×10^{-11} cm^3 s^{-1}; if the complex is assumed to break-up statistically, the implied rate of complex formation is 7.4×10^{-11} cm^3 s^{-1} (Anderson et al., 1985). The rate coefficient for vibrational relaxation of NO by O is of the same order of magnitude, 6.5×10^{-11} cm^3 s^{-1} (Fernando and Smith, 1979). Isotope exchange of ^{18}O with O_2 proceeds with a rate coefficient of 2.9×10^{-12} cm^3 s^{-1} and vibrational relaxation with a rate coefficient of 5.4×10^{-12} cm^3 s^{-1} (Anderson et al., 1985; Quack and Troe, 1977).

All of the measurements or calculations of the O–CO_2 quenching rate have been at high temperatures where the Landau–Teller model might be expected to be valid. Center (1973) found O atoms an order of magnitude more efficient at relaxing the bending mode than Ar atoms; a probability of about 1% was measured at 3000 K. Bass (1974) computed the probabilities assuming classical scattering with a hard sphere interaction and quasi-diatomic model for CO_2 for energies from 1 to 10 eV and found that when the C atom was struck by the impinging O atom, the average fraction of energy transferred to the v_2 mode was 0.447. Schatz and Redmon (1981) used quasi-classical trajectory theory to compute the cross sections for excitation of the bending mode and found a value of 1.9×10^{-15} cm^2 at 2.2 eV, whereas the excitation of the symmetric and asymmetric stretch modes were 2 and 4–5 orders of magnitude smaller, respectively. Harvey (1982) obtained cross sections 50% smaller than Schatz and Redmon (1981) with a quantum mechanical vibrational close-coupling calculation using the same potential surface. Extrapolation of these results to low temperatures using the Landau–Teller expression would give very small quenching probabilities, but is not justified. The potential used by Schatz and Redmon (1981) and Harvey (1982) is a sum of pairwise exponential repulsion terms; no portion of the potential was attractive. At center-of-mass energies larger than 1 eV, the repulsive core of the potential dominates (Bass, 1974), but at low energies, attractive forces are expected to become important.

A value for the $O-CO_2^+$ quenching rate at low energies was derived by Sharma and Nadile (1980) from measurements of the 15 μm radiation in the terrestrial atmosphere. They recommended a de-excitation rate coefficient of $6.67 \times 10^{-14} T^{0.5}$ cm^3 s^{-1}. This rate has been refined by Wintersteiner et al. (1988) to

$$1.5 \times 10^{-13} T^{0.5} + 2.32 \times 10^{-9} \exp(-76.75/T^{1/3}) \text{ cm}^3 \text{ s}^{-1}, \tag{72}$$

valid in the approximate temperature range 200–700 K. The first term in expression (72) is due to attractive forces, which dominate at low temperatures and the second is the Landau–Teller expression that accounts for the short-range repulsive forces. The potential error in this expression, about a factor of 2, are mostly due to uncertainties in the atomic O density (Wintersteiner, private communication). At 200 K, the temperature which prevails in the lower part of the Venus thermosphere, the rate coefficient would be 2×10^{-12} cm^3 s^{-1}, about 1% of gas kinetic. The rate adopted by Dickinson (1984) is less than 40% of the value in expression (72) and other recent models have assumed quenching coefficients less than 1% of the Wintersteiner et al. value. Rates on the order of 1×10^{-12} cm^3 s^{-1} appear to be consistent with the terrestrial and martian heat budgets (Dickinson et al., 1987; Roble et al., 1987; Bougher and Dickinson, 1988; Bougher et al., 1990b). The reluctance to adopt the rate derived from terrestrial 15 μm observations is not justified. Certainly, however, a low temperature laboratory measurement of this rate coefficient is badly needed, not only for models of the Venus thermosphere, but for the Earth and Mars as well.

The strong nonlinear (exponential) temperature dependence of the CO_2 15-μm cooling insures that the Venus exospheric temperature does not vary by more than ~ 65 K over the solar cycle (300–235 K) (Dickinson and Bougher, 1986) (see Section 7). In addition, the O densities themselves increase with increasing solar activity. Calculations show that the production rate of O above 100 km on Venus increases from 2.2×10^{12} to 3.3×10^{12} cm^{-2} s^{-1} from low to high solar activity (Fox, 1989b). The VTS3 model predicts an increase of O from 2.7 to 7.3% at 135 km from $F_{10.7} = 74$ to $F_{10.7} = 200$. This change in the O abundance will act to dampen the response of the thermospheric temperature to changing solar activity, increasing the cooling at higher solar fluxes. The combination of the effects of the exponential temperature dependence of the $O-CO_2$ collisional excitation rate and the change in the atomic O mixing ratio with solar activity provide the Venus thermosphere with an 'atomic oxygen thermostat', regulating CO_2 cooling, that is less effective at Mars where the O abundance is lower (Dickinson and Bougher, 1986; Bougher and Dickinson, 1988; Fox, 1989b).

The nonlinear temperature dependence of the 15 μm cooling also means that it cannot be adequately addressed by simple linear perturbation (Newtonian cooling) schemes

(Dickinson, 1973). Instead, a nonlinear temperature-dependent cooling parameterization was developed by Bougher *et al.* (1986) for use in multi-dimensional model simulations. It first requires an exact 1-D NLTE calculation of a representative reference 15-μm cooling and corresponding temperature profile. For Venus, since the day-to-night temperature variation is so large (~ 200 K), separate dayside and nightside mean reference cooling rate profiles were generated. These were later used to drive a nonlinear NLTE cool-to-space formulation for deviations from these references. The deviations depend on departures of total temperature and atomic-O concentrations from the 1-D model values. The separate reference cooling for dayside and nightside mean conditions is matched across the terminator by interpolation so that discontinuities are minimized. Such an interactive scheme is fast, yet it allows the accurate calculation of total 15-μm cooling at other temperatures at any local time over the globe. The expression for the cooling rate derived by Bougher (Equation (A.5) of Bougher *et al.* (1986)) gives approximately the correct dependence of the IR cooling on temperature, pressure, and O concentration, and reduces to reference model cooling for composition and temperature profiles of either the day- or nightside reference models. Recent Mars TGCM model studies also confirm the usefulness of this NTLE cool-to-space formulation (Bougher *et al.*, 1988c, 1990b).

The one-dimensional model results of Dickinson and Bougher (1986) showed that dayside CO_2 cooling is dominated by the 15-μm fundamental above 130 km. Near-IR heating peaks at 130 km, with 2.7-μm bands important above 120 km, and ≤ 2.0-μm bands strongly contributing over 100 to 120 km. A calculation of the dayside $v_2 = 1$ populations relative to LTE confirms that complete NLTE conditions hold above about 140 km, while LTE is a good assumption only below 110 km.

5.5. MOLECULAR AND EDDY CONDUCTION

Molecular conduction serves to transfer EUV heat deposited in the thermosphere downward to a lower level, the mesopause, where it can be radiated to space by IR active gases. On Venus, CO_2 15-μm emission dominates all others in providing this cooling near the bottom of the thermosphere. Temperatures are also affected by the amount of cooling eddy processes are assumed to produce (Hunten, 1974; Izakov, 1978). The turbulence responsible can heat the thermosphere through the dissipation of its energy; conversely, cooling can take place by eddy heat conduction in a process analogous to molecular conduction. The relative importance of turbulent heating or cooling in a given planetary thermosphere is still the subject of much debate (e.g., Hunten, 1974).

Izakov (1978) argued theoretically that conditions exist when cooling of a planetary thermosphere by turbulence predominates, and conditions when heating prevails. For Venus, eddy mixing is estimated to provide cooling (upper limit) or a compensation of heating and cooling (lower limit). Hollenbach *et al.* (1985) approached the problem from a more empirical point of view, whereby the relative roles of heating and cooling were varied to match the observed temperatures over the solar cycle. The 1-D model calculations of Dickinson (1972, 1976) and Dickinson and Bougher (1986) have all addressed the case where vertical eddy mixing gives rise only to cooling. This establishes an upper limit to the role of eddy mixing in contributing to the overall thermospheric heat budget (see Section 5.3).

Questions still remain as to the complementary roles of eddy cooling and diffusion, particularly in the context of Venus circulation models. A few key points can be gleaned from recent modeling studies. The same eddy coefficient must be used for eddy diffusion and heat conduction in the absence of theoretical arguments to the contrary (Bougher *et al.*, 1986, 1988b). However, the coefficient for eddy viscosity and diffusion may be different (Prandtl number $\neq 1$) (Mengel *et al.*, 1989). This implies that a self-consistent calculation of winds, composition, and temperatures is required to address the impact of eddy mixing in the thermosphere. The practice of calculating temperatures using eddy heat conduction independent of the corresponding mixing effect on model densities is improper. Furthermore, observed densities appear to be influenced by a combination of global wind and eddy mixing processes; i.e., smaller eddy coefficients are needed in Venus circulation models than formerly implied by 1-D (non-dynamical) models (see Section 6.6). Thus, it appears that the introduction of eddy cooling is not generally a viable method for modeling thermospheric temperature profiles until all other dynamical, radiative cooling, and heating processes have been carefully examined. The effects of small-scale mixing, although difficult to model, need to be incorporated in a self-consistent fashion. No Venus model yet devised has this capability.

6. Thermospheric Circulation and Transport

6.1. INTRODUCTION

The dynamics of Venus's atmosphere can be separated into two distinct regimes in which very different flow patterns dominate: (1) a retrograde, zonal, superrotating flow from the surface up to ~ 70 km, and (2) a mean subsolar-to-antisolar (SS–AS) solar-driven flow above ~ 95–100 km. In between (70–95 km), zonal winds are generally decreasing with altitude while thermospheric-type SS–AS winds are increasing. This is a complicated transition region in which changing zonal and meridional flow patterns may be superimposed on a possible lower return branch of the SS–AS circulation (Clancy and Muhleman, 1985b; Goldstein, 1989). A summary of these basic dynamical regions and the relevant supporting data is given in Figure 36 (Goldstein, 1989).

Solar UV–IR heating clearly drives the SS–AS circulation. An eddy viscosity appears to be required to reduce pressure-driven winds toward acceptable values (see Section 6.4). A redistribution of angular momentum by a strong cloud-top Hadley circulation maintains the retrograde zonal jets observed at mid-latitudes (Hou, 1984; Hou and Goody, 1985; Walterscheid *et al.*, 1985). Recent studies support the idea, originally proposed by Fels and Lindzen (1974), that the superrotation at cloud tops is maintained by the pumping action of thermally driven tides (Pechmann and Ingersoll, 1984; Hou, 1984; Hou and Goody, 1985). Baker and Leovy (1987) and Leovy (1987) suggest that the equatorial wind speed near Venus's cloud top is maintained by a balance between the pumping effect of the semidiurnal tide and vertical advection by the Hadley

Proposed Circulation Model for Venus

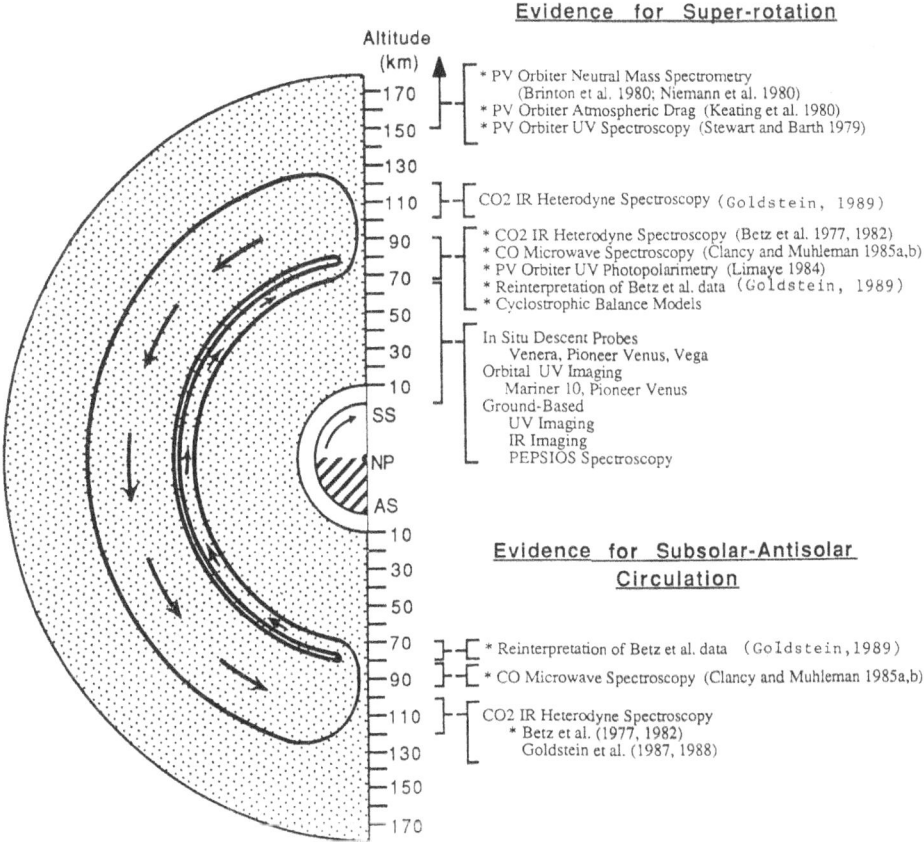

Fig. 36. Dynamical regions of the Venus mesosphere-thermosphere. Taken from Goldstein (1989).

circulation in the thermal driving region. This rapid superrotation (at all latitudes) has not yet been adequately explained.

Zonal circulation seems to exist throughout the atmosphere, although it does not dominate SS–AS flow above 100 km (Goldstein, 1989). Venus cyclostrophic balance models predict maximum zonal winds of 130 m s^{-1} near 70 km. The observed poleward warming above this level implies decreasing zonal speeds with altitude (70–90 km), becoming negligible by 90–100 km. A concomitant increase in the mean meridional component with possible 120 m s^{-1} speeds near 90–100 km has also been predicted (Taylor *et al.*, 1980). Infrared heterodyne observations detect a weak zonal retrograde component (20–30 m s^{-1}) (equatorial) at 110 ± 10 km, superimposed on strong 100–140 m s^{-1} (SS–AS) cross-terminator winds (Goldstein, 1989). Finally, a slightly larger component of retrograde zonal winds (~ 50–60 m s^{-1}) is implied above 150 km from helium and night airglow observations (Mayr et al., 1980; Stewart et al., 1980). See Section 6.5 for details of superrotation mechanisms and sensitivity studies.

There are, however, no direct or indirect measurements of wind velocity in Venus's upper atmosphere above 110 km. The circulation of most of the thermosphere must, therefore, be inferred from observations of its temperature and density structure (Mayr et al., 1980; Schubert et al., 1983; von Zahn et al., 1983) and subsequent modeling efforts (Bougher, 1985; Mayr et al., 1985; Bougher et al., 1986, 1988b, 1990a; Mengel et al., 1989).

6.2. THERMOSPHERIC CIRCULATION: SYMMETRIC VS ZONAL

The most prominant feature of the thermosphere which suggests a strong SS–AS wind system is the large contrast in temperatures and densities between day and night. Temperatures on the dayside of Venus increase from ~ 170–180 K at the 'mesopause' (100 km) level (Taylor et al., 1979, 1980; Seiff et al., 1980) to near 300–310 K above 150 km for solar maximum conditions (von Zahn et al., 1979, 1980; Niemann et al., 1979, 1980b; Hedin et al., 1983; Keating et al., 1980, 1986). The rise of temperatures above a 100 km mesopause is similar to that of the Earth's thermosphere, although the temperatures are much colder (see Figure 31). Nightside temperatures are shown to decrease from 170 K at the mesopause to values as low as 100–130 K above 150 km (Keating et al., 1979, 1980; Niemann et al., 1979, 1980b; Seiff et al., 1980; Seiff, 1982). This vertical structure is remarkably unlike the terrestrial nightside thermosphere, and has been termed a 'cryosphere' (Schubert et al., 1980). In addition, changes in temperature across the terminator are very abrupt, with the minimum value observed just beyond midnight at 02:00 LT (Mayr et al., 1980; Keating et al., 1980).

These thermospheric (above 150 km) kinetic temperatures are inferred primarily from the observed scale heights of the measured neutral densities. The diurnal variation of the heaviest species (CO_2, CO, N_2, and O) show a near symmetry about the subsolar-to-anti-solar axis (see Figure 37), with a density maximum at a fixed altitude at local noon and minimum at midnight (von Zahn et al., 1983). The slow planetary rotation coupled with dayside solar heating results in this expansion of the dayside thermosphere; i.e., the height of a given pressure surface rises (falls) with increasing (decreasing) temperatures. Nightside densities are strongly depleted with respect to dayside values, suggesting a significant contraction of the nightside thermosphere with constant pressure surfaces decreasing abruptly in altitude across the terminators. Helium and hydrogen, the two lightest species, show a diurnal density reversal, with densities higher at night than during the day (Niemann et al., 1980b; Brinton et al., 1980). This signature is consistent with their small atomic weights and correspondingly large-scale heights. These properties increase the transport efficiencies by the large scale winds for these species, just as in the atmosphere of the Earth (Mayr et al., 1978). The diurnal variation of helium at 165 km reveals a distinct (lasting) nightside bulge (see Figures 38(a) and 38(b)), with a maximum at 03:00 LT roughly 30–40 times that on the dayside (Taylor et al., 1984). Results obtained from Brinton et al. (1980) suggest a similar night–day bulge of hydrogen peaking at 05:00 LT, with a diurnal ratio of nearly 400:1 at 165 km (see Figure 25).

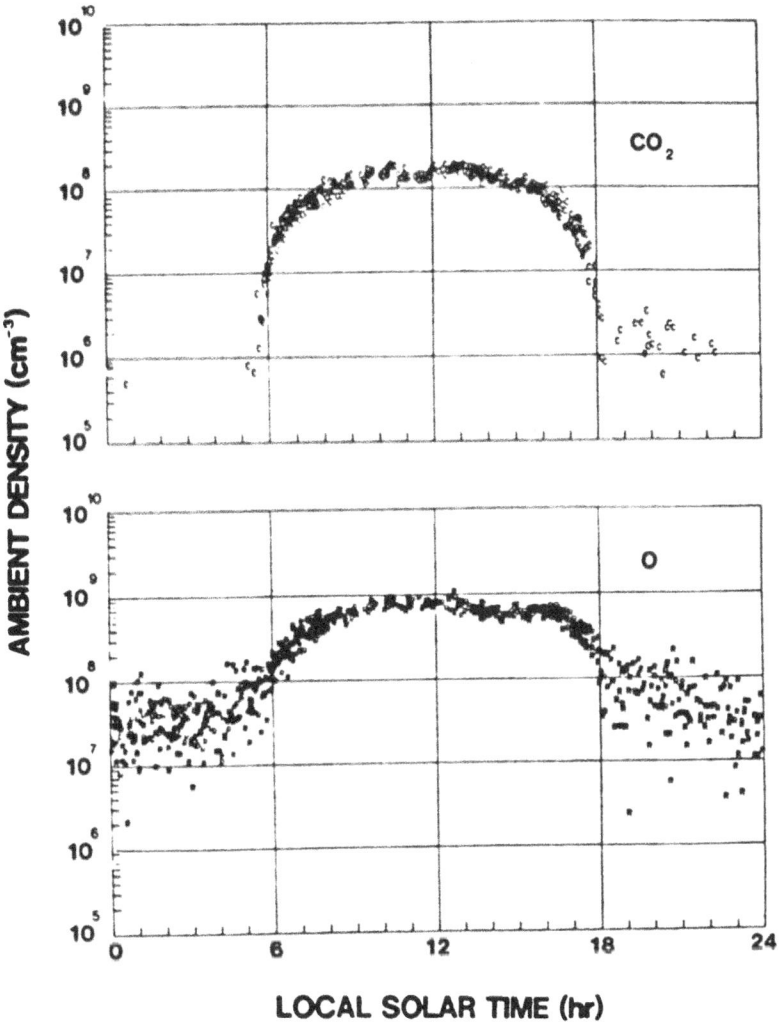

Fig. 37. Measurements of CO_2 and O densities at 170 km altitude taken by the Pioneer Venus orbiter neutral mass spectrometer (ONMS) over nearly three diurnal cycles. Taken from von Zahn *et al.* (1983).

These bulges imply a 'wind-induced diffusion' in which light constituents are preferentially transported from day-to-night, with little returning to the dayside along the return branch of the circulation (Mayr *et al.*, 1978, 1980).

This temperature difference between day and night gives rise to horizontal pressure gradients, which to first order should drive a strong subsolar-to-anti-solar (SS–AS) circulation in Venus's upper atmosphere (Dickinson and Ridley, 1977; Mayr *et al.*, 1980; Seiff, 1982). A simplified flow pattern would be axisymmetric about the Sun–Venus line, with dayside upwelling centered on the subsolar point, strong cross-terminator flow, and subsidence centered on the antisolar point (Schubert *et al.*, 1983).

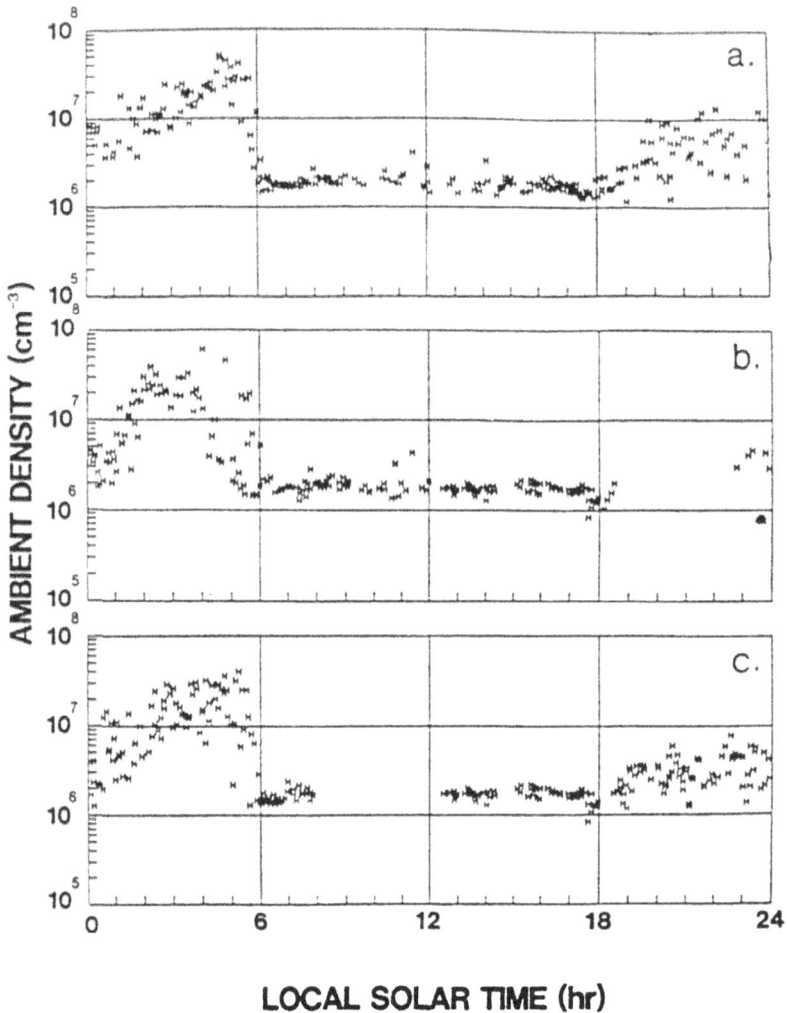

Fig. 38.   Measurements of He densities at 170 km altitude taken by the Pioneer Venus orbiter neutral mass spectrometer (ONMS) over nearly three diurnal cycles. Helium has a dawn bulge analogous to that of hydrogen, with some differences in shape. Taken from von Zahn *et al.* (1983).

Return flow, possibly between 70–90 km (Goldstein, 1989), would complete the circuit. Streamlines for transported species approximately follow constant pressure surfaces, which descend in altitude from day to night (Seiff, 1982). Supersonic velocities are estimated across the terminators for laminar flow conditions; turbulent viscosity is likely important in modifying the pressure-driven wind speeds (Seiff, 1982; Bougher *et al.*, 1986). This basic day–night flow pattern was hypothesized and studied in a series of numerical models by Dickinson and Ridley (1972, 1975, 1977) prior to Pioneer Venus measurements, and by Bougher *et al.* (1986, 1988b, 1990a), Mayr *et al.* (1980, 1985), and Mengel *et al.* (1989) afterwards.

This single symmetric cell, driven by solar UV–IR heat, must occur. However, the actual motions in Venus's thermosphere are likely a superposition of this SS–AS flow and a retrograde zonal wind component. Post-midnight maxima in the helium and hydrogen densities (Niemann *et al.*, 1979; Brinton *et al.*, 1980; Taylor *et al.*, 1984), displacement of the minimum diurnal temperature to 02:00 LT (Mayr *et al.*, 1980; Keating *et al.*, 1980), and the NO night airglow maxima at 02:00LT (Stewart *et al.*, 1980; Bougher *et al.*, 1990a) all suggest a westward superrotation of the upper atmosphere (≤ 150 km) with zonal winds speeds of 50–100 m s^{-1}. This corresponds to a rotational period of 4–8 Earth days. This modified flow pattern now includes upwelling (divergence) near the subsolar point, and subsidence (convergence) offset from midnight, yielding a nightside flow which stagnates near 02:00–03:00 LT (see Figure 39). The zonal winds add to the SS–AS flow across the evening terminator, giving stronger winds there than at dawn.

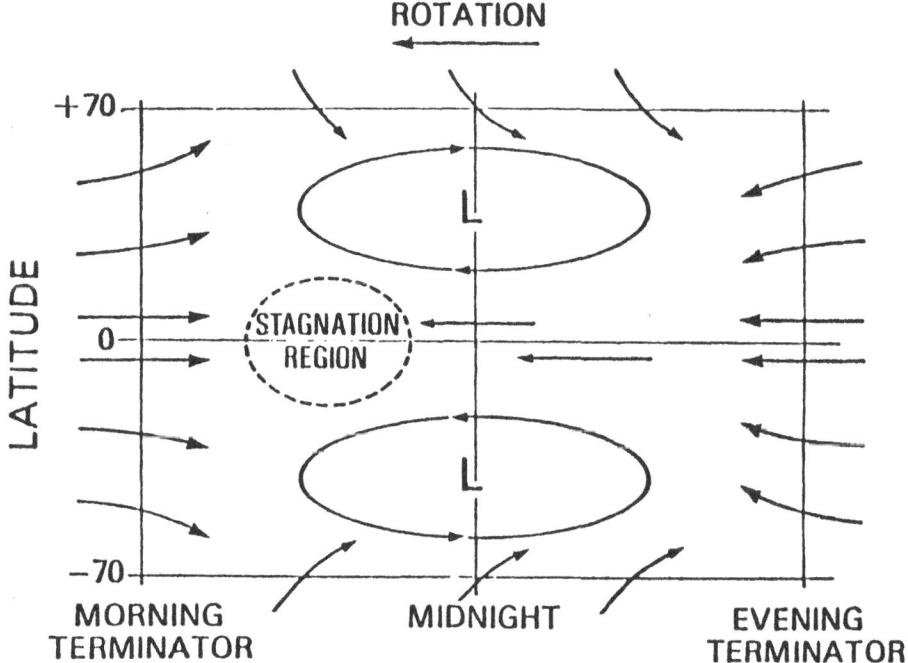

Fig. 39.   A schematic flow pattern for the nightside thermosphere of Venus. Taken from Niemann *et al.* (1980b).

Models of the Venus thermospheric circulation have been under development for nearly 18 years. Experience gained from Earth thermosphere studies initially guided the model processes and parameterizations considered. Prior to Pioneer Venus, only the SS–AS circulation pattern was generally thought to be operating in the Venus thermosphere. The Dickinson and Ridley (1972, 1975, 1977) two-dimensional model

(DRM) of the temperature, winds, and composition of the Venus thermosphere, correctly predicted the gross characteristics of this SS–AS mean circulation. Axial symmetry was assumed in choosing altitude and solar zenith angle (SZA) as the two model coordinates, in accord with the dominant SS–AS flow. Later versions of this model were fully self-consistent and used linear perturbation theory to simultaneously calculate thermal structure, winds, and CO_2, CO, and O densities above 100 km. Detailed results of the model, however, differed considerably from the Pioneer Venus observations (Schubert *et al.*, 1983; Dickinson and Bougher, 1986). Figure 40 shows

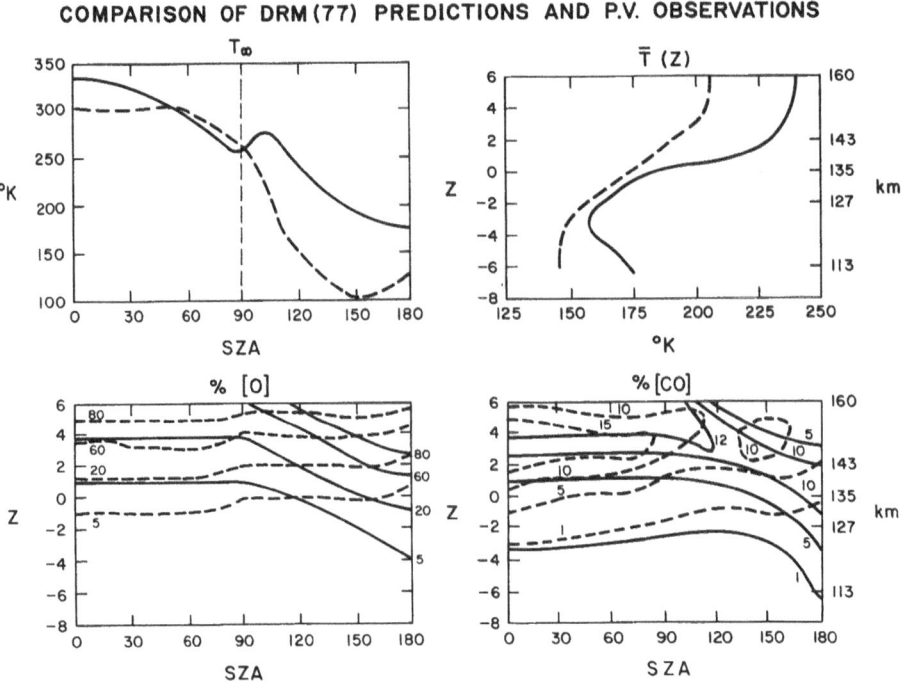

COMPARISON OF DRM (77) PREDICTIONS AND P.V. OBSERVATIONS

Fig. 40.   Comparisons of Dickinson and Ridley (1977) (DRM) model predictions and Pioneer Venus (PVO) observations. Solid lines are for the 2-D model simulations; dashed lines are from PVO observations as given by the empirical model of Hedin *et al.* (1983). Taken from Dickinson and Bougher (1986).

that the Dickinson–Ridley temperatures were somewhat too warm on the dayside and much too warm, by as much as 100 K, on the nightside. Their calculations failed to predict the nightside cryosphere entirely. Model atomic-O and CO bulges on the nightside were also predicted by the DRM, that were not observed. Correspondingly, the dayside O-mixing ratios at the F1-peak (140 km) were calculated to be 2.5–5.6%, much smaller than actually observed (17–20%) (von Zahn *et al.*, 1983). No shift in the nighttime minimum temperatures from midnight is possible using the axially-symmetric framework assumed in the DRM. Effects of zonal flow require a three-dimensional model simulation (Mayr *et al.*, 1980; Bougher, 1985; Bougher *et al.*, 1988b).

Several explanations were initially forwarded to account for these discrepancies between DRM and the data. Two groups, in particular, focused their attention on the Pioneer Venus neutral data and its global modeling. The Goddard group (cf. Mayr *et al.*, 1980, 1985) used a 3-D spectral model to simulate the observed day–night distribution of composition and temperatures. Dynamical properties of the Venus thermosphere were initially inferred using a simplified linear spectral model and Pioneer Venus ONMS measurements (Niemann *et al.*, 1979, 1980; Mayr *et al.*, 1980). A combination of parametrically prescribed strong vertical eddy diffusion ($K \leq 3 \times 10^7$ cm^2 s^{-1}) and retrograde zonal winds (5–10 day period) seemed to modify the fundamental diurnal tide sufficiently to yield a reasonable match between observed and modeled phases and magnitudes of global temperatures and densities. Strong vertical eddy diffusion was required to counteract the mass-transport effects of the large-scale winds. Non-linear processes (i.e., wind-induced diffusion), important for helium and hydrogen, were estimated to steepen the nighttime density maxima and shift them significantly toward dawn (02:00–04:00 LT). Maximum horizontal winds of ~ 200 m s^{-1} were inferred from the day–night density (temperature) variations observed by the ONMS. Strong vertical eddy diffusion, retrograde zonal winds, and reasonable day–night winds were all suggested to improve the DRM simulations.

Further work was described by Mayr *et al.* (1985), whose spectral model is more realistic in that it accounts for nonlinear processes, includes higher order tidal components, and describes the major gases in self-consistent form. Eddy diffusion and conduction were modified to vary inversely as the square root of the number density (after von Zahn *et al.*, 1980), with a maximum eddy coefficient of 6×10^7 cm^2 s^{-1}. The EUV heating efficiency was taken to be 30%. The basic diurnal contrasts in heavy (CO_2, CO, and O) and light (He) species and temperatures observed by PV were reproduced. Also, superrotating zonal winds (6-day period) were found to have little impact on the diurnal distributions of the heavy species and temperatures. However, a significant dayside depletion and nightside enhancement in O and CO densities (with respect to Hedin *et al.* (1983) values) was obtained. This occurs because the calculated solar-driven thermospheric wind system, with maximum horizontal wind speed of about 300 m s^{-1}, is too strong. The Mayr *et al.* (1985) lower thermosphere wind speeds (~ 110 m s^{-1}) are not consistent with ground-based measurements (Goldstein, 1989).

The group at NCAR updated and improved the DRM code, and subsequently adopted the finite-difference approach of the terrestrial Thermospheric General Circulation Model (TGCM) (cf. Dickinson *et al.*, 1984) modified for Venus. Bougher (1985) and Bougher *et al.* (1986) proposed that a global circulation system weaker than the DRM originally predicted would enable increased dayside O and CO densities to be maintained, while simultaneously reducing the buildup on the nightside. Furthermore, since nightside heating is maintained primarily from the circulation (i.e., adiabatic compressional heating), slower winds would also result in cooler nightside temperatures more indicative of a cryosphere. Suggestions along this line were also made by von Zahn *et al.* (1983) who noted that dayside O/CO_2 ratios increased for the series of Dickinson and Ridley (1972, 1975, 1977) models, due to progressively weaker day-to-night winds.

Schubert *et al.* (1983) also noted that dynamical processes may be required to explain the very low nightside thermospheric temperatures. Specifically, low temperatures must be maintained in the face of (1) advection and convection of dayside heat across the terminator, (2) compressional heating due to nightside subsidence, and (3) upward molecular heat conduction from the relatively warm 100 km level.

This weakened circulation hypothesis seems to give the most self-consistent explanation for all the available Pioneer Venus composition, temperature, and airglow data. An updated version of the DRM model was constructed by Bougher *et al.* (1986) to examine how SS–AS winds, eddy diffusion/conduction, and strong CO_2 15-μm cooling affect the day–night contrasts in Venus densities and temperatures. It was learned that symmetrized empirical model fields of Hedin *et al.* (1983) could be simulated largely by adding a wave-drag mechanism, resulting from turbulent dissipation effects on the horizontal winds, which weakened the day-to-night flow. Eddy viscosity and Rayleigh friction were parameterized within the horizontal momentum equation to mimic the wave-mean-flow interaction responsible (see Section 6.4). Maximum terminator winds of ≤ 230 m s^{-1} were found to be consistent with observed global composition and temperature fields. Such a reduction of winds by nearly a factor of two (from those of DRM) permitted less O and CO to be carried to the nightside. The dayside F1-peak O-mixing ratios were calculated to be $\sim 15\%$. The increased isolation of the day and nightside thermospheres also serves to enhance the diurnal temperature contrast, giving midnight exospheric temperatures of about 115 K, characteristic of the cold temperatures inferred from density data. This occurs because slower winds support less adiabatic compressional heating, the dominant source of nightside heating (Dickinson and Ridley, 1977). It was also shown that drag could not be used simultaneously with strong eddy conduction to derive the cold nightside temperatures. Slower winds are sufficient to maintain cold nightside temperatures, and are also consistent with the observed diurnal density contrasts. Winds calculated near 110 km match Earth-based measured values (100–140 m s^{-1}) quite well (Goldstein, 1989). Figures 41(a–d) illustrate contour plots from the standard NCAR 2-D model simulation (Bougher *et al.*, 1986) for solar maximum conditions ($F_{10.7} = 200$). Notice that constant pressure surfaces decrease in altitude from the day to the nightside, in agreement with the 'collapsed' nightside cryosphere observed.

Superrotation effects on thermospheric densities and temperatures can only be examined in a three-dimensional coupled chemical-dynamical model. A fully self-consistent code coupling large-scale and sub-grid scale effects, including wave-drag and superrotation, is very difficult to formulate. A prelude to such a code was developed by Bougher *et al.* (1988b), based on the NCAR terrestrial TGCM. The general framework of this Earth code (Dickinson *et al.*, 1984) was used in concert with new inputs and physical parameterizations to examine Venus thermospheric processes. The Venus TGCM (VTGCM) is internally self-consistent in that it calculates global distributions of CO_2, CO, and O that are consistent with the three-dimensional day–night temperature contrasts and corresponding large-scale winds. The model covers a 5° by 5° latitude-longitude grid, with 24 constant log-pressure levels in the vertical, extending

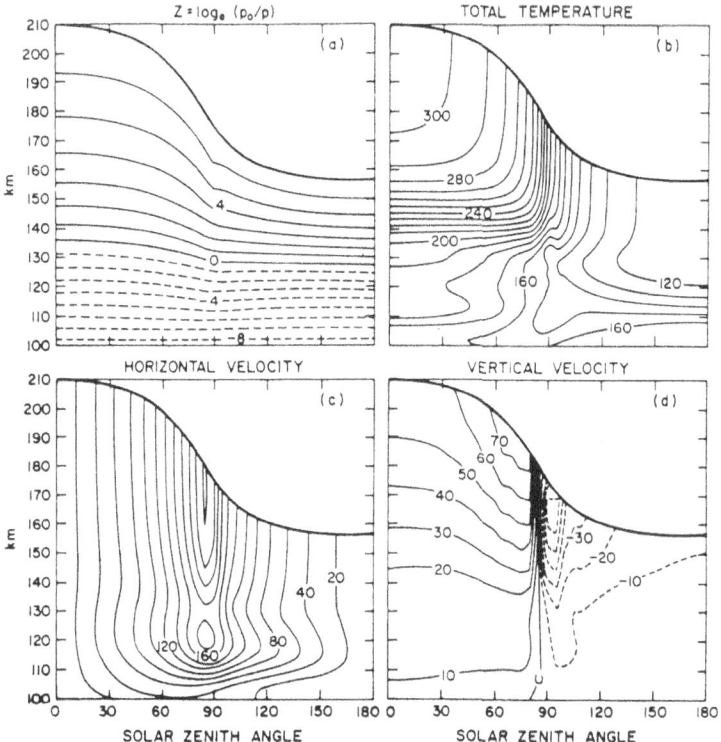

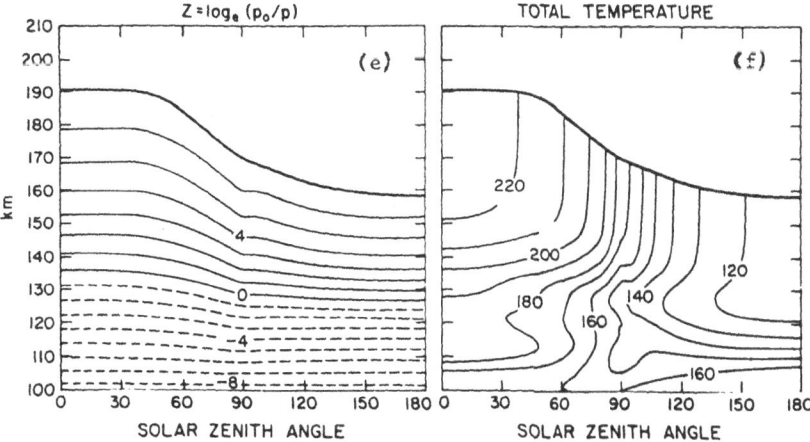

Fig. 41.   NCAR 2-D model contour plots. All frames are presented on an altitude scale for convenience. The sharp upper level cutoff ($z = 7$) corresponds to the top boundary of the model. Solar maximum simulation: (a) $z = \log_e(p_0/p)$, (b) total temperature (K), (c) horizontal velocity in m s^{-1}, (d) vertical velocity in cm s^{-1}. Solar minimum simulation: (e) $z = \log_e(p_0/p)$, (f) total temperature (K). Taken from Bougher *et al.* (1986).

from approximately 105 to 200 km at local noon. Dayside O and CO sources arise from CO_2 net dissociation; sinks are provided primarily by advection and convection of the large-scale winds. Total temperature is derived as the sum of an adopted global mean reference temperature profile (Dickinson and Bougher, 1986) and calculated perturbation temperatures. Sub-grid scale processes (i.e., eddy diffusion, viscosity, conduction, and wave-drag) are not self-consistently incorporated, but rather parameterized using standard aeronomical formulations (cf. Hunten, 1974; von Zahn et al., 1980). Rayleigh friction is used to mimic wave-drag effects on the mean flow (see Section 6.4); its incorporation is analogous to early terrestrial mesopause studies which sought to define an approximate time-scale for the required wave-breaking process (Schoeberl and Strobel, 1978).

A symmetric form of the VTGCM was first used to examine the SS–AS circulation on Venus for solar maximum conditions (Bougher et al., 1988b). Inputs and parameters comparable to the NCAR 2-D model of Bougher et al. (1986) were incorporated, including a low 10% EUV heating efficiency, an $O-CO_2$ de-excitation rate coefficient of 8×10^{-13} cm^3 s^{-1} at 300 K, a maximum nightside eddy coefficient of 1×10^7 cm^2 s^{-1}, and minimal dayside eddy diffusion and cooling ($K \leq 5 \times 10^6$ cm^2 s^{-1}). Prescribed wave-drag (Raleigh friction coefficient $\leq 10^{-4}$ s^{-1}) was primarily responsible for the relatively weak winds calculated (≤ 230 m s^{-1}), which is consistent with the observed day–night contrast in calculated densities and temperatures. Model exospheric temperatures ranged from 309 K (day) to 136 K (night), simulating the nightside cryosphere seen by Pioneer Venus instruments.

Once the VTGCM was validated (in its symmetric form) by comparison with the NCAR 2-D code, further tests were conducted to examine the asymmetric character of the circulation by incorporating prescribed superrotating retrograde zonal winds. The adopted zonal wind profile was specified to increase from zero to 60 m s^{-1} over 105 to 140 km. Such a profile is consistent with recent suggestions that zonal flow near 100–110 km is quite weak (Clancy and Muhlemann, 1985; Goldstein, 1989), while zonal flow above 140 km is significant (Mayr et al., 1980, 1985). Results show that the major (heavy) species and temperatures are not greatly affected by this superrotation. However, the exospheric temperature minimum shifts to 02:00 LT, which is consistent with data and with the convergence of the horizontal winds after midnight. Figures 42(a) and 42(b) show pressure levels slices (at an average of 162 km) of VTGCM total temperatures, horizontal winds, and vertical winds. Also, the VTGCM nightside temperatures appear to be larger in the exosphere than in the lower thermosphere (see Figure 43); the NCAR 2-D model, formulated explicity for total temperatures, does not show this behavior. This discrepancy may be due to the difficulty of generating a representative global mean temperature profile for use with VTGCM perturbation values to give total temperatures (Bougher et al., 1988b).

A more sensitive tracer of the thermospheric winds is required to further constrain the circulation. The incorporation of minor (light) species, such as helium, can provide such a tracer of the SS–AS winds and the superimposed zonal wind component (Bougher et al., 1988b; Mengel et al., 1989). The magnitude of the nightside helium bulge

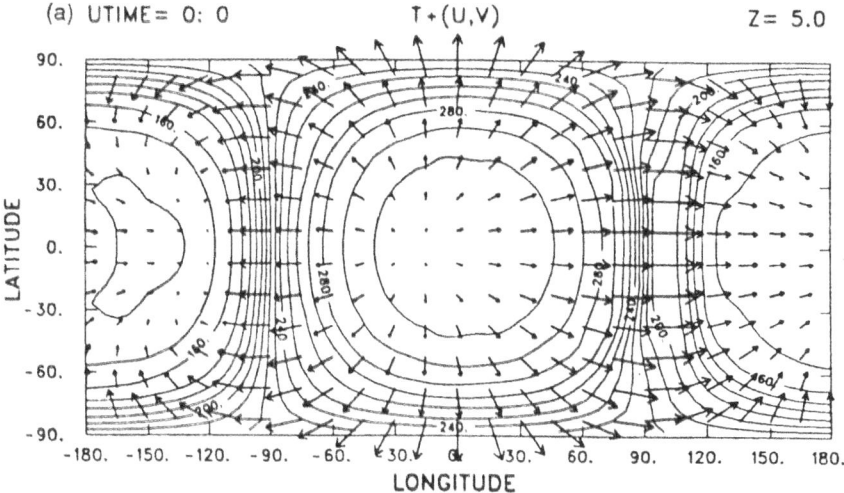

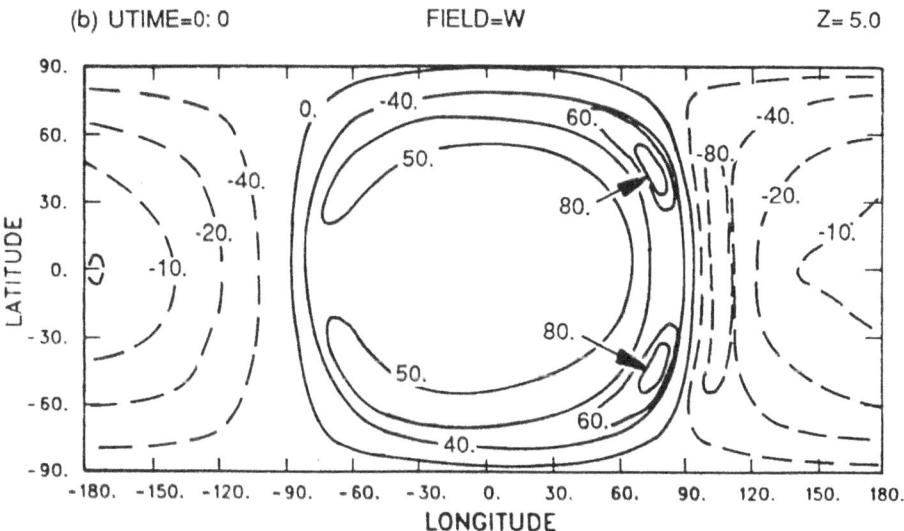

Fig. 42.   VTGCM superrotating model pressure level slices (local time vs latitude) at $z = 5$ (or roughly 162 km average altitude): (a) $T + (u, v)$ – superimposed total temperature and horizontal wind vectors, (b) w – vertical velocity in cm s^{-1}. The maximum horizontal wind vector in (a) corresponds to 230 m s^{-1}. Taken from Bougher *et al.* (1988b).

should be very sensitive to the overall strength of the circulation and the amount of global eddy diffusion. Likewise, the local-time position of the bulge is a measure of the degree of asymmetry of the otherwise symmetric SS–AS flow. Mengel *et al.* (1989) presented calculations of helium that were used to fine tune the Mayr *et al.* (1985) 3-D spectral model circulation and eddy diffusion formulation. This improved model

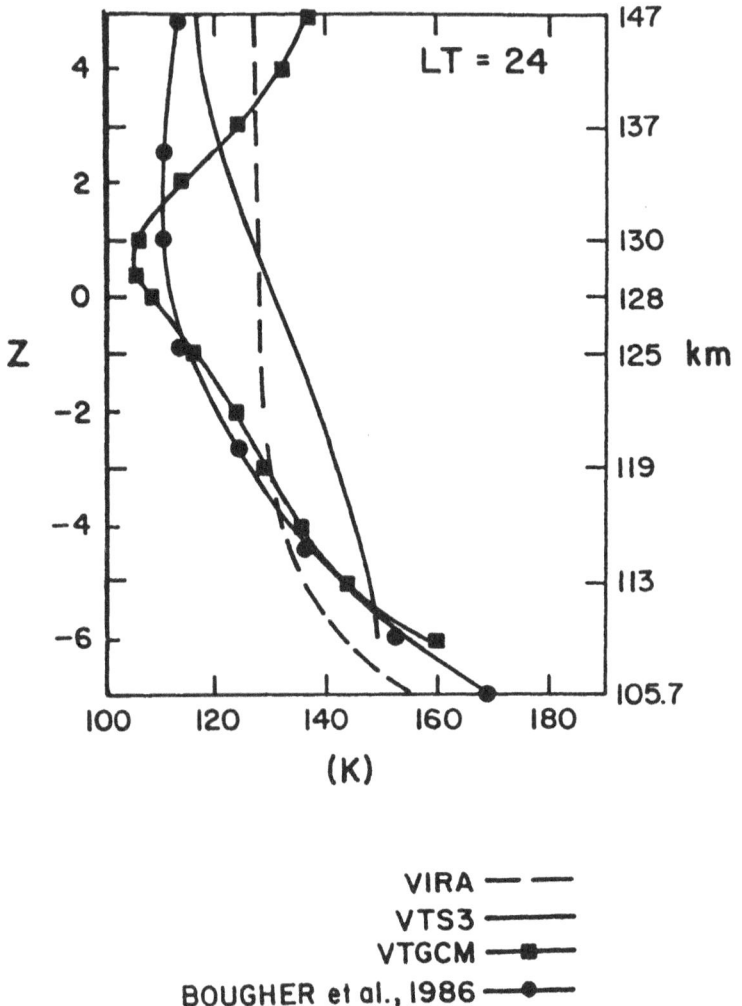

Fig. 43.   VTGCM symmetric model midnight temperature profile above 100 km. Calculated profiles are compared to the corresponding empirical model profiles of Hedin *et al.* (1983) (VTS3) and Keating *et al.* (1986) (VIRA). The NCAR 2-D model profile of Bougher *et al.* (1986) is also shown. Taken from Bougher *et al.* (1988b).

accounted for nonlinear processes and superrotation through a solid-shell approxima-tion. They assumed a 6-day superrotation period and used the EUV heating rates of Fox (1988) with an efficiency close to 20%. An altitude dependent Rayleigh friction scheme (see Section 6.4) was also adopted to weaken the pressure driven winds.

Standard spectral-model profiles for noon, midnight, and global mean temperatures are illustrated in Figure 44. The global mean exospheric temperature of 220 K is consis-tent with data, and lies between 325 K (noon) and 155 K (midnight) model values for extreme solar maximum conditions. Superrotation was found to have relatively little

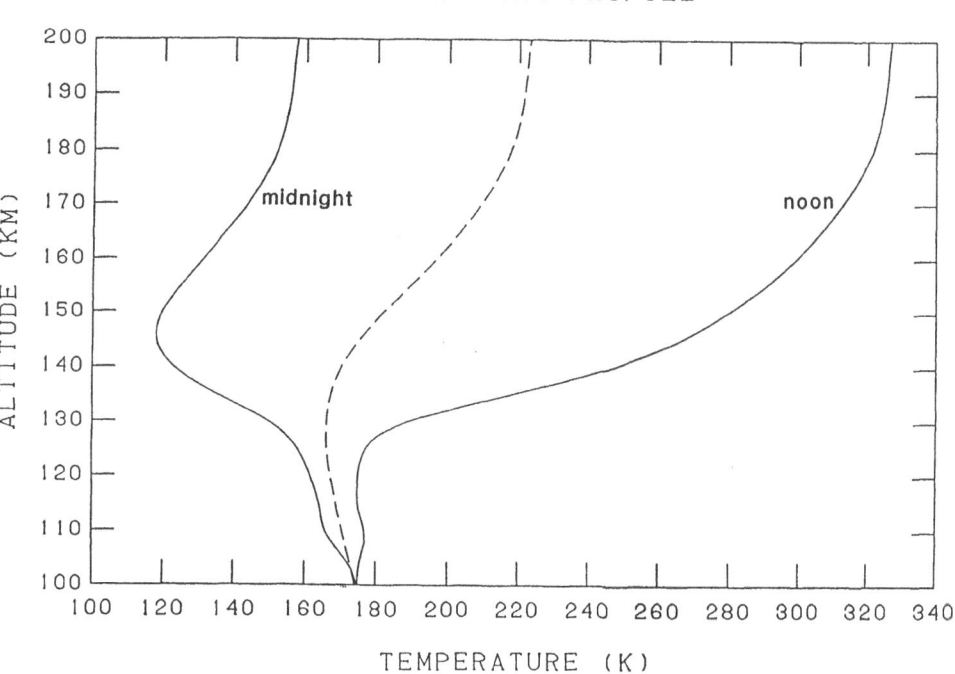

Fig. 44.   Equatorial temperature: height profiles for the standard GSFC spectral model case. The solid lines are the noon and midnight temperatures. The dashed line is the globally averaged model profile. Extreme solar maximum conditions ($F10.7 = 238$). Taken from Mengel *et al.* (1989).

effect on the temperature contrast between day and night (see Figure 45(a)); however, it did produce a shift in the temperature minimum to 02:00 LT, in accord with observations. This dawn-dusk asymmetry in the flow also enabled the nightside helium bulge to peak at 05:00 LT (see Figure 45(b)). Its night–day ratio at 170 km was calculated to be 45:1, in good agreement with the 30:1 value seen at 165 km over the three diurnal periods analyzed by Taylor *et al.* (1984). Results show that the longer the superrotation period, the larger is the day-to-night buildup in the helium density, and the shorter is the time delay in the peak density after midnight. Figure 46 schematically shows the combined effects of the SS–AS flow plus the superrotating winds on the diurnal He density distribution. Both the horizontal and the vertical He distributions are also strongly dependent on the eddy diffusion coefficient. The data are best fitted by a SS–AS circulation of ~ 100 m s^{-1}, a moderate height-dependent global mean eddy diffusion ($K \leq 3 \times 10^7$ cm^2 s^{-1}), and a superrotation period of 6 Earth days. In this model, the Prandtl number, which is defined as the ratio of the kinematic viscosity to the thermal diffusivity, and indicates the relative efficiency of mass and heat transport by turbulence, was assumed to be unity. Strong superrotating zonal winds appear to reduce (smear out) the magnitude of the nightside helium bulge from what SS–AS wind-induced diffusion would otherwise produce.

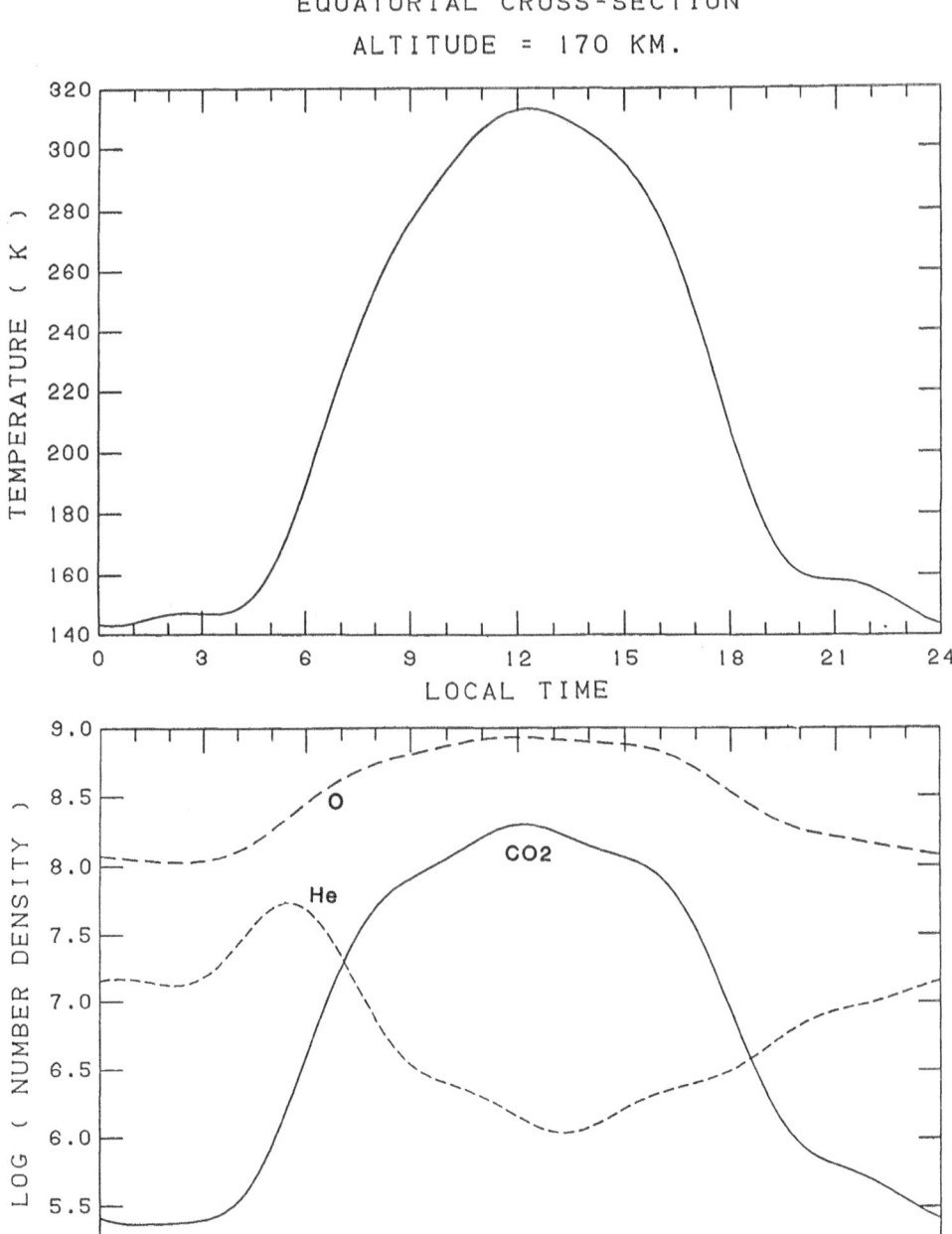

Fig. 45. Equatorial output fields vs local time at an altitude of 170 km for the standard GSFC spectral model. (a) Temperature; and (b) CO_2, O, and He number densities. Extreme solar maximum conditions ($F10.7 = 238$). Taken from Mengel *et al.* (1989).

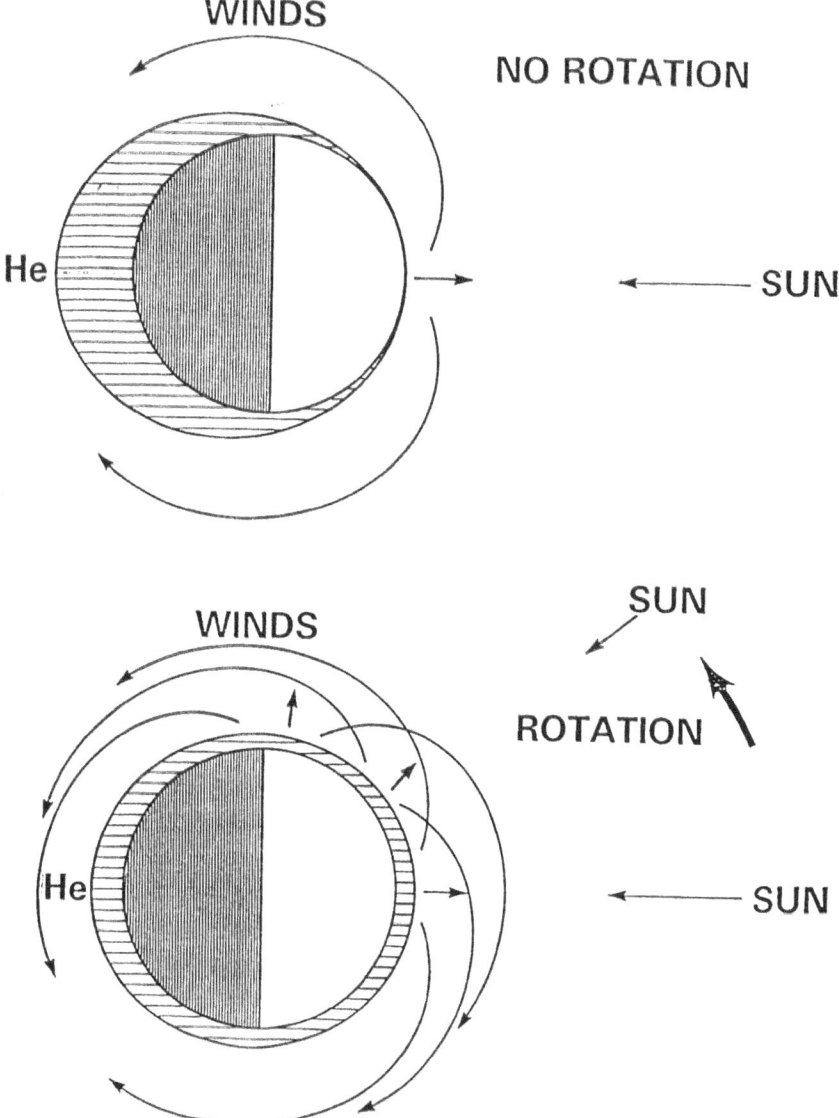

Fig. 46.   Schematic illustrating the effect of atmospheric superrotation on the magnitude and phase of the diurnal variations of helium in the Venus thermosphere. Taken from Mengel *et al.* (1989).

An additional tracer of the thermospheric circulation is given by the observed NO nightglow, which is shown to be produced by radiative recombination of N and O atoms transported from their source on the dayside to the nightside (Bougher *et al.*, 1990a). This work, described in Section 3.10, shows how the maintenance of this nightglow constrains the magnitude of the SS–AS winds (≤ 150 m s^{-1}) and global mean eddy diffusion ($K \leq 2 \times 10^7$ cm^2 s^{-1}) in the face of moderate zonal winds (~ 55 m s^{-1})

having a period of 8 days. This nightglow provides a means to fine tune the VTGCM circulation presented by Bougher *et al.* (1988b).

It is apparent that differences in the mean state of the solar maximum Venus thermospheric circulation (and small-scale processes) still exist in the most recent model simulations (Bougher *et al.*, 1990a; Mengel *et al.*, 1989). However, the major tunable parameters have been identified: (1) zonal superrotation period, (2) the magnitude of the maximum global mean eddy diffusion coefficient, and (3) the magnitude of the peak Rayleigh friction coefficient (see Section 6.4) which is used to modify SS–AS wind speeds. A reasonable range of peak SS–AS wind speeds is ~ 100–200 m s^{-1}, yielding a close match of model fields to available temperature, density, and airglow data.

6.3. Departures from radiative equilibrium: adiabatic effects

Radiative heating and cooling processes can be augmented by dynamical effects in planetary atmospheres, thereby maintaining temperatures that depart from radiative equilibrium values. This is particularly the case for Venus's nightside, in which a given planet-fixed meridian experiences darkness for as much as 117 Earth days. The night-side temperatures would be expected to be extremely cold without some sort of day-to-night heat redistribution. The NCAR dynamical models of Dickinson and Ridley (1977) and Bougher *et al.* (1986, 1988b) have all presented heat budgets thought to maintain day and night temperatures. Little difference in mean dayside heat balances is estimated between 1-D and 3-D models; i.e., adiabatic cooling (due to upwelling/divergent flow) is negligible (Bougher *et al.*, 1986).

Thermospheric winds, however, have a controlling effect on the maintenance of nightside temperatures. Weakened SS–AS winds have produced a significant change in the nightside heat budget that has enabled nightside cryospheric temperatures to be explained (Bougher *et al.*, 1986, 1988b). Figure 47 shows heating and cooling rates at 120 SZA for both the NCAR 2-D and VTGCM model calculations. Notice that adiabatic heating (from subsiding/convergent winds) largely balances CO_2 15-μm cooling below 140 km. Above, molecular conduction is offset by this same compressional heating. The magnitude of this heating is reduced by about a factor of three from the Dickinson and Ridley (1977) model, permitting the observed cold nightside temperatures (Niemann *et al.*, 1979, 1980b; Keating *et al.*, 1980) to be reproduced. Early ideas that invoked nightside eddy heat conduction to give cold nightside temperatures (cf. Niemann *et al.*, 1980b; Schubert *et al.*, 1983; Gordiets and Kulikov, 1985) appear to be inappropriate. Observed nightside temperatures (and the diurnal density contrasts) can be maintained by weakened thermospheric winds alone (Bougher *et al.*, 1986).

6.4. Drag processes

It is evident by the observed day–night density and temperature contrasts that some type of deceleration mechanism is necessary to slow the otherwise pressure-driven neutral winds (Seiff, 1982; Bougher *et al.*, 1986). Early in the Pioneer Venus mission, thought was given to ion-drag and its possible influence on neutral winds (Mayr *et al.*, 1980).

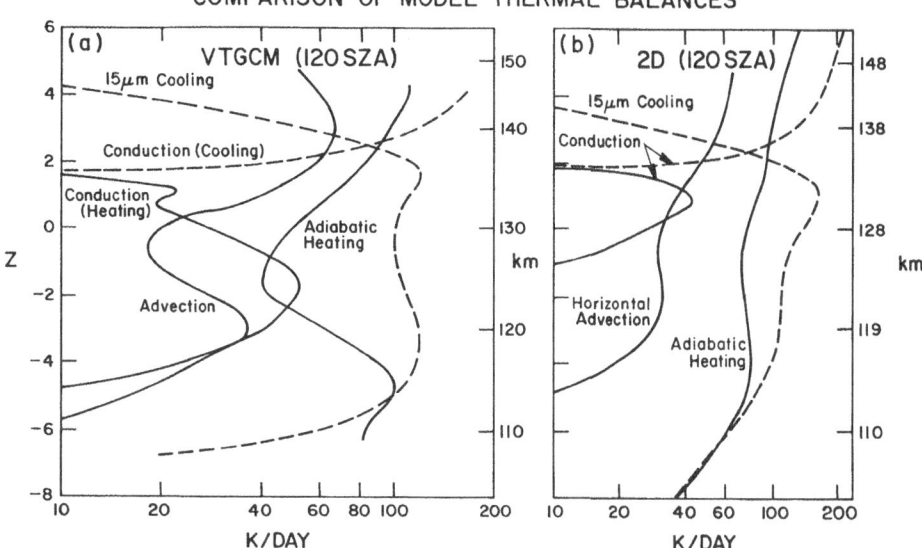

Fig. 47. A comparison of nightside mean heat balances for the NCAR VTGCM and 2-D models at 120 SZA. Units in K Earth day^{-1}. (a) VTGCM symmetric model heat balances at the equator, and (b) 2-D model heat balances. Taken from Bougher *et al.* (1988b).

Since Venus does not have a significant intrinsic magnetic field (Russell *et al.*, 1979), ions are not generally confined to field lines, thereby eliminating a 'grid' for neutrals to encounter. Rather, large horizontal ion-drift velocities, measured by the PV ion mass spectrometer (Taylor *et al.*, 1980) and the retarding potential analyzer (Knudsen *et al.*, 1980) (on the order of ≤ 1 km s^{-1} below 200 km), may carry along neutrals to the nightside. This effect was estimated to be negligible since ion-neutral collision frequencies below 200 km are small (Mayr *et al.*, 1980). The situation may be different during times of high solar wind dynamic pressure, when the region below the minimum ionopause (~ 220 km) may be penetrated by interplanetary magnetic fields (Luhmann *et al.*, 1987). Under these conditions, temporary fields may generate a possible ion-neutral drag for ion velocities in excess of neutral velocities. A definitive analysis awaits a 3-D coupled ionosphere/thermosphere calculation of global winds.

Alternatively, there is ample evidence for turbulence or wave perturbations in the upper atmosphere of Venus (Seiff *et al.*, 1980; Seiff, 1982; von Zahn *et al.*, 1983; Kasprzak *et al.*, 1988; Mayr *et al.*, 1988). All available observations, however, are snapshots of wave structure; horizontal (vertical) wavelengths in the range of 100–600 km (5–15 km) have been observed in the major neutral densities. There seems to be more wave activity during nighttime than during the day, with considerable activity occurring in the vicinity of the predawn and postdusk terminators. Comparison with the model of Mayr *et al.* (1988) shows that the observed perturbations can be interpreted as being due to gravity waves propagating upward from source regions in the lower thermosphere (130 km), or below 80 km. Most important, the data imply that significant wave energy arrives in the thermosphere from lower altitudes on a continuing basis. The effect of such waves on the thermospheric densities, temperatures, and winds is signifi-cant. Specifically, nightside neutral densities vary by as much as $\pm 50\%$ from day-to-day, likely due to upward propagating waves (Hedin et al., 1983; Kasprzak et al., 1988).

Eddy conduction, eddy or wave drag, and eddy diffusion are typically parameterized in planetary models to mimic turbulent effects on temperatures, densities, and winds. For the Earth, it has been shown that the heat and momentum budgets near the mesopause cannot both be simultaneously satisfied without a strong zonal drag force (Lindzen, 1981; Holton, 1982, 1983). Such a frictional force is required to simulate the reversal of the latitudinal temperature gradient observed at the mesopause, with the lowest temperatures at the summer pole (due to adiabatic cooling from upwelling) and the warmest at the winter pole (due to adiabatic heating from subsidence). A modified meridional circulation is responsible. The underlying mechanism has long been thought to result from the unstable breakdown of tides or gravity waves near the terrestrial mesopause. Waves generated in the troposphere are assumed to attain sufficient ampli-tudes in the mesosphere to produce superadiabatic lapse rates, giving rise to convective overturning and turbulence. This causes attenuation of further growth and a conver-gence of the vertical momentum flux, resulting in a net zonal drag force (Lindzen, 1981; Holton, 1982, 1983).

Rayleigh friction is an *ad hoc* scheme commonly used to mimic this 'wave-drag' force *in-lieu* of self-consistently calculating momentum stress terms based on wave properties and the large-scale flow. Specifically, Rayleigh friction is formulated as a damping term added to the horizontal equation of motion which is proportional to the horizontal velocity. The constant of proportionality is assigned a fixed peak value (s^{-1}) at some reference level, and prescribed to vary as a function of altitude (or pressure) below. The shape of the profile is usually chosen to be that of the corresponding eddy coefficient used in the model study.

Models of this terrestrial wave drag progressed from crude Rayleigh friction schemes (Schoeberl and Strobel, 1978), to current wave drag models patterned after Lindzen (1981) and Holton (1982), which describe self-consistent methods for parameterizing the stress and diffusion due to gravity wave breaking in terms of the mean zonal wind, static stability, and specific wave parameters. A height-dependent Rayleigh friction is a first choice when little is known about wave parameters and their source. The initial approach for Venus has been guided by these terrestrial mesopause wave-drag schemes and the underlying wave-breaking mechanism as well. Bougher et al. (1986) assumed that small-scale gravity waves are responsible for the observed wavelike features in the altitude profiles of temperature and density (Seiff et al., 1980; Schubert, 1983). A deceleration of the SS–AS winds is proposed to become important at a level where wave-breaking maximizes; i.e., also where eddy diffusion peaks on the dayside (135–140 km). A profile for drag is prescribed that increases with height up to this level, consistent with a vertical propagating spectrum of waves each of which breaks at a different level. A Rayleigh friction time-scale of $\leq 1 \times 10^{-4}$ s^{-1} is specified. This leaves

SS–AS winds below 110 km largely unaffected, while those near 150 km are reduced from Dickinson and Ridley (1977) values by a factor to two.

The day–night contrasts in observed Venus thermospheric temperatures and densities are closely matched by Bougher et al. (1986) using this wave-drag mechanism (see Section 6.2). It is, however, somewhat unsatisfying that an *ad hoc* formulation, rather than a self-consistent scheme, for wave-induced drag was used. Nevertheless, Rayleigh friction is the best that can be used until further wave parameters (and measured thermospheric winds) are available. The Bougher et al. (1986, 1988b) and Mengel et al. (1989) benchmark models, all of which rely on Rayleigh friction, serve as starting points for later studies.

6.5. THERMOSPHERIC SUPERROTATION

The observed asymmetries in the helium (and hydrogen) densities, as well as the observed NO nightglow intensity (Niemann et al., 1980b; Brinton et al., 1980; Stewart et al., 1980), suggest that the Venus thermosphere superrotates (above 150 km) with a period of 5 to 10 days (Mayr et al., 1980, 1985). The origin of this superrotation is not known. Thermospheric superrotation could be a remnant of the cloud-top 4-day zonal winds observed in the lower atmosphere (cf. Schubert and Walterscheid, 1984). A decrease in the superrotation rate between the cloud tops (4-day period) and the thermosphere is related to the temperature reversal observed above 70 km (where it increases toward the poles). In principle, thermospheric superrotation could also be generated *in situ*, presumably through zonally symmetric pressure gradients producing a geostrophic or cyclostrophic balance or by nonlinear interactions involving the diurnal tides (Mengel et al., 1989).

All circulation model studies to date have chosen to circumvent this question of the origin of this zonal flow, in order to examine its effects on the thermospheric structure (Mayr et al., 1985; Bougher et al., 1988b; Mengel et al., 1989). Mayr et al. (1985) and Mengel et al. (1989) both specify a solid-body rotation in which SS–AS winds are calculated with respect to a rotating reference frame. This has the effect of adding zonal wind speeds of ~ 75 m s^{-1} at the equator. A considerable shift in the nightside helium bulge (to 05:00 LT) is produced. However, any altitude dependence of a zonal wind profile cannot be given by assuming solid-body rotation. Bougher et al. (1988b) prescribed an altitude dependent zonal wind profile increasing from zero to 60 m s^{-1} over 105 to 140 km. This vertical variation is consistent with weak zonal winds at 100–110 km (Clancy and Muhlemann, 1985b; Goldstein, 1989) and strong zonal winds above 150 km (Mayr et al., 1980). Since tracer species are thought to generally follow constant pressure surfaces (streamlines) in their path from the day to the nightside, zonal wind shear could be important in modifying the nightside distribution of helium densities and NO nightglow intensities. A self-consistent model formulation (and mechanism) needs to be developed to explain Venus thermospheric superrotation. These initial sensitivity studies serve as useful starting points for future work.

6.6. CONCEPT OF HOMOPAUSE REVISITED

One-dimensional coupled continuity-diffusion models are a reasonable first step in modeling the vertical density profiles observed in the Venus dayside thermosphere. Von Zahn *et al.* (1980) developed a morningside model at 60 SZA for examination of BNMS densities above 130 km. The success of the model fit to observations is in large part due to the chosen estimate of the altitude dependance of the eddy coefficient profile used for eddy diffusion. The analytical representation of the eddy coefficient profile was taken from Lindzen (1971) as $K = A/\sqrt{N}$, where N is total number density, and A is a tuning factor chosen to be 1.4×10^{13} cm$^{0.5}$ s^{-1}. The helium data taken by the BNMS were used to constrain the eddy profile and provide this specific factor. The upper level cutoff was set at $K = 5 \times 10^8$ cm^2 s^{-1}, despite the fact that unique values above the homopause (where eddy and molecular diffusion are equal) are not easily determined or meaningful. Homopause altitudes were established at 136, 130, and 134 km for N_2, He, and mean composition, respectively. A similar approach was taken by Massie *et al.* (1983), who incorporated detailed chemistry and von Zahn *et al.* (1980) eddy diffusion coefficients ($K \leq 5 \times 10^7$ cm^2 s^{-1}) in an improved dayside model extending down to 100 km. A nightside mean model was also constructed assuming dayside fluxes of O- and N-atoms as upper boundary conditions for the nightside equations. One dimensional models of this type are subject to several limitations, the principal one of which concerns the neglected or improperly parameterized effects of large-scale winds on the calculated densities (Bougher, 1985).

By contrast, Venus circulation model studies of Bougher *et al.* (1990a) and Mengel *et al.* (1989) suggest that vertical small-scale eddy diffusion is only a partial contributor to the maintenance of observed day and nightside thermospheric densities. Eddy coefficients required are about 2–3 times smaller than values used by previous one-dimensional composition models (von Zahn *et al.*, 1980; Stewart *et al.*, 1980; Massie *et al.*, 1983). This implies that the vertical eddy coefficient dependence derived by von Zahn *et al.* (1980) within a one-dimensional model is not solely a signature of small-scale vertical mixing. Rather, it provides a reasonable description of compositional variations brought about by a combination of large-scale horizontal and vertical winds and small-scale mixing. The conventional homopause, a well-defined level where molecular diffusion and eddy mixing processes balance, might better be described as a statistically average level where gravitational separation is offset by global wind plus small-scale mixing effects. The changing Venus thermospheric circulation has a modest influence on this level at any time. In general, one-dimensional eddy mixing models are valuable since they quantify the magnitude of the total dynamical effect expected; however, no information on the relative contribution of winds or turbulence can be obtained (von Zahn *et al.*, 1980; Bougher *et al.*, 1986, 1990a).

Current Venus modeling efforts (Bougher *et al.*, 1990a; Mengel *et al.*, 1989) disagree as to the magnitude of this residual eddy diffusion required to supplement global circulation in maintaining observed vertical and diurnal density distributions at solar maximum. The Bougher *et al.* (1990a) VTGCM cannot tolerate global eddy diffusion exceeding $2 \times 10^7 \, cm^2 \, s^{-1}$ and still yield atomic-N and O densities that reasonably match Hedin *et al.* (1983) empirical model values. Furthermore, the altitude of the NO nightglow peak volume emission rate (115 ± 2 km) is a very sensitive indicator of the nightside eddy diffusion possible within this global circulation model. Global winds for this code reach a maximum of ~ 100–$200 \, m \, s^{-1}$ across the terminators for a super-position of SS–AS and prescribed zonal winds having a period of 8 days. Conversely, the Mengel *et al.* (1989) spectral model uses eddy diffusion reaching $3 \times 10^7 \, cm^2 \, s^{-1}$, in concert with maximum horizontal winds of $\sim 100 \, m \, s^{-1}$, to obtain observed diurnal helium distributions. Superrotating winds are included with a 6-day period.

The discrepancies in eddy, wind (wave-drag), and superrotation parameters in these two models reflect the difficulty in obtaining a unique solution without further observations, namely measurements of thermospheric winds. The effectiveness of the circulation in building up the nightside densities of atomic-O (and lighter species) decreases as the eddy coefficient increases. Likewise, the magnitude of the circulation regulates the amount that can be transported from day to night. Furthermore, superrotating winds not only shift the local time position of the nightside helium, hydrogen and NO airglow bulges, but they also serve to 'smear out' the magnitudes of these bulges. Therefore, future progress in Venus thermospheric modeling will likely require either remote or *in situ* wind measurements for convergence on a unique combination of wind, eddy, and superrotation parameters.

7. Solar Cycle Variation

Very little thermospheric data exists for Venus at solar minimum. A range of exospheric temperatures (275 ± 50 K) was obtained from Mariner 5 and 10 Lα data (Anderson, 1976; Takacs *et al.*, 1980) taken in October 1967 and February 1974, respectively. The latter corresponds to a near minimum in the solar cycle ($F_{10.7} = 68$). The Venera 9 and 10 orbiters also visited the planet in October 1975 through February 1976. Remote Lα (Bertaux *et al.*, 1978) and visible airglow (Krasnopolsky *et al.*, 1976) observations were made ($F_{10.7} = 75$). No definitive estimate of neutral temperatures could, however, be derived. Lastly, the Pioneer Venus orbiter took *in situ* neutral data for nearly three diurnal periods (1978–1980). Periapsis altitude was actively controlled to remain within an altitude range of about 145–160 km until June, 1980. Afterward, this altitude was allowed to rise to a maximum mean value of about 2200 km in 1986. Presently, periapsis is moving downward toward an entry and burn-up in 1992. The only *in situ* neutral/ion observations available are for near solar maximum conditions ($F_{10.7} = 150$–240).

Nevertheless, remote information regarding the ionosphere/thermosphere at solar minimum was obtained from various radio occultation measurements (Kliore and Mullen, 1988). Most of this data in fact comes from the Pioneer Venus Orbiter, although a few profiles come from Venera 9/10 and Mariner 10 (Kliore and Mullen, 1987; Kim *et al.*, 1989). Photochemical equilibrium conditions hold for the major dayside ion densities below ~ 180 km (Cravens *et al.*, 1981; Kim *et al.*, 1989). Therefore, the most useful electron density measurements for inferring neutral temperatures are just above the ion peak (≥ 140 km). A simple model (Kliore and Mullen, 1987) suggests that neutral temperatures during 1984–1986 at 150 km (55–75 SZA) are about 215 K, nearly 60 K lower than values at solar maximum (275 K). Mars derived neutral temperatures near its dayside ionospheric peak, also based on radio occultation electron density data, similarly show a rather weak dependence on $F_{10.7}$ (Bauer and Hantsch, 1989). Solar cycle exospheric temperatures, however, vary by a larger amount than ionospheric peak values, especially for Mars (Bougher and Dickinson, 1988; Bougher *et al.*, 1988a).

Bauer and Taylor (1981) sought to explain the weak solar cycle response of the Venus thermosphere compared to that of the Earth as arising from the stronger temperature dependence of the conductivity of CO_2 ($K \sim T^{1.23}$) compared to that of O ($K \sim T^{0.71}$). Dickinson and Bougher (1986) showed, however, that conduction is the dominant cooling mechanism only above 160 km on Venus, where O is the dominant constituent.

The Venus electron/ion density calculations of Kim *et al.* (1989) indicate that the Hedin *et al.* (1983) model predicts the neutral densities reasonably well for dayside solar minimum conditions; i.e., the resulting ion peak is given at the observed altitude (139–140 km). The empirical models of Hedin *et al.* (1983) and Keating *et al.* (1986), based primarily on *in situ* solar maximum data, predict a solar cycle exospheric temperature variation ($F_{10.7}$ = 200 to 70) of about 55 to 70 K (310 to 255 K and 310 to 240 K, respectively). The well monitored 27-day exospheric temperature variation of ± 11 K (cf. Keating and Bougher, 1987) also suggests that a larger, non-negligible, solar cycle variation exists. However, without additional Venus *in situ* density (or temperature) measurements, a firm confirmation of this rather weak thermospheric response to the solar cycle is still lacking! Figure 48 illustrates a collection of possible dayside Venus solar minimum neutral temperature profiles. The only data on the plot comes from the PV–ORO radio occultation measurements of Kliore and Mullen (1987).

The true test of a useful predictive model is its ability to reliably simulate the observed fields under a wide range of conditions. The NCAR 1-D (Dickinson and Bougher, 1986), 2-D (Bougher *et al.*, 1986) and VTGCM (Bougher *et al.*, 1990a) models have been modified to predict the Venus solar minimum thermospheric circulation (dynamical models) and structure. The primary solar cycle change occurs in the EUV fluxes shortward of 105 nm; i.e., a reduction of peak EUV heating and net CO_2 dissociation is given of approximately a factor of 3. This is consistent with a 21% smaller peak ion density (55–75 SZA) for solar minimum than solar maximum (Kliore and Mullen, 1987). Indirectly, through cooler temperatures, CO_2 15-µm cooling is also reduced, since it is strongly nonlinearly temperature dependent. The EUV fluxes of Heroux and Hinteregger (1978) were chosen, corresponding to an $F_{10.7}$ = 74.

Dickinson and Bougher (1986) confirmed that the Venus exospheric temperatures should be relatively insensitive to changes in the solar flux; dayside values were shown to change by less than ~ 65 K over the solar cycle (300 to 235 K). The strong nonlinear temperature dependence of the CO_2 cooling and the variation of atomic oxygen densities with solar activity ('the atomic O thermostat') effectively buffer against solar perturbations, thereby reducing the temperature variation from that which would occur if a balance were present between peak EUV heating and molecular conduction. Corre-

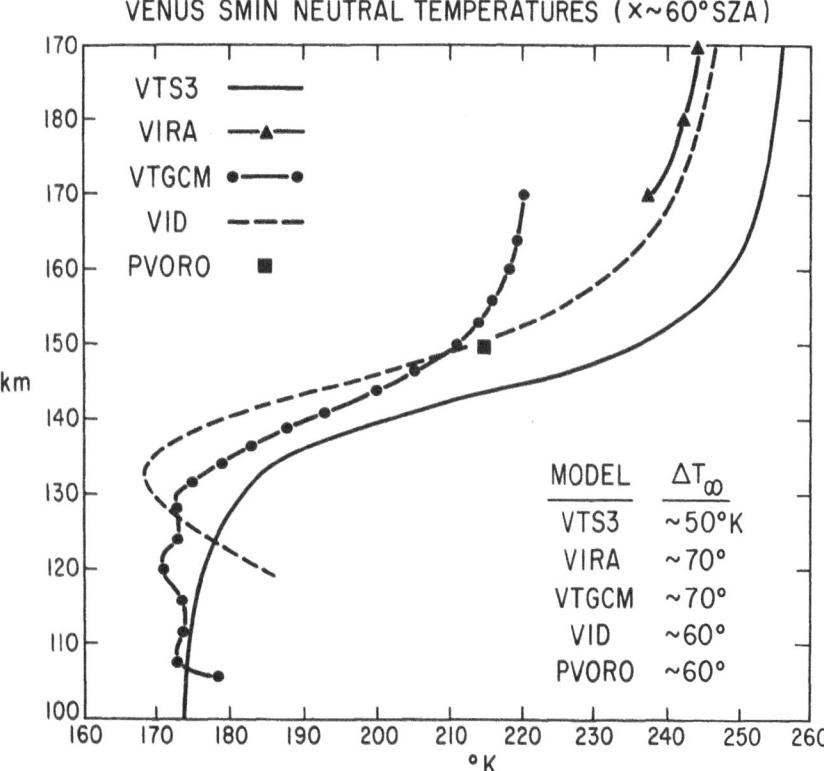

Fig. 48.   Venus dayside solar minimum neutral temperature profiles (60 SZA). VTS3 and VIRA profiles are only reasonable extrapolations from near solar maximum data. The one true data point is inferred from ORO measured electron density profiles just above the ion peak (150 km). The NCAR 1-D (V1D) and VTGCM model predictions are also shown for comparison.

sponding nightside mean calculations suggested that little change in temperatures with solar activity should be expected, so that global mean values drop by only 30 K from solar maximum.

Hydrodynamic effects were carefully examined in subsequent Venus circulation models for an improved description of solar cycle response. Figures 41(e–f) illustrate NCAR 2-D solar minimum contour plots, for comparison to corresponding solar maximum fields of Figures 41(a) and 41(b). The dayside exospheric temperatures are reduced from 300 K to ~ 230 K, giving a dayside solar cycle temperature variation of at most 70 K. These newly calculated dayside solar minimum temperatures are somewhat low, yet the values fall within the range of exospheric temperatures estimated from Mariner 5 and 10 Lα data (Anderson, 1976; Takacs et al., 1980). Also, values at 150 km (over 55–75 SZA) are close to 215 K, quite similar to those inferred from 1984–1986 radio occultation data (Kliore and Mullen, 1987). Adiabatic cooling serves to reduce dayside temperatures slightly below previous radiative equilibrium model values (Dickinson and Bougher, 1986). The solar driven winds are also 20–30 m s^{-1} weaker; this slight change causes the nightside thermospheric major densities and temperatures to remain largely the same. Both the NCAR 2-D and VTGCM solar minimum models suggest that the Venus thermospheric major densities and temperatures are relatively insensitive to the factor of 3 reduction in EUV inputs over the solar cycle (Bougher et al., 1986, 1990a).

This does not appear to be the case for tracer species and the UV nightglow. Odd-nitrogen studies using the VTGCM (see Section 3.10) predict a variation of nearly a factor of 3 in the $NO(0, 1)$ δ-band intensities over the solar cycle (see Figures 20(a) and 20(b)). This sensitivity is due to the net change in dayside production of N-atoms, which varies strongly with solar cycle. A detailed analysis of recently obtained PV–OUVS data taken from 1984–1986 will provide an opportunity to validate this VTGCM solar cycle nightglow variation. A reasonable match of the model and OUVS imaged dark-disk intensity variation over the solar cycle would be a substantial validation of the odd-nitrogen chemistry adopted and the solar cycle response predicted by the VTGCM thermospheric winds. Also, since the NO nightglow is a good tracer of the Venus thermospheric circulation, its monitoring over the solar cycle should provide a remote means to study the variation of the winds once the dayside N-production is calculated.

Acknowledgements

We would like to thank A. I. F. Stewart and L. P. Paxton for helpful suggestions. This work has been supported in part by grants from the National Aeronautics and Space Administration NAGW–665 and NAG2–523 to the Research Foundation of the State University of New York and NAG2–519 to the University of Arizona.

References

Abreu, V. J., Eastes, R. W., Yee, J. H., Solomon, S. C., and Chakrabarti, S.: 1986, 'Ultraviolet Nightglow Production Near the Magnetic Equator by Neutral Particle Precipitation', J. Geophys. Res. 91, 11365.

Adams, W. S. and Dunham, T.: 1932, 'Absorption Bands in the Infrared Spectrum of Venus', Publ. Astron. Soc. Pacific 44, 243.

Ajello, J. M.: 1971, 'Emission Cross Section of CO_2 by Electron Impact in the Interval 1260–4500 Å', J. Chem. Phys. 55, 3169.

Alge, E., Adams, N. G., and Smith, D.: 1987, 'Measurements of the Dissociative Recombination Coefficients of O_2^+, NO^+, and NH_4^+ in the Temperature Range 200–600 K', J. Phys. B16, 1433.

Ali, A. A., Ogryzlo, E. A., Shen, Y. Q., and Wassel, P. T.: 1986, 'The Formation of $O_2(a^1\Delta_g)$ in Homogeneous and Heterogeneous Atom Recombination', Can. J. Phys. 64, 1614.

Allen, D. C., Scragg, T., and Simpson, C. J. S. M.: 1980, 'Low Temperature Fluorescence Studies of the Deactivation of the Bend-Stretch Manifold of CO_2', Chemical Phys. 51, 279.

Anderson, D. E.: 1975, 'The Mariner 5 Ultraviolet Photometer Experiment: Analysis of Rayleigh-Scattering and 1304-Å Radiation from Venus', J. Geophys. Res. 80, 3063.

Anderson, D. E.: 1976, 'The Mariner 5 Ultraviolet Spectrometer Experiment: Analysis of Hydrogen Lyman-α Data', J. Geophys. Res. 81, 1213.

Anderson, S. M., Klein, F. S., and Kaufman, F.: 1985, 'Kinetics of the Isotope Exchange Reaction of ^{18}O with NO and O_2 at 298 K', J. Chem. Phys. 83, 1648.

Atkinson, R. and Welge, K. H.: 1972, 'Temperature Dependence of $O(^1S)$ Deactivation by CO_2, O_2, N_2, and Ar', J. Chem. Phys. 57, 3689.

Atreya, S. K., Pollack, J. B., and Matthews, M. S.: 1989, *Origin and Evolution of Planetary and Satellite Atmospheres,* Univ. of Arizona Press, Tucson.

Baker, N. L. and Leovy, C. B.: 1987, 'Zonal Winds Near Venus' Cloud Top Level: A Model Study of the Interaction between the Zonal Mean Circulation and the Semidiurnal Tide', *Icarus* **69**, 202.

Banks, P. M. and Kockarts, G.: 1973, *Aeronomy,* Academic Press, New York.

Barker, E. S.: 1975, 'Observations of Venus Water Vapour over the Disc of Venus', *Icarus* **25**, 268.

Barth, C. A.: 1964, 'Three-Body Reactions', *Ann. Geophys.* **20**, 183.

Barth, C. A.: 1968, 'Interpretation of Mariner 5 Lyman Alpha Measurements', *J. Atmospheric Sci.* **25**, 564.

Barth, C. A.: 1969, 'Planetary Ultraviolet Spectroscopy', *Appl. Optics* **8**, 1295.

Barth, C. A., Pierce, J. B., Kelly, K. K., Wallace, L., and Fastie, W. G.: 1967, 'Ultraviolet Emission Observed Near Venus from Mariner V', *Science* **158**, 1675.

Barth, C. A., Wallace, L., and Pearce, J. B.: 1968, 'Mariner 5 Measurement of Lyman-alpha Radiation Near Venus', *J. Geophys. Res.* **73**, 2541.

Barth, C. A., Hord, C. W., Pearce, J. B., Kelly, K. K., Anderson, G. P., and Stewart, A. I.: 1971, 'Mariner 6 and 7 Ultraviolet Spectrometer Experiment: Upper Atmospheric Data', *J. Geophys. Res.* **76**, 2213.

Barth, C. A., Hord, C. W., Stewart, A. I., and Lane, A. L.: 1972, 'Mariner 9 Ultraviolet Spectrometer Experiment: Initial Results', *Science* **175**, 309.

Bass, J. N.: 1974, 'Translation to Vibration Energy Transfer in O + NH_3 and O + CO_2 Collisions', *J. Chem. Phys.* **60**, 2913.

Bates, D. R.: 1981, 'The Green Light of the Night Sky', *Planetary Space Sci.* **29**, 1061.

Bates, D. R.: 1988a, 'Transition Probabilities of the Bands of the Oxygen Systems of the Nightglow', *Planetary Space Sci.* **36**, 869.

Bates, D. R.: 1988b, 'Excitation and Quenching of the Oxygen Bands in the Nightglow', *Planetary Space Sci.* **36**, 875.

Bates, D. R.: 1988c, 'Excitation of 557.7 nm Line in Nightglow', *Planetary Space Sci.* **36**, 883.

Bates, D. R.: 1989, 'Oxygen Band System Transition Arrays', *Planetary Space Sci.* **37**, 881.

Bates, D. R. and Zipf, E. C.: 1980, 'The $O(^1S)$ Quantum Yield from O_2^+ Dissociative Recombination', *Planetary Space Sci.* **28**, 1081.

Bauer, S. J. and Hantsch, M. H.: 1989, 'Solar Cycle Variation of the Upper Atmosphere Temperature of Mars', *Geophys. Res. Letters* **16**, 373.

Bauer, S. J. and Taylor, H. A.: 1981, 'Modulation of Venus Ion Densities Associated with Solar Variations', *Geophys. Res. Letters* **8**, 840.

Bauer, S. H., Caballero, J. F., Curtis, R., and Wiesenfeld, J. R.: 1987, 'Vibrational Relaxation Rates of $CO_2(001)$ with Various Collision Partners for $T < 300$ K', *J. Geophys. Chem.* **91**, 1778.

Baulch, D. L., Cox, R. A., Hampson, R. F., Kerr, J. A., Troe, J., and Watson, R. T.: 1980, 'Evaluated Kinetic and Photochemical Data for Atmospheric Chemistry', *J. Phys. Chem. Ref. Data* **9**, 295.

Belton, M. J. S., Hunten, D. M., and Goody, R. M.: 1968, in J. C. Brandt and M. C. McElroy (eds.), 'Quantitative Spectroscopy of Venus in the Region 8000–11 000 Å', *The Atmospheres of Venus and Mars,* Gordon and Breach, New York, pp. 69–98.

Bernstein, R. B. and Levine, R. D.: 1975, 'Role of Energy in Reactive Scattering: An Information-Theoretic Approach', *Adv. At. Mol. Phys.* **11**, 215.

Bertaux, J. L. and Clarke, J. T.: 1989a, 'Deuterium Content of the Venus Atmosphere', *Nature* **338**, 567.

Bertaux, J. L. and Clarke, J. T.: 1989b, 'Reply to Donahue, T. M., Deuterium on Venus', *Nature* **340**, 514.

Bertaux, J. L., Blamont, J., Marcelin, M., Kurt, V. G., Romanova, N. N., Smirnov, A. S.: 1978, 'Lyman-alpha Observations of Venera 9 and 10. I. The Nonthermal Hydrogen Population in the Exosphere of Venus', *Planetary Space Sci.* **26**, 817.

Bertaux, J. L., Blamont, J. E., Lupine, V. M., Kurt, V. G., Romanova, N. N., and Smirnov, A. S.: 1981, 'Venera 11 and Venera 12 Observations of EUV Emission from the Upper Atmosphere of Venus', *Planetary Space Sci.* **29**, 149.

Bertaux, J. L., Lepine, V. M., Kurt, V. G., and Smirnov, A. S.: 1982, 'Altitude Profile of H in the Atmosphere of Venus from Lyman-α Observations of Venera 11 and Venera 12 and Origin of the Hot Exospheric Component', *Icarus* **52**, 221.

Bertaux, J. L., Chassefiere, E., and Kurt, V. G.: 1985, 'Venus EUV Measurements of Hydrogen and Helium from Venera 11 and Venera 12', *Adv. Space Res.* **5**, 119.

Biermann, L., Brosowski, B., and Schmidt, H. U.: 1967, 'Interaction of the Solar Wind with a Comet', *Solar Phys.* **1**, 254.

Bougher, S. W.: 1980, 'The Ultraviolet Night Airglow of Venus – Morphology and Implications', MS Thesis, University of Colorado, Boulder.

Bougher, S. W.: 1985, 'Venus Thermospheric Circulation', PhD. Thesis, University of Michigan, Ann Arbor.

Bougher, S. W. and Dickinson, R. E.: 1988, 'Mars Mesosphere and Thermosphere. I. Global Mean Heat Budget and Thermal Structure', *J. Geophys. Res.* **93**, 7325.

Bougher, S. W., Dickinson, R. E., Ridley, E. C., Roble, R. G., Nagy, A. F., and Cravens, T. E.: 1986, 'Venus Mesosphere and Thermosphere. II. Global Circulation, Temperature, and Density Variations', *Icarus* **68**, 284.

Bougher, S. W., Roble, R. G., and Dickinson, R. E.: 1988a, 'The Thermospheres of Venus and Mars: A Comparison of Global Structure and Winds Using the NCAR–TGCM', *Bull. Am. Astron. Soc.* **20**, 851.

Bougher, S. W., Dickinson, R. E., Ridley, E. C., and Roble, R. G.: 1988b, 'Venus Mesosphere and Thermosphere. III. Three-Dimensional General Circulation with Coupled Dynamics and Composition', *Icarus* **73**, 545.

Bougher, S. W., Dickinson, R. E., Roble, R. G., and Ridley, E. C.: 1988c, 'Mars Thermospheric General Circulation Model: Calculations for the Arrival of Phobos at Mars', *Geophys. Res. Letters* **15**, 1511.

Bougher, S. W., Gerard, J. C., Stewart, A. I. F., and Fesen, C. G.: 1990a, 'The Venus Nitric Oxide Night Airglow: Model Calculations Based on the Venus Thermospheric General Circulation Model', *J. Geophys. Res.* **95**, 6271.

Bougher, S. W., Roble, R. G., Ridley, E. C., and Dickinson, R. E.: 1990b, 'The Mars Thermosphere: 2 General Circulation with Coupled Dynamics and Composition', *J. Geophys. Res.* **95**, 14811.

Braun, W., Bass, A. M., Davis, D. D., and Simmons, J. J.: 1969, 'Flash Photolysis of Carbon Suboxide: Absolute Rate Constants for Reactions of $C(^3P)$ and $C(^1D)$ with H_2, N_2, CO, NO, O_2 and CH_4', *Proc. Roy. Soc. London Ser.* **A312**, 417.

Breus, T. K., Gringauz, K. I., and Verigin, M. I.: 1985, 'On the Properties and Origin of the Venus Ionosphere', *Adv. Space Res.* **5**, 145.

Brinton, H. C., Taylor, H. A., Jr., Niemann, H. B., Mayr, H. G., Nagy, A. F., Cravens, T. E., and Strobel, D. F.: 1980, 'Venus Nighttime Hydrogen Bulge', *Geophys. Res. Letters* **7**, 865.

Broadfoot, A. L., Kumar, S., Belton, M. J. S., and McElroy, M. B.: 1974, 'Ultraviolet Observations of Venus from Mariner 10: Preliminary Results', *Science* **183**, 1315.

Broadfoot, A. L., Clapp, S. S., and Stuart, F. E.: 1977, 'Mariner 10 Ultraviolet Spectrometer: Airglow Experiment', *Space Sci. Instr.* **3**, 199.

Buchwald, M. I. and Bauer, S. H.: 1972, 'Vibrational Relaxation in CO_2 with Selected Collision Partners', *J. Phys. Chem.* **76**, 3108.

Burnett, T. and Rountree, S. P.: 1979, 'Differential and Total Cross Sections for Electron Impact Ionization of Atomic Oxygen', *Phys. Rev.* **A20**, 1468.

Carlson, R. W. and Judge, D. L.: 1973, 'Rocket Observations of the Extreme Ultraviolet Dayglow', *Planetary Space Sci.* **21**, 879.

Center, R. E.: 1973, 'Vibrational Relaxation of CO_2 by O Atoms', *J. Chem. Phys.* **59**, 3523.

Chamberlain, J. W.: 1963, 'Planetary Coronae and Atmospheric Evaporation', *Planetary Space Sci.* **11**, 901.

Chamberlain, J. W.: 1977, 'Charge Exchange in a Planetary Corona: Its Effect on the Distribution and Escape of Hydrogen', *J. Geophys. Res.* **82**, 1.

Chamberlain, J. W. and Hunten, D. M.: 1987, *Theory of Planetary Atmospheres: An Introduction to Their Physics and Chemistry*, Academic Press, New York, pp. 45–48.

Chapman, S.: 1931, 'Some Phenomena of the Upper Atmosphere', *Proc. Roy. Soc. London* **A132**, 353.

Chassefiere, E., Bertaux, J. L., Kurt, V. G., and Smirnov, A. S.: 1986, 'Venus EUV Measurements of Helium at 58.4 nm from Venera 11 and Venera 12 and Implications for the Outgassing History', *Planetary Space Sci.* **34**, 585.

Choo, Y. C. and Ming-Tuan Leu: 1985, 'Determination of $O_2(^1\Sigma_g^+)$ and $O_2(^1\Delta_g)$ yields in $Cl + O_2$ and $Cl + O_3$ Reactions', *J. Phys. Chem.* **89**, 4832.

Chu, J. O., Flynn, G. W., and Weston, R. E.: 1983, 'Spectral Distribution of Vibrational States Produced by Collisions with Fast Hydrogen Atoms from Laser Photolysis of HBr', *J. Chem. Phys.* **78**, 2990.

Clancy, R. T. and Muhleman, D. O.: 1985a, 'Diurnal CO Variations in the Venus Mesosphere from CO Microwave Spectra', *Icarus* **64**, 157.

Clancy, R. T. and Muhleman, D. O.: 1985b, 'Chemical-Dynamical Models of the Venus Mesosphere Based Upon Diurnal Microwave CO Variations', *Icarus* **64**, 183.

Cleary, D. D.: 1986, 'Daytime High Latitude Rocket Observations of the NO γ, δ, and ε Bands', *J. Geophys. Res.* **91**, 11337.

Connes, P., Connes, J., Benedict, W. S., and Kaplan, L. D.: 1967, 'Traces of HCL and HF in the Atmosphere of Venus', *Astrophys. J.* **152**, 731.

Connes, P., Connes, J., Kaplan, L. D., and Benedict, W. S.: 1968, 'Carbon Monoxide in the Venus Atmosphere', *Astrophys. J.* **152**, 731.

Connes, J., Connes, P., and Maillard, J. P.: 1969, *Atlas des spectres dans le proche infrarouge de Venus, Mars, Jupiter et Saturn*, Centre National de la Recherche Scientific, Paris.

Connes, P., Noxon, J. F., Traub, W. A., and Carleton, N. P.: 1979, '$O_2(^1\Delta)$ Emission in the Day and Night Airglow of Venus', *Astrophys. J.* **233**, L29.

Conway, R. R.: 1981, 'Spectroscopy of the Cameron Bands in the Mars Airglow', *J. Geophys. Res.* **86**, 4767.

Cook, G. R., Metzger, P. H., and Ogawa, M.: 1965, 'Photoionization and Absorption Coefficients of CO in the 600 to 1000 Å Region', *Can. J. Phys.* **43**, 1706.

Cooper, D. L., Yee, J. H., and Dalgarno, A.: 1984, 'Energy Transfer in Oxygen-Hydrogen Collisions', *Planetary Space Sci.* **32**, 825.

Cravens, T. E., Gombosi, T. I., and Nagy, A. F.: 1980, 'Hot Hydrogen in the Exosphere of Venus', *Nature (London)* **283**, 178.

Cravens, T. E., Kliore, A. J., Kozyra, J. U., and Nagy, A. F.: 1981, 'The Ionospheric Peak on the Venus Dayside', *J. Geophys. Res.* **86**, 11323.

Cravens, T. E., Brace, L. H., Taylor, H. A., Russell, C. T., Knudsen, W. L., Miller, K. L., Barnes, A., Mihalov, J. D., Scarf, F. L., Quenon, S. J., and Nagy, A. F.: 1982, 'Disappearing Ionospheres on the Nightside of Venus', *Icarus* **51**, 271.

Cravens, T. E, Crawford, S. L., Nagy, A. F., and Gombosi, T. I.: 1983, 'A Two-Dimensional Model of the Ionosphere of Venus', *J. Geophys. Res.* **88**, 5595.

Dalgarno, A. and Degges, T. C.: 1971, in C. Sagan, T. Owen, and H. J. Smith (eds.), 'CO_2^+ Dayglow on Mars and Venus', *Planetary Atmospheres*, D. Reidel Publ. Co., Dordrecht, Holland, pp. 337–345.

Degen, V.: 1981, 'Vibrational Enhancement and the Excitation of N_2^+ and the First Negative System in the High Altitude Red Aurora and the Dayside Cusp', *J. Geophys. Res.* **86**, 11, 1372.

Del Genio, A. D., Schubert, G., and Strauss, J. M.: 1979, 'Acoustic Gravity Waves in the Thermosphere of Venus', *Icarus* **39**, 401.

DeMore, W. B. and Yung, Y. L.: 1982, 'Catalytic Processes in the Atmospheres of the Earth and Venus', *Science* **217**, 12099.

DeMore, W. B., Leu, M.-T., Smith, R. H., and Yung, Y. L.: 1985, 'Laboratory Studies on the Reactions between Chlorine, Sulfur Dioxide and Oxygen: Implications for the Venus Stratosphere', *Icarus* **63**, 347.

Dickinson, R. E.: 1972, 'Infrared Radiative Heating and Cooling in the Venusian Mesosphere. I. Global Mean Radiative Equilibrium', *J. Atmospheric Sci.* **29**, 1531.

Dickinson, R. E.: 1973, 'Method for Parameterization for Infrared Cooling between Altitudes of 30 and 70 Kilometers', *J. Geophys. Res.* **78**, 4451.

Dickinson, R. E.: 1976, 'Venus Mesosphere and Thermosphere Temperature Structure: Global Mean Radiative and Conductive Equilibrium', *Icarus* **27**, 479.

Dickinson, R. E.: 1984, 'Infrared Radiative Cooling in the Mesosphere and Lower Thermosphere', *J. Atmospheric Terrest. Phys.* **46**, 995.

Dickinson, R. E. and Bougher, S. W.: 1986, 'Venus Mesosphere and Thermosphere. I. Heat Budget and Thermal Structure', *J. Geophys. Res.* **91**, 70.

Dickinson, R. E. and Ridley, E. C.: 1972, 'A Numerical Solution for the Composition of a Steady Subsolar-to-Antisolar Circulation with Application to Venus', *J. Atmospheric Sci.* **29**, 1557.

Dickinson, R. E. and Ridley, E. C.: 1975, 'A Numerical Model for the Dynamics and Composition of the Venusian Thermosphere', *J. Atmospheric Sci.* **32**, 1219.

Dickinson, R. E. and Ridley, E. C.: 1977, 'Venus Mesosphere and Thermosphere Temperature Structure. II. Day-Night Variations', *Icarus* **30**, 163.

Dickinson, R. E., Ridley, E. C., and Roble, R. G.: 1984, 'Thermospheric General Circulation with Coupled Dynamics and Composition', *J. Atmospheric Sci.* **41**, 205.

Dickinson, R. E., Roble, R. G., and Bougher, S. W.: 1987, 'Radiative Cooling in the NLTE Region of the Mesosphere and Lower Thermosphere – Global Energy Balance', *Adv. Space Res.* **7**, 10, 5.

Doering, J. P. and Gulcicek, E. E.: 1988, 'Absolute Differential and Integral Electron Excitation Cross

Sections for Atomic Oxygen: 7. The $^3P \to {}^1D$ and $^3P \to {}^1S$ Transitions from 4.0 to 30 eV', *J. Geophys. Res.* (in press).

Doering, J. P. and Gulcicek, E. E.: 1989, 'Absolute Differential and Integral Electron Excitation Cross Sections for Atomic Oxygen', *J. Geophys. Res.* **94**, 2733.

Doering, J. P. and Vaughan, S. O.: 1986, 'Absolute Experimental Differential and Integral Cross Sections for Atomic Oxygen. 1. The $({}^3P \to {}^3S^0)$ Transition (1304 Å) at 100 eV', *J. Geophys. Res.* **91**, 3279.

Doering, J. P., Gulcicek, E. E., and Vaughan, S. O.: 1985, 'Electron Impact Measurements of Oscillator Strengths for Dipole-Allowed Transitions of Atomic Oxygen', *J. Geophys. Res.* **90**, 5279.

Donahue, T. M.: 1968, 'The Upper Atmosphere of Venus: A Review', *J. Atmospheric Sci.* **25**, 568.

Donahue, T. M.: 1989, 'Deuterium on Venus', *Nature* **340**, 513.

Donahue, T. M. and Pollack, J. B.: 1983, in D. M. Hunten, L. Colin, T. M. Donahue, and V. I. Moroz (eds.), 'Origin and Evolution of Venus's Atmospheric Structure', *Venus*, Univ. of Arizona Press, Tucson.

Donahue, T. M., Hoffman, J. H., Hodges, R. R., and Watson, A. J.: 1982, 'Venus was Wet: A Measurement of the Ratio of Deuterium to Hydrogen', *Science* **216**, 630.

Doschek, G. A., Behring, W. E., and Feldman, U.: 1974, 'The Widths of the Solar He I and He II Lines at 584, 538, and 304 Å', *Astrophys. J.* **190**, L141.

Durrance, S. T.: 1981, 'The Carbon Monoxide Fourth Positive Bands in the Venus Dayglow. 1. Synthetic Spectra', *J. Geophys. Res.* **86**, 9115.

Durrance, S. T., Barth, C. A., and Stewart, A. I. F.: 1988, 'Pioneer Venus Observations of the Venus Dayglow Spectrum 1250–1430 Å', *Geophys. Res. Letters* **7**, 222.

Durrance, S. T., Conway, R. R., Barth, C. A., and Lane, A. L.: 1981, 'IUE High Resolution Observation of the Venus Dayglow Spectrum 1280–1380 Å', *Geophys. Res. Letters* **8**, 111.

Eckstrom, D. J.: 1973, 'Vibrational Relaxation of Shock-Related N_2 by Atomic Oxygen Using the IR Tracer Method', *J. Chem. Phys.* **59**, 2787.

Erdman, P. W. and Zipf, E. C.: 1983, 'Electron-Impact Excitation of the Cameron System $(a^3\Pi \to X^1\Sigma)$ of CO', *Planetary Space Sci.* **31**, 317.

Ewing, G.: 1978, 'The Role of van der Waals Molecules in Vibrational Relaxation Processes', *Chem. Phys.* **29**, 253.

Fehsenfeld, F. C., Dunkin, D. B., and Ferguson, E. E.: 1970, 'Rate Constants for the Reaction of CO_2^+ with O, O_2, and NO; N_2^+ with O and NO; and O_2^+ with NO', *Planetary Space Sci.* **18**, 1267.

Feldman, P. D., Moos, H. W., Clarke, J. T., and Lane, A. L.: 1979, 'Identification of the UV Nightglow from Venus', *Nature* **279**, 221.

Feldman, P. D., Anderson, D. E., Meier, R. R., and Gentieu, E. P.: 1981, 'The Ultraviolet Dayglow. 4. The Spectrum and Excitation of Singly Ionized Oxygen', *J. Geophys. Res.* **86**, 3583.

Fels, S. B. and Lindzen, R. S.: 1974, 'The Interaction of Thermally Driven Excited Gravity Waves with Mean Flows', *Geophys. Fluid Dyn.* **6**, 149.

Fernando, R. P. and Smith, I. W. M.: 1979, 'Vibrational Relaxation of NO by Atomic Oxygen', *Chem. Phys. Letters* **66**, 218.

Fesen, C. G., Gerard, J. C., and Rusch, D. W.: 1989, 'Rapid Deactivation of $N(^2D)$ by O: Impact on Thermospheric and Mesospheric Odd Nitrogen', *J. Geophys. Res.* **94**, 5419.

Fink, U., Larson, H. P., Kuiper, G. P., and Poppen, R. F.: 1972, 'Water Vapor in the Atmosphere of Venus', *Icarus* **17**, 617.

Flynn, G. W. and Weston R. E.: 1986, 'Hot Atoms Revisited: Laser Photolysis and Product Detection', *Ann. Rev. Phys. Chem.* **37**, 551.

Fox, J. L.: 1978, 'The Upper Atmospheres of Mars and Venus', Ph.D. Thesis, Harvard University.

Fox, J. L.: 1982a, 'The Chemistry of Metastable Species in the Venusian Ionosphere', *Icarus* **51**, 248. Erratum: 1985, *J. Geophys. Res.* **90**, 11106.

Fox, J. L.: 1982b, 'Atomic Carbon in the Atmosphere of Venus', *J. Geophys. Res.* **87**, 9211.

Fox, J. L.: 1985, 'The O_2^+ Vibrational Distribution in the Venusian Ionosphere', *Adv. Space Res.* **5**, 165.

Fox, J. L.: 1986a, 'The O_2^+ Vibrational Distribution in the Dayside Ionosphere', *Planetary Space Sci.* **34**, 1252.

Fox, J. L.: 1986b, 'Models for Aurora and Airglow Emissions from Other Planetary Atmospheres', *Can. J. Phys.* **64**, 1631.

Fox, J. L.: 1988, 'Heating Efficiencies in the Thermosphere of Venus Reconsidered', *Planetary Space Sci.* **36**, 37.

Fox, J. L.: 1989a, in J. B. A. Mitchell and S. L. Guberman (eds.), 'Dissociative Recombination in Aeronomy', *Dissociative Recombination: Theory, Experiment and Applications*, World Scientific, Singapore.

Fox, J. L.: 1989b, 'The Neutral Thermospheres of Mars and Venus', *EOS (Trans. Am. Geophys. Union)* **70**, 387.

Fox, J. L.: 1990a, 'The Red and Green Lines of Atomic Oxygen in the Nightglow of Venus', *Adv. Space Res.* **10**(5), 31.

Fox, J. L.: 1990b, *The Production of Hot Oxygen at the Exobases of the Terrestrial Planets* (in preparation).

Fox, J. L. and Black, J. H.: 1989, 'Photodissociation of CO in the Thermosphere of Venus', *Geophys. Res. Letters* **16**, 291.

Fox, J. L. and Dalgarno, A.: 1979, 'Ionization, Luminosity and Heating of the Upper Atmosphere of Mars', *J. Geophys. Res.* **84**, 7315.

Fox, J. L. and Dalgarno, A.: 1981, 'Ionization, Luminosity, and Heating of the Upper Atmosphere of Venus', *J. Geophys. Res.* **86**, 629.

Fox, J. L. and Dalgarno, A.: 1983, 'Nitrogen Escape from Mars', *J. Geophys. Res.* **88**, 9027.

Fox, J. L. and Dalgarno, A.: 1985, 'The Vibrational Distribution of N_2^+ in the Terrestrial Ionosphere', *J. Geophys. Res.* **90**, 7557.

Fox, J. L. and Stewart, A. I. F.: 1990, 'The Venus Ultraviolet Aurora: a Soft Electron Source', *J. Geophys. Res.* (submitted).

Frederick, J. E. and Rusch, D. W.: 1977, 'On the Chemistry of Metastable Atomic Nitrogen in the F Region Deduced from Simultaneous Measurements of the 5200 Å Airglow and Atmospheric Composition', *J. Geophys. Res.* **82**, 3509.

Freund, R. S.: 1971, 'Dissociation of CO_2 by Electron Impact with the Formation of Metastable $CO(a^3\Pi)$ and $O(^5S)$', *J. Chem. Phys.* **55**, 3569.

Gel'man, B. G., Zolotukhin, V. G., Lamonov, N. I., Levchuk, B. V., Lipatov, A. N., Mukhin, L. M., Nenarokov, D. F., Rotin, V. A., and Okhotnikov, B. P.: 1979, 'Analysis of Chemical Composition of Venus Atmosphere by Gas Chromatography', *Kossm. Issled.* **17**, 708; 1980, *Cosmic Res.* (Engl. transl.) **17**, 585.

Gerard, J. C., Stewart, A. I. F., and Bougher, S. W.: 1981, 'The Altitude Distribution of the Venus Ultraviolet Nightglow and Implication on Vertical Transport', *Geophys. Res. Letters* **8**, 633.

Gerard, J. C., Denye, E. J., and Lerho, H.: 1988, 'Sources and Distribution of Odd Nitrogen in the Venus Daytime Thermosphere', *Icarus* **75**, 171.

Goldstein, J.: 1989, 'Absolute Wind Speed Measurements in the Lower Thermosphere of Venus Using Infrared Heterodyne Spectroscopy', Ph.D. Thesis, University of Pennsylvania, Philadelphia.

Gordiets, B. F. and Kulikov, Y. N.: 1985, 'On the Mechanisms of Cooling of the Nightside Thermosphere of Venus', *Adv. Space Res.* **5**(9), 113.

Gordiets, B. F., Kulikov, Yu. N., Markov, M. N., and Marov, M. Ya.: 1982, 'Numerical Modelling of the Thermospheric Heat Budget', *J. Geophys. Res.* **87**, 4504.

Gordon, R. J.: 1981, 'A Metastable Complex Model for Vibrational Relaxation', *J. Chem. Phys.* **74**, 1676.

Grechnev, K. V., Istomin, V. G., Ozerov, L. N., and Klimovitskii, V. G.: 1979, 'The Mass Spectrometer from Venera 11 and 12', *Kossm. Issled.* **17**, 697; 1980, *Cosmic Res.* (Engl. transl.) **17**, 575.

Greer, R. G. H., Murtagh, D. P., McDade, I. C., Dickinson, P. H. G., Thomas, L., Jenkins, D. B., Stegman, J., Llewellyn, E. J., Witt, G., MacKinnon, D. J., and Williams, E. R.: 1986, 'ETON 1: A Data Base Pertinent to the Study of Energy Transfer in the Oxygen Nightglow', *Planetary Space Sci.* **34**, 771.

Gringauz, K. I., Verigin, M. I., Breus, T. K., and Gombosi, T.: 1979, 'The Interaction of Electrons in the Optical Umbra of Venus with the Planetary Atmosphere – The Origin of the Nighttime Ionosphere', *J. Geophys. Res.* **84**, 2123.

Grinspoon, D. H. and Lewis, J. S.: 1988, 'Cometary Water on Venus: Implications of Stochastic Impacts', *Icarus* **74**, 21.

Guberman, S. L.: 1987, 'The Production of $O(^1S)$ in Dissociative Recombination of O_2^+', *Nature* **327**, 408.

Guberman, S. L.: 1988, 'The Production of $O(^1D)$ from Dissociative Recombination of O_2^+', *Planetary Space Sci.* **36**, 47.

Gulcicek, E. E. and Doering, J. P.: 1988, 'Absolute Differential and Integral Cross Sections for Atomic Oxygen. 5. Revised Values for the $^3P \rightarrow {}^3S^0$ (1304 Å) and $^3P \rightarrow {}^3D^0$ (989 Å) Transitions Below 30 eV', *J. Geophys. Res.* **93**, 5879.

Gulcicek, E. E., Doering, J. P., and Vaughan, S. O.: 1988, 'Absolute Differential and Integral Electron Excitation Cross Sections for Atomic Oxygen. 6. The $^3P \rightarrow {}^3P$ and $^3P \rightarrow {}^5P$ Transitions from 13.87 eV to 100 eV', *J. Geophys. Res.* **93**, 5885.

Gutcheck, R. A. and Zipf, E. C.: 1973, 'Excitation of the CO Fourth Positive System by the Dissociative Recombination of CO_2^+ Ions', *J. Geophys. Res.* **78**, 5429.

Hartle, R. E. and Taylor, H. A.: 1983, 'Identification of Deuterium Ions in the Ionosphere of Venus', *Geophys. Res. Letters* **10**, 965.

Harvey, N. M.: 1982, 'A Quantum Mechanical Investigation of Vibrational Energy Transfer in $O(^3P)$ + CO_2 Collisions', *Chem. Phys. Letters* **88**, 553.

Hedin, A. E., Niemann, H. B., Kasprzak, W. T., and Seiff, A.: 1983, 'Global Empirical Model of the Venus Thermosphere', *J. Geophys. Res.* **88**, 73. Erratum: 1983, *J. Geophys. Res.* **88**, 6352.

Henry, R. J. W. and McElroy, M. B.: 1968, in J. C. Brandt and M. B. McElroy (eds.), 'Photoelectrons in Planetary Atmospheres', *The Atmospheres of Venus and Mars*, Gordon and Breach Publishers, New York.

Heroux, L. and Hinteregger, H. E.: 1978, 'Aeronomical Reference Spectrum for Solar UV Below 2000 Å', *J. Geophys. Res.* **83**, 5305.

Herzberg, G.: 1950, *Spectra of Diatomic Molecules*, Van Nostrand Reinhold Co., New York.

Hodges, R. R. and Tinsley, B. A.: 1981, 'Charge Exchange in the Venus Thermosphere as a Source of Hot Exospheric Hydrogen', *J. Geophys. Res.* **86**, 7649.

Hodges, R. R. and Tinsley, B. A.: 1986, 'The Influence of Charge Exchange on the Velocity Distribution of Hydrogen in the Venus Exosphere', *J. Geophys. Res.* **91**, 13649.

Hoffman, J. H., Hodges, R. R., Donahue, T. M., and McElroy, M. B.: 1980a, 'Composition of the Venus Lower Atmosphere from the Pioneer Venus Mass Spectrometer', *J. Geophys. Res.* **85**, 7882.

Hoffman, J. H., Oyama, V. I., and von Zahn, U.: 1980b, 'Measurements of the Venus Lower Atmosphere Composition: A Comparison of Results', *J. Geophys. Res.* **85**, 7871.

Hogan, J. S. and Stewart, R. W.: 1969, 'Exospheric Temperatures on Mars and Venus', *J. Atmospheric Sci.* **26**, 332.

Hollenbach, D. J., Prasad, S. S., and Whitten, R. C.: 1985, 'The Thermal Structure of the Dayside Upper Atmosphere of Venus Above 125 km', *Icarus* **64**, 205.

Holton, J. R.: 1982, 'The Role of Gravity Wave Induced Drag and Diffusion in the Momentum Budget of the Mesosphere', *J. Atmospheric Sci.* **39**, 791.

Holton, J. R.: 1983, 'The Influence of Wave Breaking on the General Circulation of the Middle Atmosphere', *J. Atmospheric Sci.* **40**, 2497.

Hou, A.: 1984, 'Axisymmetric Circulations Forced by Heat and Momentum Sources: A Simple Model Applicable to the Venus Atmosphere', *J. Atmospheric Sci.* **41**, 3437.

Hou, A. and Goody, R. M.: 1985, 'Diagnostic Requirements for the Superrotation of Venus', *J. Atmospheric Sci.* **42**, 413.

Hunten, D. M.: 1973, 'The Escape of H_2 from Titan', *J. Atmospheric Sci.* **30**, 726.

Hunten, D. M.: 1973, 'The Escape of Light Gases from Planetary Atmospheres', *J. Atmospheric Sci.* **30**, 1481.

Hunten, D. M.: 1974, 'Energetics of Thermospheric Eddy Transport', *J. Geophys. Res.* **79**, 2533.

Hunten, D. M.: 1982, 'Thermal and Nonthermal Escape Mechanisms for Terrestrial Bodies', *Planetary Space Sci.* **30**, 773.

Hunten, D. M. and Donahue, T. M.: 1976, 'Hydrogen Loss from the Terrestrial Planets', *Ann. Rev. Earth Planetary Sci.* **4**, 265.

Hunten, D. M. and McElroy, M. B.: 1970, 'Production and Escape of Hydrogen on Mars', *J. Geophys. Res.* **75**, 5989.

Hunten, D. M., Pepin, R. O., and Walker, J. C. G.: 1987, 'Mass Fractionation in Hydrodynamic Escape', *Icarus* **69**, 532.

Hunten, D. M., Donahue, T. M., Walker, J. C. G., and Kasting, J. F.: 1989, in S. Atreya, J. Pollack, and M. Matthews (eds.), 'Escape of Atmospheres and Loss of Water', *Origin and Evolution of Planetary and Satellite Atmospheres*, Univ. of Arizona Press, Tucson, pp. 396–422.

Ingersoll, A. P.: 1969, 'The Runaway Greenhouse: A History of Water on Venus', *J. Atmospheric Sci.* **26**, 1191, 1969.

Inoue, G. and Tsuchiya, S.: 1975, 'Vibrational Relaxation of $CO_2(00^01)$ in CO_2, He, Ne, and Ar in the Temperature Range of 300–140 K', *J. Phys. Soc. Japan* **38**, 870.

Istomin, V. G., Grechnev, K. V., Kochnev, V. A., and Ozerov, L. N.: 1979, 'Composition of Venus Lower Atmosphere from Mass-Spectrometer Data', *Kossm. Issled.* **17**, 703; 1980, *Cosmic Res.* (Engl. transl.) **17**, 581.

Izakov, M. N.: 1978, 'Effect of Turbulence on the Thermal Regime of Planetary Thermospheres', *Kossm. Issled.* **16**, 403 (in Russian).

James, T. C.: 1971, 'Transition Moments, Franck–Condon Factors, and Lifetimes of Forbidden Transitions. Calculation of the Intensity of the Cameron System of CO', *J. Chem. Phys.* **55**, 4118.

Jeans, J. H.: 1925, *The Dynamical Theory of Gases*, Cambridge Univ. Press, Cambridge.

Johnson, C. E.: 1972, 'Lifetime of $CO(a^3\Pi)$ Following Electron Impact Dissociation of CO_2', *J. Chem. Phys.* **57**, 576.

Julienne, P. S., Davis, J., and Oran, E.: 1974, 'Oxygen Recombination in the Tropical Nightglow', *J. Geophys. Res.* **79**, 2540.

Jusinski, L. E. and Slanger, T. G.: 1987, 'Determination of the Rate Coefficient for Quenching $N(^2D)$ by $O(^3P)$', *EOS Trans. AGU* **68**, 1389.

Jusinski, L. E., Black, G., and Slanger, T. G.: 1988, 'Resonance-Enhanced Multi-Photon Ionization Measurement of $N(^2D)$ Quenching by $O(^3P)$', *J. Phys. Chem.* **92**, 5977.

Kakar, R. K., Waters, J. W., and Wilson, W. J.: 1976, 'Venus: Microwave Detection of Carbon Monoxide', *Science* **191**, 379.

Kasprzak, W. T., Hedin, A. E., Niemann, H. B., and Spencer, N. W.: 1980, 'Atomic Nitrogen in the Upper Atmosphere of Venus', *Geophys. Res. Letters* **7**, 106.

Kasprzak, W. T., Hedin, A. E., Mayr, H. G., and Niemann, H. B.: 1988, 'Wavelike Perturbations Observed in the Neutral Thermosphere of Venus', *J. Geophys. Res.* **93**, 11237.

Kassel, T.: 1976, 'Scattering of Solar Lyman Alpha by the (14, 0) Fourth Positive System of CO', *J. Geophys. Res.* **81**, 1411.

Kasting, J. F. and Pollack, J. B.: 1983, 'Loss of Water from Venus. I. Hydrodynamic Escape of Hydrogen', *Icarus* **53**, 479.

Kasting, J. F. and Toon, O. B.: 1989, in S. Atreya, J. Pollack, and M. Matthews (eds.), 'Climate Evolution on the Terrestrial Planets', *Origin and Evolution of Planetary and Satellite Atmospheres*, Univ. of Arizona Press, Tucson, pp. 423–449.

Kasting, J. F., Pollack, J. B., and Ackerman, T. P.: 1984, 'Response of the Earth's Atmosphere to Increases in Solar Flux and Implications for Loss of Water from Venus', *Icarus* **57**, 335.

Keating, G. M.: 1977, 'Pioneer Venus Experiment Descriptions', *Space Sci. Rev.* **20**, 520.

Keating, G. M. and Bougher, S. W.: 1987, 'Neutral Upper Atmospheres of Venus and Mars', *Adv. Space Res.* **7**, 12, 57.

Keating, G. M., Taylor, F. W., Nicholson, J. Y., and Hinson, E. W.: 1979, 'Short-Term Cyclic Variations and Diurnal Variations of the Venus Upper Atmosphere', *Science* **205**, 62.

Keating, G. M., Nicholson, J. Y., and Lake, L. R.: 1980, 'Venus Upper Atmosphere Structure', *J. Geophys. Res.* **85**, 7941.

Keating, G. M., Bertaux, J. L., Bougher, S. W., Cravens, T. E., Dickinson, R. E., Hedin, A. E., Krasnopol'sky, V. A., Nagy, A. F., Nicholson, J. Y., III, Paxton, L. J., and von Zahn, U.: 1986, 'Models of Venus Neutral Upper Atmosphere: Structure and Composition', *Adv. Space Res.* **5**, 117.

Keldysh, M. V.: 1977, 'Venus Exploration with the Venera 9 and Venera 10 Spacecraft', *Icarus* **30**, 605.

Kenner, R. D. and Ogryzlo, E. A.: 1983, 'Quenching of $O_2(c^1\Sigma_u^-)v = 0$ by $O(^3P)$, $O_2(a^1\Delta_g)$, and Other Gases', *Can. J. Chem.* **61**, 921.

Kenner, R. D., Ogryzlo, E. A., and Turley, S.: 1979, 'On the Excitation of the Night Airglow on Earth, Venus and Mars', *J. Photochem.* **10**, 199.

Kim, J., Nagy, A. F., Cravens, T. E., and Kliore, A. J.: 1988, 'Solar Cycle Variations of the Electron Densities Near the Ionospheric Peak of Venus', *J. Geophys. Res.* **94**, 11997.

Kirby, K., Constantinides, E. R., Babeu, S., Oppenheimer, M., and Victor, G. A.: 1979, 'Photoionization and Photoabsorption Cross Sections of He, O, N_2, and O_2 for Aeronomic Calculations', *At. Data Nucl. Data Tables* **23**, 68.

Kley, D., Lawrence, G. M., and Stone, E. C.: 1977, 'The Yield of $N(^2D)$ atoms in the Dissociative Recombination of NO^+', *J. Chem. Phys.* **66**, 4157.

Kliore, A., Cain, D. L., Levy, G. S., Fjeldbo, G., and Rasool, S. I.: 1969, 'Structure of the Atmosphere of Venus Derived from Mariner V S-band Measurements', *Space Res.* **IX**, 712.

Kliore, A., Patel, I. R., Nagy, A. F., Cravens, T. E., and Gombosi, T.: 1979, 'Initial Observations of the Nightside Ionosphere of Venus from Pioneer Venus Orbiter Radio Occultations', *Science* **205**, 99.

Kliore, A. J. and Mullen, L.: 1987, 'Solar Cycle Influence on the Topside Plasma Scale Height of the Venus Dayside Ionosphere', *Bull. Am. Astron. Soc.* **19**, 846.

Kliore, A. J. and Mullen, L.: 1988, 'The Long-Term Behavior of the Main Peak of the Dayside Ionosphere of Venus During Solar Cycle 21 and Its Implications on the Effect of the Solar Cycle Upon the Electron Temperature in the Main Peak Region', *J. Geophys. Res.* **94**, 13339.

Knudsen, W. C.: 1973, 'Escape of ^{4}He and Fast O Atoms from Mars and Inferences on the ^{4}He Mixing Ratio', *J. Geophys. Res.* **78**, 8049.

Knudsen, W. C.: 1988, 'Solar Cycle Changes in the Morphology of the Venus Ionosphere', *J. Geophys. Res.* **93**, 8756.

Knudsen, W. C. and Miller, K. L.: 1985, 'Pioneer Venus Superthermal Electron Flux Measurements in the Venus Umbra', *J. Geophys. Res.* **90**, 2695.

Knudsen, W. C., Spenner, K., Whitten, R. C., Spreiter, J. R., Miller, K. L., and Novak, V.: 1979, 'Thermal Structure and Energy Influx into the Nightside Ionosphere', *Science* **205**, 105.

Knudsen, W. C., Spenner, K., Miller, K. L., and Novak, V.: 1980, 'Transport of Ionospheric O$^+$ Ions Across the Venus Terminator and Implications', *J. Geophys. Res.* **85**, 7803.

Knudsen, W. C., Miller, K. L., and Spenner, K.: 1986, *J. Geophys. Res.* **91**, 11936.

Krasnopol'sky, V. A.: 1978, 'Threshold Estimates of the Content of Some Substances in the Atmosphere of Mars and Venus from the Results of Spectroscopy of the Twilight Glow on the Mars 5, Venera 9 and Venera 10 Satellites', *Kossm. Issled.* **16**, 895; 1979, *Cosmic Res.* **16**, 713.

Krasnopol'sky, V. A.: 1979, 'Nightside Ionosphere of Venus', *Planetary Space Sci.* **27**, 1403.

Krasnopol'sky, V. A.: 1981, 'Excitation of Oxygen Emission in the Night Airglow of the Terrestrial Planets', *Planetary Space Sci.* **29**, 925.

Krasnopol'sky, V. A.: 1982a, '5577-Å Airglow and Electron Fluxes in the Nighttime Atmosphere of Venus', *Kossm. Issled.* **20**, 742; 1983, *Cosmic Res.* (Engl. transl) **20**, 530.

Krasnopol'sky, V. A.: 1982b, 'Atomic Carbon in the Atmospheres of Mars and Venus', *Kossm. Issled* **20**, 595; 1983, *Cosmic Res.* (Engl. transl.) **20**, 430.

Krasnopol'sky, V. A.: 1983a, in D. M. Hunten, L. Colin, T. M. Donahue, and V. I. Moroz (eds.), 'Venus Spectroscopy in the 3000–8000 Å Region by Veneras 9 and 10', *Venus*, Univ. of Arizona Press, Tucson, pp. 459–483.

Krasnopol'sky, V. A.: 1983b, 'Lightning and Nitric Oxide on Venus', *Planetary Space Sci.* **31**, 1363.

Krasnopol'sky, V. A.: 1985, 'Total Injection of Water Vapor Into the Venus Atmosphere', *Icarus* **62**, 221.

Krasnopol'sky, V. A.: 1986a, *Photochemistry of the Atmospheres of Mars and Venus*, Springer-Verlag, New York.

Krasnopol'sky, V. A.: 1986b, 'Oxygen Emission in the Night Airglow of the Earth, Venus and Mars', *Planetary Spoace Sci.* **34**, 511.

Krasnopol'sky, V. A. and Parshev, V. A.: 1981, 'Chemical Composition of the Atmosphere of Venus', *Nature* **282**, 610.

Krasnopol'sky, V. A. and Parshev, V. A.: 1983, in D. M. Hunten, L. Colin, T. M. Donahue, and V. I. Moroz (eds.), 'Photochemistry of the Venus Atmosphere', *Venus*, Univ. of Arizona Press, Tucson, p. 431.

Krasnopol'sky, V. A. and Tomashova, G. V.: 1980, 'Venus Nightglow Variations', *Cosmic Res.* **18**, 766.

Krasnopol'sky, V. A., Krys'ko, A. A., Rogachev, V. N., and Parshev, V. A.: 1976, 'Spectroscopy of the Night-Sky Luminescence of Venus from the Interplanetary Spacecraft Venera 9 and Venera 10', *Kossm. Issled.* **14**, 789; 1977, *Cosmic Res.* (Engl. transl.) **14**, 687.

Kumar, S. and Broadfoot, A. L.: 1975, 'Helium 584 Å Airglow Emission from Venus: Mariner 10 Observations', *Geophys. Res. Letters* **2**, 357.

Kumar, S. and Hunten, D. M.: 1974, 'Venus: An Ionospheric Model with an Exospheric Temperature of 350 K', *J. Geophys. Res.* **79**, 2529.

Kumar, S. and Taylor, H. A.: 1985, 'Deuterium on Venus: Model Comparisons with Pioneer Venus Observations of the Predawn Bulge Ionosphere', *Icarus* **62**, 494.

Kumar, S., Hunten, D. M., and Broadfoot, A. L.: 1978, 'Non-Thermal Hydrogen in the Venus Exosphere: the Ionospheric Source and the Hydrogen Budget', *Planetary Space Sci.* **26**, 1063.

Kumar, S., Hunten, D. M., and Taylor, H. A.: 1981, 'H$_2$ Abundance in the Atmosphere of Venus', *Geophys. Res. Letters* **8**, 237.

Kumar, S., Hunten, D. M., and Pollack, J. P.: 1983a, 'Nonthermal Escape of Hydrogen and Deuterium from Venus and Implications for Loss of Water', *Icarus* **55**, 369.

Kumar, S., Chakrabarti, S., Paresce, F., and Bowyer, S.: 1983b, 'The O$^+$ 834-Å Dayglow: Satellite Observations and Interpretation with a Radiation Transfer Model', *J. Geophys. Res.* **88**, 9271.

Kurt, V. G., Dostovalow, S. B., and Sheffer, E. K.: 1968, 'The Venus Far Ultraviolet Observations with Venera 4', *J. Atmospheric Sci.* **25**, 668.

Kurt, V. G., Romanova, N. N., Smirnov, A. S., Bertaux, J. L., and Blamont, J. L.: 1979, 'Venus Ultraviolet Radiation in the Wavelength Region from 300 Å to 1657 Å from Venera 11 and Venera 12 Data (Preliminary Results)', *Kossm. Issled.* **17**, 772; 1980, *Cosmic Res.* (Engl. transl.) **17**, 638.

Kurt, V. G., Smirnov, A. S., Titarchuk, L. G., Bertaux, J. L., and Lepine, V. M.: 1983, 'Scattering of Resonance Lines in the Upper Atmosphere of Venus According to Venera 11 and Venera 12 Ultraviolet Data', *Kossm. Issled.* **21**, 545; 1984, *Cosmic Res.* (Engl. transl.) **21**, 443.

Kuz'min, A. D.: 1970, 'The Atmosphere of the Planet Venus', *Radio Sci.* **5**, 339.

Kuz'min, A. D., Nanmov, A. P., Smirnova, T. V., and Vetekhnooskaya, U. N.: 1971, 'Lower Atmosphere of Venus from Radio Astronomical and Space Measurements', *Space Res.* **11**, 141.

Landau, L. and Teller, E.: 1936, 'Zur Theorie der Schalldispersion', *Physik. Z. Sowjetunion* **10**, 34.

Lawrence, G. M.: 1971, 'Quenching and Radiation Rates of $CO(a^3\Pi)$', *Chem. Phys. Letters* **9**, 575.

Lawrence, G. M., Barth, C. A., and Argabright, V.: 1977, 'Excitation of the Venus Night Airglow', *Science* **195**, 573.

Leach, S., Devoret, M., and Eland, J. H. D.: 1978a, 'Fluorescence Quantum Yields of Isotope CO_2^+ Ions', *Chem. Phys.* **33**, 113.

Leach, S., Stannard, P. R., and Gelbart, W. M.: 1978b, 'Interelectronic-State Perturbation Effects on Photoelectron Spectra and Emission Quantum Yields: Application to CO_2^+', *Mol. Phys.* **36**, 1119.

LeCompte, M. A., Paxton, L. J., and Stewart, A. I. F.: 1989, 'Analysis and Interpretation of Observations of Airglow at 297 nm in the Venus Thermosphere', *J. Geophys. Res.* **94**, 208.

Lee, L. C. and Judge, D. L.: 1972, 'Cross Sections for Production of $CO_2^+ [A^2\Pi_u; B^2\Sigma_u^+ \to X^2\Pi_g]$, Fluorescence by Vacuum Ultraviolet Radiation', *J. Chem. Phys.* **57**, 4433.

Leovy, C. B.: 1987, 'Zonal Winds Near Venus' Cloud Top Level: An Analytic Model of the Equatorial Wind Speed', *Icarus* **69**, 193.

Letzelter, C., Eidelsberg, M., Rostas, F., Breton, J., and Thieblemont, B.: 1987, 'Photoabsorption and Photodissociation Cross Sections of CO Between 88.5 and 115 nm', *Chem. Phys.* **114**, 273.

Leu, M.-T. and Yung, Y. L.: 1987, 'Determination of $O_2(a^1\Delta_g)$ and $O_2(b^1\Sigma_g^+)$ Yields in the Reaction $O + ClO \to Cl + O_2$: Implications for Photochemistry in the Atmosphere of Venus', *Geophys. Res. Letters* **14**, 949.

Levine, R. D. and Bernstein, R. B.: 1974, 'Energy Disposal and Energy Consumption in Elementary Chemical Reactions: An Information Theoretic Approach', *Accts. Chem. Res.* **7**, 393.

Lewis, J. S. and Prinn, R. G.: 1984, *Planets and Their Atmospheres: Origin and Evolution*, Academic Press, New York.

Lin, H.-M., Seaver, M., Tang, K. Y., Knight, A. E. W., and Parmenter, C. S.: 1979, 'The Role of Intermolecular Potential Well Depths in Collision-Induced State Changes', *J. Chem. Phys.* **70**, 5442.

Lindzen, R. S.: 1981, 'Turbulence and Stress Owing to Gravity Wave and Tidal Breakdown', *J. Geophys. Res.* **86**, 9707.

Link, R., Gladstone, G. R., Chakrabarti, S., and McConnell, J. C.: 1988, 'A Reanalysis of Rocket Measurements of the Ultraviolet Dayglow', *J. Geophys. Res.* **93**, 14631.

Liu, S. C. and Donahue, T. M.: 1974a, 'The Aeronomy of Hydrogen in the Atmosphere of Earth', *J. Atmospheric Sci.* **31**, 1118.

Liu, S. C. and Donahue, T. M.: 1974b, 'Mesospheric Hydrogen Related to Exospheric Escape Mechanisms', *J. Atmospheric Sci.* **31**, 1466.

Llewellyn, E. J., Solheim, B. H., Witt, G., Stegman, J., and Greer, R. G. H.: 1980, 'On the Excitation of Oxygen Emissions in the Night Airglow of the Terrestrial Planets', *J. Photochem.* **12**, 179.

Lopez-Gonzales, J., Lopez-Moreno, J., Lopez-Valverde, M. A., and Rodrigo, R.: 1989, 'Behaviour of the O_2 Infrared Atmospheric (0–0) Band in the Middle Atmosphere During Evening Twilight and at Night', *Planetary Space Sci.* **37**, 61.

Lopez-Moreno, J. J., Rodrigo, R., Moreno, F., Lopez-Puertas, M., and Molina, A.: 1988, 'Rocket Measurements of O_2 Infrared Atmospheric System in the Nightglow', *Planetary Space Sci.* **36**, 459.

Lowell, P.: 1894, 'Mars', *Pop. Astron.* **2**, 154.

Luhmann, J. G. and Kozrya, J. U.: 1990, 'Dayside Pick-Up Oxygen Ion Precipitation at Venus and Mars: Spatial Distribution, Energy Deposition and Consequences', *J. Geophys. Res.* (in press).

Marmo, F. and Engelman, A.: 1970, 'Carbon Atoms in the Upper Atmosphere of Venus', *Icarus* **12**, 128.

McDonald, G. J. G.: 1963, 'The Escape of Helium from the Earth's Atmosphere', *Rev. Geophys.* **1**, 305.

Massie, S. T., Hunten, D. M., and Sowell, D. T.: 1983, 'Day and Night Models of the Venus Thermosphere', *J. Geophys. Res.* **88**, 3955.

Mayr, H. G., Harris, I., and Spencer, N. W.: 1978, 'Some Properties of the Upper Atmosphere Dynamics', *Rev. Geophys. Space Phys.* **16**, 539.

Mayr, H. G., Harris, I., Niemann, H. B., Brinton, H. C., Spencer, N. W., Taylor, H. A., Jr., Hartle, R. E., Hoegy, W. R., and Hunten, D. M.: 1980, 'Dynamic Properties of the Thermosphere Inferred from Pioneer Venus Mass Spectrometer Measurements', *J. Geophys. Res.* **85**, 7841.

Mayr, H. G., Harris, I., Stevens-Rayburn, D. R., Niemann, H. B., Taylor, H. A., Jr., and Hartle, R. E.: 1985, 'On the Diurnal Variations in the Temperature and Composition: A Three-Dimensional Model with Superrotation', *Adv. Space Res.* **5**, 9, 109.

Mayr, H. G., Harris, I., Kasprzak, W. T., Dube, M., and Varosi, F.: 1988, 'Gravity Waves in the Upper Atmosphere of Venus', *J. Geophys. Res.* **93**, 11247.

McClelland, G. M., Saenger, K. L., Valentini, J. J., and Herschbach, D. R.: 1979, 'Vibratinal and Rotational Relaxation of Iodine in Seeded Supersonic Beams', *J. Phys. Chem.* **83**, 947.

McCoy, R. P.: 1983, 'Thermospheric Odd Nitrogen, 1. NO, N(^4S), and O(^3P) Densities from Rocket Measurements of NO δ and γ Bands, the O_2 Herzberg I Bands', *J. Geophys. Res.* **88**, 3197.

McDade, I. C., Llewellyn, E. J., Greer, R. G. H., and Murtagh, D. P.: 1987, 'ETON 6: A Rocket Measurement of the O_2 Infrared Atmospheric (0–0) Band in the Nightglow', *Planetary Space Sci.* **35**, 1541.

McElroy, M. B.: 1968, 'The Upper Atmosphere of Venus', *J. Geophys. Res.* **73**, 1513.

McElroy, M. B.: 1969, 'Structure of the Venus and Mars Atmospheres', *J. Geophys. Res.* **74**, 29.

McElroy, M. B.: 1970, 'Ionization Processes in the Atmospheres of Venus and Mars', *Ann. Geophys.* **26**, 643.

McElroy, M. B. and McConnell, J. C.: 1971, 'Atomic Carbon in the Atmospheres of Mars and Venus', *J. Geophys. Res.* **76**, 6674.

McElroy, M. B. and Hunten, D. M.: 1969, 'The Ratio of Deuterium to Hydrogen in the Venus Atmosphere', *J. Geophys. Res.* **74**, 115.

McElroy, M. B., Prather, M. J., and Rodriguez, J. M.: 1982a, 'Escape of Hydrogen from Venus', *Science* **215**, 1614.

McElroy, M. B., Prather, M. J., and Rodriguez, J. M.: 1982b, 'Loss of Oxygen from Venus', *Geophys. Res. Letters* **9**, 649.

McNeal, R. J., Whitson, M. E., and Cook, G. R.: 1974, 'Temperature Dependence of the Quenching of Vibrationally Excited Nitrogen by Atomic Oxygen', *J. Geophys. Res.* **79**, 1527.

Meier, R. R. and Anderson, D. E.: 1983, 'Determination of Atmospheric Composition and Temperature from the UV Airglow', *Planetary Space Sci.* **31**, 967.

Meier, R. R., Anderson, D. E., and Stewart, A. I. F.: 1983, 'Atomic Oxygen Emissions Observed from Pioneer Venus', *Geophys. Res. Letters* **10**, 214.

Meier, R. R., Conway, R. R., Anderson, D. E., Feldman, P. D., Eastes, R. W., Gentieu, E. P., and Christensen, A. B.: 1985, 'The Ultraviolet Dayglow at Solar Maximum. 3. Photoelectron Excited Emissions of N_2 and O', *J. Geophys. Res.* **90**, 6608.

Mengel, J. G., Mayr, H. G., Harris, I., and Stevens-Rayburn, D. R.: 1989, 'Non-Linear Three-Dimensional Spectral Model of the Venusian Thermosphere with Superrotation. II. Temperature, Composition, and Winds', *Planetary Space Sci.* **37**, 707.

Miller, K. L., Knudsen, W. C., Spenner, K., Whitten, R. C., and Novak, V.: 1980, 'Solar Zenith Angle Dependence of Ionospheric Ion and Electron Temperatures and Density on Venus', *J. Geophys. Res.* **85**, 7759.

Miller, S. M., Fell, C. P., and Steinfeld, J. I.: 1988, 'Rate Constants for Quenching of N* by O(^3P)', *EOS (Trans. Am. Geophys. Union)* **69**, 1347.

Moore, C. B., Wood, R. E., Hu, B. L., and Yardley, J. T.: 1967, 'Vibrational Energy Transfer in CO_2 Lasers', *J. Chem. Phys.* **46**, 4222.

Moos, H. W. and Rottman, G. J.: 1971, 'OI and HI emissions from the Upper Atmosphere of Venus', *Astrophys. J.* **169**, L127.

Moos, H. W., Fastie, W. G., and Bottema, M.: 1969, 'Rocket Measurement of Ultraviolet Spectra of Venus and Jupiter Between 1200 and 1800 Å', *Astrophys. J.* **155**, 887.

Moroz, V. I.: 1963, 'The Infrared Spectrum of Venus (1–2.5 μ)', *Soviet Astron.–AJ* **7**, 109.

Moroz, V. I.: 1968, 'The CO_2 Bands and Some Optical Properties of the Atmosphere over Venus', *Soviet Astron.–AJ* **11**, 653.

Moroz, V. I., Golovin, Yu. M., Ekhonomov, A. P., Moshkin, B. E., Parfent'ev, N. A., and San'ko, N. F.: 1980, *Nature* **284**, 243.

Mukhin, L. M., Gel'man, B. G., Lamonov, N. I., Mel'nikov, V. V., Nenarokov, D. F., Okhotnikov, B. P., Rotin, V. A., and Khoklov, V. N.: 1983, 'Gas-Chromatograph Analysis of the Chemical Composition of the Atmosphere of Venus by the Landers of Venera 13 and Venera 14 Spacecraft', *Kossm. Issled.* **21**, 225; 1983, *Cosmic Res.* **21**, 168.

Mul, P. M. and McGowan, J. W.: 1979, 'Temperature Dependence of Dissociative Recombination for Atmospheric Ions NO^+, O_2^+, N_2^+', *J. Phys.* **B12**, 1591.

Nagy, A. F., Kim, J., and Cravens, T. E.: 1990, 'Hot Hydrogen and Oxygen Atoms in the Upper Atmospheres of Venus and Mars', *Ann. Geophys.* **8**, 251.

Nagy, A. F. and Cravens, T. E.: 1988, 'Hot Oxygen Atoms in the Upper Atmosphere of Venus and Mars', *Geophys. Res. Letters* **15**, 433.

Nagy, A. F., Cravens, T. E., Yee, J.-H., and Stewart, A. I. F.: 1981, 'Hot Oxygen Atoms in the Upper Atmosphere of Venus', *Geophys. Res. Letters* **8**, 629.

Nakata, R. S., Watanabe, K., and Matsunaga, F. M.: 1965, 'Absorption and Photoionization Coefficients of CO_2 in the Region 580–1970 Å', *Sci. Light* **14**, 54.

Niemann, H. B., Hartle, R. E., Hedin, A. E., Kasprzak, W. T., Spencer, N. W., Hunten, D. M., and Carrigan, G. R.: 1979, 'Venus Upper Atmosphere Neutral Gas Composition: First Observations of the Diurnal Variations', *Science* **205**, 54.

Niemann, H. B., Booth, J. R., Cooley, J. E., Hartle, R. E., Kasprzak, W. T., Spencer, N. W., and Way, S. H.: 1980a, 'Pioneer Venus Orbiter Neutral Mass Spectrometer Experiment', *IEEE Trans. Geosci. Remote Sensing* **GE-18**, 60.

Niemann, H. B., Kasprzak, W. T., Hedin, A. E., Hunten, D. M., and Spencer, N. W.: 1980b, 'Mass Spectrometer Measurements of the Neutral Gas Composition of the Thermosphere and Exosphere of Venus', *J. Geophys. Res.* **85**, 7817.

Ogawa, H. S. and Judge, D. L.: 1986, 'Absolute Solar Flux Measurement Shortward of 575 Å', *J. Geophys. Res.* **91**, 7089.

Ogawa, H. S., Phillips, E., and Judge, D. L.: 1984, 'Line Width of the Solar EUV He I Resonance Emissions at 584 and 537 Å', *J. Geophys. Res.* **89**, 7537.

Oyama, V. I., Carle, G. C., Woeller, F., Pollack, J. B., Reynolds, R. T., and Craig, R. A.: 1980, 'Pioneer Venus Gas Chromatography of the Lower Atmosphere of Venus', *J. Geophys. Res.* **85**, 7891.

Parmenter, C. S. and Seaver, M.: 1979, 'The Temperature Dependence of Collision-Induced State Changes', *Chem. Phys. Letters* **67**, 279.

Partridge, H., Bauschlicher, and Langhoff, S. R.: 1990, *J. Chem. Phys.* (in press).

Paxton, L. J.: 1983, 'Atomic Carbon in the Venus Thermosphere: Observatins and Theory', Ph.D Thesis, University of Colorado, Boulder.

Paxton, L. J.: 1985, 'Pioneer Venus Orbiter Ultraviolet Spectrometer Limb Observations: Analysis and Interpretation of the 166- and 165-nm Data', *J. Geophys. Res.* **90**, 5089.

Paxton, L. J.: 1988, 'CO_2^+ and N_2^+ in the Venus Ionosphere', *J. Geophys. Res.* **93**, 8473.

Paxton, L. J. and Meier, R. R.: 1986, 'Reanalysis of Pioneer Venus Orbiter Ultraviolet Spectrometer Data: OI 1304 Intensities and Atomic Oxygen Densities', *Geophys. Res. Letters* **13**, 229.

Paxton, L. J., Anderson, D. E., and Stewart, A. I. F.: 1985, 'The Pioneer Venus Orbiter Ultraviolet Spectrometer: Analysis of Hydrogen Lyman Alpha Data', *Adv. Space Res.* **5**, 129.

Paxton, L. J., Anderson, D. E., and Stewart, A. I. F.: 1988, 'Analysis of Pioneer Venus Orbiter Ultraviolet Spectrometer Lyman α Data from Near the Subsolar Region', *J. Geophys. Res.* **93**, 1766; 1988, erratum in *J. Geophys. Res.* **93**, 11551.

Pechmann, J. B. and Ingersoll, A. P.: 1984, 'Thermal Tides in the Atmosphere of Venus: Comparison of Model Results with Observations', *J. Atmospheric Sci.* **41**, 3290.

Phillips, J. L., Stewart, A. I. F., and Luhmann, J. G.: 1986, 'The Venus Ultraviolet Aurora: Observations at 130.4 nm', *Geophys. Res. Letters* **13**, 1047.

Piper, L. G.: 1989, 'The Rate Coefficients for Quenching $N(^2D)$ by $O(^3P)$', *J. Chem. Phys.* **91**, 3516.

Piper, L. G., Donahue, M. E., and Rawlins, W. T.: 1987, 'Rate Coefficients for $N(^2D)$ Reactions', *J. Phys. Chem.* **91**, 3883.

Porter, H. S., Silverman, S. M., and Tuan, T. F.: 1974, 'On the Behavior of Airglow Under the Influence of Gravity Waves', *J. Geophys. Res.* **79**, 3827.

Poulson, L. L., Billing, G. D., and Steinfeld, J. I.: 1978, 'Temperature Dependence of HF Vibrational Relaxation', *J. Chem. Phys.* **68**, 5121.

Prather, M. J. and McElroy, M. B.: 1983, 'Helium on Venus: Implications for Uranium and Thorium', *Science* **228**, 410.

Prinn, R. G.: 1985, in J. S. Levine (ed.), 'The Photochemistry of the Atmosphere of Venus', *The Photochemistry of Atmospheres: Earth the Other Planets and Comets*, Academic Press, New York, 1985.

Quack, M. and Troe, T.: 1977, 'Vibrational Relaxation of Diatomic Molecules in Complex Forming Collisions with Reactive Atoms', *Ber. Bunsenges. Physik. Chem.* **81**, 160.

Queffelec, J. L., Rowe, B. R., Vallee, F., Gomet, J. C., and Morlais, M.: 1989, 'The Yield of Metastable Atoms Through Dissociative Recombination of O_2^+ Ions with Electrons', *J. Chem. Phys.* **91**, 5335.

Revercomb, H. E., Sromovsky, L. A., Suomi, V. E., and Boese, R. W.: 1985, 'Thermal Net Flux Measurements on the Pioneer Venus Entry Probes', *Adv. Space Res.* **5**, 81.

Roble, R. G., Dickinson, R. E., and Ridley, E. C.: 1982, 'Global Circulation and Temperature Structure of the Thermosphere with High Latitude Plasma Convection', *J. Geophys. Res.* **87**, 1599.

Roble, R. G., Ridley, E. C., and Dickinson, R. E.: 1987, 'On the Global Mean Structure of the Thermosphere', *J. Geophys. Res.* **92**, 8745.

Rodriguez, J. M., Prather, M. J., and McElroy, M. B.: 1984, 'Hydrogen on Venus: Exospheric Distribution and Escape', *Planetary Space Sci.* **32**, 1235.

Rohrbaugh, R. P. and Nisbet, J. S.: 1973, 'Effect of Energetic Oxygen Atoms on Neutral Density Models', *J. Geophys. Res.* **78**, 6768.

Rosser, W. A., Wood, A. D., and Gerry, E. T.: 1969, 'Deactivation of Vibrationally Excited Carbon Dioxide (v_3) by Collisions with Carbon Dioxide or with Nitrogen', *J. Chem. Phys.* **50**, 4996.

Rottman, G. J.: 1981, 'Rocket Measurements of the Solar Spectral Irradiance During Solar Minimum, 1972–1977', *J. Geophys. Res.* **86**, 6697.

Rottman, G. J. and Moos, H. W.: 1973, 'The Ultraviolet (1200–1900 Å) Spectrum of Venus', *J. Geophys. Res.* **78**, 8033.

Rusch, D. W. and Cravens, T. E.: 1979, 'A Model of the Neutral and Ion Nitrogen Chemistry in the Daytime Thermosphere of Venus', *Geophys. Res. Letters* **6**, 791.

Russell, C. T., Elphic, R. C., and Slavin, J. A.: 1979, 'Initial Pioneer Venus Magnetic Field Results: Dayside Observations', *Science* **203**, 745.

Samson, J. A. R. and Gardner, J. L.: 1973, 'Fluorescence Excitation and Photoelectrons Spectra of CO_2 Induced by Vacuum Ultraviolet Radiation Between 185 and 716 Å', *J. Geophys. Res.* **78**, 3663.

Saxon, R. P. and Liu, B.: 1977, '*Ab Initio* Configuration Interaction Study of the Valence States of O_2', *J. Chem. Phys.* **67**, 5432.

Schatz, G. C. and Redmon, M. J.: 1981, 'A Quasi-Classical Trajectory Study of Collisional Excitation in $O(^3P) + CO_2$', *Chem. Phys.* **58**, 195.

Schloerb, F. P., Robinson, S. E., and Irvine, W. M.: 1980, 'Observations of CO in the Stratosphere of Venus via its $J = 0 \rightarrow 1$ Rotational Transition', *Icarus* **43**, 121.

Schoeberl, M. R. and Strobel, D. F.: 1978, 'The Zonally Averaged Circulation of the Middle Atmosphere', *J. Atmospheric Sci.* **35**, 577.

Schofield, J. T., Taylor, F. W., and McCleese, D. J.: 1982, 'The Global Distribution of Water Vapor in the Middle Atmosphere of Venus', *Icarus* **52**, 263.

Schubert, G.: 1983, in D. M. Hunten, L. Colin, T. M. Donahue, and V. I. Moroz (eds.), 'General Circulation and the Dynamical State of the Venus Atmosphere', *Venus*, Univ. of Arizona Press, Tucson.

Schubert, G. and Walterscheid, R. L.: 1984, 'Propagation of Small-Scale Acoustic-Gravity Waves in the Venus Atmosphere', *J. Atmospheric Sci.* **41**, 1202.

Schubert, G., Covey, C., Del Genio, A., Elson, L. S., Keating, G., Seiff, A., Young, R. E., Apt, J., Counselman III, C. C., Kliore, A. J., Limaye, S. S., Revercomb, H. E., Sromovsky, L. A., Suomi, V. E., Taylor, F., Woo, R., and von Zahn, U.: 1980, 'Structure and Circulation of the Venus Atmosphere', *J. Geophys. Res.* **85**, 8007.

Schwartz, R. N., Slawsky, Z. I., and Herzfeld, K. F.: 1952, 'Calculation of Vibrational Relaxation Times in Gases', *J. Chem. Phys.* **20**, 1591.

Seaton, M. J. and Osterbrock, D. E.: 1957, 'Relative O II Intensities in Gaseous Nebulae', *Astrophys. J.* **66**, 125.

Seiff, A.: 1982, 'Dynamical Implications of the Observed Thermal Contrasts in Venus's Upper Atmosphere', *Icarus* **51**, 574.

Seiff, A., Kirk, D. B., Young, R. E., Blanchard, R. C., Findlay, J. T., Kelly, G. M., and Sommer, S. C.: 1980, 'Measurements of Thermal Structure and Thermal Contrasts in the Atmosphere of Venus and Related Dynamical Observations: Results from the Four Pioneer Venus Probes', *J. Geophys. Res.* **85**, 7903.

Sharma, R. D. and Brau, C. A.: 1969, 'Energy Transfer in Near-Resonant Molecular Collisions Due to Long-Range Forces with Application to Transfer of Vibrational Energy from v_3 Mode of CO_2 to N_2', *J. Chem. Phys.* **50**, 924.

Sharma, R. D. and Nadile, R. M.: 1980, 'Carbon Dioxide (v_2) Radiance Results Using a New Non-Equilibrium Model', *EOS* **61**, 17, 322.

Sharma, R. D. and Nadile, R. M.: 1981, *Carbon Dioxide (v_2) Radiance Results Using a New Non-Equilibrium Model*, AFGL Tech. Rep. 81-0064, Air Force Geophys. Lab., Hanscom Air Force Base, Bedford, Mass.

Simpson, C. J. S. M., Gait, P. D., and Simmie, J. M.: 1977, 'The Vibrational Deactivation of the Bending Mode of CO_2 and by N_2', *Chem. Phys. Letters* **47**, 133.

Sinton, W. M.: 1963, 'Infrared Observations of Venus', *Mem. Soc. Roy. Sci. Liège* **7**, 300.

Slanger, T. G.: 1978, 'Generation of $O_2(c^1\Sigma_g^-, C^3\Delta_u, A^3\Sigma_g^-)$ Emission in the Terrestrial Nightglow', *J. Chem. Phys.* **69**, 4779.

Slanger, T. G. and Black, G.: 1976, '$O(^1S)$ Production from Oxygen Atom Recombination', *J. Chem. Phys.* **64**, 3767.

Slanger, T. G. and Heustis, D. L.: 1981, '$O_2(c^1\Sigma_u^- \rightarrow X^3\Sigma_g^+)$ Emission in the Terrestrial Nightglow', *J. Geophys. Res.* **86**, 3551.

Smith, I. W. M.: 1984, 'The Role of Electronically Excited States in Recombination Reactions', *Int. J. Chem. Kinet.* **16**, 423.

Spenner, K., Knudsen, W. C., Whitten, R. C., Michelson, P. F., Miller, K. L., and Novak, V.: 1981, 'On the Maintenance of the Venus Nightside Ionosphere: Electron Precipitation and Plasma Transport', *J. Geophys.* **86**, 9170.

Stair, A. T., Jr., Sharma, R. D., Nadile, R. M., Baker, D. J., and Grieder, W. F.: 1985, 'Observations of Limb Radiance with Cryogenic Spectral Infrared Rocket Experiment', *J. Geophys. Res.* **90**, 9763.

Stewart, A. I.: 1969, paper presented at *The Tucson Conference on the Atmosphere of Venus.* Quoted in McElroy and Hunten (1969).

Stewart, A. I. F.: 1980, 'Design and Operation of the Pioneer Venus Orbiter Ultraviolet Spectrometer', *IEEE Trans. Geosci. Remote Sensing* **GE-18**, 65.

Stewart, A. I.: 1972, 'Mariner 6 and 7 Ultraviolet Spectrometer Experiment: Implications of CO_2^+, CO, and O Airglow', *J. Geophys. Res.* **77**, 54.

Stewart, A. I. and Barth, C. A.: 1979, 'Ultraviolet Night Airglow of Venus', *Science* **205**, 59.

Stewart, A. I., Anderson, D. E., Esposito, L. W., and Barth, C. A.: 1979, 'Ultraviolet Spectroscopy of Venus: Initial Results from the Pioneer Venus Orbiter', *Science* **203**, 777.

Stewart, A. I. F., Gerard, J. C., Rusch, D. W., and Bougher, S. W.: 1980, 'Morphology of the Venus Ultraviolet Night Airglow', *J. Geophys. Res.* **85**, 7861.

Stewart, R. W.: 1968, 'Interpretation of Mariner 5 and Venera 4 Data on the Upper Atmosphere of Venus', *J. Atmospheric Sci.* **25**, 578.

Stewart, R. W. and Hogan, J. S.: 1969, 'Empirical Determination of Heating Efficiencies in the Mars and Venus Atmospheres', *J. Atmospheric Sci.* **26**, 330.

Stone, E. J. and Zipf, E. C.: 1974, 'Electron Impact Excitation of the $^3S^0$ and $^5S^0$ States of Atomic Oxygen', *J. Chem. Phys.* **60**, 4237.

Stoney, G. J.: 1868, 'On the Physical Constitution of the Sun and Stars', *Proc. Roy. Soc. London* **17**, 1.

Stoney, G. J.: 1898, 'Of Atmospheres Upon Planets and Satellites', *Trans. Roy. Soc. Dublin* **6**, 305; reprinted in 1898: *Astrophys. J.* **7**, 25.

Strickland, D. J.: 1973, 'The OI 1304- and 1356-Å Emissions from the Atmosphere of Venus', *J. Geophys. Res.* **78**, 2827.

Strickland, D. J., Thomas, G. E., and Sparks, P. R.: 1972, 'Mariner 6 and 7 Ultraviolet Spectrometer Experiment: Analysis of the OI 1304- and 1356-Å Emissions', *J. Geophys. Res.* **77**, 4052.

Surkov, Yu. A., Shchelgov, O. P., Ryvkin, M. L., Sheinin, D. M., and Davydov, N. A.: 1988, 'Water Vapor Distribution in the Middle and Lower Venusian Atmosphere', *Kossm. Issled.* **25**, 678; 1988: *Cosmic Res.* **25**, 517.

Sze, N. F. and McElroy, M. B.: 1975, 'Some Problems in Venus Aeronomy', *Planetary Space Sci.* **23**, 763.

Takacs, P. Z., Broadfoot, A. L., Smith, G. R., and Kumar, S.: 1980, 'Mariner 10 Observations of Hydrogen Lyman-α Emission from the Venus Exosphere: Evidence of Complex Structure', *Planetary Space Sci.* **28**, 687.

Taylor, F. W., Diner, D. J., Elson, L. S., Hanner, M. S., McCleese, D. J., Martonchik, J. V., Reichley, P. E., Houghton, J. T., Delderfield, J., Schofield, T., Bradley, S. E., and Ingersoll, A. P.: 1979, 'Infrared Remote Sounding of the Middle Atmosphere of Venus from Pioneer Venus Orbiter', *Science* **203**, 779.

Taylor, F. W., Beer, R., Chahive, M. T., Diner, D. J., Elson, L. S., Haskins, R. D., McCleese, D. J., Martonchik, J. V., Reichley, P. E., Bradley, S. P., Delderfield, J., Schofield, J. T., Farmer, C. B., Froidevaux, L., Leung, J., Cofley, M. T., and Gille, J. C.: 1980, 'Structure and Meteorology of the Middle Atmosphere of Venus: Infrared Remote Sensing from the Pioneer Orbiter', *J. Geophys. Res.* **85**, 7963.

Taylor, H. A., Jr., Brinton, H. C., Bauer, S. J., Hartle, R. E., Cloutier, P. A., and Daniell, R. E., Jr.: 1980, 'Global Observations of the Composition and Dynamics of the Ionosphere of Venus: Implications for Solar Wind Interaction', *J. Geophys. Res.* **85**, 7765.

Taylor, H. A., Jr., Brinton, H. C., Niemann, H. B., Mayr, H. G., Hartle, R. E., Barnes, A., and Larson, J.: 1984, 'Interannual and Short-Term Variations of the Venus Nighttime Hydrogen Bulge', *J. Geophys. Res.* **89**, 10669.

Taylor, H. A., Brinton, H., Niemann, H. B., Mayr, H. G., Hartle, R., Barnes, A., and Larson, J.: 1985, '*In Situ* Results on the Variation of Neutral Atmospheric Hydrogen at Venus', *Adv. Space Res.* **5**, 125.

Taylor, M. J., Hapgood, M. A., and Rothwell, P.: 1987, 'Observations of Gravity Wave Propagation in the O I (555.7 nm), Na (589.2 nm) and the Near Infrared OH Nightglow Emissions', *Planetary Space Sci.* **35**, 413.

Thomas, R. J.: 1981, 'Analysis of Atomic Oxygen Emissions in the Airglow of the Terrestrial Planets', *J. Geophys. Res.* **86**, 206.

Tobiska, W. K., Barth, C. A., and Culp, R. D.: 1988, 'Use of Solar Lyman-α and 1–8 Å X-Rays as EUV Flux Indices', *A.I.A.A.* (preprint).

Torr, M. R. and Torr, D. G.: 1985, 'Ionization Frequencies for Solar Cycle 21: Revised', *J. Geophys. Res.* **90**, 6675.

Torr, M. R., Torr, D. G., Ong, R. A., and Hinteregger, H. E.: 1979, 'Ionization Frequencies for Major Thermospheric Constituents as a Function of Solar Cycle 21', *Geophys. Res. Letters* **6**, 771.

Torr, M. R., Torr, D. G., and Hinteregger, H. E.: 1980, 'Solar Flux Variability in the Schumann–Runge Continuum as a Function of Solar Cycle 21', *J. Geophys. Res.* **85**, 6063.

Traub, W. A. and Carleton, N. P.: 1974, in A. Woszczyk and C. Iwaniszewska (eds.), 'Observations of O_2, H_2O, and HD in Planetary Atmospheres', *Exploration of Planetary Atmospheres*, D. Reidel Publ. Co., Dordrecht, Holland, p. 223.

Trauger, J. T. and Lunine, J. I.: 1983, 'Spectroscopy of Molecular Oxygen in the Atmospheres of Venus and Mars', *Icarus* **55**, 272.

Ustinov, E. A. and Moroz, V. I.: 1978, 'Refinement of the H_2O Content Value in the Atmosphere of Venus with Narrow-Band Photometry Data from Venera 9 and Venera 10', *Kossm. Issled.* **16**, 127; 1978: *Cosmic Res.* (Engl. transl.) **16**, 98.

van Dishoeck, È. F. and Black, J. H.: 1988, 'The Photodissociation and Chemistry of Interstellar CO', *Astrophys. J.* **334**, 771.

Vaughan, S. O. and Doering, J. P.: 1986, 'Absolute Experimental Differential and Integral Electron Excitation Cross Sections for Atomic Oxygen. 2. The $(^3P \rightarrow {}^3S^0)$ Transition (1304 Å) from 16.5 to 200 eV with Comparison to Atomic Hydrogen', *J. Geophys. Res.* **91**, 13755.

Vinogradov, A. P., Surkov, U. A., and Florensky, C. P.: 1968, 'The Chemical Composition of the Venus Atmosphere Based on the Data of the Interplanetary Station Venera 4', *J. Atmospheric Sci.* **25**, 535.

Vinogradov, A. P., Surkov, Yu. A., Andreichikov, B. M., Kalinkina, O. M., and Grechischeva, I. M.: 1971, in C. Sagan, T. C. Owen, and H. J. Smith (eds.), 'The Chemical Composition of the Atmosphere of Venus', *Planetary Atmospheres*, D. Reidel Publ. Co., Dordrecht, Holland.

von Zahn, U.: 1977, 'Bus Neutral Mass Spectrometer (BNMS)', *Space Sci. Rev.* **20**, 451.

von Zahn, U. and Moroz, V. I.: 1986, 'Composition of the Venus Atmosphere Below 100 km Altitude', *Adv. Space Res.* **5**, 173.

von Zahn, U., Krankowsky, D., Mauersberger, K., Nier, A. O., and Hunten, D. M.: 1979, 'Venus Thermosphere: *In Situ* Composition Measurements, the Temperature Profile, and the Homopause Altitude', *Science* **203**, 768.

von Zahn, U., Fricke, K. H., Hunten, D. M., Krankowsky, D., Mauersberger, K., and Nier, A. O.: 1980, 'The Upper Atmosphere of Venus During Morning Conditions', *J. Geophys. Res.* **85**, 7892.

von Zahn, U., Kumar, S., Niemann, H., and Prinn, R.: 1983, in D. M. Hunten, L. Colin, T. M. Donahue, and V. I. Moroz (eds.), 'Composition of the Venus Atmosphere', *Venus*, Univ. of Arizona Press, Tucson.

Walker, J. C. G.: 1977, *Evolution of the Atmosphere*, MacMillan, New York.

Walker, J. C. G.: 1982, 'The Earliest Atmosphere of the Earth', *Precambrian Res.* **17**, 147.

Wallace, L.: 1969, 'Analysis of the Lyman-Alpha Observations of Venus Made from Mariner 5', *J. Geophys. Res.* **74**, 115.

Wallace, L., Stuart, F. E., Nagel, R. H., and Larson, D. M.: 1971, 'A Search for Deuterium on Venus', *Astrophys. J.* **168**, L29.

Wallis, M. K.: 1972, 'Comet-Like Interaction of Venus with the Solar Wind. I', *Cosmic Electrodyn.* **3**, 45.

Wallis, M. K.: 1978, 'Exospheric Density and Escape Fluxes of Atomic Isotopes on Venus and Mars', *Planetary Space Sci.* **26**, 949.

Wallis, M. K.: 1982, 'Comet-Like Interaction of Venus with the Solar Wind. III. The Atomic Oxygen Corona', *Geophys. Res. Letters* **9**, 427.

Walterscheid, R. L., Schubert, G., Newman, M., and Kliore, A. J.: 1985, 'Zonal Winds and the Angular Momentum Balance of Venus' Atmosphere Within and Above the Clouds', *J. Atmospheric Sci.* **42**, 1982.

Waterston, J. J.: 1846, *Proc. Roy. Soc. London* **5**, 604.

Waterston, J. J.: 1892, *Phil. Trans. Roy. Soc. (London)* **A183**, 1.

Watson, A. J., Donahue, T. M., and Walker, J. C. G.: 1981, 'The Dynamics of Rapidly Escaping Atmosphere: Applications to the Evolution of Earth and Venus', *Icarus* **48**, 150.

Watson, A. J., Donahue, T. M., and Kuhn, W. R.: 1984, 'Temperatures in a Runaway Greenhouse on the Evolving Venus: Implications for Water Loss', *Earth Planetary Sci. Letters* **68**, 1.

Wauchop, T. S. and Broida, H. P.: 1972, 'Lifetime and Quenching of $CO(a^2\Pi)$, Produced by Recombination of CO_2 Ions in a Helium Afterglow', *J. Chem. Phys.* **56**, 330.

Wells, W. C., Borst, W. L., and Zipf, E. C.: 1972, 'Production of $CO(a^3\Pi)$ and Other Metastable Fragments by Electron Impact Dissociation of CO_2', *Planetary Space Sci.* **31**, 317.

Wilson, W. J., Klein, M. J., Kakar, R. K., Gulkis, S., Olsen, E. T., and Ho, P. T. P.: 1981, 'Venus. I. Carbon Monoxide Distribution and Molecular-Line Searches', *Icarus* **45**, 624.

Wintersteiner, P. P., Sharma, R. D., Winick, J. R., and Picard, R.: 1988, 'Determination of $CO_2(v_2)$ Vibrational Temperature Using a New Line-by-Line Radiative Transfer Algorithm', *EOS (Trans. Am. Geophys. Union)* **69**, 1346.

Wraight, P. C.: 1982, 'Association of Atomic Oxygen and Airglow Excitation Mechanisms', *Planetary Space Sci.* **30**, 251.

Yardley, J. T.: 1980, *Introduction to Molecular Energy Transfer*, Academic Press, New York.

Yatteau, J. H.: 1983, 'Some Issues Related to the Evolution of Planetary Atmospheres', Ph.D. Thesis, Harvard University, Cambridge, MA.

Yee, J.-H. and Hays, P. B.: 1980, 'The Oxygen Polar Corona', *J. Geophys. Res.* **85**, 1795.

Yee, J.-H. and Killeen, T. L.: 1986, 'Thermospheric Production of $O(^1S)$ by Dissociative Recombination of Vibrationally Excited O_2^+', *Planetary Space Sci.* **34**, 1101.

Yee, J.-H., Meriwether, J. W., and Hays, P. B.: 1980, 'Detection of a Corona of Fast Oxygen Atoms During Solar Maximum', *J. Geophys. Res.* **85**, 3396.

Yoshino, K., Stark, G., Smith, P. L., Parkinson, W. H., and Ito, K.: 1988, 'High Resolution Spectra and Photoabsorption Coefficients for Carbon Monoxide Absorption Bands Between 94.0 and 100.4 nm', *J. Physique* **49**, C1–37.

Young, L. D. G.: 1972, 'High Resolution Spectra of Venus', *Icarus* **17**, 632.

Yung, Y. L. and DeMore, W. B.: 1982, 'Photochemistry of the Stratosphere of Venus: Implications for Atmospheric Evolution', *Icarus* **51**, 199.

Zahnle, K. L. and Kasting, J. F.: 1986, 'Mass Fractionation During Transonic Escape and Implications for Loss of Water from Mars and Venus', *Icarus* **68**, 462.

Zipf, E. C.: 1970, 'The Dissociative Recombination of O_2^+ Ions into Specifically Identified States', *Bull. Am. Phys. Soc.* **15**, 498.

Zipf, E. C.: 1986, 'On the Direct and Dissociative Excitation of the $O(3s^3S^0)$ State by Electron Impact on Atomic and Molecular Oxygen', *J. Phys.* **B19**, 2199.

Zipf, E. C. and Erdman, P. W.: 1985, 'Electron Impact Excitation of Atomic Oxygen: Revised Cross Sections', *J. Geophys. Res.* **90**, 11087.

Zipf, E. C. and Stone, E. J.: 1971, 'Photoelectron Excitation of Atomic-Oxygen Resonance Radiation in the Terrestrial Nightglow', *J. Geophys. Res.* **26**, 6865.

The manufacturer's authorised representative in the EU is Springer
Nature Customer Service Centre GmbH, Europaplatz 3, 69115 Heidelberg,
Germany. If you have any concerns regarding our products, please
contact ProductSafety@springernature.com

Printed and bound by CPI Group (UK) Ltd, Croydon, CR0 4YY

23/04/2026

02095624-0009